Anthony J. Zartycki 3rd

from Mercedes T. Davis

THE MOLECULAR UNIVERSE

IAU SYMPOSIUM No. 280

COVER ILLUSTRATION:

The far infrared spectrum of the Orion Nebula, observed by the Herschel Space Observatory, superimposed on an image obtained by the Spitzer telescope.

The Orion Nebula is known to be one of the most prolific chemical factories in space, although the full extent of its chemistry and the pathways for molecule formation are not well understood. This extremely rich spectrum, obtained with the HIFI spectrometer in January 2010, covers the range 1.059 to 1.115 THz. Among the molecules that can be identified in this spectrum (marked on the corresponding lines) are water, carbon monoxide, formaldehyde, methanol, dimethyl ether, hydrogen cyanide, sulphur oxide, sulphur dioxide and their isotope analogues. It is expected that many new organic molecules will also be identified. Identification of the many spectral features visible in the Orion spectrum with transitions of particular molecular species requires sophisticated molecular spectroscopy databases, which collect the results from many years of theoretical and laboratory spectroscopy work.

HERSCHEL is an ESA space observatory with science instruments provided by European-led Principal Investigator consortia, with important participation from NASA. The high resolution spectrometer HIFI was designed and built by a nationally-funded consortium led by SRON Netherlands Institute for Space Research. The consortium includes institutes from France, Germany, USA, Canada, Ireland, Italy, Poland, Russia, Spain, Sweden, Switzerland, and Taiwan.

Credit of the image: ESA, NASA, Spitzer, HIFI consortium, E. Bergin and the HEXOS team.

INTERNATIONAL ASTRONOMICAL UNION

UNION ASTRONOMIQUE INTERNATIONALE

International Astronomical Union

IAU

THE MOLECULAR UNIVERSE

PROCEEDINGS OF THE 280th SYMPOSIUM OF THE INTERNATIONAL ASTRONOMICAL UNION HELD IN TOLEDO, SPAIN MAY 30 - JUNE 3, 2011

Edited by

JOSÉ CERNICHARO
Centro de Astrobiología, INTA-CSIC, Torrejón de Ardoz, Madrid, SPAIN

and

RAFAEL BACHILLER
Observatorio Astronómico Nacional, IGN, Madrid, SPAIN

CAMBRIDGE
UNIVERSITY PRESS

CAMBRIDGE UNIVERSITY PRESS
The Edinburgh Building, Cambridge CB2 2RU, UnitedKingdom
40 West 20th Street, New York, NY 10011–4211, USA
10 Stamford Road, Oakleigh, Melbourne 3166, Australia

First published 2011

Printed in the United Kingdom at the University Press, Cambridge

Typeset in System LaTeX 2_ε

A catalogue record for this book is available from the British Library

Library of Congress Cataloguing in Publication data

This journal issue has been printed on FSC-certified paper and cover board. FSC is an independent, non-governmental, not-for-profit organization established to promote the responsible management of the worlds forests. Please see www.fsc.org for information.

ISBN 9781107019805 hardback
ISSN 1743-9213

Table of Contents

Preface

Astrochemistry is experiencing a true golden age. The role of molecules in astronomy is becoming so important that it is no longer an exaggeration to refer to a sizeable portion of the universe as "The Molecular Universe", the title that has been adopted for this symposium.

Recent successes in the field result from advances in observational, laboratory, theoretical, and modeling work.

To illustrate the recent observational advances, it can be mentioned that molecules other than CO have now been detected up to redshifts of more than 6. On the scale of entire galaxies, molecular maps are revealing striking spatial differences reflecting different physical processes (photodissociation, shocks, high temperature chemistry, etc.). On a local scale, hot cores, regions of complex molecular growth in the vicinity of high- and low-mass young stellar objects, have been observed to be far more complex and heterogeneous than previously thought. Negatively charged molecular anions have been found for the first time in a variety of sources, including the cold core TMC-1 and the envelopes of evolved stars. Millimeter-wave interferometers and infrared telescopes (both ground-based and in space) are revealing a surprisingly rich chemistry in the planet-forming zones of circumstellar disks and, at the same time, are providing a complete inventory of ices in star- and planet-forming regions. The study of exoplanetary atmospheres has begun, and evidence exists for methane, carbon dioxide, and water. The inventory of molecules in cometary atmospheres continues to grow, with interesting variations found among comets of different origin.

The laboratory studies, which were traditionally focused on gas-phase processes, are now significantly shifting emphasis toward gas-grain interactions. Surface science experiments are being applied to chemical and physical processes on analogs of interstellar dust particles. For example, the efficiency of photodesorption has been measured for the first time for simple ices, and the photodissociation of methanol ice along with impurities supports new models of complex molecule formation on ices. The complicated formation of water ice on surfaces at low temperature, relevant to the interpretation of Herschel data, has been studied, as has the thermal evaporation of mixed ices. The formation of molecular hydrogen on high-temperature grains, a complex process, has been thoroughly investigated. Cometary samples returned from the Stardust mission as well as meteoritic and IDP samples are studied in the laboratory with increasingly sophisticated instruments. Much effort has gone into the study of gas-phase rotational spectra at frequencies as high as 2 THz; an indispensable work for the identification of spectral lines to be detected with Herschel, SOFIA and ALMA. Laboratory studies of PAHs continue to be essential to interpret the wealth and variety of infrared features detected in the interstellar medium.

From the theoretical point of view, the chemical processes that occur on dust grains are normally treated in models by so-called rate equations. In some instances, these equations are inaccurate, and more computationally intensive methods, known as stochastic approaches, must be used. Much progress has been made in creating models that combine gas-phase chemistry, which is treated by rate equations, and surface chemistry, which is treated stochastically with new methods. Quantum chemical potentials, used sometimes with classical dynamics and sometimes with quantum mechanical dynamics, have been employed to determine accurate inelastic scattering cross sections, needed in the analysis of molecular spectral intensities.

Chemical simulations are improving in a number of ways. The major networks of gas-phase reactions now include processes involving anions, which act as catalysts for the production of larger neutral species. The formation of complex molecules in hot cores is being studied with a new approach: complex gas-phase species are produced on granular surfaces as they are heating up in the presence of a young stellar object; the rising temperature allows molecules and radicals produced by photodissociation to diffuse more readily around grains and eventually evaporate. Models of protoplanetary disks are being challenged to reproduce the mid-infrared detection of water, acetylene (HCCH), HCN, and other species. The inner disk regions require that chemical networks be able to operate at temperatures as high as 1000 K. Models of photon-dominated regions (PDRs) have reached new levels of sophistication and are now available for general distribution. Hydrodynamics is being applied to more chemical models in an attempt to merge realistic dynamics and chemistry.

We finally note that molecular databases, coupled with radiative transfer programs, continue to be developed for general distribution and provide astronomers with a variety of tools to make line intensity predictions. These tools can either be combined with the output of chemo-dynamical models or used in a stand-alone mode to analyze molecular observations in terms of physical conditions and molecular abundances. Such tools are indeed becoming increasingly important in the preparation of observations for ALMA, JWST and ELTs, whether applied to the smallest scales in nearby protoplanetary disks or to the most distant galaxies.

The above examples illustrate the broad variety and high interest of hot topics in current Astrochemistry. The symposium was so intense, and the number of oral and poster contributions so high, that only the review and invited talks have found place in this volume. The abstracts of all contributions were published in a booklet produced by the Local Organizing Committee and were posted at the SAO/NASA Astrophysics Data System (ADS). The nearly 400 contributions (including oral presentations and posters) which are not present in this volume are listed here as an Appendix, and have been made public in the Internet web site of the conference *(http://cab.inta-csic.es/molecular_universe/)* forming an excellent and indispensable complement to this book.

We very much hope that this book will make investigators aware of the many and exciting new progresses achieved by Astrochemistry on the last 6 years. and we also hope to acquaint them about the explosive growth in the subject to be coming in the next few years, when ALMA, JWST and the ELTs will come along, opening a fully new era both in Astrophysics and Astrochemistry.

José Cernicharo and Rafael Bachiller
Editors
Madrid, July 31, 2011

THE ORGANIZING COMMITTEE

Scientific

E. van Dishoeck (Chair, The Netherlands)
Y. Aikawa (Japan)
G. A. Blake (USA)
J. Cernicharo (Spain)
M. Guélin (France)
S. Kwok (China)
K. Menten (Germany)
F. Salama (USA)
A. Sternberg (Israel)
E. Herbst (Vice-chair, USA)
J. Black (Sweden)
P. Caselli (United Kingdom)
G. Garay (Chile)
U. Jørgensen (Denmark)
J. Maier (Switzerland)
T. Millar (United Kingdom)
I. Sims (France)

Local

J. Cernicharo (Chair, CAB, CSIC-INTA)
M. Castellanos (CAB)
J. Albaladejo (UCLM)
A. Fernández-Clavero (CAB)
S. García-Burillo (OAN)
J. R. Goicoechea (CAB)
A. Sáiz-López (CIAC)
R. Bachiller (Vice-chair, OAN, IGN)
P. de Vicente (OAN)
B. Cabañas (UCLM)
A. Fuente (OAN)
G. Muñoz-Caro (CAB)
J. R. Pardo (CAB)
M. R. de Armas (OAN)

Acknowledgements

The symposium is sponsored and supported by the IAU Division VI (Interstellar Matter) and IAU Commissions No. 14 (Atomic and Molecular Data), No. 15 (Comets and Minor Planets), No. 16 (Physical Studies of Planets and Satellites), No. 28 (Galaxies), and No. 40 (Radio Astronomy).

The Local Organizing Committee operated under the auspices of the
Centro de Astrobiología (CAB) & Observatorio Astronómico Nacional (OAN)

Funding by
International Astronomical Union (IAU),
Ministerio de Ciencia y Tecnología (MICINN),
Consejo Superior de Investigaciones Científicas (CSIC),
Instituto Nacional de Técnica Aeoroespacial (INTA),
Instituto Geográfico Nacional (IGN),
Universidad Castilla-La Mancha (UCLM),
and
MICINN-CONSOLIDER-INGENIO program
(grant ASTROMOL CSD2009–00038)
is gratefully acknowledged.

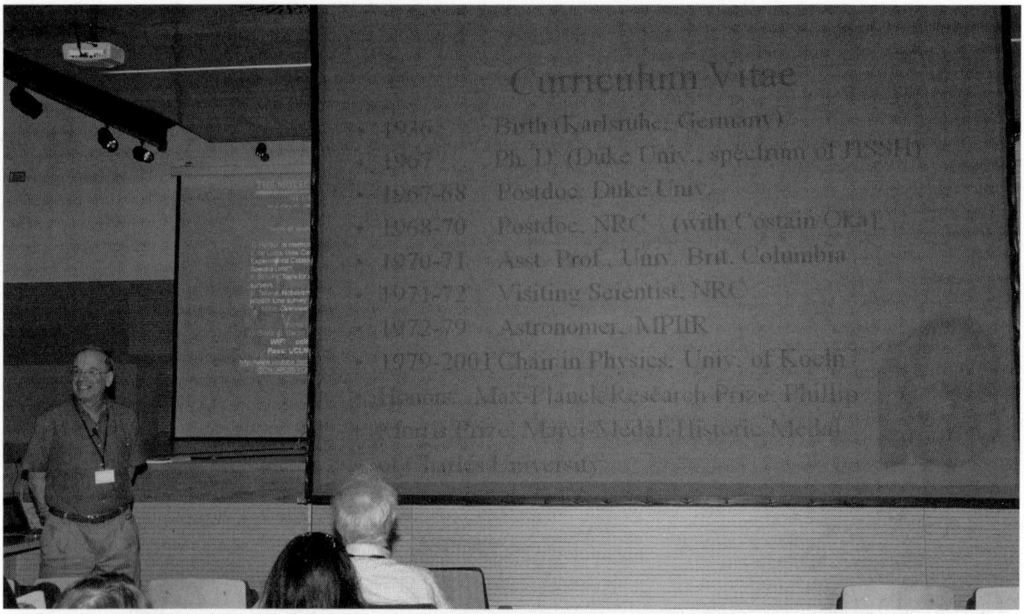
Curriculum Vitae
Birth (Karlsruhe, Germany)
1967 Ph.D. (Duke Univ.,
1967-68 Postdoc. Duke Univ.
1968-70 Postdoc. NRC (with Costain Oka)
1970-71 Asst. Prof., Univ. Brit. Columbia
1971-72 Visiting Scientist, NRC
1972-79 Astronomer, MPIfR
1979-2001 Chair in Physics, Univ. of Koeln
Max-Planck Research Prize

WIFI
Pass:
radical.
A + e →

Super-Earths & Life
A interdisciplinary puzzle
Biomarkers of Habitable
Worlds

REGISTRATION
(P-Z)
REGISTRATION
(P-Z)
THE MOLECULAR UNIVERSE
THE MOLECULAR UNIVERSE

REGISTRATION
(A-G)

Participants

Kinsuk **Acharyya**, S N Bose National Centre for Basic Sciences, India	acharyya@bose.res.in
Gilles **Adande**, University of Arizona, U.S.A.	gadande@email.arizona.edu
Marcelino **Agúndez**, Observatoire de Paris, France	marcelino.agundez@obspm.fr
Yuri **Aikawa**, Kobe Univ., Japan	aikawa@kobe-u.ac.jp
Eiji **Akiyama**, Ibaraki University, Japan	09nd401y@mcs.ibaraki.ac.jp
José **Albaladejo Pérez**, Universidad de Castilla-La Mancha, Spain	Jose.Albaladejo@uclm.es
Tobias **Albertsson**, MPIA, Heidelberg, Germany	albertsson@mpia.de
Isabel **Aleman**, IAG-USP, University of Sao Paulo, Brazil	isabel@astro.iag.usp.br
Conel **Alexander**, Carnegie Institution of Washington. Dept. Terrestrial Magnet, U.S.A.	alexander@dtm.ciw.edu
José **Alonso**, Dpto. Química-Física e Inorgánica. Fac. Ciencias, U. Valladolid, Spain	jlalonso@qf.uva.es
Hector **Alvaro Galué**, FOM Rijnhuizen, Netherlands	alvaro@rijnhuizen.nl
Diana **Andrade**, Universidade do Vale do Paraíba, Brazil	dianaufrj@gmail.com
Carina **Arasa**, Leiden Institute of Chemistry, Leiden University, Netherlands	arasa@strw.leidenuniv.nl
Hector **Arce**, Yale University, U.S.A.	hector.arce@yale.edu
Giambattista **Aresu**, Kapteyn Astronomical Institute, Netherlands	giambattista.aresu@gmail.com
Gisela **B. Esplugues**, Centro de Astrobiología, Spain	espluguesbg@cab.inta-csic.es
Rafael **Bachiller**, Observatorio Astronómico Nacional, Spain	r.bachiller@oan.es
Bernabe **Ballesteros**, Universidad de Castilla - La Mancha, Spain	bernabe.ballesteros@uclm.es
Jeanette **Bast**, Leiden Observatory, Netherlands	bast@strw.leidenuniv.nl
Tom **Bell**, Centro de Astrobiología, Spain	tab@cab.inta-csic.es
Arnold **Benz**, ETH Zurich, Switzerland	benz@astro.phys.ethz.ch
Alicia **Berciano Alba**, ASTRON, Netherlands	berciano@astron.nl
Astrid **Bergeat**, Institut des Sciences Moléculaires / Université Bordeaux 1, France	a.bergeat@ism.u-bordeaux1.fr
Edwin **Bergin**, University of Michigan, U.S.A.	ebergin@umich.edu
Olivier **Berné**, Leiden Observatory, Netherlands	berne@strw.leidenuniv.nl
Mathieu **Bertin**, LPMAA Université Pierre et Marie Curie, France	mathieu.bertin@upmc.fr
Daniela **Bettoni**, INAF - Osservatorio Astronomico di Padova, Italy	daniela.bettoni@oapd.inaf.it
Ludovic **Biennier**, Institut de Physique, Univ. de Rennes 1 - CNR, France	ludovic.biennier@univ-rennes1.fr
Veronica **Bierbaum**, University of Colorado, U.S.A.	veronica.bierbaum@colorado.edu
Thomas **Bisbas**, University College London, U.K.	tb@star.ucl.ac.uk
Chiara **Biscaro**, Basel University, Switzerland	chiara.biscaro@unibas.ch
Luca **Bizzocchi**, Centro de Astronomia e Astrofisica da Universidade de Lisboa, Portugal	bizzocchi@oal.ul.pt
John **Black**, Chalmers University of Technology, Sweden	John.Black@chalmers.se
Geoffrey **Blake**, Division of Geological & Planetary Sciences California Inst, U.S.A.	gab@gps.caltech.edu
Dominique **Bockelée-Morvan**, LESIA, Observatoire de Paris, France	dominique.bockelee@obspm.fr
Anna **Boluda**, Freelance journalist	anna.boluda@gmail.com
Sylvain **Bontemps**, Observatoire de Bordeaux, France	bontemps@obs.u-bordeaux1.fr
Jean-Baptiste **Bossa**, Leiden Observatory, Netherlands	bossa@strw.leidenuniv.nl
Sandrine **Bottinelli**, IRAP / Univ. Toulouse, France	sandrine.bottinelli@cesr.fr
Stefano **Bovino**, Department of Chemistry, University of Rome Sapienza, Italy	s.bovino@caspur.it
Rogier **Braakman**, Santa Fe Institute, U.S.A.	rogier@santafe.edu
Wendy **Brown**, University College London, U.K.	w.a.brown@ucl.ac.uk
Joanna **Brown**, Harvard-Smithsonian Center for Astrophysics, U.S.A.	joannabrown@cfa.harvard.edu
Simon **Bruderer**, MPE Garching, Germany	simonbruderer@gmail.com
Christof **Buchbender**, Instituto de Radio Astronomía Milimétrica , Spain	buchbend@iram.es
Gemma **Busquet**, INAF-Istituto di Fisica dello Spazio Interplanetario, Italy	gemma.busquet@ifsi-roma.inaf.it
Denis **Buechel**, 1. Physical Institute: University of Cologne, Germany	buechel@ph1.uni-koeln.de
Sylvie **Cabrit**, LERMA - Observatoire de Paris, France	sylvie.cabrit@obspm.fr
Jan **Cami**, The University of Western Ontario, Canada	jcami@uwo.ca
Alessandra **Candian**, University of Nottingham, U.K.	alessandra.candian@gmail.com
André **Canosa**, Institut de Physique de Rennes, France	andre.canosa@univ-rennes1.fr
Cristina **Cappa**, Facultad de Ciencias Astronómicas y Geofísicas, Argentina	ccappa@fcaglp.unlp.edu.ar
Fabio **Carelli**, University of Rome 'Sapienza', Italy Department of Chemistry, Italy	f.carelli@caspur.it
Paul **Carroll**, Emory University, U.S.A.	pbcarro@emory.edu
Miguel **Carvajal**, Universidad de Huelva, Spain	miguel.carvajal@dfa.uhu.es
Paola **Caselli**, University of Leeds, U.K.	p.caselli@leeds.ac.uk
Andrew **Cassidy**, Department of Physics and Astronomy, Aarhus University, Denmark	amc@phys.au.dk
Marcelo **Castellanos Beltrán**, Centro de Astrobiología (CSIC-INTA), Spain	mcastell@cab.inta-csic.es
Jose **Cernicharo**, CAB. INTA-CSIC, Spain	jcernicharo@cab.inta-csic.es
Qiang **Chang**, The Ohio state University, U.S.A.	chang.442@osu.edu
German **Chaparro Molano**, Kapteyn Astronomical Institute, Netherlands	chaparro@astro.rug.nl
Steven **Charnley**, NASA Goddard Space Flight Center, U.S.A.	Steven.B.Charnley@nasa.gov
Luis **Chavarria**, LAB/OASU, France	chavarria@obs.u-bordeaux1.fr
Jo-Hsin **Chen**, Jet Propulsion Laboratory / Caltech, U.S.A.	Jo-Hsin.Chen@jpl.nasa.gov
Isabelle **Cherchneff**, Basel University, Switzerland	isabelle.cherchneff@unibas.ch
Jean **Chiar**, SETI Institute, U.S.A.	jchiar@seti.org
Thierry **Chiavassa**, Aix-Marseille University, France	thierry.chiavassa@univ-provence.fr
Yunhee **Choi**, Kapteyn Astronomical Institute / SRON, Netherlands	y.choi@astro.rug.nl
Helen **Christie**, University College London, U.K.	hc@star.ucl.ac.uk
Lauren **Cleeves**, University of Michigan, U.S.A.	cleeves@umich.edu
Claudia **Comito**, Max-Planck-Institut fuer Radioastronomie, Germany	ccomito@mpifr.de
Cesar **Contreras**, NASA-Ames Research Center, U.S.A.	cesar.contreras@nasa.gov
Amanda **Cook**, NASA Ames Research Center (ORAU), U.S.A.	amanda.m.cook@nasa.gov
Carla Maria **Coppola**, Physics & Astronomy Dept, Univ. Bari, Italy	carla.coppola@chimica.uniba.it
Martin **Cordiner**, NASA Goddard Space Flight Center, U.S.A.	martin.cordiner@nasa.gov
Diane **Cormier**, Service d'Astrophysique, CEA Saclay, France, France	diane.cormier@cea.fr
Anne **Coupeaud**, IRAP, France	anne.coupeaud@cesr.fr
Audrey **Coutens**, IRAP, France	coutens@cesr.fr
Nick **Cox**, Institute of Astronomy, K.U.Leuven, Belgium	nick@ster.kuleuven.be
Kyle **Crabtree**, University of Illinois, U.S.A.	kcrabtr2@illinois.edu
Nathan **Crockett**, University of Michigan, U.S.A.	ncrocket@umich.edu
Gustavo **Cruz Díaz**, Centro de Astrobiología, INTA-CSIC, Spain	gcruz2012@gmail.com
Attila **Csaszar**, Institute of Chemistry, Eotvos University, Hungary	csaszar@chem.elte.hu
Sara **Cuadrado Prado**, Centro de Astrobiología, INTA-CSIC, Spain	quiscp@gmail.com
charles **Cunningham**, Herzberg Institute of Astrophysics, Canada	charles.cunningham@nrc.ca
Herma **Cuppen**, Inst. for Molecules and Materials, Radboud University, Netherlands	hcuppen@science.ru.nl
Steven **Cuylle**, Leiden Observatory, Netherlands	cuylle@strw.leidenuniv.nl
Alexander **Dalgarno**, Harvard University, U.S.A.	adalgarno@cfa.harvard.edu
Fabien **Daniel**, Centro de Astrobiología, INTA-CSIC, Spain	fabien128@gmail.com
Ankan **Das**, Indian Centre for Space Physics, India	ankan.das@gmail.com
Thijs **de Graauw**, Joint ALMA Observatory, Chile	tdegraau@alma.cl

Massimo **De Luca**, LERMA - CNRS - ENS France, France massimo.de.luca@lra.ens.fr
Frank **De Lucia**, Ohio State University, U.S.A. fcd@mps.ohio-state.edu
Pablo **de Vicente**, Observatorio Astronómico Nacional, Spain pablo.devicente@oan.es
Bernard **de Vries**, Instituut voor Sterrenkunde, Leuven, Belgium, Belgium ben@ster.kuleuven.be
Leen **Decin**, Instituut voor Sterrenkunde, Leuven, Belgium, Belgium Leen.Decin@ster.kuleuven.be
Carolin **Dedes**, Institute for Astronomy ETH Zurich, Switzerland carolin.dedes@astro.phys.ethz.ch
Rainer **Dietsche**, University of Basel, Switzerland rainer.dietsche@unibas.ch
Francois **Dulieu**, LERMA. Observatoire de Paris et Univ. Cergy Pontoise, France francois.dulieu@obspm.fr
Anne **Dutrey**, Laboratoire d'Astrophysique de Bordeaux , France anne.dutrey@obs.u-bordeaux1.fr
Yves **Ellinger**, UPMC, France ellinger@lct.jussieu.fr
Pierre **Encrenaz**, Observatoire de Paris, France pierre.encrenaz@obspm.fr
Christian **Endres**, University of Cologne, Germany endres@ph1.uni-koeln.de
Sonia **Erattupuzha Joseph**, University of Basel, Switzerland sonia.erattupuzha@unibas.ch
Antonio **Escobar**, Centro de Astrobiología, INTA-CSIC, Spain jimenezea@inta.es
Jarken **Esimbek**, Urumqi Observatory, National Astronomical Observatories, CAS, China jarken@uao.ac.cn
Aaseef **Esmail**, University of Basel, Switzerland aaseef.esmail@unibas.ch
Mireya **Etxaluze**, Harvard-Smithsonian Center for Astrophysics, U.S.A. metxaluz@cfa.harvard.edu
Edith **Falgarone**, LERMA/LRA. Ecole Normale Supérieure, France edith.falgarone@lra.ens.fr
Cecile **Favre**, Department of Physics and Astronomy, Aarhus University, Denmark favre@phys.au.dk
Edith **Fayolle**, Leiden Observatory, Netherlands fayolle@strw.leidenuniv.nl
Nicole **Feautrier**, LERMA, Paris Observatory, France Nicole.Feautrier@obspm.fr
Gleb **Fedoseev**, Leiden Observatory, Netherlands fedoseev@strw.leidenuniv.nl
Paul **Feldman**, Johns Hopkins University, U.S.A. pdf@pha.jhu.edu
Alicia **Fernández Clavero**, Centro de Astrobiología, INTA-CSIC, Spain fernandezca@cab.inta-csic.es
Paola **Fiadino**, Centro de Astrobiología, INTA-CSIC, Spain fiadinop@cab.inta-csic.es
David **Field**, University of Aarhus, Denmark dfield@phys.au.dk
Jean-Hugues **Fillion**, LPMAA, Université Pierre et Marie Curie, France jean-hugues.fillion@upmc.fr
Nicolas **Flagey**, Jet Propulsion Laboratory, U.S.A. nflagey@jpl.nasa.gov
Francesco **Fontani**, ESO & IRAM, France fontani@iram.fr
Kevin **France** , University of Colorado, U.S.A. kevin.france@colorado.edu
Pau **Frau**, Institut de Ciències de l'Espai, IEEC-CSIC, Spain frau@ice.cat
Douglas **Friedel**, University of Illinois, U.S.A. friedel@astro.illinois.edu
Emil **Friis**, Institute of Physics and Astronomy, Aarhus University, Denmark ef06@phys.au.dk
Asunción **Fuente**, Observatorio Astronómico NAcional, Spain a.fuente@oan.es
Kenji **Furuya**, Kobe university, Japan furuya@stu.kobe-u.ac.jp
Giuseppe **Galletta**, Universitá di Padova, Italy, Italy giuseppe.galletta@unipd.it
Marina **Galvagni**, Institute for Theoretical Physics, University of Zurich, Switzerland galva@physik.uzh.ch
Oscar **Gálvez**, Instituto de Estructura de la Materia, CSIC, Spain ogalvez@iem.cfmac.csic.es
Yu **Gao**, Purple Mountain Observatory, China yugao@pmo.ac.cn
Guido **Garay**, Universidad de Chile, Chile guido@das.uchile.cl
Santiago **García-Burillo**, Observatorio Astronómico Nacional, Spain s.gburillo@oan.es
Anibal **García-Hernandez**, Instituto de Astrofisica de Canarias (IAC), Spain agarcia@iac.es
Pedro **García-Lario**, Herschel Science Centre ESAC/ESA, Spain Pedro.Garcia-Lario@sciops.esa.int
Iryna **Garkusha**, University of Basel, Switzerland iryna.garkusha@unibas.ch
Robin **Garrod**, Cornell University, U.S.A. rgarrod@astro.cornell.edu
Lisseth **Gavilan**, Observatoire de Paris et Université de Cergy-Pontoise, Peru lisseth.gavilan@obspm.fr
Thomas **Geballe**, Gemini Observatory, U.S.A. tgeballe@gemini.edu
Wolf **Geppert**, Stockholm University, Sweden wgeppert@hotmail.com
Maryvonne **Gerin**, LERMA, France maryvonne.gerin@ens.fr
Erika **Gibb**, University of Missouri - St. Louis, U.S.A. gibbe@umsl.edu
Thomas **Giesen**, I. Physikalisches Institut, University of Cologne, Germany giesen@ph1.uni-koeln.de
Simon **Glover**, ITA/ZAH, Heidelberg, Germany sglover@ita.uni-heidelberg.de
David **Gobrecht**, University of Basel, Switzerland Dave.Gobrecht@stud.unibas.ch
Benjamin **Godard**, Centro de Astrobiología, INTA-CSIC, Madrid, Spain, Spain bgodard@cab.inta-csic.es
Marie **Godard**, Institut d'Astrophysique Spatiale, France marie.godard@ias.u-psud.fr
Javier **Goicoechea**, Centro de Astrobiología, INTA-CSIC, Madrid, Spain, Spain jr.goicoechea@cab.inta-csic.es
Paul **Goldsmith**, JPL, California Inst. of Technolog, U.S.A. Paul.F.Goldsmith@jpl.nasa.gov
Beatriz **González García**, ESAC, ESA, Villafranca del Castillo, Spain beatriz.gonzalez@sciops.esa.int
Fedor **Goumans**, Leiden University, Netherlands fedor.goumans@gmail.com
Pierre **Gratier**, Institut de RadioAstronomie Millimetrique (IRAM), France gratier@iram.fr
Michel **Guélin**, IRAM, France michel.guelin@wanadoo.fr
Stephane **Guilloteau** Laboratoire d'Astrophysique de Bordeaux , France guilloteau@obs.u-bordeaux1.fr
Varun **Gupta**, University of Basel, Switzerland varun.gupta@unibas.ch
Antoine **Gusdorf**, Max Planck Institute for Radioastronomy, Germany agusdorf@mpifr-bonn.mpg.de
Joseph **Guss**, University of Leiden, Netherlands guss@strw.leidenuniv.nl
Viviana **Guzmán**, LERMA-ENS, France viviana.guzman@lra.ens.fr
Lizette **Guzmán-Ramirez**, JBCA, University of Manchester, U.K. lizette.ramirez@postgrad.manchester.ac.u
Emilie **Habart**, Institut d'Astrophysique Spatiale, France emilie.habart@ias.u-psud.fr
Tetsuya **Hama**, Institute of Low Temperature Science, Hokkaido Univ., Japan hama@lowtem.hokudai.ac.jp
Mark **Hammonds**, University of Nottingham, U.K. pcxmh3@nottingham.ac.uk
Nanase **Harada**, The Ohio State University, U.S.A. nanaseh@mps.ohio-state.edu
Daniel **Harsono**, Leiden Observatory, Netherlands harsono@strw.leidenuniv.nl
Tatsuhiko **Hasegawa**, Academia Sinica Inst. of Astron. and Astronphys., Taiwan hasegawa@asiaa.sinica.edu.tw
Eric **Hébrard**, Laboratoire d'Astrophysique de Bordeaux, France hebrard@obs.u-bordeaux1.fr
Jonathan **Henshaw**, University of Leeds, U.K. phy5jh@leeds.ac.uk
Eric **Herbst**, The Ohio State University, U.S.A. herbst.6@osu.edu
Gregory **Herczeg**, Max-Planck-Institut für Extraterrestriche Physik, Germany gregoryh@mpe.mpg.de
Fabrice **Herpin**, LAB/OASU, France herpin@obs.u-bordeaux1.fr
Cinthya **Herrera**, Institut d'Astrophysique Spatiale, France cherrera@ias.u-psud.fr
Víctor **Herrero**, Instituto de Estructura de la Materia, CSIC, Spain vherrero@iem.cfmac.csic.es
Hiroshi **Hidaka**, Institute of Low Temperature Science, Hokkaido University, Japan hidaka@lowtem.hokudai.ac.jp
Ugo **Hincelin**, Laboratoire d'Astrophysique, Univ. Bordeaux, France Ugo.Hincelin@obs.u-bordeaux1.fr
Michiel **Hogerheijde**, Leiden Observatory, Netherlands michiel@strw.leidenuniv.nl
Sacha **Hony**, AIM/CEA-Saclay, France sacha.hony@cea.fr
Jiri **Horacek**, Charles University in Prague Institute of Theoretical Physics, Czech Republic jiho@matfyz.cz
Liv **Hornekaer**, Aarhus University, Dept. Physics and Astronomy, Denmark liv@phys.au.dk
Subhon **Ibadov**, Institute of Astrophysics TAS, Tajikstan ibadovsu@yandex.ru
John **Ilee**, The University Of Leeds, U.K. pyjdi@leeds.ac.uk
Katharina **Immer**, Max-Planck-Institut für Radioastronomie; CfA, U.S.A. kimmer@mpifr-bonn.mpg.de
Natalia **Inostroza**, CSCIC - Daza Valdés, Spain natalia.inostrozapino@gmail.com
Sergio **Ioppolo**, Leiden Observatory, Leiden University, Netherlands ioppolo@strw.leidenuniv.nl
William **Irvine**, University of Massachusetts, U.S.A. irvine@astro.umass.edu

Karoliina **Isokoski**, Leiden Observatory, Netherlands	isokoski@strw.leidenuniv.nl
Donald **Jennings**, Goddard Space Flight Center, U.S.A.	donald.e.jennings@nasa.gov
Elena **Jiménez**, Universidad de Castilla-La Mancha, Spain	Elena.Jimenez@uclm.es
Izaskun **Jimenez-Serra**, Harvard-Smithsonian CfA, U.S.A.	ijimenez-serra@cfa.harvard.edu
Doug **Johnstone**, NRC Canada, Herzberg Inst. of Astrophysics, Canada	douglas.johnstone@nrc-cnrc.gc.ca
Jes **Jørgensen**, Centre for Star and Planet Formation, Univ. Copenhagen, Denmark	jeskj@nbi.dk
Eric **Josselin**, Lab. Univers & Particules de Montpellier - Université, France	eric.josselin@univ-montp2.fr
Pavol **Jusko**, Charles University in Prague, Czech Republic	pavol.jusko@gmail.com
Sergei **Kalenskii**, Astro Space Center, Lebedev Physical Institute, Russia	kalensky@asc.rssi.ru
Lisa **Kaltenegger**, MPIA, Germany	lkaltenegger@cfa.harvard.edu
Juris **Kalvans**, Ventspils University College & University of Latvia, Latvia	kalvans@lu.lv
Mikhel **Kama**, University of Amsterdam, The Netherlands	M.Kama@uva.nl
Inga **Kamp**, Kapteyn Institute, University of Groningen, Netherlands	kamp@astro.rug.nl
Agata **Karska**, Max-Planck-Institut für extraterrestrische Physik, Germany	agata.karska@gmail.com
Michael **Kaufman**, San Jose State University, U.S.A.	Michael.Kaufman@sjsu.edu
Maja **Kazmierczak**, Nicolaus Copernicus University, Poland, Poland	maja.kazmierczak@gmail.com
Yeghis **keheyan**, CNR, Italy	yeghis.keheyan@uniroma1.it
Theo **Khouri**, Universiteit van Amsterdam, Netherlands	theokhouri@yahoo.com.br
Hyo Jeong **Kim**, The University of Texas at Austin, U.S.A.	hyojeong@astro.as.utexas.edu
Stephen **Klippenstein**, Argonne National Laboratory, U.S.A.	sjk@anl.gov
Lars **Kluge**, I. Physikalisches Institut, University of Cologne, Germany	kluge@ph1.uni-koeln.de
Kirsten **Knudsen**, Chalmers University of Technology, Sweden	kraiberg@chalmers.se
Monika **Koerber**, I. Physikalisches Institut, University of Cologne, Germany	koerber@ph1.uni-koeln.de
Akira **Kouchi**, Institute of Low Temperature Science, Hokkadio University, Japan	kouchi@lowtem.hokudai.ac.jp
Jacek **Krelowski**, Torun CfA, Copernicus University, Torun, Poland	jacek@astri.uni.torun.pl
Lars **Kristensen**, Leiden Observatory, Netherlands	kristensen@strw.leidenuniv.nl
Jay **Kroll**, Emory University, U.S.A.	jay.kroll3@gmail.com
Yi-Jehng **Kuan**, National Taiwan Normal University, Taiwan	kuan@ntnu.edu.tw
Sun **Kwok**, The University of Hong Kong, China	sunkwok@hku.hk
Jacob **Laas**, Emory University, U.S.A.	jlaas@emory.edu
Fred **Lahuis**, SRON Netherlands Institute for Space Research, Netherlands	F.Lahuis@sron.nl
Julien **Lambert**, Lab. Univers & Particules, Montpellier Univ., France	julien.lambert@univ-montp2.fr
Thanja **Lamberts**, Leiden Observatory, University of Leiden, Netherlands	lamberts@strw.leidenuniv.nl
William **Langer**, Jet Propulsion Laboratory, Caltech, U.S.A.	William.Langer@jpl.nasa.gov
Grigorii **Larionov**, Astro Space Center of Lebedev Physical Institute., Russia	grigorii99@mail.ru
Franck **Le Petit**, LUTH - Paris Observatory, France	Franck.LePetit@obspm.fr
Sébastien **Le Picard**, Inst. de Physique, Univ. Rennes 1, France	sebastien.le-picard@univ-rennes1.fr
Dae-Hee **Lee**, Korea Astronomy and Space science Institute, South Korea	dhlee@kasi.re.kr
Jeong-Eun **Lee**, Kyung Hee University, South Korea	jelee.jeongeunlee@gmail.com
Bertrand **Lefloch**, IPAG, France	lefloch@obs.ujf-grenoble.fr
Jean Louis **Lemaire**, Obs. Paris & Univ. Cergy-Pontoise, France	jean-louis.lemaire@obspm.fr
Pierre **Lesaffre**, CNRS,LERMA/ENS/LRA, France	pierre.lesaffre@ens.fr
Silvia **Leurini**, MPIfR, Germany	sleurini@mpifr.de
Francois **Levrier**, LERMA-ENS, France	francois.levrier@ens.fr
Xiaohu **Li**, Leiden Observatory, Netherlands	lixiaohu@dicp.ac.cn
Johan **Lindberg**, Centre Star & Planet Formation Univ. Copenhaguen, Denmark	jolindbe@gmail.com
Harold **Linnartz**, Leiden Obs., Sackler Lab. for Astrophysics, Netherlands	linnartz@strw.leidenuniv.nl
Dariusz **Lis**, California Institute of Technology, U.S.A.	dcl@caltech.edu
Sheng-Yuan **Liu**, Academia Sinica, Inst. of Astron. and Astrophys., Taiwan	syliu@asiaa.sinica.edu.tw
Alexandra **Lockwood**, California Institute of Technology, U.S.A.	alock@gps.caltech.edu
Edo **Loenen**, Leiden Observatory, Netherlands	loenen@strw.leidenuniv.nl
Jean-Christophe **Loison**, ISM, Université de Bordeaux 1-CNRS, France	jc.loison@ism.u-bordeaux1.fr
Robin **Lombaert**, Instituut voor Sterrenkunde KULeuven, Belgium	robinl@ster.kuleuven.be
Savino **Longo**, Dept Chemistry Univ. - IMIP CNR Bari, Italy	savino.longo@ba.imip.cnr.it
Juan **López**, Universidad de Valladolid, Spain	jclopez@qf.uva.es
Ana **López-Sepulcre**, IPAG - Grenoble, France	sepulcre@arcetri.astro.it
Jonathan **Lunine**, University of Arizona. Dept. of Planetary Sciences, U.S.A.	jlunine@lpl.arizona.edu
Roxana **Lupu**, University of Pennsylvania, U.S.A.	lroxana@sas.upenn.edu
James **Lyons**, UCLA, U.S.A.	jimlyons@ucla.edu
Suzanne **Madden**, CEA, Saclay SAp, France	suzanne.madden@cea.fr
José **Madiedo**, University of Huelva, Spain	madiedo@uhu.es
Jean-Pierre **Maillard**, Institut d'Astrophysique de Paris, France	maillard@iap.fr
Liton **Majumdar**, Indian Center for Space Physics, India	liton.icsp@gmail.com
Margot **Mandy**, University of Northern British Columbia, Canada	mandy@unbc.ca
Nuria **Marcelino**, National Radio Astronomy Observatory, U.S.A.	nmarceli@nrao.edu
Sergio **Martín**, European Southern Observatory, Chile	smartin@eso.org
Jesus **Martín-Pintado**, Centro de Astrobiologia, INTA-CSIC, Spain	jmartin@cab.inta-csic.es
Belén **Maté**, Instituto de Estructura de la Materia, CSIC, Spain	bmate@iem.cfmac.csic.es
Christopher **Materese**, SETI, U.S.A.	christopher.k.materese@nasa.gov
Rainer **Mauersberger**, Joint ALMA Observatory, Chile	rmauersb@alma.cl
Carolyn **McCoey**, University of Waterloo, Canada	cmccoey@astro.uwaterloo.ca
Daniel **McElroy**, Queen's University Belfast, Northern Ireland	dmcelroy03@qub.ac.uk
Brett **McGuire**, Emory University, U.S.A.	brett.mcguire@emory.edu
Alan **McLoughlin**, Queen's University Belfast, U.K.	amcloughlin929@qub.ac.uk
Gwendolyn **Meeus**, Universidad Autonoma de Madrid, Spain, Spain	gwen@gmeeus.be
Tom **Megeath**, University of Toledo, U.S.A.	tommegeath@gmail.com
Rowin **Meijerink**, Leiden Observatory, Netherlands	meijerink@strw.leidenuniv.nl
Gary **Melnick**, Harvard-Smithsonian Center for Astrophysics, U.S.A.	gmelnick@cfa.harvard.edu
Vito **Mennella**, INAF-Osservatorio Astronomico di Capodimonte, Italy	mennella@na.astro.it
Karl **Menten**, Max-Planck-Institut fuer Radioastronomie, Germany	kmenten@mpifr.de
Pablo **Merino**, Centro de Astrobiologia INTA-CSIC, Spain	merinomp@cab.inta-csic.es
Elisabetta **Micelotta**, Goddard Space Flight Center, U.S.A.	elisabetta.micelotta@nasa.gov
Stefanie **Milam**, NASA Goddard Space Flight Center, U.S.A.	stefanie.n.milam@nasa.gov
Tom **Millar**, Queen's University Belfast, U.K.	Tom.Millar@qub.ac.uk
Faviola **Molina**, Inst. Theoretical Astrophysics, Heidelberg Univ., Germany	fmolina@ita.uni-heidelberg.de
Raquel **Monje**, California Institute of Technology , U.S.A.	raquel@caltech.edu
Julien **Montillaud**, IRAP - Toulouse, France, France	julien.montillaud@cesr.fr
Thierry **Montmerle**, Institut d'Astrophysique de Paris, France	montmerle@iap.fr
Bhaswati **Mookerjea**, Tata Institute of Fundamental Research , India	bhaswati@tifr.res.in

Oscar **Morata**, ASIAA, Taiwan omorata@asiaa.sinica.edu.tw
Stefanie **Muehle**, Joint Institute for VLBI in Europe, Netherlands muehle@jive.nl
Sebastien **Muller**, Onsala Space Observatory, Sweden mullers@chalmers.se
Guillermo **Muñoz Caro**, Centro de Astrobiología, , INTA-CSIC, Spain munozcg@cab.inta-csic.es
Andra **Muntean**, Queen's University Belfast, U.K. e.muntean@qub.ac.uk
Holger **Müller**, I. Physikalisches Institut Universität zu Köln, Germany hspm@ph1.uni-koeln.de
Zsofia **Nagy**, Kapteyn Astronomical Institute & SRON, Netherlands nagy@astro.rug.nl
Adam **Nagy**, University of Basel, Switzerland adam.nagy@unibas.ch
Binukumar **Nair**, The Open University, UK, U.K. b.g.nair@open.ac.uk
Tac **Nakajima**, National Astronomical Observatory of Japan, Japan nakajima@nro.nao.ac.jp
David **Neufeld**, Johns Hopkins University, U.S.A. neufeld@pha.jhu.edu
Hideko **Nomura**, Kyoto University, Japan nomura@kusastro.kyoto-u.ac.jp
Michel **Nuevo**, NASA Ames Research Center, SETI, U.S.A. michel.nuevo-1@nasa.gov
Yasuhiro **Oba**, Institute of Low Temperature Science, Hokkaido Univ., Japan oba@lowtem.hokudai.ac.jp
Karin **Oberg**, Harvard-Smithsonian Center for Astrophysics, U.S.A. koberg@cfa.harvard.edu
Angela **Occhiogrosso**, University College of London, U.K. angela.occhiogrosso.10@ucl.ac.uk
Hans **Olofsson**, Onsala Space Observatory, Sweden hans.olofsson@chalmers.se
Volker **Ossenkopf**, I. Physikalisches Institut der Universität zu Köln, Germany ossk@ph1.uni-koeln.de
Susana **Pacheco Vázquez**, Institut Planétologie et Astrophys., Grenoble, France pachecos@ujf-grenoble.fr
Laurent **Pagani**, LERMA & UMR8112 Observatoire de Paris & CNRS, France laurent.pagani@obspm.fr
Aina **Palau**, Institut de Ciències de l'Espai, CSIC-IEEC, Spain palau@ieec.uab.es
MariaElisabetta **Palumbo**, INAF - Osservatorio Astrofisico di Catania, Italy mepalumbo@oact.inaf.it
Juan **Pardo**, Centro de Astrobiología, INTA-CSIC, Spain jrpardo@cab.inta-csic.es
Andrey **Paska**, The University of Manchester, U.K. andrey.paska@postgrad.manchester.ac.uk
Francoise **Pauzat**, CNRS/UPMC, France pauzat@lct.jussieu.fr
John **Pearson**, Jet Propulsion Laboratory, California Institute of Technolog, U.S.A. John.C.Pearson@jpl.nasa.gov
Els **Peeters**, University of Western Ontario & SETI Institute, Canada epeeters@uwo.ca
Yezhe **Pei**, Ohio State University, U.S.A. pei.10@buckeyemail.osu.edu
Ruisheng **Peng**, Caltech Submillimeter Observatory, U.S.A. peng@submm.caltech.edu
Tzu-Cheng **Peng**, Laboratoire d'Astrophysique de Bordeaux, France Tzu-Cheng.Peng@obs.u-bordeaux1.fr
Juan-Pablo **Perez-Beaupuits**, Max-Planck Institut für Radioastronomie, Germany jp@mpifr.de
Francisco **Pérez-Bernal**, Universidad de Huelva, Spain francisco.perez@dfaie.uhu.es
Amélie **Pernet**, UPMC, France pernet@lct.jussieu.fr
Magnus **Persson**, Centre for Star and Planet Formation, Copenhagen University, Denmark magnusp@snm.ku.dk
Carina **Persson**, Onsala space observatory, Chalmers University of Technology, Sweden carina.persson@chalmers.se
Dawn **Peterson**, Harvard-Smithsonian Center for Astrophysics, U.S.A. dpeterson@cfa.harvard.edu
Andreea **Petric**, Caltech, U.S.A. ap@astro.caltech.edu
Annemieke **Petrignani**, Leiden Observatory, Netherlands a.petrignani@umail.leidenuniv.nl
Thomas **Phillips**, California Institute of Technology, U.S.A. tgp@phobos.caltech.edu
Göran **Pilbratt**, ESA Astrophysics Missions Div., Research and Scientific Sup, Netherlands gpilbratt@rssd.esa.int
Paolo **Pilleri**, Observatorio Astronómico Nacional & Centro de Astrobiologia , Spain paolo.pilleri@gmail.com
Sergio **Pilling**, UNIVAP, Brazil sergiopilling@yahoo.com.br
Julien **Pilme**, Laboratoire de Chimie Théorique. UPMC, France pilme@lct.jussieu.fr
Nuria **Pinol Ferrer**, Department of Astronomy, Stockholm University, Sweden npi@astro.su.se
Ciro **Pinto**, SRON - Netherlands Institute for Space Research, Netherlands c.pinto@sron.nl
Claire **Pirim**, LADIR UMR 7075/CNRS UPMC, France pirim@spmol.jussieu.fr
Rene **Plume**, University of Calgary, Canada plume@ras.ucalgary.ca
Linda **Podio**, Kapteyn Astronomical Institute, Netherlands podio@astro.rug.nl
Thomas **Posch**, University of Wien, Austria, thomas.posch@univie.ac.at
Fabrizio **Puletti**, University College London, U.K. uccafpu@ucl.ac.uk
Guillermo **Quintana-Lacaci**, Instituto de Radioastonomía Milimétrica, Spain quintana@iram.es
Mary **Radhuber**, Emory University, U.S.A. mradhub@emory.edu
Ranjini **Raghunandan**, University of Basel, Switzerland r.raghunandan@unibas.ch
Mark **Rawlings**, Joint ALMA Observatory, Chile mrawling@alma.cl
Pilar **Redondo**, Universidad de Valladolid, Spain predondo@qf.uva.es
Miguel **Requena Torres**, Max-Planck-Institute fuer Radioastronomie, Germany mrequena@mpifr-bonn.mpg.de
Anita **Richards**, UK ALMA Regional Centre, JBCA, University of Manchester, U.K. amsr@jb.man.ac.uk
Paul **Rimmer**, Ohio State University, U.S.A. pbrimmer@mps.ohio-state.edu
Denise **Riquelme**, Instituto de Radioastonomía Milimétrica, Spain riquelme@iram.es
Víctor M. **Rivilla**, Centro de Astrobiología INTA-CSIC, Spain rivilla@cab.inta-csic.es
Ricardo **Rizzo**, Centro de Astrobiología, INTA-CSIC, Spain ricardo@cab.inta-csic.es
Julia **Roberts**, Centro de Astrobiología, INTA-CSIC, Spain robertsj@cab.inta-csic.es
Arturo **Rodríguez Franco**, Centro de Astrobiología, INTA-CSIC, Spain rodriguezfa@cab.inta-csic.es
Evelyne **Roueff**, Observatoire de Paris, France evelyne.roueff@obspm.fr
Gaël **Rouillé**, Universität Jena - Max-Planck-Institut für Astronomie, Germany gael.rouille@uni-jena.de
Georgij **Rudnitskij**, Sternberg Astronomical Institute, Moscow State University, Russia gmr@sai.msu.ru
Nicola **Sacchi**, INAF-IFSI, Italy nicola.sacchi@ifsi-roma.inaf.it
Alfonso **Sáiz López**, Lab. Ciencias de la Atmósfera y el Clima, Spain a.saiz-lopez@ciac.jccm-csic.es
Nami **Sakai**, The University of Tokyo, Japan nami@taurus.phys.s.u-tokyo.ac.jp
Takeshi **Sakai**, The University of Tokyo, Japan sakai@ioa.s.u-tokyo.ac.jp
Farid **Salama**, NASA-Ames Research Center, U.S.A. Farid.Salama@nasa.gov
Dinalva **Sales**, UFRGS - Universidade Federal do Rio Grande do Sul, Brazil dinalva.aires@ufrgs.br
Francisco **Salgado**, Leiden Observatory, Netherlands salgado@strw.leidenuniv.nl
Colette **Salyk**, University of Texas at Austin, McDonald Observatory, U.S.A. csalyk@astro.as.utexas.edu
Irene **San José García**, Leiden Observatory, Netherlands sanjose@strw.leidenuniv.nl
Carmen **Sánchez Contreras**, Astrobiology Center, INTA-CSIC, Spain csanchez@cab.inta-csic.es
Alvaro **Sanchez-Monge**, INAF-Osservatorio Astrofisico di Arcetri, Italy asanchez@am.ub.es
Rogelio **Sánchez Verdasco**, Centro de Astrobiología, INTA-CSIC, Spain sanchezvr@cab.inta-csic.es
Scott **Sandford**, NASA-Ames Research Center, U.S.A. Scott.A.Sandford@nasa.gov
Gina **Santangelo**, INAF - Osservatorio Astronomico di Roma, Italy gina.santangelo@oa-roma.inaf.it
Arkaprabha **Sarangi**, Basel University, Switzerland arkaprabha.sarangi@unibas.ch
Peter **Sarre**, The University of Nottingham , U.K. Peter.Sarre@Nottingham.ac.uk
Peter **Schilke**, I. Physikalisches Institut der Universität zu Köln, Germany schilke@ph1.uni-koeln.de
Eric **Schindhelm**, University of Colorado, U.S.A. eric.schindhelm@colorado.edu
Stephan **Schlemmer**, I. Physikalisches Institut, Universität zu Köln, Germany schlemmer@ph1.uni-koeln.de
Miroslaw **Schmidt**, Nicolaus Copernicus Astronomical Center, Poland schmidt@ncac.torun.pl
Nicola **Schneider**, CEA Saclay, France nschneid@cea.fr
Dmitry **Semenov**, Max Planck Institute of Astronomy, Heidelberg, Germany semenov@mpia.de
Maria Luisa **Senent Diez**, Instituto de Estructura de la Materia, CSIC, Spain senent@iem.cfmac.csic.es
Bethmini **Senevirathne**, University of Gothenburg, Sweden bethmini@chem.gu.se
Sachindev **Shenoy**, NASA Ames Research Center, U.S.A. sachindev.s.shenoy@nasa.gov

Takashi **Shimonishi**, Dept. Astronomy, Graduate School of Science, Japan shimonishi@astron.s.u-tokyo.ac.jp
Ivar **Shmeld**, Ventspils University College, VIRAC, Latvia ivarss@venta.lv
Ian **Sims**, Universite de Rennes1, France ian.sims@univ-rennes1.fr
Howard **Smith**, Harvard-Smithsonian Cente for Astrophysics, U.S.A. hsmith@cfa.harvard.edu
Ian **Smith**, University of Cambridge, U.K. i.w.m.smith@bham.ac.uk
Rachel **Smith**, University of California, Los Angeles, U.S.A. rsmith@ess.ucla.edu
Theodore **Snow**, University of Colorado, U.S.A. tsnow@casa.colorado.edu
Amiel **Sternberg**, Tel Aviv University, Israel amiel@astro.tau.ac.il
Yu-Nung **Su**, Academia Sinica, Institute of Astronomy and Astrophysics, Taiwan ynsu@asiaa.sinica.edu.tw
Ewelina **Szymanska**, The Open University , U.K. e.szymanska@open.ac.uk
Marian **Szymczak**, Torun Centre for Astronomy, Nicolaus Copernicus University, Poland msz@astro.uni.torun.pl
Mario **Tafalla**, Observatorio Astronómico Nacional, Spain m.tafalla@oan.es
Shuro **Takano**, Nobeyama Radio Observatory, National Astronomical Observator, Japan stakano@nro.nao.ac.jp
Dahbia **Talbi**, LUPM CNRS/Université de Montpellier II, France dahbia.talbi@univ-montp2.fr
Jonathan **Tan**, University of Florida, U.S.A. jt@astro.ufl.edu
Vianney **Taquet**, Institut de Planétologie et d'Astrophysique de Grenoble, France vianney.taquet@obs.ujf-grenoble.fr
Konstantinos **Tassis**, Jet Propulsion Laboratory / Caltech, U.S.A. ktassis@jpl.nasa.gov
Belén **Tercero**, Centro de Astrobiología, INTA-CSIC, Spain terceromb@cab.inta-csic.es
David **Teyssier**, ESAC, Spain dteyssier@sciops.esa.int
Patrice **Theule**, University of Provence, France patrice.theule@univ-provence.fr
Wing-Fai **Thi**, Institut de Planetologie et d'Astrophysique de Grenoble, France thiw@obs.ujf-grenoble.fr
Sven **Thorwirth**, I. Physikalisches Institut Universität zu Köln, Germany sthorwirth@ph1.uni-koeln.de
John **Thrower**, Aarhus University, Denmark thrower@phys.au.dk
Alexander **Tielens**, Leiden Observatory, Netherlands tielens@strw.leidenuniv.nl
Giovanna **Tinetti**, University College London. Dept. of Physics and Astronomy , U.K. g.tinetti@ucl.ac.uk
Samuel **Tisi**, Department of Physics and Astronomy, University of Waterloo, Canada stisi@sciborg.uwaterloo.ca
Thomas **Townsend**, Universidad de Castilla- La Mancha, Spain thomasmtownsend@gmail.com
Josep **Trigo Rodríguez**, Institute of Space Sciences, CSIC-IEEC, Spain trigo@ieec.uab.es
Matthew **Troutman**, University of Missouri - St. Louis, U.S.A. troutmanm@umsl.edu
Antonio **Usero**, Observatorio Astronómico Nacional , Spain a.usero@oan.es
Nathalie **Vaeck**, Université Libre de Bruxelles, Belgium nvaeck@ulb.ac.be
Nienke **van der Marel**, Leiden Observatory, Netherlands nmarel@strw.leidenuniv.nl
Floris **van der Tak**, SRON, Netherlands vdtak@sron.nl
Ewine **van Dishoeck**, Leiden Observatory/MPE, Netherlands ewine@strw.leidenuniv.nl
Peter **van Hoof**, Royal Observatory of Belgium, Belgium p.vanhoof@oma.be
Huib Jan **van Langevelde**, JIVE & Sterrewacht Leiden, Netherlands langevelde@jive.nl
Magda **Vasta**, INAF OA Arcetri - Florence, Italy mvasta@arcetri.astro.it
Charlotte **Vastel**, IRAP/Université de Toulouse, France vastel@cesr.fr
Anton **Vasyunin**, The Ohio State University, U.S.A. vasyunin@mps.ohio-state.edu
Tatiana **Vasyunina**, The Ohio State University, U.S.A. vasyunina@mpia.de
Luis **Velilla Prieto**, CAB, Astrobiology Center, INTA-CSIC., Spain lvelilla@cab.inta-csic.es
Thangasamy **Velusamy**, Jet Propulsion Laboratory, Caltech, U.S.A. velusamy@jpl.nasa.gov
Gianfranco **Vidali**, Syracuse University, U.S.A. gvidali@syr.edu
Ruud **Visser**, University of Michigan, U.S.A. visserr@umich.edu
Serena **Viti**, Department of Physics and Astronomy, University College Lond, U.K. sv@star.ucl.ac.uk
Stéphane **Vranckx**, Free University of Brussels, Belgium stvranck@ulb.ac.be
Valentine **Wakelam**, Labo. d'Astrophysique, Univ. Bordeaux, France wakelam@obs.u-bordeaux1.fr
Malcolm **Walmsley**, INAF-Osservatorio di Arcetri, Italy walmsley@arcetri.astro.it
Catherine **Walsh**, Queen's University Belfast, U.K. catherine.walsh@qub.ac.uk
Adam **Walters**, IRAP, France walters@cesr.fr
Susanne **Wampfler**, Inst. for Astronomy, ETH Zurich, Switzerland wampfler@astro.phys.ethz.ch
Kuo-Song **Wang**, Leiden Observatory, Netherlands kswang@strw.leidenuniv.nl
Naoki **Watanabe**, Inst. of Low Temperature Science, Hokkaido Univ., Japan watanabe@lowtem.hokudai.ac.jp
Yoshimasa **Watanabe**, The University of Tokyo, Japan nabe@taurus.phys.s.u-tokyo.ac.jp
Susanna **Widicus Weaver**, Emory University, U.S.A. susanna.widicus.weaver@emory.edu
Dmitri **Wiebe**, Institute of Astronomy of the RAS, Russia dwiebe@inasan.ru
Eva **Wirström**, NASA Goddard Space Flight Center, U.S.A. eva.s.wirstrom@nasa.gov
Markus **Wittkowski**, ESO, Germany mwittkow@eso.org
Mark **Wolfire**, University of Maryland, U.S.A. mwolfire@astro.umd.edu
Paul **Woods**, University College London, U.K. dr.paul.woods@gmail.com
Bohan **Wu**, University of Basel, Switzerland bohan.wu@unibas.ch
Ronin **Wu**, Service d'Astrophysique CEA Saclay, France ronin.wu@cea.fr
Friedrich **Wyrowski**, Max Planck Institute for Radioastronomy, Bonn, Germany wyrowski@mpifr-bonn.mpg.de
Satoshi **Yamamoto**, Department of Physics, The University of Tokyo, Japan yamamoto@phys.s.u-tokyo.ac.jp
Umut **Yildiz**, Leiden Observatory, Netherlands yildiz@strw.leidenuniv.nl
Walter **Yvart**, LERMA, France walter.yvart@obspm.fr
Laimons **Zacs**, University of Latvia, Latvia zacs@latnet.lv
Natalia R. **Zelmanovitch**, Centro de Astrobiología, INTA-CSIC, Spain n.zelman@cab.inta-csic.es
Junfeng **Zhen**, Sackler laboratory for astrophysics, leiden observatory, uni, Netherlands zhen@strw.leidenuniv.nl
Jianjun **zhou**, Urumqi Observatory, National Astronomical Observatories, CAS, China zhoujj@uao.ac.cn
Igor **Zinchenko**, Institute of Applied Physics, Russian Academy of Sciences, Russia zin@appl.sci-nnov.ru
Emilie-Laure **Zins**, LADIR Université Pierre et Marie Curie, France zins@spmol.jussieu.fr

Address by the Scientific Organizing Committee

Over 435 participants from 31 countries gathered in Toledo, Spain to attend IAU Symposium 280, entitled The Molecular Universe, which took place from 30 May to 3 June 2011 at the Technological Campus of the University of Castilla-La Mancha in Toledo, Spain. This is the main worldwide conference in the field of astrochemistry, held every ~5 years, and covering all areas in which molecules are found, from Solar system to the highest redshift galaxies. This breadth of topics sets the IAU symposia series apart from other meetings in the field.

The Local Organizing Committee, chaired by J. Cernicharo and R. Bachiller, organized both the scientific and structural aspects of the meeting very well. Almost all possible logistic problems were handled amicably by M. Castellanos together with the other members of the LOC. The cultural mecca that is Toledo added a sense of awe and excitement to the symposium. The large size of the meeting did not interfere with the proceedings in any way. A large number of questions were asked of speakers, who, given their relative youth and diversity, brought many different viewpoints to the discussions. The three dedicated 2.5-hr. poster sessions were very well attended and enriched the experience of the participants. Informal conversations, held at intermissions from the speaking program, and during the poster sessions, were many and spirited. The large number of younger scientists at the meeting was quite impressive, and confirmed that the field of astrochemistry is entering a period of rapid growth led by new and exceedingly powerful telescopes.

The scientific organization of the symposium was undertaken by a very active IAU Working Group on Astrochemistry, under the sponsorship of IAU Commission 34 (Division VI), with co-sponsorship provided by Division VI (Interstellar Matter), Division VIII (Galaxies and the Universe), Division X (Radio Astronomy) as well as Commissions 51 (Bio-astronomy), 36 (Theory of Stellar Atmospheres), and 14 (Atomic and Molecular Data). There have now been six IAU symposia on astrochemistry, starting with the one held in India (1985; IAU Symposium 120). Later meetings in the series were held in Brazil (1991; IAU Symposium 150), the Netherlands (1996; IAU Symposium 178); South Korea (1999; IAU Symposium 197), and California, USA (2005; IAU Symposium 231). Each symposium has been larger than its predecessor, showing that astrochemistry is becoming a larger and more diverse community.

The scientific program of the symposium was divided into three parts: invited and review talks, contributed talks, and poster presentations. The Scientific Organizing Committee, which was the Working Group on Astrochemistry, democratically elected the speakers who gave contributed talks among the many applicants. Overall, there were 42 invited and review talks, 32 contributed talks, and nearly 360 posters. In the oral program were three sessions on new results from the Herschel Space Observatory labeled Herschel hot results, as well as a panel discussion entitled On to ALMA. The panel members adjudicated a contest in which young investigators competed to win a prize for the best and next best projects for ALMA with the constraint of at most 10 hours observing time. There were three large poster sessions, and awards were given to the best posters in each of the three from personal funds by E. van Dishoeck. During the third poster session, there were also computer demonstrations of databases.

After brief words of welcome by E. van Dishoeck, chair of the SOC, and J. Cernicharo, chair of the LOC, the 4.5-day oral program started with a general introduction on the molecular universe by A. Tielens, which was followed by a session on star formation. This field has become broader since the last astrochemistry symposium, and observational talks concerning stages of both low-mass and high-mass star formation were given, as was a theoretical talk on a new class of models that combine hydrodynamics with chemical simulations in the formation of protostellar cores. This session was followed by the first session of hot results from Herschel, which emphasized observations of water vapor, molecules in protostellar shocks, and a wide spectral survey toward Orion KL.

Astrochemistry certainly extends to planetary studies, including solar system objects. A session on these objects was held on the first day of the meeting, starting with a review talk on the chemistry of the solar system, including the origin of water on Earth, which was followed by talks on comets, meteorites, and the atmospheres of Titan and Saturn. The power of sample return missions to solar system bodies was emphasized. The second day of the meeting started with a session on evolved stars, in which supernova chemistry was also discussed. Talks on the molecular evolution from AGB stars to planetary nebulae, the role of time-dependent anionic chemistry (involving negatively-charge molecules) in IRC+10216, and the detection of fullerenes in assorted environments rounded out the session. Complex molecules are well known in IRC+10216 and other selected circumstellar sources, so this session merged well with the next one on star formation and complex molecules. Here observations of complex molecules were discussed in a variety of objects, along with current gas-grain simulations as well as possible future simulations involving the use of stochastic methods to improve the surface chemistry occurring in granular icy mantles.

Astrochemistry is based on the laboratory and theoretical study of basic atomic and molecular processes, and two sessions were held on this subject. The first concerned gas-phase processes, where a review talk was given on gas-phase reactions as a function of temperature, followed by a talk concerning the theory of low-temperature reactions, and one on experimental studies on the rates of reactions involving anions and how they relate to the observations of such species in various sources. The second day of the meeting ended with another Herschel hot topic session, highlighted by the report of an unambiguous detection of molecular oxygen in the interstellar medium.

The topic of protoplanetary disks occupied the first group of speakers on Wednesday, with talks on the phenomenal developments in observations at a variety of wavelengths ranging from the millimeter to the far-UV and an emphasis on interferometry. Modeling was also discussed, as was the chemical history of molecules from the hot core to the disk stage. Another session on basic molecular processes followed, this one emphasizing surface processes in the laboratory and in space. Much progress has been made during the last decade in this field, but there is still a great need for further laboratory studies before robust interstellar chemical simulations including surface processes can be constructed.

Although most of astrochemistry still revolves around galactic sources, the field of extragalactic astrochemistry will receive a big boost with the onset of ALMA observations. So, it was quite appropriate to have a session on extragalactic astrochemistry, which was held on Thursday morning. This field was understood to include the early universe, so talks on early chemistry were included along with a talk on extragalactic line surveys. It is impressive to see spectra of extragalactic sources with similar complexity to those found in galactic star-forming regions three decades ago! Next in line was the explosive topic of exoplanets and their atmospheres, which will occupy more and more astrochemists as more is learned about planetary atmospheres. Talks on observations, atmospheric models and their chemistry, as well as biomarkers of habitable worlds were included. The inclusion of astrobiology is a sign that this field is gaining importance and certainly overlaps with areas of astrochemistry such as the formation of complex molecules. The final session on Thursday concerned the tools of analysis and databases. Starting with a brief memorial to the late astronomer and astrochemist Gisbert Winnewisser, this session included talks on how to reduce the problem of unidentified lines in hot cores, on various tools for analysis of spectral surveys, on a legacy line survey from the Nobeyama telescope, and on databases and their uses. The session ended with the panel discussion discussed previously.

The last day of the symposium started with a session nominally on diffuse clouds and photon-dominated regions (PDRs). The role of turbulence in diffuse clouds was discussed, as was a controversial candidate for a carrier of several diffuse interstellar bands (H_2C_3). A talk on both PDRs and XDRs (X-ray dominated regions) was given as was a more general talk on diffuse interstellar bands. A number of aspects of the PAH hypothesis were touched upon. Finally, the complex nature of the central molecular zone of our galaxy, as seen through the infra-red spectrum of H_3^+, was explored. Next came the third of the Herschel hot topic sessions, which included talks on observations of diffuse clouds in the spiral arms of the Milky Way, carbon chemistry in translucent clouds, and the detection of C_3 in envelopes of star-forming regions. The detection of the reactive ions OH^+ and H_2O^+ in a variety of sources was an exceptionally interesting topic. The oral program was concluded with an exceedingly thoughtful summary of the field, past, present, and future, by J. Black.

The reader of this volume will find a cornucopia of riches concerning the state of astrochemistry before the fundamental changes that will occur when observations using the ALMA interferometer add greatly to our knowledge of sources throughout the universe. By the time of the next astrochemical symposium, much progress will have been made and the field will have grown both in size and, we trust, in understanding.

Eric Herbst
Secretary, Scientific Organizing Committee
Toledo, June 2011

The Molecular Universe
Proceedings IAU Symposium No. 280, 2011
José Cernicharo & Rafael Bachiller, eds.

doi:10.1017/S1743921311024811

In Memoriam: Gisbert Winnewisser

Eric Herbst[1], Pierre Encrenaz[2], and Rafael Bachiller[3]
[1]University of Virginia and Ohio State University, USA
[2]Observatoire de Paris, France
[3]Observatorio Astronómico Nacional, Spain

Gisbert Winnewisser, emeritus Professor of Physics at the University of Köln, passed away in March 2011 after a long illness. His dedication to molecular spectroscopy in the laboratory and in the interstellar medium, coupled with his very influential voice for molecular science will be extremely difficult to replace.

Gisbert Winnewisser

Born in Karlsruhe on September 7, 1936, Gisbert moved with his family to a farm in the Black Forest where they lived until 1946. Gisberts brother Manfred, his twin sister Ingrid, and Gisbert lived simply and independently with their mother, after their father died near Moscow. This tragic event marked Gisbert deeply. The family went back to Karlsruhe in 1946 where Gisbert graduated from gymnasium in 1956. He then became a physics student at the Technical University of Karlsruhe and graduated with a Master of Science degree in 1963. His brother Manfred had then become a post-doctoral fellow in the laboratory of Walter Gordy, a well-known molecular spectroscopist whose book on the subject, co-authored with Cook, is still known as a bible in the subject. Gisbert decided to follow the same career path, and became a PhD student under Professor Gordy at Duke University in Durham, North Carolina where he stayed until 1968. In the interim, he married Helga ter Jung in 1964 in Karlsruhe, and the couple stayed in Durham where their son Carsten was born in 1966. Gisbert defended his PhD thesis in 1967, the topic of which is the rotational-torsional spectrum of the internal rotor molecule HSSH, an intriguing molecule on which he kept working for decades, along with similar species such as HOOH and HSOH.

Gisbert and his family then moved to Ottawa, where Gisbert had been appointed a postdoctoral associate in the world- famous spectroscopic laboratory of Gerhard Herzberg. He worked there until 1972 with a short stay in Vancouver at the University of British Columbia. This was the time of the first observations of interstellar molecules: H_2O, NH_3, H_2CO, CO, CN, OCS, etc. Gisbert realized what could be done in this new field with his knowledge of spectroscopy, and he accepted a permanent position at the Max Planck Institute for Radioastronomy in Bonn as a molecular radio astronomer. He spent thousands of hours observing the interstellar medium and discovered formic acid, HNC, and vinyl cyanide, as well as isotopomers of HCCCN. He also participated in the earliest studies of TMC1, now the best-known cold interstellar cloud core. Gisbert also collaborated with his brother Manfred, then professor of physics at Justus Liebig University in Giessen, and his sister-in-law Brenda in assorted spectroscopic projects. He started to defend the idea of a large submillimeter antenna in space as a member of ESA's Astrophysics Working Group.

In 1979, he was offered and accepted a chair in Physics at the University of Köln, and he became Director of the I. Physikalishes Institute. Gisbert made of this Institute one of the worlds leading centers for high-resolution molecular spectroscopy, developing new instrumentation for ground and space applications, including acousto-optical detectors, local oscillators, SIS junctions, and new submillimeter spectrometers. His laboratory provided unique equipment for the satellites SWAS, Herschel, and for ASTRO (the Antarctic Submillimeter Telescope and Remote Observatory) . The transfer of carcinotron techniques from Russia to his laboratory, where it was used with his OROTRON spectrometer, kept alive this technology.

Gisbert built a submillimeter telescope in Köln and then moved it to the Gornergrat, near Zermatt in southern Switzerland. The telescope, known as KOSMA, helped to educate many students in high-frequency radio astronomy. Indeed, the number of spectroscopy and astronomy students who received and continue to receive their degrees at the I. Physikalisches Institute is exceedingly impressive.

Gisbert was a major figure on the international scene and helped to promote interstellar molecular astronomy. The series of Zermatt conferences on the latest discoveries in the interstellar medium and in the laboratory attracted many well-known and younger spectroscopists and molecular astronomers. He also co-founded the LEA Hires program (Laboratoire Européen Associé for High Resolution Spectroscopy), a European association of spectroscopy laboratories in Germany, France, Belgium and Spain, specializing in complementary techniques and theories.

Gisbert is the author or coauthor of more than 400 publications, and has received many international awards, including the Honorary Medal of the University of Helsinki (1985), the Max Planck Research Prize (1993), the Philip Morris Prize (1994), the Honorary Medal of the Ioannes Marcus Marci Spectroscopic Society (1996), and the Historical Medal of the Charles University in Prague (2000). NASA awarded him a Group Achievement Award for his contribution to the SWAS satellite.

On behalf of all Gisbert Winnewisser's students, colleagues, associates, coauthors, and friends, we dedicate this volume to his memory.

The Molecular Universe
Proceedings IAU Symposium No. 280, 2011
José Cernicharo & Rafael Bachiller, eds.

doi:10.1017/S1743921311024823

The Molecular Universe

A. G. G. M. Tielens[1]

[1]Leiden Observatory, Leiden University,
PO Box 9513, NL-2300RA, Leiden, the Netherlands
email: tielens@strw.leidenuniv.nl

Abstract. Over the last 20 years, we have discovered that we live in a molecular Universe: A Universe with a rich and varied organic inventory; A Universe where molecules are abundant and widespread; A Universe where molecules play a central role in key processes that dominate the structure and evolution of galaxies; A Universe where molecules provide convenient thermometers and barometers to probe local physical conditions; A Universe where molecules can work together to form such complex species as you and me. Understanding the origin and evolution of interstellar and circumstellar molecules is thus key to understanding the Universe around us and our place in it and has become a fundamental goal of modern astrophysics. This review focuses on the organic inventory and the chemical processes that may play a role in stablishing molecular complexity in regions of planet formation.

1. Introduction

The field of astrochemistry is heavily driven by new observational tools that have become available over the last 20 years; in particular, space based missions that have opened up the IR and submillimeter window at an ever accelerating pace. Our progress in understanding the Molecular Universe is greatly aided by close collaborations between astronomers, molecular physicists, astrochemists, spectroscopists, and physical chemists who work together in loosely organized networks. In this review, I will sketch the progress that we have made over the last 20 years and outline some of the challenges that are facing us. The focus will be on understanding the unique and complex organic inventory of regions of star and planet formation that may well represent the prebiotic roots to life. Astrochemistry is much broader that this and key questions such as "What is the role of molecules in the evolution of the Universe ?" and "How can we use molecules to study the Universe ?" are by necessity not addressed here.

Over the last twenty years, we have seen a major shift in our thinking on life in the Universe. First, ground-based Doppler and transient studies have lead to the discovery of $\sim$ 1700 exoplanets (and counting). Sixty of these resemble the Earth and at the end of the Kepler mission, we can expect to know some 200 terrestrial planets. Of the 1700 suspected planets, some 50 are in the habitable zone of their star and 5 of those are Earth-like. This census is very much incomplete and we are only starting to address the general architecture of planetary systems and how this compares to the Solar system, but already it is clear that planetary systems are common. Second, in biology, it has become clear that life can adapt itself to very extreme conditions. Life can thrive at very low and at very high temperatures, under very alkaline, very acidic, and very saline conditions, in very arid regions (and wet regions, duh): Extremophiles are everywhere and consequently the habitable zone of planetary systems may be quite large. Liquid water – often taken as the defining factor of the habitable zone – may also be more widespread than previously thought and moons such as Europa may have an "'internal" ocean in which (primitive) life can be hypothesized. In the Solar system, the presence of liquid water has been established – albeit in the past – on Mars and some meteoritic parent bodies. Hence, conditions suitable for life may be widespread in the Universe. Third, the oldest preserved cells date back some 2.5 billion years and a number of possible, or perhaps even likely, indicators for the existence of microbial fossils have been found in rocks that date back to some 3.5 billion years ago. That latter date is very close to the age of the late heavy bombardment when the interaction between Jupiter and Saturn reordered the Solar system. The impacts associated with this bombardment would have sterilized the Earth (over and over again) and so this timeline would leave only $\sim$ 200 Myr for the emergence of the first cells on Earth. In all, these new developments have completed the Copernican/Darwinian revolution and the consensus is now that life is widespread in the Universe.

With this shift in paradigm, astrochemistry and its relationship to the prebiotic roots of life have emerged as key areas of research in astronomy. This can be taken at two levels: The first question is: "What are the main reservoirs of carbon in regions of planet formation ?" Or alternatively, "How and in what form was carbon delivered to planets in the habitable zone ?" Obviously, if most of the carbon is in the form of very volatile species (e.g., CO, CH_4), then delivery has to be by cometary or icy-asteroidal bodies presumably during a "late heavy bombardment" phase in the assemblage of planetary systems. Alternatively, most of the carbon may be locked up in macromolecular moieties and incorporation into planets may then occur "in-situ" during the growth of planetary embryo's. These different reservoirs result thus in a very different history and very different carbon content of newly formed planets. The second and deeper level that we can examine in the relationship between astrochemistry and the origin of life is: "Does astrochemistry "prime" regions of planet formation for the emergence of life ?" Or phrased alternatively: "Does astrochemistry jump start life's chemistry by delivering the right molecular building blocks (e.g., amino acids) that can be readily assembled in any Darwinian pond ?" In this scenario, astrochemistry takes the place of the Urey-Miller experiment†, which actually may not have been very relevant for conditions on the early Earth. We can identify a number of scenarios differing in the importance of astrochemistry for the origin of life: Did the Universe "deliver" 1) carbon which got converted into life's building blocks on the early Earth (Urey-Miller); 2) a primed organic mix (e.g., H_2CO, HCN) which then were converted into e.g., amino acids on the early Earth; 3) specific molecules with well defined biological functions (e.g., amino acids, lipids) which were immediately assembled into cells after being dropped into Darwin's little pond.

The focus of this review is on the main reservoirs of molecular carbon in regions of star and planet formation – CO and PAHs – and the astrochemical processes that "release" this carbon and make it available for the formation of species that might be of interest from a prebiotic point of view. Much carbon may be in the form of soot particles, solid kerogen-like macromolecular structures or otherwise frozen out of active chemistry and these reservoirs are of no concern to this review except to note that if these are the reservoirs life grew from, astrochemistry has little prebiotic relevance.

2. Overview

Much of the carbon in regions of star and planet formation is locked up in very stable species that do not readily partake in chemical reactions. The story of interstellar organic

† In the Urey-Miller experiment, the discharge converts simple molecular feedstock in formaldehyde and cyanide that are subsequently converted into amino acids through Strecker synthesis in water.

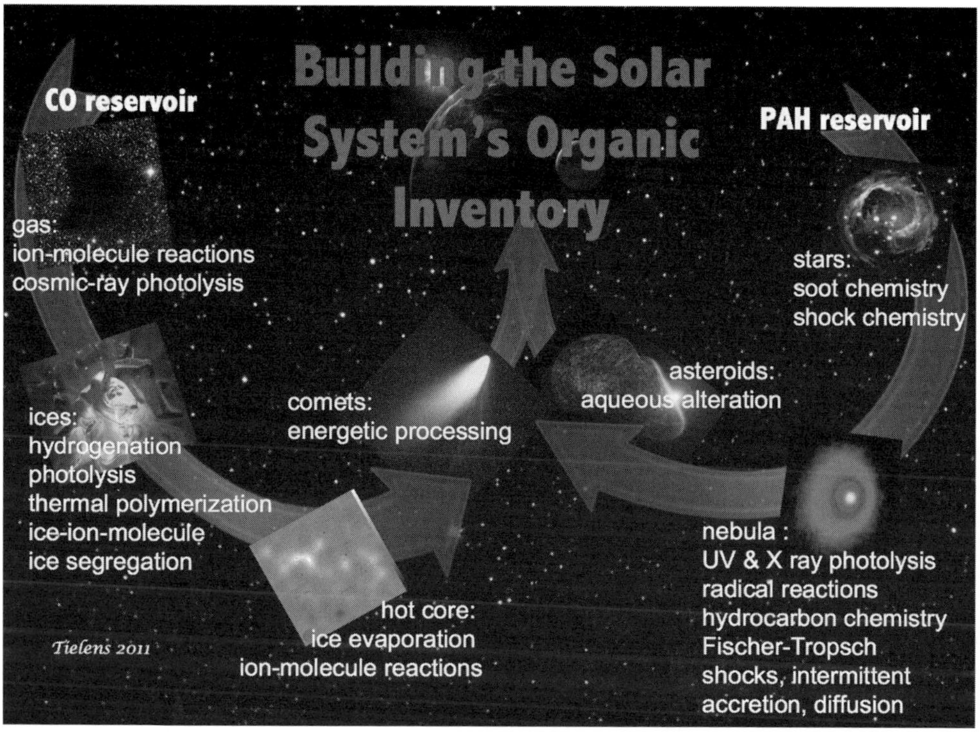

Figure 1. The organic inventory of the Solar System derives from a vast array of processes acting in a wide range of environments. Globally, two independent routes can be recognized. The first one builds up complex species from small radicals and starts with CO in dark clouds. Ion-molecule reactions, grain surface chemistry, and photoprocessing converts the main gaseous reservoir, CO, into complex species. The other route breaks down very complex species (e.g., PAHs) injected into the interstellar medium by stars into smaller and smaller species. Eventually, the species produced by either of these two chemical routes can become part of planetesimals and cometesimals in a protoplanetary disk environment which deliver this organic inventory to the nascent planets in the habitable zone.

chemistry is therefore by necessity a story that starts with breaking the carbon out of these species. The two main molecular reservoirs are CO and Polycyclic Aromatic Hydrocarbon (PAH) molecules. Figure 1 illustrates relevant chemical routes. In dark clouds, ion molecule chemistry converts gaseous carbon into predominantly carbon monoxide. After accretion onto grains, hydrogenation reactions convert this molecule efficiently into formaldehyde and methanol as well as traces of other species in an icy grain mantle. Energetic processing of these ices by UV photons and cosmic rays can convert these molecules into more complex species, while heating by the protostar can initiate thermal polymerization reactions. The newly formed protostar will sublime these ices and fast ions in shocks can sputter some molecules into the gas phase. Subsequent ion-neutral and neutral-neutral reactions in the warm dense gas of the hot core surrounding the protostar may convert, in particular, methanol into more complex species such as dimethyl ether and methyl formate. These can then accrete again into ices and other planetesimals.

The other route starts with the formation of PAHs and other large, complex molecules in stellar ejecta through chemical processes akin to soot chemistry. The injected PAH family is further processed in the interstellar medium by energetic photons and particles, weeding out the less stable members. Further processing can occur in the photosphere of the protoplanetary disk by stellar UV and X-ray photons. Furthermore, in the inner, hot

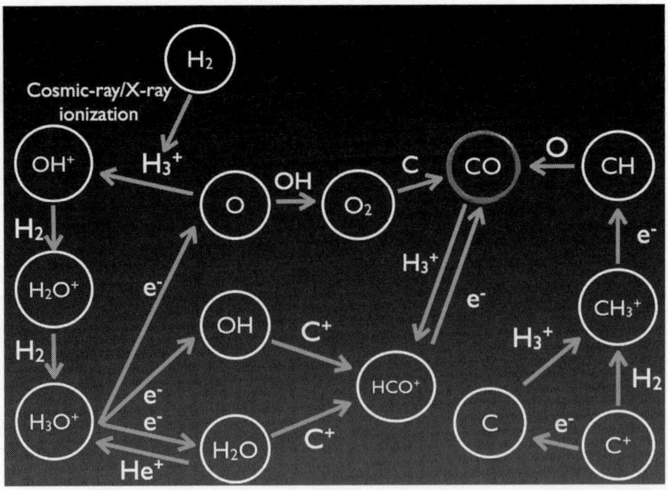

Figure 2. First chemical steps in ion-molecule route convert C predominantly into CO in dark clouds (Herbst & Klemperer 1973; Watson 1973). Figure adapted from Bergin (2011).

and dense (mid-plane) regions, H and OH radicals will breakdown these complex species into acetylene, methane and carbon monoxide and carbon dioxide. Turbulence will diffuse these species and their daughter products – formed through neutral-neutral reactions – throughout the protoplanetary disk. Catalytic Fischer-Tropsch reactions on iron grains may also be of importance. Eventually, the species produced in these two chemical routes can become part of planetesimals and cometesimals where further processing may occur before this organic inventory is delivered to nascent planets.

3. The CO reservoir

3.1. *Ion-molecule chemistry in dark clouds*

Gas phase chemistry in dark clouds is predominantly ion-molecule chemistry where cosmic ray ionization of H_2 forms H_2^+ which reacts on with H_2 to form protonated molecular hydrogen, H_3^+ (fig. 2). The proton is then efficiently handed over to other atoms and molecules until recombination occurs to a small molecule or radical. The upshot of it all is that ion-molecule chemistry converts all elemental carbon into CO on a very short timescale; e.g., essentially for every C-atom, one H_2 has to be ionized ($\tau_{chem} \simeq \mathcal{A}_C/\zeta_{CR} \simeq 2 \times 10^6$ yr with $\mathcal{A}_C$ the carbon abundance and ζ_{CR} the cosmic ray ionization rate. Along the way, the small hydrocarbon radicals may (temporarily) form unsaturated hydrocarbon species such as acetylene and acetylenic derivatives but eventually, and largely, carbon is burned to carbon monoxide.

Molecular complexity is not one of the defining characteristics of ion-molecule chemistry. Observed abundances of even such simple (saturated) species such as CH_3OH and NH_3 cannot be explained by gas phase routes. Also, while acetylenic derivatives such as the cyanopolyynes (eg., HC_3N) are a natural exponent of ion-molecule chemistry, their absolute abundance can only be explained well if there is an efficient way to break the C out of CO. One mechanism to do so is through cosmic-ray-produced UV photons, but the effect is limited: Few of these photons dissociate CO and the C/CO abundance produced this way is only $\simeq 6 \times 10^{-3}$. Reactions with H_3^+ and carbon insertion reactions form then simple hydrocarbons such as CH_4 and C_2H_2 which can react on to other hydrocarbon species including the cyanopolyynes. Much higher abundances of acetylenic species can be produced during early times before all the carbon has been locked up in CO but it is hard to imagine that molecular clouds are typically that young. There are other ways that C can be broken out of CO (eg., shocks) but then shocks have to efficiently process clouds on a large scale. I should point out here that some key aspects are not understood; e.g., the observed variation in the abundance of these acetylenic derivatives and what makes some clouds carbon-chain-rich (Hirota *et al.* 2009) and that may hold important clues.

In summary, with qualifications, gas-phase models of interstellar clouds have been very successful. Among the key successes are explanations for the high abundance of CO, the presence of ions such as H_3^+ and HCO^+, the presence of radicals such as OH and CH, the presence of isomers such as HNC/HCN, the rich array of acetylenic derivatives present, including the cyano-polyynes, and the high observed deuteration levels in species such as DCO^+. Qualitatively, the agreement between gas-phase models and the observations is very impressive. Quantitatively, however, the agreement is wanting. The most extensive reaction networks contain some 4000 reactions among 400 species, and reach an order of magnitude agreement with observed abundances for some 80% of the species. Partly, this is because our chemical knowledge is imperfect: less than 20% of the reactions included have actually been studied in the laboratory. The remainder are educated guesses. It is not even known whether all the important chemical routes are included and whether perhaps some routes have been included that are inhibited at low temperature. This is one area of research where we should focus our attention through close collaborations between astrochemical modelers and chemists (Wakelam *et al.* 2010).

For the story of this review, the bottom line is that in terms of molecular complexity gas phase ion-molecule chemistry is in essence a dead-end, locking the carbon away in a very unreactive species that, from a prebiotic point of view, is unimportant. And this view is confirmed by observations with simple-organics/CO ratios of $\simeq 10^{-3}$ and complex-organics/CO ratios of $\simeq 10^{-5}$.

3.2. *Grain surface chemistry and interstellar ices*

Grain surface chemistry involves predominantly hydrogenation and oxidation reactions and leads to the formation of simple saturated species; e.g., H_2O, CO, CO_2, CH_3OH, H_2CO, CH_4, and NH_3. Focusing here on hydrogenation, atomic hydrogen is mobile, has a long residence time and because it is so light can tunnel through appreciable activation barriers (Tielens & Hagen 1982). Hence, H will "collide" often with coreactants on a grain surface and is very reactive. Reactions of H with CO and H_2CO are particularly interesting here, since this transforms CO first into formaldehyde and then methanol (Fig. 3). Over the past 10 years, these theoretical ideas have been extensively investigated in the laboratory and the viability of this activation route of CO is now well established (Hiraoka *et al.* 1998; Hidaka *et al.* 2004; Fuchs *et al.* 2009; Ioppolo *et al.* 2008). This route can also lead to more complex species because the radical intermediaries, HCO and CH_3O, may react with accreted C (and O) atoms rather than H atoms (Fig. 3). The mobility of C and O atoms on a grain surface are not well known and matrix studies yield puzzling results (Tielens & Allamandola 2011), but if these atoms are mobile a rich chemistry can be initiated. Possible routes towards larger organic compounds are indicated in Figure 3. However, unlike the initial steps towards H_2CO and CH_3OH, these routes towards further chemical complexity have not been tested experimentally and isomerization at intermediate steps may conceivably lead to other products instead. I do want to point out that, while the precise chemical routes are not yet well characterized, they are expected to be chemical specific and isomers can be expected to have very different abundances.

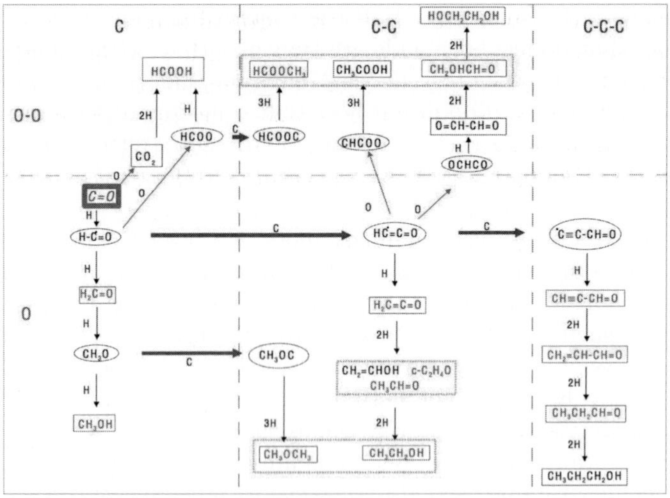

Figure 3. CO accreted on a grain surface is activated through reaction with H. Eventually these H-reactions will form H_2CO and CH_3OH (black arrows). Intermediate radicals such as HCO and CH_3O can react with other atoms leading to larger organic molecules (red and blue arrows) and possible routes towards, among others, HCOOH, CH_3CH_2OH, and CH_3OCH_3 are indicated. Figure taken from Requena-Torres *et al.* (2008), adapted from Tielens & Hagen (1982); Charnley & Rodgers (2009).

Grain surface chemistry models have been very successful in qualitatively explaining the molecular inventory of interstellar ices: the high degree of saturation, the importance of methanol, and the presence of hydrogenated and oxidized species. In addition, the high deuteration of formaldehyde and methanol, including the presence of doubly and triply deuterated species, in hot cores is often – and probably rightfully so – cited in support of the importance of grain surface chemistry. Quantitatively, grain surface chemistry models suffer from a lack of detailed reaction rates and branching ratios but we are now reaching the point where quantitative laboratory studies become feasible and, by the next astrochemistry symposium, we should strive to have reliable, quantitative models in hand.

From the point of view of this review, grain surface chemistry is an efficient way of "activating" the C in CO. Observationally, the ratio of CH_3OH/CO can reach values as high as 1 in interstellar ices. As an aside, the observed strong variation in the methanol abundance in ices is not well understood theoretically. The formation of more complex organic species on grain surfaces is, however, limited by the accretion of atomic C (c.f., Fig. 3) and hence is set by the release of C from CO through cosmic-ray-produced UV photons in the gas phase (c.f., § 3.1) and only traces of complex species (complex-organics/CO$\sim 10^{-3}$) are expected in ice mantles through these processes. Unfortunately, absorption spectroscopy does not allow direct detection of trace species in interstellar ices and this important aspect of grain surface chemistry can only be tested indirectly through e.g., studies of the composition of hot cores where ices have been released into the gas phase (c.f., § 3.3).

3.3. *The chemistry of hot cores*

Hot cores are regions of warm (> 100 K), dense ($> 10^6$ cm^{-3}) gas near massive protostars that are characterized by the presence of high abundances of hydrogenated species, in particular H_2O, NH_3, H_2CO, and CH_3OH (Blake *et al.* 1987; Wright *et al.* 1996). In recent years, regions with similar characteristics as hot cores – but much smaller and therefore labeled hot corino's – have also been discovered around solar-type protostars (Bottinelli *et al.* 2004). Because of the similarity to the composition of interstellar ices, it is now generally accepted that these species reflect the sublimation of interstellar ices by the newly formed star when the dust reaches a temperature of $\sim$ 100 K. The observed high abundance of deuterated species (e.g., HDCO, D_2CO, CH_2DOH, CHD_2OH, CD_3OH, CH_3OD, and NH_2D; Parise *et al.* (2006); Ceccarelli *et al.* (2001)) is also taken as evidence for the importance of ice sublimation in these regions as high deuteration is characteristic for regions of $\sim$ 10 K, e.g., much colder than hot cores. In this view, ices are formed at low temperatures during the prestellar core phase of the molecular cloud and, after the formation of the protostar, the sublimated gas has not yet had the time to "equilibrate" with the gas (e.g., $\tau_{chem} \sim 10^5$ yr). Sputtering in shocks may also be important in releasing these molecules into the gas phase.

Hot cores also contain a multitude of more complex species such as ethyl alcohol, formic acid, dimethyl ether, methyl formate, methyl cyanide, and ethyl cyanide. The origin of these species is less obvious. Initially, like H_2O and CH_3OH, they were thought to represent grain-surface products and putative chemical routes can be identified for species such as ethanol, dimethyl ether, and methyl formate (c.f., Fig. 3; Tielens & Hagen (1982); Charnley & Rodgers (2009)). However, the absence of quantitative reaction rates and branching ratios hampers detailed model efforts. At the same time, the near-impossibility of detecting trace abundances of these species in interstellar ices severly limit observational approaches to assessing the importance of grain surface chemistry routes. In general, though, it can be expected that these species will have very low abundances given the low accretion rate of atomic C. Moreover, the presence of the nitrogen bearing species is difficult to understand in this scenario. As a result, a number of different chemical routes towards molecular complexity have been proposed. These are discussed in the next three subsections.

3.3.1. *Chemistry driven by evaporating ices*

The injection of H_2O, CH_3OH, and NH_3 from the ices into the gas phase can drive a rich chemistry in these warm dense surroundings. Charnley *et al.* (1992) realized that methanol may play a key role in driving molecular complexity because protonated methanol will transfer CH_4^+ groups (rather than H^+). Chemical routes towards the complex (oxide) species observed in hot cores can then be formulated. For example, the following route is available for dimethyl ether,

$$H_3^+ \xrightarrow{H_2O} H_3O^+ \xrightarrow{CH_3OH} CH_3OH_2^+ \xrightarrow{CH_3OH} CH_3OCH_4^+ \xrightarrow{e} CH_3OCH_3 \qquad (3.1)$$

For hot cores rich in N-bearing species, NH_3 is the precursor ice molecule which is channeled to HCN and then to CN in the gas. The reaction of cyanogen with specific hydrocarbons (also presumed to come from ices) leads then to the suite of N-bearing species observed (Charnley *et al.* 1992; Caselli *et al.* 1993). The timescale for the formation of these complex organics is some 3×10^4 yr and after some 10^5 yr gas phase chemistry will have driven all the sublimated organics to CO.

Recent experiments have called these chemical routes into question. First, formation of some of the key intermediaries is inhibited (Horn *et al.* 2004). Second, dissociative electron recombination leads to fragmentation (Geppert *et al.* 20006). In addition, the inferred chemical clock is difficult to reconcile with the rapid dynamical timescales associated with hot corinos.

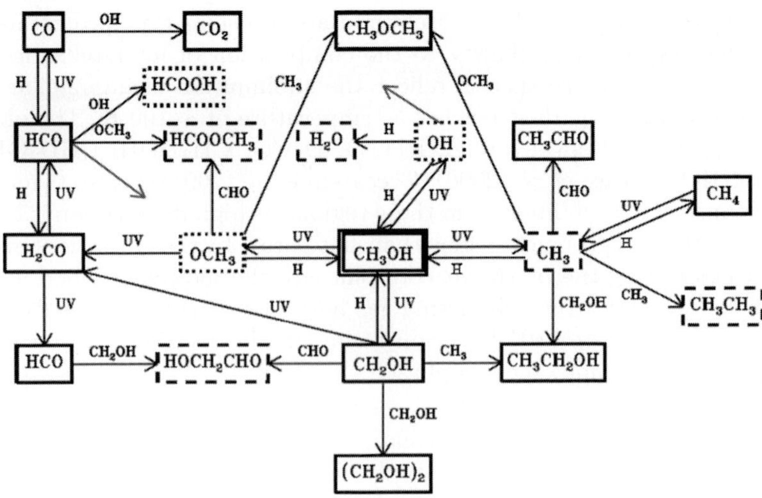

Figure 4. UV photolysis of ices containing H_2O and CH_3OH produces radicls which can react on to form more complex organic molecules. This figure illustrates a number of relevant chemical routes identified based upon laboratory experiments. Figure taken from Öberg *et al.* (2009).

3.3.2. *Energetic processing of interstellar ices*

UV photolysis and/or ion bombardment of interstellar ices will produce (hot) radicals – specifically H & OH – and some of these will react with other molecules while others will rattle around until they lose their excess energy and become trapped in the matrix. Upon warm up, the trapped radicals will diffuse and can recombine with each other, forming more complex species (Fig. 4). Recent laboratory studies have focused on putting these ideas on a firm footing identifying the chemical routes involved (Öberg *et al.* 2009; Hudson *et al.* 2008). Astronomical models have been developed, incorporating ice photochemistry under astrophysically relevant conditions (Garrod *et al.* 2008). These ideas are well worth pursuing further as a number of critical issues are examined. First, this photolysis is unlikely to be very chemical specific. Rather, starting with realistic interstellar ice mixtures, a large number of radicals can be produced which can recombine to a wide variety of species and isomers. It is unclear whether this chemical diversity can be reconciled with the astronomical record which points towards chemical specificity. Second, prolonged exposure to UV or particle radiation will lead to the production of macromolecular structures. Observationally, there is no evidence for such "organic goop" in the interstellar dust record. Also, this removes carbon from the complex organic inventory that is the focus of this review. The presence of NH_3 and H_2CO in the ices may provide an efficient polymerization route for the conversion of molecules to solid materials during warm up (Schutte *et al.* 1993) and this complication has to be evaluated. Finally, HCO is a key, initial product of the photolysis of ices containin CO but the telltale absorption signature of this molecule has never been observed, questioning the relevance of these chemical routes.

3.3.3. *Ions and interstellar ices*

Chemical schemes based upon CH_3 transfer in ice-loaded environments – as originally proposed by Charnley *et al.* (1992) – are very elegant and appealing but the experimental evidence argues against it. However, a variant of this may be of relevance for the composition of hot cores. Specifically, it has become abundantly clear in recent years that

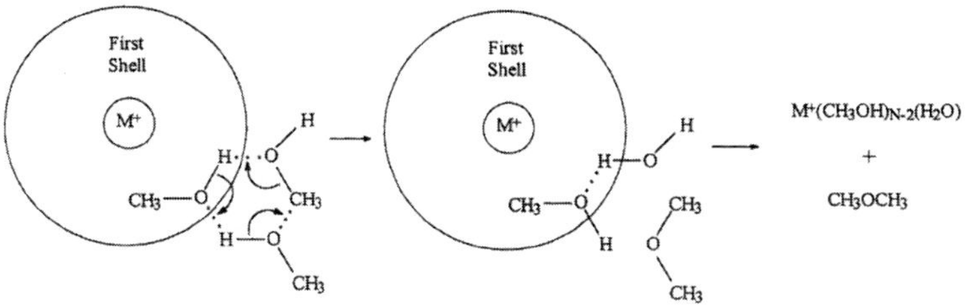

Figure 5. Possible chemical route towards molecular complexity in methanol-rich interstellar ices. The circle around the central (alkali) M^+ ion represents the first solvation shell of methanol. Methanol molecules in the second solvation shell and hydrogen bonded to the first are (stereochemically) in position to react. In this way, two methanol species are converted into dimethyl ether and water ($M^+(CH_3OH)_N \rightarrow M^+(CH_3OH)_{N-2}H_2O + CH_3OCH_3$). Figure taken from Selegue & Lisy (1994).

charge can be an important aspect for interstellar ices. First, the 4.62 μm band is now generally attributed to the OCN^- ion (Demyk *et al.* 1998). Second, any PAHs trapped in these ices can act as electron acceptors, localizing the charge and preventing electron diffusion and recombination (Bouwman *et al.* 2011). Likewise, trapped atoms such as Na and K can be efficient electron donors in interstellar ices. Third, recent studies on dipole alignment of molecular films (Balog *et al.* 2009) also point towards the importance of electric fields for the evolution of interstellar ices. Accepting that interstellar ices are charged and that these charges are localized, new interesting chemical routes akin to ion-molecule gas phase reactions but now in the solid state open up. Specifically, figure 5 illustates a possible route towards dimethyl ether, based upon molecular beams experiments on methanol droplets (Selegue & Lisy 1994). So, the scenario would be: typical interstellar ices containing CH_3OH and other species (eg., H_2O) including traces of alkali metals are charged by visible and UV photons. Upon warm up by a nearby protostar to $\sim$ 80 K, the methanol segregates out from the water – known to be important in some interstellar sightlines (Boogert *et al.* 2000; Dartois *et al.* 1999) – and sterochemistry can ensue. The resulting more-weakly-bound dimethyl ether molecule can then readily sublimate. While there is good experimental support for this reaction scheme in droplets, the importance of methyl transfer without fragmentation in ices (on astrophysical timescales) is unknown and this whole scheme is highly speculative. Regardless, this discussion points towards issues that we have ignored in the past and that may actually be very important: charging of ices, stereochemistry, segregation, molecular solubility in water/methanol, and solid-state ion-molecule chemistry.

4. The PAH reservoir

Large ($\sim$ 50 – 100 C-atom) PAH molecules are known to be abundant and ubiquitous in the interstellar medium. These molecules lock up $\sim$ 5 % of the elemental C and hence represent a carbon reservoir which is comparable to that of gaseous CO. PAH species are often thought to form in the ejecta from C-rich AGB stars as critical intermediaries or byproducts of the dust formation process and detailed chemical routes based upon terrestrial soot studies have been proposed (Frenklach & Feigelson 1989; Cherchneff *et al.* 1992). Alternatively, PAHs may result from shattering of carbonaceous materials in grain-grain collisions in shocks in the interstellar medium itself (Jones *et al.* 1996). In the ISM, these molecular species will be further processed by UV and X-ray photons and

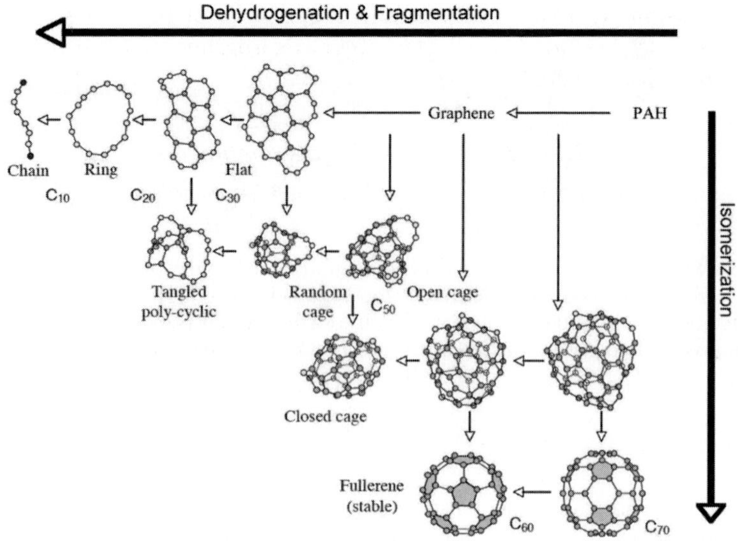

Figure 6. Chemical evolution of PAHs in the interstellar medium under the influence of UV photons combines the effects of dehydrogenation and fragmentation with those of isomerization. Fully hydrogenated PAHs – injected by stars into the ISM (top right side) – are photolyzed near bright stars, leading to complete H-loss (e.g., the weakest link) and the formation of graphene. Further fragmentation may lead to the formation of flats, rings, and chains. However, this process competes with isomerization to various types of stable intermediaries such as cages and fullerenes. Figure taken from Berné & Tielens (2011).

by energetic particles in shock waves, hot gas or cosmic rays (Micelotta *et al.* 2010a,b, 2011), before becoming incorporated into regions of star and planet formation.

4.1. *PAH processing in the ISM*

Figure 6 schematically illustrates the chemical evolution of PAHs in space. Upon excitation, several fragmentation channels will compete with each other (and with IR fluorescence). Experiments show that loss of H dominates (Jochims *et al.* 1994; Joblin 2003). For typical interstellar conditions and relevant PAH sizes, there is then a kinetic equilibrium between fragmentation and H-addition reactions. However, close to a star, H-loss will dominate and PAHs are quickly stripped of all their H's leading to the formation of graphene structures (Le Page *et al.* 2001; Berné & Tielens 2011). These can then fragment futher by losing C-atoms, forming rings, chains and small hydrocarbon radicals (Fig. 6). This fragmentation will compete with isomerization of graphene to cages and fullerene structures. The latter are much more stable to fragmentation than PAHs and can survive very close to a star. Given the calculated rapid destruction timescale of interstellar PAHs, it seems unavoidable that chemical growth of PAHs has to take place in the ISM. At large distances from stars, PAHs are in kinetic equilibrium with the H-flux and phenyl-like intermediaries can be abundant (Le Page *et al.* 2001). Laboratory studies show that such radicals can react with hydrocarbon species even at low tempertures and it is conceivable that this chemical growth process drives PAHs to ever larger sizes (Parker *et al.* 2011). The small hydrocarbon radicals "released" in the fragmentation may also set off a rich and interesting chemistry that may well contribute to the rich inventory of hydrocarbon radicals in photodissociation regions (Pety *et al.* 2005). Finally, while the focus here has been on fragmentation and chemistry driven by UV photon, similar fragmentation channels are available when PAHs are excited by energetic ions in shocks or hot plasmas (Postma *et al.* 2010). The details of these chemical routes have not yet been explored and dedicated experimental programs will be required to assess their viability (cf., Joblin (2003)).

4.2. *PAH processing in protostellar disks*

The photochemical PAH processing outlined in § 4.1 is also very relevant for protoplanetary disks. Stellar UV photons may process PAHs in disk photospheres while X-ray photons from low mass T Tauri stars may penetrate deeply into the disk. X-rays will leave PAHs much more highly excited than the (non-ionizing) UV photons available in PDRs and hence may be particularly efficient in driving this type of photochemistry. PAHs can also be chemically processed by e.g., H and OH radicals in warm gas in the inner disk and early accretion phases (Kress *et al.* 2010). Figure 7 shows the calculated chemical timescale for PAH loss as a function of the temperature of the gas. The adopted pressure is typical for models of protoplanetary disks but the results are not very sensitive to this. PAHs are stable when the dynamical timescale of the system is shorter than this chemical timescale and typically this is the case for temperatures less than 1000 K (Kress *et al.* 2010). As this figure indicates, PAHs will be quickly destroyed ($< 10^6$ yr) in the inner regions ($<$ 5 AU) of protoplanetary disks. The processes that created the chondrules in the Solar system likely also drove an interesting chemistry characterized by higher temperatures but shorter timescales. Finally, we note that the gas in the photosphere of the inner disk is also very warm but the PAHs themselves are not (except immediately after UV photon absorption). Hence whether this radical chemistry can play a role in the disk photosphere will need to be assessed.

Small hydrocarbon released by the destruction of PAHs (e.g., C_2H_2) will be partly oxidized to CO (and CO_2) and partly channeled to CH_4 and this is an effective way to drive the CH_4/CO ratio to values more appropriate of "chemical equilibrium" at the temperatures and pressures of the disks. Small, catacondensed PAHs are more susceptible to attack by radicals than larger, pericondensed PAHs, resulting in faster chemical timescales for the former as compared to the latter. Temperatures and pressures are very conducive to chemical growth of PAHs as well and a fraction of the C_2H_2 released by the radical attack on catacondensed PAHs will flow along these routes. However, eventually, even the more stable PAHs will typically be destroyed by the radicals. Finally, we mention that if the diffusion timescale in the disk is much faster than the oxidation timescale, the hydrocarbons produced this way (CH_4, C_2H_2, and their radicals) will spread throughout the disk and be available for a rich organic chemistry (Gail 2002).

5. Implications

As emphasized in this review and illustrated in figure 1, the organic reservoir of the habitable zone of protoplanetary disks reflects a variety of chemical processes that release C from the otherwise unreactive reservoirs, CO and PAHs. For CO, hydrogenation on grain surfaces is key to activating the carbon. Subsequent ion-molecule chemistry either in the gas phase or in the solid state combined with warm up or photolysis & warm up may then form a wide array of complex organics. For PAHs, photolysis can lead to fullerenes, cages & hydrocarbon radicals, while warm radical chemistry destroys PAHs, forming methane, acetylene & stable PAHs (temporarily), and (after diffusion) acetylenic derivatives. All of these processes likely played a role in setting the organic inventory of the Solar system and may have contributed to life's chemical start on the Earth.

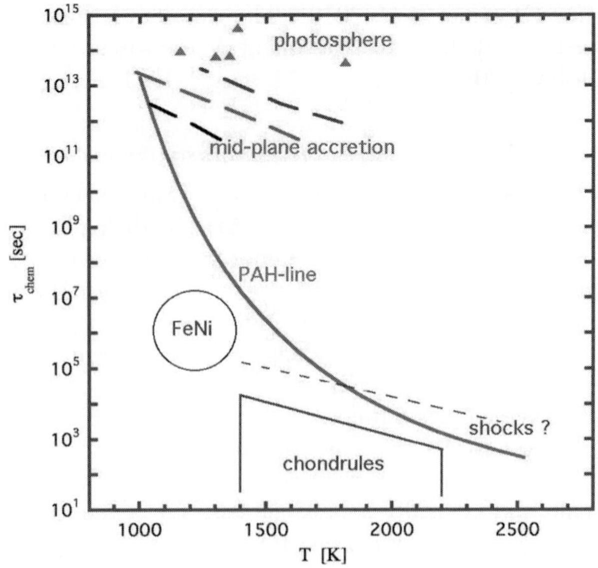

Figure 7. Chemical timescale for the breakdown of PAHs through chemical attack by H and OH radicals as a function of the temperature of the gas (Kress *et al.* 2010). At a given temperature, PAHs will be stable for timescales below the line marked "PAH-line". The black, green, and blue dashed lines indicate calculated temperatures in the mid plane at 1, 2, and 5 AU (between $10^4 - 10^6$ yr) for a typical model for the evolution of the early Solar nebula (Bell *et al.* 1997). The box labeled chondrules gives the temperature-time regime experienced by chondrules in meteorites as derived from their texture (Connolly *et al.* 2006). The line labeled shocks is an indication that the processes producing chondrules may have lead to more extreme conditions without leaving traces in the meteorites. The circle indicates typical conditions derived from analysis of meteoritic FeNi (Meibom *et al.* 2000). The red triangles mark the temperatures derived from K-band CO observations (Najita *et al.* 2003) of a few T-Tauri stars for which accurate isochrone ages have been determined (Bertout *et al.* 2007).

5.1. *Chemical inventory of protoplanetary disks*

Infrared studies have revealed an interesting array of very abundant molecules in the warm gas of protoplanetary disks and hot cores including CH_4, H_2O, C_2H_2, HCN, and HNCO (Najita *et al.* 2003; Lahuis *et al.* 2007; Knez *et al.* 2009). Some of these are readily made in the warm dense gas while others are possibly ice-release-products. However, C_2H_2 and HCN are not typical surface chemistry products and the observed abundances in these regions are much higher than in prestellar cores, so they are not freeze out products either. Acetylene might be produced by the breakdown of PAHs (§ 4.2) in the inner disk followed by rapid diffusion throughout the disk (Kress *et al.* 2010). Hydrogen cyanide also reflects a high temperature chemistry and may be produced in the same regions and perhaps even through the same processes if PAHs do contain N (Hudgins *et al.* 2005). The observed high abundances of species such as C_2H_6 in comets (Mumma *et al.* 2003) may reflect diffusion of PAH-breakdown-products (hydrocarbon radicals) and their reactions throughout the disk.

5.2. *Chemical inventory of comets*

Comets are generally thought to contain a pristine record of the prestellar core and/or the quiescent outskirts of the protosolar nebula. However, it is now clear that some components – crystalline silicates, CAI fragments – may well have been formed elsewhere in the Solar nebula and then diffused outwards. The chemical inventory of comets

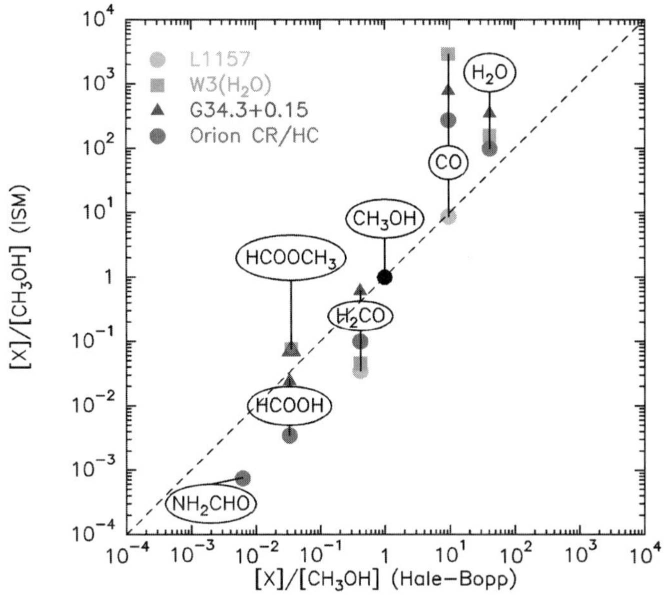

Figure 8. Comparison of the chemical composition of comets and hot cores. The commonality in chemical inventory and abundances of complex organic molecules points towards similarity in chemical processing involved. Figure taken from Bockelée-Morvan *et al.* (2000).

resembles that of hot cores with species such as methanol, formaldehyde, formic acid, methyl formate, and formamide. Even the observed abundances are quite similar (Fig. 8; Bockelée-Morvan *et al.* (2000). This resemblance of the organic inventory in composition and abundance may well point towards a shared chemical history. If cometary organics form by aggregating interstellar ice grains, then this would point towards an in situ origin for these complex organics – e.g., energetic processing of ices –, since more complex scenarios involving sublimation, gas phase reactions, macroscopic diffusion, and distillation would face difficulties preserving the relative abundances of these species.

5.3. *Chemical inventory of meteorites*

Carbonaceous meteorites are among the most primitive objects of the solar system. They exhibit significant carbon contents mostly occuring as insoluble organic matter. The macromolecular structure of this material resembles that of terrestrial kerogen†. In addition, carbonaceous meteorites contain numerous soluble organic compounds. Many of these have familiar biochemical counterparts (e.g., amino acids, fatty acids, purines, pyrimidines, and sugars; Cronin & Chang (1993); Botta & Bada (2002) but the soluble organic inventory is very rich and includes alcohols, aldehydes, amides, amines, mono- and dicarboxylic acids, aliphatic and aromatic hydrocarbons, heterocyclic aromatics, hydroxy acids, ketones, phosphonic and sulfonic acids, sulfides and ethers. Concentrations of the major representatives of these classes vary widely from less than 10 parts per million (amines) to tens of parts per million (amino acids) to hundreds of parts per million (carboxylic acids). The total soluble fraction can be as extreme as 0.3 (eg., Murchison) in CM carbonaceous meteorites.

In contrast to the Earth's biosphere, meteoritic biomolecules are a diverse and random mixture and this lack of specificity points towards an origin in abiotic processes. Both the soluble and the insoluble organic components of meteorites show an anomalous isotopic

† That should not be take to imply a biological origin for this material.

composition in H, C, and N. In particular, the enrichment in D is evidence for formation at the low temperatures characteristics for the cold interiors of dense clouds rather than the warm nebular gas available in the asteroidal formation zone. Hence, it is generally accepted that the initial reservoir was inherited from the molecular cloud from which the Solar system formed and/or the outskirts of the Solar nebula. However, this record has largely been obliterated. In particular, the coexistence of soluble organics with clay minerals suggest active chemistry during the aqueous (and thermal) alteration of the parent body, such as hydrolysis of nitriles and saturation of double bonds, and plausible reaction routes have been formulated (e.g., Strecker synthesis) (Peltzer & Bada 1978). Similar processes likely played a role on the early Earth as well.

6. Summary

As outlined in this short review, the observed organic inventory of regions of planet formation as well as that of Solar system bodies is very rich and diverse, reflecting a wide variety of chemical processes contributing as material sojourns from their stellar "birth" sites to their planetary "resting" places. The two main molecular reservoirs of carbon in molecular clouds – CO and PAHs – are converted into a variety of more complex organic molecules through ion-molecule and grain surface chemistry in one case and UV photochemistry and radical chemistry in the other case (Fig. 1). Three different types of "scenarios" can be formulated differing in the intrinsic role of astrochemistry for the early emergence of life: 1) Delivery of carbon which is converted on the early Earth to the required building blocks (Urey-Miller hypothesis). 2) Delivery of a primed organic mixture (H_2CO, HCN) which were converted with the 'Strecker' mechanism into amino acids and the like on the Early Earth ("hamburger helper" model). 3) Delivery of the specific molecules necessary to jump start life ("Universal Darwinian pond"). On asteroids, process 2 seems to have take place. If asteroids delivered the organic inventory to Earth, then process 3 may well be key. Moreover, some laboratory studies show that processed interstellar ices – and therefore perhaps comets could as well – may well have delivered the essential ingredients which transformed into the first cells in water (Deamer *et al.* 2002). So, while the extent of astrochemistry's contribution to the origin of life is still to be discovered, it has become clear over the last decade that chemistry and biology are as important for understanding the Universe and our place in it as physics.

References

Balog, R., Cicman, P., Jones, N. C., & Field, D., 2009, *Phys Rev Let*, 201, 73003
Bell, K. R., Cassen, P. M., Klahr, H. H., & Henning, T. 1997, *ApJ*, 486, 372
Bergin, E. A., 2011, in *The Interstellar Medium*, ed: (Springer Verlag, Berlin), in press
Berné, O. & Tielens, A. G. G.. M., 2011, submitted
Bertout, C., Siess, L., & Cabrit, S. 2007, *A & A*, 473, L21
Blake, G. A., Sutton, E. C., Masson, C. R., & Phillips, T. G. 1987, *ApJ*, 315, 621
Bockelée-Morvan, D., *et al.* 2000, *A & A*, 353, 1101
Boogert, A. C. A., *et al.* 2000, *A & A*, 353, 349
Botta, O. & Bada, J. L., 2002, *Surveys in Geophysics*, 23, 411
Bottinelli, S. & *et al.*, 2004, *ApJ*, 617, L69
Bouwman, J., Cuppen, H. M., Steglich, M., Allamandola, L. J., & Linnartz, H. 2011, *A & A*, 529, A46
Caselli, P., Hasegawa, T. I., & Herbst, E. 1993, *ApJ*, 408, 548
Ceccarelli, C., Loinard, L., Castets, A., Tielens, A. G. G. M., Caux, E., Lefloch, B., & Vastel, C. 2001, *A & A*, 372, 998

Charnley, S. B., Tielens, A. G. G. M., & Millar, T. J. 1992, *ApJ*, 399, L71

Charnley, S. B. & Rodgers, S. B. 2009, *Bioastronomy 2007: Molecules, Microbes and Extraterrestrial Life*, 420, 29

Cherchneff, I., Barker, J. R., & Tielens, A. G. G. M. 1992, *ApJ*, 401, 269

Connolly, H. C., Desch, S. J., Ash, R. D., & Jones, R. H., 2006, in *Meteorites and the Early Solar System II* (University of Arizona Press), p383

Cronin, J. R. & Chang, S., 1993, in *The Chemistry of Lifes Origin*, eds. J.M. Greenberg, C.X. Mendoza-Gmez, and V. Pirronello, (Kluwer: Dordrecht), p209

Dartois, E., Demyk, K., d'Hendecourt, L., & Ehrenfreund, P. 1999, *A & A*, 351, 1066

Deamer, D., Dworkin, J. P., Sandford, S. A., Bernstein, M. P., & Allamandola, L. J., 2002, *Astrobiology*, 2, 371

Demyk, K., Dartois, E., D'Hendecourt, L., Jourdain de Muizon, M., Heras, A. M., & Breitfellner, M. 1998, *A & A*, 339, 553

Frenklach, M. & Feigelson, E. D. 1989, *ApJ*, 341, 372

Fuchs, G. W., Cuppen, H. M., Ioppolo, S., *et al.*, 2009, *A & A*, 505, 629

Gail, H.-P. 2002, *A & A*, 390, 253

Garrod, R. T., Weaver, S. L. W., & Herbst, E. 2008, *ApJ*, 682, 283

Geppert, W., *et al.*, 2006, *Faraday discussions*, 133, 177

Herbst, E. & Klemperer, W. 1973, *ApJ*, 185, 505

Hidaka, H., Watanabe, N., Shiraki, T., *et al.*, 2004, *ApJ*, 614, 1124

Hiraoka, K., Miyagoshi, T., Takayama, T., *et al.*, 1998, *ApJ*, 498, 710

Hirota, T., Ohishi, M., & Yamamoto, S. 2009, *ApJ*, 699, 585

Horn, A., *et al.*, 2004, *ApJ*, 611, 605

Hudgins, D. M., Bauschlicher, C. W., Jr., & Allamandola, L. J. 2005, *ApJ*, 632, 316

Hudson, R. L., Palumbo, M. E., Strazzulla, G., Moore, M. H., Cooper, J. F., & Sturner, S. J., 2008, in *The Solar System Beyond Neptune*, eds: M. A. Barucci, H. Boehnhardt, D. P. Cruikshank, and A. Morbidelli, (University of Arizona Press, Tucson), p. 507

Ioppolo, S., Cuppen, H. M., Romanzin, C., *et al.*, 2008, *ApJ*, 686, 1474

Joblin, C., 2003, SF2A-2003: Semaine de lAstrophysique Francaise, 175

Jochims, H. W., Ruhl, E., Baumgartel, H., Tobita, S., & Leach, S. 1994,*ApJ*, 420, 307

Jones, A. P., Tielens, A. G. G.. M., & Hollenbach, D. J., 1996, *ApJ*, 469, 740

Knez, C., Lacy, J. H., Evans, N. J., II, van Dishoeck, E. F., & Richter, M. J. 2009, *ApJ*, 696, 471

Kress, M. E., Tielens, A. G. G. M., & Frenklach, M. 2010, *Advances in Space Research*, 46, 44

Lahuis, F., *et al.* 2007, *ApJ*, 659, 296

Le Page, V., Snow, T. P., & Bierbaum, V. M. 2001, *ApJS*, 132, 233

Meibom, A., Desch, S. J., Krot, A. N., Cuzzi, J. N., Petaev, M. I., Wilson, L., & Keil, K. 2000, *Science*, 288, 839

Micelotta, E. R., Jones, A. P., & Tielens, A. G. G. M. 2010, *A & A*, 510, A36

Micelotta, E. R., Jones, A. P., & Tielens, A. G. G. M. 2010, *A & A*, 510, A37

Micelotta, E. R., Jones, A. P., & Tielens, A. G. G. M. 2010, *A & A*, 526, 52

Mumma, M. J., Disanti, M. A., dello Russo, N., Magee-Sauer, K., Gibb, E., & Novak, R. 2003, *Advances in Space Research*, 31, 2563

Najita, J., Carr, J. S., & Mathieu, R. D. 2003, *ApJ*, 589, 931

Öberg, K. I., Garrod, R. T., van Dishoeck, E. F., & Linnartz, H. 2009, *A & A*, 504, 891

Parise, B., Ceccarelli, C., Tielens, A. G. G. M., Castets, A., Caux, E., Lefloch, B., & Maret, S. 2006, *A & A*, 453, 949

Perker, D., *et al.*, 2011, submitted

Peltzer, E. T. & Bada, J. L., 1978, *Nature*, 272, 443

Pety, J., Teyssier, D., Fossé, D., Gerin, M., Roueff, E., Abergel, A., Habart, E., & Cernicharo, J. 2005, *A & A*, 435, 885

Postma, J., Bari, S., Hoekstra, R., Tielens, A. G. G.. M., & Schlathter, T., 2010, *ApJ*, 708, 435

Requena-Torres, M. A., Martín-Pintado, J., Martín, S., & Morris, M. R. 2008, *ApJ*, 672, 352

Schutte, W. A., Allamandola, L. J., & Sandford, S. A. 1993, *Icarus*, 104, 118

Selegue, T. J. & List, J. M., 1994, *J Am Chem Soc*, 116, 4874

Tielens, A. G. G. M. & Hagen, W. 1982, *A & A*, 114, 245
Tielens, A. G. G. M. & Allamandola, L. J., 2011, *Physics and Chemistry at Low Temperatures*, ed. L. Khriachtchev, (Pan Stanford, Singapore), p341
Wakelam, V., *et al.*, 2010, *Space Sci Rev*, 156, 13
Watson, W. D. 1973, *ApJ*, 183, L17
Wright, M. C. H., Plambeck, R. L., & Wilner, D. J. 1996, *ApJ*, 469, 216

The Molecular Universe
Proceedings IAU Symposium No. 280, 2011
José Cernicharo & Rafael Bachiller, eds.

doi:10.1017/S1743921311024835

Observational Studies of Pre-Stellar Cores and Infrared Dark Clouds

Paola Caselli[1]

[1]School of Physics and Astronomy, University of Leeds, Leeds LS2 9JT, UK
email: p.caselli@leeds.ac.uk

Abstract. Stars like our Sun and planets like our Earth form in dense regions within interstellar molecular clouds, called pre-stellar cores (PSCs). PSCs provide the initial conditions in the process of star and planet formation. In the past 15 years, detailed observations of (low-mass) PSCs in nearby molecular cloud complexes have allowed us to find that they are cold (T < 10 K) and quiescent (molecular line widths are close to thermal), with a chemistry profoundly affected by molecular freeze-out onto dust grains. In these conditions, deuterated molecules flourish, becoming the best tools to unveil the PSC physical and chemical structure. Despite their apparent simplicity, PSCs still offer puzzles to solve and they are far from being completely understood. For example, what is happening to the gas and dust in their nuclei (the future stellar cradles) is still a mystery that awaits for ALMA. Other important questions are: how do different environments and external conditions affect the PSC physical/chemical structure? Are PSCs in high-mass star forming regions similar to the well-known low-mass PSCs? Here I review observational and theoretical work on PSCs in nearby molecular cloud complexes and the ongoing search and study of massive PSCs embedded in infrared dark clouds (IRDCs), which host the initial conditions for stellar cluster and high-mass star formation.

Keywords. astrochemistry; line: profiles; molecular data; molecular processes; radiative transfer; stars: formation; ISM: clouds, dust, extinction, kinematics and dynamics, molecules.

1. Introduction

For about 30 years we have known that stars like our Sun form within the dense regions of interstellar molecular clouds (e.g. Benson & Myers 1989; Beichman *et al.* 1986), where submicron sized dust grains and molecules such as H_2, CO and more complex organic species coexist and interact (Herbst & van Dishoeck 2009). These regions are called dense cores (Myers *et al.* 1983) and represent the initial conditions of the star formation process (e.g. Shu *et al.* 1987). In fact, observations in quiescent regions, such as the Pipe nebula (Alves *et al.* 2007; Rathborne *et al.* 2009) and the Aquila Rift (André *et al.* 2010; Könyves *et al.* 2010) indicate that the stellar mass function is the reflection of the core mass function (CMF). If so, processes controlling the formation of the cores sets the initial mass function (IMF; Salpeter 1955). Observations of dense cores and their interpretation are therefore fundamental to put constraints on star and planet formation theory as well as on simulations of molecular cloud formation and evolution. How can we study the initial conditions of star and planet formation by looking at dense cores?

The initial conditions are soon lost as dense cores become associated with protostars, which drive powerful outflows and alter the physical and chemical properties of the parent cloud (e.g. Arce *et al.* 2007, 2010; Foster *et al.* 2009; Friesen *et al.* 2010a; Pineda *et al.* 2011). Thus one first needs to identify dense cores on the verge of star formation, the so-called pre-stellar cores (PSCs). PSCs are typically cold (temperature $\leqslant$10 K) and have volume densities $> \simeq 2\times10^{-19}$ g cm^{-3}, or number densities $>5\times10^4$ H_2 molecules per cc (e.g. Ward-Thompson *et al.* 1994; Crapsi *et al.* 2005). In such environments, the dust mostly emits sub-millimetre radiation (peaking around 300μm), and the rotational transitions of molecules such as CO emit in the readily accessible millimetre spectrum. While CO is the most abundant molecule after H_2, there are a host of other observable molecules that provide additional information. For example, ammonia (NH_3; e.g. Goodman *et al.* 1993) and dyazenilium (N_2H^+; Caselli *et al.* 1995, 2002a) both have lower optical depths and are better tracers of dense gas. Both molecules have hyperfine spectra useful in determining the gas temperature and column densities. Over the past 15 years, observations of dust, CO, NH_3, and N_2H^+ have established the basic structure of the cores: nearly hydrostatic equilibrium and heated from the outside by starlight. More recent studies have been able to analyze observations within this basic framework to deduce some of the physical and chemical processes in molecular clouds such as internal wave motions, UV heating, molecular depletion and desorption between gas and solid phases, and chemical fractionation. In Sect. 2 below we discuss this further. The more recent study of PSCs in high-mass star forming regions is described in Sect. 3.

2. Low-mass PSCs

The state of the art in the PSCs field is summarized below. More comprehensive reviews can be found in Di Francesco *et al.* (2007) and Bergin & Tafalla (2007).

2.1. *Molecular freeze-out*

First identified by Ward-Thompson *et al.* (1994) as centrally concentrated cores in the dust millimetre continuum emission, PSCs are characterized by relatively large central number densities (a few $\times\,10^6$ cm^{-3}), low central kinetic temperature (T$\,\simeq\,$6 K; Crapsi *et al.* 2007), and kinematic features which can be reproduced by gravitational collapse motions (e.g. Caselli *et al.* 2002b; van der Tak *et al.* 2005). CO molecules (typically used to derive cloud masses, assuming a canonical CO/H_2 abundance ratio of $\simeq 10^{-4}$ cm^{-3}, as measured in nearby molecular clouds; e.g. Frerking *et al.* 1987, Pineda *et al.* 2008) do not trace the dust peak at all (Willacy *et al.* 1998; Caselli *et al.* 1999; Bacmann *et al.* 2002; Pagani *et al.* 2007). To reproduce observations toward the PSC L1544, about 90% of CO molecules should leave the gas phase, on average along the line of sight, and over 99% of them must deplete in the core nucleus (Caselli *et al.* 1999). The only easy way to get rid of gas-phase molecules in quiescent and cold regions is to deposit them on the surface of dust particles, forming thick icy mantles (e.g. Leger 1983; Lacy *et al.* 1994; Ossenkopf & Henning 1994; Ormel *et al.* 2009; Chiar *et al.* 2011).

Unlike CO and other C-bearing species (Tafalla *et al.* 2006), N-bearing molecules do not significantly freeze-out in PSC centers (Hily-Blant *et al.* 2010): only a factor $\simeq$2 abundance drop for N_2H^+ (Caselli *et al.* 2002c, Bergin *et al.* 2002) and hints of abundance rise for NH_3 (Tafalla *et al.* 2004). This is a puzzle in regions where CO is heavily depleted, given that the binding energy of N_2 (the precursor of N_2H^+ and NH_3) onto bare grains and icy surfaces is similar to the CO binding energy (Öberg *et al.* 2005) and the sticking coefficient is the same for the two species ($\simeq$1; Bisschop *et al.* 2006). Also, NH_3 is significantly more strongly bound on grain mantles than CO. A possible explanation may be the slow conversion of atomic nitrogen into N_2, because of the slow neutral-neutral reactions involved in the process (Hily-Blant *et al.* 2010), the possibly low sticking probability of N ($\simeq$0.1; Flower *et al.* 2006) and the fact that the freeze-out of CO decreases the destruction rate of N_2H^+ and molecular ions important for the formation of NH_3 (such as NH^+; see also Aikawa *et al.* 2005, 2008). Indeed, Maret *et al.* (2006) found evidence of a low fraction of N_2 ($\simeq 10^{-6}$ with respect to H nuclei) and a consequent high N/N_2 ratio toward the starless core B68, although some N has to be trapped on the surface of dust grains to avoid a too efficient production of N_2. Interestingly, Hily-Blant *et al.* (2010) also require a low fractional abundance of atomic nitrogen in the gas phase ($\leqslant 10^{-6}$), to reproduce the observed abundance ratio of N-bearing species tracing the CO-depleted zone of five starless cores (but uncertainties on the rate coefficients of key neutral-neutral reactions prevent unambiguous conclusions). In any case, in the central few hundred AU of a PSC (the PSC nucleus), where the volume density is $>10^6\,\mathrm{cm^{-3}}$, molecules freeze-out onto dust grains with time scales $\leqslant 10^3/[n(\mathrm{H_2})/10^6\,\mathrm{cm^{-3}}]\,\mathrm{yr}$ (assuming a gas-to-dust mass ratio of 100 and a dust grain radius of 0.1 μm; e.g. van Dishoeck *et al.* 1993, Flower *et al.* 2005), and no high density tracer is expected to survive in the gas phase (except for those molecules composed by the volatile light elements H and D, such as H_3^+ and its deuterated isotopologues; see Sect. 2.2). An example of this is provided by the recent interferometric observations of N_2H^+ (1-0) toward Oph B (Friesen *et al.* 2010b), where no emission was found toward dust continuum peaks within the core.

Keto & Caselli (2008, 2010) used the basic temperature-density model to predict line intensities and profiles for different values of the cosmic ray ionization rate and the reverse process to depletion, molecular desorption, which is dependent on cosmic rays. Detailed comparison with observed molecular lines put limits on the rate of cosmic-ray ionization ($\simeq 1\times10^{-17}\,\mathrm{s^{-1}}$) and desorption (about 30 times larger than the value deduced by Hasegawa & Herbst 1993 and in agreement with more recent studies by Bringa & Johnson 2004 and Herbst & Cuppen 2006). Moreover, to reproduce the low temperature observed toward the core nucleus ($\simeq$6 K) the dust opacity has to increase by a factor of about 4 compared to the standard dust opacities of Ossenkopf & Henning (1994). This is consistent with fluffy dust, suggesting that dust grains have started to coagulate (see also Ormel *et al.* 2009), with a consequent reduction of the molecular freeze-out rate because of the smaller surface area available for adsorption (Flower *et al.* 2005).

2.2. *Deuterium fractionation*

Deuterated molecules are highly enhanced in PSC nuclei. The reason for this is briefly outlined here. In cold and dark clouds (such as PSCs), H_2 molecules are mainly destroyed by cosmic rays, which produce H_2^+. As soon as H_2^+ is formed, it reacts with another H_2 molecule, forming one of the most important molecular ions in astrochemistry: H_3^+. This molecular ion can easily cede a proton to atoms and other species, starting astrochemical processes and increasing the molecular complexity in the interstellar medium. H_3^+ can also exchange its proton with a deuteron, through the reaction: $H_3^+ + HD \rightarrow H_2D^+ + H_2 + \Delta E$ ($\Delta E = 230$ K; Millar *et al.* 1989). If the gas temperature is below 20 K, such as in PSCs, the above exothermic reaction can only proceed from left to right, enhancing the H_2D^+/H_3^+ abundance ratio. This is reflected in species (such as DCO^+ and N_2D^+), which form in reactions of neutrals (CO and N_2, respectively) with H_2D^+. The ratio between deuterated and non-deuterated species (e.g. N_2D^+/N_2H^+) is called D-fraction (R_D) and its values can be as large as a few %, about 3 orders of magnitude larger than the cosmic D/H ratio ($\simeq 10^{-5}$; e.g. Linsky *et al.* 2006).

R_D increases even farther (to about 13 orders of magnitude in multiply deuterated species; e.g. Ceccarelli *et al.* 2007) if the abundance of the main destruction partners of H_3^+ and H_2D^+ (mainly CO and O) is reduced, e.g. via freeze-out. This was predicted by Dalgarno & Lepp (1984) and observationally proved toward several starless cores and PSCs (Caselli *et al.* 2002c; Bacmann *et al.* 2003). Crapsi *et al.* (2005) found criteria to identify PSCs among starless cores by measuring R_D and studying line profiles: PSCs have (i) high R_D ($>10\%$); (ii) large amount of CO freeze-out ($>99\%$); (iii) large central density ($>5\times10^5\,\mathrm{cm^{-3}}$) and centrally concentrated density profile; (iv) kinematics consistent with central gravitational infall. This is important because PSCs, the future stellar cradles, are the best candidates where to find the initial conditions in the process of star formation. Starless cores with lower densities and shallower density profiles may oscillate around an equilibrium configuration (e.g. Lada *et al.* 2003) or expand and disperse.

The large amount of molecular freeze-out and the consequent boost in the D-fraction explain why PSCs are the strongest emitters of H_2D^+ and D_2H^+ in the Galaxy (Caselli *et al.* 2003, 2008; Vastel *et al.* 2004). To reproduce these observations, chemical models had to include multiply deuterated species (Roberts *et al.* 2003). Being bright (main beam brightness temperatures $\simeq$1 K), the ortho-$H_2D^+(1_{1,0}$-$1_{1,1})$ line is an ideal tracer of PSC nuclei, even in those regions where most of the heavier species are gone from the gas phase (the "complete depletion" zone; Walmsley *et al.* 2004). The emission of ortho-H_2D^+ extends over sizes $\simeq$5,000 AU toward L1544 (Vastel *et al.* 2006), similar to the para-$D_2H^+(1_{1,0}$-$1_{0,1})$ line (Parise *et al.* 2011). This is the size of the "deuteration zone", where CO is mostly frozen onto dust grains and bright lines of N_2D^+ and deuterated ammonia can be observed (e.g. Crapsi *et al.* 2005, 2007). Caselli *et al.* (2008) extended the search of ortho-H_2D^+ in a sample of PSCs and cores with embedded protostars. Variations in the amount of R_D and ortho-H_2D^+ column densities were correlated with physical parameters and found that large uncertainties in the interpretation were present, due to our poor knowledge of the cosmic-ray ionization rate (Dalgarno 2006), the dust grain size distribution (Ormel *et al.* 2009), the H_2 ortho-to-para ratio (Troscompt *et al.* 2009; Pagani *et al.* 2009) and the abundance of atomic oxygen in molecular clouds and dense cores (Melnick & Bergin 2005). Caselli *et al.* (2008) suggested that an important source of the scatter observed in the correlation plots is probably due to variations in the ortho-to-para ratio of H_2D^+, from >0.5 for PSCs, with gas temperatures <10 K, to 0.03 for protostellar cores, consistent with chemical models of dense cores with temperature $\simeq$15 K gas (Flower *et al.* 2004; see also Emprechtinger *et al.* 2009 and Friesen *et al.* 2010a for further discussion on the variation of deuterium fractionation toward protostellar objects).

2.3. *Physical Structure*

Measurements of dust and gas temperature profiles (Crapsi *et al.* 2007; Pagani *et al.* 2007), as well as estimates of their density structure, kinematics, chemical composition and ionization fraction (e.g. Caselli *et al.* 2002b; Bergin *et al.* 2002; Schnee *et al.* 2007; Keto & Caselli 2008, 2010) provided the following conclusions: (i) PSCs are mainly thermally supported (with some additional energy furnished by waves and subsonic turbulence; e.g. Caselli *et al.* 2002a; Lada *et al.* 2008). (ii) Subsonic oscillations and large scale wave motions are important to PSC internal energy (Keto & Field 2005; Broderick *et al.* 2008); (iii) PCSs are heated from outside by starlight and throughout by cosmic rays (e.g. Evans *et al.* 2001; Zucconi *et al.* 2001); (iv) main coolants are molecular line radiation (in particular CO) in the outer regions and dust in the central nuclei (e.g. Goldsmith 2001; Galli *et al.* 2002); (v) their density structure is set by the Lane-Emden equation (e.g. Alves *et al.* 2001); (vi) the molecular abundance across PSCs is set by freeze-out/desorption processes in the core centers and by photodissociation in the outskirts (Pavlyuchenkov *et al.* 2008; Keto & Caselli 2008); (vii) PSCs are thermally supercritical cloud cores (Keto & Caselli 2008), with infall velocities of $\simeq$0.1 km s^{-1} peaking at $\simeq$1000 AU from the core center (Keto & Caselli 2010).

High angular resolution observations (probing linear sizes of $\simeq$800 AU) using the Very Large Array (VLA) and the Plateau de Bure Interferometer (PdBI) have shown that the specific angular momentum of the PSC L1544 drops from $\simeq 5\times10^{21}$ cm^2 s^{-1} at sizes of 0.04 (traced by NH_3) to $\simeq 3.5\times10^{20}$ cm^2 s^{-1} at sizes of 0.001 pc (traced by NH_2D; Crapsi *et al.* 2007). While the former value is typical of starless cores (e.g. Ohashi *et al.* 1999), the smaller obtained toward the core nucleus resembles values measured toward protostellar envelopes and it is consistent with what found in more evolved objects (Belloche *et al.* 2002). ALMA will allow us to dive into the central few thousands AU and study the physical conditions and kinematics of the core nucleus, the future protostar and protoplatenary disk cradle.

Figure 1 shows a schematic summary of the physical and chemical structure of a representative PSC based on recent observational and theoretical work. The panel is divided in five regions with different gray scales, representing various chemical zones: (i) the "complete depletion" zone ($r \leqslant 1000$ AU, $A_V \geqslant 60$ mag), not yet discovered but predicted by chemical models (e.g. Walmsley *et al.* 2004). Here, the main tracers of the gas are expected to be the ortho-H_2D^+ ($1_{1,0}$-$1_{1,1}$) and para-D_2H^+ ($1_{1,0}$-$1_{0,1}$), both observable with ALMA. (ii) The "depletion zone" ($1000 < r \leqslant 5{,}000$ AU, $A_V \geqslant 20$ mag), where the peak of deuterated molecules such as N_2D^+, D_2CO and deuterated ammonia are found (e.g. Caselli *et al.* 2002a; Bacmann *et al.* 2003; Crapsi *et al.* 2007). Other non-deuterated molecules more resistant than CO to the freeze-out have also been observed in the same region: CN, HCN, HNC (Hily-Blant *et al.* 2010, see also Tafalla *et al.* 2006). (iii) The "CO freeze-out zone" ($r \leqslant 7{,}000$ AU, $A_V \geqslant 5$ mag) where the majority of CO molecules (>99%, Caselli *et al.* 1999) are locked onto icy dust grain mantles. (iv) The "dark-cloud chemistry zone" ($7{,}000 < r \leqslant 15{,}000$ AU, $A_V \geqslant 4$ mag), traced by CO molecules and higher density tracers chemically linked to CO, such as HCO^+. Because carbon atoms are probably not completely locked into CO molecules (due to the presence of interstellar UV photons), carbon-chain species such as CCS and CCH (Tafalla *et al.* 2006; Padovani *et al.* 2009) can also be found here. (v) The "photodissociation region" (PDR; e.g. Kaufman *et al.* 1999), where FUV ($6\,\mathrm{eV} < h\nu < 13.6\,\mathrm{eV}$) photons play a crucial role in the chemistry, dissociating molecules and maintaining most of the carbon and oxygen in atomic form in the gas phase. The detailed structure of PSCs of course depends on the environment, so that the size and physical/chemical properties of the various PSC layers are expected to change in different molecular cloud complexes or in more massive star forming regions. Understanding how external conditions affect the structure of PSCs is crucial to shed light on the star formation process in different environments.

2.4. *The water content*

Oxygen is the third most abundant element in the Universe and it has a crucial role in the chemical and physical structure of molecular clouds, where large fractions are locked into CO and H_2O molecules. Unfortunately, large uncertainties in the chemistry of oxygen are still present due to the difficulty of observing H_2O and O in molecular clouds. In dark ($A_V > 4$ mag) clouds, H_2O is thought to be mostly in solid form (Hollenbach *et al.* 2009), in agreement with the stringent upper limits on the ortho-H_2O($1_{1,0}$-$1_{0,1}$) obtained toward dark clouds and dense cores ($x(H_2O \equiv N(H_2O)/N(H_2) < 3\times10^{-8}$, 8×10^{-9} and 7×10^{-9} toward B68, Oph D and Cha-MMS1, respectively; Bergin & Snell 2002; Klotz *et al.* 2008). These limits are however difficult to interpret, given that radiative transfer modeling has shown that line trapping and absorption of the dust continuum can heavily affect the water emission from dense interstellar clouds (Poelman *et al.* 2007). Moreover, a significant fraction of atomic oxygen can still be maintained in the gas phase, thanks to non-thermal (e.g. cosmic-ray induced) desorption processes (Caselli *et al.* 2002c), as in fact measured with ISO (Infrared Space Observatory) toward two molecular clouds (Caux *et al.* 1999; Vastel *et al.* 2000).

The superior angular resolution and sensitivity of the Herschel Space Observatory (Pilbratt *et al.* 2010) has recently furnished the first column density measure of ortho-H_2O

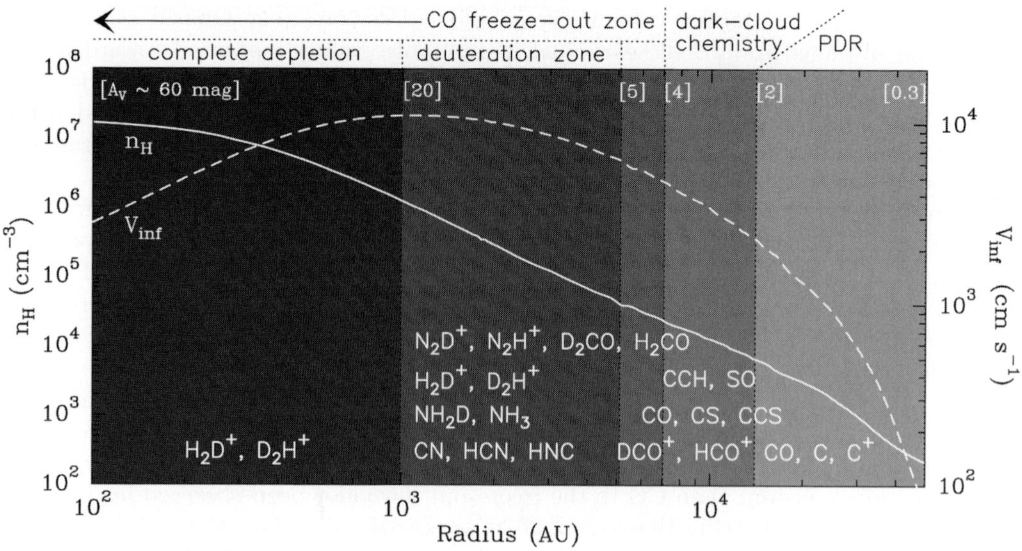

Figure 1. Schematic representation of the physical and chemical structure of a PSC in a low–mass star forming region, based on detailed observations and hydrodynamic/radiative transfer modeling of the PSC L1544, isolated PSC embedded in the Taurus molecular cloud complex. The white curves represent the hydrogen nuclei volume density ($n_{\rm H}$; solid, left labels) and the infall velocity ($V_{\rm inf}$; dashed, right labels) profiles as derived by Keto & Caselli (2010). The density is predicted to be 2×10^7 cm^{-3} within the central 500 AU. The infall velocity has a peak of $\simeq$0.1 km s^{-1} around 1,000 AU. The panel is divided in five regions with different gray scales, representing various chemical zones (see text).

toward the PSC L1544 (Caselli *et al.* 2010, see Figure 2), within the WISH (Water In Star-forming regions with Herschel, van Dishoeck *et al.*2011) guaranteed time programme. No water line was detected toward the starless core B68 (from which a stringent upper limit of the water abundance was deduced: $x(\text{ortho} - \text{H}_2\text{O}) < 1.3\times10^{-9}$). The water detection toward L1544 has been possible because L1544, unlike B68, is centrally concentrated and thus has a relatively bright FIR nucleus (about 5 Jy at 557 GHz within the Herschel beam of 40″). In fact, the ortho-$\text{H}_2\text{O}(1_{1,0}\text{-}1_{0,1})$ line has been detected in *absorption* at a 5σ level (Figure 2). The column density is $N(\text{ortho} - \text{H}_2\text{O}) \simeq 8\times10^{12}$ cm^{-2}. Using the H_2 column density deduced by the 1.3 mm dust continuum emission (Ward-Thompson *et al.* 1999; Figure 2) within the Herschel beam, Caselli *et al.* (2010) found $x(\text{ortho} - \text{H}_2\text{O}) \simeq 2\times10^{-10}$ toward the central few thousand AUs, where CO is highly frozen onto dust grains. This value is close to that measured in the envelope of massive young stellar objects (van der Tak *et al.* 2010). Radiative transfer models of collapsing Bonnor-Ebert spheres with the inclusion of simplified chemistry based on the Hollenbach *et al.* (2009) model, were able to reproduce the observations assuming a peak abundance of water vapor of $\simeq10^{-8}$ at 0.1 pc from the PSC center. Higher gas phase abundance of water are ruled out because the predicted absorption is too strong. These observations have been recently repeated and the S/N significantly improved so that a more detailed analysis and comparison with chemical models is now possible and will soon be presented (Caselli *et al.* 2011, in prep.).

Figure 2 shows the ortho-$\text{H}_2\text{O}(1_{1,0}\text{-}1_{0,1})$ spectrum toward L1544 as presented in Caselli *et al.* (2010), where a comparison with the CO(1-0) profiles is also given. It is interesting

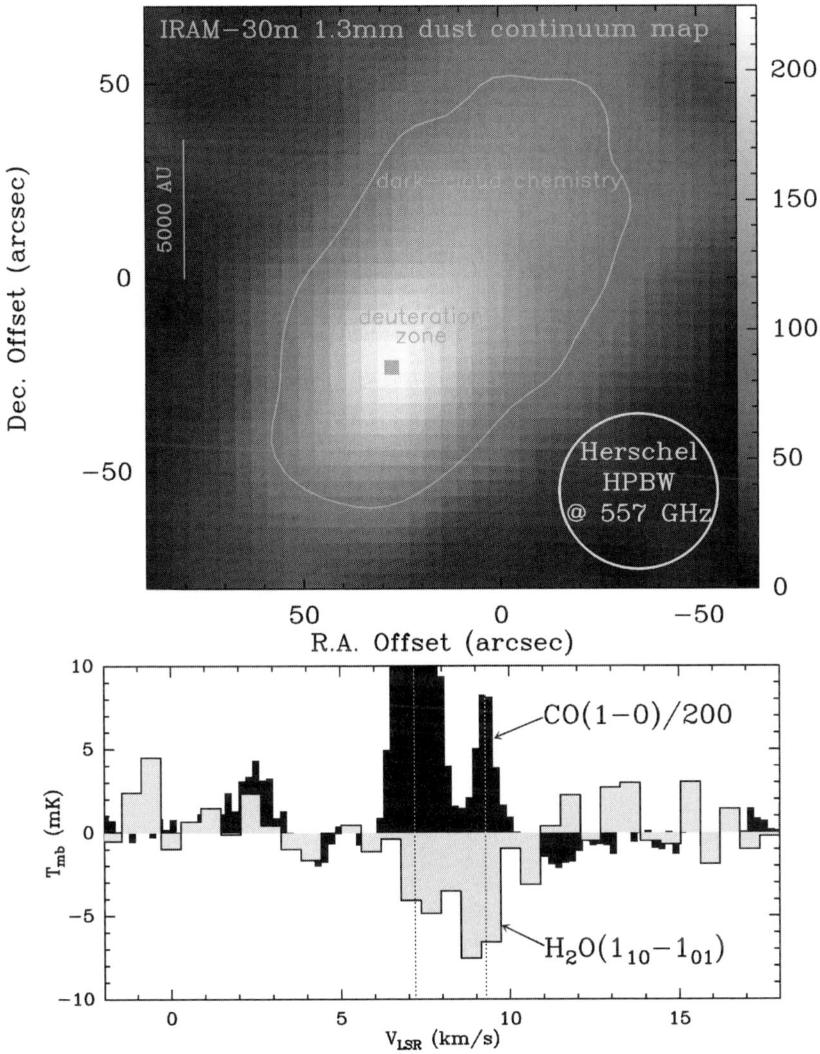

Figure 2. (**Top panel**) The 1.3 mm dust continuum emission map obtained at the IRAM-30m antenna by Ward-Thompson *et al.* (1999), smoothed at 22″ resolution to improve the sensitivity. The wedge on the right is in units of mJy/22″-beam. The contours represent the 50% and 70% of the dust continuum peak (225 mJy/beam). They approximately coincide with the size of the "dark-cloud chemistry" zone and "deuteration zone", respectively (see Figure 1). The square locate the dust peak position, observed by Herschel (RA(J2000) = $05^h\ 04^m\ 17.21^s$, Dec(J2000) = $25^{\rm deg}\ 10'\ 42.8''$). The Herschel beam at 557 GHz is the circle at the bottom-right corner. (**Bottom panel**) The feature in absorption is the ortho-$H_2O(1_{1,0}$-$1_{0,1})$ line observed against the 11 mK continuum emission of the dust at the same frequency (from Caselli *et al.* 2010). The black histogram is the CO(1-0) line (scaled down by a factor of 200) observed with the Five College Radio Astronomy Observatory (FCRAO).

to see that the water absorption is present in the same velocity range of the CO(1-0) emission line, including the faint CO component redshifted by about 2 km s^{-1} away from the centroid velocity of the L1544 dense gas tracers. These are the first observations of water vapor in dark clouds and they are crucial to shed light on the oxygen chemistry and the formation of water vapor in quiescent regions.

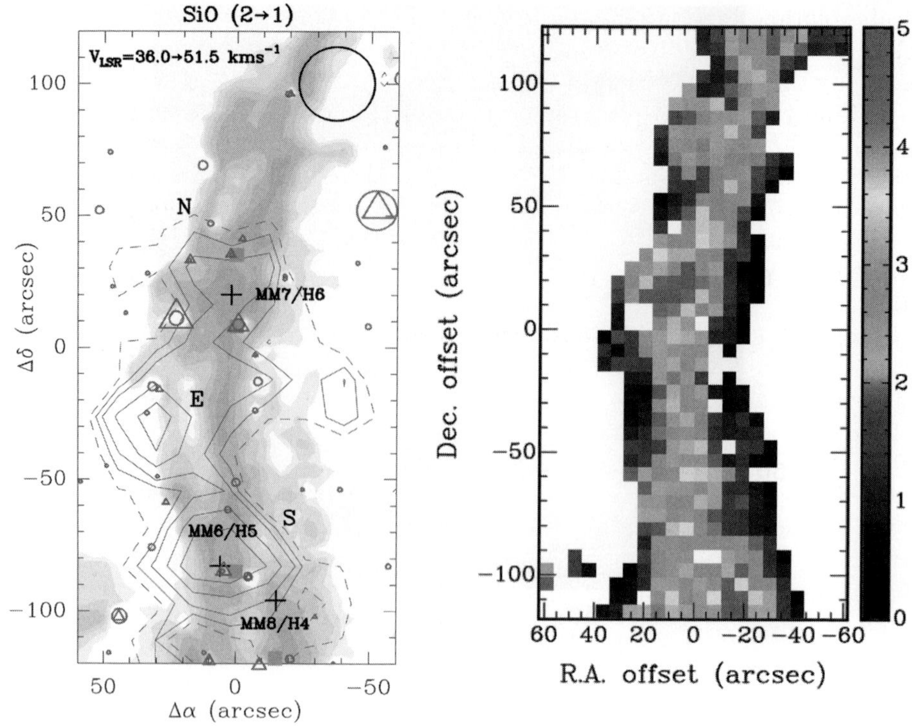

Figure 3. (**Left panel**) Integrated intensity map of the SiO (2-1) line towards G035.39-00.33 (blue contours), overlapped on the mass surface density map of Butler & Tan (2009, grey scale). The contour levels of the SiO emission are 10 (2σ; dashed contour), 15, 20, 30 and 40 mK km s^{-1}. For the mass surface density map, contours are 0.014 (2σ), 0.021, 0.035, 0.049, 0.07, 0.105 and 0.14 g cm^{-2}. Crosses indicate the cores found by Rathborne *et al.* (2006). Red open circles, red open triangles and green squares show the location of the 8- and 24-μm sources and 4.5-μm extended emission, respectively. The marker sizes used for the 8- and 24-μm sources have been scaled by the source flux. The SiO (2-1) beam size is plotted at the upper-right corner (adapted from Jimenez-Serra *et al.* 2010). (**Right panel**) CO depletion factor map (from Hernandez *et al.* 2011) of G035.39-00.33. The abundance of CO decreases by factors 3-5 in the highest extinction regions, indicating the presence of cold dust ($T < 20$ K) and consequent CO freeze-out.

3. Massive PSCs

The majority of stars form in clusters within giant molecular clouds (GMCs; e.g. Lada & Lada 2003), so it is crucial to understand how stellar clusters form. However, the initial conditions in the process of high-mass star and stellar cluster formation are still highly debated. Understanding what causes a particular region of a GMC to attain high densities on the way to forming a star cluster is a major open question in the fields of star and galaxy formation. There are two main classes of theories: spontaneous gravitational instability perhaps regulated by photo-ionization (McKee 1989) or turbulence (Krumholz & McKee 2005) and triggering by some external process, such as HII regions (Deharveng *et al.* 2005), supernova blastwaves (Palous *et al.* 1994), turbulent motions (MacLow & Klessen 2004) or GMC collisions (Tan 2000). Detailed studies of the chemical/physical structure and kinematics of massive PSCs (MPSCs) and comparison with the properties of low-mass PSCs (see Sect. 2) are sorely needed.

The search and study of MPSCs in high-mass star forming regions is however at its infancy, given that it is challenged by the MPSC short time scales, large distances and their proximity to active sites of high-mass star and cluster formation, which can affect their dynamical evolution. For these reasons, the initial conditions in the process of high-mass and stellar cluster formation require the search for relatively quiescent massive clouds and then study the embedded stellar population content to identify starless regions, possibly away from the reach of powerful outflows driven by massive young stellar objects (MYSO) embedded in nearby clumps. The best candidates to host *pristine* MPSCs are the so-called infrared dark clouds (IRDCs; e.g. Beuther *et al.* 2007; Zhang *et al.* 2009), in particular those with a relatively large fraction of quiescent gas (i.e. low star formation activity) and filamentary structures, which are predicted to be the initial morphologies of GMCs in several classes of dynamical formation models (e.g. van Loo *et al.* 2007; Heitsch *et al.* 2009).

IRDCs are high extinction regions viewed against the diffuse mid-IR Galactic background (Pérault *et al.* 1996; Egan *et al.* 1998; Peretto & Fuller 2009). They have very high-mass surface densities that are similar to known regions of massive star and star cluster formation (Tan 2005). They are cold (temperatures between 8 and 20 K; Pillai *et al.* 2006; Peretto *et al.* 2010; Stamatellos *et al.* 2010; Devine *et al.* 2011; Ragan *et al.* 2011), host clumps with densities $\simeq 10^4$-10^6 cm^{-3} (Teyssier *et al.* 2002; Butler & Tan 2009; Swift 2009; Zhang *et al.* 2011), show chemical compositions similar to low-mass PSCs (Vasyunina *et al.* 2011), including large deuterium fractions R_D (Pillai *et al.* 2007; Chen *et al.* 2010). In fact, Fontani *et al.* (2011) have found significantly larger N_2D^+/N_2H^+ column density ratios toward a sample of high mass starless cores (HMSCs) embedded in cold IRDCs (between 0.3 and 0.7) than the rest of their sample, which includes HMSCs embedded in more active high-mass star forming regions as well as clumps associated with high-mass protostellar objects and ultra-compact HII (UCHII) regions. Thus, R_D measurements are powerful tools to identify MPSCs among the IRDC starless cores, which are now readily found by Herschel on a galactic scale (e.g. Molinari *et al.* 2010; Wilcock *et al.* 2010; Henning *et al.* 2010).

Detailed mapping and analysis has been recently done (and more it is under way) of a highly filamentary IRDC: G035.39-00.33. This region (i) has one of the most extreme filamentary structure in the Rathborne *et al.* (2006) sample of IRDCs studied in the mm continuum emission; (ii) shows extended quiescent regions with high-mass surface density and relatively low star formation activity (as can be seen from the lack of a large population of sources in the Spitzer 24 μm image, see Figure 3, left panel); (iii) is relatively nearby (D = 2.9 kpc). This IRDC overlaps with Spitzer-GLIMPSE data, and is cloud H in the sample of 10 clouds (A-J) studied by Butler & Tan (2009) using extinction mapping in the 8μm IRAC band. Hernandez & Tan (2011) have studied the kinematics of the gas associated with this cloud using ^{13}CO(1-0) from the Galactic Ring Survey (GRS; Jackson *et al.* 2006) and found evidence for disturbed kinematics, perhaps indicating the filament has formed very recently. Widespread SiO(2-1) emission has been found along the filament, suggestive of a large scale shock caused by the recent collision of molecular filaments, which may have produced the IRDC (Jimenez-Serra *et al.* 2010; Figure 3, left panel). Hernandez *et al.* (2011) have measured signicant levels of CO freeze-out (factor of 5; see Figure 3, right panel), indicative of the presence of cold and dense gas. Henshaw *et al.* (2011, in prep) found that the main starless core in IRDC is located at the position where different filaments appear to merge, indicating a recent formation via cloud-cloud collision. The presence of extended SiO emission and CO freeze-out, plus the evidence of filament merging (Henshaw *et al.* 2011, in prep.), further suggest that this region is dynamically young, so it is an ideal target where to study the initial conditions in the process of high-mass star and stellar cluster formation. Higher angular resolution observations are under way to connect the large scale structure of IRDC with the embedded PSCs and clumps. This will allow us to make quantitative comparisons with models of giant molecular cloud (GMC) formation and evolution (e.g. Heitsch *et al.* 2009, 2011; Butler *et al.*, in prep.), and achieve a deeper understanding of the physical processes involved in the earliest phases of high-mass and cluster forming regions.

Acknowledgements. I am grateful to many collaborators & friends who made all this work possible and very enjoyable, in particular Luca Bizzocchi, Tyler Bourke, Michael Butler, Cecilia Ceccarelli, Antonio Crapsi, Francesco Fontani, Jonathan Henshaw, Audra Hernandez, Itaskun Jiménez-Serra, Eric Keto, Phil Myers, Jaime Pineda, Scott Schnee, Mario Tafalla, Jonathan Tan, Floris van der Tak, Charlotte Vastel, Malcolm Walmsley. A special thank to Audra for her help with Figure 3, and to Mario, Jonathan and Malcolm for their careful reading of the manuscript and important suggestions.

References

Aikawa, Y., Herbst, E., Roberts, H., & Caselli, P. 2005, *ApJ*, 620, 330
Aikawa, Y., Wakelam, V., Garrod, R. T., & Herbst, E. 2008, *ApJ*, 674, 984
Alves, J. F., Lada, C. J., & Lada, E. A. 2001, *Nature*, 409, 159
Alves, J., Lombardi, M., & Lada, C. J. 2007, *A&A*, 462, L17
André, P., *et al.*, 2010, *A&A*, 518, L102
Arce, H. G., Borkin, M. A., Goodman, A. A., Pineda, J. E., & Halle, M. W. 2010, *ApJ*, 715, 1170
Arce, H. G., Shepherd, D., Gueth, F., Lee, C.-F., Bachiller, R., Rosen, A., & Beuther, H. 2007, *Protostars and Planets V*, 245
Bacmann, A., Lefloch, B., Ceccarelli, C., Castets, A., Steinacker, J., & Loinard, L. 2002, *A&A*, 389, L6
Bacmann, A., Lefloch, B., Ceccarelli, C., Steinacker, J., Castets, A., & Loinard, L. 2003, *ApJ*, 585, L55
Beichman, C. A., Myers, P. C., Emerson, J. P., Harris, S., Mathieu, R., Benson, P. J., & Jennings, R. E. 1986, *ApJ*, 307, 337
Belloche, A., André, P., Despois, D., & Blinder, S. 2002, *A&A*, 393, 927
Benson, P. J. & Myers, P. C. 1989, *ApJS*, 71, 89
Bergin, E. A., Alves, J., Huard, T., & Lada, C. J. 2002, *ApJ*, 570, L101
Bergin, E. A. & Snell, R. L. 2002, *ApJ*, 581, L105
Bergin, E. A. & Tafalla, M. 2007, *ARAA*, 45, 339
Beuther, H., Churchwell, E. B., McKee, C. F., & Tan, J. C. 2007, *Protostars and Planets V*, 165
Bisschop, S. E., Fraser, H. J., Öberg, K. I., van Dishoeck, E. F., & Schlemmer, S. 2006, *A&A*, 449, 1297
Bringa, E. M. & Johnson, R. E. 2004, *ApJ*, 603, 159
Broderick, A. E., Narayan, R., Keto, E., & Lada, C. J. 2008, *ApJ*, 682, 1095
Butler, M. J. & Tan, J. C. 2009, *ApJ*, 696, 484
Caselli, P., Benson, P. J., Myers, P. C., & Tafalla, M. 2002a, *ApJ*, 572, 238
Caselli, P., *et al.*, 2010, *A&A*, 521, L29
Caselli, P., Myers, P. C., & Thaddeus, P. 1995, *ApJ*, 455, L77
Caselli, P., van der Tak, F. F. S., Ceccarelli, C., & Bacmann, A. 2003, *A&A*, 403, L37
Caselli, P., Vastel, C., Ceccarelli, C., van der Tak, F. F. S., Crapsi, A., & Bacmann, A. 2008, *A&A*, 492, 703
Caselli, P., Walmsley, C. M., Tafalla, M., Dore, L., & Myers, P. C. 1999, *ApJ*, 523, L165
Caselli, P., Walmsley, C. M., Zucconi, A., Tafalla, M., Dore, L., & Myers, P. C. 2002b, *ApJ*, 565, 331
Caselli, P., Walmsley, C. M., Zucconi, A., Tafalla, M., Dore, L., & Myers, P. C. 2002c, *ApJ*, 565, 344
Caux, E., *et al.*, 1999, *A&A*, 347, L1
Ceccarelli, C., Caselli, P., Herbst, E., Tielens, A. G. G. M., & Caux, E. 2007, *Protostars and Planets V*, 47

Chen, H.-R., Liu, S.-Y., Su, Y.-N., & Zhang, Q. 2010, *ApJ*, 713, L50
Chiar, J. E., *et al.*, 2011, *ApJ*, 731, 9
Crapsi, A., Caselli, P., Walmsley, C. M., Myers, P. C., Tafalla, M., Lee, C. W., & Bourke, T. L. 2005, *ApJ*, 619, 379
Crapsi, A., Caselli, P., Walmsley, M. C., & Tafalla, M. 2007, *A&A*, 470, 221
Dalgarno, A. 2006, *Proceedings of the National Academy of Science*, 103, 12269
Dalgarno, A. & Lepp, S. 1984, *ApJ*, 287, L47
Devine, K. E., Chandler, C. J., Brogan, C., Churchwell, E., Indebetouw, R., Shirley, Y., & Borg, K. J. 2011, *ApJ*, 733, 44
Di Francesco, J., Evans, N. J., II, Caselli, P., Myers, P. C., Shirley, Y., Aikawa, Y., & Tafalla, M. 2007, *Protostars and Planets V*, 17
Deharveng, L., Zavagno, A., & Caplan, J. 2005, *A&A*, 433, 565
Egan, M. P., Shipman, R. F., Price, S. D., Carey, S. J., Clark, F. O., & Cohen, M. 1998, *ApJ*, 494, L199
Emprechtinger, M., Caselli, P., Volgenau, N. H., Stutzki, J., & Wiedner, M. C. 2009, *A&A*, 493, 89
Evans, N. J., II, Rawlings, J. M. C., Shirley, Y. L., & Mundy, L. G. 2001, *ApJ*, 557, 193
Flower, D. R., Pineau des Forêts, G., & Walmsley, C. M. 2004, *A&A*, 427, 887
Flower, D. R., Pineau Des Forêts, G., & Walmsley, C. M. 2005, *A&A*, 436, 933
Flower, D. R., Pineau Des Forêts, G., & Walmsley, C. M. 2006, *A&A*, 456, 215
Fontani, F., *et al.*, 2011, A&A, 529, L7
Foster, J. B., Rosolowsky, E. W., Kauffmann, J., Pineda, J. E., Borkin, M. A., Caselli, P., Myers, P. C., & Goodman, A. A. 2009, *ApJ*, 696, 298
Friesen, R. K., Di Francesco, J., Myers, P. C., Belloche, A., Shirley, Y. L., Bourke, T. L., & André, P. 2010a, *ApJ*, 718, 666
Friesen, R. K., Di Francesco, J., Shimajiri, Y., & Takakuwa, S. 2010b, *ApJ*, 708, 1002
Frerking, M. A., Langer, W. D., & Wilson, R. W. 1987, *ApJ*, 313, 320
Galli, D., Walmsley, M., & Gonçalves, J. 2002, *A&A*, 394, 275
Goldsmith, P. F. 2001, *ApJ*, 557, 736
Goodman, A. A., Benson, P. J., Fuller, G. A., & Myers, P. C. 1993, *ApJ*, 406, 528
Hasegawa, T. I. & Herbst, E. 1993, *MNRAS*, 261, 83
Heitsch, F., Ballesteros-Paredes, J., & Hartmann, L. 2009, *ApJ*, 704, 1735
Heitsch, F., Naab, T., & Walch, S. 2011, *MNRAS*, 415, 271
Henning, T., Linz, H., Krause, O., Ragan, S., Beuther, H., Launhardt, R., Nielbock, M., & Vasyunina, T. 2010, *A&A*, 518, L95
Herbst, E. & Cuppen, H. M. 2006, *Proceedings of the National Academy of Science*, 103, 12257
Herbst, E. & van Dishoeck, E. F. 2009, *ARAA*, 47, 427
Hernandez, A. K. & Tan, J. C. 2011, *ApJ*, 730, 44
Hernandez, A. K., Tan, J. C., Caselli, P., Butler, M. J., Jimenez-Serra, I., Fontani, F., & Barnes, P. 2011, *ApJ*, in press (arXiv:1106.2499)
Hily-Blant, P., Walmsley, M., Pineau Des Forêts, G., & Flower, D. 2010, *A&A*, 513, A41
Hollenbach, D., Kaufman, M. J., Bergin, E. A., & Melnick, G. J. 2009, *ApJ*, 690, 1497
Jiménez-Serra, I., Caselli, P., Tan, J. C., Hernandez, A. K., Fontani, F., Butler, M. J., & van Loo, S. 2010, *MNRAS*, 406, 187
Kaufman, M. J., Wolfire, M. G., Hollenbach, D. J., & Luhman, M. L. 1999, *ApJ*, 527, 795
Keto, E. & Caselli, P. 2008, *ApJ*, 683, 238
Keto, E. & Caselli, P. 2010, *MNRAS*, 402, 1625
Keto, E. & Field, G. 2005, *ApJ*, 635, 1151
Klotz, A., Harju, J., Ristorcelli, I., Juvela, M., Boudet, N., & Haikala, L. K. 2008, *A&A*, 492, 767
Könyves, V., *et al.*, 2010, *A&A*, 518, L106
Krumholz, M. R. & McKee, C. F. 2005, *ApJ*, 630, 250
Lacy, J. H., Baas, F., Allamandola, L. J., van de Bult, C. E. P., Persson, S. E., McGregor, P. J., Lonsdale, C. J., & Geballe, T. R. 1984, *ApJ*, 276, 533
Lada, C. J., Bergin, E. A., Alves, J. F., & Huard, T. L. 2003, *ApJ*, 586, 286

Lada, C. J., Muench, A. A., Rathborne, J., Alves, J. F., & Lombardi, M. 2008, *ApJ*, 672, 410
Lada, C. J. & Lada, E. A. 2003, *ARAA*, 41, 57
Leger, A. 1983, *A&A*, 123, 271
Linsky, J. L., *et al.*, 2006, *ApJ*, 647, 1106
Mac Low, M.-M. & Klessen, R. S. 2004, *Reviews of Modern Physics*, 76, 125
Maret, S., Bergin, E. A., & Lada, C. J. 2006, *Nature*, 442, 425
McKee, C. F. 1989, *ApJ*, 345, 782
Melnick, G. J. & Bergin, E. A. 2005, *Advances in Space Research*, 36, 1027
Millar, T. J., Bennett, A., & Herbst, E. 1989, *ApJ*, 340, 906
Molinari, S., *et al.*, 2010, *A&A*, 518, L100
Myers, P. C., Linke, R. A., & Benson, P. J. 1983, *ApJ*, 264, 517
Öberg, K. I., van Broekhuizen, F., Fraser, H. J., Bisschop, S. E., van Dishoeck, E. F., & Schlemmer, S. 2005, *ApJ*, 621, L33
Ohashi, N., Lee, S. W., Wilner, D. J., & Hayashi, M. 1999, *ApJ*, 518, L41
Ormel, C. W., Paszun, D., Dominik, C., & Tielens, A. G. G. M. 2009, *A&A*, 502, 845
Ossenkopf, V. & Henning, T. 1994, *A&A*, 291, 943
Padovani, M., Walmsley, C. M., Tafalla, M., Galli, D., & Müller, H. S. P. 2009, *A&A*, 505, 1199
Pagani, L., Bacmann, A., Cabrit, S., & Vastel, C. 2007, *A&A*, 467, 179
Pagani, L., *et al.*, 2009, *A&A*, 494, 623
Palous, J., Tenorio-Tagle, G., & Franco, J. 1994, *MNRAS*, 270, 75
Parise, B., Belloche, A., Du, F., Güsten, R., & Menten, K. M. 2011, *A&A*, 526, A31
Pavlyuchenkov, Y., Wiebe, D., Shustov, B., Henning, T., Launhardt, R., & Semenov, D. 2008, *ApJ*, 689, 335
Perault, M., *et al.*, 1996, *A&A*, 315, L165
Peretto, N. & Fuller, G. A. 2009, *A&A*, 505, 405
Peretto, N., *et al.*, 2010, *A&A*, 518, L98
Pilbratt, G. L., *et al.*, 2010, *A&A*, 518, L1
Pillai, T., Wyrowski, F., Carey, S. J., & Menten, K. M. 2006, *A&A*, 450, 569
Pillai, T., Wyrowski, F., Hatchell, J., Gibb, A. G., & Thompson, M. A. 2007, *A&A*, 467, 207
Pineda, J. E., Caselli, P., & Goodman, A. A. 2008, *ApJ*, 679, 481
Pineda, J. E., Goodman, A. A., Arce, H. G., Caselli, P., Longmore, S., & Corder, S. 2011, *ApJ*, in press (arXiv:1106.5474)
Poelman, D. R., Spaans, M., & Tielens, A. G. G. M. 2007, *A&A*, 464, 1023
Ragan, S., Bergin, E., & Wilner, D. 2011, *ApJ*, in press (arXiv:1105.4182)
Rathborne, J. M., Jackson, J. M., & Simon, R. 2006, *ApJ*, 641, 389
Rathborne, J. M., Lada, C. J., Muench, A. A., Alves, J. F., Kainulainen, J., & Lombardi, M. 2009, *ApJ*, 699, 742
Roberts, H., Herbst, E., & Millar, T. J. 2003, *ApJ*, 591, L41
Salpeter, E. E. 1955, *ApJ*, 121, 161
Schnee, S., Caselli, P., Goodman, A., Arce, H. G., Ballesteros-Paredes, J., & Kuchibhotla, K. 2007, *ApJ*, 671, 1839
Shu, F. H., Adams, F. C., & Lizano, S. 1987, *ARAA*, 25, 23
Stamatellos, D., Griffin, M. J., Kirk, J. M., Molinari, S., Sibthorpe, B., Ward-Thompson, D., Whitworth, A. P., & Wilcock, L. A. 2010, *MNRAS*, 409, 12
Swift, J. J. 2009, *ApJ*, 705, 1456
Tafalla, M., Myers, P. C., Caselli, P., & Walmsley, C. M. 2004, *A&A*, 416, 191
Tafalla, M., Santiago-García, J., Myers, P. C., Caselli, P., Walmsley, C. M., & Crapsi, A. 2006, *A&A*, 455, 577
Tan, J. C. 2000, *ApJ*, 536, 173
Tan, J. C. 2005, *Cores to Clusters: Star Formation with Next Generation Telescopes*, 87
Teyssier, D., Hennebelle, P., & Pérault, M. 2002, *A&A*, 382, 624
Troscompt, N., Faure, A., Maret, S., Ceccarelli, C., Hily-Blant, P., & Wiesenfeld, L. 2009, *A&A*, 506, 1243
van der Tak, F. F. S., Caselli, P., & Ceccarelli, C. 2005, *A&A*, 439, 195

van der Tak, F. F. S., *et al.*, 2010, *A&A*, 518, L107
van Dishoeck, E. F., Blake, G. A., Draine, B. T., & Lunine, J. I. 1993, *Protostars and Planets III*, 163
van Dishoeck, E. F., *et al.*, 2011, *PASP*, 123, 138
van Loo, S., Falle, S. A. E. G., Hartquist, T. W., & Moore, T. J. T. 2007, *A&A*, 471, 213
Vastel, C., Caselli, P., Ceccarelli, C., Phillips, T., Wiedner, M. C., Peng, R., Houde, M., & Dominik, C. 2006, *ApJ*, 645, 1198
Vasyunina, T., Linz, H., Henning, T., Zinchenko, I., Beuther, H., & Voronkov, M. 2011, *A&A*, 527, A88
Vastel, C., Caux, E., Ceccarelli, C., Castets, A., Gry, C., & Baluteau, J. P. 2000, *A&A*, 357, 994
Vastel, C., Phillips, T. G., & Yoshida, H. 2004, *ApJ*, 606, L127
Walmsley, C. M., Flower, D. R., & Pineau des Forêts, G. 2004, *A&A*, 418, 1035
Ward-Thompson, D., Motte, F., & Andre, P. 1999, *MNRAS*, 305, 143
Ward-Thompson, D., Scott, P. F., Hills, R. E., & Andre, P. 1994, *MNRAS*, 268, 276
Wilcock, L. A., *et al.*, 2011, *A&A*, 526, A159
Willacy, K., Langer, W. D., & Velusamy, T. 1998, *ApJ*, 507, L171
Zhang, Q., Wang, Y., Pillai, T., & Rathborne, J. 2009, *ApJ*, 696, 268
Zhang, S. B., Yang, J., Xu, Y., Pandian, J. D., Menten, K. M., & Henkel, C. 2011, *ApJS*, 193, 10
Zucconi, A., Walmsley, C. M., & Galli, D. 2001, *A&A*, 376, 650

Discussion

MELNICK: A note of caution... You refer to the abundance of H_2O, $x(H_2O)$, in several contexts relating to pre-stellar cores. Because we believe that the gas-phase H_2O resides primarily in a relatively narrow zone near the cloud surface, dividing the inferred H_2O column density with that of any tracer that samples the bulk of the cloud volume, such as ^{13}CO or dust, will result in an underestimate of $x(H_2O)$.

CASELLI: This is true if models are indeed a good representation of reality. However, unlike model predictions, our recent Herschel observations toward L1544 suggest that H_2O must be present toward the center of the pre-stellar core (to produce the emission observed in the high sensitivity HRS spectrum) and not much toward the outskirts (to avoid total absorption). For L1544, we use the dust continuum emission to derive the H_2 column density and then compute $x(H_2O)$. Chemical/dynamical models are then used to derive the abundance profile (Keto & Caselli 2010, with a chemistry based on Hollenbach *et al.* 2009). Our recent observations are consistent with an almost constant value of $x(H_2O)$ (10^{-10}-10^{-9}) across the core.

GOLDSMITH: How does the kinematic model explain the line profile of water observed in L1544?

CASELLI: What we see is the typical infall asymmetry also seen in other molecular transitions such as HCO^+ (3-2) and CS(2-1), see e.g. Mardones *et al.* (1997) and Gregersen *et al.* (1997). The blue peak is coming from a region close to the center and behind the core nucleus. The gas in front of the core nucleus is redshifted because of the infall and the water line (with high critical density) is heavily self-absorbed. The absorption feature is below the continuum level set by dust emission at the frequency of the water line.

VAN DER TAK: Detection of para-H_2O towards L1544 may not be harder than ortho-H_2O because of the stronger dust continuum.

CASELLI: This is a very interesting suggestion and we will explore this more (although preliminary calculations showed integration times quite large, it would be extremely important to uniquely measure the ortho/para ratio of H_2O in a quiescent source).

SEMENOV: The problems of chemical models of pre-stellar cores can in fact be the problem with physical models. For example, fractal structure of PSCs will enable much needed FUV-evaporation of ices in cloud cores. This may explain the need of having high H_2O abundances in the centers of PSCs.

CASELLI: This could indeed be the case, although so far VLA and PdBI data does not show evidence of fractal structure. Besides, the scenario you propose may also enhance the abundance of water in the envelope, and this is against what we observe. However, ALMA will shed light on the small scale structure of PSCs and definitely help to more quantitatively answer your question.

The Molecular Universe
Proceedings IAU Symposium No. 280, 2011
José Cernicharo & Rafael Bachiller, eds.

doi:10.1017/S1743921311024847

Hydrodynamical-Chemical Models from Prestellar Cores to Protostellar Cores

Yuri Aikawa[1], **Kenji Furuya**[1], **Valentine Wakelam**[2], **Frank Hersant**[2], **Tomoaki Matsumoto**[3], **Kazuya Saigo**[4], **Kengo Tomida**[4], **Koji Tomisaka**[4], **Robin Garrod**[5] **and Eric Herbst**[6]

[1]Department of Earth and Planetary Sciences, Kobe University, Japan
email: aikawa@kobe-u.ac.jp

[2]Université Bordeaux, France, [3]Housei University, Japan, [4]National Astronomical Observatory of Japan, [5]Cornell University, USA, [6]Ohio State University

Abstract. We investigate the molecular evolution in star forming cores from dense cloud cores ($n_H \sim 10^4$ cm^{-3}, $T \sim 10$ K) to protostellar cores. A detailed gas-grain reaction network is solved in infalling fluid parcels in 1-D radiation hydrodynamic model. Large organic molecules are mainly formed via grain-surface reaction at $T \sim$ several 10 K and sublimated to the gas-phase at ~ 100 K, while carbon-chain species are formed at a few 10 K after the sublimation of CH_4 ice. The former accounts for the high abundance of large organic molecules in hot corinos such as IRAS16293, and the latter accounts for the carbon chain species observed toward L1527. The relative abundance of carbon chain species and large organic species would depend on the collapse time scale and/or temperature in the dense core stage. The large organic molecules and carbon chains in the protostellar cores are heavily deuterated; although they are formed in the warm temperatures, their ingredients have high D/H ratios, which are set in the cold core phase and isothermal collapse phase. HCOOH is formed by the gas-phase reaction of OH with the sublimated H_2CO, and is further enriched in Deuterium due to the exothermic exchange reaction of OH + D $\rightarrow$ OD + H.

In the fluid parcels of the 1-D collapse model, warm temperature $T \sim$ several 10 K lasts for only $\sim 10^4$ yr, and the fluid parcels fall to the central star in ~ 100 yr after the temperature of the parcel rises to $T \geqslant 100$ K. These timescales are determined by the size of the warm region and infall ($\sim$ free-fall) velocity: $r_{\rm warm}/t_{\rm ff}$. In reality, circum stellar disk is formed, in which fluid parcels stay for a longer timescale than the infall timescale. We investigate the molecular evolution in the disk by simply assuming that a fluid parcel stays at a constant temperature and density (i.e. a fixed disk radius) for $10^4 - 10^5$ yrs. We found that some organic species which are underestimated in our 1-D collapse model, such as CH_3OCH_3 and $HCOOCH_3$, become abundant in the disk. We also found that these disk species have very high D/H ratio as well, since their ingredients are highly deuterated.

Finally we investigate molecular evolution in a 3D hydrodynamic simulation of star forming core. We found CH_3OH are abundant in the vicinity of the first core. The abundances of large organic species are determined mainly by the local temperature (sublimation), because of the short lifetime of the first core and the efficient mass accretion via angular momentum transfer.

Keywords. stars: formation, ISM: clouds, ISM: abundances

---

1. Introduction: Hydro-Chemical Models

In star-forming cores, temperature and density vary temporally and spatially. Major chemical processes, which determine the molecular abundances in these cores, also change according to the physical conditions. In prestellar cores with high densities and low temperatures, various molecules are depleted onto grains, where they are hydrogenated by grain-surface reactions. In protostellar phase, on the other hand, molecules are desorbed to the gas phase in the central high-temperature regions. The molecular abundances in the star-forming cores thus change dramatically in the course of the star formation.

It is also well established that molecular clouds are generally not in chemical equilibrium; chemical timescale is comparable or longer than the dynamical timescale of the clouds. For example, the adsorption timescale of gaseous species onto grain surfaces in cold prestellar cores is

$$\pi a^2 \sqrt{\frac{8kT}{\pi m}} S n_{\rm grain} \sim 1 \times 10^6 \left(\frac{10^4 {\rm cm}^{-3}}{n_{\rm H}}\right) [{\rm yr}], \tag{1.1}$$

where a ($\sim 10^{-5}$ cm) and $n_{\rm grain}$ are the size and number density of the grain particles, respectively, and $S(\sim 1.0)$ is the sticking probability. It is comparable to the free-fall timescale of a cold core

$$t_{\rm ff} = \sqrt{\frac{3\pi}{32G\rho}} \sim 4.3 \times 10^5 \left(\frac{10^4 {\rm cm}^{-3}}{n_{\rm H}}\right)^{1/2} [{\rm yr}]. \tag{1.2}$$

We thus need hydro-chemical models in order to investigate molecular abundances and their spatial distribution in star forming cores theoretically.

In the hydro chemical models of star-forming cores, we solve the chemical reaction network (i.e. rate equations) in Lagrange fluid parcels. If the hydrodynamic models are calculated in the Eulerian coordinate, we follow the trajectories of Lagrangian parcels to get the temporal variation of physical parameters in the fluid parcels. The Lagrangian coordinate is more convenient than the Eulerian coordinate to understand the non-equilibrium chemistry, in which molecular abundances are determined not only by the local physical conditions but also by the temporal variation of physical parameters in the past.

In this article, we present hydro-chemical models of low-mass star forming cores. The models start from cold dense cores and evolve to protostellar cores. We first show a spherical core model, in which we calculate molecular abundances and D/H ratios in the gas phase and ice mantles. Then we briefly discuss the chemistry in circum stellar disks. Finally we present preliminary results of a 3-D hydro-chemical model of star-forming cores.

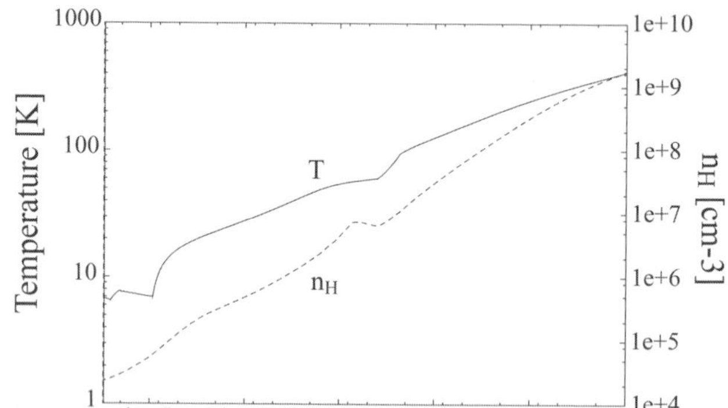

Figure 1. Temporal variation of density and temperature in the fluid parcel which falls from $r \sim 1 \times 10^4$ AU to 2.5 AU.

2. Spherical model

2.1. *Radiation hydrodynamic model*

Hydrodynamic evolution of star-forming cores have been extensively studied by numerical calculations for decades (Klein *et al.* 2007 and references therein). There are also well known semi-analytical models such as Larson-Penston collapse (Larson 1969; Penston 1969) and inside-out collapse model (Shu 1977). But many of these models assume isothermal conditions or polytropes, and thus do not solve the energy equation for temperatures, which is critically important for chemistry. For hydro-chemical models we need radiation hydrodynamic models in which energy equations are solved.

Here we adopt the 1D (spherical) radiation hydrodynamic model of a low-mass star formation by Masunaga *et al.* (1998) and Masunaga & Inutsuka (2000) (Aikawa *et al.* 2008). Initially, the central density of the dense prestellar core is $\rho = 1.4 \times 10^{-19}$ g cm^{-3}, which corresponds to a number density of hydrogen nuclei $n_{\rm H} \sim 6 \times 10^4$ cm^{-3}. The contraction is almost isothermal as long as the cooling rate overwhelms the compressional heating, but eventually the latter dominates and the temperature rises in the central region. The model goes through the first core phase and the second collapse, and follows the evolution of the protostar up to the age of 9.3×10^4 yr.

Figure 1 shows the temporal variation of density and temperature in the fluid parcel which is initially at 1×10^4 AU and falls to 2.5 AU at the final step of the model. Naturally, the density and temperature increase with time. It is well known that the free-fall timescale of a isothermal sphere is proportional to $\rho^{-1/2}$. While the core temperature rises in the central region, the infall velocity accelerates toward the center until the fluid parcel hits the surface of central hydrostatic core (first core or protostar). Hence the temporal variation of density and temperature accelerate as well. In order to highlight the rapid rise near and at the final stage, the horizontal axis in Figure 1 is set to be the logarithm of $t(\text{final}) - t$, where $t(\text{final})$ corresponds to 9.3×10^4 yr after the birth of the protostar.

Temporal variation of temperature is critically important for chemistry. Garrod & Herbst (2006) show that the grain-surface reactions of heavy-element species are efficient in a limited temperature range: several 10 K. Once the temperature gets higher than $\sim$ 100 K, most volatile species are sublimated to the gas phase, where they start the gas-phase reactions. In Figure 1, we can easily read the timescale of these warm phases; the fluid parcel stays at the temperature of several 10 K for slightly less than 10^4 yr. After the fluid parcel gets waremr than 100 K, it falls to the central star in a few 100 yr. Although these timescales slightly depend on the model parameters such as the initial mass of the core, they do not significantly difference from the value in Figure 1; the timescales are approximately equal to $r_{\rm warm}/v_{\rm infall}$, where $r_{\rm warm}$ is the size of the warm region, and the infall velocity $v_{\rm infall}$ is close to the free-fall velocity.

We solve the rate equations of the gas-grain reaction network of Garrod & Herbst (2006) along the temporal variation of the physical parameters in Figure 1. We have also extended the original model of Garrod & Herbst (2006) to include deuterated species referring to Millar *et al.* (1989) and Roberts *et al.* (2004) in order to investigate molecular D/H ratios. Here we assume that the dense core is initially supported by turbulence and/or magnetic fields; we calculate the reaction network model under the cold cloud condition (T=10K, $n_{\rm H} \sim 10^4$ cm^{-3}) for 1×10^6 yr before we adopt the temporal variation of physical parameters in Figure 1. By performing similar calculations for various infalling fluid parcels, we can obtain radial distribution of molecular abundances in the protostellar core.

2.2. *Molecular Abundances in the Protostellar Core*

Figure 2 shows the radial distribution of physical parameters and molecular abundances in the protostellar core of age 9.3×10^4 yr. Many molecules (e.g. NH_3, H_2O and large organic molecules) are sublimated to the gas-phase at $r \sim 100$ AU, where temperature reaches ~ 100 K. The ionization degree decreases toward the center, because it is proportional to $n_{\rm H}^{-1/2}$. In the outer part of the envelope ($r \gtrsim 1000$ AU) HCO^+ is the dominant positive ion. In the inner regions, on the other hand, molecules with higher proton affinities than CO are sublimated. There, the dominant ions are H_3CO^+ and NH_4^+.

Methanol and H_2CO ices are abundant out to 10^4 AU, which indicates that they are formed even in the molecular cloud conditions ($T \sim 10$ K) (Figure 2*c*). Other large organic molecules (both in the gas phase and ice mantle), on the other hand, are abundant only inside a few 1000 AU; they are formed by the grain surfaces reactions among heavy element species at warm temperature ($T \sim 40$ K) and the gas-phase reactions of sublimated species such as H_2CO (Garrod & Herbst 2006; Aikawa *et al.* 2008). The large organic species are sublimated to be observed as hot corinos (e.g. Ceccarelli. *et al.* 2007)

Figure 2(d) shows the distribution of carbon chain species. They increases toward the center at $r \sim 1000$ AU, where CH_4 is sublimated to the gas phase and partly converted to these carbon chains via gas-phase reactions (Sakai *et al.* 2008; Aikawa *et al.* 2008; Hassel *et al.* 2009). It is observed as warm carbon chain chemistry (WCCC) (see Sakai *et al.* in this volume).

It is worth noticing that both hot corino species and WCCC species are formed in our model. In observations, on the other hand, the carbon chains are not abundant in Hot Corinos, and the large organic molecules are not abundant in WCCC objects (Sakai *et al.* 2009). It should depend on the initial ice abundance of CH_3OH and CH_4, which are formed in the cold prestellar phase (Sakai *et al.* 2009). Aikawa *et al.* (2005) investigated

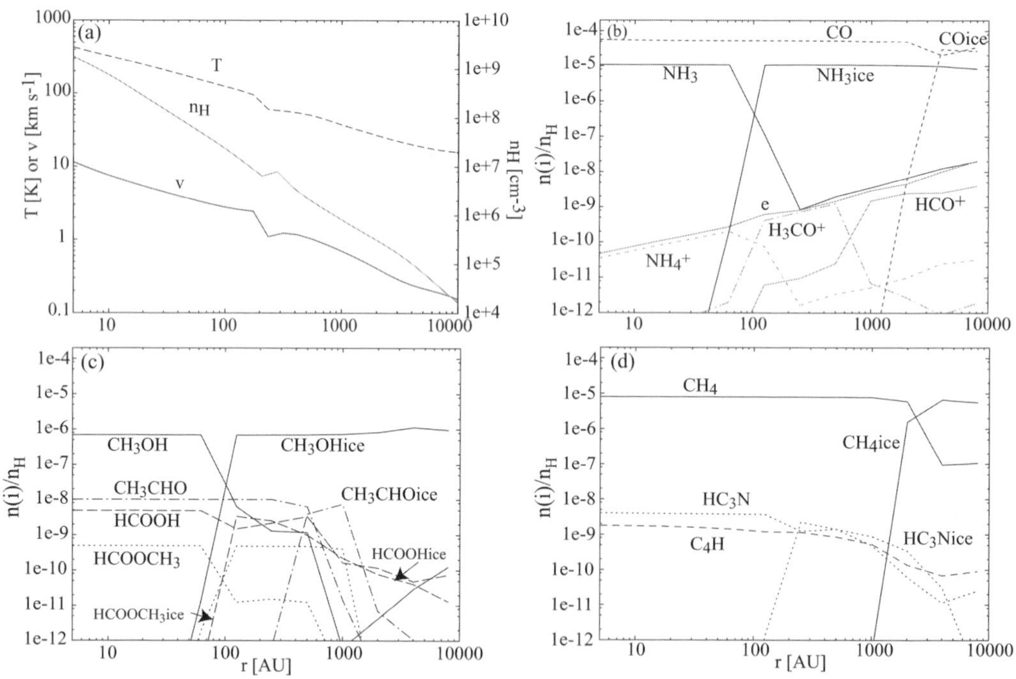

Figure 2. Distribution of (a) physical parameters and (b-d) molecular abundances in the protostellar core of age 9.3×10^4 yr.

gas-grain chemistry in isothermal (cold) collapse phase and shows that CH_3OH is more abundant than CH_4 ice in the slow collapse model starting from near equilibrium conditions, while CH_4 ice is more abundant in the model of faster collapse with a smaller pressure/gravity ratio. The abundance of CH_3OH ice and CH_4 ice may also depend on the temperature in the cold phase, i.e. whether it is slightly higher or lower than 10 K.

2.3. *D/H ratios*

Figure 3 shows the temporal variation of deuterated species and their normal isotope counterparts in the fluid parcel falling to 2.5 AU. In the cold phase, molecules are enriched with Deuterium due to the exothermic exchange reactions such as $H_3^+ + HD \rightarrow H_2D^+ + H_2$. Although such exothermic exchange reactions are limited to several species (Millar *et al.* 1989), the high D/H ratio propagates to other molecules via gas-phase reactions and grain-surface reactions. For example, H_2D^+ gives its deuteron to other molecules and atoms via ion-molecule reactions. H_2D^+ also recombines to produce D atom, which are adsorbed onto grains and hydrogenate the ice mantle species. Once the core gets warm, H_2D^+ is rapidly destroyed by the endothermic exchange reaction with H_2. But deuterated neutral species are not easily destroyed (ex. CH_3D in Figure 3*a*). Large organic molecules and warm carbon chains, which are formed at warm temperatures, also have high D/H ratios, because they are formed from molecules with high D/H ratios. In our model, DCOOH/HCOOH ratio is further enhanced at t(final)-$t \sim$ a few 10^3 yr. HCOOH is formed by the reaction of sublimated H_2CO with OH, which is deuterated by the exchange reaction OH + D $\rightarrow$ OD + H with high exothermicity ($\Delta E \sim 810$ K.)

3. Circumstellar disk

So far, we adopted a spherical symmetric model of star-forming cores. In the spherical model, fluid parcels fall to the central star in a short timescale ($\sim$ 100 yr), once they enter the region of $T \geqslant 100$ K (Figure 1). But in reality, dense cores are rotating ($\omega \sim 10^{-14}$ s^{-1}), and the spherical symmetry breaks in the central regions as the collapse proceeds. Because of the angular momentum conservation, the centrifugal force increases for infalling fluid parcels and eventually balances with the gravity at the centrifugal radius

$$r_{\rm cent} = \frac{(r^2\omega)^2_{\rm init}}{GM} \sim 400 \left(\frac{r}{0.1\rm pc}\right)^4 \left(\frac{\omega}{1\times 10^{-14}\rm s^{-1}}\right)^2 \left(\frac{1M_\odot}{M}\right) \rm AU \qquad (3.1)$$

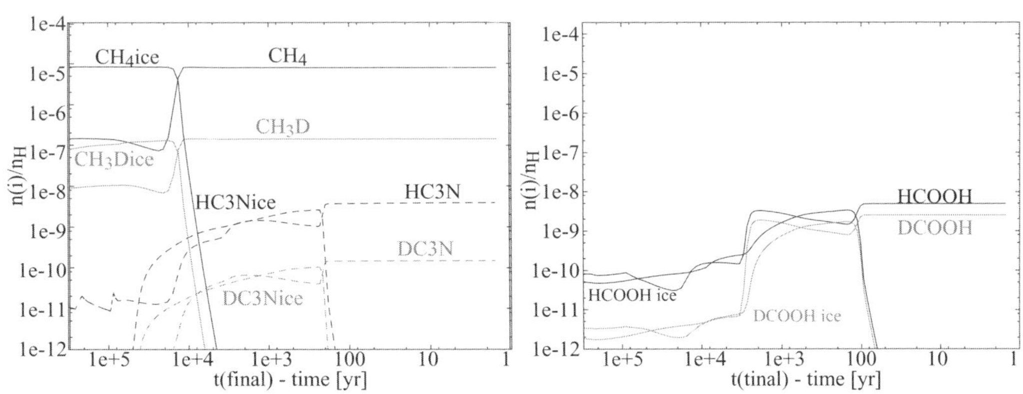

Figure 3. Temporal variation of deuterated molecules and thier normal isotope counterparts in the fluid parcel falling to $r = 2.5$ AU. The horizon axis is the same as that of Figure 1.

to form a circumstellar disk. It means that the fluid parcels stay in the warm dense regions for a longer time scale than assumed in the spherical model. In order to investigate the chemistry in this prolonged warm phase, we performed a pseudo time dependent model (i.e. fixed density and temperature). As the initial abundances, we adopt the molecular abundances at 2.5 AU in final step of the spherical collapse model.

Figure 4 shows the temporal variation of molecular abundances at (a) $T = 260$ K and (b) $T = 40$ K. The density is set to be $n_{\rm H} = 4 \times 10^8$ cm^{-3}. We can see that significant amount of gaseous CH_3OCH_3 and $HCOOCH_3$ ice are formed in $10^4 - 10^5$ yr via gas-phase reactions and grain-surface reactions, respectively. It should be noted that the abundances of these species are underestimated in our spherical model ($n(CH_3OCH_3)/n_{\rm H} = 2\times10^{-11}$ and $n(HCOOCH_3)/n_{\rm H} = 5 \times 10^{-10}$ inside $r \sim 100$ AU) compared with the estimated abundances in the observations: $n(CH_3OCH_3)/n_{\rm H} \sim 10^{-7}$ and $n(HCOOCH_3)/n_{\rm H} \sim 10^{-9} - 10^{-7}$ (Cazaux *et al.* 2003; Chandler *et al.* 2005; Kuan *et al.* 2004). It indicates that hot corinos are composed of the envelope and the circumstellar disk, and some species (e.g. H_2CO and CH_3CHO) are abundant in the envelope, while others (e.g. CH_3OCH_3) are abundant in the disk regions. Observations with a high spatial resolution is needed to confirm this prediction.

It should also be noted that the D/H ratio of CH_3OCH_3 are a few %, inspite of its formation at 260 K (Figure 4*a*). Again, the high D/H ratio is a relic of the low-temperature chemistry; CH_3OCH_3 is formed from molecules (e.g. CH_3OH) with high D/H ratios, which are set in the cold phase.

4. 3D model of the first core formation

4.1. *the physical model and trajectories*

In order to understand the chemistry in star-forming cores with disks more quantitatively, it is necessary to adopt 2D or 3D models. Semi-analytical 2D model is presented in Visser *et al.* in this volume. Here we show preliminary results of chemistry in a 3D model, in which a cold dense core collapses to form the first core (Furuya *et al.* in prep). The first core is a transient hydrostatic core of H_2; after several 10^3 yrs, H_2 dissociates and the core collapses to form a protostar. In spite its short life time, it plays important roles in the star formation: (i) given a large amount of angular momentum, it can fragment to form a binary system (Matsumoto & Hanawa 2003), (ii) it drives an outflow (Tomisaka

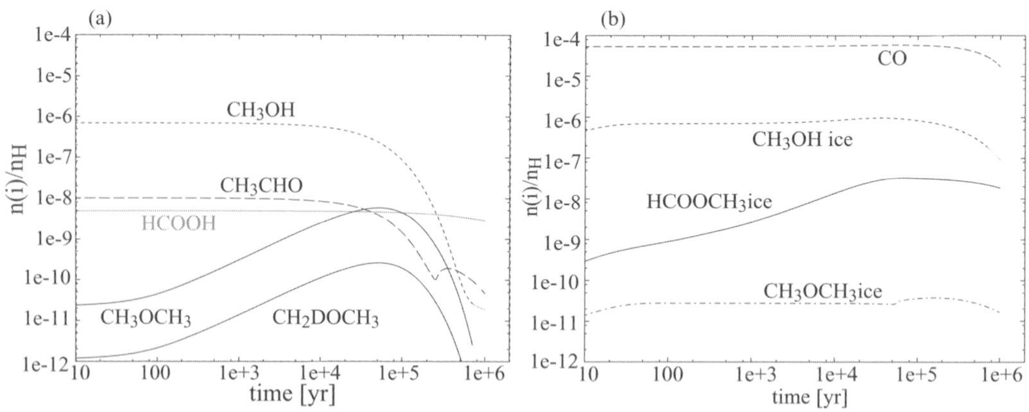

Figure 4. Temporal variation of molecular abundances at fixed temperatures and density after the fluid parcel enters the disk. The temperature is (a) $T = 260$ K and (b) $T = 40$ K, while the density is set to $n_{\rm H} = 4 \times 10^8$ cm^{-3}.

2002) and (iii) outer part of the first core evolves to the protoplanetary disks (Machida *et al.* 2010).

We adopt the 3D radiation hydrodynamic model by Tomida *et al.* (2010); the model is based on the 3D nested grid code of Matsumoto & Hanawa (2003) and developed to include the radiation transfer with the gray flux-limited diffusion approximation. The central dense region is divided to finer grids to resolve the structure; the finest grid corresponds to 0.08 AU, while the size of the model core as a whole extends to ~ 0.1 pc. Figure 5(a) shows trajectories of three fluid parcels. In the envelope region, the trajectories are almost free-fall, until the parcels enter the disk in the central region. The temporal variation of physical parameters in the fluid parcels are shown in Figure 5(b).

Since the hydrodynamic model contains a wide range of density and temperature, we combine several chemical network models to follow the chemistry in the parcels. At low temperatures ($T \leqslant 100$ K), we adopt the gas-grain model of Garrod & Herbst (2006), while at the higher temperatures we adopt Harada *et al.* (2010). We also implemented collisional dissociations and three-body reactions, which can be important at high temperatures and high densities, respectively, referring to Willacy *et al.* (1998). Rate coefficients and activation barriers of some collisional dissociations are updated referring to NIST-Janaf Thermochemical Tables of the heat of formation. Grain charge is solved referring to Umebayashi (1983); while at low densities ($n_{\rm H} \leqslant 10^{12}$ cm^{-3}) most of the grain particles are negatively charged and can be an important recombination partner of positive ions, neutral grains are dominant at higher densities.

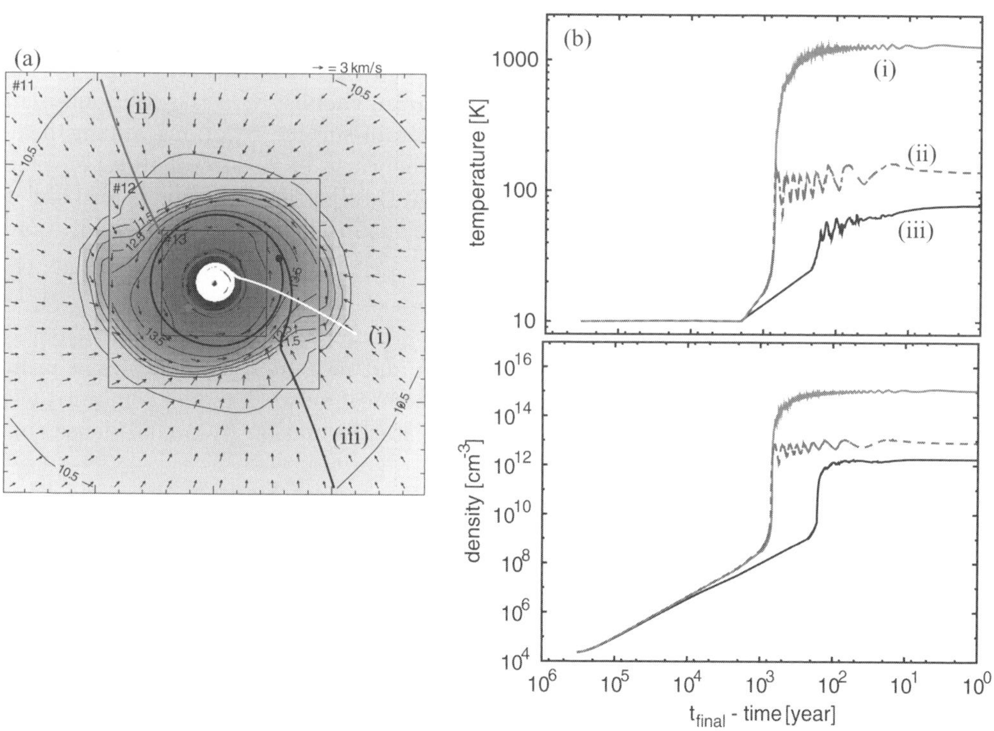

Figure 5. (a) Trajectories of the representative three fluid parcels in the 3D radiation hydrodynamic model. The counters depict the density distribution in the $z = 0$ plane. (b) Temporal variation of temperature and density in the three fluid parcels.

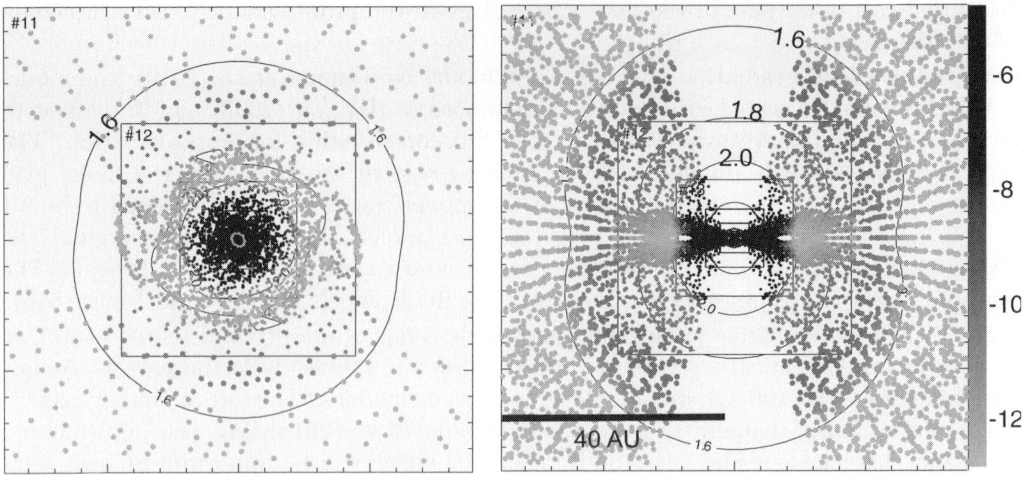

Figure 6. Spatial distribution of CH_3OH in the 3D first core model (gray dots). The line contours depict the temperature distribution in the $z = 0$ plane (left) and $y = 0$ plane (right).

4.2. *Molecular abundances*

In total we followed about 5000 fluid parcels and solved the gas-grain chemistry in each parcel. Figure 6 shows the abundances of gaseous CH_3OH in the x-z plane and the x-y plane. In the x-z plane, the CH_3OH distribution is almost spherical, simply following the temperature distribution. Our result indicates that CH_3OH can be a good tracer of the infalling envelope and the surface of the first core. In the x-y plane, on the other hand, the abundance decreases at the arms of the gravitational instability, where the density is slightly higher than the inter-arm region. In the zeroth approximation, the molecular abundances are determined by the gas-grain chemistry in the cold phase and sublimation at the temporal local temperature. Grain-surface reactions among heavy element molecules are limited, because the size of the warm ($T \sim$ several 10 K) envelope is small in the first core stage. Gas-phase reactions in the $T \geqslant 100$ K region is also limited by the short life time of the first core and the efficient mass accretion via gravitational instability.

5. Summary

We have investigated the molecular abundances and molecular D/H ratios in protostellar cores by calculating the detailed gas-grain chemical network model combined with the 1D and 3D radiation hydrodynamic models of star formation from cold dense cores to protostars. Our finding are as follows.

• While H_2CO and CH_3OH can be formed in molecular cloud conditions ($T \sim 10$ K), other large organic species such as CH_3CHO are formed mainly by the grain-surface reactions and gas-phase reactions at $T \sim$ several 10 K. In the infalling fluid parcels, such warm temperature lasts for $r_{\rm warm}/v_{\rm infall} \sim 10^4$ yr. Warm carbon chains are also formed from sublimated CH_4 at $T \geqslant 25$ K.

• While the spherical symmetry (1D model) is a good approximation in the envelope, a disk-like structures appears at $r \leqslant r_{\rm cent}$. Duration of the warm temperature chemistry can be longer than mentioned above once the fluid parcels enter the disk region. The prolonged warm chemistry are favorable in producing some large organic species such as CH_3OCH_3 and $HCOOCH_3$, which are underestimated in our spherical model. It indicates that the envelope and the disk are characterized by different molecules in hot corinos.

The Molecular Universe
Proceedings IAU Symposium No. 280, 2011
José Cernicharo & Rafael Bachiller, eds.

doi:10.1017/S1743921311024859

Observations of Complex Molecules in Low-Mass Protostars

Nami Sakai and Satoshi Yamamoto

Department of Physics and Research Center for the Early Universe, The University of Tokyo, Bunkyo-ku, Tokyo 113-0033, Japan
email: nami@phys.s.u-tokyo.ac.jp

Abstract. Low-mass star forming regions are rich inventories of complex organic molecules. Furthermore, they show significant chemical diversity even among sources in a similar physical evolutionary stage (i.e. Class 0 sources). One distinct case is the hot corino chemistry characterized by rich existence of saturated complex organic molecules such as $HCOOCH_3$ and C_2H_5CN, whereas the other is the warm carbon-chain chemistry (WCCC) characterized by extraordinary richness of unsaturated complex organic molecules such as carbon-chain molecules. We here summarize these observational achievements during the last decade, and present a unified picture of carbon chemistry in low-mass protostellar cores. The chemical diversity most likely originates from the source-to-source difference in chemical compositions of grain mantles. In particular, the gas-phase abundance of CH_4 evaporated from grain mantles is thought to be a key factor for appearance of WCCC. The origin of the diversity and its evolution to protopranetary disks are discussed.

Keywords. ISM: molecules, ISM: abundances, stars: formation

1. Introduction

So far more than 160 interstellar molecules have been identified mainly by radioastronomical observations, and about 1/3 of them are 'complex' molecules having 6 atoms or more (c.f. Herbst & van Dishoeck 2009). This clearly indicates the high chemical complexity of interstellar medium, which would ultimately be related to an origin of rich substances in our solar system. Hence, a detailed understanding of the chemical composition and its evolution in star forming regions is a central issue for astrochemistry.

Complex organic molecules (COMs) are classified into two general groups, saturated COMs and unsaturated COMs. Saturated COMs are fully or almost fully hydrogenated molecules such as CH_3OH, $HCOOCH_3$, $(CH_3)_2O$, and C_2H_5CN. These molecules are known to exist abundantly in hot cores in high-mass star forming regions. Since the gas-phase chemical reactions cannot produce them efficiently, they are thought to be formed by grain surface reactions (e.g., Charnley *et al.* 1997; Watanabe & Kouchi 2002; Garrod & Herbst 2006; Herbst & van Dishoeck 2009). The produced molecules stay on grain mantles in the cold phase, and they are librated into the gas phase after the birth of high-mass stars. Since 1980s, the hot core chemistry has extensively been studied to account for the observational results in various high-mass star forming regions and external galaxies (e.g., Blake *et al.* 1987; Millar *et al.* 1991; Nomura & Millar 2004; Herbst & van Dishoeck 2009). On the other hand, unsaturated COMs are represented by so called carbon-chain molecules having rich multiple bonds. They are known to exist abundantly and preferentially in young starless cores such as TMC-1. In fact, carbon-chain molecules are dominant organic molecules in TMC-1, according to the spectral line survey by Kaifu *et al.* (2004). Carbon-chain molecules are thought to be produced in the gas phase starting from C^+ and C, and hence, they are only abundant in young starless cores, where C^+ and C have not been converted to CO completely (Suzuki *et al.* 1992; Benson *et al.* 1998).

In contrast to the extensive astrochemical studies on high-mass star forming regions and cold starless cores, chemistry of low-mass star forming regions was sparsely studied until 1990s. A pioneering study toward this direction is the spectral line survey of the low-mass Class 0 protostar, IRAS 16293-2422, by Blake *et al.* (1994) and van Dishoeck *et al.* (1995). They studied the chemical composition of this source in an unbiased way for the first time. Although they found rich CH_3OH and SO_2, which can be recognized as a signature of the hot-core like chemistry, they could not detect the lines of larger COMs such as $HCOOCH_3$, $(CH_3)_2O$, and C_2H_5CN at that time, probably because of insufficient sensitivities in their observations. The chemical composition of low-mass star forming regions is of fundamental importance, because it can be regarded as the initial condition for chemical evolution to protoplanetary disks and planets. Thanks to great improvements of sensitivity in millimeter- and submillimeter-wave observations, such studies have extensively been carried out in 2000s, resulting in the following discoveries.

2. Hot Corino Chemistry and Warm Carbon-Chain Chemistry

First discovery was the detection of large saturated COMs such as $HCOOCH_3$, $(CH_3)_2O$, and C_2H_5CN in the low-mass star-forming region, IRAS16293-2422 (Cazaux *et al.* 2003), revealing that the hot-core like chemistry is also occurring even in a low-mass source. Furthermore, the interferometric observations showed that these molecules are distributed within a few 100 AU region (at temperature of $\sim$ 100 K) around the protostar (Bottinelli *et al.* 2004b; Kuan *et al.* 2004). These results along with subsequent detections of the saturated COMs in other low-mass star forming regions clearly demonstrated that fairly complex organic molecules are already formed in the protostellar-core phase, which would be brought into protoplanetary disks. Then, it was thought that they may be related to an origin of pre-solar organic molecules found in meteorites (e.g., Cronin & Pizzarello 1997 and references therein). Just as in the case of the hot core chemistry in high-mass star forming regions, saturated COMs are formed on dust grains during cold and warm-up phases (Garrod & Herbst 2006), and are liberated into the gas phase in a small hot region (a few 100 AU) around the protostar, called hot corino. Now four Class 0 protostars (IRAS 16293-2422, NGC1333 IRAS4A, IRAS4B, and IRAS2A) are known to harbor hot corinos (Cazaux *et al.* 2003; Bottinelli *et al.* 2004a; Jørgensen *et al.* 2005; Sakai *et al.* 2006; Bottinelli *et al.* 2007).

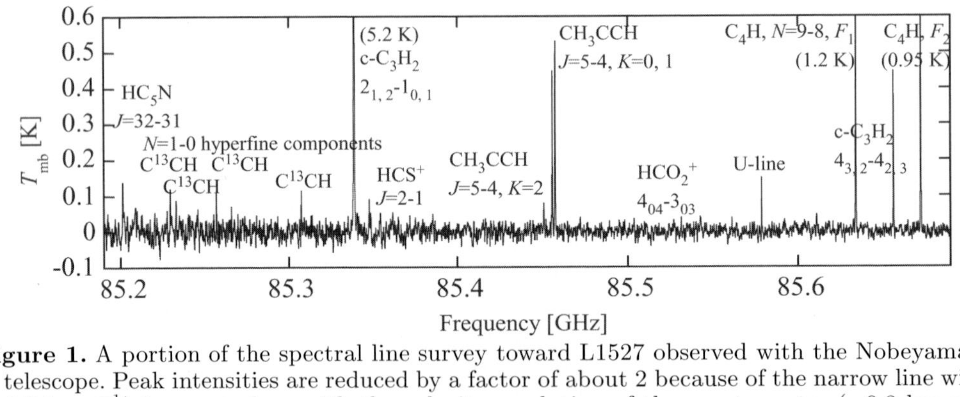

Figure 1. A portion of the spectral line survey toward L1527 observed with the Nobeyama 45 m telescope. Peak intensities are reduced by a factor of about 2 because of the narrow line width ($\sim$0.5 km s^{-1}) in comparison with the velocity resolution of the spectrometer ($\sim$0.8 km s^{-1}). In addition to the bright lines of C_4H and c-C_3H_2, even the very high excitation line of HC_5N ($J = 32 - 31$, $E_u = 67$ K) can be seen.

On the other hand, the extraordinary richness of unsaturated COMs such as C_4H, C_4H_2, and HC_5N was found in the lukewarm region around the low-mass Class 0 protostar, L1527 (Sakai *et al.* 2008a). For instance, the high excitation lines of C_4H ($N = 9-8$), C_2H ($N = 3-2$) were detected with very high intensities (Figure 1). Emissions of various long carbon-chain molecules and their anions like C_4H^- and C_6H^- were also detected (e.g. Sakai *et al.* 2007, 2008b; Agúndez *et al.* (2008)) (Figure 2), while the lines of saturated COMs were not (Sakai *et al.* 2008c). Hence, the protostellar core L1527 has a completely different chemical composition from that of hot corinos. A small mapping observation of C_4H with the Nobeyama 45 m telescope shows that the distribution is concentrated toward the protostar in a few thousand AU. More importantly, the line width becomes broader toward the protostar position. It is therefore highly likely that carbon-chain molecules do exist even in the gas infalling to the protostar. This result is striking, since it had been thought that carbon-chain molecules are abundant only in young starless cores and deficient in the evolved starless cores and star forming cores (e.g., Suzuki *et al.* 1992; Benson *et al.* 1998). To account for these results, a new production scheme of carbon-chain molecules triggered by evaporation of CH_4 from ice mantles was proposed (Sakai *et al.* 2008a). It is called warm carbon-chain chemistry (WCCC). The sublimation temperature of CH_4 is 25 K (Collings *et al.* 2004), being much lower than

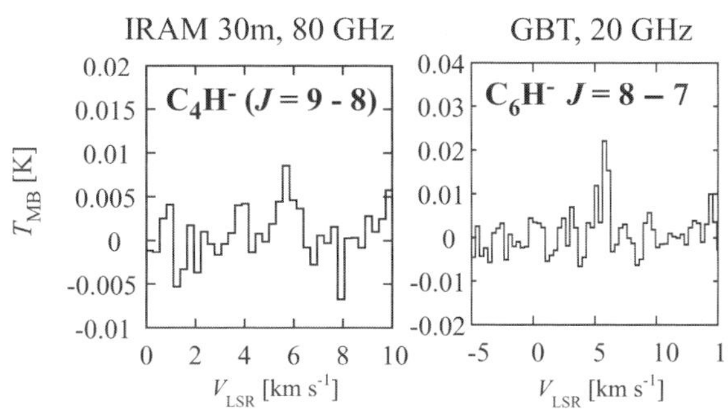

Figure 2. Spectral line profiles of carbon-chain anions observed toward L1527 (Sakai *et al.* 2007, 2008b).

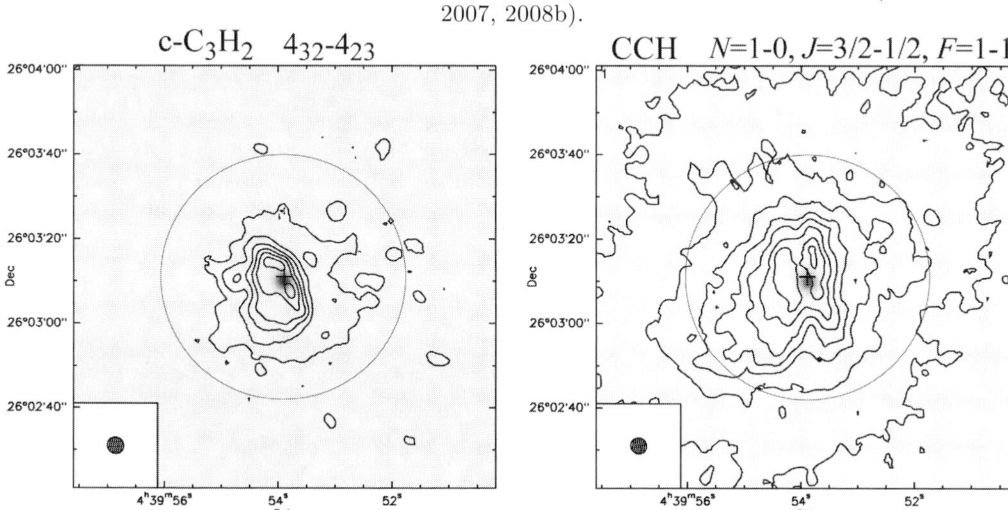

Figure 3. Integrated intensity maps of c-C_3H_2 and CCH observed toward L1527 with PdBI (Sakai *et al.* 2010. The maps are prepared by combining the single dish (IRAM 30 m) data to recover the extended component. The circles represent a field of view of PdBI.

that of H_2O. Hence, CH_4 is evaporated in a warm region extending to a few thousand AU, and reacts with C^+ in the gas phase to produce various carbon-chain molecules. The basic mechanism of WCCC was independently confirmed by chemical model calculations (Aikawa *et al.* 2008; Hassel *et al.* 2008; Harada & Herbst 2008).

An important question is then whether such peculiar chemistry is specific to L1527 or not. With this in mind, we searched for a similar source among many low-mass protostars, and finally found the second WCCC source, IRAS 15398-3359, in Lupus molecular cloud 1 ($d = 150$ pc) (Sakai *et al.* 2009a). Its spectral features resemble those in L1527, and various carbon-chain molecules are as abundant as in L1527. Furthermore, the distribution of the high excitation line of CCH ($N = 4 - 3$) shows central concentration around the protostar. According to the Spitzer observations, a deep infrared absorption feature of solid CH_4 is seen toward this source (Öberg *et al.* 2008), implying a favorable condition for WCCC. The discovery of the WCCC source outside the Taurus region suggests that WCCC can no longer be ascribed to a peculiarity of a particular region, but should be regarded as one of characteristic chemical features of low-mass protostellar cores.

Recently, we conducted an interferometric observation of carbon-chain molecules and their related molecules toward L1527 with PdBI, and confirmed that carbon-chain molecules are distributed within a few thousand AU around the protostar, as shown in Figure 3 (Sakai *et al.* 2010). The distributions show a steep increase inward of a radius of 500-1000 AU from the protostar, implying an abundance jump by an order of magnitude. The radius corresponds to the dust temperature of 20-30 K, which is just around the evaporation temperature of CH_4. This result further supports the WCCC in L1527. More interestingly, a small dip (500 AU scale) can be seen toward the protostar position. We carried out a simple radiative transfer calculation by employing the density and temperature structure of the L1527 protostellar core reported by Jørgensen *et al.* (2002), and found that the abundance of the carbon-chain molecules decreases in the innermost part by a factor of 10. Nevertheless, carbon-chain molecules are thought to survive partly in the gas infalling to the protostar, since the blue- and red-shifted components by $\sim$1 km s^{-1} are centrally concentrated (See Figures 2-4 of Sakai *et al.* (2008a)). Hence, it is likely that a portion of carbon-chain molecules produced in the Class 0 phase will be delivered to the protostellar disk.

3. Origin and Future of the Chemical Diversity

So far, it has been thought that the chemical compositions of low-mass star forming regions are more or less uniform, and hence, the chemical evolution from a protostellar disk to a protoplanetary disk has no significant variations. However, the above observational

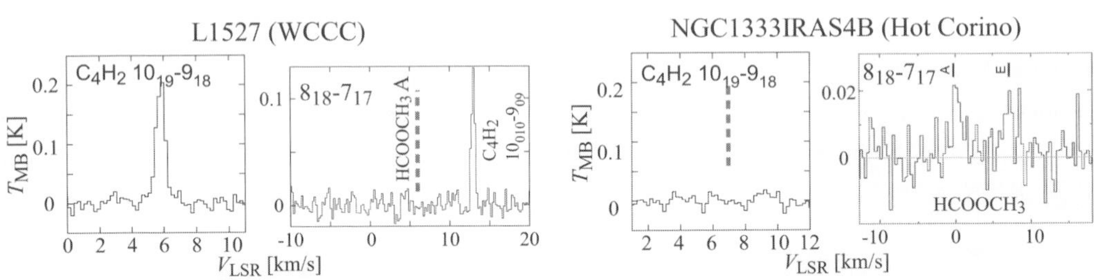

Figure 4. Chemical diversity of the Class 0 sources. In the WCCC source L1527, the carbon-chain molecules are abundant, whereas the complex organic molecules tend to be deficient. In the hot corino source NGC1333 IRAS4B, the situation is opposite. The C_4H_2 and $HCOOCH_3$ spectra in L1527 are taken from Sakai *et al.* (2008a) and Sakai *et al.* (2008c), respectively. The $HCOOCH_3$ spectrum in NGC1333 IRAS4B is from Sakai *et al.* (2006).

results clearly demonstrate diversity of chemical compositions in low-mass protostellar cores. At present, we know the hot corino and WCCC sources as the two extreme cases. These two types of sources show exclusive nature in chemical compositions. In hot corinos, unsaturated COMs such as carbon-chain molecules are generally deficient, whereas in WCCC sources, saturated COMS such as $HCOOCH_3$ are deficient (Figure 4). This implies that the chemical evolution to the protoplanetary disk is not a single pathway, and the chemical compositions of protoplanetary disks could be different from source to source.

3.1. *Carbon Chemistry around Low-Mass Protostar*

To consider the origin of the chemical diversity, we first summarize the chemical processes which would occur during the temperature raise due to protostellar activities. We start from a cold ($T < 10$ K) dense starless core, where most of the molecules except for H_2 are depleted on dust grains and no carbon chemistry is occurring. After the birth of the protostar, the temperature starts to rise. We can define the following three major steps in carbon-chemistry.

(1) Restart of carbon-chemistry ($T \sim 20$ K): At about 20 K, CO is evaporated from grain mantles, which restarts the basic carbon-chemistry by the $He^+ + CO \rightarrow C^+ + O + He$ reaction. The produced C^+ reacts with H_2 to form CH_2^+, and then makes various organic molecules. This is essentially the same as the conventional carbon chemistry in young starless cores (e.g., Herbst 1983). Since the $C^+ + H_2$ reaction is a slow radiative-association process, the efficiency to produce organic molecules is not very high.

(2) Warm carbon-chain chemistry ($T \sim 25$ K): When the temperature is raised up to 25 K, CH_4 can be evaporated. This realizes a temporary carbon-rich condition in the gas phase. Since C^+ already exists in stage (1), CH_4 reacts with C^+ to produce carbon-chain molecules. For this process to be effective, the fractional abundance of CH_4 relative to H_2 in the gas phase should be higher than several times 10^{-7}. In this case, CH_4 can be a major reactant for C^+ in place of OH and other molecules. Note that H_2O is still frozen on dust grains, and does not contribute to the C^+ destruction in this stage.

(3) Hot corino chemistry ($T = 70 - 100$ K): The ice mantle will be evaporated at this temperature. Complex organic molecules trapped in the ice mantle will also be released into the gas phase, which realizes the hot corino chemistry. At the same time, the $C^+ + CH_4$ reaction will be suppressed, because overwhelming H_2O molecules act as the main destructor of C^+. The $C^+ + H_2O$ reaction just reproduces CO through HCO^+. This would result in the deficiency of carbon-chain molecules.

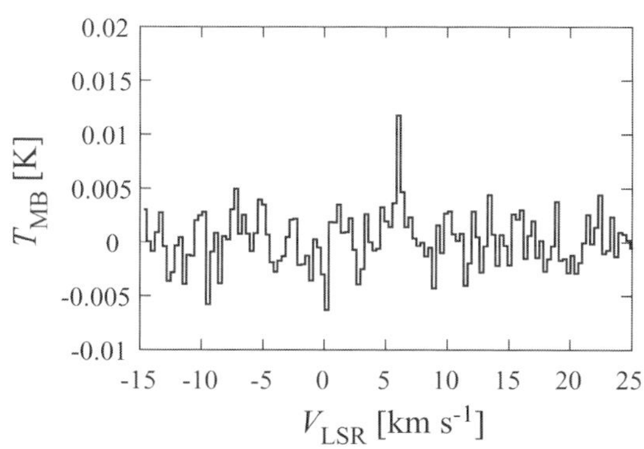

Figure 5. Spectral line profile of CH_3D ($J_K = 1_0 - 0_0$) observed toward L1527 with HHST.

According to the above picture, the WCCC becomes effective only when the gas-phase CH_4 abundance is higher than several times 10^{-7}. In this case, the chemical evolution will proceed in order of (1), (2), and (3). If the CH_4 abundance is not enough for WCCC, the evolution will proceed from (1) to (3) directly by skipping stage (2). Hence, the gas-phase CH_4 abundance is a key factor for appearance of WCCC. In fact, we have recently detected the rotational spectral line of CH_3D toward the prototypical WCCC source, L1527, and have found that the derived CH_4 abundance ($f = (4.3 - 15.2) \times 10^{-6}$) indeed fulfills this criterion (Figure 5). A significant spread in the CH_4/H_2O abundance ratio in the solid phase is seen in infrared observations of low-mass protostars (Öberg *et al.* 2008). Considering that the difference of the CH_4 abundance would be amplified in abundances of carbon-chain molecules through their formation processes, it is highly likely that the abundance of CH_4 supplied from grain mantles would control the chemical composition in the low-mass protostars. In the present stage, it is not certain whether the WCCC sources have a hot corino activity in the closest vicinity of the protostar. However, the hot corino activity should be weak even if it does exist, since the CH_3OH lines ($J_K = 7_K - 6_K$) are much weaker in the WCCC sources than in the hot corino sources (Jørgensen *et al.* 2007; Sakai *et al.* in prep) and large saturated COMs are not detected in L1527. This situation is schematically shown in Figure 6.

From the above considerations, the chemical diversity would not originate from a simple evolutionary effect, but the different chemical compositions of grain mantles are needed to be considered. This is consistent with the fact that the hot corino sources are not always more evolved than the WCCC sources.

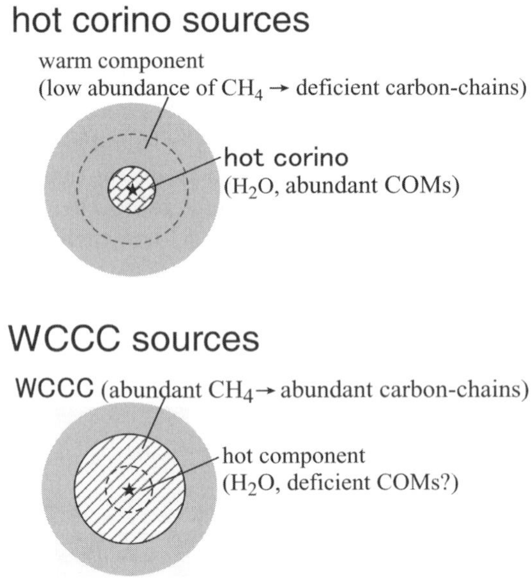

Figure 6. Physical appearance of the hot corino sources and the WCCC sources. In the hot corino sources, carbon-chain molecules are deficient over the protostellar core, whereas saturated COMs are abundant in the vicinity of the protostar (a few hundred AU). In the WCCC sources, carbon-chain molecules are distributed abundantly in a lukewarm region near the protostar (a few thousand AU). On the other hand, large saturated COMs are not detected, indicating that the hot corino activity is very weak.

3.2. *Origin of the Different Mantle Compositions*

As seen in the above unified picture of carbon chemistry in low-mass star forming regions, hot corino chemistry and the WCCC would reflect the different chemical compositions of grain mantles. This difference would reflect the past history during the starless core phase, when the grain mantle had grown up. One important factor is the duration time of the starless-core phase (Sakai *et al.* 2009a) (Figure 7). It could depend on large-scale environmental effects including magnetic field strength and turbulent energy input. The WCCC prefers the short starless-core phase. In this case, more carbon atoms are depleted directly onto dust grains before being converted to CO in the gas phase, which makes more CH_4 on dust grains by hydrogenation. In contrast, the hot corino chemistry requires the long starless-core phase. In this case, CO is mainly depleted onto dust grains to form saturated COMs, whereas CH_4 is hardly produced.

The short starless-core phase for the WCCC sources is also supported by the deuterium fractionation ratios. Sakai *et al.* (2009b) observed many deuterated carbon-chain molecules, and found that the deuterium fractionation ratios are moderate (2-7 %). The CH_2DOH/CH_3OH ratio is derived to be lower than 3 %, which is much lower than that found in hot corinos (Parise *et al.* 2006). The observed ratios mean less degree of the CO depletion, which is consistent with the short time scale of the starless-core phase.

The two cases shown in Figure 7 are the extreme ones, and the intermediate case would be possible, depending on the length of the starless-core phase. In this way, the continuous chemical variation shown in our C_4H survey (Sakai *et al.* 2009a) can naturally be interpreted. If this is the case, we would be able to obtain novel information on the past contracting process of the parent core from what type of chemistry is now occurring near the protostar. We think that this can be a new approach to understanding physical processes in the early stages of star formation.

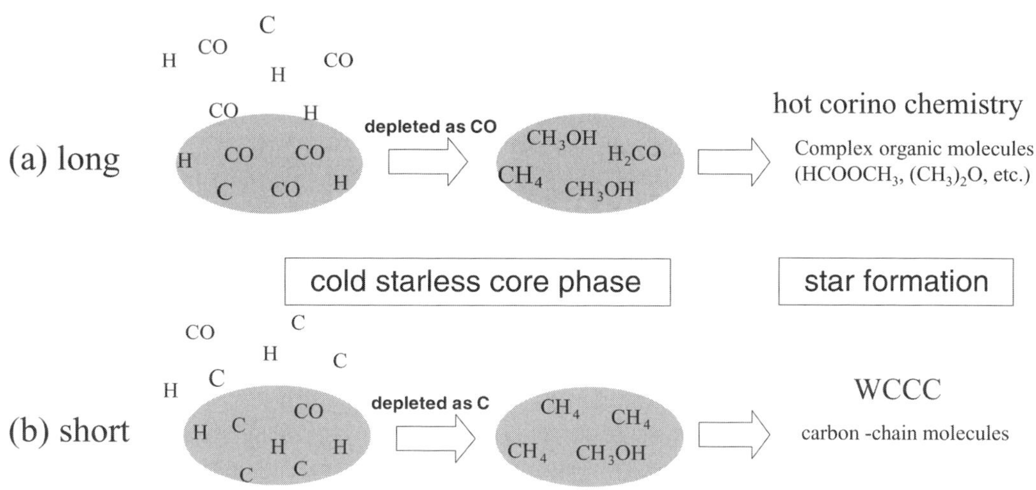

Figure 7. Origin of chemical variation. The two extreme cases are shown. The case (a) represents the chemical evolution for a long starless core phase. Most of the C atoms are converted to the CO molecules in the gas phase, which then deplete onto dust grains. They then form CH_3OH, H_2CO, and more complex organic molecules through hydrogenation reactions and radical reactions on the grain surfaces. On the other hand, the case (b) corresponds to a short starless-core phase in which a certain amount of the C atoms deplete directly onto the dust grains, and then form CH_4 effectively through hydrogenation. This realizes WCCC.

4. Summary and Implications

During the last decade, chemical compositions of low-mass star-forming regions have been investigated extensively, as summarized above. Now, we recognize the significant chemical diversity even among sources in a similar physical evolutionary stage. Nevertheless, we still have many unknowns on the chemical evolution of protostellar cores, some of which are highlighted below:
(1) We do not know whether the hot corino activity exists inside the WCCC sources, as mentioned above. According to the unified picture, the hot corino chemistry would occur in the vicinity of the protostar even for the WCCC sources. However, the hot corino activity is known to be weak in these sources. This means that saturated COMs are not well produced in the WCCC sources for some reasons. To address this problem, we have to study the weak hot corino activity in the WCCC sources with sensitive and high-resolution observations. Furthermore, it is necessary to understand the formation mechanism of saturated COMs on grain mantles in more detail.
(2) How unique are the WCCC and the hot corino chemistry among the low-mass star forming regions? We need to investigate chemical compositions of more low-mass protostellar cores for statistical arguments.
(3) How is the chemical diversity of the Class 0 sources brought into the later stages? Although the relation between hot corinos and protostellar disks is still unknown, saturated COMs are acturally distributed within a few 100 AU region around the protostar. On the other hand, carbon-chain molecules also exist in the closest vicinity of the protostar ($<$ 500 AU) in the WCCC sources (Sakai *et al.* 2010). Thus, the chemical variation would be delivered to the protoplanetary disks. Its detailed processes are needed to be clarified.

Here we stress that the above questions are only major ones. We still have a big gap between chemistry of star-forming regions and that of protoplanetary disks. In the next decade, a dramatic progress is expected in this area with the full operation of ALMA.

Finally, we mention unique importance of COMs in tracing the chemical evolution. Observations of 'fundamental' molecules such as CO, HCN and HCO^+ are essential to study the current physical structures and kinematics of star-forming regions. Their relatively high abundances ensure bright emissions in many sources to make such a business easy. However, fundamental species appear everywhere, and it is difficult to extract chemical characteristic of each source from their observations. On the other hand, COMs are much more sensitive to chemical and physical environments and are suitable for chemical characterization. A drawback is faintness of the lines, and hence, sensitive observations are necessary. If we can overcome this for instance with ALMA, complex molecules will gives us fruitful information, not "complex" information.

Acknowledgements

The authors are grateful to Yuri Aikawa, Michael Burton, Cecilia Ceccarelli, Eric Herbst, Tomoya Hirota, Jes Jørgensen, Nagayoshi Ohashi, Takeshi Sakai, and Yancy Shirley for their stimulating discussions. This study is supported by Grant-in-Aids from Ministry of Education, Culture, Sports, Science, and Technologies (21224002 and 21740135).

References

Agúndez., M. Cernicharo, J., Guélin, M., Gerin, M., McCarthy, M. C., & Thaddeus, P. 2008, *A&A*, 478, L19

Aikawa, Y., Wakelam, V., Garrod, R. T., & Herbst, E. 2008, *ApJ*, 674, 984

Benson, P. J., Caselli, P., & Myers, P. C. 1998, *ApJ*, 506, 743
Blake, G. A., Sutton, E. C., Masson, C. R., & Phillips, T. G. 1987, *ApJ*, 315, 621
Blake, G. A., van Dischoeck, E. F., Jansen, D. J., Groesbeck, T. D., & Mundy, L. G. 1994, *ApJ*, 428, 680
Bottinelli, S., *et al.*, 2004, *ApJ*, 615, 354
Bottinelli, S., *et al.*, 2004, *ApJ*, 617, L69
Bottinelli, S., Ceccarelli, C., Williams, J. P., & Lefloch, B. 2007, *A&A*, 463, 601
Cazaux, S., *et al.*, 2003, *ApJ*, 593, L51
Charnley, S. B., Tielens, A. G. G. M., & Rodgers, S. D. 1997, *ApJ*, 482, L203
Collings, M. P., *et al.*, 2004, *MNRAS*, 354, 1133
Cronin, J. R. & Pizzarello, S. 1997, *Sci.*, 275, 951
Garrod, R. T. & Herbst, E. 2006, *A&A*, 457, 927
Harada, N. & Herbst, E. 2008, *ApJ*, 685, 272
Hassel, G. E., Herbst, E., & Garrod, R. T. 2008, *ApJ*, 681, 1385
Herbst, E. 1983, *ApJS*, 53, 41
Herbst, E. & van Dishoeck, E. F. 2009, *ARAA*, 47, 427
Jørgensen, J. K., Schöier, F. L., & van Dishoeck, E. F. 2002, *A&A*, 389, 908
Jørgensen, J. K., Bourke, T. L., Myers, P. C., Schöier, F. L., van Dishoeck, E. F., & Wilner, D. J. 2005, *ApJ*, 632, 973
Jørgensen, J. K., *et al.*, 2007, *ApJ*, 659, 479
Kaifu, N., *et al.*, 2004, *PASJ*, 56, 69
Kuan, Y.-J., *et al.*, 2004, *ApJ*, 616, L27
Millar, T. J., Herbst, E., & Charnley, S. B. 1991, *ApJ*, 369, 147
Nomura, H. & Millar, T. J. 2004, *A&A*, 414, 409
Öberg, K. 2008, *ApJ*, 678, 1032
Parise, B., *et al.*, 2006, *A&A*, 453, 949
Sakai, N., Sakai, T., & Yamamoto, S. 2006, *PASJ*, 58, L15
Sakai, N., Sakai, T., Osamura, Y., & Yamamoto, S. 2007, *ApJ*, 667, L65
Sakai, N., Sakai, T., Hirota, T., & Yamamoto, S. 2008a, *ApJ*, 672, 371
Sakai, N., Sakai, T., & Yamamoto, S. 2008b, *ApJ*, 673, L71
Sakai, N., Sakai, T., & Yamamoto, S. 2008c, *Ap&SS*, 313, 153
Sakai, N., Sakai, T., Hirota, T., Burton, M., & Yamamoto, S. 2009a, *ApJ*, 697, 769
Sakai, N., Sakai, T., Hirota, T., & Yamamoto, S. 2009b, *ApJ*, 702, 1025
Sakai, N., Sakai, T., Hirota, T., & Yamamoto, S. 2010, *ApJ*, 722, 1633
Suzuki, H., *et al.*, 1992, *ApJ*, 392, 551
van Dishoeck, E. F., Blake, G. A., Jansen, D. J., & Groesbeck, T. D. 1995, *ApJ*, 447, 760
Watanabe, N. & Kouchi, A. 2002, *ApJ*, 571, L73

Discussion

Doug Johnstone: You suggest that the difference between Hot Corino Sources and WCCC sources is due to the speed with which they collapse but how important might the environment of the core be? For exampe, a core at the edge of a cloud may live in a higher radiation field?

Nami Sakai: I agree that environmental effects are important. We should concretely identify what types of environmental effects discriminate between the hot corino chemistry and WCCC. One important factor would be the large scale star-formation activity in the region where the protostar resides. Although the number of the sample is limited, a region harboring hot corinos is usually a formation site of star clusters, whereas a region harboring WCCC sources is relatively quiescent. Such a difference in the large-scale environment may affect the contracting speed of the clouds. The position of the cores in a parent cloud could also be a possible cause, as you mention. More statistics are essential.

E. F. van Dishoeck: It is not clear to me why you so easily dismiss the information on the "spread" in observed CH_4 ice abundances toward pre-stellar cores and protostars, since these data should probe close to the sublimation temperature of CH_4 ice.

Nami Sakai: We think that the CH_4 ice abundance would strongly depend on the density distribution and the temperature structure, and hence, we have directly measured the gas-phase CH_4 abundance in a warm region to test our WCCC scenario. However, I agree that we should more carefully make use of the CH_4 ice abundances reported so far, because a factor of difference in the CH_4 abundance would cause a big difference in the abundances of complex species.

Martin Cordiner: Are there any other observational tracers to test the idea a rapid freeze-out produces elevated ice CH_4 abundances?

Nami Sakai: The molecular D/H ratios would be an independent tracer. The relatively low fractions observed in L1527 suggest the freeze-out of gaseous species in early stages of chemical evolution of its parent cloud.

S. Viti: TPD experiments (e.g. Collings *et al.* 2004, Viti *et al.* 2004) show that not all the CO and CH_4 evaporate at 20 K and 25 K respectively. In fact you should have about 40 % of CH_4 still on the grains up to ~70 K. How does this agree your conclusion?

Nami Sakai: Thank you very much for your comments. It would not affect seriously on our conclusion. Even if that portion (~40 %) of the CH_4 ice still remains on grains and is released into the gas phase in the hot (> 70 K) region, carbon-chain production would not be effective, because H_2O becomes the main destructor of C^+ in such a hot region. However, your point will be important for detailed modeling.

The Molecular Universe
Proceedings IAU Symposium No. 280, 2011
José Cernicharo & Rafael Bachiller, eds.

doi:10.1017/S1743921311024860

Interferometric Studies of Low-Mass Protostars

Jes K. Jørgensen[1]

[1]Niels Bohr Institute & Centre for Star and Planet Formation, University of Copenhagen
Juliane Maries Vej 30 DK-2100, Copenhagen Ø, Denmark
email: jeskj@nbi.dk

Abstract. With the advances in high angular resolution (sub)millimeter observations of low-mass protostars, windows of opportunities are opening up for very detailed studies of the molecular structure of star forming regions on wide range of spatial scales. Deeply embedded protostars provide an important laboratory to study the chemistry of star formation – providing the link between dense regions in molecular clouds from which stars are formed, i.e., the initial conditions and the end product in terms of, e.g., disk and planet formation. High angular resolution observations at (sub)millimeter wavelengths provide an important tool for studying the chemical composition of such low-mass protostars. They for example constrain the spatial molecular abundance variations – and can thereby identify which species are useful tracers of different components of the protostars at different evolutionary stages. In this review I discuss the possibilities and limitations of using high angular resolution (sub)millimeter interferometric observations for studying the chemical evolution of low-mass protostars – with a particular keen eye toward near-future ALMA observations.

Keywords. astrochemistry — stars: formation — planetary systems: protoplanetary disks — ISM: abundances — ISM: molecules — radiative transfer — techniques: high angular resolution — techniques: interferometric

1. Introduction

Our standard picture suggests that low-mass stars form from the gravitational collapse of dense molecular cloud cores. In the earliest stages, the newly formed young star is deeply embedded in its infalling envelope of cold gas and dust obscuring its radiation. Eventually this envelope is dissipated revealing the pre-main sequence star surrounded by the circumstellar disk. From an astrochemical point of view, one of the prime goals is to understand how dust and gas evolves during this process – from the cold molecular clouds until it arrives in a planet-forming disk.

In this context, studies of young stars during their embedded (Class 0 and I) stages are of prime importance for our understanding of the formation of solar-type stars and their disks. On the one hand, they make it possible to probe the star formation process right after the collapse has occurred and thereby reveal the physical and chemical relationship between the protostar and the environment in which it forms. On the other hand, the initial conditions for the subsequent evolution of the young star are determined during these embedded stages. This is when the star accretes the bulk of its mass and a circumstellar disk forms. The physical and chemical structure of this disk may determine whether planets are formed around a given young star later in its evolution.

The evolution of the gas and dust during these stages is a complex process. It involves following the chemistry through environments with significantly varying physical properties (densities, temperatures); properties that are changing on time-scales that are relatively short or comparable to, e.g., the time-scales for molecules to freeze-out on dust

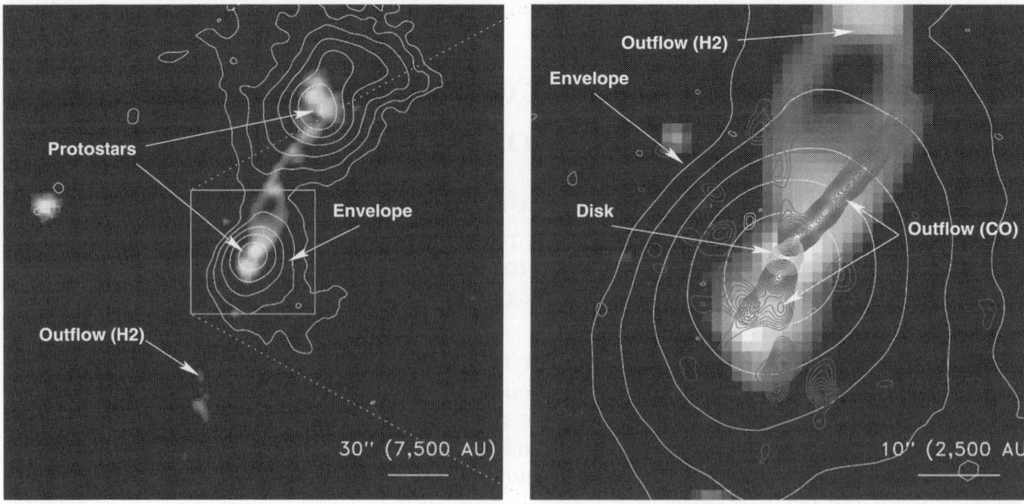

Figure 1. Observations of embedded protostars in the L1448 region with the Class 0 protostar L1448-C in the center of the picture. **The envelope** is revealed by its cold dust probed by JCMT/SCUBA 850 μm continuum observations. **The outflow** is traced by Spitzer observations of shocked H_2 at temperatures of 500–1000 K (green color in background image) and by SMA observations of CO $J = 2-1$ emission at temperatures of 20–50 K (red and dark blue contours in right panel). **The disk** is revealed by SMA 850 μm continuum observations (light-blue contours in right panel). *Spitzer/IRAC observations from Jørgensen et al. (2006) (background image; blue is 3.6 μm; green 4.5 μm and red 8.0 μm) and SMA and JCMT observations from Jørgensen et al. (2007).*

grains, for grain-surface reactions in the dust grains' ice mantles or for gas-phase chemistry in the warm gas in the envelope. For a typical molecular cloud core, densities range from just $10^3 - 10^4$ molecules cm^{-3} in the molecular cloud stage to $10^7 - 10^8$ cm^{-3} in the inner envelopes of low-mass protostars, and potentially even higher in their disks. Likewise, temperatures typically range from around 10 K in the outer envelopes to a few hundred K in the inner envelope or surface layers of the circumstellar disks – and can locally increase even further through shocks related to the accretion or caused by the ubiquitous outflows.

To fully understand the chemical evolution of low-mass protostars we therefore need to observe a wide range of spatial scales with observational techniques sensitive to these large variations in physical conditions. (Sub)millimeter interferometric observations provide such a unique opportunity to probe the warm and dense material in the innermost regions of protostellar systems – and to image the radial variations in the chemistry related to the changes in the underlying physical structures of the protostars and their environments.

2. Continuum studies of low-mass protostars

Continuum emission at (sub)millimeter wavelengths is dominated by thermal radiation from the dust grains in the envelopes and thus measure the combined effects of increasing temperatures and column densities toward the center of the protostellar cores. On larger (> 1000 AU) scales, the temperature and density distributions of the protostellar envelopes can be determined from radiative transfer modeling of single-dish submillimeter continuum images and mid- and far-infrared spectral energy distributions of the young stellar objects (e.g., Jørgensen *et al.* 2002; Schöier *et al.* 2002; Shirley *et al.* 2002). Such radiative transfer models show that the bulk of the mass in typical infalling envelopes

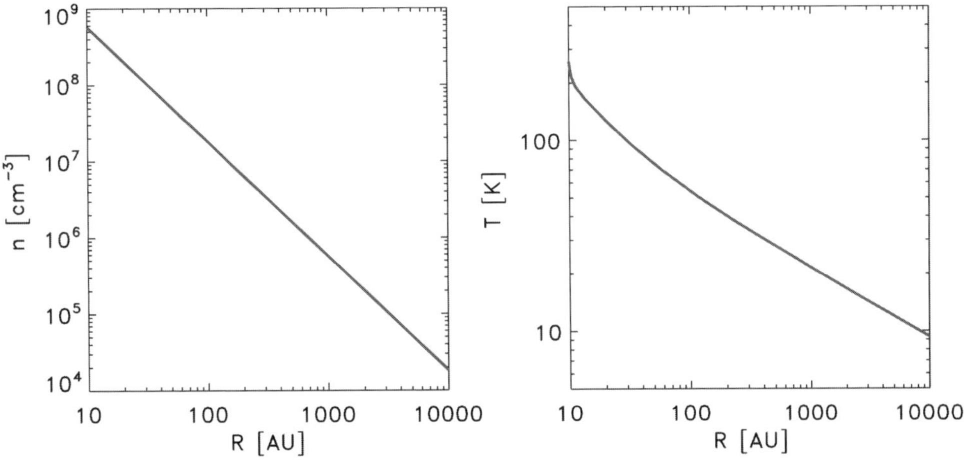

Figure 2. Density and temperature profile for 1 $M_\odot$ envelope around 5 $L_\odot$ protostar. The envelope is assumed to have a power-law density profile, $n \propto r^{-1.5}$, corresponding to material in free-fall toward the center, an inner radius of 10 AU and an outer radius of 10,000 AU. The temperature is calculated self-consistently using the DUSTY radiative transfer code (Ivezić *et al.* 1999; see also Jørgensen *et al.* 2002).

is located on large scales and that gas and dust heated to 20 K or more are located in the inner 1000 AU (5″ at a typical distance of 200 pc for nearby low-mass star forming regions).

These types of profiles provide a good first order description of the physical conditions in the protostellar envelopes necessary to solve for excitation of molecules, predict line strengths and, e.g., calculate synthetic spectra that can be compared to observations to constrain molecular abundances or kinematics of low-mass protostars. However, it is worth noting that these are dependent on a number of underlying assumptions about among others *(i)* simplified structures of the sources, *(ii)* dust properties – in particular opacities that may vary as function of density, temperature and more, *(iii)* some degree of coupling between the dust and gas – e.g., in terms of the thermal balance.

2.1. *Extrapolation of envelope structures to small radii*

In recent years significant efforts have been put into characterizing the inner few hundred AU of embedded protostars through continuum observations at (sub)millimeter wavelengths. One of the expected outcomes of the collapse of rotating protostellar cores is the formation of a circumstellar accretion disk. Theoretically the growth of this disk is expected to be closely coupled to the rotation speed and infall rate in the initial core (Terebey *et al.* 1984) – as well as whether magnetic fields introduce differential rotation (e.g., Basu 1998).

At (sub)millimeter wavelengths the continuum emission from such disks is likely dominated by the thermal radiation from their dust grains, which is becoming stronger at shorter wavelengths with $F_\nu \propto \nu^{2-3}$ or steeper. (Sub)millimeter interferometric observations of the continuum emission toward embedded protostars coupled to detailed radiative transfer models similar to those described above (e.g., Jørgensen *et al.* 2004a, 2005a, 2009; Brinch *et al.* 2007; Chiang *et al.* 2008; Lommen *et al.* 2008; Enoch *et al.* 2009, 2011) show evidence for compact continuum emission on $\sim 1''$ scales (inner ~ 100 AU) that cannot be explained by typical infalling envelopes (Fig. 3). If these compact continuum components indeed reflect the presence of circumstellar disks on ~ 100 AU scales, this would argue in favor of a scenario in which disks are formed early, and grow rapidly to sizes comparable to those around more evolved sources (e.g., Jørgensen *et al.* 2009; Enoch *et al.* 2011). But the compact continuum emission can also be explained by, e.g., magnetic pseudo-disks (e.g., Brinch *et al.* 2009; Maury *et al.* 2010), i.e., flattened disk-like structures that extend out to 500–1000 AU scales, but are not rotationally supported (Galli & Shu 1993a,b; Hennebelle & Ciardi 2009), or by magnetic accretion shock walls (Chiang *et al.* 2008).

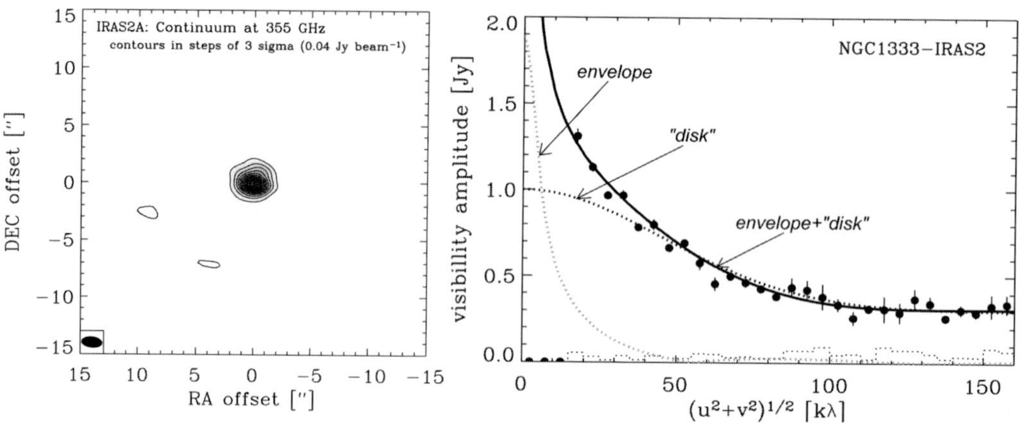

Figure 3. SMA 850 μm continuum observations from the deeply embedded (Class 0) protostar NGC1333-IRAS2A (Jørgensen *et al.* 2005a). **Left:** image at 355 GHz. **Right:** visibility amplitude as function of projected baselines probing scales from about 20″ to 1″ (2000 AU to 100 AU radius). The results of dust radiative transfer modeling of the sources are also shown: a model with a pure envelope (explaining the larger-scale SCUBA emission and single-dish line observations of the collapsing core on scales from a few thousand to 10,000 AU) cannot account for most of the emission seen with the SMA, but with a compact continuum component ("disk") introduced the emission is well-reproduced on all scales.

In more evolved protostars, some of these compact continuum components can unambiguously be attributed to Keplerian disks through resolved line observations revealing their kinematical structures (e.g., Brinch *et al.* 2007; Lommen *et al.* 2008; Jørgensen *et al.* 2009). However, for the more embedded sources with envelope masses larger than about 0.1 $M_{\odot}$, the key (sub)millimeter transitions of species such as CO, HCO^+, CS, and HCN, traditionally used for studying low-mass protostars, become optically thick on few hundred AU scales and are thus not able to probe the region where rotation starts to dominate over infall. This lack of kinematical information makes the interpretation of the compact continuum structures unclear. One option is instead to target the isotopologues of the same molecular species: in fact, for one of the best studied Class 0 protostars, NGC 1333-IRAS2A, high angular SMA observations of the $H^{13}CN$ isotopologue actually suggest that the kinematics of the envelope is dominated by infall rather than rotation down to few hundred AU scales (Brinch *et al.* 2009) – perhaps evidence for the pseudo-disk interpretation. Generally, however, the current interferometers have lacked the sensitivity to image the emission from low surface brightness emission from these isotopologues and such studies are consequently sparse.

No matter the physical nature of the compact continuum components, the main issue raised by these observations is that the profiles of low-mass protostars remain poorly constrained on $\sim$ 100 AU scales. They illustrate that extrapolations of density profiles established on a few thousand AU scales are problematic and absolute abundances of molecular species measured relative to such density profiles – either through high excitation transitions or in resolved observations – should be taken with care when used for interpretations of the chemical structures of protostars. Another solution is to consider relative abundances of different molecular species that can be resolved spatially, although for this differences in the chemical structures of the studied molecules (e.g., radial variations) should be taken into account. Comparing, e.g., the (sub)millimeter rotational transitions from molecules such as CO that are thought to desorb at temperatures of 20–30 K to molecules only thought to be present in much warmer gas will likewise be strongly model dependent.

3. Interferometric observations of the gas chemistry of low-mass protostars

As for the continuum observations there are definite benefits of going to the shorter submillimeter wavelengths when studying low-mass protostars. Interpretation of lower excitation lines are complicated by the fact that a large fraction of their emission may arise in the outer cold regions of the envelopes, where the lines can also become optically thick. As an example, Fig. 4 compares the radius where the $\tau = 1$ surface is encountered for continuum and HCO^+ 3–2 line emission (calculated from the outside inwards) in spherical, free-falling envelopes (density profile, $n \propto r^{-1.5}$) and a singular isothermal spheres ($n \propto r^{-2}$) of differing masses. As shown in the figure, for typical low-mass protostars the continuum emission remains optically thin down to scales of a few AU even for steep density profiles and massive envelopes. In contrast, the line emission of a commonly used tracer such as HCO^+ 3–2 becomes optically thick at radii > 100 AU even for envelopes of a few 0.1 $M_\odot$ mass. This means that the continuum optical thickness will not obscure the line radiation and that it, at least from this perspective, should be possible to find molecular species tracing the center regions of these protostars. However, many of the more common molecular species traditionally targeted using (sub)millimeter interferometry are not sensitive to what is going on the smallest scales – for example, making observations of the kinematics of the inner rotating and infalling regions of the protostellar envelopes and disks difficult (see also Jørgensen *et al.* 2009).

3.1. *Radial chemical variations in envelopes: large to intermediate scales*

As noted above, the chemistry in protostellar environments is thought to evolve on time-scales similar to the dynamical time-scales for the protostellar envelopes. For example, freeze-out of molecules such as CO onto dust grains is expected to occur over time-scales of $10^4 - 10^5$ years at the densities characterizing the outer ($T < 20 - 30$ K) envelope (e.g., Jørgensen *et al.* 2005b). As the material is falling-in toward the center of the cores, the molecular species will evaporate back into the gas-phase as the temperature due to heating by the central protostar exceeds the desorption temperature for the ices (typically 20–30 K for CO, 90–100 K for water containing ices etc.).

Based on line radiative transfer modeling of single-dish observations of protostellar envelopes Jørgensen *et al.* (2004b, 2005b) suggested that molecules such as CO and related species would show characteristic "drop abundance" profiles. That is, an outer envelope with a small degree of depletion because of the low densities and associated long time-scales for freeze-out, an intermediate region where the density is high enough that significant depletion can occur, but the temperature is low enough to avoid thermal desorption and an inner region where the molecules desorb back into the gas-phase due to higher temperatures (see, e.g., middle panel in right column of Fig. 5). The validity of

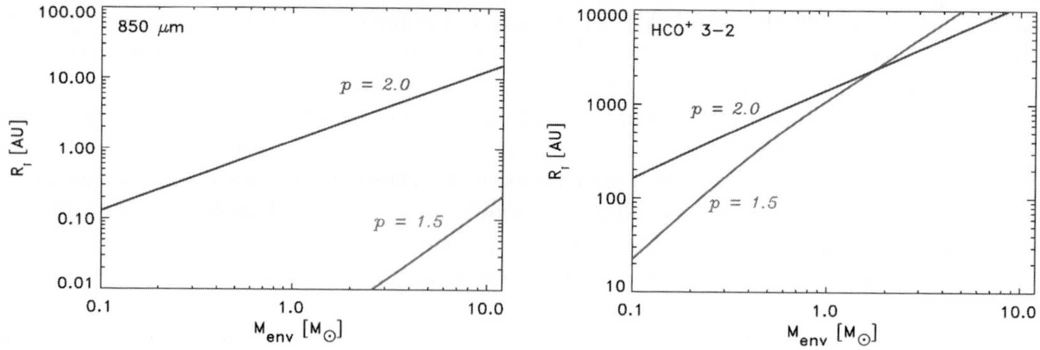

Figure 4. Radius where the $\tau = 1$ surface is reached for emission from continuum at 850 μm (left) and the HCO^+ 3–2 line (right) as function of envelope mass. Note the differences in scales of the Y-axes. In each plot two curves are shown for envelopes with density profiles of $n \propto r^{-1.5}$ and $n \propto r^{-2}$, respectively. For HCO^+ the optical depth was calculated using the Radex escape probability code (van der Tak *et al.* 2007) assuming a temperature of 25 K, density of 5×10^5, HCO^+ abundance with respect to H_2 of 10^{-9} and line width of 3 km s^{-1} typical values observed for the envelopes of low-mass protostars. The optical depth is only weakly dependent on the density and temperature in this regime, but naturally increases with increasing abundance and decreasing line width.

this scenario (in particular, the presence of the "intermediate" and "inner" regions") can be tested by direct imaging of different molecular species and by constraining their radial abundance variations. Fig. 5, as an example, compares the radial profiles (distribution of visibility amplitudes as function of projected baselines) for three molecular species, SO, $H^{13}CN$ and CH_3OH, toward the Class 0 protostar, NGC 1333-IRAS2A (compare also to Fig. 3). The three species show significantly different emission morphologies: SO is well reproduced by a constant abundance throughout the envelope and any abundance enhancement on small-scales can be ruled out. The emission of $H^{13}CN$ cannot be explained by a constant abundance, but rather require a drop in abundance similar to the profiles described above. CH_3OH finally shows compact emission that can best be described as originating in either the circumstellar disk or in the inner warm envelope – or "hot corino".

The freeze-out and desorption of molecules also play an important role in regulating the gas-phase chemistry in different regions of the protostellar envelopes. Perhaps the simplest example is the chemistry of the key molecular ion N_2H^+. N_2H^+ is strongly regulated by the presence or absence of CO: when CO is present in the gas-phase it acts as the main destroyer of N_2H^+, lowering its abundance significantly driving nitrogen into N_2. However, when CO is starting to freeze-out, the abundance of N_2H^+ is enhanced (e.g., Jørgensen *et al.* 2004b, and references therein). Again, this and similar effects can be tested through imaging of emission from CO and N_2H^+ at millimeter wavelengths. Fig. 6 shows one example for the L483 protostellar cores (Jørgensen 2004); similar morphologies have been observed toward Barnard 1c (Matthews *et al.* 2006) and IRAS 16293-2422 (Jørgensen *et al.* 2011).

3.2. *The innermost regions: hot corinos and disks*

One of the most interesting recent developments in the astrochemistry of low-mass protostars has undoubtedly been the detections on complex organics in low-mass protostars (Cazaux *et al.* 2003; Bottinelli *et al.* 2004; Sakai *et al.* 2006). A parallel has been proposed between these "hot corinos" and the high-mass protostellar "hot cores" related to the evaporation of icy-mantles on scales where the dust and gas is heated to

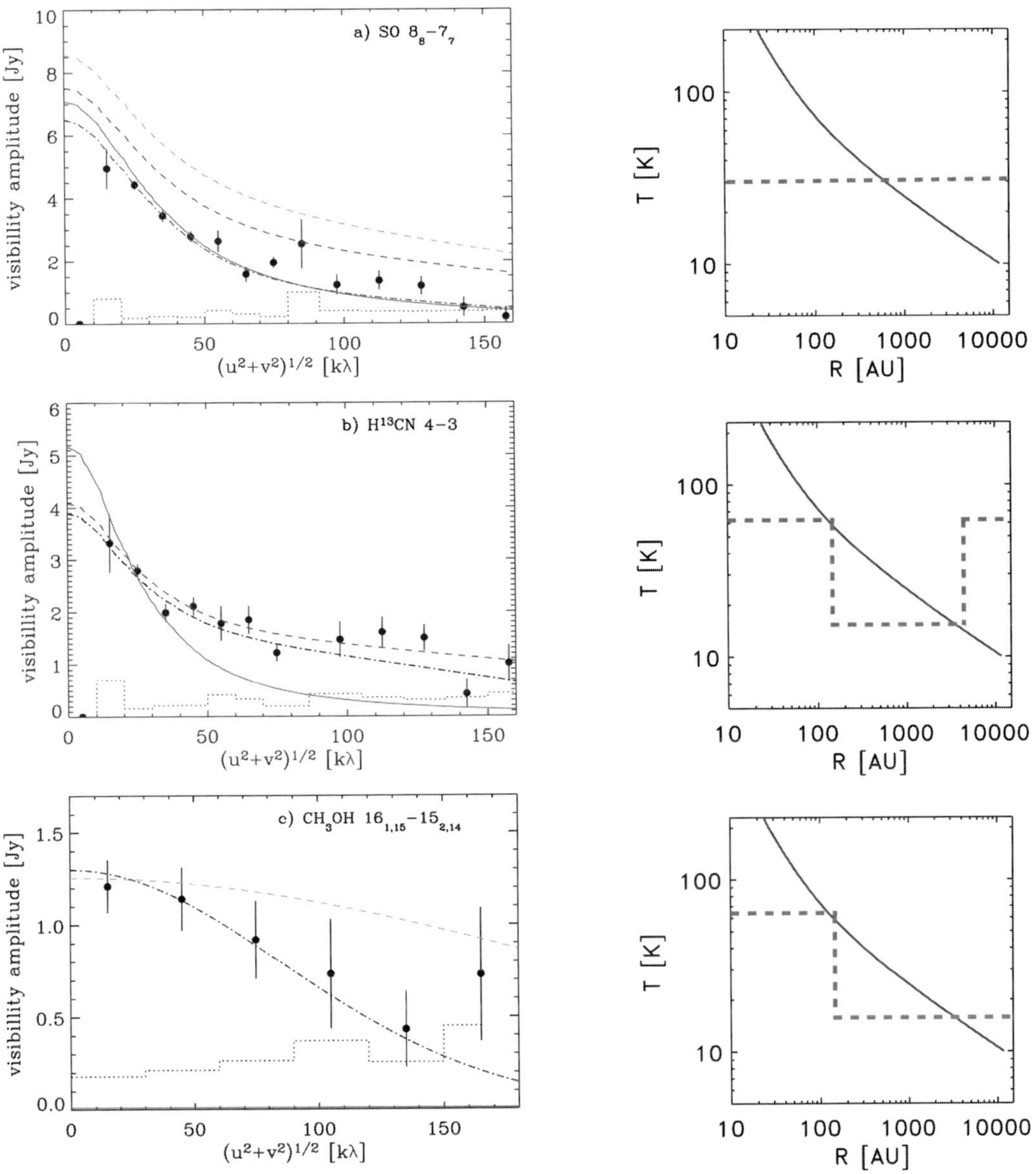

Figure 5. Abundance profiles for three different molecular species (SO, $H^{13}CN$ and CH_3OH in the top, middle and bottom panels, respectively) toward the Class 0 protostar NGC 1333-IRAS2A imaged with the SMA (Jørgensen *et al.* 2005a). The **left column** shows the visibility amplitudes (averaged in the spectral domain over the line profile) as function of projected baseline length. The error bars indicate 1σ errors and the dotted histogram the zero-expectation level (i.e., the anticipated amplitude in the absence of source emission). In each panel the solid line indicates the best fit model for the given species from Jørgensen *et al.* (2004b, 2005c) (constant abundance for SO and $H^{13}CN$ and jump abundance for CH_3OH). the dashed-dotted lines indicate envelope models with an additional disk component with a Gaussian brightness distribution (FWHM of 200 AU). The dashed lines indicate additional models: for SO models with abundance enhancements in the inner regions by factors 10 (dark) and 100 (light). For $H^{13}CN$ models with a drop abundance profile and for CH_3OH a model with a jump in abundance corresponding to a hot corino. The **right column** shows the corresponding abundance profiles for the three species (dashed lines) and the temperature profile for NGC 1333-IRAS2A (solid line). For further details see Jørgensen *et al.* (2005a).

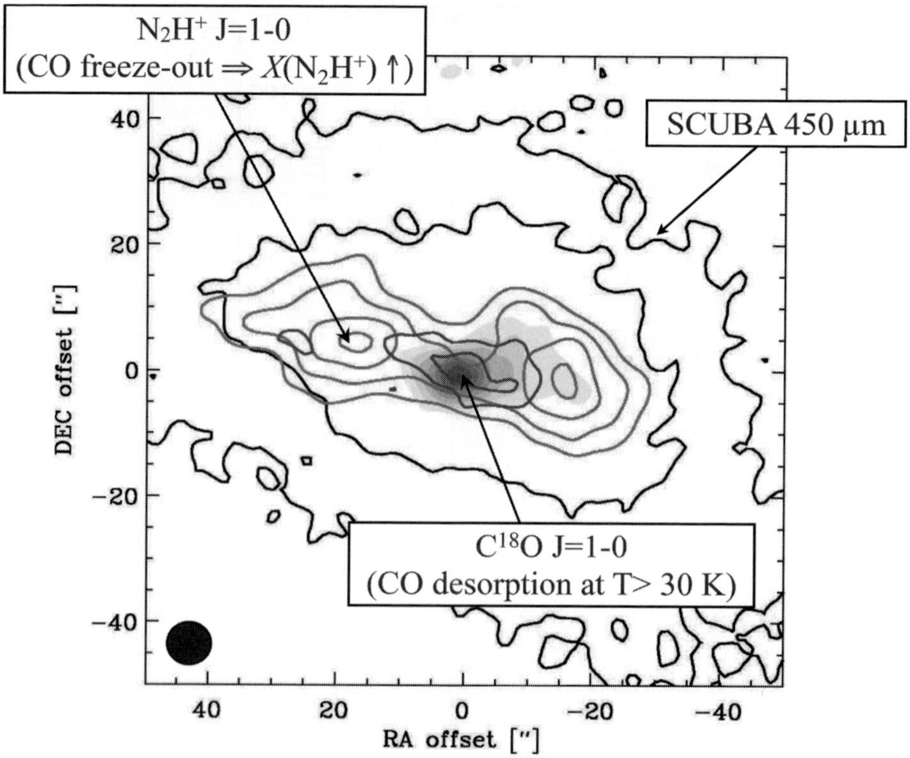

Figure 6. Morphology of the L483 protostellar core in 450 μm continuum from JCMT/SCUBA observations as well as N_2H^+ 1–0 and $C^{18}O$ 1–0 from observations with the Owens Valley Radio Observatory Millimeter Array (for further details see Jørgensen 2004). The 450 μm observations show the overall (asymmetric) distribution of dust in the core. The $C^{18}O$ emission is only present at the center where CO evaporates of the dust grains at temperatures of 30 K, whereas N_2H^+ predominantly is present at larger scales where CO is frozen out. Adapted from Jørgensen (2004).

temperatures above 100 K by the radiation from the newly formed protostars. Until now, four of these "hot corinos" have been suggested to be present toward the low-mass protostars, IRAS 16293-2422, NGC 1333-IRAS2A, NGC 1333-IRAS4A and NGC 1333-IRAS4B, based on single-dish observations of the before mentioned complex organics and through modeling of H_2CO and CH_3OH observations (e.g., Ceccarelli *et al.* 2000; Schöier *et al.* 2002; Maret *et al.* 2005). The hot corinos themselves, however, are severely diluted in single-dish beams ($\lesssim$ 100 AU sizes, compared to typical single-dish telescope beams of 1000–2000 AU). The interpretation of single-dish observations and comparison to models therefore relies on extrapolation of the density and temperature distribution from observations on larger scales that as pointed out are highly uncertain on the $\lesssim$ 100 AU scales where the hot corino activity may take place. There are therefore good reasons to push to high angular resolution.

Studies targeting lines of CH_3OH and H_2CO have highlighted the problems in separating the emission from outflows and possible hot corinos (Jørgensen *et al.* 2007; Schöier *et al.* 2004). The main, medium excitation ($E_u = 50$ K and above) transitions of H_2CO and CH_3OH in the 300 GHz windows are strongly affected by compact outflow emission up to 10″ scales. For the more complex organic molecules, the picture at first glance appears simpler: for the four well-studied hot corinos images of the complex organic molecules show relatively compact emission with extents comparable to the typical

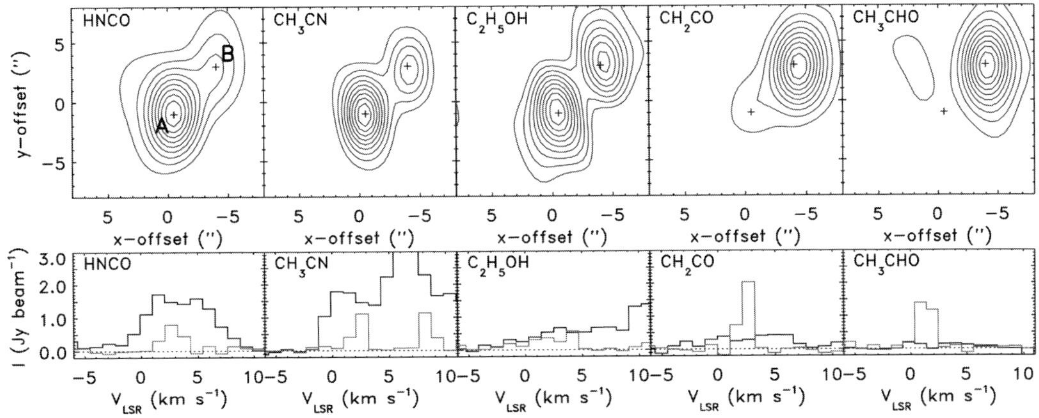

Figure 7. Observations of complex organics toward the IRAS 16293-2422 binary system (Bisschop *et al.* 2008; Jørgensen *et al.* 2011).

synthesized beam sizes of about 1″ (200 AU). As an example, high angular resolution interferometer observations of IRAS 16293-2422 (Bottinelli *et al.* 2004; Kuan *et al.* 2004; Bisschop *et al.* 2008), show that line emission of the organic molecules peaks close to the position of two compact dust continuum peaks indicating the locations of the central protostars (Fig. 7) – although the relative strengths of the complex organics differ between the two sources. However, low surface brightness emission from many other species, not necessarily associated with the hot corinos, at first glance also shows similar compact morphologies with deconvolved sizes comparable to the current interferometers' typical angular resolution (Jørgensen *et al.* 2007, 2011). Also, as pointed out by Chandler *et al.* (2005) from their subarcsecond observations of IRAS 16293-2422, there are small offsets of order 0.1 ″ between the location of the dust and the peaks of the complex organics, which is now also seen for the three other hot corino sources (Persson *et al.* 2011). Chandler *et al.* (2005) argued that these offsets (and similar offsets for high excitation transitions of non-organic molecules) reflected the action of shocks related to the protostellar jets rather than the passive (thermal) heating by the central protostars.

Another possibility, namely that many of these species reside in the circumstellar disks, has been prompted by recent subarcsecond millimeter wavelength resolution observations of a line of the water isotopologue, $H_2^{18}O$, at 203.4 GHz (1.4 mm) using the IRAM Plateau de Bure Interferometer (PdBI; Jørgensen & van Dishoeck) as well Spitzer Space Telescope detections of water at mid-infrared wavelengths (Watson *et al.* 2007) toward the deeply embedded protostar NGC 1333-IRAS4B. Because of the high densities required to excite the mid-infrared lines, the Spitzer Space Telescope detections were interpreted as signs of an ongoing accretion shock in the circumstellar disk. The IRAM PdBI observations confirmed that the water has its origin in the inner 25 AU (radius) of the disk predicted on basis of models of the Spitzer observations – and furthermore showed a tentative velocity gradient perpendicular to the propagation direction of the outflow probed by high angular resolution CO and maser observations. Similar high angular resolution ground-based detections of the $H_2^{18}O$ 203.4 GHz transition toward three other hot corino candidates (Persson *et al.* 2011; and M. Persson *et al.*, in prep.), suggests that this source is not unique. The simultaneous detections of complex organic species on similar scales supports the basic idea that the release of complex organics is prompted by the evaporation of water-rich ice mantles in these systems.

3.3. *Chemistry of 100 AU regions*

No matter the physical origin of the complex organic molecules and water, the interferometric observations show that they are present on few hundred scales. In their SMA study of IRAS 16293-2422 Bisschop *et al.* (2008) found strong chemical differentiation of the observed complex organic molecules between the two components in the binary system (Fig. 7): the nitrogen-bearing species HNCO and CH_3CN were detected toward both components in the binary but much stronger in the "A" source where the opposite was found to be the case for the oxygen-bearing species, C_2H_5OH, CH_2CO and CH_3CHO – the latter only seen toward the "B" source. Likewise, Persson *et al.* (2011) used millimeter interferometric observations of the $H_2^{18}O$ isotopologue to trace the abundances of a few complex organics relative to water in the warm inner regions of three other low-mass hot corinos in NGC 1333. They also found differences in the abundances of, e.g., C_2H_5CN and CH_3OCH_3 relative to water: while the latter is relatively constant between the three different sources, the C_2H_5CN to water abundance ratio varies by orders of magnitude. Similar differences between oxygen- and nitrogen-bearing species have also been seen in other types of star-forming regions (e.g., the sample of high-mass stars surveyed by Bisschop *et al.* 2007). As discussed by Bisschop *et al.* (2008), these differences can be explained through differences in their formation on grain surfaces and, e.g., the initial ice compositions. Differences between the oxygen bearing species also suggest that simple successive hydrogenation reactions on the grain surfaces are insufficient to explain the observed gas-phase abundances.

The $H_2^{18}O$ observations also makes it possible to revisit a long-standing discussion concerning the astrochemistry of water – namely what is the deuteration of water. It is thought that the amount of deuterated relative to non-deuterated water is established in these icy grain mantles. Determining exactly when this ratio is established and how it varies, is an important for understanding, e.g., whether water undergoes significant processing in warm regions of protostellar envelopes or disks.

Multi-transition HDO and H_2O single-dish observations from ground- and space-based observations have previously been used to estimate the HDO/H_2O abundance ratios in low-mass protostellar envelopes (e.g., Stark *et al.* 2004; Parise *et al.* 2005; Liu *et al.* 2011). Despite the advantages offered by the ranges of excitation conditions probed by the multi-transitions observations, these studies are still complicated by the relatively large beam sizes of the ground-based single-dish and space observatories and are therefore strongly dependent on exact line radiative transfer models utilized for the interpretation. In particular, the H_2O abundance is highly uncertain: even recent Herschel observations are predominantly sensitive to the impact of outflows and shocks on larger scales and only provide upper limits on the water abundance in the inner protostellar envelopes (e.g., Kristensen *et al.* 2010), although recent Herschel observations of $H_2^{18}O$ toward low-mass protostars (e.g., R. Visser *et al.*, in prep) may alleviate some of these problems.

Ground-based interferometric observations make it possible to directly measure the HDO/H_2O abundance ratio in the warm gas close to the central protostars. A non-detection of the HDO $3_{12} - 2_{21}$ transition at 225.9 GHz toward NGC 1333-IRAS4B from the SMA together with the detected 203.4 GHz transition of $H_2^{18}O$ for example put a strong 3σ upper limit of 6×10^{-4} to the HDO/H_2O ratio on scales of a few hundred AU, thus implying that the ratio is not enhanced above the cometary value. A detection of both the $H_2^{18}O$ 203.4 GHz and HDO 225.9 GHz transition toward the "A" component in IRAS 16293-2422 (M. Persson *et al.*, in prep.) likewise suggests an HDO/H_2O abundance ratio of 1–2×10^{-3} – only slightly enhanced above the cometary value. Higher sensitivity (multi-transition) $H_2^{18}O$ and HDO observations with ALMA for more sources are needed to settle this issue. However, to target the 203.4 GHz $H_2^{18}O$ transition with this sensitivity the full complement of ALMA band-5 receivers will be necessary.

4. Outlook: on to ALMA

Clearly many of the above issues will be addressed with the coming of the Atacama Large Millimeter Array that at the writing of this proceedings paper just had its deadline for "cycle 0" proposals. The main features of ALMA are:

• *Its high sensitivity* due to the excellent site, state-of-the-art technology and, in particular, large collecting area. This will allow studies of rare molecular species, isotopologues as well as faint lines that selectively probe specific components in the protostellar environments.

• *Its high angular resolution* with baselines out to 16 km will resolve structures down to few AU scales in the centers of protostars, including the forming circumstellar disks and their physical and chemical structures.

• *Its imaging capabilities* with the large number of antennae that will provide broad (u, v) coverage even in relatively short observations and thereby not be as affected by spatial filtering (resolving out of large-scale smooth structures) or dynamical range as current interferometers. This will in particular be key for imaging the few hundred AU scales of protostellar envelopes understanding, e.g., the interactions between the protostellar outflows and envelopes – or putting hard constraints on the radial variations of molecular species.

Of course, designing feasible programs fully utilizing the promise of ALMA requires some thought. For example, the highest angular resolution will only be achievable for the brightest lines – some of which will be optically thick on large scales and thus provide limited information. Thus, particular care needs to be taken to pick the optimal balance between achievable angular resolution/sensitivity and the posed questions. Also, there are many reasons to expect that the protostars will show significantly more complex structures than what is captured by for example traditional models for protostellar cores, pushing the need to even more sophisticated modeling tools. Still, this is of course a challenge in itself, and for anyone up to it, it is clear that ALMA will provide great possibilities for comprehensive molecular imaging studies of protostellar systems (as well as many other types of sources) that will keep astrochemists busy for years to come.

References

Basu, S. 1998, *ApJ*, 509, 229

Bisschop, S. E., Jørgensen, J. K., Bourke, T. L., Bottinelli, S., & van Dishoeck, E. F. 2008, *A&A*, 488, 959

Bisschop, S. E., Jørgensen, J. K., van Dishoeck, E. F., & de Wachter, E. B. M. 2007, *A&A*, 465, 913

Bottinelli, S., Ceccarelli, C., Neri, R., *et al.* 2004, *ApJ*, 617, L69

Brinch, C., Crapsi, A., Jørgensen, J. K., Hogerheijde, M. R., & Hill, T. 2007, *A&A*, 475, 915

Brinch, C., Jørgensen, J. K., & Hogerheijde, M. R. 2009, *A&A*, 502, 199

Cazaux, S., Tielens, A. G. G. M., Ceccarelli, C., *et al.*, 2003, *ApJ*, 593, L51

Ceccarelli, C., Loinard, L., Castets, A., Tielens, A. G. G. M., & Caux, E. 2000, *A&A*, 357, L9

Chandler, C. J., Brogan, C. L., Shirley, Y. L., & Loinard, L. 2005, *ApJ*, 632, 371

Chiang, H.-F., Looney, L. W., Tassis, K., Mundy, L. G., & Mouschovias, T. C. 2008, *ApJ*, 680, 474

Enoch, M. L., Corder, S., Duchene, G., *et al.*, 2011, ArXiv e-prints

Enoch, M. L., Corder, S., Dunham, M. M., & Duchêne, G. 2009, *ApJ*, 707, 103

Galli, D. & Shu, F. H. 1993a, *ApJ*, 417, 220
Galli, D. & Shu, F. H. 1993b, *ApJ*, 417, 243
Hennebelle, P. & Ciardi, A. 2009, *A&A*, 506, L29
Ivezić, Ž., Nenkova, M., & Elitzur, M. 1999, User Manual for DUSTY, University of Kentucky Internal Report
Jørgensen, J. K. 2004, *A&A*, 424, 589
Jørgensen, J. K., Bourke, T. L., Myers, P. C., *et al.*, 2007, *ApJ*, 659, 479
Jørgensen, J. K., Bourke, T. L., Myers, P. C., *et al.*, 2005a, *ApJ*, 632, 973
Jørgensen, J. K., Bourke, T. L., Nguyen Luong, Q., & Takakuwa, S. 2011, å, submitted
Jørgensen, J. K., Harvey, P. M., Evans, N. J. I., *et al.*, 2006, *ApJ*, 645, 1246
Jørgensen, J. K., Hogerheijde, M. R., van Dishoeck, E. F., Blake, G. A., & Schöier, F. L. 2004a, *A&A*, 413, 993
Jørgensen, J. K., Schöier, F. L., & van Dishoeck, E. F. 2002, *A&A*, 389, 908
Jørgensen, J. K., Schöier, F. L., & van Dishoeck, E. F. 2004b, *A&A*, 416, 603
Jørgensen, J. K., Schöier, F. L., & van Dishoeck, E. F. 2005b, *A&A*, 435, 177
Jørgensen, J. K., Schöier, F. L., & van Dishoeck, E. F. 2005c, *A&A*, 437, 501
Jørgensen, J. K. & van Dishoeck, E. F. 2010, *ApJ*, 710, L72
Jørgensen, J. K., van Dishoeck, E. F., Visser, R., *et al.*, 2009, *A&A*, 507, 861
Kristensen, L. E., Visser, R., van Dishoeck, E. F., *et al.*, 2010, ArXiv e-prints
Kuan, Y., Huang, H., Charnley, S. B., *et al.*, 2004, *ApJ*, 616, L27
Liu, F., Parise, B., Kristensen, L., *et al.*, 2011, *A&A*, 527, A19
Lommen, D., Jørgensen, J. K., van Dishoeck, E. F., & Crapsi, A. 2008, *A&A*, 481, 141
Maret, S., Ceccarelli, C., Tielens, A. G. G. M., *et al.*, 2005, *A&A*, 442, 527
Matthews, B. C., Hogerheijde, M. R., Jørgensen, J. K., & Bergin, E. A. 2006, *ApJ*, 652, 1374
Maury, A. J., André, P., Hennebelle, P., *et al.*, 2010, *A&A*, 512, A40
Parise, B., Caux, E., Castets, A., *et al.*, 2005, *A&A*, 431, 547
Persson, M., Jørgensen, J. K., & van Dishoeck, E. F. 2011, *A&A*, submitted
Sakai, N., Sakai, T., & Yamamoto, S. 2006, *PASJ*, 58, L15
Schöier, F. L., Jørgensen, J. K., van Dishoeck, E. F., & Blake, G. A. 2002, *A&A*, 390, 1001
Schöier, F. L., Jørgensen, J. K., van Dishoeck, E. F., & Blake, G. A. 2004, *A&A*, 418, 185
Shirley, Y. L., Evans, N. J., & Rawlings, J. M. C. 2002, *ApJ*, 575, 337
Stark, R., Sandell, G., Beck, S. C., *et al.*, 2004, *ApJ*, 608, 341
Terebey, S., Shu, F. H., & Cassen, P. 1984, *ApJ*, 286, 529
van der Tak, F. F. S., Black, J. H., Schöier, F. L., Jansen, D. J., & van Dishoeck, E. F. 2007, *A&A*, 468, 627
Watson, D. M., Bohac, C. J., Hull, C., *et al.*, 2007, *Nature*, 448, 1026

Discussion

MAUERSBERGER: H_2O has a complex spectrum and the lines you were observing are high in the excitation ladder. How much are your abundance estimations influenced by masing?

JØRGENSEN: This is an important issue indeed. In the specific cases of the IRAM Plateau de Bure observations, we actually have evidence that the $H_2^{18}O$ transition is not severely by masing – for example that the widths of the observed $H_2^{18}O$ lines are very close to those of other species. Also, species-to-species variations in the line strengths are much smaller than the source-to-source variations.

The Molecular Universe
Proceedings IAU Symposium No. 280, 2011
José Cernicharo & Rafael Bachiller, eds.

doi:10.1017/S1743921311024872

Ices in Starless and Starforming Cores

Karin I. Öberg[1], A. C. Adwin Boogert[2], Klaus M. Pontoppidan[3], Saskia van den Broek[4], Ewine F. van Dishoeck[4], Sandrine Bottinelli[4,5], Geoffrey A. Blake[6] and Neal J. Evans II[7]

[1]Harvard-Smithsonian Center for Astrophysics, 60 Garden St, Cambridge, MA 02139, USA
[2]IPAC, NASA Herschel Science Center, Caltech, Pasadena, CA 91125, USA
[3]Space Telescope Science Institute, 3700 San Martin Drive, Baltimore, MD 21218, USA
[4]Leiden Observatory, Leiden University, PO Box 9513, 2300 RA Leiden, the Netherlands
[5]CESR, CNRS-UMR 5187, 9 ave. Colonel Roche, BP 4346, 31028 Toulouse Cedex 4, France
[6]Caltech, Division of Geological and Planetary Sciences, Pasadena, CA 91125, USA
[7]Dep. of Astronomy, UT Austin, 1 University Station C1400, Austin, TX 78712, USA

Abstract. Icy grain mantles are commonly observed through infrared spectroscopy toward dense clouds, cloud cores, protostellar envelopes and protoplanetary disks. Up to 80% of the available oxygen, carbon and nitrogen are found in such ices; the most common ice constituents – H_2O, CO_2 and CO – are second in abundance only to H_2 in many star forming regions. In addition to being a molecular reservoir, ice chemistry is responsible for much of the chemical evolution from H_2O to complex, prebiotic molecules. Combining the exisiting ISO, Spitzer, VLT and Keck ice data results in a large sample of ice sources (~80) that span all stages of star formation and a large range of protostellar luminosities (<0.1–10^5 $L_\odot$). Here we summarize the different techniques that have been applied to mine this ice data set on information on typical ice compositions in different environments and what this implies about how ices form and evolve during star and planet formation. The focus is on how to maximize the use of empirical constraints from ice observations, followed by the application of information from experiments and models. This strategy is used to identify ice bands and to constrain which ices form early during cloud formation, which form later in the prestellar core and which require protostellar heat and/or UV radiation to form. The utility of statistical tests, survival analysis and ice maps is highlighted; the latter directly reveals that the prestellar ice formation takes place in two phases, associated with H_2O and CO ice formation, respectively, and that most protostellar ice variation can be explained by differences in the prestellar CO ice formation stage. Finally, special attention is paid to the difficulty of observing complex ices directly and how gas observations, experiments and models help in constraining this ice chemistry stage.

Keywords. astrochemistry, astrobiology, line: identification, line: profiles, molecular processes, circumstellar matter, ISM: molecules, infrared: ISM

1. Introduction

The evolution from starless to star-forming cores and further on to (extra)solar systems is accompanied by a rich chemistry, which will affect planet and planetesimal compositions. Much of this chemical evolution takes place on interstellar grain surfaces, in icy mantles. Ices are generally observed in the cold and dense interstellar medium that are the nurseries of stars and are also abundant in our own solar system in comets and other outer solar system bodies. Such icy bodies are proposed to have delivered volatiles to the young earth. Understanding their link to the chemistry observed during star formation is key to constrain the amount of organic material delivered to Earth as well as to exo-planets.

The first ices were detected in the interstellar medium almost 40 years ago by Gillett & Forrest (1973). H_2O and CO were established early on as common ice constituents with abundances reaching 10^{-4} n_{H_2}. This makes ices the most common molecules after

H_2 in many star forming regions. CO_2 is the third major ice constituent. Because of abundant atmospheric CO_2, it was only detected in the ISM after the launch of IRAS by D'Hendecourt & Jourdain de Muizon(1989). Following IRAS, the *Infrared Space Observatory* (ISO) and the *Spitzer Space Telescope* have, together with ground-based efforts, continued the ice exploration and interstellar ices. In addition to the major ice constituents, H_2O, CO and CO_2 are observed to often contain smaller amounts of CH_3OH, CH_4, NH_3 and OCN^-/XCN (see Boogert & Ehrenfreund (2004) for a review).

The simple ices observed in the ISM may become the source of complex molecules as described by Charnley *et al.* (1992) and Garrod *et al.* (2008) and determining ice abundances and production channels in star forming regions is of great interest for studies of prebiotic chemistry. Ices can form from direct freeze-out of molecules from the gas phase, through atomic addition on grain and ice surfaces, and through energetic processing of already existing ices. At the low temperatures (<20 K) prevalent in cloud cores, atoms and molecules will stick to the grain upon impact. In this stage grains are well protected from external UV radiation and atomic addition reactions probably dominate. Tielens & Hagen (1982) worked out in detail how the atoms and molecules can be hydrogenated and oxygenated to form all commonly observed ices, except for CO, which condenses directly from the gas-phase. In this scheme, H_2O forms from O+H, CH_3OH from CO+H etc. CH_3OH can also form from UV processing of H_2O:CH_4 ice mixtures, however, and there has been some debate on which formation pathway dominates. Gibb *et al.* (2004) and others have also suggested that at least some CO_2 and XCN form from thermal or UV processing, mainly based on high-mass protostellar studies with *ISO*.

Spitzer and complimentary ground-based surveys have more recently allowed us to observe ices toward large numbers of low-mass protostars and cloud cores, which have provided new observational constraints on how ice formation and evolution depend on the local environment. The aim of this contribution is to review the constraints supplied by ice observations toward both *ISO* and *Spitzer* targets, on the identifications of ice bands and of the main formation paths of the identified carriers. The figures are taken from Öberg *et al.* (2011, subm. to ApJ) unless otherwise noted.

2. Ice observations

Ices are generally observed in absorption at mid-IR wavelengths and can therefore only be observed in lines of sight with infrared background sources. Protostars are excellent IR sources and ices in protostellar envelopes are the most extensively studied (Fig. 1). Early ice observations were exclusively ground based and thus limited to observe ice bands that coincide with atmospheric windows. IRAS and especially *ISO* opened up the full infrared spectral region, resulting in the first complete ice inventories toward bright protostars and a few cloud lines of sight. The *ISO* ice results are described in Gibb *et al.* (2004) and these still constitute the largest high-mass protostellar sample.

Compared to previous space missions, *Spitzer*'s high sensitivity enabled observations of ices toward low-mass protostars and also toward previously inaccessible background stars. The largest sample was obtained within the *c2d* program, defined in Evans *et al.* (2003), which included IRS spectra of 50 low-mass protostars. The different ice bands were analyzed in a series of papers by Boogert *et al.* (2008), Pontoppidan *et al.* (2008), Öberg *et al.* (2008) and Bottinelli *et al.* (2010). Additional *Spitzer* observations exist on ices toward background sources, looking through molecular clouds at a range of extinctions (reported by Bergin *et al.* (2005), Knez *et al.* (2005), Pontoppidan (2006), Whittet *et al.* (2009) and Boogert *et al.* (2011)) and toward protostars in the high-UV environment of IC 1396A by Reach *et al.* (2009). Because the *Spitzer* spectrometer had a lower cutoff at

~5 μm, complete ice inventories can only be obtained by adding complementary ground-based spectroscopy to cover the strong 3 μm H_2O, the 4.65 μm CO, the 4.60-4.62 μm XCN, and the 3.53 μm CH_3OH features and this was done using VLT and Keck. Results are summarized in Öberg *et al.* (2011, subm. to ApJ).

The combined *Spitzer*, ISO, VLT and Keck ice sample spans the full range of observed luminosities from the low-luminosity object L1014 IRS ($<0.1L_\odot$) to the 10^5 $L_\odot$ massive protostar W33A (Fig. 1). The low-mass young stellar objects (YSOs) are located in Perseus, Taurus, Serpens, Ophiuchus and Corona Australis, as well as a number of nearby isolated dense cores, and thus cover star-forming regions with different histories and physical environments. The sample also represents a wide range of cloud, cloud core and YSO evolutionary stages, from envelope-dominated Class 0 sources to disks.

3. Ice identifications

Figure 1 shows ice spectra toward protostars with the identifications of different bands to H_2O, CO, CO_2, CH_3OH, NH_3, CH_4 and OCN^- ice marked. van Broekhuizen *et al.* (2005) found that the observed XCN band at 4.62 μm consists of two components. Based on laboratory spectroscopy from van Broekhuizen *et al.* (2004), at least one of the components is due to OCN^-. OCN^- in a separate ice matrix is probably the carrier of the second XCN band component (centered at 2175 cm^{-1}), but other identifications have also been suggested by e.g. Fraser *et al.* (2005).

In general the frequency of a solid state molecular vibration depends on the bonding environment of the vibrating molecule. This results in that the positions and shapes of ice spectroscopic features depend on the ice mixture(s). Fig. 2 from Bottinelli *et al.* (2010) demonstrates how the CH_3OH FWHM and band center depend on the ice mixture in laboratory experiments and how this can be used to constrain that protostellar CH_3OH ices are not present in H_2O-dominated ices.

Some ice bands, such as CO and CO_2, cannot be fitted by a single laboratory ice mixture, but seem to trace molecules in two or more ice phases, often referred to as polar

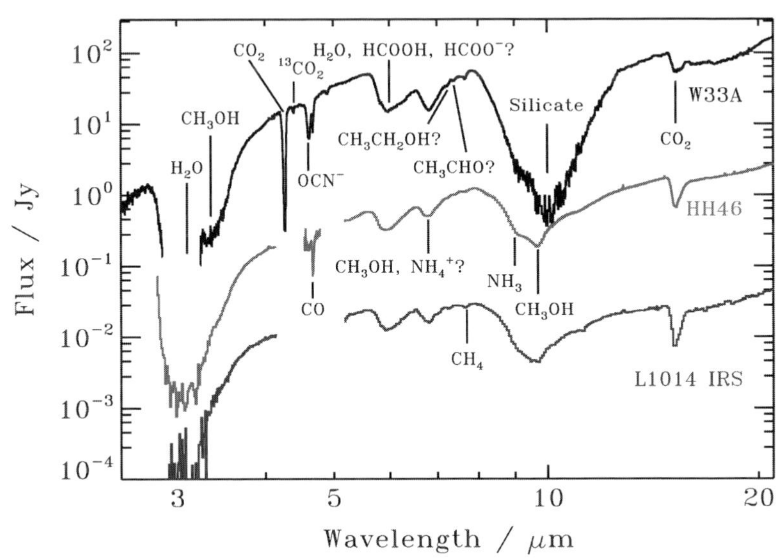

Figure 1. Ice spectra toward the protostars W33A (10^5 $L_\odot$), HH46 (12 $L_\odot$) and L1014 IRS (0.09 $L_\odot$) from Gibb *et al.* (2000), Boogert *et al.* (2004) and Boogert *et al.* (2008). The 3 μm portions of the spectra have been binned to increase the S/N.

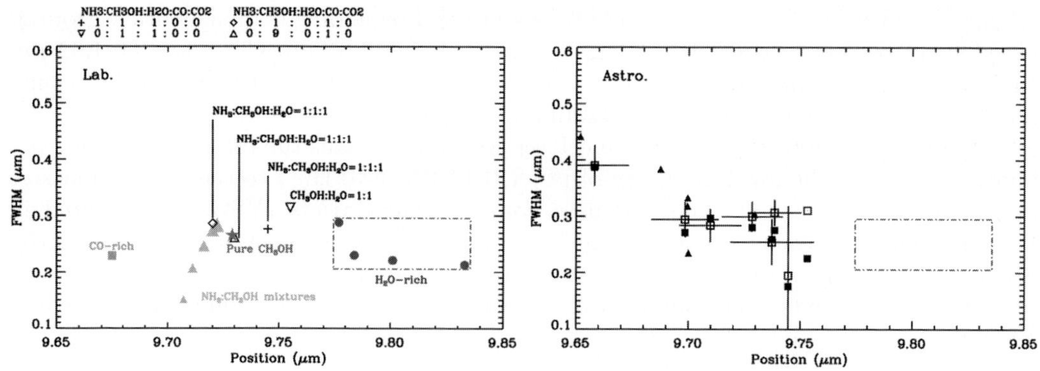

Figure 2. Figure from Bottinelli *et al.* (2010) showing the FWHM and peak position of the CH_3OH feature in laboratory mixtures at 15 K (left panel) and in Spitzer *c2d* spectra (right panel). In the right panel an increasing symbol size is indicative of increasing CH_3OH content. In both panels, the dash-dot polygons delimitate the parameter space of FWHM and peak positions corresponding to H_2O-rich mixtures.

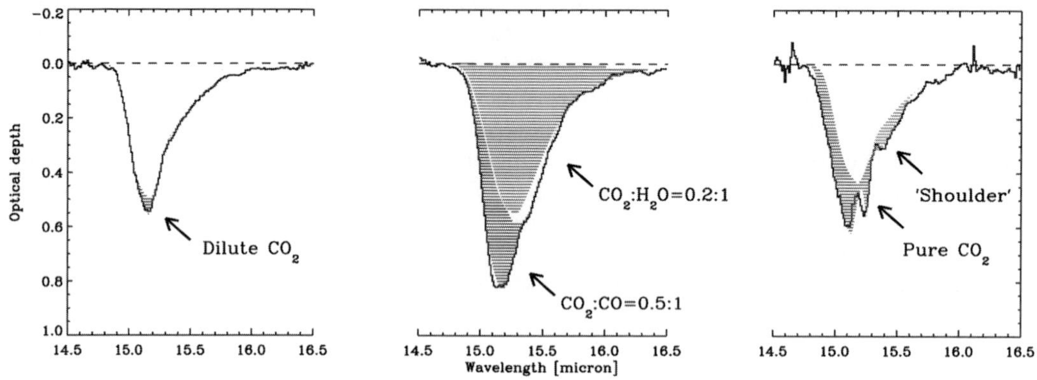

Figure 3. Sketch of the five different components used to fit the CO_2 band on top of IRS 51, SVS 4-5, and RNO 91 spectra from Pontoppidan *et al.* (2008)

and apolar. Traditionally the observed spectra have been directly compared to a superposition of laboratory pure and mixed ice spectra to constrain ice components, see e.g. Merrill *et al.* (1976), Gibb *et al.* (2004) and Zasowski *et al.* (2009). The constraints are often degenerate, however, since ice spectral features vary with ice composition, temperature and radiation processing. To address this, observed ice bands are now commonly decomposed phenomenologically and then the components are compared to laboratory spectra, see Tielens *et al.* (1991), Pendleton *et al.* (1999), Keane *et al.* (2001) and Boogert *et al.* (2008). Because all observed spectra are decomposed into the same, small number of components, this method provides information on the sample as a whole, i.e. it directly shows which parts of the spectral profile are ubiquitous and which are environment dependent. Figure 3 from Pontoppidan *et al.* (2008) demonstrates how the derived CO_2 ice components can be identified with pure CO_2 ice, CO_2 mixed with CO (CO_2:CO), CO_2 mixed with H_2O ice (CO_2:H_2O) and a shoulder associated with CO_2 mixed with CH_3OH. Pontoppidan *et al.* (2003) showed that the CO ice band can be similarly decomposed into three components corresponding to pure CO ice, CO:CO_2 and CO:H_2O or CO:CH_3OH.

Figure 1 also shows tentative identifications to HCOOH and NH_4^+ in a complex band between 5 and 7 μm. To constrain the carriers of this band Boogert *et al.* (2008) decomposed it into five different components (C1–C5) after subtraction of the contribution

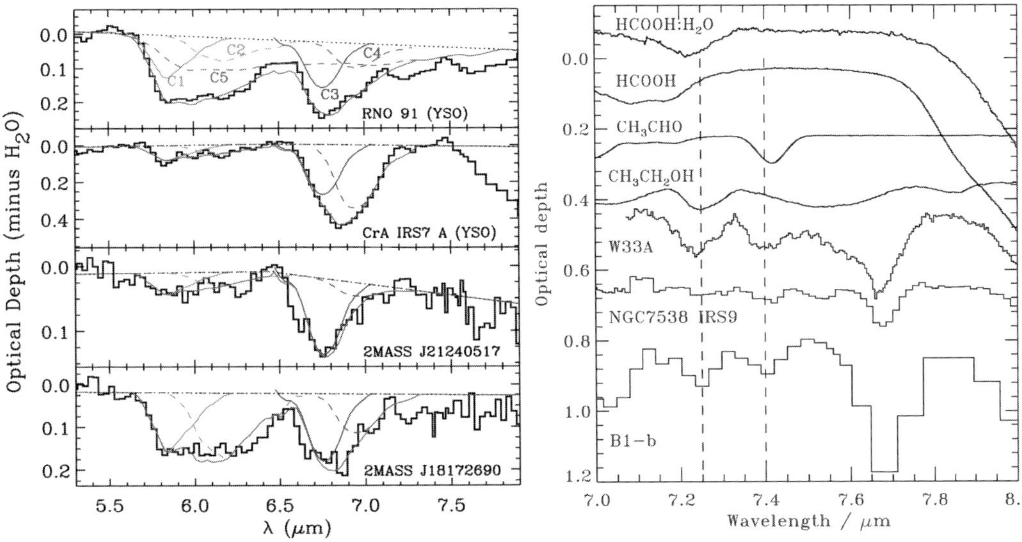

Figure 4. *Left panel:* Decomposition of the absorption features in the 5-8 μm spectra of two background stars (bottom panels), compared to two YSOs. The sum of the components is indicated with a thick red line. Figure from Boogert *et al.* (2011). *Right panel:* Spectra toward W33A, NGC7538 IRS9 and B1-b, following subtraction of a local spline continuum, plotted together with laboratory spectra of HCOOH, CH_3CHO and CH_3CH_2OH ices. The dashed lines mark the 7.25 and 7.40 μm features usually assigned to HCOOH and $HCOO^-$.

from H_2O ice (Fig. 4). From comparison with laboratory ice spectra, C1 has been identified with HCOOH and H_2CO, C2 with $HCOO^-$ and NH_3, C3 with NH_4^+ and CH_3OH, C4 with NH_4^+ and C5 with warm H_2O and anions. The 7.25 μm has been used to derive HCOOH ice abundances by e.g. Schutte *et al.* (1999), Gibb *et al.* (2004) and Boogert *et al.* (2008). The right panel of Figure 4 shows that CH_3CH_2OH is an equally plausible carrier, however. The presence of a 7.25 μm feature is evidence for the formation of complex ices, but derivations of HCOOH abundances from this band is not possible without higher resolution spectra that can distinguish between the two carriers.

4. Ice abundance studies

Following ice identification, ice column densities are determined by dividing the integrated optical depth of the ice feature by the band strength. Ice abundances are expressed with respect to H_2O ice, because H_2O ice is predicted to form early during star formation, is the most abundant ice, and has a high sublimation point. Determining ice abundances is important to predict the chemical evolution in different environments and to constrain the C, O and N budget. In addition to these direct uses, the following sections demonstrate how ice abundance statistics can be used to put constraints on how ices form. Comparison of ice distributions between clouds and protostars constrain which ices require protostellar heating or UV processing to form. Ice abundance variations constrain which ices co-form with H_2O and which ices form in a separate stage. Ice correlations provide further constraints on co-formation, and anti-correlations can be used to demonstrate competitive formation. In these investigations it is key to quantify how significant differences and correlations really are. Finally, ice maps provide direct evidence of how ice formation depends on the position of a source in a cloud or a core.

Table 1. Recommended ice abundances toward low- and high-mass protostars and clouds.

Ice feature	Low-mass	High-mass	background
H_2O	100	100	100
CO	29	13	31
CO_2	29	13	38
CH_3OH	3	4	4
NH_3	5	5	–
CH_4	5	2	–
XCN	0.3	0.6	–
pure CO	21	3	–
CO:H_2O	13	10	–
CO:CO_2	2	0.3	–
pure CO_2	2	2	–
CO_2:H_2O	20	9	24
CO_2:CO	5	5	6
CO_2 shoulder	0.8	1	–
OCN^-	0.2	0.6	–
2175 cm^{-1}	0.2	0.1	–
C1 (HCOOH + H_2CO)[a]	2.5	2.1	2.8
C2 ($HCOO^-$+NH_3)[a]	1.1	1.3	2.5
C3 (NH_4^+ + CH_3OH)[a]	4.3	4.3	3.7
C4 (NH_4^+)[a]	2.3	4.3	2.1
C5 (warm H_2O + anions)[a]	0.9	1.4	–

[a]Because of multiple band carriers, the reported number is $\tau_{\rm peak}/(N_{\rm H_2O} \times 10^{-18})$.

4.1. *Typical abundances*

Table 1 lists the representative ice abundances found toward low-mass protostars from the *c2d* survey, high-mass protostars from Gibb *et al.* (2004)and background stars from Knez *et al.* (2005) and Boogert *et al.* (2011). The reported median abundances are calculated using the Kaplan-Meier (KM) estimate of the survival function. This is a non-parametric procedure that takes into account the constraints provided by upper limits. When calculating the KM estimate, the detections and upper limits are ordered from low to high and the upper limits are given the values of the nearest lower detections. For example in a sample of four detections of 0.5, 1, 3 and 4 and an upper limit of <2, the upper limit is treated as a detection of 1. The KM estimate and how to apply statistical tests on it is reviewed by Feigelson & Nelson (1985). Compared to medians based on only detected ice abundances, the medians calculated from the KM estimate provide more accurate descriptions of ice populations with significant upper limits. The difference is significant, e.g. the CH_3OH and NH_3 medians toward high-mass protostars are 40-70% lower when taking into account the upper limits, resulting in similar abundances toward low- and high-mass YSOs and clouds of 3–5%. The CO and CO_2 abundances vary between the different environments, but are always second in abundance only to H_2O.

4.2. *The C, O and N budget*

The amount of C, O and N that are typically bound up in ice mantles is important for the life cycle of the elements in the interstellar medium. Przybilla *et al.* (2008) found that the total C, O and N abundances in the solar neighborhood are 2.1, 5.8 and 0.58×10^{-4} per hydrogen nuclei, respectively. The fractional C, O and N abundance in ices with respect to these total C, O and N abundances can be calculated from the presented ice abundances with respect to H_2O ice together with a median H_2O abundance with respect to hydrogen

Table 2. The percentages of C, O and N bound up in protostellar ices.

	C_{ice} / C_{total}	O_{ice} / O_{total}	N_{ice} / N_{total}	C_{ice} / C_{vol}	O_{ice} / O_{vol}
Low-mass median	15%	16%	10%	27%	34%
Low-mass max	46%	29%	35%	83%	61%
High-mass median	8%	12%	12%	14%	25%
High-mass max	18%	17%	22%	32%	36%

of $5 \times 10^{-5} n_H$ from Pontoppidan *et al.* (2004) and Boogert & Ehrenfreund (2004). The resulting median percentages of C, O and N in ices are 8–16% with respect to the total elemental abundances. When the C and O bound up in grains is subtracted, these values increase to 14–34% (Table 2). For the most ice-rich sources in the sample, the amount of O and C in ice approaches the abundances of volatile O and C, indicative of that ice formation only stops when the gas-phase is completely depleted of these elements.

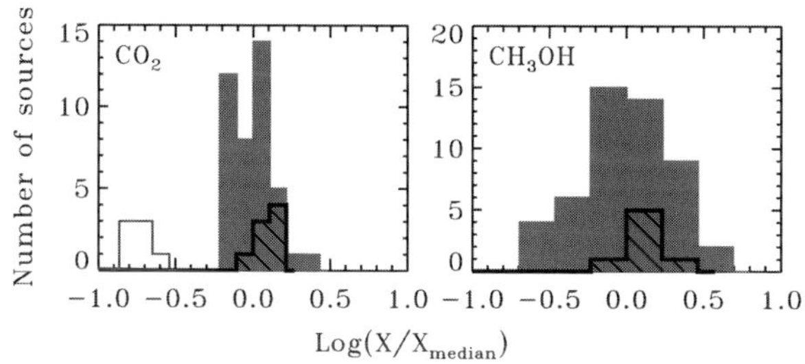

Figure 5. Histograms of CO_2 and CH_3OH toward low-mass protostars (red fill), our background star sample (black line-fill) and Taurus background stars (blue contours).

4.3. *Cloud versus protostellar ices*

Comparisons between protostars and background stars provide direct limits on which ices form before the protostar turns on and starts to heat and irradiate its surroundings. Cook *et al.* (2011) showed that in Taurus the CO_2 ice fraction in clouds is lower compared to low-mass protostellar envelopes, and proposed that at least some of the CO_2 observed toward protostars forms from energetic processing. Similarly, CH_3OH ice has been proposed to be a product of protostellar UV ice-processing by e.g. Gibb *et al.* (2004). Figure 5 shows that the CO_2 and CH_3OH ice abundances toward protostars and background stars from Boogert *et al.* (2011) are similar, and that the Taurus CO_2 ice abundances from Whittet *et al.* (2007) are poor templates for ice abundances toward average cloud cores.

Boogert *et al.* (2011) show that there is also no statistically significant difference in the C1-4 components between the protostars and background stars, but the C5 component is absent in cloud sources.The only other feature that is present exclusively toward protostars is pure CO_2 ice, which is proposed to form from thermally induced ice segregation or distillation. The protostar thus seems to have a very limited impact on the observable ice composition. This entails that protostellar ice abundances mainly reflect the ice formation process in the previous cold cloud phase.

4.4. *Protostellar ice abundance variations*

Ice abundance variations between different sources depend on how sensitive the ice formation and destruction pathways are to the local environment. Since ice abundances are with respect to H_2O, a small variation of a species indicates co-formation with H_2O, while large ice abundance variations are indicative of different formation and/or destruction dependencies than H_2O ice. Figure 6 shows that the CO_2, CH_4 and NH_3 abundance distributions are narrow, indicative of that these ices form together with H_2O, while the CO, OCN^- and CH_3OH distributions are broad. The ice abundances around high-mass protostars are generally lower compared to low-mass protostars, indicating that heat and UV mainly destroy simple ices. The large CH_3OH, CO and OCN^- abundance variations must then originate from a separate prestellar ice formation stage compared to the stage associated with H_2O formation. This is consistent with observations that most prestellar and protostellar CO_2 is present in a H_2O rich ice, while CH_3OH and CO ices are not. Figure 6 shows the distributions of the individual CO and CO_2 components. The pure CO and CO_2 ice components have the broadest distributions, consistent with their expected dependence on the protostellar envelope temperature for ice evaporation and segregation. All other CO and CO_2 component distributions, except for CO_2:H_2O, are broad as well, indicative of a different history compared to the H_2O-rich ice phase.

4.5. *Ice maps*

The spatial extent of ices can be mapped in clouds and cloud cores if enough background sources are available. Mapping of ice abundances in dense clouds by Bergin *et al.* (2005) has revealed a similar formation threshold of H_2O and CO_2 at A_V=4 mag. CO is only observed at higher extinctions. The H_2O-rich ice thus forms first.

Pontoppidan *et al.* (2008) shows that toward the Oph-F core, where protostars are used as background sources to probe the cloud core ices, CO ice abundances increase dramatically toward the core (Fig. 7). Pontoppidan (2006) interpreted this as catastrophic freeze-out of CO in the pre-stellar stage once a certain density and temperature is reached. Figure 7 shows that the order of magnitude increase in CO ice with respect to H_2O toward the core is accompanied by an increase in CO_2:CO. In contrast, the CO_2:H_2O ice component is almost constant. CO:H_2O and XCN band are the only other species that

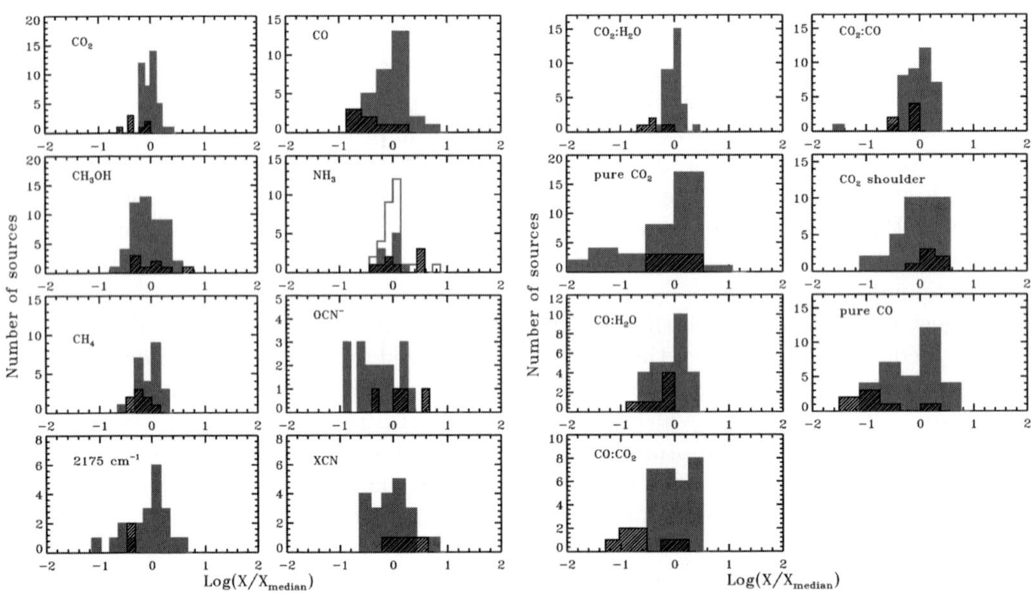

Figure 6. Histograms of ice and ice component abundances toward low-mass (red) and high-mass (black) protostars. The abundances are log-transformed and normalized to the median detected low-mass protostellar ice abundance. Note the difference in spread between species.

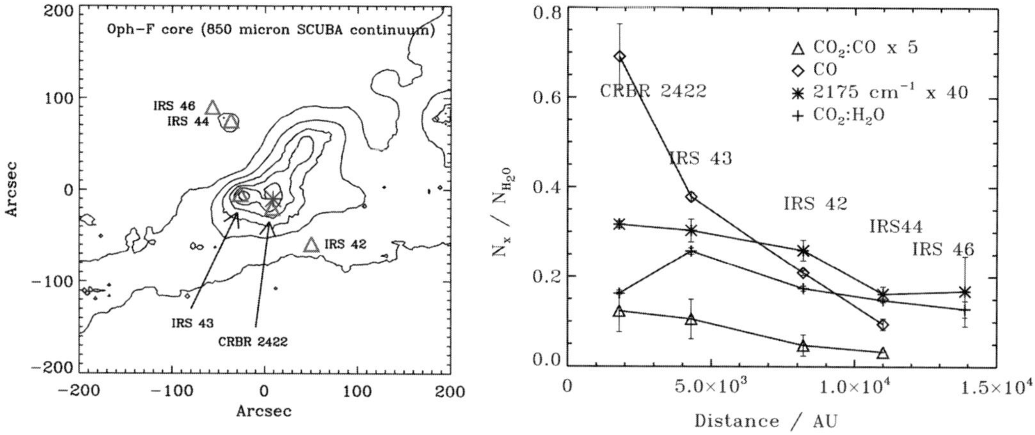

Figure 7. *Left panel:* The locations of the sources used to construct the ice map plotted on the SCUBA 850 μm map from Pontoppidan (2006) with respect to the central core (star). *Right panel:* Ice abundances at different distances from the Oph-F core with respect to H_2O ice. The 2175 cm^{-1} abundances are scaled by 40 and the CO_2:CO abundances by 5 for clarity.

increase monotonically toward the densest part of Oph-F core. These bands thus appear directly related to CO freeze-out, indicating that their broad abundance distributions in Fig. 6 are due to their dependence on the prestellar CO freeze-out level. No trend with distance to the core center is seen or expected for ices that form with H_2O (e.g. CH_4 and NH_3), or of species dependent on protostellar heating (e.g. pure CO_2 ice) or of components with potentially multiple carriers, such as the C1-5 bands. CH_3OH is only detected toward one of the sources and no trend can be extracted.

4.6. *Ice abundance correlations*

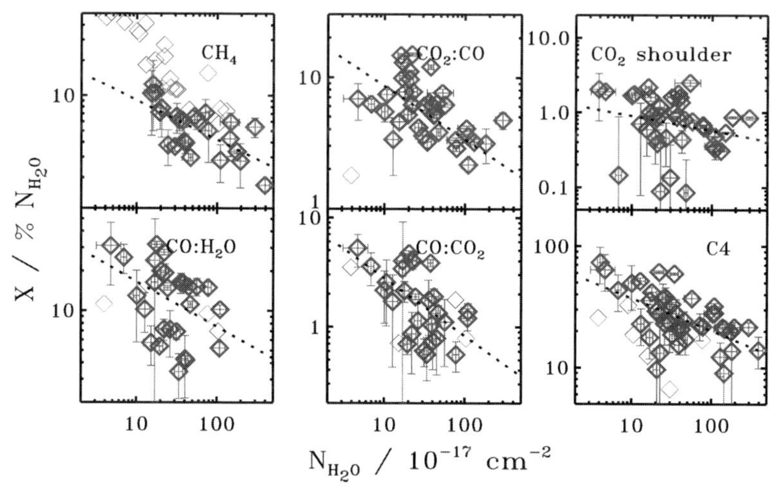

Figure 8. Statistically significant correlations between ice abundances w.r.t. H_2O (N_X / % N_{H_2O}) and the H_2O ice column density for low-mass protostars, with upper limits plotted with thin symbols. The significance of the correlations was measured using Spearman's rank correlation test, which makes no assumptions about the type of correlation, while the dotted line shows the best log-log fit to the data to guide the eye.

Ice maps are extremely useful, but only possible in a few lines of sight with fortuitous placement of background sources. Ice correlation tests is another tool used to constrain the ice formation history, used by e.g. Boogert *et al.* (2008) to investigate which ice components depend on ice heating to form. Figure 8 shows an example of such a correlation study using the *c2d* low-mass protostellar sample. Among the investigated ices, the abundance w.r.t H_2O of CH_4, CO_2:CO, the CO_2 shoulder, CO:H_2O, CO:CO_2 and C4 are anti-correlated with the H_2O column density at 97–99% confidence with the rank correlation test introduced by Spearman (1904). The CO-related anti-correlations are consistent with the §4.5 results that H_2O-rich and a CO-rich ice phases form separately in clouds and cloud cores. The anti-correlations are expected if there is competition between the formation of these two phases; i.e. the more oxygen that is bound up in H_2O ice, the less may be available to form CO and thus ices that depend on CO freeze-out. Still in each individual core CO ice abundance and H_2O ice column are expected to correlate because of increasing CO freeze-out toward the densest part of the cloud core. This accounts for some of the scatter in the plots. The CH_4–H_2O trend is strongest for low H_2O column densities and may be due to more efficient CH_4 formation during the earliest H_2O formation stage when then C/CO ratio is high.

5. An ice evolution scenario

Building on the presented results above as well as additional experiments, models and observations, this section discusses what the current constraints on the ice evolution from clouds to protostars.

5.1. *Ice formation in clouds and cloud cores*

From ice maps of cloud edges, the first ices to form are H_2O dominated with large concentrations of CO_2 (Fig. 9). CH_4 formation is the most efficient at the earliest stages of H_2O ice formation, when the C/CO ratio in the gas phase is still high (§4.6). The narrow abundance distribution of NH_3 is evidence for that it too forms with H_2O. This ice formation stage depend on hydrogenation of atoms, O+H, C+H and N+H; CO_2 can form efficiently from CO+OH at low temperatures, based on experiments by Oba *et al.* (2010), Ioppolo *et al.* (2011) and Noble *et al.* (2011). Most CO that is frozen out in this phase must be converted into CO_2 since CO_2:H_2O is more abundant than CO:H_2O. From experiments by Hiraoka *et al.* (1998), H preferentially reacts with atoms compared to CO, explaining why no CH_3OH forms in this ice formation phase.

At some point the CO/O ratio becomes large enough that most frozen-out CO is no longer converted into CO_2 and H_2O is no longer the most efficiently formed ice. We mark this as the breaking point between early H_2O dominated and late CO-driven pre-stellar ice formation. The ices that form in this later stage are not expected to co-vary with H_2O, but rather depend on the CO-freeze-out efficiency. The latter is both temperature and density dependent and may vary greatly dependent on cloud structure, cloud collapse time scales and local radiation environment. This explains why many identified ice and ice components that are definitely present during the pre-stellar stage have abundances that vary by orders of magnitude with respect to H_2O. In addition, the anti-correlation between H_2O ice and the abundance of several CO-ice components suggest that the time available for this CO-based chemistry decreases when a longer time is spent in the H_2O-ice formation stage, where O is converted into H_2O and CO_2:H_2O ice.

CO_2:CO, CO:H_2O and the XCN band all correlate with CO freeze-out in the Oph-F ice map. CO_2 may still form from OH+CO at this stage, but a high CO/O ratio results in sparse H_2O formation compared to CO and thus in a CO-rich CO_2 ice. This is also the outcome from recent modeling by Garrod & Pauly (2011). OCN^- may form from CO+NH, followed by proton transfer, and the XCN-CO freeze-out correlation supports that OCN^- is responsible for the entire XCN band. Cuppen *et al.* (2009) have used recent experiments to show that CH_3OH and H_2CO can form from hydrogenation of CO under ISM conditions and CH_3OH abundance should also depend on CO freeze-out. In addition to atomic processing, the ices are continuously exposed to cosmic rays and cosmic-ray induced UV photons – Shen *et al.* (2004) have estimated the relative doses. This may both affect the simple ice abundances and result in some cold formation of more complex molecules.

5.2. *Thermal and UV processing of ices around protostars*

Though much of the observed ice evolution can be explained by processes during the prestellar stage, some ice features are formed or destroyed by the protostar. Based on the observations of pure CO_2 by Pontoppidan *et al.* (2008) and experiments by e.g. Öberg *et al.* (2009b), evaporation of CO from CO_2:CO ice and segregation of H_2O:CO_2 ices are efficient above 20–30 K in protostellar envelopes. Pure CO_2 is also observed by Kim. *et al.* (2011) around extremely low-luminosity protostars, however, and may there be a remnant of past outbursts that heated up the envelope beyond its low-accretion temperature, since ice segregation is an irreversible process. Evans *et al.* (2009) and Dunham *et al.* (2010) have invoked episodic accretion to explain the luminosity problem of low-mass YSOs, i.e. the observation that most YSOs are significantly less luminous than predicted by protostellar evolutionary models with steady accretion.

The observed lower CO and CH_4 ice abundances toward high-mass protostars compared to low-mass YSOs can also be explained by protostellar ice heating since these are very volatile ices. From experiments by e.g. Sandford & Allamandola (1990), Collings *et al.* (2004) and Fayolle *et al.* (2011), some CO and CH_4 can be trapped in H_2O ices,

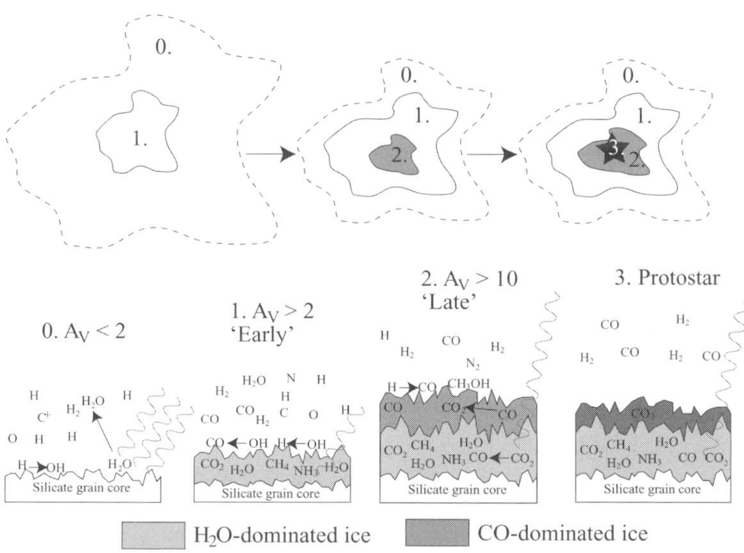

Figure 9. The key ice reactions during cloud and star formation, where 0. corresponds to cloud edges where UV photons can penetrate, 1. the early stages of dense cloud formation, 2. the later formation of cloud cores and 3. the protostellar envelope. Ice formation begins in 1. with hydrogenation of atoms resulting in a H_2O dominated ice with a high CO_2 concentration (CO_2:H_2O). At later times CO freezes out catastrophically, resulting in a second layer where CO_2 formation continues (CO_2:CO) and CH_3OH formation begins. Small amounts of ice surface chemistry products are maintained in gas-phase due to non-thermal desorption. In the protostellar stage desorption of CO starts at 20 K and H_2O:CO_2 segregation at 30 K.

however, which is important for predicting ice abundances in moderately heated regions, such as the comet forming zones in protoplanetary disks.

From experiments by e.g. Öberg *et al.* (2009a), moderate ice heating to 20–50 K also enables the movement of radicals within the ice. Radicals are produced in ices by photodissociation of the condensed molecules. Models by Garrod *et al.* (2008) predict that especially CH_3OH photodissociation followed by radical diffusion result in complex ice formation and protostellar envelopes are thus expected to be efficient factories of complex organics. UV radiation of CO:CH_3OH ice mixtures produce a more HCO-rich complex chemistry, which may dominate at lower temperatures when CO is still frozen out. The ice feature at 7.25 μm identified with to HCOOH or CH_3CH_2OH provides direct evidence for this complex ice chemistry. Most complex ice features cannot be directly observed, however, because of band confusion. Instead most constraints on this ice evolutionary phase comes from millimeter observations of desorbed ice chemistry products.

Complex organic molecules such as CH_3CH_2OH and $HCOOCH_3$ are observed in the inner regions around high-mass protostars, in so called hot cores, at abundances that cannot be reproduced by gas phase chemistry. Furthermore, Bisschop *et al.* (2007) found that they correlate with the ice chemistry product CH_3OH. These observational pieces of evidence, together with the existence of a plausible ice production pathway of complex molecules, strongly support a scenario in which complex ices are produced in the outer envelope and then desorb into the gas-phase as icy grains flow in towards the protostar.

IRAS 16293-2422, observed by e.g. van Dishoeck *et al.* (1995) and Cazaux *et al.* (2003) is a low-mass equivalent to the high-mass hot cores, and there are at least five other low-mass protostars where complex molecules such as $HCOOCH_3$ and CH_3OCH_3 are detected by Bottinelli *et al.* (2004), Jørgensen *et al.* (2005) and van der Marel *et al.* (2011). Arce *et al.* (2008) also detected complex molecules in a low-mass outflow and Öberg *et al.* (2010) recently detected $HCOOCH_3$ in a cold and quiescent part of a low-mass star forming region that probably has been illuminated by UV through a protostellar outflow cavity. Combining these observational results, Öberg *et al.* (2011b, subm. to ApJ) find that $HCOOCH_3$ generally dominates the complex chemistry toward outflows and low-mass protostellar envelopes, consistent with experiments on CO:CH_3OH ice photochemistry below 25 K. In contrast, observations of high-mass and resolved low-mass hot cores contain more CH_3OCH_3 or C_2H_5OH than $HCOOCH_3$. This is evidence for a sequential formation of complex molecules, starting with HCO-rich molecules as long as CO ice is abundant, followed by $CH_{3/2}$-rich molecules at higher ice temperatures. Gas phase observations can thus provide information on the ice chemistry, but interpretations of these more indirect constraints rely on a combination of experimental and model results.

6. Conclusions

Thanks to a combination of ground-based and space telescope ice observations and years of laboratory ice spectroscopy and ice chemistry experiments we have good constraints on the typical ice abundances and ice compositions in a range of interstellar and circumstellar environments. A key result coming out of these observations is that most ices up to CH_3OH in complexity have already formed by the time the protostar turns on. Ice abundance variations must therefore be mainly due to variations in the prestellar ice formation processes. Variations in CO freeze-out and the subsequent CO-based chemistry is a likely main contributor to the observed differences in ice abundances in different lines of sight, but other processes may be important as well. Thermal and UV processing of ices by the protostar is important for the continued evolution of the ice, leading up to the formation of hot core molecules and the complex organics found in comets. To further constrain the ice evolution from cloud cores to comets will require a combination of high-resolution IR studies to detect complex molecules directly in the ice, millimeter observations toward protostars and disks to image desorbed ice chemistry products, and more detailed experiments and models of both the early ice formation stages and of the reactions that convert simple ices into more complex ones.

References

Arce, H. G., Santiago-García, J., Jørgensen, *et al.* 2008, *ApJl*, 681, L21
Bergin, E. A., Melnick, G. J., Gerakines, P. A., *et al.* 2005, *ApJl*, 627, L33
Bisschop, S. E., Jørgensen, J. K., van Dishoeck, E. F., *et al.* 2007b, *A&A*, 465, 913
Boogert, A. C. A. & Ehrenfreund, P. 2004, in ASP Conf. Ser. 309: Astrophysics of Dust, 547–572
Boogert, A. C. A., Huard, T. L., Cook, A. M., *et al.* 2011, *ApJ*, 729, 92
Boogert, A. C. A., Pontoppidan, K. M., Knez, C., *et al.* 2008, *ApJ*, 678, 985
Boogert, A. C. A., Pontoppidan, K. M., Lahuis, F., *et al.* 2004, *ApJs*, 154, 359
Bottinelli, S., Adwin Boogert, A. C., Bouwman, J., *et al.* 2010, *ApJ*, 718, 1100
Bottinelli, S., Ceccarelli, C., Neri, R., *et al.* 2004, *ApJl*, 617, L69
Cazaux, S., Tielens, A. G. G. M., Ceccarelli, C., *et al.* 2003, *ApJl*, 593, L51
Charnley, S. B. 2004, *Advances in Space Res.*, 33, 23
Charnley, S. B., Tielens, A. G. G. M., & Millar, T. J. 1992, *ApJl*, 399, L71
Collings, M. P., Anderson, M. A., Chen, R., *et al.* 2004, *MNRAS*, 354, 1133
Cook, A. M., Whittet, D. C. B., Shenoy, S. S., *et al.* 2011, *ApJ*, 730, 124
Cuppen, H. M., van Dishoeck, E. F., Herbst, E., & Tielens, A. G. G. M. 2009, *A&A*, 508, 275
D'Hendecourt, L. B. & Jourdain de Muizon, M. 1989, *A&A*, 223, L5
Dunham, M. M., Evans, N. J., Bourke, T. L., *et al.* 2010, *ApJ*, 721, 995
Evans, N. J., Dunham, M. M., Jørgensen, J. K., *et al.* 2009, *ApJs*, 181, 321
Evans, II, N. J., Allen, L. E., Blake, G. A., *et al.* 2003, *PASP*, 115, 965
Fayolle, E. C., Öberg, K. I., Cuppen, H. M., Visser, R., & Linnartz, H. 2011, *A&A*, 529, A74+
Feigelson, E. D. & Nelson, P. I. 1985, *ApJ*, 293, 192
Fraser, H. J., Bisschop, S. E., Pontoppidan, K. M., *et al.* 2005, *MNRAS*, 356, 1283
Garrod, R. T. & Pauly, T. 2011, ArXiv e-prints
Garrod, R. T., Weaver, S. L. W., & Herbst, E. 2008, *ApJ*, 682, 283
Gibb, E. L., Whittet, D. C. B., Boogert, A. C. A., & Tielens, A. G. G. M. 2004, *ApJs*, 151, 35
Gibb, E. L., Whittet, D. C. B., Schutte, W. A., *et al.* 2000, *ApJ*, 536, 347
Gillett, F. C. & Forrest, W. J. 1973, *ApJ*, 179, 483
Hiraoka, K., Miyagoshi, T., Takayama, T., Yamamoto, K., & Kihara, Y. 1998, *ApJ*, 498, 710
Ioppolo, S., van Boheemen, Y., Cuppen, H. M., *et al.* H. 2011, *MNRAS*, 238
Jørgensen, J. K., Bourke, T. L., Myers, P. C., *et al.* 2005, *ApJ*, 632, 973
Keane, J. V., Boonman, A. M. S., Tielens, A. G. G. M., *et al.* 2001, *A&A*, 376, L5
Kim., H. J., Evans, II, N. J., Dunham, M. M., & Lee, J. E. 2011, in IAU Symposium, 280, 216
Knez, C., Boogert, A. C. A., Pontoppidan, K. M., *et al.* 2005, *ApJl*, 635, L145
Merrill, K. M., Russell, R. W., & Soifer, B. T. 1976, *ApJ*, 207, 763
Oba, Y., Watanabe, N., Kouchi, A., Hama, T., & Pirronello, V. 2010, *ApJl*, 712, L174
Öberg, K. I., Boogert, A. C. A., Pontoppidan, K. M., *et al.* 2008, *ApJ*, 678, 1032
Öberg, K. I., Bottinelli, S., Jørgensen, J. K., & van Dishoeck, E. F. 2010, *ApJ*, 716, 825
Öberg, K. I., Fayolle, E. C., Cuppen, H. M., *et al.* H. 2009a, *A&A*, 505, 183
Öberg, K. I., Garrod, R. T., van Dishoeck, E. F., & Linnartz, H. 2009, *A&A*, 504, 891
Noble, J. A., Dulieu, F., Congiu, E., & Fraser, H. J. 2011, *ApJ*, 735, 121
Pendleton, Y. J., Tielens, A. G. G. M., Tokunaga, A. T., *et al.* 1999, *ApJ*, 513, 294
Pontoppidan, K. M. 2006, *A&A*, 453, L47
Pontoppidan, K. M., Boogert, A. C. A., Fraser, H. J., *et al.* 2008, *ApJ*, 678, 1005
Pontoppidan, K. M., Dartois, E., van Dishoeck, E. F., *et al.* 2003a, *A&A*, 404, L17
Pontoppidan, K. M., Fraser, H. J., Dartois, E., *et al.* 2003b, *A&A*, 408, 981

Pontoppidan, K. M., van Dishoeck, E. F., & Dartois, E. 2004, *A&A*, 426, 925
Przybilla, N., Nieva, M., & Butler, K. 2008, *ApJl*, 688, L103
Reach, W. T., Faied, D., Rho, J., *et al.* 2009, *ApJ*, 690, 683
Sandford, S. A. & Allamandola, L. J. 1990, *ApJ*, 355, 357
Schutte, W. A., Boogert, A. C. A., Tielens, A. G. G. M., *et al.* 1999, *A&A*, 343, 966
Shen, C. J., Greenberg, J. M., Schutte, W. A., & van Dishoeck, E. F. 2004, *A&A*, 415, 203
Spearman, C. 1904, *The American Journal of Psychology*, 15, pp. 72
Tielens, A. G. G. M. & Hagen, W. 1982, *A&A*, 114, 245
Tielens, A. G. G. M., Tokunaga, A. T., Geballe, T. R., & Baas, F. 1991, *ApJ*, 381, 181
van der Marel, N., Öberg, K. I., Kristensen, L., *et al.* 2011, IAU Symposium 280, 365
van Broekhuizen, F. A., Keane, J. V., & Schutte, W. A. 2004, *A&A*, 415, 425
van Broekhuizen, F. A., Pontoppidan, K. M., Fraser, H. J., *et al.* 2005, *A&A*, 441, 249
van Dishoeck, E. F., Blake, G. A., Jansen, D. J., & Groesbeck, T. D. 1995, *ApJ*, 447, 760
Whittet, D. C. B., Cook, A. M., Chiar, J. E., *et al.* 2009, *ApJ*, 695, 94
Whittet, D. C. B., Shenoy, S. S., Bergin, E. A., *et al.* 2007, *ApJ*, 655, 332
Zasowski, G., Kemper, F., Watson, D. M., *et al.* 2009, *ApJ*, 694, 459

Discussion

AMANDA COOK: At least once in your talk you referred to a pure CO_2 component in the ice. Since it seems like CO_2 and H_2O forms in tandem in molecular clouds, how would you explain the presence of pure CO_2 ice?

ÖBERG: Pure CO_2 ice can form from evaporation of CO at 20 K from the CO_2:CO ice layer that is a result of late CO_2 formation in cloud cores, or from segregation of the H_2O:CO_2 ice layer starting at 30 K. Pure CO_2 is therefore only expected in environments where ices have been heated to at least 20–30 K.

NAME SAKAI: What is the origin of the difference of molecular abundances in the ice between low-mass and high-mass sources? Are the ice abundances in high-mass cases measured in pre-stellar cores?

ÖBERG: The high-mass sources are all protostars. Low CO and CH_4 ice abundances toward high-mass YSOs are probably due to that the protostars have heated most of the envelopes, resulting in evaporation of volatile ices. The difference in CO_2 ice abundances may be due to either CO_2 evaporation in the high-mass protostellar envelope, or due to less CO_2 formation in the high-mass pre-stellar phase because of different collapse time scales, and density and temperature structures in low- and high-mass prestellar cores

JACOB LAAS: The recent studies by Gerakines *et al.* have found that NH_3 and other N-bearing species can have large effects on certain line profiles in ices. Could this have any effect on the spectral signatures near 7μm that you say are strong indicators of complex organic molecules?

ÖBERG: NH_3 acts as a base in ices, transforming e.g. HCOOH into $HCOO^-$. If the two bands at 7.2–7.5 μm are due (partially) to HCOOH and $HCOO^-$, NH_3 may regulate the relative intensities of the two bands. If the bands are instead due to CH_3CH_2OH and CH_3CHO, NH_3 should not have a large effect on the band characteristics.

The Molecular Universe
Proceedings IAU Symposium No. 280, 2011
José Cernicharo & Rafael Bachiller, eds.

doi:10.1017/S1743921311024884

Models of Hot Cores with Complex Molecules

Susanna L. Widicus Weaver[1], **Robin T. Garrod**[2], **Jacob C. Laas**[1], **and Eric Herbst**[3]

[1]Department of Chemistry, Emory University
1515 Dickey Drive, Atlanta, GA, 30322, USA
email: `susanna.widicus.weaver@emory.edu`

[2]Department of Astronomy, Cornell University
304 Space Sciences Building, Ithaca, NY, 14853, USA
email: `rgarrod@astro.cornell.edu`

[3]Departments of Physics, Chemistry and Astronomy, The Ohio State University
Physics Research Building, 191 West Woodruff Avenue, Columbus, OH, 43210
email: `herbst.6@osu.edu`

Abstract. Recent models of hot cores have incorporated previously-uninvestigated chemical pathways that lead to the formation of complex organic molecules (COMs; i.e. species containing six or more atoms). In addition to the gas-phase ion-molecule reactions long thought to dominate the organic chemistry in these regions, these models now include photodissociation-driven grain surface reaction pathways that can also lead to COMs. Here, simple grain surface ice species photodissociate to form small radicals such as OH, CH_3, CH_2OH, CH_3O, HCO, and NH_2. These species become mobile at temperatures above 30 K during the warm-up phase of star formation. Radical-radical addition reactions on grain surfaces can then form an array of COMs that are ejected into the gas phase at higher temperatures. Photodissociation experiments on pure and mixed ices also show that these complex molecules can indeed form from simple species. The molecules predicted to form from this type of chemistry reasonably match the organic inventory observed in high mass hot cores such as Sgr B2(N) and Orion-KL. However, the relative abundances of the observed molecules differ from the predicted values, and also differ between sources. Given this disparity, it remains unclear whether grain surface chemistry governed by photodissociation is the dominant mechanism for the formation of COMs, or whether other unexplored gas-phase reaction pathways could also contribute significantly to their formation. The influence that the physical conditions of the source have on the chemical inventory also remains unclear. Here we overview the chemical pathways for COM formation in hot cores. We also present new modeling results that begin to narrow down the possible routes for production of COMs based on the observed relative abundances of methyl formate ($HCOOCH_3$) and its $C_2H_4O_2$ structural isomers.

Keywords. astrochemistry, molecular processes, methods: numerical, stars: formation, ISM: abundances, ISM: clouds, ISM: molecules

1. Introduction

Over 160 molecules have now been detected in interstellar and circumstellar environments, and a large fraction of these molecules are classified as complex organic molecules, or COMs (i.e. those organic species containing six or more atoms; Herbst & van Dishoeck 2009). The majority of COMs have been discovered in the warm environments of hot cores, where it was presumed that gas-phase ion-molecule reactions were the predominant formation mechanism (Charnley 2001). However, theoretical and experimental investigations of the possible gas-phase formation pathways for the highly-abundant COM methyl formate ($HCOOCH_3$) showed that its formation through these types of reactions cannot explain its observed abundance (Horn *et al.* 2004). In addition, methyl formate has been detected in regions where the timescales are too short for sufficient build-up of COMs through gas-phase reactions. Such regions include shocked regions in the Galactic Center (Requena-Torres *et al.* 2006), low-mass hot cores (Ceccarelli 2005), and cold envelopes around embedded protostars (Öberg *et al.* 2010). These discoveries call into question the gas-phase formation mechanisms previously proposed for methyl formate and other related COMs.

Recent astrochemical modeling studies (Garrod & Herbst 2006; Garrod *et al.* 2008) have invoked grain surface radical-radical reactions to explain the high abundance of methyl formate in hot cores. Additional studies (Laas *et al.* 2011) have further investigated the relative abundance of methyl formate to its $C_2H_4O_2$ isomers acetic acid (CH_3COOH) and glycolaldehyde ($HOCH_2CHO$). An overview of these hot core chemical models and associated results is presented here.

2. Chemical Models of Hot Cores

2.1. *Previous Hot Core Models*

Previous models of hot cores assumed limits on interstellar chemical reactions based on the characteristic temperatures of interstellar clouds. The assumption was long made that only simple single-atom addition reactions could occur on grain surfaces, and that the more complex chemistry occurred through gas-phase ion-molecule reactions in warmer regions of star formation. Indeed, at 10 K, only single atoms are mobile on an ice surface, limiting the possible condensed-phase chemistry to the formation of simple molecules such as water and methanol. The physical model assumed in these previous studies for the evolution of the hot core was a single temperature jump, where the simpler molecules trapped in ices in cold clouds are injected into the gas phase through thermal evaporation from the grains. A complex network of gas-phase ion-molecule chemistry was then invoked to explain the formation of COMs in hot cores. See the review of Charnley (2001) and references therein for a more comprehensive overview of these earlier hot core chemical models.

2.2. *New Gas/Grain Hot Core Models*

2.2.1. *Grain Surface Chemistry During the Warm-up Phase of Star Formation*

In the studies of Garrod & Herbst (2006) and Garrod *et al.* (2008), a second-order power law was used to model the warm-up phase of a hot core. Such an approach had been suggested by Viti & Williams (1999) to address the time-dependent evaporation of the ice mantles. This change in physical model allowed the employment of a more complex network of chemical reactions on grain surfaces at temperatures that fell within the range between the two extremes of a dense cloud (10 K) and a hot core (200 K). As in the case of the previous models, the single-atom addition reactions on grain surfaces are the most efficient reactions at colder temperatures. However, molecular species become mobile on ice surfaces above $\sim$30 K, enabling an extensive chemical reaction network to be employed for grain surface chemistry.

The assumption is made in the model of Garrod *et al.* (2008) that the simple molecules formed in colder regions dissociate in the ice matrix through either direct or cosmic-ray induced photodissociation to form OH, CH_3, CH_3O, CH_2OH, HCO, and NH_2 radicals through the following processes:

$$H_2O \rightarrow H + OH \tag{2.1}$$

$$\begin{aligned} CH_3OH \quad &\rightarrow CH_3 + OH \\ &\rightarrow CH_2OH + H \\ &\rightarrow CH_3O + H \end{aligned} \tag{2.2}$$

$$H_2CO \rightarrow H + HCO \tag{2.3}$$

$$NH_3 \rightarrow NH_2 + H \tag{2.4}$$

The radicals formed from this photodissociation become mobile at temperatures >30 K. More complex organics such as methyl formate and its structural isomers are formed through the combination reactions of these mobile radicals during the warm-up phase of star formation. The grain surface reactions leading to methyl formate and glycolaldehyde are:

$$CH_3O + HCO \rightarrow HCOOCH_3 \tag{2.5}$$

$$CH_2OH + HCO \rightarrow HOCH_2CHO \tag{2.6}$$

The grain surface chemistry that leads to acetic acid is more complicated, and a sequence of three reactions is required to explain its formation:

$$\begin{aligned} CH_3 + HCO \quad &\rightarrow CH_3CHO \\ CH_3CHO + H \quad &\rightarrow H_2 + CH_3CO \\ CH_3CO + OH \quad &\rightarrow CH_3COOH \end{aligned} \tag{2.7}$$

This type of grain surface chemistry can build up significant chemical complexity in the envelopes around young stars. The molecules formed from this grain surface chemistry then evaporate at warmer temperatures. The molecules released from the ices undergo subsequent ion-molecule gas-phase reactions to form even more complex organic molecules.

The results of the Garrod *et al.* (2008) gas/grain model are shown in Figure 1. Here, the peak abundances predicted by the model are compared to the observed abundances in the Sagittarius B2(N-LMH) source. It is clear that the addition of radical-radical grain surface reactions has improved the agreement between the models and observations considerably, as significant abundances are predicted for a variety of COMs that have been detected in the Sgr B2(N-LMH) hot core. Despite this improvement, the relative abundances of many COMs, and in particular methyl formate and its structural isomers, are not consistent with the Sgr observations. Some results of this study made it clear that additional studies examining COM formation in these sources are required. More sophisticated hydrodynamics models, such as those employed by Aikawa *et al.* (2008), may also be needed to properly model the complex physical environment of a hot core.

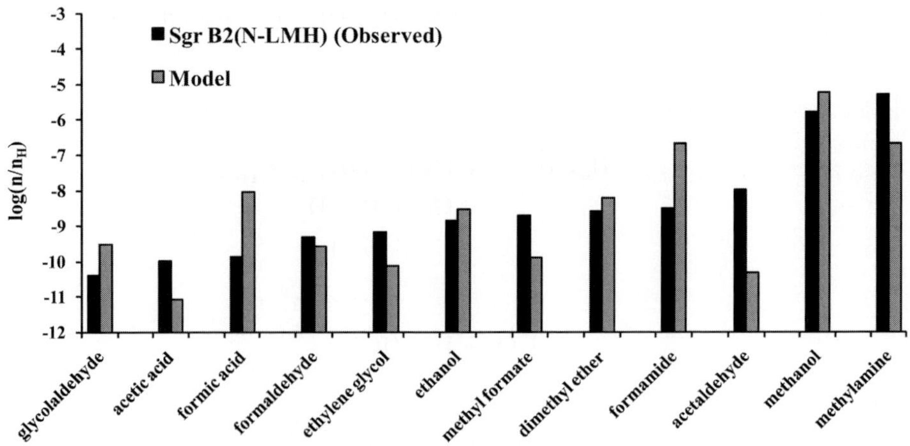

Figure 1. Results of initial gas/grain hot core model as adapted from Garrod *et al.* (2008). This model incorporates photodissociation-induced grain surface reactions to form complex organic molecules. Comparison is made between predicted peak abundance for each molecule in the model and the observational results for the Sgr B2(N-LMH) hot core.

2.2.2. *Testing the Influence of Methanol Photodissociation Branching Ratios*

Examination of Reactions 2.5 - 2.7 reveals that the photodissociation of methanol plays a significant role in the production of COMs through grain surface radical-radical reactions. Methanol is the most abundant organic molecule in interstellar ices and is therefore the dominant source of the CH_3, CH_2OH, and CH_3O radicals on grain surfaces. Methanol photodissociation is therefore the primary route to several of the major organic functional groups that make up COMs. Methanol photodissociation branching ratios therefore strongly influence structural isomerism in interstellar clouds, and are particularly crucial for methyl formate and its structural isomers.

There is little agreement in the literature as to the values for either the condensed-phase or gas-phase methanol photodissociation branching ratios. A laboratory study by Öberg *et al.* (2009) of methanol photodissociation in interstellar ice analogs has made the first step toward determining the information required for astrochemical models. The experiment monitors the stable species that form from reactions between the photodissociation products, rather than directly monitoring the photodissociation products themselves. Therefore, the reaction rates and branching ratios derived from these ice photodissociation measurements are difficult to employ in astrochemical models because determination of branching ratios is also dependent on chemical modeling of the ice chemistry. The model employed by Öberg *et al.* (2009) does measure the time-dependent abundance of CH_2OH and H_2CO, but not other methanol photolysis products.

Reports of gas-phase methanol photodissociation studies are extensive in the literature, but there is very little agreement as to the branching ratios for the dissociation pathways. This disagreement arises primarily from the methods used to monitor product formation. Hagege *et al.* (1968) determined the primary processes at 189.9 nm to be the three processes included in Reaction 2.2 above, plus the additional route

$$CH_3OH \rightarrow CH_2O + H_2 \tag{2.8}$$

with relative ratios of $CH_2OH + CH_3O{:}CH_3{:}CH_2O = 0.75{:}0.05{:}0.20$. These conclusions were based on measurements of rates of various reactions and an analysis of a kinetics model; these values have remained the most accepted photodissociation branching ratios to date. Thus, CH_2OH and CH_3O are thought to be the primary photodissociation products for methanol. A more recent gas-phase study by Satyapal *et al.* (1989) agrees qualitatively with the conclusions by Hagege *et al.* However, improved photodissociation branching ratios have not been reported in the literature because no experimental technique used for such studies distinguishes between the CH_2OH and CH_3O species in measurement.

The gas/grain chemical network of Garrod *et al.* (2008) was the first to distinguish between CH_2OH and CH_3O, and assumed a methanol photodissociation branching ratio of 60% for the CH_3 + OH channel, with the remainder split between the CH_3O and CH_2OH channels. The recent work of Laas *et al.* (2011) has explored the extreme limits of these photodissociation branching ratios, with trials being conducted for each of the three limits of the methanol photodissociation mechanism. In each trial, one channel (i.e. CH_2OH, CH_3O, or CH_3) was set to a branching ratio of 90%, and the other two channels set to 5% each. The results of the Laas *et al.* (2011) study are shown in Figure 2, focusing specifically on methyl formate and its structural isomers. It was found that changes to the gas-phase photodissociation branching ratios for methanol had little effect on COM chemistry, as would be expected given that the major production routes for most COMs are grain-surface reactions. In contrast, changes to the grain surface photodissociation branching ratios for methanol significantly influence the relative abundances of COMs, particularly methyl formate and its structural isomers. As can be seen in Figure 2, the set of branching ratios favoring the CH_3O channel give the best match to the Sgr B2(N-LMH) methyl formate abundance. However, the branching ratios favoring the CH_3 channel give a better match to observations for acetic acid and glycolaldehyde. Additional laboratory and observational studies are required to determine the values appropriate for interstellar ices.

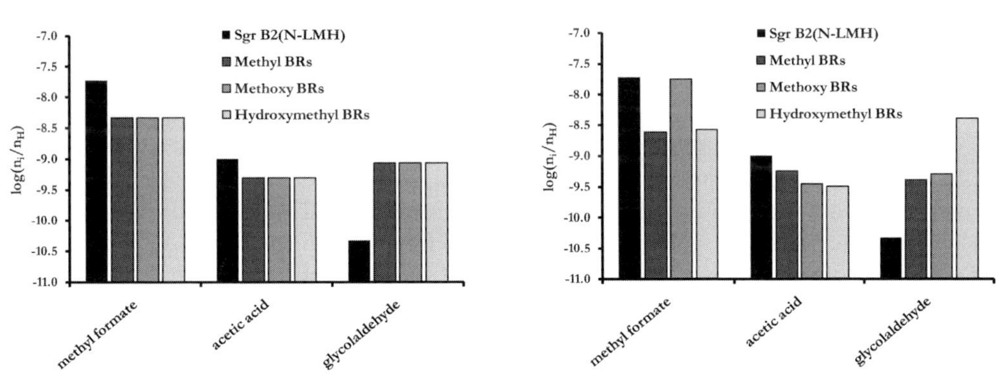

Figure 2. Peak abundances predicted by modeling studies where the methanol photodissociation branching ratios were adjusted using the gas/grain network of Garrod *et al.* (2008), as adapted from the results of Laas *et al.* (2011). The results from adjustment of the gas-phase methanol photodissociation branching ratios are shown in the left panel; the results from adjustment of the grain surface methanol photodissociation branching ratios are shown in the right panel. The trials are labeled by the name of the major photodissociation product, such that in the case of the "Methoxy BRs" trial, the CH_3O branching ratio was set to 0.9, while the CH_2OH and CH_3 channels were each set to a branching ratio value of 0.05.

In addition to investigating the methanol photodissociation branching ratios, the Laas *et al.* (2011) study examined the influence of warm-up timescale on COM chemistry. The results of this analysis are shown in Figure 3 for three warm-up timescales. These results show that a longer warm-up timescale favors production of COMs through grain surface chemistry. This trend arises because the cloud spends a longer time in the temperature window that favors radical-radical grain surface chemistry. Longer warm-up timescales are required to build up significant abundances of the more complex species such as acetic acid that require multiple grain surface reactions for formation.

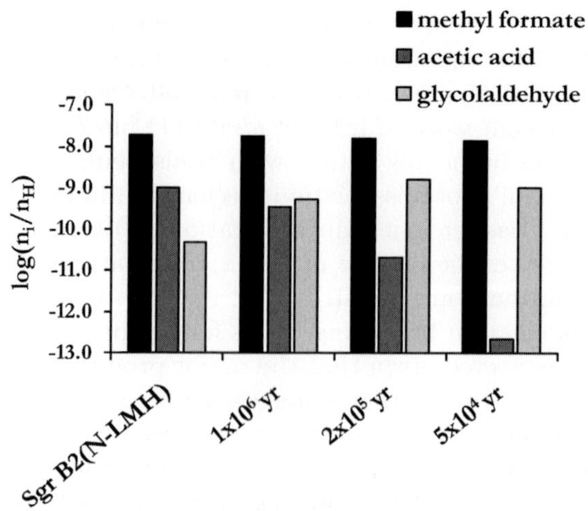

Figure 3. Results of modeling studies investigating the influence of warm-up timescale on the relative peak abundances of methyl formate and its structural isomers acetic acid and glycolaldehyde, as adapted from Laas *et al.* (2011).

2.2.3. *Testing Additional Gas Phase Routes to Methyl Formate*

Although the Laas *et al.* (2011) model gives better agreement with the observations for methyl formate over the gas/grain model of Garrod *et al.* (2008), it is clear that the model still needs improvement before it will match observations. New observational information has led us to investigate additional gas-phase formation pathways for methyl formate. Recent interferometric studies (Neill *et al.* 2011) have probed the spatial distributions of methyl formate, methanol, formic acid, and dimethyl ether in the Orion-KL star forming region. These observations show that methyl formate is spatially correlated with both methanol and dimethyl ether, but is spatially anti-correlated with formic acid. Pate and coworkers postulated that this anti-correlation might indicate a gas-phase chemical mechanism that forms methyl formate at the expense of formic acid (Neill *et al.* 2011). They propose two additional gas-phase formation pathways for methyl formate that have not been considered in previous astrochemical models. One such pathway is the Fischer esterification reaction between protonated formic acid and methanol to form protonated methyl formate:

$$HCOOH_2^+ + CH_3OH \rightarrow HC(OH)_2OCH_3^+ + H_2O \quad (2.9)$$

Calculations show that this reaction has an activation energy of ~3100 K (Neill *et al.* 2011), and is therefore unlikely to be an efficient formation mechanism for methyl formate under interstellar conditions. However, the methyl cation transfer reaction between protonated methanol and formic acid

$$HCOOH + CH_3OH_2^+ \rightarrow HC(OH)_2OCH_3^+ + H_2O \tag{2.10}$$

was found to have no activation energy when proceeding to the *trans* conformer of protonated methyl formate (Neill *et al.* 2011). The activation energy for the reaction leading to *cis*-protonated methyl formate was calculated to be ~1320 K (Neill *et al.* 2011). The two conformers of methyl formate are structurally distinct, with the orientation of the -CH_3O group relative to the -CHO group being the distinguishing structural difference. The *cis* conformer is energetically more stable, and is therefore the conformer commonly detected in hot cores. The *cis* conformer was calculated to be 2880 K lower in energy than the *trans* species (Neill *et al.* 2011). The protonated versions of the *cis* and *trans* stereoisomers have similar structures to their neutral counterparts, but in this case the *trans* species is lower in energy by 2880 K (Neill *et al.* 2011).

These reactions were added to the gas/grain network of Laas *et al.* (2011) and combined with the methoxy branching ratio trial to test their influence on the formation of methyl formate in hot cores. The results are shown in Figure 4. It was found that these reactions had no effect on the predicted peak abundance of *cis*-methyl formate, which is the conformer that has been detected in the interstellar medium. However, a significant amount of *trans*-methyl formate was predicted to form at the higher temperatures of hot cores, with fractional abundances relative to H approaching 10^{-11} at temperatures above 100 K. Pate and coworkers have now obtained the laboratory spectrum of *trans*-methyl formate, and report its preliminary detection in the Sgr B2(N-LMH) hot core (Neill *et al.* 2011), in agreement with the model prediction. It should be noted here that this detection is not in Orion-KL, where the anti-correlation between formic acid and methyl formate emission has been observed in imaging studies. Additional observations are necessary to examine the spatial distributions of these molecules in Sgr B2 and search for *trans*-methyl formate in Orion-KL.

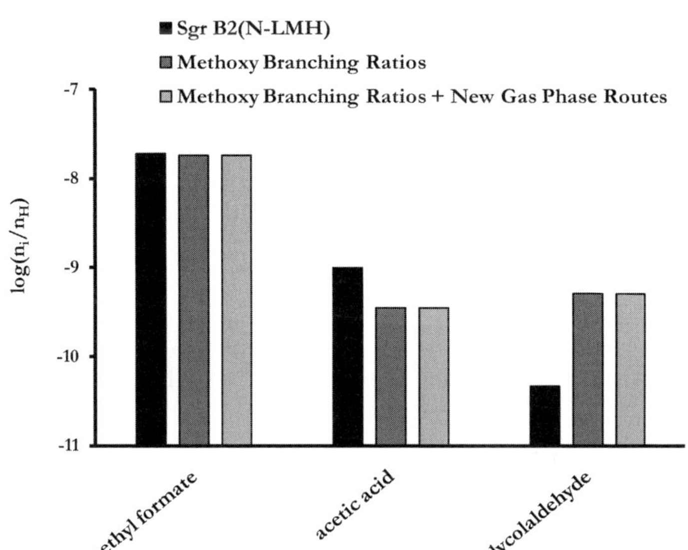

Figure 4. Results of modeling studies investigating the influence of two new gas-phase formation routes for methyl formate, as adapted from Laas *et al.* (2011).

The *trans* conformer of methyl formate may serve as an excellent tracer of gas-phase chemistry in star-forming regions. Its predominant formation mechanism is the electron recombination reaction of protonated *trans*-methyl formate, which only forms from the gas-phase methyl cation transfer reaction between protonated methanol and formic acid. The assumption is made that *trans*-methyl formate forms exclusively through this gas-phase reaction pathway in hot cores, and no grain surface chemistry contributes directly to its abundance. The basis for this assumption is that the two methyl formate conformers are not distinguished on the grain surface, and any methyl formate liberated from grains will be deposited into the gas phase with the expected thermal distribution based on the relative energies of the two conformers. It is not clear if this assumption is valid, and additional investigation of stereochemical effects should be considered in future work. However, identification of additional molecules that can serve as tracers of grain surface or gas phase chemistry would be an important step forward in understanding the chemical mechanisms at play in interstellar clouds.

3. Conclusions

Previous interstellar chemical models, in which gas-phase ion-molecule reactions were assumed to dominate the formation of complex molecules, could not satisfactorily explain the observed abundances of COMs in hot cores. New gas/grain models that incorporate a warm-up model for star formation show that significant abundances of COMs are predicted to form from photodissociation-induced radical-radical reactions on grain surfaces. Additional studies have investigated the formation of methyl formate and its structural isomers acetic acid and glycolaldehyde, and found that the branching ratios for methanol photodissociation on grains strongly influence the relative abundances of these COMs. The timescale for warm-up also greatly influences COM production, with longer warm-up timescales favoring more complex chemistry. Future work will examine the possibility for a higher level of COM complexity being reached through grain surface reactions.

In addition to these studies involving grain surface chemistry, recent interferometric observations indicate an anti-correlation between methyl formate and formic acid in the Orion-KL region, prompting studies of additional gas-phase formation mechanisms for methyl formate that involve a formic acid precursor. These studies show that the addition of two new pathways to methyl formate through its protonated precursor do not influence the predicted peak abundance of the *cis* conformer of methyl formate (which is the one detected in interstellar clouds). These two reactions are the Fischer esterification reaction between protonated formic acid and methanol and the methyl cation transfer reaction between protonated methanol and formic acid. Despite the inefficient formation of *cis*-methyl formate through these processes, the methyl cation transfer reaction does have a barrierless reaction pathway leading to the *trans* stereoisomer of protonated methyl formate. A large abundance of *trans*-methyl formate is predicted at temperatures above 100 K in hot cores. Follow-up laboratory and observational studies have confirmed the presence of this methyl formate conformer in Sgr B2(N-LMH). This conformer may serve as an excellent tracer of gas phase chemistry in interstellar clouds because there are no grain-surface reactions that directly form *trans*-methyl formate. Full quantitative understanding of the chemical mechanisms in interstellar clouds requires identification of additional molecules that can serve as tracers of grain surface or gas phase chemistry. Additional studies will examine other such barrierless chemical pathways involving known COMs. Observational studies that enable comparison to sources other than Orion-KL and Sgr B2(N) can also provide crucial information to constrain models of COM chemistry in hot cores.

References

Aikawa, Y., Wakelam, V., Garrod, R. T., & Herbst, E. 2008, *ApJ*, 674, 984

Charnley, S. 2001, in: F. Giovanelli (ed.), *The Bridge Between the Big Bang & Biology*, (Rome: Cons. Naz. d. Ricerche), p. 139

Ceccarelli, C. 2006, in: D. C. Lis, G. A. Blake, & E. Herbst (eds.), *Astrochemistry: Recent Successes and Current Challenges*, Proc. IAU Symposium 231 (Cambridge, UK: Cambridge Univ. Press), p. 1

Garrod, R. T. & Herbst, E. 2006, *A&A*, 457, 927

Garrod, R. T., Widicus Weaver, S. L., & Herbst, E. 2008, *ApJ*, 682, 283

Hagege, J., Roberge, P. C., & Vermeil, C. 1968, *Trans. Faraday Soc.*, 64, 3288

Herbst, E. & van Dishoeck, E. F. 2009, *ARAA*, 47, 427

Horn, A., Møllendal, H., Sekiguchi, O., Uggerud, E., Roberts, H., Herbst, E., Viggiano, A. A., & Fridgen, T. D. 2004, *ApJ*, 611, 605

Laas, J. C., Garrod, R. T., Herbst, E., & Widicus Weaver, S. L. 2011, *ApJ*, 728, 71

Neill, J. L., Steber, A. L., Muckle, M. T., Zaleski, D. T., Lattanzi, V., Spezzano, S., McCarthy, M. C., Remijan, A. J., Friedel, D. N., Widicus Weaver, S. L., & Pate, B. H. 2011, *J. Phys. Chem. A*, 115, 6472

Öberg, K. I., Garrod, R. T., van Dishoeck, E. F., & Linnartz, H. 2009, *A&A*, 504, 891

Öberg, K. I., Bottinelli, S., Jørgensen, J. K., & van Dishoeck, E. F. 2010, *ApJ*, 716, 825

Requena-Torres, M. A., Martín-Pintado, J., Rodríguez-Franco, A., Martín, S., Rodríguez-Fernández, N. J., & de Vincente, P. 2006, *A&A*, 455, 971

Satyapal, S., Park, J., Bersohn, R., & Katz, B. 1989, *J. Phys. Chem.*, 91, 6873

Viti, S. & William, D. A. 1999, *MNRAS*, 305, 755

Discussion

FIELD: What is the role of very low energy electrons in initiating chemistry in ice mantles?

WIDICUS WEAVER: High-energy electrons (i.e. cosmic rays) are considered in the grain surface chemistry included in these models. However, we do not consider the influence of low-energy electrons here.

GUÉLIN: Sgr B2 is a distant source and current millimeter-wave observations have a low spatial resolution and mix-up different types of sources (hot cores, cold clouds, ...). You should really compare your model results with observations of well resolved sources, such as Orion KL hot cores. There, it seems that acetic acid is hardly or not detected, whereas methyl formate is very strong (e.g. C. Favre *et al.* *A&A* 2011).

WIDICUS WEAVER: Sgr B2 is not the most ideal source for comparison to these models because it is clear that both the chemical and physical environment here is very complex and unusual. Acetic acid detections in other sources are limited, and glycolaldehyde has only been detected in two sources. My group is conducting observations of a variety of sources targeting COMs so that we have additional sources to which we can compare the model predictions.

The Molecular Universe
Proceedings IAU Symposium No. 280, 2011
José Cernicharo & Rafael Bachiller, eds.

doi:10.1017/S1743921311024896

Molecules in Bipolar Outflows

Mario Tafalla[1] **and Rafael Bachiller**[1]

[1]Observatorio Astronómico Nacional (IGN)
Alfonso XII 3, 28014 Madrid, Spain
email: m.tafalla@oan.es, r.bachiller@oan.es

Abstract. Bipolar outflows constitute some of the best laboratories to study shock chemistry in the interstellar medium. A number of molecular species have their abundance enhanced by several orders of magnitude in the outflow gas, likely as a combined result of dust mantle disruption and high temperature gas chemistry, and therefore become sensitive indicators of the physical changes taking place in the shock. Identifying these species and understanding their chemical behavior is therefore of high interest both to chemical studies and to our understanding of the star-formation process. Here we review some of the recent progress in the study of the molecular composition of bipolar outflows, with emphasis in the tracers most relevant for shock chemistry. As we discuss, there has been rapid progress both in characterizing the molecular composition of certain outflows as well as in modeling the chemical processes likely involved. However, a number of limitations still affect our understanding of outflow chemistry. These include a very limited statistical approach in the observations and a dependence of the models on plane-parallel shocks, which cannot reproduce the observed wing morphology of the lines. We finish our contribution by discussing the chemistry of the so-called extremely high velocity component, which seems different from the rest of the outflow and may originate in the wind from the very vicinity of the protostar.

Keywords. ISM: molecules, ISM: jets and outflows, stars: formation, radio lines: ISM

1. Introduction

Bipolar outflows are one of the most studied phenomena of the star-formation process. They result from the supersonic acceleration of gas in two opposite directions by a newly formed star, and were first identified more than three decades ago with radio observations (Snell, Loren, & Plambeck 1980). Since their discovery, bipolar outflows have been identified around protostars of nearly all masses, from below the brown dwarf limit to the precursors of the ultra-compact HII regions, and in environments as different as isolated globules and cluster-forming regions. This ubiquity of the outflow phenomenon suggest that bipolar outflow formation is a necessary element of the physics of star formation, likely associated to the need for the gas to lose angular momentum in its process of forming a highly compact object (see Arce *et al.* 2007 for a recent review).

Interest on the chemistry of bipolar outflows goes back in time as far as the interest on their physics. Outflow emission originates from ambient gas that has been accelerated by a supersonic wind, and therefore has been shock-processed. As a result, the study of outflow chemistry is unavoidably intertwined to the study of shock chemistry in the interstellar gas. Early questions on this chemistry were raised as soon as high velocity molecular gas was observed, and were related to the survival or shock-production of the CO molecules seen at velocities of tens of km s^{-1} in the wings of outflow spectra (Kwan & Scoville 1976) and the up to a hundred km s^{-1} velocities seen in some water masers

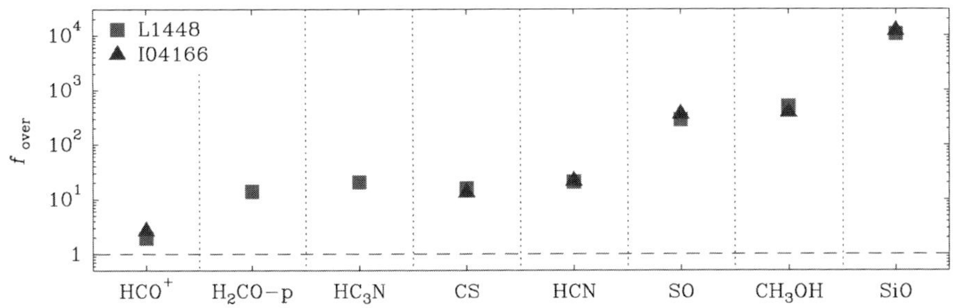

Figure 1. Overabundance factors with respect to dense core values for a number of molecules towards the outflows from L1448-mm and IRAS 04166+2706. Data from Tafalla *et al.* (2010).

(Morris 1976). Prompted by these observations, the first (and still relevant) models of molecule formation in shocks soon appeared (Hollenbach & McKee 1979).

With time, additional chemical processes have been identified in the gas accelerated by outflows. Current interest includes the enhancement in the abundance of high-density gas tracers like SiO and CH_3OH, which is seen toward a number of outflow sources powered both by high and low-mass young stellar objects (Bachiller 1996). These abundance enhancements are believed to result from the erosion of dust grains via sputtering and grain-grain collisions, perhaps combined with a number of gas-phase chemical reactions favored by the temperature increase caused by the shock (see van Dishoeck & Blake 1998 for a review).

The amount of the abundance enhancement depends on the species under consideration, and typically ranges from factors of a few for species like HCO^+ to several orders of magnitude for species like CH_3OH and SiO (Fig. 1). Not all outflows show the same degree of molecular richness, however, and this apparent selectivity of outflow chemistry has lead to the creation of special category of outflows, the so-called "chemically active" outflows (Bachiller *et al.* 2001).

2. Chemically active outflows

Although easily distinguishable when observed in species like CH_3OH and SiO, the chemically active outflows do not stand apart from the rest when observed in CO (the standard outflow tracer), either by their line strength, velocity extent, or spatial collimation. It is well established that the chemically active phase corresponds to an early period in the outflow development, as the driving sources of this family of outflows tend to be Class 0 sources. The exact physical cause of the chemical richness is however unclear. One possibility is that the youngest outflows must encounter denser envelopes along their paths, and therefore must naturally give rise to stronger shocks. Another possibility is that early chemical activity is associated with the higher energy known to characterize the earliest outflow phases (Bontemps *et al.* 1996). In either way, the lack of chemical richness in the more evolved outflows indicates that chemical activity is highly transient, and that any signature of its presence must disappear quickly in the protostellar life (near the transition between Class 0 and Class I, Tafalla *et al.* 2000). Rapid depletion of the enhanced species via freeze out onto cold dust grains is the most likely cause of this effect (Bergin *et al.* 1998).

Among the group of chemically active outflows, the one in the L1157 dark cloud stands out for the strength of its lines in species like SiO and CH_3OH. As a result, it has been

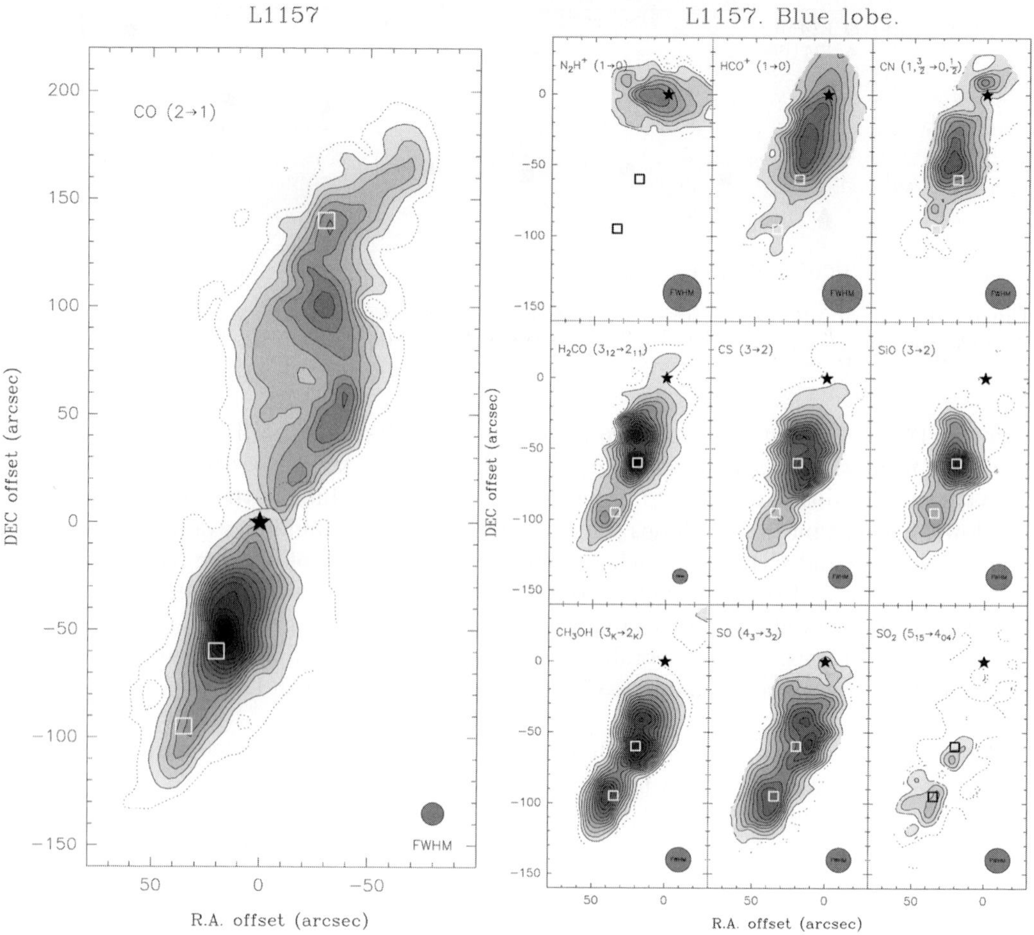

Figure 2. Left: map of CO(2-1) emission from the L1157 outflow. **Right:** maps of integrated intensity for different molecular species towards the blue (southern) lobe of the L1157 outflow. Note the different extension and peak position of each species. Figures from Bachiller *et al.* (2001).

the subject of a very intense observing campaign, including the study using the Herschel Space Observatory presented by Lefloch (2011, this volume). The L1157 outflow was first identified by Umemoto *et al.* (1992) due to its strong CO lines. Its chemical activity was recognized very early on (Mikami *et al.* 1992), and has been characterized in great detail by Bachiller & Pérez-Gutiérrez (1997) and Bachiller *et al.* (2001). The chemical activity of L1157 is specially prominent towards several regions often denoted as B1 and B2 (for blue lobe) and R (for red lobe), and are most likely associated with regions where the ambient cloud is being strongly shocked by the outflow wind. Interferometric observations of these regions reveal a complex pattern of spatial distributions that indicate a very fragmented small-scale structure (Benedettini *et al.* 2007). As illustration of this chemical richness, we show in Fig. 2 (right panel) a series of maps of the southern lobe of the L1157 outflow in a number of molecules that are selectively enhanced in different parts of the outflow (especially B1 and B2).

3. Recent progress trancing outflow gas with molecules

It is not possible to review here with detail the rapid progress made in the study of molecules in outflows during the last few years. In this section we present a selection of studies that have been carried out since the last IAU Astrochemistry meeting in 2005. We concentrate on low-mass outflow studies and in molecules that seem specially relevant to our understanding of shock chemistry.

3.1. H_2O

Probably the most significant progress since the Asilomar meeting concerns the study of the H_2O molecule, thanks to the coming on line of a series of satellite observatories: the Submillimeter Wave Astronomy Observatory (SWAS), Odin, the Spitzer Space Telescope, and the Herschel Space Observatory.

Since the early discovery of its maser emission in outflows, water has been considered a sensitive outflow tracer. The maser emission is so bright that it can be easily detected across the galaxy, and the abundance of water required to produce the emission must have been significantly enhanced by shocks (Elitzur & de Jong 1978, Kaufman & Neufeld 1996, Cernicharo *et al.* 1996). The thermal emission of water, on the other hand, has taken longer to become a recognized tracer of outflow activity, especially in low-mass outflows. Early observations with the Infrared Space Observatory (ISO) revealed thermal water emission from a number of bipolar outflows (Liseau *et al.* 1996, Nisini *et al.* 1999, Giannini *et al.* 2001). These ISO spectra, however, did not resolve spectrally even the broadest outflow profiles, a fact that was later overcome by the SWAS and Odin telescopes, which observed the $H_2O(1_{10}–1_{01})$ transition with a resolution better than 1 km s^{-1}. These instruments, however, had a limited angular resolution, approximately 4 arcminutes for SWAS and half that size for Odin, that only allowed to study global averages of the water emission over the whole outflow, or at least over each of their lobes. Compilations of this work have been presented by Franklin *et al.* (2008) (SWAS) and Bjerkeli *et al.* (2009) (Odin).

In the past two years, the Herschel Space Observatory (HSO) has started to provide water data with spatial resolution comparable to that of the large radio telescopes from the ground and, thanks to its Heterodyne Instrument for the Far Infrared (HIFI), also with velocity resolutions below 1 km s^{-1}. Two HSO Guaranteed Time Key Programmes have a strong component of low-mass outflow studies, the Chemical Herschel Surveys of Star Forming Regions (CHESS) and Water in Star-forming regions with Herschel (WISH, see van Dishoeck *et al.* 2011), while Herschel/HIFI Observations of EXtraOrdinary Sources (HEXOS) covers Orion. As a number of contributions in this volume by members of these teams detail the first results of these programs, here we will only present a brief description of the new exciting outflow science that the HSO is providing, and we refer the reader to the contributions by Bertrand Lefloch (CHESS), Lars Kristensen (WISH), and Nathan Crockett (HEXOS) for further details.

It is safe to say that we are still exploring the tip of the iceberg in our research of water emission from outflows. Much work still needs to be done to understand the excitation conditions of the H_2O emitting gas, and this understanding is critical to extract the full potential of the water data. Preliminary work on water excitation in outflows is presented in this meeting by the poster contributions of Vasta *et al.* and Santangelo *et al.*, which summarize the first results of a multi-transition analysis of water towards the very young outflows in L1157 and L1448. A more limited treatment in terms of water lines, but more extensive in terms of outflow sources, comes from the so-called "water outflow survey" carried out also as part of the WISH project, and illustrated in Fig. 3. This outflow survey

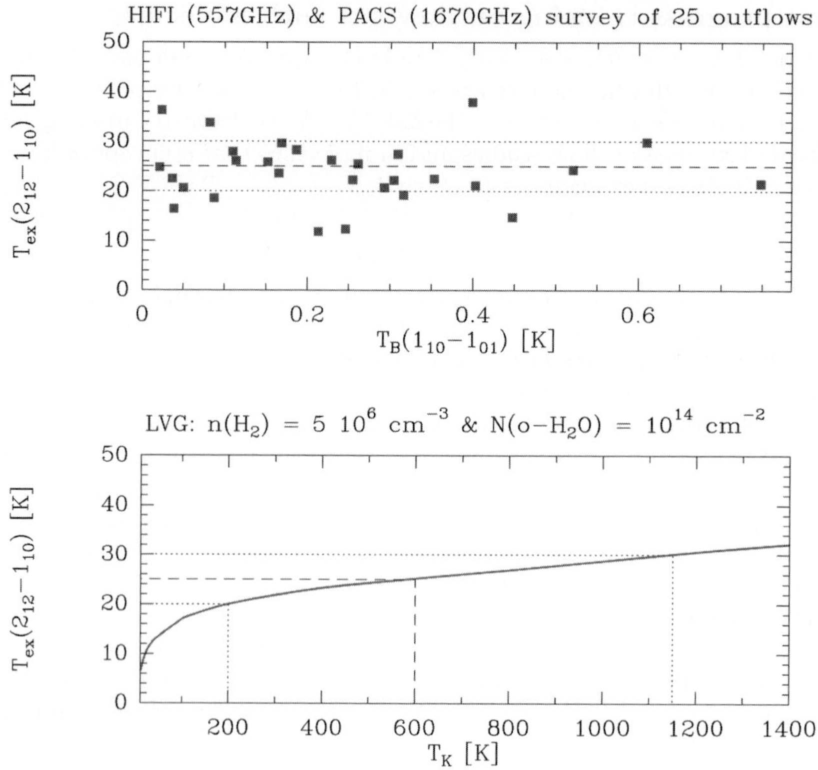

Figure 3. Top: excitation temperature between the first two excited levels of o-H_2O as a function of peak intensity in the 557 GHz H_2O line, as measured in a sample of 25 outflows within the WISH program. Note the constant value at a level of about 25 K (optically thin conditions have been assumed). **Bottom:** LVG radiative transfer results for representative physical conditions of the emitting gas ($n(H_2) = 5\ 10^6\ cm^{-3}$ and $N(o\text{-}H_2O) = 10^{14}\ cm^{-2}$). As can be seen, the excitation temperatures derived in the top panel require gas kinetic temperatures in the range 200-1100 K. From Tafalla *et al.* (2011), in preparation.

has explored about 25 different objects by observing in each of them two positions using two ortho-water lines, the 1_{10}–1_{01} at 557 GHz with HIFI and the 2_{12}–1_{01} at 1670 GHz with PACS. As seen in Fig. 3, the data from this survey shows a clear trend for the ratio of these two lines (when convolved to the same angular resolution) to be approximately constant independent on the intensity of the 557 GHz line. This ratio, assuming optically thin conditions (tested with an LVG code), is proportional to the excitation temperature between the first two excited levels of ortho-water ($T_{ex}(2_{12} - 1_{10})$), and as Fig. 3 (top panel) shows is approximately constant over the whole sample at a level of approximately 25 ± 5 K. Such a low level of scatter suggests that the bulk of the water emission we observe in outflows (at least from the lowest energy levels) arises from a rather narrow range of physical conditions.

Although a $T_{ex}(2_{12} - 1_{10})$ value of around 25 K may seem a relatively low temperature, it is important to remember that the large Einstein A coefficients of water make this molecule hard to become excited collisionally. To estimate the kinetic temperature of the gas responsible for this T_{ex} of 25 K, we need to estimate the balance between collisional excitation and radiative de-excitation (both spontaneous and induced). We have done this using the large velocity gradient (LVG) approximation together with the recently calculated collision rates by Faure *et al.* (2007) (as provided by the LAMDA data base, Schöier *et al.* 2005), and assuming values for the gas volume density and H_2O column densities in line with those found by multi-transition analysis like those in the posters by Vasta *et al.* and Santangelo *et al.* As shown in Fig. 3 (bottom panel), $T_{ex}(2_{12} - 1_{10})$ values of around 25 K require relatively high gas temperatures, in the vicinity of 500 K. This result adds to existing evidence that the gas responsible for the observed water emission from outflows corresponds to a component warmer than that typically observed with low-J CO transitions.

3.2. H_2

H_2 emission from outflows has been observed widely since its discovery in Orion, and reveals the presence of gas heated to temperatures up to a few thousand Kelvin (Gautier *et al.* 1976). Most of the H_2 work has so far concentrated on the v=1–0 S(1) line at 2.12 μm, which is easily observable from the ground. A testimony of the widespread nature of this emission in outflows is its use to define the recently created category of Molecular Hydrogen emission-line Objects (MHOs), which are analogous to the optically-based Herbig-Haro objects but include more deeply embedded line-emitting regions thanks to its use of NIR wavelengths (Davis *et al.* 2010). At the time of writing this article, the MHO on-line catalog (`http://www.jach.hawaii.edu/UKIRT/MHCat/`) is growing steadily and contains more than 1200 entries.

Although some of the pure rotational transitions of H_2 are also observable from the ground (e.g., Burton *et al.* 1989), sampling the full H_2 spectrum, and therefore characterizing the population of its energy levels, requires the use of space-based telescopes. ISO observations in the late 1990's carried out the first complete observations of the H_2 spectrum in Orion, and revealed a distribution of excitation temperatures that ranged from about 600 K in the lowest levels to more than 3000 K in the highest levels (Rosenthal *et al.* 2000).

More recently, the Infrared Spectrograph (IRS) on board of the Spitzer Space Telescope has covered the 5-33 μm wavelength range, and therefore has provided access to the lowest pure rotational lines of H_2. The high sensitivity of this instrument has allowed the systematic mapping of rotational emission from low mass outflows. Neufeld *et al.* (2006), for example, observed the S(0)-S(7) transitions towards the HH 7-11 and HH 54 outflows, and found that the ortho-to-para ratio is spatially variable, and presents values significantly below the expected equilibrium value of 3 for the observed gas temperature (which ranges between 400 and 1200 K). Such non-equilibrium ortho-to-para ratios can be understood as being a remnant of the gas pre-shock conditions, where the gas temperature was significantly lower ($\sim$ 50 K).

A more systematic study of the H_2 emission from low mass outflows using Spitzer observations comes from the series of papers by Neufeld *et al.* (2009) and Nisini *et al.* (2010), and the poster by Giannini *et al.* in this meeting, which concentrate on the outflows from L1157, L1448, BHR71, NGC 2071, and VLA 1623. Despite their differences in evolutionary stage and central source luminosity, all these outflows present a number of similar trends. One of them is the need for a multiplicity of temperatures to explain the H_2 emission, and which these authors fit with a distribution of mass with temperature that follows a power law T^{-b} with b in the range 2.3-3.3. In addition, the ortho-to-para ratio in all objects with the exception of L1448 is below the equilibrium value of 3, and suggest the presence of an activation energy in the para-to-ortho conversion. Similar results were obtained by Maret *et al.* (2009) in their study of the outflows in the NGC 1333 region.

3.3. *Complex Molecules*

There has also been substantial progress in the study of "complex molecules" in outflows over the last few years. These organic species are usually defined as containing 6 or more atoms, and are found in a number of different environments that range from the Orion hot core to the cold core of TMC-1 (see Herbst & van Dishoeck 2009 for a recent review).

In outflows, the most widely observed complex molecule is methanol (CH_3OH), whose abundance is often enhanced by several orders of magnitude, most likely due to its release from dust grain mantles (see Fig. 1). A recent result concerning methanol comes from the study of its high energy transitions by Codella *et al.* (2010), who used the Herschel Space Telescope to observe L1157-B1 as part of the CHESS project. These authors found that the excitation of the high-J lines (up to J=13) is characterized by a rotation temperature of 106 K, which is almost one order of magnitude higher than the rotation temperature of 12 K derived from the low-J lines (Bachiller *et al.* 1995). Given the excitation conditions of CH_3OH, a rotational temperature of more than 100 K implies a gas kinetic temperature of at least 200 K, and such a warm component represents a link between the low-excitation gas seen in most molecular tracers and the hotter gas traced by H_2 emission. (See also Nisini *et al.* 2007 for a similar result obtained from SiO observations.)

Molecules larger than CH_3OH have also been recently identified in L1157-B1. Arce *et al.* (2008) carried out deep integrations with the IRAM 30m telescope towards this position and detected for the first time in an outflow $HCOOCH_3$ (methyl formate), CH_3CN (methyl cyanide), HCOOH (formic acid), and C_2H_5OH (ethanol). The detection of these molecules shows that outflow shocks can lead to a very rich chemistry, that in low-mass sources was previously restricted to the hot corino regions in the vicinity of some protostars like IRAS 16293-2422 (Cazaux *et al.* 2003).

Additional detection of complex molecules in the L1157-B1 region comes from a 3mm spectral line survey being carried out with the Nobeyama 45m telescope, and whose first results have been presented by Sugimura *et al.* (2011). This survey has already produced new detections of complex molecules in this chemically active outflow: CH_3CHO (acetaldehyde) and the mono-deuterated variety of methanol CH_2DOH. From the first analysis of the line survey, Sugimura *et al.* (2011) find that the relative abundance of the complex molecules with respect to methanol is significantly lower in L1157 than in the prototype hot corino IRAS 16293-2422. These authors suggest that such a difference reveals a difference in the chemical processing of the dust in the cloud regions sampled by the outflow shock compared to the vicinity of the protostar sampled in the hot corino phase. Detection of more complex molecules is possible given the on-going nature of the survey.

Another recent detection towards the L1157-B1 position is that of HNCO (isocyanic acid) by Rodríguez-Fernández *et al.* (2010). Although not a complex molecule, HNCO is a well-known organic tracer of warm environments, like hot cores, the Galactic center, and even extragalactic sources, and the good correlation of its distribution with that of methanol in IC 342 had led to the suggestion that this species is shock-sensitive (Meier & Turner 2005). The detection of HNCO in L1157-B1 by Rodríguez-Fernández *et al.* (2010) not only reinforces this suggestion, but illustrates the important role of relatively simple regions like L1157-B1, which is far enough from the YSO to be considered a "pure shock" region (without stellar heating), and therefore provides a clean template to recognize shock chemistry in more complex environments.

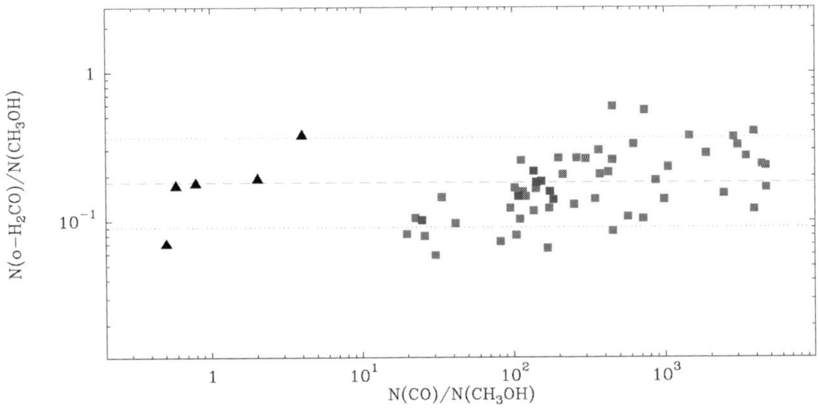

Figure 4. Abundance ratio between H_2CO and CH_3OH in the outflow gas phase (squares) and in the ices of dust grains (triangles). The red squares represent values for a sample of low-mass outflows from Santiago-García *et al.* (in preparation), and the blue squares are values towards high-mass star forming regions from Bisschop *et al.* 2007. The black triangles are ice values from Gibb *et al.* (2004). The good match between gas-phase and solid-phase values reinforces the idea that the abundance enhancement of H_2CO and CH_3OH in outflows results from the release of these molecules from the mantles.

4. Do we understand outflow chemistry?

A meeting like this one is an appropriate occasion to look critically at the state of our understanding of outflow chemistry in a more global way than usually done in research papers. A look of the literature reveals a generalized consensus in the belief that shock chemistry can explain most (or all) abundance anomalies observed in the outflow gas (apart from the EHV component discussed in the following section). This consensus, however, seems based more on the lack of observational counter-examples and alternative models than in the existence of a quantitative proof that a single shock chemistry model can explain simultaneously the set of abundances observed in an object like L1157-B1.

This incomplete state of affairs mostly results from a too-focused approach in outflow-chemistry work. From the observational point of view, and this paper testifies it, most of this work has been dedicated to a single object, the L1157 outflow, and more exactly, to the B1 position of this outflow. Although L1157 is sometimes called "prototypical" in terms of outflow chemistry, the lack of objects with comparable chemical richness and complexity suggest that L1157 is probably extreme, if not for its abundances, at least for the column densities responsible for its very bright lines.

Clearly, more general studies of outflow chemistry are needed, not only to understand how typical L1157 is, but to reconstruct the full cycle of outflow chemical activity represented by other very young outflows. An important limitation for any statistical approach is the lack of good samples. This is partly the result of the small number of outflows that so far have been explored in molecular lines other than CO, but also because of the need for full maps to identify chemical hot spots in the flows. Despite the limitations, some work has been done in this direction. Fig. 4 shows an example of how an statistical approach can further illuminate the origin of the observed abundances. This plot represents the CH_3OH/H_2CO ratio for the sample of low-mass outflows studied by Santiago-García *et al.* (in preparation), and shows that this gas-phase ratio remains approximately constant over a relatively large range of individual abundances.

The (gas-phase) ratio, in addition, matches closely the solid-phase ratio derived from ice mantles in a different sample of objects, and this good match supports strongly the interpretation that the CH_3OH and H_2CO abundance enhancements in outflows result from the release of these species from the mantles of grains due to sputtering in a shock.

While observations tend to concentrate on one object at a time, chemical models tend to solve one species at a time, and this makes it difficult to asses whether a single set of physical conditions can explain the variety of observations, even for a single object like L1157. A more important limitation of the chemical models is their reliance on plane-parallel shock models. In a plane parallel shock, there is a single velocity for the shock, and all the material suffers the same type of acceleration. As a result, the predicted emerging spectrum is characterized by a single spike at the velocity of the shock and a rapidly declining tail of pre-shocked material towards low velocities (e.g., Figs. 8 in both Gusdorf *et al.* 2008 and Flower & Pineau Des Forêts 2010). This shape of the spectrum is the opposite to what it is observed in a typical outflow wing, which has most of the material moving at low velocities and a minority of the gas moving at top speeds (see next section for a discussion of the different nature of the spiky extremely high velocity regime). As the optical depth of the emission is proportional to the amount of material per unit velocity, the compression of shocked gas into a small range of velocities predicted by the plane parallel model leads to a substantial overestimate of the optical depth. By comparing in the figures mentioned above the width of the spikes with the full velocity extent of the spectrum, we estimate that the optical depth overestimate resulting from the artificial limited range of velocities in the emission must range from factors of a few to more than order of magnitude. Thus, comparing observations with the predictions from chemical models must be done with care when the model predicts a non-negligible of optical depth in the emission.

The systematic reliance on plane-parallel models in the analysis of outflows is more a reflection of our poor understanding of the internal kinematics in the outflow gas than a conviction in the model providing the correct geometry. In fact, the observation of a wing feature in the emission spectrum of almost every outflow indicates that a multiple-velocity description is needed, and that most of the gas in the outflow moves at intrinsically low velocities. How to incorporate this description to chemical models is not yet clear, although a first step may be to parametrize the mix of velocities with a power-law description like that used in the multi-temperature analysis of the H_2 data by Neufeld *et al.* (2006) and described above. Using such a more realistic description of the outflow velocity field will clearly bring a better agreement between the predicted and observed spectra. More interestingly, a multiple-velocity description of the shocked gas will allow exploring the changes in the gas composition as a function of the velocity that a number of observations are starting to uncover (Codella *et al.* 2010, Tafalla *et al.* 2010).

5. The EHV gas, a different chemical component?

So far we have referred to the "outflow emission" as the one appearing in the spectra forming a wing where the intensity decreases steadily with velocity. Although such type of wing emission is the most characteristic signature of outflow gas, a few very young Class 0 objects show in their spectra an additional outflow component that looks like a close-to-gaussian secondary peak. This component, often referred to as the extremely

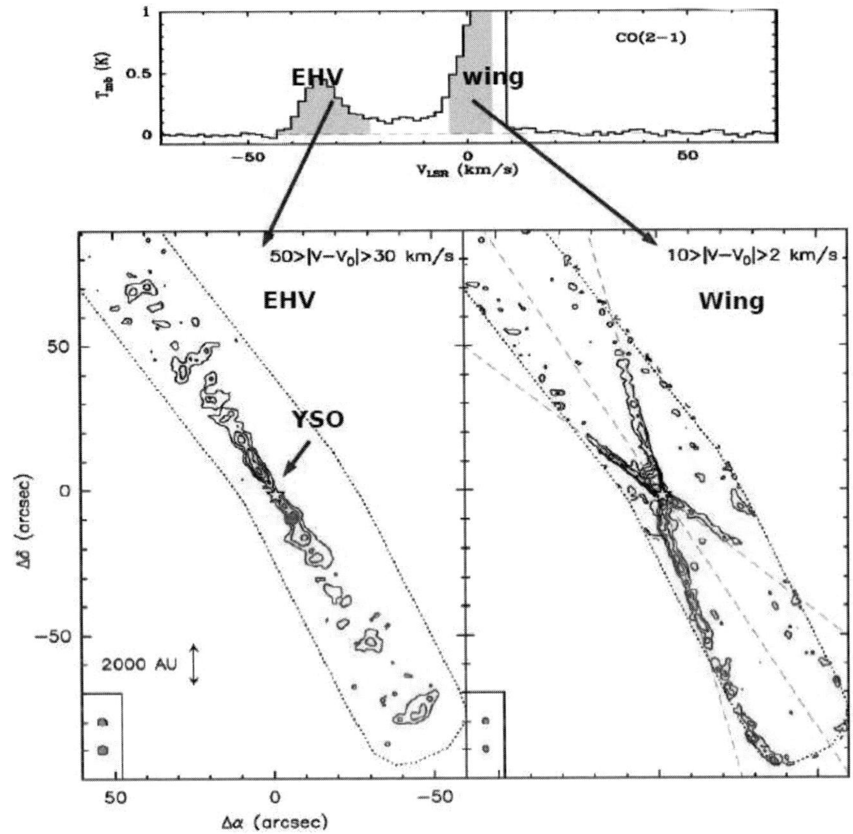

Figure 5. Average blue lobe spectrum and velocity-integrated maps of the CO(2-1) emission from the IRAS 04166+2706 outflow illustrating the different spatial distribution and spectral signature of the wing and the EHV components. Data from Santiago-García *et al.* (2009).

high velocity (EHV) gas given its high speed (tens of km s^{-1}), was first discovered in the outflow powered by L1448-mm (Bachiller *et al.* 1990), and by now it has been observed in a (reduced) number of outflows.

One of the outflows where an EHV component has been recently identified is the one powered by IRAS 04166+2706 (IRAS 04166 hereafter, see Tafalla *et al.* 2004). In Fig. 5 we present the CO(2-1) emission from this outflow as observed by Santiago-García *et al.* (2009) using the Plateau de Bure Interferometer. The figure shows that the EHV emission appears distinct from the wing emission both in spectral signature and spatial distribution. As can be seen, the (lower-velocity) wing component forms in the maps an X-shaped structure centered on the IRAS source and suggestive of tracing the walls of a pair of evacuated cavities (a fact confirmed by the association of the northern blue cavity with an IR cometary nebula). The EHV emission, on the other hand, appears in the maps as a jet-like feature that runs along the middle of the cavities and consists of a collection of discrete peaks. As shown by Santiago-García *et al.* (2009), the peaks of EHV emission are located so symmetrically from the IRAS source that each has a counterpart on the other side less than 2″ away from its expected position. Such a level of symmetry

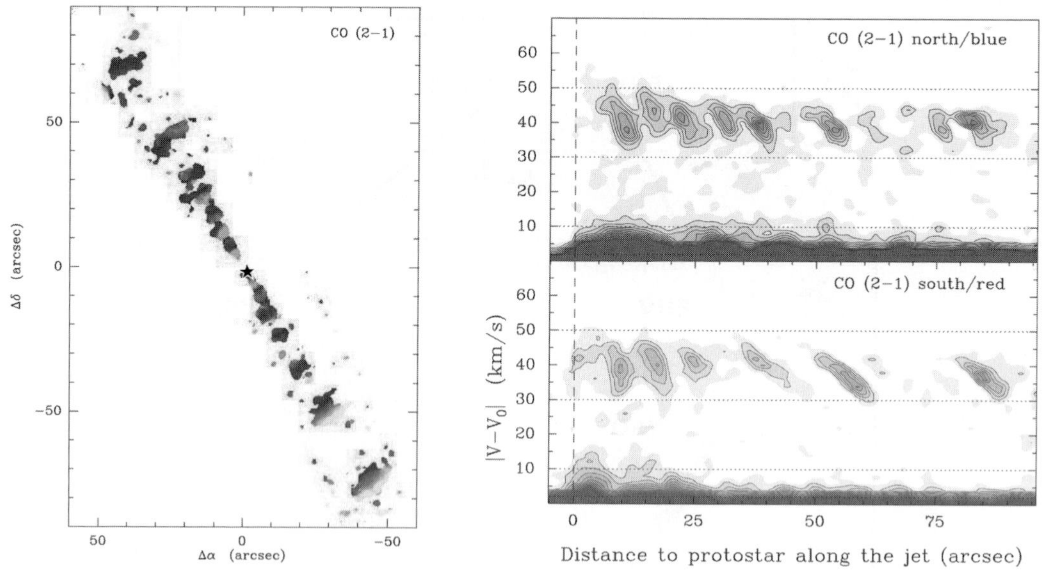

Figure 6. Left: First momentum plot of the CO(2-1) EHV emission towards the outflow from IRAS 04166+2706 illustrating the presence of velocity oscillations. **Right:** Position-velocity diagrams of CO(2-1) along the outflow axis. Note the sawtooth pattern in the EHV regime. From Santiago-García *et al.* (2009).

suggests that the EHV peaks arise from events that took place near the central source and that have since propagated outwards with the flow.

A clue to the origin of the EHV peaks comes from a pattern of oscillations in their velocity field. As illustrated in Fig. 6, the oscillations form a sawtooth pattern in the position-velocity diagram indicative of a combination of a close-to-constant mean velocity together with strong velocity gradients inside each EHV peak. The sense of these gradients is the same in all EHV peaks, and indicates that the upstream gas moves faster than the downstream gas. This sawtooth velocity pattern in the PV diagram is in striking agreement with the predictions for a set of internal shocks in a jet modeled by Stone & Norman (1993) to simulate the pattern of knots seen in optical jets (see their Figure 16). Thus, the combination of a symmetric and fragmented structure together with a sawtooth velocity pattern suggests that the EHV emission in IRAS 04166 arises not from ambient accelerated gas (like the rest of the outflow), but from the internal shocks in a pulsating jet.

If the EHV gas represents jet material, it must be coming from the protostar or its nearmost vicinity, and we can therefore expect its chemical composition to differ substantially from that of the wing gas, which consists of ambient gas that has been shock-accelerated, and therefore has had a very different thermal history. To test this possible difference between these two components of the outflow, Tafalla *et al.* (2010) have carried out the first molecular survey of EHV gas by making deep integrations towards two outflow positions known to have bright EHV and wing components, one in the L1448 outflow and the other in the IRAS 04166 outflow. Previous to this survey, only CO and SiO had been detected in the EHV gas of any outflow, despite the large number of species detected in the wing component of outflows like that of L1157 (see above). Thanks to the new deep integrations, carried out with the IRAM 30m telescope, and some of them lasting several hours, the number of species observed in the EHV gas has more than doubled, with clear

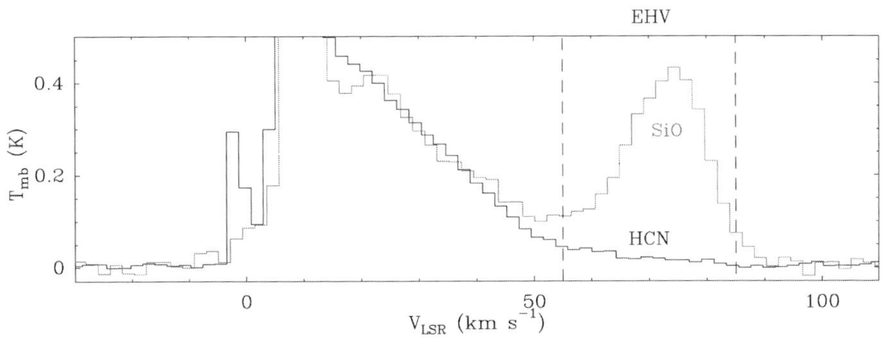

Figure 7. Comparison between SiO(2-1) and HCN(1-0) spectra towards the red lobe of the L1448-mm outflow. The SiO line has been scaled to match the HCN intensity in the wing, in order to better illustrate the sudden drop of HCN/SiO abundance in the EHV component. From Tafalla *et al.* (2010).

detections of SO, CH_3OH, and H_2CO. Possible detections of HCO^+ and CS have also resulted from these deep integrations, although their status remains unclear because of possible contamination with emission from the wing outflow component, which extends at a low level up to velocities comparable to those of the EHV gas. (See also Kristensen *et al.* 2011 for the recent detection of H_2O in the EHV gas of L1448.)

Probably the most interesting result from the EHV molecular survey of Tafalla *et al.* (2010) is the evidence for a change in the carbon-to-oxygen ratio between the wing and the EHV components. This is illustrated in Fig. 7 with a superposition of the SiO(2–0) and HCN(1-0) spectra towards the target position in the L1448 outflow (the IRAS 04166 target position presents a similar behavior but weaker lines). As can be seen, when scaled appropriately, the SiO(2-1) and HCN(1-0) spectra track each other in the (red) wing regime over several tens of km s^{-1}, suggesting that the HCN/SiO abundance ratio remains constant over this range of velocities. In contrast with this smooth behavior in the wing regime, the intensity ratio between HCN and SiO drops by more than one order of magnitude in the EHV regime. This intensity drop is unlikely to result from a difference in excitation between the two velocity regimes, as multi-transition analysis of SiO, SO, and CH_3OH show only small changes in the excitation temperature of these molecules between the wing and EHV gas. The most likely cause of the sudden drop in the HCN intensity towards the EHV regime is a similar, order-of-magnitude drop in the abundance of this species towards this outflow component. A similar drop in the abundance of CS is also seen in the data, suggesting that the HCN drop is not a peculiarity of this molecule but a common feature of C-bearing species. Indeed, all molecules with clear detection in the EHV component have so far been O-bearing (CO, SiO, SO, CH_3OH, H_2CO, and H_2O).

The finding of chemical differences between the EHV and wing outflow components in both L1448 and IRAS 04166 opens a new tool to explore the still mysterious mechanism of outflow acceleration. If the EHV gas truly arises from a jet, it must have originated in the innermost vicinity of the central object, and therefore must carry information about the physical conditions in the outflow acceleration region, which is of a few AU or even a few stellar radii, depending on the model, and therefore inaccessible to current instrumentation. To fully exploit the potential of this information, however, a new generation of chemical models is needed. The chemistry of protostellar winds was originally studied by Glassgold *et al.* (1991), but little progress has been done on this issue in the twenty years past since this pioneering work. It is reassuring to see that there is a renewed interest in the topic, as illustrated by the poster contribution in this meeting of Yvart *et al.* Clearly more modeling and observational progress is needed in this new and exciting line of study of molecules in outflows. Let's hope we can all see its results in the next IAU Astrochemistry Symposium a few years from now.

References

Arce, H. G., Shepherd, D., Gueth, F., Lee, C.-F., Bachiller, R., Rosen, A., & Beuther, H. 2007, in *Protostars and Planets V*, (Tucson: Univ. of Arizona Press, Tucson), p. 245

Arce, H. G., Santiago-García, J., Jørgensen, J. K., Tafalla, M., & Bachiller, R. 2008, *ApJL*, 681, L21

Bachiller, R., Cernicharo, J., Martín-Pintado, J., Tafalla, M., & Lazareff, B. 1990, *A&A*, 231, 174

Bachiller, R., Liechti, S., Walmsley, C. M., & Colomer, F. 1995, *A&A*, 295, L51

Bachiller, R. 1996, *ARAA*, 34, 111

Bachiller, R. & Perez Gutierrez, M. 1997, *ApJL*, 487, L93

Bachiller, R., Pérez Gutiérrez, M., Kumar, M. S. N., & Tafalla, M. 2001, *A&A*, 372, 899

Benedettini, M., Viti, S., Codella, C., Bachiller, R., Gueth, F., Beltrán, M. T., Dutrey, A., & Guilloteau, S. 2007, *MNRAS* 381, 1127

Bergin, E. A., Neufeld, D. A., & Melnick, G. J. 1998, *ApJ*, 499, 777

Bisschop, S. E., Jørgensen, J. K., van Dishoeck, E. F., & de Wachter, E. B. M. 2007, *A&A*, 465, 913

Bjerkeli, P., *et al.* 2009, *A&A*, 507, 1455

Bontemps, S., André, P., Terebey, S., & Cabrit, S. 1996, *A&A*, 311, 858

Burton, M. G., Brand, P. W. J. L., Geballe, T. R., & Webster, A. S. 1989, *MNRAS*, 236, 409

Cazaux, S., Tielens, A. G. G. M., Ceccarelli, C., Castets, A., Wakelam, V., Caux, E., Parise, B., & Teyssier, D. 2003, *ApJL*, 593, L51

Cernicharo, J., Bachiller, R., & González-Alfonso, E. 1996, *A&A*, 305, L5

Codella, C., *et al.* 2010, *A&A*, 518, L112

Davis, C. J., Gell, R., Khanzadyan, T., Smith, M. D., & Jenness, T. 2010, *A&A*, 511, A24

Elitzur, M. & de Jong, T. 1978, *ApJ*, 67, 323

Faure, A., Crimier, N., Ceccarelli, C., Valiron, P., Wiesenfeld, L., & Dubernet, M. L. 2007, *A&A*, 472, 1029

Flower, D. R. & Pineau Des Forêts, G. 2010, *MNRAS*, 406, 1745

Franklin, J., Snell, R. L., Kaufman, M. J., Melnick, G. J., Neufeld, D. A., Hollenbach, D. J., & Bergin, E. A. 2008, *ApJ*, 674, 1015

Gautier, T. N., III, Fink, U., Larson, H. P., & Treffers, R. R. 1976, *ApJL*, 207, L129

Giannini, T., Nisini, B., & Lorenzetti, D. 2001, *ApJ*, 555, 40

Gibb, E. L., Whittet, D. C. B., Boogert, A. C. A., & Tielens, A. G. G. M. 2004, *ApJS*, 151, 35

Glassgold, A. E., Mamon, G. A., & Huggins, P. J. 1991, *ApJ*, 373, 254

Gusdorf, A., Pineau Des Forêts, G., Cabrit, S., & Flower, D. R. 2008, *A&A*, 490, 695

Herbst, E. & van Dishoeck, E. F. 2009, *ARAA*, 47, 427

Hollenbach, D. & McKee, C. F. 1979, *ApJS*, 41, 555

Kaufman, M. J. & Neufeld, D. A. 1996, *ApJ*, 456, 611

Kristensen, L. E., van Dishoeck, E. F., Tafalla, M., Bachiller, R., Nisini, B., Liseau, R., & Yıldız, U. A. 2011, *A&A*, 531, L1

Kwan, J. & Scoville, N. 1976, *ApJL*, 210, L39

Liseau, R., *et al.* 1996, *A&A*, 315, L181

Maret, S., *et al.* 2009, *ApJ*, 698, 1244

Meier, D. S. & Turner, J. L. 2005, *ApJ*, 618, 259

Mikami, H., Umemoto, T., Yamamoto, S., & Saito, S. 1992, *ApJL*, 392, L87
Morris, M. 1976, *ApJ*, 210, 100
Nisini, B., *et al.* 1999, *A&A*, 350, 529
Nisini, B., Codella, C., Giannini, T., Santiago Garcia, J., Richer, J. S., Bachiller, R., & Tafalla, M. 2007, *A&A*, 462, 163
Nisini, B., Giannini, T., Neufeld, D. A., Yuan, Y., Antoniucci, S., Bergin, E. A., & Melnick, G. J. 2010, *ApJ*, 724, 69
Neufeld, D. A., *et al.* 2006, *ApJ*, 649, 816
Neufeld, D. A., *et al.* 2009, *ApJ*, 706, 170
Rodríguez-Fernández, N. J., Tafalla, M., Gueth, F., & Bachiller, R. 2010, *A&A*, 516, A98
Rosenthal, D., Bertoldi, F., & Drapatz, S. 2000, *A&A*, 356, 705
Santiago-García, J., Tafalla, M., Johnstone, D., & Bachiller, R. 2009, *A&A*, 495, 169
Schöier, F. L., van der Tak, F. F. S., van Dishoeck, E. F., & Black, J. H. 2005, *A&A*, 432, 369
Snell, R. L., Loren, R. B., & Plambeck, R. L. 1980, *ApJL*, 239, L17
Stone, J. M. & Norman, M. L. 1993, *ApJ*, 413, 210
Sugimura, M., *et al.* 2011, *PASJ*, 63, 459
Tafalla, M., Myers, P. C., Mardones, D., & Bachiller, R. 2000, *A&A*, 359, 967
Tafalla, M., Santiago, J., Johnstone, D., & Bachiller, R. 2004, *A&A*, 423, L21
Tafalla, M., Santiago-García, J., Hacar, A., & Bachiller, R. 2010, *A&A*, 522, A91
Umemoto, T., Iwata, T., Fukui, Y., Mikami, H., Yamamoto, S., Kameya, O., & Hirano, N. 1992, *ApJL*, 392, L83
van Dishoeck, E. F. & Blake, G. A. 1998, *ARAA*, 36, 317
van Dishoeck, E. F., *et al.* 2011, *PASP* 123, 138

Discussion

STERNBERG: Could you comment on whether it might be possible to study angular momentum loss using these molecular diagnostics?

TAFALLA: Rotation in molecular jets has been searched for, and there are a few possible detections in the literature. Our observations on the jet from IRAS 04166, however, suggest that the kinematics of the gas in the jet is dominated by lateral ejection in the internal shocks, and I suspect that this faster motion is going to overwhelm any rotation signature that may be present.

ELLINGER: You have shown similar abundances of HCN in three different objects. Do you observe the same thing for HNC?

TAFALLA: Bachiller & Pérez Gutiérrez (1997) estimated an HNC/HCN ratio of 0.1 towards L1157-B1. In the fast wing of L1448, the non detection of HNC in Tafalla *et al.* (2010) sets an upper limit that is about one order of magnitude lower.

MENTEN: What do we know about deuterium enhancement in the outflow gas, for example for methanol?

TAFALLA: Sugimura *et al.* (2011) have recently presented the first detection of mono deuterated methanol towards L1157-B1, and derive an CH_2DOH/CH_3OH abundance ratio in the range 0.013-0.029, which is significantly lower than that derived for IRAS 16293 but similar to the one measured in Orion KL.

VISSER: You showed that the HCN/SiO abundance ratio is much lower in the EHV component than it is in the wing, and you interpreted this as a lower elemental C/O ratio in the EHV gas. Could the low HCN/SiO ratio also indicate a low elemental N/O ratio?

TAFALLA: It could be. Still, we find a similar drop in the CS/SiO ratio between the wing and the EHV gas, although we have lower signal-to-noise in the CS data. This seems to point more to a deficit of carbon than of nitrogen, although we are dealing here with a very limited set of observations.

The Molecular Universe
Proceedings IAU Symposium No. 280, 2011
José Cernicharo & Rafael Bachiller, eds.

doi:10.1017/S1743921311024902

Millimetre/Sub-millimetre Observations of Circumstellar Disks

Anne Dutrey[1]

[1]Laboratoire d'Astrophysique de Bordeaux, OASU, CNRS UMR 5804, 2 rue de l'observatoire, BP89, F-33271 Floirac Cedex, France
email: Anne.Dutrey@obs.u-bordeaux1.fr

Abstract. TTauri disks located in nearby star-forming regions (e.g. Taurus-Auriga at 140 pc) are thought to be the site of planet formation, since proto-planetary disks orbiting around active (still accreting) TTauri stars should contain, in many cases, enough gas to form giant gaseous planets. As such, circumstellar disks are ideal laboratories to study planet formation, provided the gas and dust observations have enough sensitivity and resolving power. I will focus in these proceedings, on recent results of molecular observations which unveil the physical conditions of gas disks and reveal the weakness of our current understanding and modeling.

Keywords. astrochemistry, stars: formation, planetary systems: protoplanetary disks

1. Introduction

Since the early nineties, optical, near infrared (NIR) and millimetre/submillimetre (mm/submm) observations of low and intermediate mass stars ($\sim$ 1 Myr), such as TTauri and Herbig Ae stars, have shown that these young stars are surrounded by circumstellar disks. These proto-planetary disks, residuals from the molecular cores which have formed the stars, still contain enough gas and dust to form planets. Although the dust emission in the optical and in NIR (scattered light emission) is the easiest tracer of proto-planetary disks, H_2 molecules (which does not deplete on dust grains) represent more than 70 % of the total mass in disks and remain difficult to observe because of the lack of a permanent dipole moment. Some direct H_2 detections are possible but the observed lines mainly trace the warm gas located in the inner disks (R< 10-20 AU), while in many cases, disks are known to extend out to several 100 AU. These cool ($\sim$ 10-50 K) outer regions, which may even contain most of the mass, can only be characterized by molecules having rotational lines from low energy levels and detectable with large millimeter and sub-millimeter interferometers. CO is the most abundant molecule (even if it can deplete on dust grains) after H_2 and its lowest rotational lines are easily excited (and thermalized) by collisions with H_2. Current heterodyne mm/submm arrays begin to reach the sensitivity, resolving power and spectral resolution needed to map cold molecular disks and millimeter spectro-imaging have revealed, for example, that disks are in Keplerian rotation. The SMA has recently opened the sub-millimeter window, while the IRAM Plateau de Bure interferomter (PdBI) routinely provides images of 0.6" resolution at 1.3 mm. Therefore, recent mm/submm observations have provided many new direct constraints on physical and chemical structure of disks which could not be addressed by disk continuum observations. With ALMA entering in operation, our current knowledge on planet forming regions in disks will likely significantly improve in the next years.

In this review, I will focus on recent molecular results which provide new constraints on the gas chemistry and physics in disks. In particular, I will present studies obtained with the IRAM array in the context of the CID ("Chemistry in Disks") project, a

European collaboration dealing with the observation and modeling of gas disks to retrieve the physical and chemical conditions leading to planet formation. † After a general presentation of the state of art of molecular gas disks, I will present our new results and discuss how they can change our current thinking about disk chemistry.

2. Current Model for Gas and Dust Disks

For disks of about 1 Myr old, the mass and dynamics are dominated by the gas and the gas-to-dust ratio is supposed to be of the order of 100 but there is no accurate measurements of this quantity yet. Another major unknown in disks is the exact radial and vertical distribution of the gas and its evolution with time. This is particularly critical in the inner regions ($R < 20 - 50$ AU) where planet formation should occur.

So far, molecular surveys are sensitivity limited to the most abundant molecules, namely those found in cold molecular clouds. In addition to CO, ^{13}CO and $C^{18}O$, only HCO^+, $H^{13}CO^+$, DCO^+, CS, HCN, HNC, DCN, CN, H_2CO, N_2H^+ and C_2H have been firmly detected. The study of these molecules can however improve our knowledege on the gas disk density provided the excitation conditions are properly understood and modeled (Dutrey *et al.* (1997)). As soon as the full ALMA array will be operational, the detection rate of molecules may significantly increase, revealing the complexity of the disk molecular chemistry and its vertical stratification.

2.1. *Current Gas Disk Structure*

Chemistry is an essential agent in the shaping of the gas disks. Molecular (and atomic) abundances also play a key role in determining the cooling rate of the gas, and thereby the thermal balance of the disk. It is then essential to understand where in the disk the molecules are located.

With the exception of the very inner disk ($R \leqslant 10$ AU) where accretion can be a source of heat, disks are mostly heated by the stellar UV radiation and X rays, the interstellar UV field and the cosmic rays. This leads to the current scheme where both the gas and dust have vertical and radial temperature variations, the upper layers being superheated by the stellar UV field (Chiang & Goldreich (1997)). Given this temperature and the vertical and radial density distributions (the gas is expected to be in hydrostatic equilibrium), one can vertically defines three different "chemical" zones. The disk upper layer acts like a PDR formed of ions and then neutral gas, deeper in the disk, there is the warm molecular layer and then the cold mid-plane, mostly devoid of molecules, with the exception of H_2. The mid-plane temperature drops significantly below the freeze out temperature of CO, with a chemistry close to that of cold dense cores. This standard model is illustrated in Fig.1.

2.2. *Measuring temperature and densities from molecular lines Observations*

An important source of uncertainty in the analysis of molecular maps comes from the existence of non-LTE effects in the line excitation. A proper analysis of any line detection in disks requires then adequate handling of the line formation process, and in particular of the disk kinematics. Failure to do so can result in mis-interpretation of spectral lines, as shown by Guilloteau *et al.* (2006) for HDO in DM Tau. Modeling the surface density and temperature are usually done assuming power law radial dependencies with in some cases an exponential decay to express the viscous spread out of the surface density (see for example, Guilloteau *et al.* (2011)) .

† The CID consortium is an international collaboration (Dutrey *et al.* 2007b) between researchers from IRAM, MPIA and LAB which have been recently extended to american and taiwanese collaborators.

Fortunately, the first levels of the CO rotational lines (J=2-1 and J=1-0) are thermalized and hence allow a direct measurement of the gas temperature and surface density depending on the line opacity. For a thermalized transition, the brightness temperature $T_B(r)$ which is measured by the interferometer, is $\sim T_K$ (kinetic temperature) in the optically thick case and is a function of $\Sigma(r), T_K(r)$ ($\Sigma(r)$ being the surface density) depending on the transition for a partially optically thin line. It roughly scales as $\Sigma(r)/T_K(r)$ at high temperatures (see Dutrey *et al.* (2007a) for a discussion).

Due to different opacities, the ^{12}CO, ^{13}CO J=1-0 and J=2-1 sample different disk layers. In the case of DM Tau, the pioneer CO study of Dartois *et al.* (2003) confirmed predictions from models of disks externally heated by their central star (e.g. D'Alessio *et al.* (1999)). Close to the mid-plane, the temperature appears significantly cooler ($\sim$ 13 K) than at the CO disk surface ($\sim$ 30 K at 100 AU). This happens in the region of the disk where the dust is still optically thick to the stellar radiation while it is already optically thin to its own emission, around $r \sim 50 - 200$ AU in the DM Tau case. Beyond $r \geqslant 200$ AU where the dust becomes optically thin to both processes, the temperature profile appears vertically isothermal (Dartois *et al.* (2003)). The fact that a significant fraction of the DM Tau disk has a temperature apparently below the CO freeze out point (17 K) has been interpreted in the framework of the turbulence where vertical mixing should mix the vertical molecular layers. This result also suggests that photo-desorption rates are not properly known (Hersant *et al.* (2009)).

Using the IRAM array, Pietu *et al.* (2007) refine the study of DM Tau by measuring both the temperature and the surface density of CO and HCO^+ in DM Tau, LkCa15 and MWC480. Fig. 2 shows the results of the χ^2 minimization performed on the resolved observations (LTE conditions are reached for the observed transitions) in DM Tau. The right scale presents the signal-to-noise ratio obtained on the minimized parameter versus radius. At high opacity (e.g. ^{12}CO J=2-1), the temperature can be measured in a significant disk fraction while the estimate of the molecular surface density is only possible in the outer optically thin region. This is the opposite for an almost optically thin

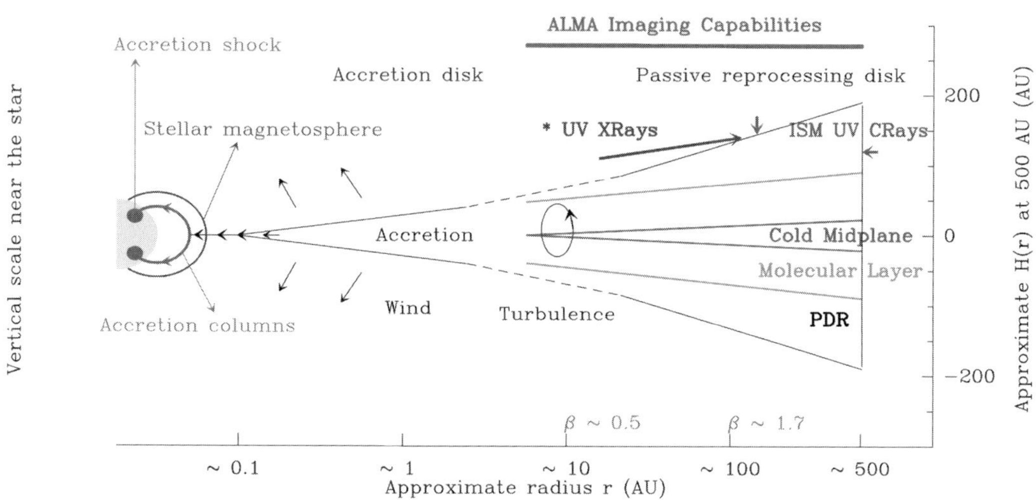

Figure 1. A schematic view of a protoplanetary disk orbiting a TTauri star and located in a nearby star-forming region (150 pc). Current models predict three layers from the mid-plane to the top: 1) the cold mid-plane ($\sim$ 10 K) where most molecules are frozen on grains, 2) the intermediate warm ($\sim$ 20-40 K) molecular layer and 3) the Photon dominated (PDR) upper layer. Following the recent results from Guilloteau *et al.* (2011), the expected value of the dust spectral index β versus radius in the mm range is given.

transition such as ^{13}CO J=1-0. This method will remain robust using ALMA and the exceptional sensitivity and resolving power of the new array will also allow observers to refine the accuracy on the radial and vertical dependencies of the temperature and density. Moreover, such a method can also be extended to higher CO transitions which would sample the disk atmosphere revealing the impact of X-rays and UV irradiation on the disk (Qi *et al.* (2006)).

Finally since CO observations are usually not sensitivity limited, spectro-imaging of CO provides invaluable information on the disk structure and geometry. In a recent paper, Dutrey *et al.* (2008) have found that the Keplerian gas disk of GM Aur was innerly truncated at $\sim$ 20 AU by studying the CO velocity pattern observed with the IRAM array. In a Keplerian velocity field, the presence of an inner cavity of radius r_{in} would suppress line emission at velocities above $V_{r_{in}} = \sqrt{\frac{G.M_*}{r_{in}}}.sin(i)$ compared to the systemic velocity (i being the inclination angle). Such an analysis is only possible on data with both high spectroscopic resolution (e.g. 0.1 km/s) and signal-to-noise ratio. This will likely become very common using ALMA to study the gas content of inner disks and constrain photo-evaporation processes or possible tidal interactions in the planet forming zone.

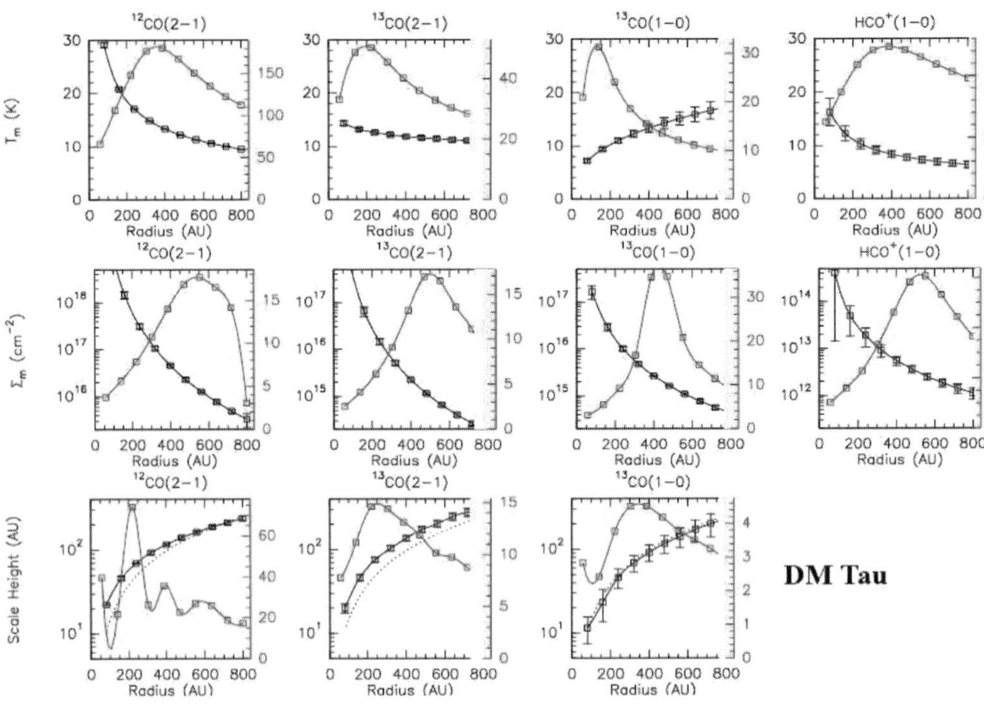

Figure 2. From Pietu *et al.* (2007). Analysis of IRAM PdBI observations of the DM Tau gas disk. Top panel: Kinetic temperature versus radius, Medium panel: Molecular surface densities versus radius, Top panel: scale height versus radius. These physical parameters are respectively derived from left to right (by decreasing opacity): ^{12}CO J=2-1, ^{13}CO J=2-1, ^{13}CO J=1-0 and HCO$^+$ J=1-0. The black scale corresponds to the parameter scale, the red one to the S/N ratio on the derived parameter. The kinetic temperature is better determined in the inner part of the disk which is optically thick, while the surface density is better determined in the optically thin outer part. The scale height is more difficult to constrain but CO results are in agrement with an hydrostatic regime.

3. New Results derived from recent mm/submm molecular Surveys

Fig.3 illustrates the quality of data which is now obtained with current arrays. Such a quality is required to properly measured physical gas disk parameters and begin quantitative comparisons with surface density predictions from chemical models.

3.1. *Global Properties*

In the recent years, spectro-imaging of molecular disks have also open to investigations some important ingredients which play a role both in the chemistry and in the physics of disks. Current arrays allow observers to derive averaged values on the whole disk but ALMA will provide in many cases radial dependencies on these key parameters.

3.1.1. *Turbulence*

Most measurements of the turbulence in disks are indirect and rely on the α prescription through modeling of the spectral energy distribution (SED), or radial density profile (Guilloteau *et al.* (2011)). Dartois *et al.* (2003) then Pietu *et al.* (2007) made the first direct measurement of the random component of the gas velocity field from spatially and spectrally resolved molecular observations. Their analysis of the CO maps on DM Tau and LkCa15 revealed that in outer TTauri disks the averaged value of the turbulence correspond to largely subsonic motions.

3.1.2. *Ionization*

Only a few molecular ions have been observed in disks: HCO^+, DCO^+ and N_2H^+ (Pietu *et al.* (2007), Dutrey *et al.* (2007b), Qi *et al.* 2008). The most abundant molecular ion is HCO^+. Dutrey *et al.* (2007b) found a N_2H^+/HCO^+ ratio of $\sim 0.02 - 0.05$ in the DM Tau and LkCa15 disks, a value similar to that found in cold dark clouds. Deriving an ionization fraction from the current data remain delicate. Using CO and HCO^+ observations from the SMA, Qi *et al.* (2008) have reported an ionization fraction of $X(e^-) = n(e^-)/n(H_2) \simeq 10^{-7}$ in TW Hydra.

3.1.3. *Deuteration*

Since H_2D^+ is not yet detected, DCO^+ and DCN remain the only deuterated molecular species observed in disks (e.g. DM Tau: Guilloteau *et al.* (2006) and TW Hydra: Qi *et al.* (2008)). Searching for deuterated species in disks clearly requests sensitivity and ALMA (Chapillon *et al.* (2011a)).

3.2. *The layered chemical Model*

3.2.1. *Towards a colder molecular layer ?*

After CO, other molecules such as CCH (Henning *et al.* (2010)) or CN and HCN (Chapillon *et al.* 2011b) have been studied in DM Tau, LkCa15 and MWC 480. In all cases, the chemical modeling reproduces the magnitude of the observed column densities within a factor 3-10 but fails to reproduce its radial dependencies. Surprisingly, the analysis of the DM Tau and LkCa15 disks gives excitation temperatures around 10 K or even below ($\sim$ 6 K) beyond a radius of 100 AU. For MWC480, Chapillon *et al.* 2011 have derived from their study, an excitation temperature for the CN J=2-1 line which is around 30 K, in agreement both with the kinetic temperature derived by Pietu *et al.* (2007) from their CO analysis, and chemical predictions for the molecular layer. Since MWC480 is an Herbig Ae star of spectral type A4, the disk is also expected to be warmer and molecules may remain in the gas phase.

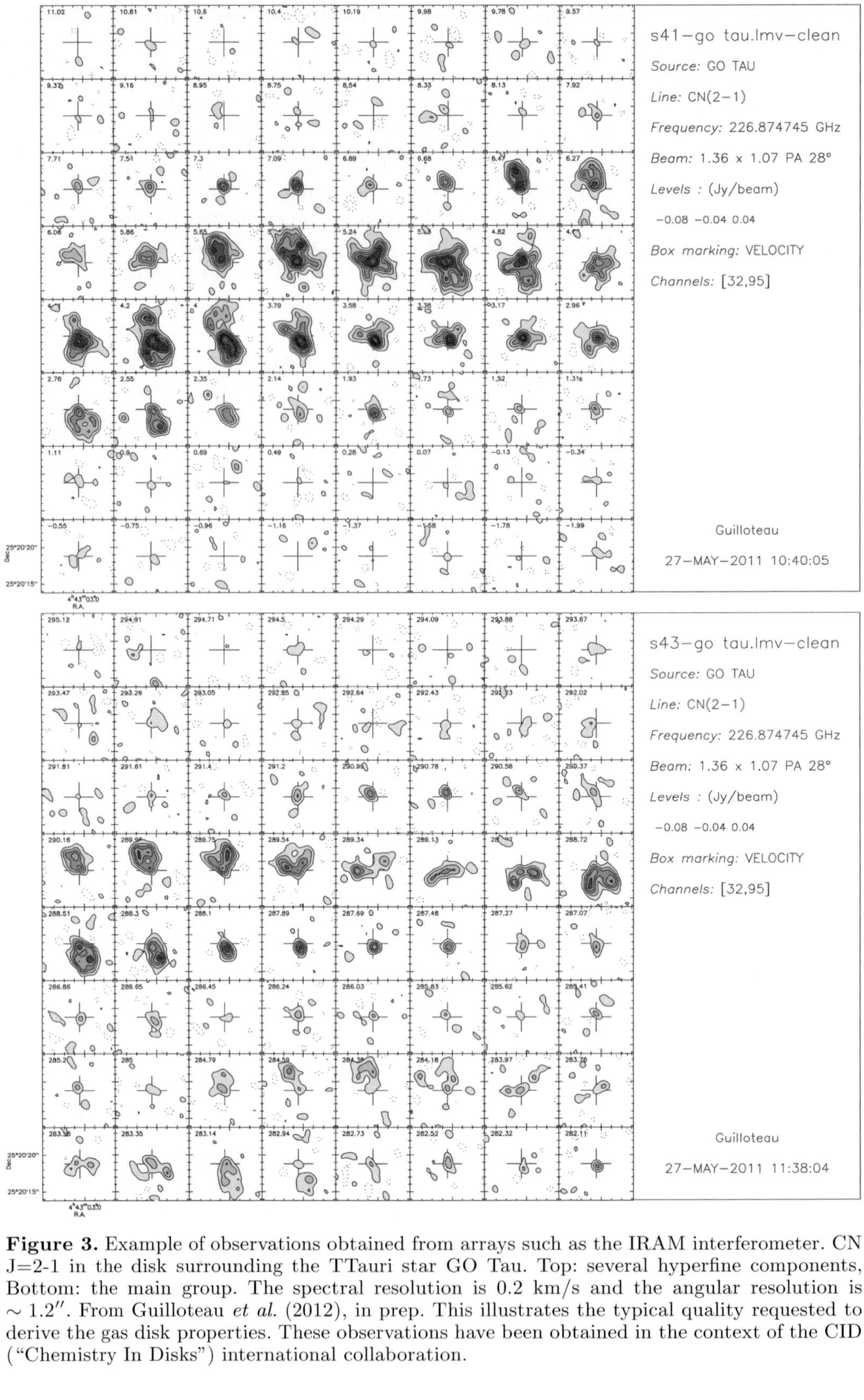

Figure 3. Example of observations obtained from arrays such as the IRAM interferometer. CN J=2-1 in the disk surrounding the TTauri star GO Tau. Top: several hyperfine components, Bottom: the main group. The spectral resolution is 0.2 km/s and the angular resolution is $\sim 1.2''$. From Guilloteau *et al.* (2012), in prep. This illustrates the typical quality requested to derive the gas disk properties. These observations have been obtained in the context of the CID ("Chemistry In Disks") international collaboration.

To explain the apparently low temperatures observed in the TTauri disks, several possibilities can be invoked. First of all, using expected physical conditions, the observed transitions (CCH 1-0, 2-1, HCN 1-0 and CN 2-1) should be thermalized at least in a large fraction of these disks. Furthermore sub-thermal excitation is not sufficient to explain these low temperatures which are also similar to those derived from thermalized CO J-0 and J=2-1 transitions (Pietu *et al.* (2007)). Turbulence may also play a role by allowing vertical mixing from warmer upper molecular layers (Semenov *et al.* (2006), Aikawa (2007)). In the case of DM Tau, Chapillon *et al.* 2011b have investigated the possibility to have a lower gas-to-dust ratio (by about a factor $\sim$ 6). In that case sub-thermal excitation remains possible but the predicted column densities from HCN and CN are low compared to the observed ones. Finally, another possibility would be that "cold" chemistry for molecules such as CN or CCH and in particular photo-desorption rates are poorly known (Öberg *et al.* (2009), Hersant *et al.* (2009)).

Waiting from more accurate observations with ALMA, these apparently low temperatures point toward a colder molecular layer. This strongly suggests that our current knowledge of the disk physics (thermal and density structure, role of dust settling, gas-to-dust-ratio estimate, vertical and radial mixing, etc...) and chemistry is more limited than commonly believed.

3.2.2. *Characterization of the Mid-Plane*

Tracing the cold mid-plane is challenging since most molecules should be frozen onto grains. The best candidate to trace this depleted zone is the ion H_2D^+ because it is formed in gas phase at low temperature and in a so cold environment, it cannot be chemically destroyed by molecules such as N_2 and CO which are stick onto grains. In a recent paper, Chapillon *et al.* (2011a), have observed the H_2D^+ $1_{1,1} - 1_{1,0}$ transition with APEX in DM Tau and TW Hya. No detection was obtained but their upper limit was about 3 times better than the previous values. In their chemical analysis (using the gas-grain chemical model developed by Parise et collaborators 2011), they investigated several disk models (varying the density, temperature and outer radius), grain sizes (0.1, 1 and 10 μm), CO abundances (10^{-4},10^{-5} and 10^{-6}) and rate of cosmic rays ionization (10^{-17}, 3.10^{-17} and 10^{-16} s^{-1}). Their data only firmly exclude the case with both a low CO abundance and small grains. If the CO abundance is $\geqslant 10^{-5}$, any grain size can fit the data. This study clearly shows that H_2D^+ is harder to detect than expected and that a significant sensitivity improvement is also needed to get constraining results on the physics and chemistry of the cold depleted mid-plane of TTauri disks.

3.3. *Impacts of the Dust Properties on Gas Phase Chemistry*

The dust is a major actor in these environments through several aspects and all chemical models are severely limited in their treatment of the influence of dust on the chemistry. The photoelectric effect on grains can affect the gas temperature, especially in the upper layers if the dust has already settled towards the disk mid-plane. Grain growth and dust settling directly affect the dust opacity, influencing the penetration of radiation through the medium and its thermal balance. Grain growth also affects the grain surface chemistry, by reducing the available surface area (per unit mass) but can also change the thermal balance in discs because the dust temperature is expected to depend on grain size. This has never been investigated by astro-chemical models and may significantly influence the gas disk structure because of the existence of important vertical and radial temperature gradients in disks, associated with strong grain growth. From a CO analysis of PdBI data obtained on the disks of CQ Tau and MWC758, Chapillon *et al.* (2008), reported a strong CO depletion ($\sim$ 100) which cannot be explained by depletion on grains since the disks are warm ($\sim 30 - 100$ K at 100 AU). Interestingly, they suggest that CO may have been removed from the gas phase during the cold pre-stellar phase by adsorption on small grains. During the warmer proto-stellar phase, grain growth also occurs and CO may stay locked on larger grains which remain colder than small grains during reheating.

3.3.1. *Gas-to-Dust Ratio*

Dust settling along the mid-plane implies that the gas-to-dust ratio changes with the altitude in the disk. Recent observations of dust disks also suggest that the gas-to-dust ratio changes with radius. Indeed, Guilloteau *et al.* (2011) have reported the first evidences for radial segregation on grain size in several disks observed with the IRAM array at 2.8 and 1.3mm. They found that large grains are preferentially located in inner disks (within $R < 50 - 70$ AU) while the dust spectral index in the outer part of the disks suggest that these grains are small. As an example, Fig.4 presents the dust spectral index versus radius in the DL Tau disk. Such a result has lots of implications on the disk physics, it implies in particular that the gas-to-dust ratio and extinction curve in the UV are changing with radius. Future chemical models will have to take into account such changes in order to properly reproduce the physical conditions (e.g. UV illumination and penetration) in disks.

Measuring the gas-to-dust ratio is a difficult task both from the modeling and observational points of view. Following their CO interferometric study of the CQ Tau disk (Chapillon *et al.* (2008), Chapillon *et al.* (2010)) have performed using APEX a deep

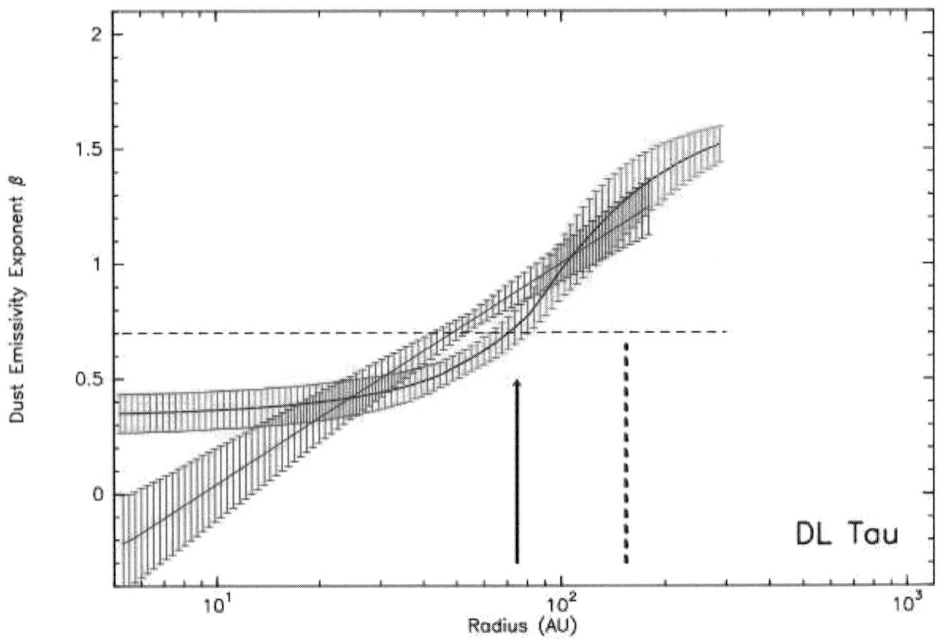

Figure 4. From Guilloteau *et al.* (2011). Derived dust emissivity (spectral index) versus radius in the dust disk surrounding the TTauri DL Tau. The spectral index has been derived from the resolved mm observations obtained with the IRAM array. There is a clear change of the dust properties within the radius. In the inner disk (R$< 50 - 60$ AU), the dust is dominated by large particles with a β value of ~ 0.5 while in the outer disk the dust appears ISM-like with a β around ~ 1.7. Red and Green curves: two functionalities have been checked for the dust radial dependencies and the trend is the same in both cases.

search for CI but failed to detect it. Their goal was to quantify the reservoir of Carbon. Using the PDR Meudon code, they simultaneously modeled the CO, CI and CII column densities and compared the result to the observed column densities of CO and CI (upper limit). Their study strongly suggests that only a low gas-to-dust ratio ($\lesssim 10$) in presence of a weak UV field can explain the presence of CO and the non detection of CI. Therefore CQ Tau would be a disk in a transition phase between a proto-planetary gas-rich disk and a gas-free debris disk. Even though many unknowns remain, such a method seems promising to quantify the gas-to-dust ratio in disks where the physical structure is basically understood from resolved interferometric observations.

3.3.2. *Importance of Gas-Grains Interactions revealed by Sulfur Chemistry*

Protoplanetary disks are particular because they present strong density and temperature gradients on short spatial scales. For example, at 100 AU, the mid-plane temperature is expected to be around 7-12 K while 3 scale-heights above (about 30 AU), the temperature should be warm enough to allow the presence of a molecular layer. The chemical coupling between gas and dust through grain surface reactions may be a major ingredient of disk chemistry, this being emphasized by the presence of vertical turbulence since many molecules can remain trapped onto grains in the mid-plane.

In a recent paper, Dutrey *et al.* (2011) made a deep search for SO and H_2S using the 30-m radiotelescope in the disks surrounding GO Tau, DM Tau, LkCa15 and MWC480. They failed to detect these molecules but they analyzed the data together with the existing CS observations (obtained with IRAM facilities, CID project). They compared the molecular column densities derived from the observations with chemical predictions made using Nautilus (Hersant *et al.* (2009)), a gas-grain chemistry code dedicated to disks. For each disk, the density and temperature profiles have been taken from previous analysis (e.g. Pietu *et al.* (2007)). Fig.5 presents the ratio of the observed molecular densities (upper limits only for SO and H_2S) to predicted column densities. Their model reproduces the SO and CS column densities reasonably well, but fails to match the upper limits obtained on H_2S by at least one order of magnitude. This suggests that at the high

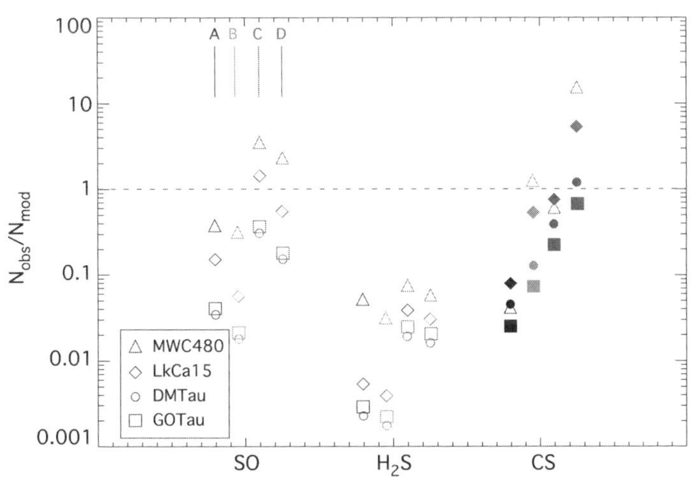

Figure 5. From Dutrey *et al.* (2011), CID paper V. Sulfur bearing observations with the IRAM 30-m radiotelescope (SO,H_2S) and PdBI (CS). Constraining upper limits have been obtained on SO and H_2S allowing us to make a quantitative analysis of the Sulfur-bearing molecules in four disks. The observations have been analyzed using the chemical code Nautilus (Hersant *et al.* (2009)).

densities and low temperatures encountered around disk mid-planes, H_2S remains locked onto the grain surfaces where it reacts to form other species preventing desorption of H_2S. Indeed, some recent experiments by Garozzo *et al.* (2010) have shown that H_2S on grains is easily destroyed by cosmic rays and lead to the formation of C_2S, SO_2 and OCS on grains. These studies also suggest that most of the sulphur may be in the form of a sulphur rich residuum, which could be polymers of sulfur or amorphous aggregates of sulfur (Wakelam *et al.* (2004)). The associated grain surface reactions are not yet incorporated in chemical models.

These recent results emphasize the need of grain surface reactions in astro-chemical models in presence of high density and low temperature associated to a UV photon source, as it is the case in protoplanetary disks.

4. Open Questions

The detection rate of new molecules has not significantly improved since 2000 but with the advent of ALMA, the observation of chemistry in circumstellar disks is beginning a new era. However, there are some trends which appear from the recent observational results. Among them, it is important to mention:

- Multi-line, multi-isotopologue CO interferometric studies allow observers and modelers to retrieve the basic density and temperature structures of disks. This method will become more and more accurate with ALMA.
- Both CO, CN, HCN and CCH molecular studies apparently reveal very low temperature in the gas phase in several TTauri disks. These observations suggest that the molecular layer is at least partly colder than predicted and that there are some ingredients still missing in our understanding of the disk physics and/or chemistry. More accurate observational data will refine these studies.
- Dust is one of the major agent shaping the disk physics and chemistry. In particular, it plays a fundamental role to control the UV disk illumination and the temperature. It also strongly influences the chemistry since a significant part of the disk is at a temperature which is below the freeze out temperature of most molecules. Therefore chemical reactions on grain surface need to be incorporated in chemical models.
- Deriving the gas-to-dust ratio remains one of the most difficult task. This would be even more complex than current models suggest because recent dust observations reveal that the gas-to-dust ratio also changes radially.
- Finally, several upper limits such as that obtained on H_2D^+ suggest that ALMA observations of disks may not be so easy and will require an adequate observational strategy.

Acknowledgements

I would like to acknowledge all CID members for a very fruitful collaboration, in particular S.Guilloteau, E.Chapillon, F.Hersant, D.Semenov, V.Pietu and T.Henning. I also thank all the KIDA team (http://kida.obs.u-bordeaux1.fr/) and V.Wakelam for providing chemical reactions as accurate as possible for astrophysics.

References

Aikawa Y. 2007 *A&A*, 656, L93

D'Alessio P., Calvet N., Hartmann L., Lizano S., & Cantó, J. 1999 *ApJ*, 527, 893

Chapillon E., Guilloteau S., Dutrey A., & Piétu V. 2008 *A&A*, 488, 565

Chapillon E., Parise B., Guilloteau S., Dutrey A., & Wakelam V. 2010 *A&A*, 520, 61

Chapillon E., Parise B., Guilloteau S., & Du F. 2011 *A&A*, in press
Chapillon E., Guilloteau S., Dutrey A., Pietu V., & Guelin M. 2011 *A&A*, in press
Chiang E. I. & Goldreich P. 1997 *ApJ*, 490, 368
Dartois E., Dutrey A., & Guilloteau S. 2003 *A&A*, 399, 773
Dutrey, A. Guilloteau S., & Guelin M. 1997 *A&A*, 317, L55
Dutrey A., Guilloteau S., & Ho P. 2007 *Protostars and Planets V, Reipurth B., Jewitt D. & Keil K. Eds*, 495
Dutrey A., Henning T., Guilloteau S., Semenov D., Piétu V., Schreyer K., Bacmann A., Launhardt R., Pety J., & Gueth F. 2007 *A&A*, 464, 615
Dutrey A., Guilloteau S., Piétu V., Chapillon E., Gueth F., Henning T., Launhardt R., Pavlyuchenkov Y., Schreyer K., & Semenov D. 2008 *A&A*, 490, L15
Dutrey A., Wakelam V., Boehler Y., Guilloteau S., Hersant F., Semenov D., Chapillon E., Henning T., Piétu V., Launhardt R., Gueth F., & Schreyer K.2011 *A&A*, in press
Garozzo M., Fulvio D., Kanuchova Z., Palumbo M. E., & Strazzulla D. 2010, *A&A*, 509, A67
Guilloteau S., Piétu, V., Dutrey A., & Guélin M. 2006 *A&A*, 448, L5
Guilloteau S., Dutrey A., Piétu V., & Boehler Y. 2011 *A&A*, X,Y
Henning T., Semenov D., Guilloteau S., Dutrey A., Hersant F., Wakelam V., Chapillon E., Launhardt R., Piétu V., & Schreyer K. 2010 *ApJ*, 714, 1511
Hersant F., Wakelam V., Dutrey A., Guilloteau S., & Herbst E. 2009 *A&A*, 493, L49
Öberg K. I., Linnartz H., Visser R., & van Dishoeck E. F. 2009 *ApJ*, 693, 1209
Piétu V., Dutrey A., & Guilloteau S. 2007 *A&A*, 467, 163
Qi C., Wilner D. J., Calvet N., Bourke T. L., Blake G. A., Hogerheijde M. R., Ho P. T. P., & Bergin E. 2008 *ApJ*, 636, L157
Qi C., Wilner D. J., Aikawa Y., Blake G. A., & Hogerheijde M. R. 2006 *ApJ*, 681, 1396
Semenov D., Wiebe D., & Henning, T. 2006 *ApJ*, 647, L57
Semenov D., Hersant F., Wakelam V., Dutrey A., Chapillon E., Guilloteau S, Henning T., Launhardt R., Piétu V., & Schreyer K. 2010 *A&A*, 522, A42
Wakelam V., Caselli P., Ceccarelli C., Herbst E., & Castets A. 2004 *A&A*, 422, 159

Discussion

E. van Dishoeck: Can you comment on the current state of determining the outer radius of the gas disk and its importance on the analysis ?

A.Dutrey: The accuracy on the determination of the gas outer radius depends, of course, on the sensitivity and angular resolution of the observations. As usual, the S/N ratio must be properly taken into account to provide relevant error bars on the measurement. As soon as the molecular disk is spectroscopically and spatially resolved, which is the case for many mm interferometric data (at least those from PdBI), the outer radius measurement is accurate enough to have a very limited influence on the other parameters (details can be found in Dutrey *et al.* (2007a), Pietu *et al.* (2007)). Moreover, there are a few examples of CO disks (e.g. DM Tau, Dartois *et al.* (2003)) which suggest that selective photo-dissociation can explain the different disk sizes observed in ^{12}CO and in ^{13}CO (the ^{13}CO disk being smaller).

The Molecular Universe
Proceedings IAU Symposium No. 280, 2011
José Cernicharo & Rafael Bachiller, eds.

doi:10.1017/S1743921311024914

Chemical Evolution of a Protoplanetary Disk

Dmitry A. Semenov[1]
[1]Max Planck Institute for Astronomy, Königstuhl 17, D–69117 Heidelberg, Germany
email: semenov@mpia.de

Abstract. In this paper we review recent progress in our understanding of the chemical evolution of protoplanetary disks. Current observational constraints and theoretical modeling on the chemical composition of gas and dust in these systems are presented. Strong variations of temperature, density, high-energy radiation intensities in these disks, both radially and vertically, result in a peculiar disk chemical structure, where a variety of processes are active. In hot, dilute and heavily irradiated atmosphere only the most photostable simple radicals and atoms and atomic ions exist, formed by gas-phase processes. Beneath the atmosphere a partly UV-shielded, warm molecular layer is located, where high-energy radiation drives rich ion-molecule and radical-radical chemistry, both in the gas phase and on dust surfaces. In a cold, dense, dark disk midplane many molecules are frozen out, forming thick icy mantles where surface chemistry is active and where complex polyatomic (organic) species are synthesized. Dynamical processes affect disk chemical composition by enriching it in abundances of complex species produced via slow surface processes, which will become detectable with ALMA.

Keywords. accretion disks, astrochemistry, diffusion, line: formation, molecular processes, protoplanetary disks, submillimeter, techniques: interferometric, turbulence

---

1. Introduction

Protoplanetary disks (PPDs) are ubiquitous around young stars (e.g., Lin & Papaloizou 1980; Lissauer 1987). Their chemical composition and physical properties regulate the efficiency and timescale of planet formation. Molecules and dust serve as heating and cooling agents of the gas, while dust grains also dominate the disk opacities. Molecular lines observed at infrared and (sub-)millimeter wavelengths are useful probes of physical conditions (temperature, density, kinematics, turbulence) in PPDs.

Over the past decade significant progress has been achieved in our understanding of disk chemical composition, both theoretically and observationally. Since molecular hydrogen is not observable when it is cold, we have to rely on other trace species to infer disk physics and chemistry (Table 1). Multi-molecule, multi-transition interferometric observations, coupled to line radiative transfer and chemical modeling, allowed to constrain disk sizes, kinematics, distribution of temperature, surface density, and molecular column densities (see reviews by Bergin *et al.* (2007) and Dutrey *et al.* (2007a)).

Observations of thermal dust emission are employed to constrain dust properties, infer disk masses, whereas IR spectroscopy is used to study mineralogical composition of dust in PPDs. Recently, with space-borne (*Spitzer*) and ground-based (Keck, VLT, Subaru) infrared telescopes, molecules have been detected in very inner zones of planet-forming systems, at $r \lesssim 1 - 10$ AU.

The conditions of planets formation in the early Solar system have been revealed by a detailed analysis of chemical and mineralogical composition of meteoritic samples and cometary dust particles (e.g., Bradley 2005). The recent *Stardust* and *Genesis* space missions have returned first samples of pristine materials, likely of cometary origin, showing a complex structure of high-temperature crystalline silicates embedded in low-temperature

Table 1. Molecular species commonly utilized to study disks

Tracer	Quantity	Midplane	Molecular layer	Atmosphere	Inner zone
^{12}CO, ^{13}CO	Temperature	mm*	mm	mm	IR
H_2	—	0	0	0	IR
NH_3	—	cm	cm	0	0
CS, H_2CO	Density	0	mm	0	IR
CCH, HCN, CN	Photochemistry	0	mm	0	IR
HCO^+	Ionization	0	mm	0	0
N_2H^+	—	mm	0	0	0
C^+	—	0	0	IR	IR
complex organics	Surface processes	IR**	IR-cm	0	IR,mm
DCO^+, DCN, H_2D^+	Deuterium fractionation	mm	mm	0	0

* – "mm/cm" and "IR" mean radio-interferometric and infrared observations, respectively.
** – Complex molecules frozen onto dust surfaces could be detected through absorption lines in infrared, while the gas-phase counterparts emit at (sub-) millimeter frequencies.

condensates (Brownlee *et al.* 2004; Flynn *et al.* 2006; Brownlee *et al.* 2008). The presence of crystalline silicates in outer regions of protoplanetary disks has also been revealed (e.g., van Boekel *et al.* 2004; Juhász *et al.* 2010). An isotopic analysis of refractory condensates in unaltered chondritic meteorites shows strong evidence that the inner part of the Solar Nebula has been almost completely mixed during the first several Myr of evolution (e.g., Boss 2004; Ciesla 2009).

These intriguing findings are partly understood in modern astrochemical models of protoplanetary disks (Aikawa & Herbst 1999; Markwick *et al.* 2002; van Zadelhoff *et al.* 2003; Tscharnuter & Gail 2007; Agúndez *et al.* 2008; Woods & Willacy 2009; Visser *et al.* 2009b; Walsh *et al.* 2010; Semenov & Wiebe 2011). The major result of the chemical modeling is that disks have a layered chemical structure due to heavy freeze-out of gas-phase molecules in the cold midplane and their photodissociation in the atmosphere. Observed column densities of CO, HCO^+, N_2H^+, CN, HCN, HNC, CS, etc. are qualitatively agree with the chemical models.

In this paper we briefly review the major observational findings and modeling predictions for the chemical composition and evolution of protoplanetary disks.

2. Observational constrains on composition of gas and ices

Apart from CO and its isotopologues, and occasionally HCO^+, DCO^+, CN, HCN, DCN, CCH, H_2CO, and CS, the molecular content of protoplanetary disks remains largely unknown (e.g., Dutrey *et al.* 1997; Kastner *et al.* 1997; Aikawa *et al.* 2003; Thi *et al.* 2004; Piétu *et al.* 2007; Qi *et al.* 2008; Henning *et al.* 2010). Molecular line data are limited in sensitivity and resolution. Thus the spatial distribution of molecular abundances is still poorly determined (e.g., Piétu *et al.* 2005; Dutrey *et al.* 2007b; Panić *et al.* 2009). Observational facilities such as the Plateau de Bure interferometer (France) and Submillimeter Array (USA) have permitted measurements of several brightest nearby disks (DM Tau, LkCa 15, AB Aur, and TW Hya). Soon, with the Atacama Large Millimeter Array entering it's full power in 2013, we will make substantial progress in our knowledge of molecular disk structure.

Typically, studies of disk physics begin with observations of bright CO lines. These lines are thermally excited at densities $\sim 10^3 - 10^4$ cm^{-3}. The ^{12}CO lines are optically thick and their intensities measure kinetic temperature in the upper disk layer (e.g., Dutrey *et al.* 1997). The lines of less abundant ^{13}CO and $C^{18}O$ are typically optically thin or partially optically thick and are sensitive to both temperature and corresponding column densities throughout the entire disk. Strong CO lines are most suitable for accurate determination of disk kinematics as well as orientation and geometry. Their measured widths indicate that turbulence in disks is subsonic, with typical velocities of about $0.05 - 0.2$ km s^{-1} (Dutrey *et al.* 2007a; Semenov *et al.* 2010a; Hughes *et al.* 2011).

It has been found that disks appear progressively larger from observations of dust continuum, onwards to $C^{18}O$, ^{13}CO, and ^{12}CO, respectively, with typical values of $300 - 1000$ AU. This is a manifestation of selective isotopic photodissociation. Most disks exhibit a vertical temperature gradient, ranging from ~ 10 K at the midplane to $\sim 50 - 100$ K in the atmosphere region, as determined by physical models (e.g., Dartois *et al.* 2003; Qi *et al.* 2006; Piétu *et al.* 2007; Isella *et al.* 2010), though several disks with large inner cavities do not show evidence for such a gradient (e.g., GM Aur and LkCa 15; Dutrey *et al.* 2008; Hughes *et al.* 2009).

By comparing intensities of 6-5 to 2-1 CO transitions it was found by Qi *et al.* (2006) that the TW Hya disk has a surface region that is superheated, so that an additional heating mechanism is required, possibly stellar X-ray radiation (Glassgold *et al.* 1997). A significant reservoir of very cold CO, HCO^+, CN and HCN gases has been found in the disk of DM Tau at temperatures $\lesssim$ 6-17 K, which cannot be explained by conventional chemical models without invoking a non-thermal desorption or transport mechanism (e.g., Semenov *et al.* 2006; Aikawa 2007; Hersant *et al.* 2009).

The second most readily observed molecular species in disks is HCO^+. The low-lying transitions of this ion are thermalized at densities of about 10^5 cm^{-3}. This is one of the most abundant charged molecule in disks, the other being C^+ (not observable at millimeter wavelengths) and H_3^+ (lacking dipole moment). The ionization degree measured from HCO^+ is $\sim 10^{-10} - 10^{-9}$ inside warm molecular disk layer (Qi *et al.* 2003; Dutrey *et al.* 2007b).

Less strong lines of C_2H, CN, and HCN are sensitive to the intensity and shape of the incident UV spectrum and are excellent tracers of photochemistry (e.g., Bergin *et al.* 2003a). Recently detected DCO^+ and DCN have abundances that are about $1 - 10\%$ of the HCO^+ and HCN densities (Qi *et al.* 2008). It remains to be verified whether such a large degree of deuteration is a heritage of cloud chemistry or produced *in situ*.

A key observational result is that molecular abundances are depleted by factors 5-100 compared to the values in the Taurus Molecular Cloud (Dutrey *et al.* 2007a). Since disks have higher densities up to $10^7 - 10^{10}$ cm^{-3} and enshrouded in stronger ionizing radiation fields, this depletion can be attributed to a combined effect of photodissociation and freeze-out.

The results from the Infrared Space Observatory and Spitzer telescope have proven the existence of a significant amount of frozen material and various types of silicates and polycyclic aromatic hydrocarbons (PAH) in disks (van den Ancker *et al.* 2000; van Dishoeck 2004; Bouwman *et al.* 2008). The PAH features at ~ 3–12μm probe the incident radiation field and density distribution of the upper disk (e.g., Habart *et al.* 2004). Recently, with space-borne (*Spitzer*) and ground-based (Keck, VLT, Subaru) infrared telescopes, molecules have been detected in very inner zones of planet-forming systems, at $r \lesssim 1 - 10$ AU. Rotational-vibrational emission lines from CO, CO_2, C_2H_2, HCN, OH, H_2O imply a rich chemistry driven by endothermic reactions or reactions with activation barriers and photoprocesses (Lahuis *et al.* 2006; Carr & Najita 2008; Salyk *et al.* 2008; Pontoppidan *et al.* 2008; Pascucci *et al.* 2009; van der Plas *et al.* 2009; Salyk *et al.* 2011). Through *ISO* and *Spitzer* infrared spectroscopy abundant ices in cold disk regions

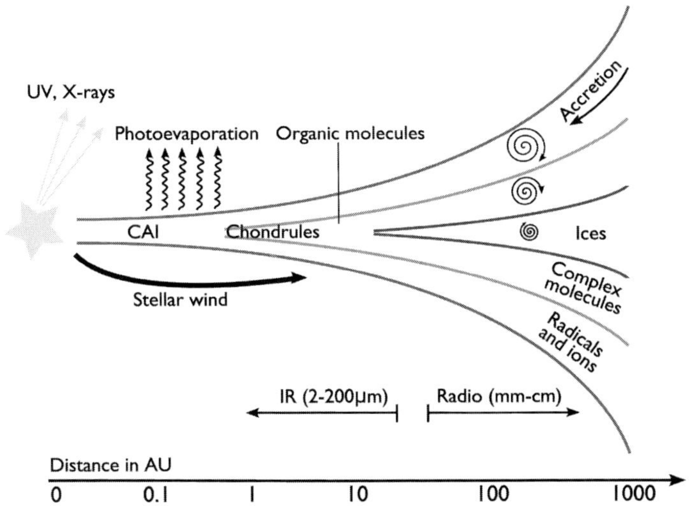

Figure 1. Physical and chemical structure of a protoplanetary disk. Timescales are rough estimates and given for a radius of about 100 AU.

consisting of water ice and substantial amounts ($\sim 1 - 30\%$) of volatile materials like CO, CO_2, NH_3, CH_4, H_2CO, and HCOOH have been detected (e.g., Pontoppidan *et al.* 2005; Terada *et al.* 2007; Zasowski *et al.* 2009).

In Table 1 the various molecules used to study protoplanetary disks are summarized.

3. Global evolutionary picture of disk chemical structure

The current scheme of the disk chemical structure is presented in Fig. 1. The disk can be divided into 4 distinct chemical regions, primarily determined by temperature. The warm "inner zone" corresponds to radii < 20 AU that are accessible with the IR instruments, while the other 3 regions dubbed "midplane", "molecular layer", and "atmosphere" represent the outer disk regions ($r > 20 - 50$ AU) that are observed with radio-interferometers.

Dense midplane is opaque to the UV and X-ray radiation and remains cold and essentially neutral ($10 - 20$ K). Chemical complexity in this region is initially reached by fast ion-molecular reactions in the gas, followed by slow accretion onto grains and surface reactions (mostly hydrogenation). Once formed, a molecule is seldom re-emitted into the gas phase. Typical chemical timescales for this region are determined by freeze-out and surface reactions, and are $> 10^5 - 10^6$ years in the outer disk. In the very inner part, the midplane is hot due to viscous accretion heating ($T \gtrsim 50$ K) and gas-grain interactions are not important, so the chemical timescale is due to gas-phase neutral-neutral reactions ($\sim$100 years). The planet-forming zone is in general in a chemical steady-state, which allows the application of simple condensation/evaporation thermodynamical chemical models.

Adjacent to the midplane lies a less dense, warmer layer ($T \sim 30 - 70$ K) that is partly shielded from stellar and interstellar UV/X- ray radiation. Young stars emit intense non-thermal UV and thermal and variable X-ray radiation fields that strongly mediate disk chemistry at intermediate heights (e.g., Bergin *et al.* 2007). UV excess in T Tauri stars is related to their chromospheric activity (e.g., Bouvier *et al.* 2007). Soft X-ray radiation

Table 2. Chemical reactions active in disks

Process	Formula	Midplane	Molecular layer	Atmosphere	Inner zone
Bond formation					
Radiative association	A + B → AB + hν	X	X	X	X
Surface formation	A + B‖gr → AB + gr	X	X	0	0
Three-body	A + B + M → AB + M	0	0	0	X
Bond destruction					
Photodissociation	AB + hν → A + B	0	X	X	X
Dissociation by CRP	AB + CRP → A + B	X	X	0	0
Dissociation by X-rays	—	0	X	X	X
Dissociative recombination	$AB^+ + e^- \rightarrow A + B$	X	X	X	X
Bond restructuring					
Neutral-neutral	A + BC → AB + C	X	X	0	X
Ion-molecule	$A^+ + BC \rightarrow AB^+ + C$	X	X	X	X
Charge transfer	$A^+ + BC \rightarrow A + BC^+$	X	X	X	X
Unchanged bond					
Photoionization	$AB + h\nu \rightarrow AB^+ + e^-$	0	X	X	X
Ionization by CRP	$AB + CRP \rightarrow AB^+ + e^-$	X	X	0	0
Ionization by X-rays	—	0	X	X	X

has likely the same origin, while harder X-rays are produced in magnetic reconnection loops or due to jets (e.g., Güdel & Nazé 2009).

These energetic photons dissociate and ionize gas and photodesorb surface species, thus enriching the gas composition and initiating a rich chemistry. Abundances of most molecules attain high concentrations in this zone, and numerous molecular lines are excited and observable emission produced. Chemistry does not reach a steady state in this region and a typical timescale, as determined by surface chemistry and gas-grain interactions, exceeds 10^5 years.

Above the intermediate layer a hot, dilute, and heavily irradiated disk atmosphere exists ($T \gtrsim 100$ K). This is a molecularly-poor region, where only simple light hydrocarbons, their ions, and other radicals such as CCH and CN are able to survive. Chemical timescales are short (~100 years) and defined by photochemical processes and dissociate recombination.

4. Gas-phase chemistry

Models of chemical evolution employ a set of chemical reactions of relevance to conditions in disks. Modern astrochemical databases include up to 600 species involved in 4-7000 reactions (Smith *et al.* 2004; Le Petit *et al.* 2006; Woodall *et al.* 2007; Wakelam 2009). Only $10-20\%$ of reaction rates have been studied in the laboratory or calculated theoretically and thus models are prone to inherent uncertainties (see, e.g., Wakelam *et al.* 2010). All reactions can be divided into 4 distinct groups that are dominant in different disk regions, see Table 2 (after van Dishoeck & Black 1988).

Apart from the very dense inner zone, all reactions in disks are two-body processes. Three-body reactions become competitive only at $\lesssim 10$ AU, where $n \gtrsim 10^{10}$ cm^{-3} (Aikawa *et al.* 1999). The main processes leading to formation of molecular bonds are slow radiative association and surface reactions. Upon collision, a collisional complex in an excited state may form, which is stabilized with a low probability by emission of a photon (e.g., Bates 1951; Williams 1972; Herbst & Klemperer 1973). For example, formation of light hydrocarbons starts with radiative association of C^+ and H_2, leading to excited CH_2^+ (Herbst 1985).

Ionized by FUV, X-ray photons or CRPs, ions and molecule drive rapid ion-neutral chemistry, which constitutes the largest fraction of astrochemical models. These reactions are exothermic, with high rate coefficients $\sim 10^{-9}$ cm^{-3} s^{-1}, which often increase toward low temperatures ($\beta < 0$) (e.g., Dalgarno & Black 1976). Ion-neutral reactions result in bond restructuring of the reactants. Some of the most important reactions of this category are protonation reactions.

Molecular ions are destroyed by dissociative recombination with electrons and negatively charged grains. These processes are especially fast at low temperatures, with typical rates of about 10^{-7} cm^{-3} s^{-1} at 10 K (Woodall *et al.* 2007). For nearly all observed species, dissociative recombination is an important formation pathway (e.g., water and hydrocarbons). Often, at later evolutionary times, $\gtrsim 10^5$ years, dissociative recombination is balanced by protonation reactions, e.g. $CO + H_3^+ \rightarrow HCO^+ + H_2$ followed by $HCO^+ + e^- \rightarrow CO + H$. Products and branching ratios for polyatomic ions are not easily obtainable (e.g., Bates & Herbst 1988; Spanel & Smith 1994).

A number of neutral-neutral reactions involving open-shell radicals can also be active in both cold outer and warm inner disk regions (van Dishoeck 1998). The typical rate coefficient for these reactions is $\sim 10^{-11}$–10^{-10} cm^{-3} s^{-1}, i.e., only about an order of magnitude lower than for the ion-molecule processes (e.g., Clary *et al.* 1994; Smith *et al.* 2004). One of the most interesting reactions of this type is formation of HCO^+ upon collision between O and CH ($\alpha_0 = 2.0 \cdot 10^{-11}$ cm^{-3} s^{-1}; Woodall *et al.* 2007).

5. Photochemistry

The photochemical reactions are overviewed in this book in the paper by van Dishoeck & Visser and only briefly discussed here.

Young T Tau stars ($T_{\rm eff} \simeq 4\,000$ K) emit intense non-thermal UV radiation that has a spectra different from the interstellar UV field (Bergin *et al.* 2003a), with a prominent Lyman$_\alpha$ line, while hot ($T_{\rm eff} \gtrsim 10\,000$ K) Herbig Ae/Be stars produce a lot of thermal UV emission. The overall intensity of the stellar UV radiation at 100 AU from the star can be as high as 500 and 10^5 for a T Tau and a Herbig Ae star, respectively, in units of the interstellar UV field (Habing 1968).

Many molecules, such as CO, H_2, and CN, are dissociated by radiation at short wavelengths ($\lambda \lesssim 1\,100$ Å), while photodissociation of other species, such as HCN, occurs at longer wavelengths (van Dishoeck & Black 1988; van Dishoeck *et al.* 2006). Some molecules (e.g., H_2 and CO) dissociate via absorption of UV photons at discrete lines to the excited (Rydberg) states, whereas other molecules are dissociated either by the continuum (e.g., CH_4) or by the continuum and lines (e.g., C_2) (van Dishoeck 1988; van Dishoeck *et al.* 2006). The high ratio of CN to HCN abundances, observed in disks (Dutrey *et al.* 1997), can be explained by more intense photodissociation of HCN if part of the UV flux comes as Ly$_\alpha$ photons (1216$\mathring{A}$; Bergin *et al.* 2003a).

Since the dissociation of the abundant H_2 and CO molecules result from photoabsorption at discrete wavelengths, isotopically selective photodissociation based on self-shielding is possible (e.g., Visser *et al.* 2009a). Two conditions for that are required: 1) dissociation via line absorption for each isotopically substituted molecule, and 2) differential photolysis that depends upon the isotopic abundances. Self-shielding occurs when the spectral lines leading to dissociation of the major isotopic species optically saturate, while the other residual lines relevant for dissociation of the minor isotopes remain transparent. Under conditions of protoplanetary disks, isotopic-selective chemistry may occur for abundant CO molecules (e.g., Dalgarno & Stephens 1970; Thiemens & Heidenreich 1983; van Dishoeck & Black 1988; Lee *et al.* 1996; Visser *et al.* 2009a).

6. Gas-grain interactions

In cold disk regions grains serve as a passive sink for heavy molecules and electrons, while providing free electrons to dissociate positive ions. In the darkest midplane zone dust grains become the dominant charged species (Semenov *et al.* 2004). At low temperatures of $\lesssim 20-50$ K, many molecules adhere to a grain with a nearly 100% sticking probability since their kinetic energy is much smaller than the binding energy (see, e.g., d'Hendecourt *et al.* 1985; Buch & Zhang 1991). This sticking probability also depends on surface properties, such as porosity and distribution of surface chemi- and physisorption sites. Chemisorption requires formation of a chemical bond between a surface species and a grain, and such a species does not evaporate easily (e.g., Cazaux *et al.* 2005).

The most effective desorption processes for disk chemistry are thermal evaporation, cosmic ray induced desorption, and photodesorption. Thermal evaporation occurs if a molecule has energy that exceeds it's binding energy. Typical binding energies of physisorbed species are about 1 000 K for light molecules like CO and N_2 (Bisschop *et al.* 2006), and much larger for heavier cyanopolyynes and carbon chains. Chemisorbed species do not desorb until very high temperatures of $100-1000$ K are reached.

In disk midplane CRPs provide energy that allows dust mantles (partly) thermally evaporate. A relativistic iron nucleus may eventually collide with a grain and impulsively heat it, releasing a portion of the volatile component (e.g., Watson & Salpeter 1972; Leger *et al.* 1985; Hartquist & Williams 1990). In less opaque disk regions, penetrating UV photons lead to photoevaporation of surface species. The probability of evaporation per one UV photon has been measured in the laboratory for some simple molecules, such as CO, H_2O, CH_4, and NH_3, and is about 10^{-6}–10^{-2} (e.g., Öberg *et al.* 2007, 2009b,a). The dilute UV radiation field produced by cosmic rays in the disk midplane (Prasad & Tarafdar 1983) can also be important for high values of the photodesorption rates. Another mechanism is the X-ray induced desorption which is efficient only for tiny grains, $\lesssim 0.05\mu$m (Najita *et al.* 2001).

7. Surface formation of complex species

The surfaces of dust grains serve as a catalyst for reactions that do not proceed efficiently in the gas phase. The most notable example is formation of molecular hydrogen, which occurs almost entirely on dust surfaces (e.g., Hollenbach & Salpeter 1971; Watson & Salpeter 1972). Tielens & Hagen (1982) have studied the chemical evolution of surface species on large grains and considered the effect of various migration rates for light atoms and heavy molecules. A typical 0.1μm amorphous silicate grain accommodates $\sim 10^6$ surface sites available for accretion. An atom or light radical, if it is not chemisorbed, may migrate over the surface from site to site by thermal hopping, when its energy exceeds the barriers for particle motions and react with other species.

There is an increasing body of evidence from laboratory measurements that complex molecules like methanol cannot be produced in the gas phase via radiative association forming a large protonated precursor followed by its dissociative recombination (Geppert *et al.* 2005). Thus, surface reactions due to thermal hopping remain the only viable formation pathway for production of complex (organic) molecules in protoplanetary disks, found also in meteoritic samples in Solar system.

8. Importance of dynamical processes for disk chemistry

While most of the chemical studies are still based on laminar disk models, evidences for mixing call for a more sophisticated treatment. Models of the early Solar nebula with radial transport by advective flows have been developed (e.g., Morfill & Völk 1984; Gail 2001, 2002; Wehrstedt & Gail 2002; Boss 2004; Keller & Gail 2004). Ilgner *et al.* (2004) for the first time modeled the influence of turbulent diffusion in the vertical direction and advection flows in the radial direction on the chemical composition of the inner disk region. They found that dynamical processes significantly affect the chemical evolution of sulfur-bearing species. Willacy *et al.* (2006) have shown that 1D vertical mixing modifies chemical composition of the outer disk region and that the mixing results better agree to observations. Semenov *et al.* (2006, Paper I) and Aikawa (2007) have found that turbulent transport allows explaining the presence of a large amount of cold ($\lesssim$ 15 K) CO gas in the disk of DM Tau. Tscharnuter & Gail (2007) have used a 2D disk chemo-hydrodynamical model and showed that in the disk midplane matter moves outward, carrying out the angular momentum, while the accretion flows toward the star are located at elevated altitudes. Consequently, gas-phase species produced by warm chemistry in the inner nebula can be steadily transported into the cold outer region and freeze out. A radial advection model has also been utilized by Nomura *et al.* (2009), who have demonstrated that inward radial transport enhances abundances of organic molecules (produced mainly on dust surfaces in cold outer regions). Hersant *et al.* (2009) have studied various mechanisms to retain gas-phase CO in very cold disk regions. They concluded that efficient photodesorption in moderately obscured disk regions ($A_{\rm V} < 5^{\rm m}$) greatly enhances gas-phase CO concentrations, while the role of vertical mixing is less important. Finally, Heinzeller *et al.* (2011) have investigated the disk chemical evolution with radial advection, vertical mixing, and vertical wind transport processes. They have found that the disk wind has a negligible effect on disk chemistry, whereas the radial accretion alters the molecular abundances in the cold midplane, and the vertical turbulent mixing affects the chemistry in the warm molecular layer.

We study the influence of turbulence mixing on the chemical evolution of protoplanetary disks in Semenov & Wiebe (2011). Mixing is important in disks since a chemical steady-state is not reached due to long timescales associated with surface chemical processes and slow evaporation of heavy molecules. Our analysis was based on the α-model of a $\sim$ 5 Myr DM Tau disk coupled to the large-scale gas-grain chemical code "ALCHEMIC" (Semenov *et al.* 2010b). To account for production of complex molecules, our chemical network was supplied with a set of surface reactions and photoprocessing of ices.

The adopted flaring disk structure is based on a 1+1D steady-state α-model similar to D'Alessio *et al.* (1999) model. The non-thermal FUV radiation field from DM Tau is represented by the scaled ISRF of Draine (1978), with the un-attenuated intensity at 100 AU of $\chi_*(100) = 410$ (e.g., Bergin *et al.* 2003b). For the X-ray luminosity of the star we adopt a value of 10^{30} erg s^{-1}, which is constrained by recent measurements with Chandra and XMM in the range of 0.3-10 keV.

The turbulence in disks is likely driven by the magnetorotational instability (MRI), which is operative even in a weakly ionized medium (e.g., Balbus & Hawley 1991). This turbulence causes anomalous viscosity that enables efficient redistribution of the angular momentum. We have followed the parametrization of Shakura & Sunyaev (1973), where

Table 3. Detectable tracers of turbulent mixing.

Steadfast	**Hypersensitive**
CO	Heavy hydrocarbons (e.g.,C_6H_6)
H_2O ice	C_2S
	C_3S
	CO_2
	O_2
	SO
	SO_2
	OCN
	Complex organics (e.g., HCOOH)

turbulent viscosity ν is related to local disk properties such as the vertical spatial scale $H(r)$, the sound speed $c_s(r, z)$, and the dimensionless parameter α:

$$\nu(r, z) = \alpha\, c_s(r, z)\, H(r). \tag{8.1}$$

From observational constraints α is $\sim 0.001 - 0.1$ (Andrews & Williams 2007; Guilloteau *et al.* 2011), so we adopt the constant value of 0.01. Consequently, the diffusion coefficient is calculated as

$$D_{\rm turb}(r, z) = \nu(r, z)/Sc, \tag{8.2}$$

where Sc is the Schmidt number describing efficiency of turbulent diffusivity (see e.g. Shakura & Sunyaev 1973; Schräpler & Henning 2004). We assume that gas-phase species and dust grains are well mixed, and transported with the same diffusion coefficient.

We show that the higher the ratio of the characteristic chemical timescale to the turbulent transport timescale for a given molecule, the higher the probability that its column density will be affected by dynamical processes. Thus, turbulent transport influences abundances of many gas-phase species and especially ices. Vertical mixing is more important as it affects the evolution of gas-phase and surface species of any kind, whereas the effect of radial mixing is pronounced mostly for the evolution of ices. The radial temperature gradient is weaker, and thus is only relevant for the evolution of polyatomic ices formed via surface reactions of heavy radicals, whereas steep vertical gradients of temperature and high-energy radiation intensity cause much sharper transition from the ice-dominated chemistry in the disk midplane to the rich gas-phase chemistry in the molecular layer.

The simple molecules that are unresponsive to dynamical transport include C_2H, C^+, CH_4, CN, CO, HCN, HNC, H_2CO, OH, as well as water and ammonia ices. The sensitive species to dynamics are carbon chains and other heavy species, in particular sulfur-bearing and complex organic molecules frozen onto the dust grains. Mixing steadily transports ice-coated grains in warmer regions, allowing more efficient surface processing due to enhanced hopping rates of heavy radicals. In warm intermediate layer these organically-rich ices evaporate, and in the inner disk they can also be photodissociated by CRP/X-ray-induced UV photons. The importance of mixing is higher in an inner, planet-forming disk zone, where thermal, density, and high-energy radiation gradients are stronger than in the outer region. Still, mixing does not completely erase the layered chemical structure of protoplanetary disks.

Several promising detectable tracers of dynamical processes in protoplanetary disks are the column density ratios of the CO_2, O_2, SO, SO_2, C_2S, C_3S to that of CO and the water ice (see Table 3). The detection of complex species (e.g., dimethyl ether, formic acid, methyl formate, etc.) in protoplanetary disks with ALMA and JWST will be a strong indication that chemical evolution of these objects is influenced by transport processes.

References

Agúndez, M., Cernicharo, J., & Goicoechea, J. R. 2008, *A&A*, 483, 831

Aikawa, Y. 2007, *ApJ*, 656, L93

Aikawa, Y. & Herbst, E. 1999, *A&A*, 351, 233

Aikawa, Y., Momose, M., Thi, W.-F., *et al.* 2003, *PASJ*, 55, 11

Aikawa, Y., Umebayashi, T., Nakano, T., & Miyama, S. M. 1999, *ApJ*, 519, 705

Andrews, S. M. & Williams, J. P. 2007, *ApJ*, 659, 705

Balbus, S. A. & Hawley, J. F. 1991, *ApJ*, 376, 214

Bates, D. R. 1951, *MNRAS*, 111, 303

Bates, D. R. & Herbst, E. 1988, in ASSL Vol. 146: Rate Coefficients in Astrochemistry, ed. T. J. Millar & D. A. Williams (Kluwer Academic Publishers, Dordrecht), 41–48

Bergin, E., Calvet, N., D'Alessio, P., & Herczeg, G. J. 2003a, *ApJ*, 591, L159

Bergin, E., Calvet, N., D'Alessio, P., & Herczeg, G. J. 2003b, *ApJ*, 591, L159

Bergin, E. A., Aikawa, Y., Blake, G. A., & van Dishoeck, E. F. 2007, in Protostars and Planets V, ed. B. Reipurth, D. Jewitt, & K. Keil, 751–766

Bisschop, S. E., Fraser, H. J., Öberg, K. I., van Dishoeck, E. F., & Schlemmer, S. 2006, *A&A*, 449, 1297

Boss, A. P. 2004, *ApJ*, 616, 1265

Bouvier, J., Alencar, S. H. P., Harries, T. J., Johns-Krull, C. M., & Romanova, M. M. 2007, Protostars and Planets V, 479

Bouwman, J., Henning, T., Hillenbrand, L. A., *et al.* 2008, *ApJ*, 683, 479

Bradley, J. P. 2005, in Meteorites, Comets and Planets: Treatise on Geochemistry, Volume 1, ed. A. M. Davis, H. D. Holland, & K. K. Turekian (Elsevier B), 689–+

Brownlee, D. E., Horz, F., Newburn, R. L., *et al.* 2004, *Science*, 304, 1764

Brownlee, D. E., Joswiak, D. J., Matrajt, G., Bradley, J. P., & Ebel, D. S. 2008, in Lunar and Planetary Institute Conference Abstracts, Vol. 39, Lunar and Planetary Institute Conference Abstracts, 1978–+

Buch, V. & Zhang, Q. 1991, *ApJ*, 379, 647

Carr, J. S. & Najita, J. R. 2008, *Science*, 319, 1504

Cazaux, S., Caselli, P., Tielens, A. G. G. M., LeBourlot, J., & Walmsley, M. 2005, Journal of Physics Conference Series, 6, 155

Ciesla, F. J. 2009, Icarus, 200, 655

Clary, D. C., Haider, N., Husain, D., & Kabir, M. 1994, *ApJ*, 422, 416

D'Alessio, P., Calvet, N., Hartmann, L., Lizano, S., & Cantó, J. 1999, *ApJ*, 527, 893

Dalgarno, A. & Black, J. H. 1976, Reports of Progress in Physics, 39, 573

Dalgarno, A. & Stephens, T. L. 1970, *ApJ*, 160, L107

Dartois, E., Dutrey, A., & Guilloteau, S. 2003, *A&A*, 399, 773

d'Hendecourt, L. B., Allamandola, L. J., & Greenberg, J. M. 1985, *A&A*, 152, 130

Draine, B. T. 1978, *ApJS*, 36, 595

Dutrey, A., Guilloteau, S., & Guelin, M. 1997, *A&A*, 317, L55

Dutrey, A., Guilloteau, S., & Ho, P. 2007a, in Protostars and Planets V, ed. B. Reipurth, D. Jewitt, & K. Keil, 495–506

Dutrey, A., Guilloteau, S., Piétu, V., *et al.* 2008, *A&A*, 490, L15

Dutrey, A., Henning, T., Guilloteau, S., *et al.* 2007b, *A&A*, 464, 615

Flynn, G. J., Bleuet, P., Borg, J., *et al.* 2006, *Science*, 314, 1731

Gail, H.-P. 2001, *A&A*, 378, 192

Gail, H.-P. 2002, *A&A*, 390, 253

Geppert, W. D., Hellberg, F., Österdahl, F., *et al.* 2005, in IAU Symposium, Vol. 231, Astrochemistry: Recent Successes and Current Challenges, ed. D. C. Lis, G. A. Blake, & E. Herbst, 117–124

Glassgold, A. E., Najita, J., & Igea, J. 1997, *ApJ*, 480, 344
Güdel, M. & Nazé, Y. 2009, *A&A Rev.*, 17, 309
Guilloteau, S., Dutrey, A., Piétu, V., & Boehler, Y. 2011, *A&A*, 529, A105+
Habart, E., Natta, A., & Krügel, E. 2004, *A&A*, 427, 179
Habing, H. J. 1968, *Bull. Astron. Inst. Netherlands*, 19, 421
Hartquist, T. W. & Williams, D. A. 1990, *MNRAS*, 247, 343
Heinzeller, D., Nomura, H., Walsh, C., & Millar, T. J. 2011, *ApJ*, 731, 115
Henning, T., Semenov, D., Guilloteau, S., *et al.* 2010, *ApJ*, 714, 1511
Herbst, E. 1985, *ApJ*, 291, 226
Herbst, E. & Klemperer, W. 1973, *ApJ*, 185, 505
Hersant, F., Wakelam, V., Dutrey, A., Guilloteau, S., & Herbst, E. 2009, *A&A*, 493, L49
Hollenbach, D. & Salpeter, E. E. 1971, *ApJ*, 163, 155
Hughes, A. M., Andrews, S. M., Espaillat, C., *et al.* 2009, *ApJ*, 698, 131
Hughes, A. M., Wilner, D. J., Andrews, S. M., Qi, C., & Hogerheijde, M. R. 2011, *ApJ*, 727, 85
Ilgner, M., Henning, T., Markwick, A. J., & Millar, T. J. 2004, *A&A*, 415, 643
Isella, A., Natta, A., Wilner, D., Carpenter, J. M., & Testi, L. 2010, *ApJ*, 725, 1735
Juhász, A., Bouwman, J., Henning, T., *et al.* 2010, *ApJ*, 721, 431
Kastner, J. H., Zuckerman, B., Weintraub, D. A., & Forveille, T. 1997, *Science*, 277, 67
Keller, C. & Gail, H. 2004, *A&A*, 415, 1177
Lahuis, F., van Dishoeck, E. F., Boogert, A. C. A., *et al.* 2006, *ApJ*, 636, L145
Le Petit, F., Nehmé, C., Le Bourlot, J., & Roueff, E. 2006, *ApJS*, 164, 506
Lee, H.-H., Herbst, E., Pineau des Forets, G., Roueff, E., & Le Bourlot, J. 1996, *A&A*, 311, 690
Leger, A., Jura, M., & Omont, A. 1985, *A&A*, 144, 147
Lin, D. N. C. & Papaloizou, J. 1980, *MNRAS*, 191, 37
Lissauer, J. J. 1987, *Icarus*, 69, 249
Markwick, A. J., Ilgner, M., Millar, T. J., & Henning, T. 2002, *A&A*, 385, 632
Morfill, G. E. & Völk, H. J. 1984, *ApJ*, 287, 371
Najita, J., Bergin, E. A., & Ullom, J. N. 2001, *ApJ*, 561, 880
Nomura, H., Aikawa, Y., Nakagawa, Y., & Millar, T. J. 2009, *A&A*, 495, 183
Öberg, K. I., Fuchs, G. W., Awad, Z., *et al.* 2007, *ApJ*, 662, L23
Öberg, K. I., Linnartz, H., Visser, R., & van Dishoeck, E. F. 2009a, *ApJ*, 693, 1209
Öberg, K. I., van Dishoeck, E. F., & Linnartz, H. 2009b, *A&A*, 496, 281
Panić, O., Hogerheijde, M. R., Wilner, D., & Qi, C. 2009, *A&A*, 501, 269
Pascucci, I., Apai, D., Luhman, K., *et al.* 2009, *ApJ*, 696, 143
Piétu, V., Dutrey, A., & Guilloteau, S. 2007, *A&A*, 467, 163
Piétu, V., Guilloteau, S., & Dutrey, A. 2005, *A&A*, 443, 945
Pontoppidan, K. M., Blake, G. A., van Dishoeck, E. F., *et al.* 2008, *ApJ*, 684, 1323
Pontoppidan, K. M., Dullemond, C. P., van Dishoeck, E. F., *et al.* 2005, *ApJ*, 622, 463
Prasad, S. S. & Tarafdar, S. P. 1983, *ApJ*, 267, 603
Qi, C., Kessler, J. E., Koerner, D. W., Sargent, A. I., & Blake, G. A. 2003, *ApJ*, 597, 986
Qi, C., Wilner, D. J., Aikawa, Y., Blake, G. A., & Hogerheijde, M. R. 2008, *ApJ*, 681, 1396
Qi, C., Wilner, D. J., Calvet, N., *et al.* 2006, *ApJ*, 636, L157
Salyk, C., Pontoppidan, K. M., Blake, G. A., *et al.* 2008, *ApJ*, 676, L49
Salyk, C., Pontoppidan, K. M., Blake, G. A., Najita, J. R., & Carr, J. S. 2011, *ApJ*, 731, 130
Schräpler, R. & Henning, T. 2004, *ApJ*, 614, 960
Semenov, D., Chakraborty, S., & Thiemens, M. 2010a, Chemical and Isotopic Evolution of the Solar Nebula and Protoplanetary Disks, ed. Apai, D. A. & Lauretta, D. S., 97–127
Semenov, D., Hersant, F., Wakelam, V., *et al.* 2010b, *A&A*, 522, A42+
Semenov, D. & Wiebe, D. 2011, ArXiv e-prints, 1104.4358
Semenov, D., Wiebe, D., & Henning, T. 2004, *A&A*, 417, 93
Semenov, D., Wiebe, D., & Henning, T. 2006, *ApJ*, 647, L57
Shakura, N. I. & Sunyaev, R. A. 1973, *A&A*, 24, 337
Smith, I. W. M., Herbst, E., & Chang, Q. 2004, *MNRAS*, 350, 323
Spanel, P. & Smith, D. 1994, *Chemical Physics Letters*, 229, 262

Terada, H., Tokunaga, A. T., Kobayashi, N., *et al.* 2007, *ApJ*, 667, 303
Thi, W.-F., van Zadelhoff, G.-J., & van Dishoeck, E. F. 2004, *A&A*, 425, 955
Thiemens, M. H. & Heidenreich, III, J. E. 1983, *Science*, 219, 1073
Tielens, A. G. G. M. & Hagen, W. 1982, *A&A*, 114, 245
Tscharnuter, W. M. & Gail, H.-P. 2007, *A&A*, 463, 369
van Boekel, R., Min, M., Leinert, C., *et al.* 2004, *Nature*, 432, 479
van den Ancker, M. E., Bouwman, J., Wesselius, P. R., *et al.* 2000, *A&A*, 357, 325
van der Plas, G., van den Ancker, M. E., Acke, B., *et al.* 2009, *A&A*, 500, 1137
van Dishoeck, E. F. 1988, in ASSL Vol. 146: Rate Coefficients in Astrochemistry, ed. T. Millar & D. Williams (Kluwer Academic Publishers, Dordrecht), 49–72
van Dishoeck, E. F. 1998, in The Molecular Astrophysics of Stars and Galaxies, ed. T. W. Hartquist & D. A. Williams (Clarendon Press, Oxford), 53–100
van Dishoeck, E. F. 2004, *ARA&A*, 42, 119
van Dishoeck, E. F. & Black, J. H. 1988, *ApJ*, 334, 771
van Dishoeck, E. F., Jonkheid, B., & van Hemert, M. C. 2006, in Faraday discussion, Vol. 133, Chemical evolution of the Universe, ed. I. R. Sims & D. A. Williams, 231–244
van Zadelhoff, G.-J., Aikawa, Y., Hogerheijde, M. R., & van Dishoeck, E. F. 2003, *A&A*, 397, 789
Visser, R., van Dishoeck, E. F., & Black, J. H. 2009a, *A&A*, 503, 323
Visser, R., van Dishoeck, E. F., Doty, S. D., & Dullemond, C. P. 2009b, *A&A*, 495, 881
Wakelam, V. 2009, in American Astronomical Society Meeting Abstracts, Vol. 214, American Astronomical Society Meeting Abstracts, 402.15–+
Wakelam, V., Smith, I. W. M., Herbst, E., *et al.* 2010, *Space Sci. Rev.*, 156, 13
Walsh, C., Millar, T. J., & Nomura, H. 2010, ArXiv e-prints
Watson, W. D. & Salpeter, E. E. 1972, *ApJ*, 174, 321
Wehrstedt, M. & Gail, H. 2002, *A&A*, 385, 181
Willacy, K., Langer, W., Allen, M., & Bryden, G. 2006, *ApJ*, 644, 1202
Williams, D. A. 1972, *Astrophys. Lett.*, 10, L17
Woodall, J., Agúndez, M., Markwick-Kemper, A. J., & Millar, T. J. 2007, *A&A*, 466, 1197
Woods, P. M. & Willacy, K. 2009, *ApJ*, 693, 1360
Zasowski, G., Kemper, F., Watson, D. M., *et al.* 2009, *ApJ*, 694, 459

Discussion

PAUL M. WOODS: There were some dynamical models recently that show that the accretion flow proceeds inwards along the disc surface and outwards along the disc midplane. Are those models credible, and how would that affect the disc chemistry?

D. SEMENOV: Turbulence is a 3D phenomenon and its magnitude varies greatly across a disk. Global 3D MHD models show that viscous transport occurs in all directions in any given disk location, albeit with different efficiencies, and that there is no a specific direction for it. According to observations, advective transport seems to be an effective mechanism that regulates overall lifetime of disks, $\sim 1 - 10$ Myr, in which most of disk matter accretes onto the central star and photoevaporates. Consequently, disk structure changes with time, as well as the FUV and X-ray penetration, and grain properties. This strongly affects chemical evolution of protoplanetary disks, and favors the use of evolutionary disk models (see R. Vissier's presentation and proceeding paper.

JAMES R. LYONS: What turbulent viscosity parameter α-values were used for slow vs. fast mixing?

D. SEMENOV: We considered Fickian diffusion, in which the diffusion coefficient $D_{\rm turb}$ (r, z) is attributed to the α-parameterized viscosity, $\alpha = 0.01$: $D_{\rm turb}(r, z) = \nu(r, z)/Sc = \alpha\, c_{\rm s}(r, z)\, H(r)/Sc$, where the characteristic spatial scale is $H(r)$ and the sound speed is

$c_s(r, z)$. The Sc the Schmidt number describing efficiency of turbulent diffusivity. The "fast" mixing model has $Sc = 1$ and the "slow" model has "Sc=100".

JAMES R. LYONS: Have you included isotopes in vertical mixing calculations?

D. SEMENOV: No.

JAMES R. LYONS: What did you mean by "isotopic homogeneity of the Solar nebula"? It is not homogeneous in oxygen isotopes.

D. SEMENOV: I've meant that the inner, 1-20 AU region of the Solar nebula has been isotopically homogenized at bulk level, though oxygen anomalies at percentage level still persist.

The Molecular Universe
Proceedings IAU Symposium No. 280, 2011
José Cernicharo & Rafael Bachiller, eds.

doi:10.1017/S1743921311024926

Infrared Observational Studies of Gas Molecules in Disks

C. Salyk[1]

[1]The University of Texas, Department of Astronomy, 1 University Station, C1400, Austin, TX 78712, USA

Abstract. There remain many fundamental unanswered questions about protoplanetary disks, including how (and if?) they form planets, how mass is transferred through the disk and onto the star, and how they ultimately disperse. Also, a major goal of protoplanetary disk studies is to understand the relationship between disk properties and the physical and chemical properties of planetary systems. IR molecular spectroscopy is a particularly powerful tool for probing the conditions and physical process in protoplanetary disks, which are too small and close to their parent stars to be imaged with ease. I will discuss the suite of infrared molecular transitions observed to date, which highlight the following three techniques of IR spectroscopy. Firstly, line shapes and strengths can be used as tracers of disk physics, including volatile condensation/evaporation, photo-processes, grain growth and turbulence. Secondly, observations of multiple molecular abundances provide constraints for disk chemical models, which may ultimately help explain the great diversity of planetary bodies. Finally, resolved line shapes and spectro-astrometry provide a means to study disk structure on extremely small size scales. Because IR observations are typically sensitive to radii of a few AU or smaller, the processes and structures being probed are relevant to the birth and growth of terrestrial and giant planets. Recent results that I will highlight include the discovery of a multitude of molecules in disks around sun-like stars (including H_2O, OH, HCN, C_2H_2 and CO_2), with detection rates that depend on stellar mass, constraints on gas mass and location in transitional disks, detection and characterization of 'snow lines', measurements of inner disk rims, and detections of inner disk asymmetries. I will also discuss how IR spectroscopy will remain relevant even with the emergence of facilities such as ALMA, as it allows us to connect the conditions in terrestrial-planet-forming regions with those in the cold outer reaches of disks, and to better construct a comprehensive understanding of the nature of protoplanetary disks.

Keywords. stars: circumstellar material, planets and satellites: formation, infrared: stars

1. Introduction

Protoplanetary disks are believed to be the birthplace of planets, and so offer us an opportunity to observe planet formation in progress, and to put our solar system in a larger astronomical context. In particular, studies of protoplanetary disks allow us to ask whether the solar system is a likely or unlikely outcome of the planet formation process. Protoplanetary disks form as the result of angular momentum conservation during the formation of young stars, and so planet formation is intimately linked with star formation. There are numerous star-forming regions within a distance of $\sim$500 pc from the sun and therefore ample opportunity to study such regions. However, investigations are hampered by the small angular separations and high contrast between protostar and disk. This is especially true for terrestrial and gas giant planet-forming regions in disks ($\lesssim$ 5 AU), which are $\lesssim 0.03''$ from their parent star at typical cloud distances ($\sim$ 150 pc).

Infrared molecular spectroscopy offers many advantages for overcoming the difficulties involved in observing planet forming regions in disks. Most importantly, in most of the infrared regime, the disk emission dominates over the stellar emission, and there is no need to block out the light of the central star. Also, since the disk has vertical and radial temperature gradients, particular molecular transitions nearly always originate in restricted regions of the disk. Thus by observing different transitions, one can obtain spatial information — even though the region is formally spatially unresolved. Finally, since the molecules are subject to Keplerian rotation, Doppler-induced line broadening can be used as a proxy for disk location, and provide further spatial information.

The use of infrared molecular lines to study protoplanetary disks can be roughly separated into the following categories. Firstly, line shapes and strengths can be used as tracers of disk physics, including volatile condensation/evaporation, photo-processes, grain growth and turbulence. Secondly, observations of multiple molecular abundances provide constraints for disk chemical models, which may ultimately help explain the great diversity of planetary bodies. Finally, resolved line shapes and spectro-astrometry provide a means to study disk structure on extremely small size scales. Until $\sim$ 2008, such studies were primarily limited to observations of rovibrational CO, but the field has since had a major breakthrough with the discovery that water and organics are commonly observed in the mid-infrared. In this review, I will discuss several molecular tracers in turn, highlighting how each has and can be used to probe particular aspects of protoplanetary disks and planet formation around low-mass (T Tauri; cTT) and mid-mass (Herbig Ae/Be; HAeBe) stars. I will particularly focus on recent observations and studies, and possible directions for future work.

2. CO rovibrational emission

2.1. *First detections*

The CO molecule is an excellent probe of inner protoplanetary disk atmospheres, due to its high abundance and high dissociation temperature. CO overtone emission was one of the first molecular transitions observed around young stars (Scoville *et al.* 1993) and found to be consistent with a disk origin (Scoville *et al.* 1983; Carr *et al.* 1989; Calvet *et al.* 1991). However, it seemed to be restricted to emission from young sources with strong outflows, with a further possible correlation with higher-mass (HAe/Be) stars (Najita *et al.* 2000). CO fundamental emission at $\sim 4.6\mu$m, on the other hand, was found to be a common feature of the spectra of disks around both T Tauri (Najita *et al.* 2000; Najita *et al.* 2003) and Herbig Ae/Be (Blake & Boogert 2004) stars. An analysis of lineshapes confirmed inner disks (radii < 0.1 AU for cTT disks and < 1 AU for HAeBe disks) as the dominant origin of the emission. These emission lines were strong evidence for a warm disk atmosphere, with a temperature inversion, likely caused by stellar irradiation.

2.2. *Transitional disks*

CO fundamental emission lines have been used to shed light on the properties and origin of transitional disks — disks with inner depletions of small dust grains (extending from the stellar surface to distances between a few and a few tens of AU; Andrews *et al.* 2011). The cause for the dust depletion is still actively being debated, with possible explanations including photoevaporation from the central star (Hollenbach *et al.* 1994; Alexander *et al.* 2007; Ercolano *et al.* 2008), grain growth, the presence of one or more stellar (Ireland & Kraus 2008) or sub-stellar (e.g., Rice *et al.* 2003) companions, or some combination of these effects. One key prediction of the photoevaporation hypothesis is that the inner disk should be cleared of gas as well as dust. The detection of CO fundamental emission from two canonical transitional disks, GM Aur and TW Hya (Rettig *et al.* 2004; Salyk *et al.* 2007), suggested the presence of gas in the dust-depleted regions, and lineshape

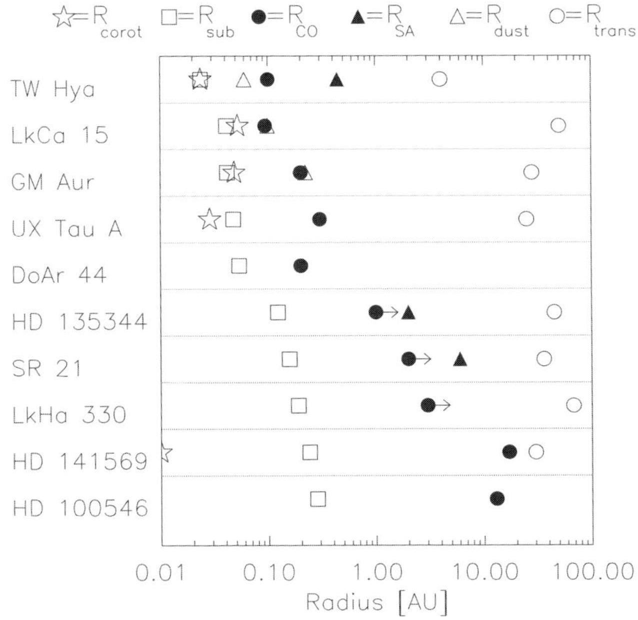

Figure 1. Comparison between $R_{\rm corot}$ (the corotation radius), $R_{\rm sub}$ (the dust sublimation radius), R_{CO} (the CO inner radius derived from lineshapes), $R_{\rm SA}$ (the CO radius derived from spectro-astrometry), $R_{\rm dust}$ (the dust inner radius derived from interferometry) and $R_{\rm trans}$ (the radius at which the disk becomes optically thick). R_{CO} for HD 141569 and HD 100546 are from Brittain *et al.* (2003) and Brittain *et al.* (2009), respectively; others are from Salyk *et al.* (2011b). ($R_{\rm SA}$ from Pontoppidan *et al.* 2008; $R_{\rm trans}$ from Marsh *et al.* 2002, Brown *et al.* 2009, Akeson *et al.* 2011 and Andrews *et al.* 2011). Figure from Salyk *et al.* (2011b).

analysis confirmed that the emission originated at radii < 1 AU (Salyk *et al.* 2007). These results have been further supported by spectro-astrometric measurements, which provide a firm upper-limit on the emission radii (Pontoppidan *et al.* 2008). These results show that some transitional disks were not formed by photoevaporation.

However, the full picture is more complex and interesting. Some transitional disks show CO fundamental emission originating inside the cleared region, but for others the emission originates at or near the wall at which the disk becomes optically thick, and for still others, there is no emission at all (Brittain *et al.* 2003; Brittain *et al.* 2009; Salyk *et al.* 2009). These results suggest that transitional disks are an inhomogeneous class of disks, with some formed by planetary companions, but others formed by stellar companions or photoevaporation (Salyk *et al.* 2009). A summary of CO emission radii and other observed radii in transitional disks can be seen in Figure 1.

2.3. *Inner disk structure*

Because the molecules in protoplanetary disks undergo (nearly) Keplerian motion, the shape of spectrally resolved emission lines reflect their radial origin. In particular, the emission line wings originate from the innermost part of the molecular disk. An analysis of lineshapes from T Tauri disks was consistent with emitting radii <0.1 AU, or similar to the stellar corotation radius (Najita *et al.* 2003) — the approximate radius at which the disk is expected to be truncated due to magnetic interaction with the host star. However, observations of CO fundamental emission from HAeBe disks hinted at a dependence of emitting location on stellar mass, as these emission lines had radial origins in the range

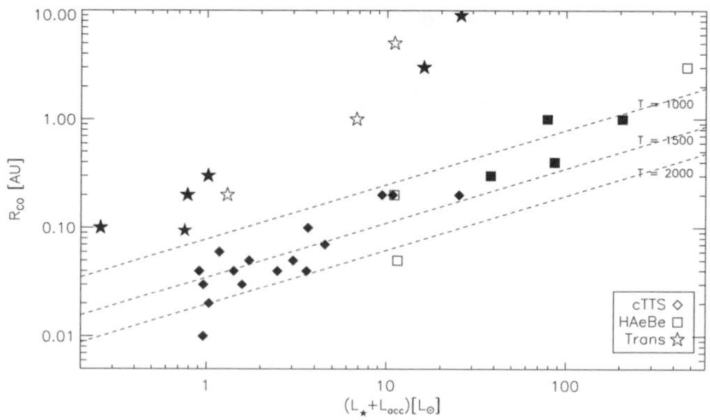

Figure 2. CO inner radius against luminosity. Diamonds are cTTS disks, squares are HAeBe disks and stars are transitional disks. Filled symbols have measured accretion rates, which are incorporated into $L_{\rm acc}$, while un-filled symbols have $L_{\rm acc} = 0$. Dashed lines show theoretical curves for dust sublimation radius as a function of luminosity. Figure from Salyk *et al.* (2011b).

of 0.5–1 AU (Blake & Boogert 2004). A more complete study, spanning several orders of magnitude in luminosity, now shows that the inner radius of the CO disk scales with stellar luminosity (see Figure 2). The observed trend is similar to a trend observed for the inner extent of dust emission, which is believed to be set by the dust sublimation temperature (see Dullemond & Monnier 2010 for a review). The CO inner radius might be expected to trace the dust inner radius, since CO can be easily photodissociated by stellar UV radiation (van Dishoeck & Black 1988). Interestingly, larger CO radii are derived for transitional disks. An exciting possibility is that this reflects dynamical truncation by a planet, though a more prosaic explanation — that the low dust densities in transitional disks allow the UV to dissociate a larger portion of the disks — is possible.

Inner disk structure is now being probed at an even deeper level with spectro-astrometry. With this technique, the centroid of an emission line is measured as a function of velocity, and can be compared with a disk model to measure emitting locations, disk inclinations and disk position angles. Along with firmly establishing an inner disk origin for CO fundamental emission from both classical and transitional protoplanetary disks, this technique has begun to reveal disk asymmetries at radii of a few AU or less, as well as possible disk warps (Pontoppidan *et al.* 2011).

2.4. *Disk winds*

Although a majority of CO fundamental emission from protoplanetary disks can be recreated with a Keplerian disk model (e.g. Najita *et al.* 2003; Salyk *et al.* 2011), a significant minority of sources show Lorentzian-like emission profiles that can not be produced with such a model (Bast *et al.* 2011; Pontoppidan *et al.* 2011). The lineshapes, measured at extremely high spectral resolution ($R \sim 10^5$) with VLT-CRIRES, require a significant low-velocity emission component (see Figure 3). The source of this component is still under debate, but a leading hypothesis is that the source is a low-velocity disk wind, as described in Figure 3 (Pontoppidan *et al.* 2011).

2.5. *Istopologues*

Observations of the isotopologues of CO allow for an investigation of fractionation processes. Taking advantage of the high spectral resolution of CRIRES and a special binary system geometry, Smith *et al.* (2009) used absorption spectroscopy to measure ^{12}C:^{13}C and C^{17}O:C^{18}O:C^{16}O ratios for the disk around VV CrA. They find mass-dependent

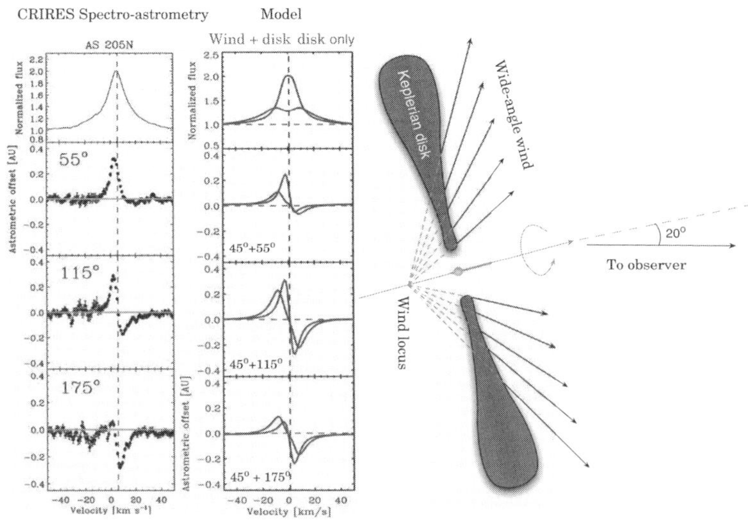

Figure 3. Sketch of a disk wind model that qualitatively reproduces the broad single-peaked CO line spectra and CRIRES spectroastrometry from highly accreting T Tauri stars — in this example, AS 205N, a T Tauri star in Ophiuchus. The left panels show the observed CO rovibrational ($v = 1 - 0$ around 4.7 μm) line spectrum and spectro-astrometry at 3 different slit position angles. The middle panels show two 2D radiative transfer models: one for a purely Keplerian disk, and one for a Keplerian disk in which a wide-angle molecular wind is launched from the disk surface. The right panel contains a sketch of the basic wind geometry.Figure from Pontoppidan *et al.* (2011).

deficits of $C^{17}O$ and $C^{18}O$, which they suggest is consistent with selective self-shielding — similar to what has been proposed to explain isotopic anomalies in the solar system. They also find a ^{12}C:^{13}C ratio nearly twice that of the expected Interstellar Medium value, and follow-up observations confirm significant cloud isotopic heterogeneity, which they attribute to ice-gas fractionation (Smith *et al.* 2010).

3. H_2

H_2 is believed to be the most abundant molecular component of circumstellar disks, but as a non-polar molecule, it remains difficult to detect via vibrational and rotational transitions. Rovibrational ($v = 1 - 0$ S(1)) emission has been detected by a number of authors (e.g. Weintraub *et al.* 2000; Bary *et al.* 2003; Ramsay Howat *et al.* 2007; Carmona *et al.* 2008), but spatially resolved observations suggest that the emission lines may be generated via outflow shocks (Beck *et al.* 2008), rather than in the quiescent disk. In a survey for pure rotational lines from disks around a variety of protostars (spanning classes I-III, with class II including both HAeBe, cTT and transitional disks), H_2 was detected only from class I sources, a single HAeBe source and a single class III source (Bitner *et al.* 2008). The lines are centered at the source velocity, and spectrally and spatially unresolved, suggesting a disk origin at 10-50 AU; however, the results of Beck *et al.* (2008) show that an outflow origin can not be ruled out. With few infrared detections of H_2, and much ambiguity about their origin, infrared H_2 lines have not turned out to be a very advantageous probe of protoplanetary disks.

4. H_2O and OH

4.1. *Detections and Origin*

Until recently, detections of water vapor in protoplanetary disks remained elusive. A single unambiguous detection of H_2O overtone emission was obtained from what turned out to be an atypical disk (Carr *et al.* 2004), probably representing a younger, more active stage of evolution than a classical T Tauri disk. In 2008, detections of water, OH and organics (the last to be discussed in §5) were detected from several more typical cTT disks with the Spitzer Space Telescope InfraRed Spectrograph (Spitzer-IRS; Carr & Najita 2008; Salyk *et al.* 2008). Ground-based high-resolution data confirmed an origin in the planet forming (few AU) region (Salyk *et al.* 2008). This exciting discovery opened up an entirely new avenue of discovery into the study of disks. In particular, as a volatile species actively involved in disk transport and chemistry, water is an ideal molecule for investigating the chemical state and evolution of disks, and condensation processes critical for the growth of giant planet cores.

As a result of a large follow-up survey of disks with Spitzer-IRS, water vapor emission is now known to be a common feature of mid-infrared spectra of cTTs disks (Pontoppidan *et al.* 2011; see also Figure 4). The richness of this dataset has only begun to be explored.

4.2. *Dependence on system parameters*

Analysis of the large database of Spitzer-IRS spectra immediately showed a striking trend. While detection rates for water and other molecules were greater than $\sim 40\%$ for cTT disks, these same emission lines were virtually absent from the spectra of HAeBe and transitional disks (Figure 4). Is this an observational bias, due to differing excitation conditions, or due to different chemistry between the two classes of sources? Radiative transfer models predict similar line/continuum ratios for water lines from both cTT and HeAeBe disks if the disks have similar structures and water abundances, so the effect does not appear to be an observational bias (Pontoppidan *et al.* 2010). The detectability of CO lines from HAeBe disks shows that line formation is possible, although CO line/continuum ratios are lower for HAeBe than cTT disks, pointing to at least some difference in line excitation conditions (Salyk *et al.* 2011a). Also, disk structure does affect the detectability of water lines amongst the cTT population, with detection rate anti-correlated with the 13–30 μm slope — a measure of disk flaring.

However, there is also intriguing evidence for chemical differences between cTT and HAeBe or transitional disks. Bethell & Bergin (2009) show that the observed water is likely produced in situ and that it can be photodissociated via UV radiation, but is capable of self-shielding. Whether or not the water survives depends on the UV flux, which can be an order of magnitude or more greater for HAeBe disks than cTT disks. Similarly, transitional disks have low dust densities, which leads to less dust shielding from UV, and at least one transitional disk, TW Hya, shows emission from highly excited OH molecules, which could be produced via photodissociation of water (Najita *et al.* 2010). Near-IR observations of water vibrational emission also point to chemical differences. While near-IR spectra of cTT disks show water/OH line/continuum ratios near 1, water is undetected in HAeBe disks, with water/OH line ratios at least an order of magnitude lower (Mandell *et al.* 2008; Fedele *et al.* 2011; Figure 5). Such a difference may have profound implications for the chemistry of planets produced in disks around stars with a variety of stellar masses.

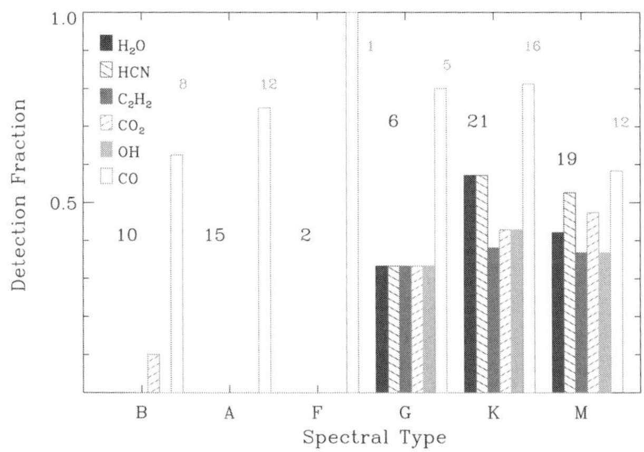

Figure 4. Detection rate as a function of spectral type for several infrared molecular tracers, with the number of each spectral type shown above. The detection rate drops dramatically for spectral types earlier than G/F, except for CO. Figure from Pontoppidan *et al.* (2010).

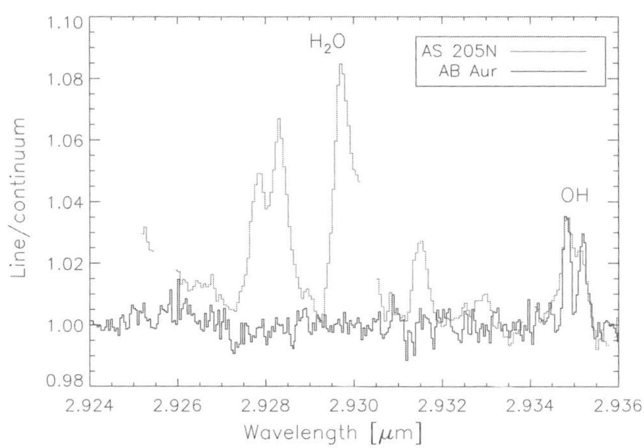

Figure 5. A comparison of water and OH emission lines observed with Keck-NIRSPEC from AS 205 N (Salyk *et al.* 2008) and AB Aur (Mandell *et al.* 2008), highlighting the radically different water/OH line ratios for cTT and HAeBe disks. The OH line/continuum ratio for AB Aur was scaled to match that of AS 205 N.

4.3. *Water photochemistry, condensation and transport*

One major goal for the study of water in protoplanetary disks is to understand its chemistry, condensation and transport, and their relation to the growth of planetesimals and the delivery of water to the proto-Earth. Infrared spectra — specifically, the relative strength of low- and high-excitation lines — can be used to determine the radial extent of water vapor, and hence the radius of water ice condensation (the so-called snow-line). This is a crucial parameter for models of planetary system formation, as ice may be a necessary ingredient to form cores large enough to contract a large gaseous envelope (as in the case of Jupiter or Saturn). A qualitative examination of the Spitzer spectra shows a relative dearth of flux from low-excitation lines, and points to a snow-line at only ~ 1 AU — well inside the canonical solar system snow-line of 5 AU (Meijerink *et al.* 2009). A more quantitative exploration of this measurement is in order, and will be greatly facilitated by the combination of Spitzer-IRS with far-IR spectra from the Herschel Space Observatory, since the relative line fluxes in the mid- and far-IR are very sensitive to the location of the snow-line.

5. Other Molecules

The detection of more complex molecules is necessary to test our understanding of disk chemistry and its relation to the great diversity of planets in our solar system. Sophisticated chemical models, incorporating dozens of molecular and atomic species (e.g. Markwick *et al.* 2002; Agundez *et al.* 2008; Willacy *et al.* 2009; Woitke *et al.* 2009) have been developed, but the observations necessary to test their conclusions are still in early stages. Infrared features of C_2H_2, HCN and CO_2 were first discovered in absorption by Lahuis *et al.* (2006) and provided evidence for high temperature chemistry. However, their detection required the uncommon geometry of a nearly edge-on disk. C_2H_2, HCN and CO_2 were finally found in emission by Carr & Najita, and more extensive surveys have confirmed that such emission is a common feature of the spectra of cTT disks (Pontoppidan *et al.* 2010; Carr *et al.* 2011).

Although a wide variety of molecular line strength ratios are observed in cTT disks (Salyk *et al.* 2011a; Carr & Najita 2011), the causes of the variation have remained elusive, as line ratios do not clearly correlate with system variables. One possible exception is the observation that the highest HCN/H_2O flux ratios are observed in the most massive disks, perhaps because planetesimals have formed, sequestered O, and created a region of Carbon-rich chemistry (Carr & Najita 2011). Additionally, Teske *et al.* (2011) note a correlation between HCN fluxes and both stellar accretion rates and X-ray luminosities. These results are all preliminary, and some care must be taken to distinguish cause and effect from covariance. Nevertheless, such studies hold much promise for the near future.

Chemical differences have also been observed between solar-like and cool stars, based on a survey with the Spitzer-IRS low-resolution modules (Pascucci *et al.* 2009). Cool stars display a significant underabundance of HCN relative to C_2H_2, which the authors propose originates from a difference in UV irradiation. Such a result is another demonstration that stellar mass can have a significant effect on disk chemistry, and possibly influence the types of planets that can be produced around any given star.

6. Future Directions

6.1. *In-depth analysis of existing datasets*

A number of rich datasets of molecular spectra have already been obtained, including datasets from Keck-NIRSPEC, VLT-CRIRES, Gemini-Michelle and Spitzer-IRS. These data have been qualitatively examined, and often studied with the use of LTE models. However, given the preponderance of recently acquired high-quality data, there is certainly great science to be derived from existing datasets. A few of the outstanding questions that may still be answered with existing datasets include: Is a low-velocity disk wind the correct explanation for the anomalous, strongly-peaked CO line profiles (Bast *et al.* 2011; Pontoppidan *et al.* 2011), and if so, what are its properties? Are molecular lines time variable, and if so, what does that tell us about the nature of the planet-forming environment, and the validity of quiescent disk models? What is the cause of variations in line ratios from organic molecules observed in Spitzer-IRS spectra? How do chemical models match up quantitatively with existing data, and what does this tell us about the relevance of various chemical processes?

6.2. *Improvements in spectral resolution and sensitivity*

Many improvements in spectral resolution, sensitivity and wavelength coverage will be made possible in the coming years via upgrades to existing facilities as well as new facilities and instruments. Combining higher spectral resolution data with the existing Spitzer archive will reduce many modeling degeneracies and make these data more powerful. Improved sensitivity (such as with the James Webb Space Telescope) will allow for the discovery of more molecular species, improving our understanding of chemical pathways and the planet formation environment. It will also allow studies to probe more distant clusters, and fainter stars, such as brown dwarfs, and to search for remnant gas in more evolved disks. Finally, high-resolution instruments with enhanced wavelength coverage will make it possible to conduct large surveys for undiscovered molecules, and achieve simultaneous observations of many species.

6.3. *The time domain*

Time variations of molecular emission features are essentially unstudied, and can easily be explored by combining new and existing datasets. One question upon which such studies would shed light is the origin and nature of the proposed low-velocity disk wind (Pontoppidan *et al.* 2011). Also, the presence of disk structure variability could have profound impacts for forming protoplanets. Finally, variability could be the result of warps and asymmetric structures caused by the interaction of the disk with embedded protoplanets.

6.4. *Spatial distribution of chemistry*

The combination of existing data with data from current and future facilities — notably the Herschel Space Observatory and ALMA — will allow us to spatially map the distribution of molecules in disks. Herschel is already allowing us to map the location of ice lines in disks with Spitzer-IRS water emission. ALMA will expand this possibility to other volatile species. Roughly speaking, near-IR spectra provide chemistry at ~ 1 AU or smaller, while Spitzer-IRS spectra provide a view of the chemistry at radii of a few AU. ALMA, on the other hand, will spatially resolve size scales of a few AU with its largest baseline, and with a sensitivity greatly exceeding that of all existing sub-millimeter or millimeter wave facilities. Thus, for the first time it will be possible to map the abundance of molecules throughout protoplanetary disks. This will bring us closer than ever before to understanding the origin of the physical and chemical diversity of planetary systems.

References

Agúndez, M., Cernicharo, J., & Goicoechea, J. R. 2008, *A&A*, 483, 831

Akeson, R. L., *et al.* 2011, *ApJ*, 728, 96

Alexander, R. D. & Armitage, P. J. 2007, *MNRAS*, 375, 500

Andrews, S. M., Wilner, D. J., Espaillat, C., Hughes, A. M., Dullemond, C. P., McClure, M. K., Qi, C., & Brown, J. M. 2011, *ApJ*, 732, 42

Bast, J. E., Brown, J. M., Herczeg, G. J., van Dishoeck, E. F., & Pontoppidan, K. M. 2011, *A&A*, 571, 119

Bethell, T. & Bergin, E. 2009, *Science*, 326, 1675

Bitner, M. A., *et al.* 2008, *ApJ*, 688, 1326

Blake, G. A. & Boogert, A. C. A. 2004, *ApJ*(Letters), 606, L73

Brittain, S. D., Rettig, T. W., Simon, T., Kulesa, C., DiSanti, M. A., & Dello Russo, N. 2003, *ApJ*, 588, 535

Brittain, S. D., Najita, J. R., & Carr, J. S. 2009, *ApJ*, 702, 85

Brown, J. M., Blake, G. A., Qi, C., Dullemond, C. P., Wilner, D. J., & Williams, J. P. 2009, *ApJ*, 704, 496

Calvet, N., Patino, A., Magris, G. C., & D'Alessio, P. 1991, *ApJ*, 380, 617

Carmona, A., *et al.* 2008, *A&A*, 477, 839

Carr, J. S. 1989, *ApJ*, 345, 522

Carr, J. S., Tokunaga, A. T., & Najita, J. 2004, *ApJ*, 603, 213

Carr, J. S. & Najita, J. R. 2008, *Science*, 319, 1504

Carr, J. S. & Najita, J. R. 2011, *ApJ*, 733, 102

Dullemond, C. P. & Monnier, J. D. 2010, *ARAA*, 48, 205

Ercolano, B., Drake, J. J., Raymond, J. C., & Clarke, C. C. 2008, *ApJ*, 688, 398

Fedele, D., Pascucci, I., Brittain, S., Kamp, I., Woitke, P., Williams, J. P., Dent, W. R. F., & Thi, W.-F. 2011, *ApJ*, 732, 106

Hollenbach, D., Johnstone, D., Lizano, S., & Shu, F. 1994, *ApJ*, 428, 654

Ireland, M. J. & Kraus, A. L. 2008, *ApJ*(Letters), 678, L59

Lahuis, F., *et al.* 2006, *ApJ*(Letters), 636, L145

Mandell, A. M., Mumma, M. J., Blake, G. A., Bonev, B. P., Villanueva, G. L., & Salyk, C. 2008, *ApJ*(Letters), 681, L25

Markwick, A. J., Ilgner, M., Millar, T. J., & Henning, T. 2002, *A&A*, 385, 632

Marsh, K. A., Silverstone, M. D., Becklin, E. E., Koerner, D. W., Werner, M. W., Weinberger, A. J., & Ressler, M. E. 2002, *ApJ*, 573, 425

Meijerink, R., Pontoppidan, K. M., Blake, G. A., Poelman, D. R., & Dullemond, C. P. 2009, *ApJ*, 704, 1471

Najita, J. R., Edwards, S., Basri, G., & Carr, J. 2000, Protostars and Planets IV, 457

Najita, J., Carr, J. S., & Mathieu, R. D. 2003, *ApJ*, 589, 931

Najita, J. R., Carr, J. S., Strom, S. E., Watson, D. M., Pascucci, I., Hollenbach, D., Gorti, U., & Keller, L. 2010, *ApJ*, 712, 274

Pascucci, I., Apai, D., Luhman, K., Henning, T., Bouwman, J., Meyer, M. R., Lahuis, F., & Natta, A. 2009, *ApJ*, 696, 143

Pontoppidan, K. M., Blake, G. A., van Dishoeck, E. F., Smette, A., Ireland, M. J., & Brown, J. 2008, *ApJ*, 684, 1323

Pontoppidan, K. M., Salyk, C., Blake, G. A., Meijerink, R., Carr, J. S., & Najita, J. 2010, *ApJ*, 720, 887

Pontoppidan, K. M., Salyk, C., Blake, G. A., & Käufl, H. U. 2010, *ApJ*(Letters), 722, L173

Pontoppidan, K. M., Blake, G. A., & Smette, A. 2011, *ApJ*, 733, 84

Ramsay Howat, S. K. & Greaves, J. S. 2007, *MNRAS*, 379, 1658

Rettig, T. W., Haywood, J., Simon, T., Brittain, S. D., & Gibb, E. 2004, *ApJ*, 616, L163

Rice, W. K. M., Wood, K., Armitage, P. J., Whitney, B. A., & Bjorkman, J. E. 2003, *MNRAS*, 342, 79

Salyk, C., Blake, G. A., Boogert, A. C. A., & Brown, J. M. 2007, *ApJ*(Letters), 655, L105

Salyk, C., Pontoppidan, K. M., Blake, G. A., Lahuis, F., van Dishoeck, E. F., & Evans, N. J., II 2008, *ApJ*(Letters), 676, L49

Salyk, C., Blake, G. A., Boogert, A. C. A., & Brown, J. M. 2009, *ApJ*, 699, 330

Salyk, C., Pontoppidan, K. M., Blake, G. A., Najita, J. R., & Carr, J. S. 2011, *ApJ*, 731, 130

Salyk, C., Blake, G. A., Boogert, A. C. A., & Brown, J. M., 2011, submitted

Scoville, N., Kleinmann, S. G., Hall, D. N. B., & Ridgway, S. T. 1983, *ApJ*, 275, 201

Smith, R. L., Pontoppidan, K. M., Young, E. D., Morris, M. R., & van Dishoeck, E. F. 2009, *ApJ*, 701, 163

Smith, R. L., Pontoppidan, K. M., Young, E. D., & Morris, M. R. 2010, Disks, Meteorites, Planetesimals, 6012

Teske, J. K., Najita, J. R., Carr, J. S., Pascucci, I., Apai, D., & Henning, T. 2011, arXiv:1104.0249

van Dishoeck, E. F. & Black, J. H. 1988, *ApJ*, 334, 771

Willacy, K. & Woods, P. M. 2009, *ApJ*, 703, 479

Woitke, P., Kamp, I., & Thi, W.-F. 2009, *A&A*, 501, 383

Discussion

LYONS: For observations from ~1 AU, how high above the midplane is being probed?

SALYK: The emitting region is very small, and most likely limited by the dust opacity. This brings to light the important point that most infrared observations probe only the upper atmosphere and do not necessarily represent the bulk chemistry. Instead, observations must be considered in the context of a full disk model.

LAHUIS: Besides being chemically sensitive, the mid-IR features are also (very) sensitive to perturbations in the inner disk. This is shown by EX-Lupi and IRS 46 C_2H_2 and HCN being very sensitive — more so than H_2O. This makes the interpretation of the mid-IR spectra even more interesting.

SALYK: Sure.

SEMENOV: What makes the lines non-Keplerian? Is it turbulence or asymmetries in inner disk regions (e.g. clumpiness, spiral arms)? Many young stars show rapid luminosity variations at optical wavelengths (~ hours – weeks timescales).

SALYK: The leading hypothesis right now is a slow molecular disk wind, although the current model is not physical, but rather a simple parametric model. The observational constraint is that the model must produce a large flux with low velocities, centered at the stellar velocity. We are looking into this further, including pursuing observations of variability.

The Molecular Universe
Proceedings IAU Symposium No. 280, 2011
José Cernicharo & Rafael Bachiller, eds.

doi:10.1017/S1743921311024938

Chemical History of Molecules in Circumstellar Disks

Ruud Visser[1,2], Ewine F. van Dishoeck[2,3] and Steven D. Doty[4]

[1]Department of Astronomy, University of Michigan, 500 Church Street, Ann Arbor, MI 48109-1042, USA
email: visserr@umich.edu

[2]Leiden Observatory, Leiden University, P.O. Box 9513, 2300 RA Leiden, the Netherlands

[3]Max-Planck-Institut für Extraterrestrische Physik, Giessenbachstrasse 1, 85748 Garching, Germany

[4]Department of Physics and Astronomy, Denison University, Granville, OH 43023, USA

Abstract. The chemical composition of a protoplanetary disk is determined not only by in situ chemical processes during the disk phase, but also by the history of the gas and dust before it accreted from the natal envelope. In order to understand the disk's chemical composition at the time of planet formation, especially in the midplane, one has to go back in time and retrace the chemistry to the molecular cloud that collapsed to form the disk and the central star. Here we present a new astrochemical model that aims to do just that. The model follows the core collapse and disk formation in two dimensions, which turns out to be a critical upgrade over older collapse models. We predict chemical stratification in the disk due to different physical conditions encountered along different streamlines. We argue that the disk-envelope accretion shock does not play a significant role for the material in the disk at the end of the collapse phase. Finally, our model suggests that complex organic species are formed on the grain surfaces at temperatures of 20 to 40 K, rather than in the gas phase in the $T > 100$ K hot corino.

Keywords. astrochemistry – stars: formation – circumstellar matter – planetary systems: protoplanetary disks – molecular processes

1. Introduction

We live on a world rich in complex organic molecules, all of which ultimately derive from a cold pre-stellar core with a much simpler chemical composition. How did the chemistry in the solar system evolve from that original simple state to its current complex state? This chapter focuses on one particular part of the entire evolutionary track: the transition from a pre-stellar core to a protoplanetary disk.

The chemical evolution in a collapsing envelope has been studied extensively with one-dimensional (1D) spherical models (e.g., Bergin & Langer 1997; Rodgers & Charnley 2003; Lee *et al.* 2004; Aikawa *et al.* 2008), but such models cannot properly describe the circumstellar disk. We have recently constructed a new collapse model that describes the envelope and the disk in two dimensions (2D) as an axisymmetric structure (Visser *et al.* 2009a, 2011; Visser & Dullemond 2010). By taking the vertical extent of the disk into account, we find that accretion from the envelope onto the disk largely takes place near the cold outer edge, rather than near the warmer inner parts as suggested by earlier models. The lower temperatures in our scenario allow for a larger fraction of pristine material to enter the disk. The 2D models also show strong evidence of layered accretion in the disk: new material arriving from the envelope always accretes on top of the older material already in the disk (Fig. 1). Another improvement in our model is that we calculate the dust temperatures throughout the disk and the envelope with a full radiative

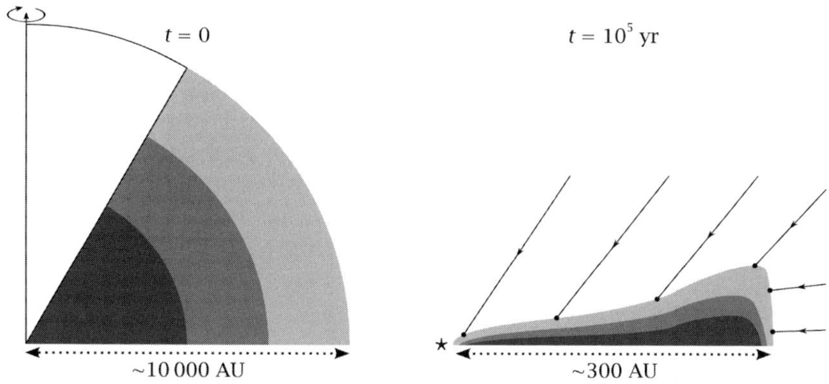

Figure 1. Schematic view of layered accretion. The outer parts of the original cloud core (left) end up as the surface layers of the circumstellar disk (right). The white part of the core is roughly the part that disappears into the outflow. The arrows illustrate the flow of material onto the disk. If the vertical extent of the disk is not taken into account properly, as happened in some older models, such streamlines would extend further in and the radii at which material accretes would be underestimated.

transfer code. The approximate temperature calculations used by some older models are not accurate enough for a reliable investigation of the chemistry.

We provide here a simple summary of our model (Sect. 2) and highlight some of the first results (Sect. 3). In particular, we focus on the chemical history of H_2O and other simple molecules as material flows from the pre-stellar core to the planet-forming disk. The discussion (Sect. 4) addresses two open questions in star formation: how strong is the disk-envelope accretion shock, and do hot corinos really exist?

2. Model description

In order to follow the chemical evolution from a pre-stellar core to a circumstellar disk, we need a physical framework to provide densities, temperatures and other quantities that may affect the chemistry. We choose to use a fast semi-analytical model for the density and velocity profiles and a full continuum radiative transfer code for the dust temperature and the radiation field. The dust temperature is particularly important for the gas-grain interactions. Hydrodynamical codes often use an approximation of the dust temperature and are therefore less suited to study the chemical evolution.

The full model procedure is explained in detail in Visser *et al.* (2009a), Visser & Dullemond (2010) and Visser *et al.* (2011). We summarise the model in this section by way of the step-wise description from Fig. 2.

2.1. *Initial conditions*

The physical and chemical evolution of the collapsing envelope depends strongly on the initial conditions. The three most prominent parameters in our model are the mass of the cloud core before the onset of collapse (M_0), the isothermal sound speed (c_s), and the core's solid-body rotation rate (Ω_0). A higher rotation rate leads to a larger and more massive disk, as do a lower sound speed and a higher core mass (Visser *et al.* 2009a). Lowering the sound speed or increasing the core mass both tend to produce a colder disk, with larger fractions of ices relative to gas-phase species. Increasing the solid-body rotation rate also leads to a colder disk. The core mass and the sound speed together determine how long it takes for the entire core to accrete onto the star and disk: $t_{\rm acc} \equiv M_0/\dot{M} \approx M_0 G/c_s^3$ (Shu 1977).

2.2. *Density and velocity*

We follow the inside-out solution of Shu (1977) for a gravitationally collapsing spherical envelope. However, because of the core's rotation, spherical symmetry is lost at the onset of collapse. The effects of rotation can be treated as a perturbation of the pure spherical solution (Cassen & Moosman 1981; Terebey *et al.* 1984). In the perturbed solution, matter no longer streams in radially, but it is deflected towards the gravitational midplane and builds up a circumstellar disk. After the disk is first formed, it evolves by ongoing accretion from the envelope and by viscous spreading to conserve angular momentum (Shakura & Sunyaev 1973; Lynden-Bell & Pringle 1974; Dullemond *et al.* 2006).

The size of the disk is initially set by the centrifugal radius ($R_c \approx (1/16)c_s^3 t^3 \Omega_0$, with t the time since the onset of collapse), because the streamlines from the envelope only hit the midplane inside of that point. However, viscous spreading can push the outer edge of the disk to several times R_c (Visser *et al.* 2009a; van Weeren *et al.* 2009). We adopt a Gaussian profile for the vertical density structure of the disk, from which we also derive the vertical velocity profile. The boundary between the disk and the envelope is conveniently defined as the surface where the thermal pressure from the disk equals the ram pressure from the infalling envelope (Visser & Dullemond 2010).

The analytical collapse solution of Terebey *et al.* (1984) does not include a bipolar outflow, so we have to put one in as an ad-hoc addition. Based on observations and theoretical predictions (Velusamy & Langer 1998; Cantó *et al.* 2008), we adopt curved cavity walls whose outflow angle grows with time; specifically, at a height z above the midplane, the size $R_{\rm cav}$ of the cavity is proportional to $t^2 z^{2/3}$ (Visser *et al.* 2011).

2.3. *Temperature and radiation field*

The chemical evolution depends strongly on the temperature, so we employ a full radiative transfer method instead of an analytical approximation for step 3 of our model. The only heating source taken into account is the central star, whose luminosity and effective temperature evolve according to Adams & Shu (1986) and Young & Evans (2005). Taking the density profiles from step 2, we compute the 2D dust temperature profiles at a range of time steps with RADMC (Dullemond & Dominik 2004). As additional output, RADMC also provides the intensity of the radiation field as function of wavelength at each point in the disk and remnant envelope, as required for the photon-driven reactions in our chemical network.

We set the gas temperature equal to the dust temperature throughout the disk and the envelope, even though this is a poor assumption in the surface of the disk and the inner parts of the envelope (Kamp & Dullemond 2004; Jonkheid *et al.* 2004). Planned updates to the model include a separate treatment of the dust and gas temperatures. The effect of the accretion shock on the temperature profiles is discussed in Sect. 4.

2.4. *Infall trajectories*

We follow the chemical evolution as material streams from the natal core towards the circumstellar disk, so we have to solve the chemistry in a Lagrangian frame. The semi-analytical nature of the density and velocity solutions offers two ways to approach the problem. The first way, followed in Visser *et al.* (2009a), consists of populating the core with a large number of points (at least 10 000) and calculating the infall streamlines in the forward direction. Alternatively, as done in Visser *et al.* (2011), we can select points of interest in the disk at the end of the collapse phase and calculate the trajectories backwards in time to their point of origin in the core. In both cases, we end up with a set of trajectories (Cartesian or polar coordinates as function of time) with corresponding physical conditions (density, temperature, UV field and extinction as function of position

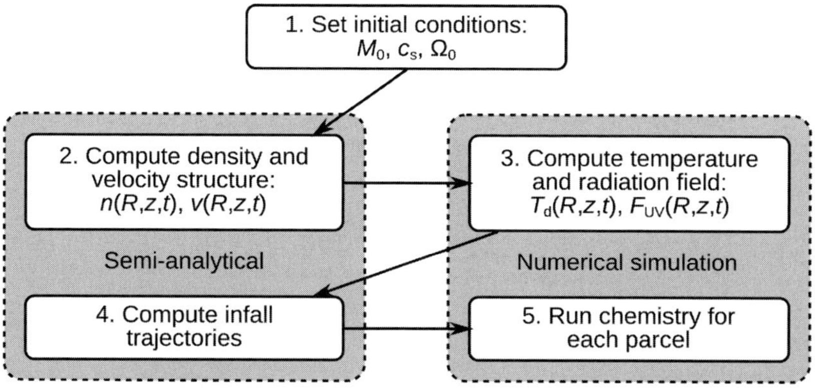

Figure 2. Step-wise summary of our 2D axisymmetric collapse model. Steps 2 and 4 are semi–analytical, while steps 3 and 5 consist of detailed numerical simulations. (Figure adapted from Visser *et al.* (2011).)

and time). These are used as input for the chemistry code. Each trajectory is treated in isolation, i.e., the material streaming along one trajectory is assumed not to interact with material streaming along other trajectories.

2.5. *Chemistry*

The basis of our chemical network is the UMIST06 database (Woodall *et al.* 2007) as modified by Bruderer *et al.* (2009), except that X-ray chemistry is not included. Full details are provided in Visser *et al.* (2011). The network includes freeze-out and evaporation (both thermal and photon-driven), but we limit the grain-surface chemistry to a handful of simple hydrogenation reactions: atomic H, C, N, O and S can react with incoming H to form H_2, CH_4, NH_3, H_2O and H_2S (Black & van Dishoeck 1987; Bergin & Langer 1997; Hollenbach *et al.* 2009). Other potentially important grain-surface reactions, such as the formation of H_2O out of O_2 (Ioppolo *et al.* 2008) or the formation of H_2CO and CH_3OH out of CO (Fuchs *et al.* 2009), have not yet been incorporated. The gas-phase network includes photodissociation and photoionisation reactions (accounting specifically for self-shielding of CO; Visser *et al.* 2009b), which dominate the chemistry close to the outflow wall and in the surface layers of the disk.

In order to set the chemical composition at the onset of collapse ($t = 0$), we evolve the initially atomic gas for a period of 1 Myr at $n_H = 8 \times 10^4$ cm^{-3} and $T_{gas} = T_{dust} = 10$ K. The resulting solid and gas-phase abundances are consistent with those observed in prestellar cores (e.g., di Francesco *et al.* 2007). They form the initial conditions for the chemical evolution along each infall trajectory.

3. Results

3.1. *Key chemical trends*

This section highlights a few key results from the standard model of Visser *et al.* (2009a, 2011), which has initial conditions $M_0 = 1.0\ M_\odot$, $c_s = 0.26$ km s^{-1} and $\Omega_0 = 10^{-14}$ s^{-1}. The chemistry is run for the full duration of the envelope accretion phase ($t_{acc} = 2.52 \times 10^5$ yr), following the 1 Myr pre-collapse phase.

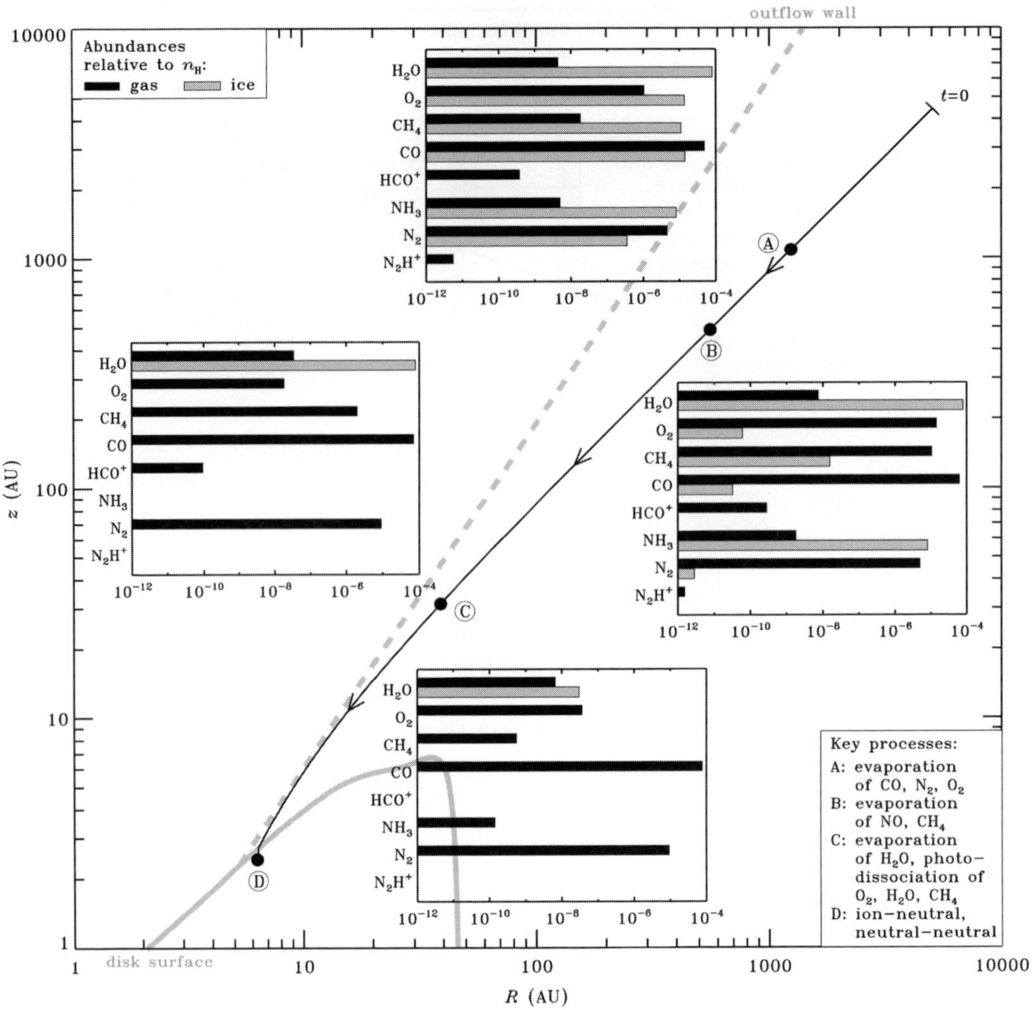

Figure 3. Schematic view of the chemical evolution along one infall trajectory (solid black curve). The solid and dashed grey lines denote the surface of the disk and the outflow wall, both at $t = t_{\rm acc}$. The abundances of some key species (black bars: gas; grey bars: ice) are plotted at four points along the trajectory (A: top panel; B: right panel; C: left panel; D: bottom panel). The key processes governing the overall chemistry at each point are listed in the bottom right. (Figure adapted from Visser *et al.* (2011).)

We first focus on the chemical evolution along one particular trajectory, starting near the edge of the original core ($R_{\rm i} = 5050$ AU, $z_{\rm i} = 4420$ AU) and ending at $t = t_{\rm acc}$ in the inner part of the disk, about 0.2 AU below the surface ($R_{\rm f} = 6.3$ AU, $z_{\rm f} = 2.4$ AU). The chemistry along this and other trajectories is governed by changes in the physical conditions. Figure 3 shows how the solid and gas-phase abundances of eight important species change along the streamline. We identify four chemical events:

• At point A, the temperature reaches a value of about 20 K and simple ices like CO, N_2 and O_2 begin to evaporate.

• The temperature increases further to 30–40 K at point B, driving CH_4 and NO off the grains.

• Moving in further, the dust is heated to the evaporation temperature of H_2O (100 K). This close to the outflow cavity wall, the gas is exposed to the stellar UV field, dissociating

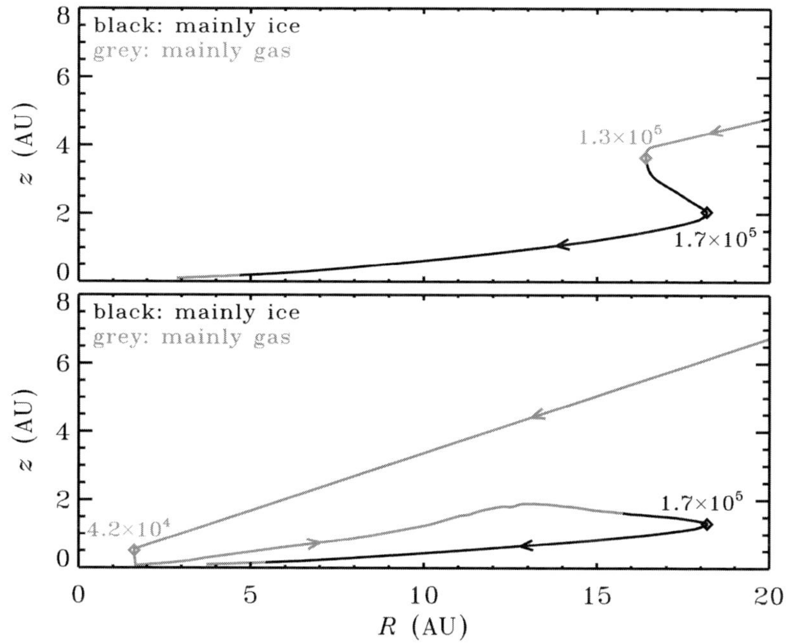

Figure 4. Examples of qualitatively different infall trajectories terminating at nearly the same point at the end of the accretion phase: $R = 2.9$ AU (top) and 3.7 AU (bottom). The black and grey portions of the trajectories indicate where most of the water is frozen out or in the gas phase. The points where the trajectories reverse direction are marked with the time in years.

the evaporating H_2O and other previously evaporated species. Our model does not include an excess UV field for the young star, so there are not enough high-energy photons to dissociate CO.

• Finally, at point D, the high density (10^{10}–10^{11} cm^{-3}) allows some of the dissociated species to reform in ion-neutral and neutral-neutral reactions.

3.2. *Chemical stratification*

Different streamlines exhibit different physical conditions. In order to fully understand the chemical evolution from every point in the pre-stellar core to every point in the circumstellar disk, we have to investigate more than just the single trajectory from Fig. 3. In this regard, it is interesting to look at where material from different parts of the core ends up in the disk.

The hydrodynamical simulations of Brinch *et al.* (2008) and our semi-analytical model (Visser *et al.* 2009a) both show the envelope to accrete onto the disk in a layered fashion (Fig. 1). Material from near the centre of the pre-stellar core is the first to become part of the disk, and material from larger radii continuously piles up on top. The thickness of the disk is small relative to its radial size (typical scale heights are 0.1–0.2R), but the outer edge is high enough that a substantial fraction of the infalling material hits there rather than on the surface (Fig. 1; see also Fig. 3 in Visser & Dullemond 2010). The infalling material has to compete with the outwardly expanding disk, resulting in some material being pushed up and over the outer edge in the hydrodynamical simulations (see also van Weeren *et al.* 2009).

Individual streamlines can exhibit large qualitative differences despite terminating at nearly the same point in the disk. Figure 4 shows two such streamlines, with termini at

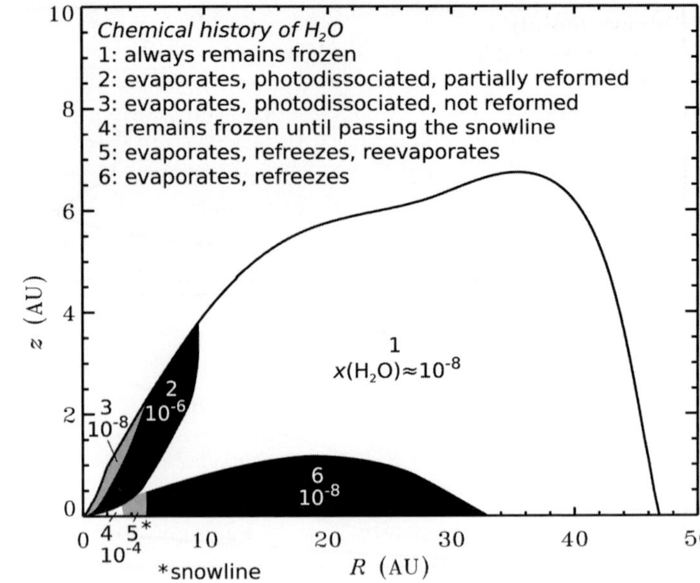

Figure 5. Schematic view of the history of H_2O gas and ice towards different parts of the disk. The typical gas-phase abundance is indicated for each zone (zones 4 and 5 both have 10^{-4}). The position of the snowline ($R = 5$ AU, between zones 5 and 6) is marked with an asterisk. Note the disproportionality of the R and z axes. (Figure adapted from Visser *et al.* (2011).)

2.9 and 3.7 AU from the star. The streamline in the bottom panel originates at $r = 1100$ AU from the centre of the pre-stellar core. It reaches the disk at a fairly early time (4.2×10^4 yr after the onset of collapse), when the disk is only a few AU large. The parcel of gas following this streamline gets swept up in the outward viscous expansion until, at $t = 1.7 \times 10^5$ yr, enough mass has accreted at larger radii that it gets pushed back in again. The streamline in the bottom panel originates further out in the natal core, at $r = 3500$ AU, and therefore accretes at a later time, $t = 1.3 \times 10^5$ yr, when the disk has grown to about 20 AU. It experiences a brief period of outward motion before moving in further.

In terms of the chemical evolution of the material flowing along these two streamlines, the key difference lies in the conditions encountered before $t = 10^5$ yr. The bottom streamline passes much closer to the star than does the other one (1.6 versus 16 AU); it is heated to more than 500 K, whereas the top parcel initially does not get warmer than 120 K. The latter is enough to drive H_2O and other non-refractory species off the grains (as indicated by the grey portion of the streamline up to 1.3×10^5 yr), but not enough to power the kind of high-temperature gas-phase chemistry that is possible at 500 K.

The different types of streamlines give rise to different chemical zones in the disk. This effect of "chemical stratification" is visualised in Fig. 5, which traces the history of H_2O gas and ice towards various parts of the disk. (Similar figures for CH_4 and NH_3 appear in Visser *et al.* (2011).) The streamlines terminating in zone 1 never encounter temperatures above 100 K, so H_2O always remains frozen on the grains. Zone 2 contains the terminus of the streamline from Fig. 3; some H_2O ends up here, but most of it is dissociated by the stellar radiation field. The same happens for material ending up in zone 3, except that no H_2O reforms at all after being destroyed. Parcels ending up in zone 4 do not encounter a strong radiation field; H_2O in these parcels stays on the grains until passing the snowline at ~5 AU. Finally, zones 5 and 6 consist largely of material that accreted at early times, following trajectories like the one in the bottom panel of Fig. 4. H_2O in these parcels undergoes a freeze-thaw cycle: it evaporates during the initial inwards part of the trajectory and freezes out again during the outwards leg. Parcels ending up in zone 5 undergo another evaporation event as they pass the snowline; material in zone 6 lies beyond the snowline, so H_2O remains frozen the rest of the time.

The chemical stratification comes with one major caveat: mixing. In our model, there is no exchange of material between adjacent parcels. In reality, however, turbulent mixing is likely to change the sizes and locations of the chemical zones, and to make the borders between the zones more diffuse (e.g., Semenov *et al.* 2006). Spectroscopic observations at 5–10 AU resolution, as possible with ALMA, are required to determine to what extent chemical stratification occurs in actual disks.

4. Discussion

4.1. *How strong is the disk-envelope accretion shock?*

When the supersonically infalling material from the envelope hits the nearly static disk, it passes through a J-type shock. However, this shock may not play as large a role as commonly believed. Neufeld & Hollenbach (1994) showed that the strength of the shock decreases with decreasing pre-shock density and decreasing shock velocity. For a given radius, the density and velocity decrease as the collapse proceeds; for a given point in time, both decrease with radius (Cassen & Moosman 1981; Terebey *et al.* 1984).

During the second half of the collapse phase (from $t = 0.5t_{\rm acc}$ to $t_{\rm acc}$), at least 80% of the infalling material accretes onto the outer half of the disk. (The exact number depends on the initial conditions.) At the end of the collapse phase, the bulk of the disk consists of material that accreted at fairly late stages and large radii. Most of the material that accreted at early times or small radii gets pushed into the star. Hence, we expect that the bulk of the gas and the dust in a T Tauri or Herbig Ae/Be disk did not pass through a strong accretion shock. Quantitatively, we showed in Visser *et al.* (2009a) that the shock has a smaller effect on the temperature history of the disk material than does the radiation from the young star. In terms of the chemical evolution, this means the desorption of H_2O and organic species is likely unaffected by the accretion shock. Species with lower binding energies, however, may be more sensitive.

4.2. *Do hot corinos exist?*

Observations of complex organic species like methyl formate ($HCOOCH_3$) and dimethyl ether (CH_3OCH_3) in low-mass protostars have been linked to the high-temperature (> 100 K) inner envelope, also known as the hot core or hot corino (Ceccarelli 2004). One possible formation pathway is powered by the evaporation of H_2O and other non-refractory species at 100 K, followed by a chain of high-temperature gas-phase reactions to transform simple carbon-bearing molecules into second-generation complex organics (Herbst & van Dishoeck 2009; see also Charnley *et al.* 1992). Alternatively, the organics may be formed abundantly on the grain surfaces through mild photochemistry at 20–40 K (Garrod & Herbst 2006; Garrod *et al.* 2008). Which scenario is more likely?

The first formation route requires the gas to spend at least several 10^3 yr at temperatures above 100 K (Charnley *et al.* 1992). In our model, however, the material in the inner envelope is essentially in freefall towards the star or the inner disk, and it only spends a few 100 yr at the required temperatures (Visser *et al.* 2009a; see also Schöier *et al.* 2002). The grain-surface pathway takes several 10^4 yr at 20–40 K (Garrod & Herbst

2006). Such temperatures are attainable both in the envelope before the material reaches the hot inner part and inside the disk (e.g., Fig. 12 in Visser *et al.* 2009a). Once formed on the grains, these first-generation organics can be returned to the gas phase by thermal evaporation in the hot corino, the hot inner disk, or the surface layers of the outer disk. They can also be liberated by non-thermal processes such as photodesorption in the disk and shocks in the outflow walls. In any scenario, the mass of the hot corino is quite limited: the amount of material wedged between the disk and the outflow cavity is at most a few per cent of the amount of material inside the disk. Hence, even if the hot corino is rich in organics, there still would not be enough of them to explain the observations.

Based on the timescales and the mass, we conclude that the observed second-generation complex organics are not associated with the hot core or hot corino. The circumstellar disk and outflow walls offer a more likely explanation: they have a larger mass reservoir and are dynamically stable enough for the observed organics to form abundantly.

5. Conclusions

We have constructed a two-dimensional axisymmetric model to simulate the collapse of a pre-stellar core into a young star and its surrounding disk. We use the model to study the chemical evolution of the gas and the dust streaming from the envelope to the disk. Comparing the streamlines to what happens in one-dimensional models, we conclude that it is imperative to study core collapse and disk formation in 2D. Our model predicts chemical stratification in the disk due to different physical conditions (density, temperature, radiation field) encountered along different streamlines. We argue that the disk-envelope accretion shock does not play a significant role for the material in the disk at the end of the collapse phase. Finally, our model suggests that complex organic species are formed on the grain surfaces at temperatures of 20 to 40 K, rather than in the gas phase in the $T > 100$ K hot corino.

References

Adams, F. C. & Shu, F. H. 1986, *ApJ*, 308, 836

Aikawa, Y., Wakelam, V., Garrod, R. T., & Herbst, E. 2008, *ApJ*, 674, 984

Bergin, E. A. & Langer, W. D. 1997, *ApJ*, 486, 316

Black, J. H. & van Dishoeck, E. F. 1987, *ApJ*, 322, 412

Brinch, C., Hogerheijde, M. R., & Richling, S. 2008, *A&A*, 489, 607

Bruderer, S., Doty, S. D., & Benz, A. O. 2009, *ApJS*, 183, 179

Cantó, J., Raga, A. C., & Williams, D. A. 2008, Rev. Mex. Astron. Astr., 44, 293

Cassen, P. & Moosman, A. 1981, *Icarus*, 48, 353

Ceccarelli, C. 2004, in *ASP Conf. Ser. 323: Star Formation in the Interstellar Medium: In Honor of David Hollenbach*, ed. D. Johnstone, F.C. Adams, D.N.C. Lin, D.A. Neufeld, & E.C. Ostriker (San Francisco: ASP), 195

di Francesco, J., Evans, II, N. J., & Caselli, *et al.* 2007, *Protostars & Planets V*, 17

Dullemond, C. P. & Dominik, C. 2004, *A&A*, 417, 159

Dullemond, C. P., Apai, D., & Walch, S. 2006, *ApJ*, 640, L67

Fuchs, G. W., Cuppen, H. M., & Ioppolo, S., *et al.* 2009, *A&A*, 505, 629

Garrod, R. T. & Herbst, E. 2006, *A&A*, 457, 927

Garrod, R. T., Weaver, S. L. W.., & Herbst, E. 2008, *ApJ*, 682, 283

Herbst, E. & van Dishoeck, E. F. 2009, *ARA&A*, 47, 427

Hollenbach, D., Kaufman, M. J., Bergin, E. A., & Melnick, G. J. 2009, *ApJ*, 690, 1497

Ioppolo, S., Cuppen, H. M., Romanzin, C., van Dishoeck, E. F., & Linnartz, H. 2008, *ApJ*, 686, 1474
Jonkheid, B., Faas, F. G. A.., van Zadelhoff, G.-J. & van Dishoeck, E. F. 2004, *A&A*, 428, 511
Kamp, I. & Dullemond, C. P. 2004, *ApJ*, 615, 991
Lee, J.-E., Bergin, E. A., & Evans, II, N. J. 2004, *ApJ*, 617, 360
Lynden-Bell, D. & Pringle, J. E. 1974, *MNRAS*, 168, 603
Neufeld, D. A. & Hollenbach, D. J. 1994, *ApJ*, 428, 170
Rodgers, S. D. & Charnley, S. B. 2003, *ApJ*, 585, 355
Schöier, F. L., Jørgensen, J. K., van Dishoeck, E. F., & Blake, G. A. 2002, *A&A*, 390, 1001
Semenov, D., Wiebe, D., & Henning, T. 2006, *ApJ*, 647, L57
Shu, F. H. 1977, *ApJ*, 214, 488
Terebey, S., Shu, F. H., & Cassen, P. 1984, *ApJ*, 286, 529
van Weeren, R. J., Brinch, C., & Hogerheijde, M. R. 2009, *A&A*, 497, 773
Velusamy, T. & Langer, W. D. 1998, *Nature*, 392, 685
Visser, R., van Dishoeck, E. F., Doty, S. D., & Dullemond, C. P. 2009a, *A&A*, 495, 881
Visser, R., van Dishoeck, E. F., & Black, J. H. 2009b, *A&A*, 503, 323
Visser, R. & Dullemond, C. P. 2010, *A&A*, 519, A28
Visser, R., Doty, S. D., & van Dishoeck, E. F. 2011, *A&A*, submitted
Woodall, J., Agúndez, M., Markwick-Kemper, A. J., & Millar, T. J. 2007, *A&A*, 466, 1197
Young, C. H. & Evans, II, N. J. 2005, *ApJ*, 627, 293

Discussion

YAMAMOTO: How do you assume the initial conditions in your model? We now know that various complex molecules exist in protostellar cores. How are they brought into the inner disk?

VISSER: We have a static pre-collapse phase of 10^6 yr to produce typical pre-stellar core abundances. Since we do not have any grain-surface chemistry other than simple hydrogenation of O, C, N and S, complex organics are not formed abundantly. However, based on the models by Garrod & Herbst (2006) and Garrod *et al.* (2008), which show that complex organics can be formed in a few 10^4 yr at $T \approx 40$ K, we predict that complex organics are formed abundantly inside the disk. We have not considered warm carbon-chain chemistry.

JOHNSTONE: How important are accretion shocks for the chemical evolution of infalling gas?

VISSER: The accretion shock is not important for the bulk of the material in the disk at the end of the accretion phase, because that material accreted at late times and large radii, and therefore did not encounter a strong shock (Neufeld & Hollenbach 1994). Material that accretes earlier encounters a stronger shock, but this material disappears into the star while the rest of the disk is formed. We refer to Sect. 4.1 for more details.

AIKAWA: What is the cause of the vortices that appear in the animation? Do they also appear in your 2D analytical model?

VISSER: The vortices are due to material piling up at the outer edge of the disk and being pushed up and over the outer edge in the 2D simulation. They do not appear in the semi-analytical model.

IRVINE: Evidence suggests that most stars, including the Sun, did not form in isolation (the Shu model). Have you considered how this might affect your results?

VISSER: No, we have not considered this in any detail. We expect that many of the results hold for circumbinary envelopes and disks, as long as the interbinary distance is smaller than 10 AU or so. If the protostar in question forms in the vicinity of an O or B star, the strong external radiation field would result in a larger fraction of atomic and ionised species at the outer edge of the envelope and the disk.

The Molecular Universe
Proceedings IAU Symposium No. 280, 2011
José Cernicharo & Rafael Bachiller, eds.

doi:10.1017/S174392131102494X

The PAH Hypothesis after 25 Years

Els Peeters[1,2]

[1]Department of Physics and Astronomy, University of Western Ontario
London, ON N6A 3K7, Canada
email: epeeters@uwo.ca

[2]SETI Institute, 189 Bernardo Ave.,
Suite 100, Mountain View, CA 94043, USA
email: epeeters@seti.org

Abstract. The infrared spectra of many galactic and extragalactic objects are dominated by emission features at 3.3, 6.2, 7.7, 8.6 and 11.2 μm. The carriers of these features remained a mystery for almost a decade, hence the bands were dubbed the unidentified infrared (UIR) bands. Since the mid-80's, the UIR bands are generally attributed to the IR fluorescence of Polycyclic Aromatic Hydrocarbon molecules (PAHs) upon absorption of UV photons – the PAH hypothesis. Here we review the progress made over the past 25 years in understanding the UIR bands and their carriers.

Keywords. infrared emission features, ISM: molecules, ISM: lines and bands, (stars:) circumstellar matter, infrared: general, astrochemistry, techniques: spectroscopic

1. Introduction

Soon after infrared (IR) detectors became available and opened up the infrared universe, broad, resolved emission features were observed centered at 8.6 and 11.2 μm (Gillet *et al.* 1973) and subsequently at 3.3, 6.2 and 7.7 μm (Russel *et al.* 1978). These emission bands always appeared together and were found in planetary nebulae, H II regions, and reflection nebulae. Since the carrier of these bands remained unidentified for almost a decade, they were dubbed the Unidentified InfraRed (UIR) bands.

Duley & Williams (1981) recognized that the UIR bands peak at wavelengths corresponding to the vibrational modes characteristic of aromatic material. This led to several proposed carriers of the UIR bands, including Hydrogenated Amorphous Carbon (HAC, e.g. Duley & Williams 1983; Borghesi *et al.* 1987), Quenched Carbon Composites (QCC, e.g. Sakata *et al.* 1984), Polycyclic Aromatic Hydrocarbons (PAHs, e.g. Puget & Léger 1984; Allamandola *et al.* 1985), coal (e.g. Papoular *et al.* 1989), nanodiamonds (Jones & d'Hendecourt 2000), Rydberg matter (Holmlid 2000) and Locally Aromatic Polycyclic Hydrocarbons (Petrie *et al.* 2003). The main arguments against several proposed carriers are (i) the UIR bands are observed in reflection nebulae, and (ii) their color temperature is independent of the distance from the illuminating star (Sellgren 1984). Indeed, in these environments, classical dust grains are too cold to emit at mid-infrared wavelengths. Instead, these observations suggest that the carrier of the UIR bands attains very high temperature upon absorption of a single UV photon and hence small species, which have a limited heat capacity, are required (Sellgren 1984).

PAH molecules fulfill this requirement. Upon absorption of a single UV photon, PAHs reach a temperature of ~1000K. They subsequently relax mainly by IR emission in their vibrational modes leading to the UIR bands. Specifically, the 3.3 μm band is due to the CH stretching mode, the 6.2 μm band to the CC stretching mode, the 7.7 μm band to coupled CC stretching and CH in-plane bending modes, the 8.6 μm band to the

CH in-plane bending mode and the 10-15 μm region to the CH out-of-plane (CH_{oop}) bending modes. PAHs contain a large fraction of the available carbon (10-20%) and play an important role in various physical and chemical processes. Thus, the PAH hypothesis was born (Puget & Léger 1984, 1989; Allamandola *et al.* 1985, 1989).

In this paper, we focus on the current status of the observational characteristics of the PAH emission bands†, their dependence on the local physical conditions and the implications for the physical and chemical characteristics of the carriers. We also highlight the current and future use of these emission bands as a diagnostic tool.

There is an extensive body of literature on PAHs and related species in astrophysical environments; we refer in particular to the reviews by Allamandola *et al.* (1989), Puget & Léger (1989), Tielens (2005, 2008), and the proceedings: "PAHs in the Universe" (Joblin & Tielens 2011).

2. The rich PAH spectrum

The Infrared Space Observatory (ISO) and the Spitzer Space Telescope showcased the richness and complexity of the astronomical PAH spectra. Beside the main bands at 3.3, 6.2, 7.7, 8.6, 11.2, and 12.7 μm, a plethora of weaker bands is observed at 3.4, 3.5, 5.25, 5.75, 6.0, 6.6, 6.9, 7.2-7.4, 8.2, 10.5, 10.8, 11.0, 12.0, 13.5, 14.2, 15.8, 16.4, 16.6, 17.0, 17.2, 17.4, and 17.8 μm (see Fig. 1; ‡). In addition, deuterated PAHs, PADs, are tentatively detected at 4.4 and 4.65 μm (Peeters *et al.* 2004a).

The PAH bands are often found on top of broad plateaus located roughly at 3.2–3.6, 6–9, 11–14 and 15–20 μm. These plateaus are part of the PAH spectrum. However, within the literature, they are treated in two different ways. One method considers them as part of the individual PAH emission bands, which are fitted by Lorentzian or Drude profiles (e.g. Boulanger *et al.* 1996; Smith *et al.* 2007; Galliano *et al.* 2008b). Alternatively, the plateaus are treated as features that are separate and distinct from the individual PAH bands. A local spline continuum is then determined to distinguish between both components (e.g. Hony *et al.* 2001, Peeters *et al.* 2002, van Diedenhoven *et al.* 2004). Neither approach is however technically correct (for a detailed discussion, see Tielens 2008). The applied method clearly determines the derived PAH band strengths and profiles. But fortunately, the overall conclusions are independent of the chosen approach (Uchida *et al.* 2000, Galliano *et al.* 2008b).

In addition to the complexity of the PAH spectrum, ISO and Spitzer revealed that the PAH bands and their carriers pervade the universe. They are observed in a large variety of astronomical environments including the (diffuse) ISM, reflection nebulae, T-Tauri stars, Herbig AeBe stars, H II regions, post-AGB stars, planetary nebulae and, more exceptionally in few AGB stars, WR stars, novae, and supernovae remnants. Moreover, they are widespread in various types of galaxies up to redshifts of ~3 (spectroscopically) and ~6 (photometrically). The PAH bands generally dominate the mid-IR spectra of these sources; a remarkable 10-15% of the total power output of galaxies is emitted in these PAH bands (Smith *et al.* 2007). Clearly, the carriers of these infrared emission bands are abundantly present throughout the universe.

† For the remaining of this paper, we refer to the UIR bands as the PAH bands.
‡ A band at 18.9 μm has been attributed to PAH emission as well. However, this band is now assigned to the fullerene C_{60} (Cami *et al.* 2010, Sellgren *et al.* 2010).

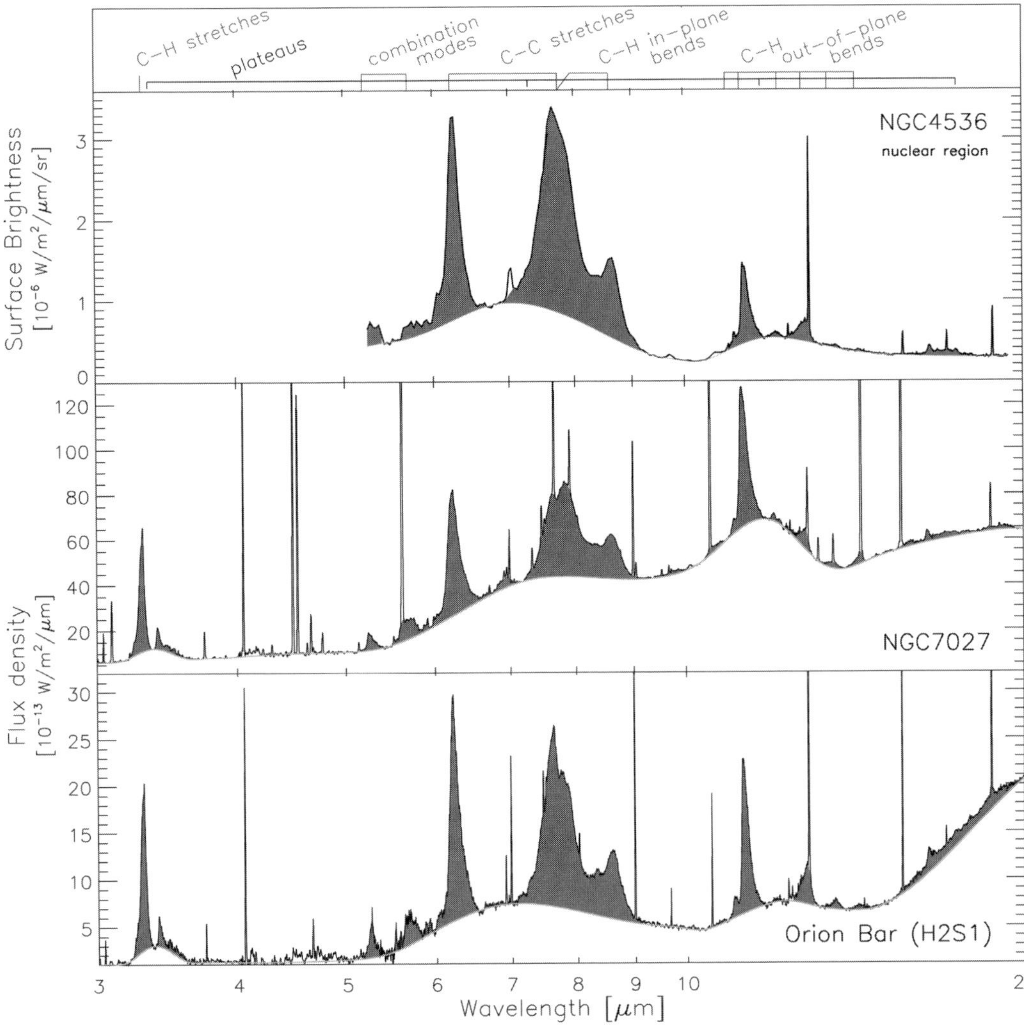

Figure 1. The ISO-SWS spectra of the planetary nebula NGC 7027 and the Photo-Dissociation Region (PDR) the Orion Bar, and the Spitzer-IRS spectrum of the nuclear region of NGC 4536, an HII-type galaxy, (SINGS legacy program; Kennicutt *et al.* 2003) illustrate the richness and variety of the PAH spectrum. Also indicated are the aromatic mode identifications of the major PAH bands. Figure adapted from Peeters 2011.

3. Band Intensities

3.1. *Observational facts*

The relative intensities of the PAH bands vary from source to source and spatially within extended sources. Some PAH band intensities show strong correlations with each other, revealing the underlying connection amongst the PAH bands (Fig. 2, e.g. Hony *et al.* 2001; Vermeij *et al.* 2002; Brandl *et al.* 2006; Sellgren *et al.* 2007, Smith *et al.* 2007; Galliano *et al.* 2008b; Acke *et al.* 2010; Bernard-Salas *et al.* 2010; Boersma *et al.* 2010; Rosenberg *et al.* 2011; Peeters *et al.* 2011). Specifically, the strength of the 3.3 μm feature correlates with that of the 11.2 μm feature. Similarly, the strength of the 6.2, 7.7, 8.6, 11.0, 12.7 and 16.4 μm features correlate well with each other. These variations are found in the Milky Way, the Magellanic Clouds and nearby galaxies.

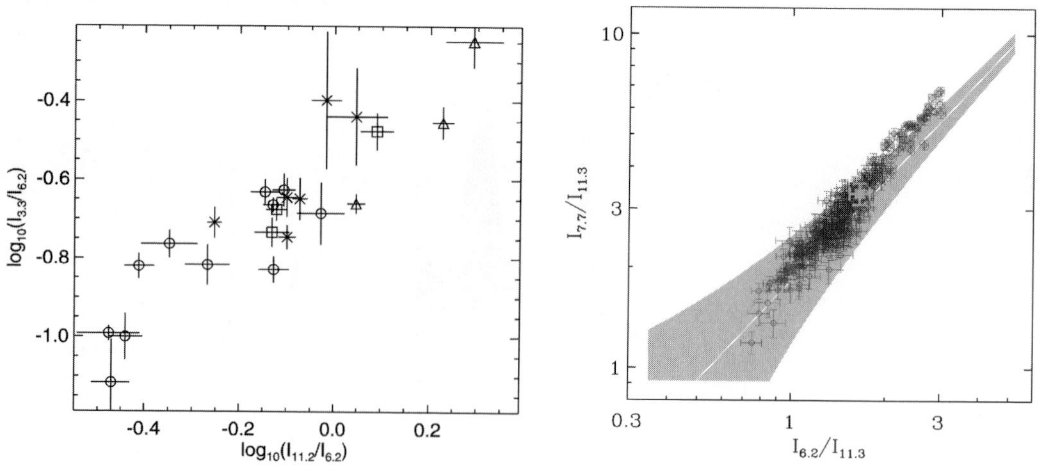

Figure 2. Left: The correlation of the 3.3 and 11.2 μm band strengths normalized to the 6.2 μm band strength. Hexagons represents H II regions, stars intermediate mass star-forming regions, squares reflection nebulae and triangles are planetary nebulae. Figure from Hony *et al.* 2001. **Right:** The variation in the CC/CH band ratios across the starburst galaxy M82. The grey filled area represent the correlation obtained for the integrated spectra of a large sample of galactic and extragalactic sources. The white (gold) symbol is the value of the global measurement over the entire galaxy. Figure from Galliano *et al.* 2008b.

Few sources exhibit peculiar PAH band strength ratios. Some T-Tauri stars and HAeBe-stars show weak or no 7.7 and 8.6 μm PAH emission (Acke & van den Ancker 2004, Geers *et al.* 2006). Similarly, several galaxies harboring an AGN show very weak emission in the 6 to 9 μm region relative to the 11.2 μm band intensity (Kaneda *et al.* 2005; Smith *et al.* 2007; Bregman *et al.* 2008). Moreover, for these galaxies, the 7.7/11.2 PAH ratio depends on the hardness of the radiation field, in contrast to galaxies with H II region or starburst-like characteristics.

In addition to variations in the relative band intensities, the total PAH intensity is also highly variable. The intensity ratio of PAHs to Very Small Grains (VSGs, determined around 15-25 μm) varies spatially across extended Galactic H II regions (e.g. Lebouteiller *et al.* 2007, Churchwell *et al.* 2009). In particular, the dust continuum at 15-25 μm dominates inside the H II region while in the Photo-Dissociation Region (PDR), the PAHs emission is strongest. Similarly, the PAH/VSG ratio in galaxies depends on the hardness of the radiation field. Furthermore, the ratio of the PAH intensity to the total dust emission declines with decreasing metallicity, and hence is in addition to a decrease in total dust content in low metallicity environments (e.g. Madden *et al.* 2006; Brandl *et al.* 2006; Engelbracht *et al.* 2006; Galliano *et al.* 2008a; Gordon *et al.* 2008; Calzetti 2011 and references therein). The underlying cause of this relation is still unclear; the two main scenarios are a less efficient PAH formation process (metallicity effect) or in an increased PAH processing (modification and/or destruction of PAHs by the hard radiation field).

3.2. *Implications*

Laboratory and theoretical spectroscopy of PAHs and related species reveal that their intrinsic spectra are set by charge, size, molecular (edge) structure and, temperature. Consequently, the observed variety in astronomical PAH spectra reflects the properties of the present PAHs. Here we give two examples.

Laboratory and theoretical studies on PAHs clearly show the large influence of the PAH charge state on their IR spectra, in particular on the band intensities (e.g. Hudgins

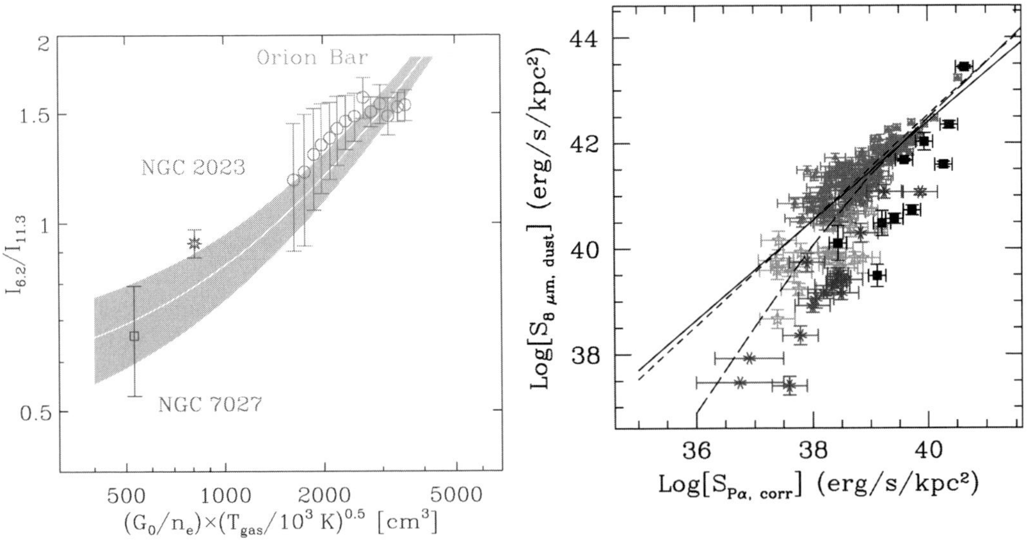

Figure 3. Left: Empirical calibration of the 6.2/11.2 band strength ratio as a function of the ionization parameter, $G_0 T^{1/2}/n_e$, where G_0 is the UV radiation field, T the gas temperature and n_e the electron density. Figure from Galliano *et al.* 2008b. **Right:** The surface luminosity density at 8 μm (stellar-continuum subtracted) as a function of the surface luminosity density of the extinction corrected Pα line for low metallicity starburst galaxies (black; Engelbracht *et al.* 2005) and H II knots. The latter are divided in 3 metallicity bins: high (red filled triangles), intermediate (green stars) and low (blue asterisks). Also shown are the best linear fit through the high metallicity data points (continuous line) and the best linear fit with unity slope (dashed line). Low metallicity regions and starburst galaxies exhibit a depressed 8 μm emission due to a deficiency of PAH emission. Figure from Calzetti 2011.

& Allamandola 2004 and reference therein). Neutral PAHs emit strongly at 3.3 and 11.2 μm while ionized PAHs show dominant emission in the 5-10 μm region. Furthermore, in the 10-15 μm region, the 11.0 μm PAH band is attributed to ionized PAHs and the 11.2 μm band to neutral PAHs. This is consistent with the observed variations in the intensities of the main PAH bands: the degree of ionization dominates the observed range in the intensity ratios of the 6.2, 7.7 or 8.6 μm PAH band and the 3.3 or 11.2 μm PAH band (Joblin *et al.* 1996; Hony *et al.* 2001; Galliano *et al.* 2008b).

The observed relative intensities of the CH_{oop} bands (10-15 μm region) unveil the molecular edge structure of the PAHs (Hony *et al.* 2001; Bauschlicher *et al.* 2008, 2009). Specifically, the peak wavelengths of the CH_{oop} bending modes depend strongly on the number of adjacent peripheral C-atoms bonded to an H-atom (Bellamy 1958; Hony *et al.* 2001). The dominance of the 11.2 μm band in planetary nebulae then indicates PAHs characterized by very compact molecular structures with long smooth edges. In contrast, interstellar PAHs have more corners, either because they are on average smaller or they are more irregular larger species. Similarly, the peak position of the CH stretching mode depends on the molecular edge structure and suggests that the edge structure of astronomical PAHs is remarkably regular and smooth (Bauschlicher *et al.* 2009).

3.3. *PAH toolbox*

The local physical conditions determine the ionization fraction of the PAHs, which in turn specifies the (6.2, 7.7, 8.6)/11.2 PAH intensity ratio. Hence, if the variations in the physical conditions are known (i.e. determined from other diagnostics such as e.g. PDR models), an empirical calibration of the PAH bands with the local physical conditions can be established (Fig. 3, Galliano *et al.* 2008b, Berné *et al.* 2009). Given that the PAH bands are omnipresent and easily detectable, they then become a powerful diagnostic tool for the local physical conditions.

Given that PAHs are excited by UV radiation, the PAH emission bands are very bright in massive star-forming regions and so are a powerful tracer of star formation throughout the universe. They are utilized to determine the contribution of star formation to the total power budget of IR emission from galactic nuclei (e.g. Genzel *et al.* 1998; Lutz *et al.* 1998; Peeters *et al.* 2004b; Smith *et al.* 2007). In addition, they are extensively being explored as a quantitative probe for the star formation rates of galaxies (Fig. 3, Calzetti 2011 and references therein). This is not that straightforward because the PAH emission depends on metallicity and, PAHs are also excited by the older stellar population, unrelated to star formation.

Finally, the presence of PAHs is used to distinguish between shocked gas and PDRs (van den Ancker *et al.* 2000) and to determine redshifts in distant galaxies (e.g. Yan *et al.* 2007).

4. Band profiles

4.1. *Observational facts*

The profiles of the PAH emission bands range from fairly symmetric (e.g. the 3.3 and 8.6 μm bands) to highly asymmetric with a steep blue rise and a red-shaded wing (e.g., 3.4, 5.25, 6.2, 11.2 μm; Barker *et al.* 1987; Roche *et al.* 1996; Verstraete *et al.* 2001; Pech *et al.* 2002; Boersma *et al.* 2008). In contrast, the 12.7 μm band profile has blue-shaded wing with a steep red decline. Additionally, the 7.7 μm complex consists out of at least 2 components: one peaking at $\sim$ 7.6 μm and the second between 7.8-8 μm (e.g. Bregman 1989). Further substructure may be present near 7.2-7.5 μm and at 8.2 μm (e.g. Moutou *et al.* 1999; Peeters *et al.* 2002).

The band profiles for the main PAH bands show clear variability (e.g. Bregman 1989; Cohen *et al.* 1989; Tokunaga *et al.* 1991; Peeters *et al.* 2002; van Diedenhoven *et al.* 2004). They are classified in three classes A, B and C, which is primarily based on the peak position of the bands (see Fig. 4; Peeters *et al.* 2002; van Diedenhoven *et al.* 2004). For the 7.7 μm complex, this results in a band profile with either a dominant 7.6 μm component (class A) or a dominant component peaking between 7.8 and 8 μm (class B) or a very broad band peaking at $\sim$ 8.2 μm with a weak to absent 8.6 μm PAH band (class C). Within this classification, class A and C show little variation while large differences in peak position and profile are present within class B. In addition, few objects have band profiles that encompasses 2 classes (AB: Van Kerckhoven 2002; Peeters *et al.* 2002, Boersma *et al.* 2008; BC: Sloan *et al.* 2007). Hence, the band profiles seem to span a continuous distribution with class A and C as the two extremes.

These band profile variations are observed in the Milky Way, and recently also in the Magellanic Clouds (Bernard-Salas *et al.* 2009, Matsuura *et al.* 2010). They are also observed within (some) extended objects (e.g. Bregman & Temi 2005, Song *et al.* 2007). Remarkably, the 7.7 μm complex varies within reflection nebulae (Bregman & Temi 2005); however, no changes are seen within H II regions (Galliano *et al.* 2008b). Similarly, the classes depend on object type and hence environment (Peeters *et al.* 2002, van Diedenhoven *et al.* 2004). Class A is primarily observed in H II regions, reflection nebulae, the ISM and galaxies but includes few post-AGB stars and planetary nebulae; class B is found in post-AGB stars, most planetary nebulae and isolated Herbig AeBe stars; and class C is primarily observed towards post-AGB stars but includes few Herbig AeBe

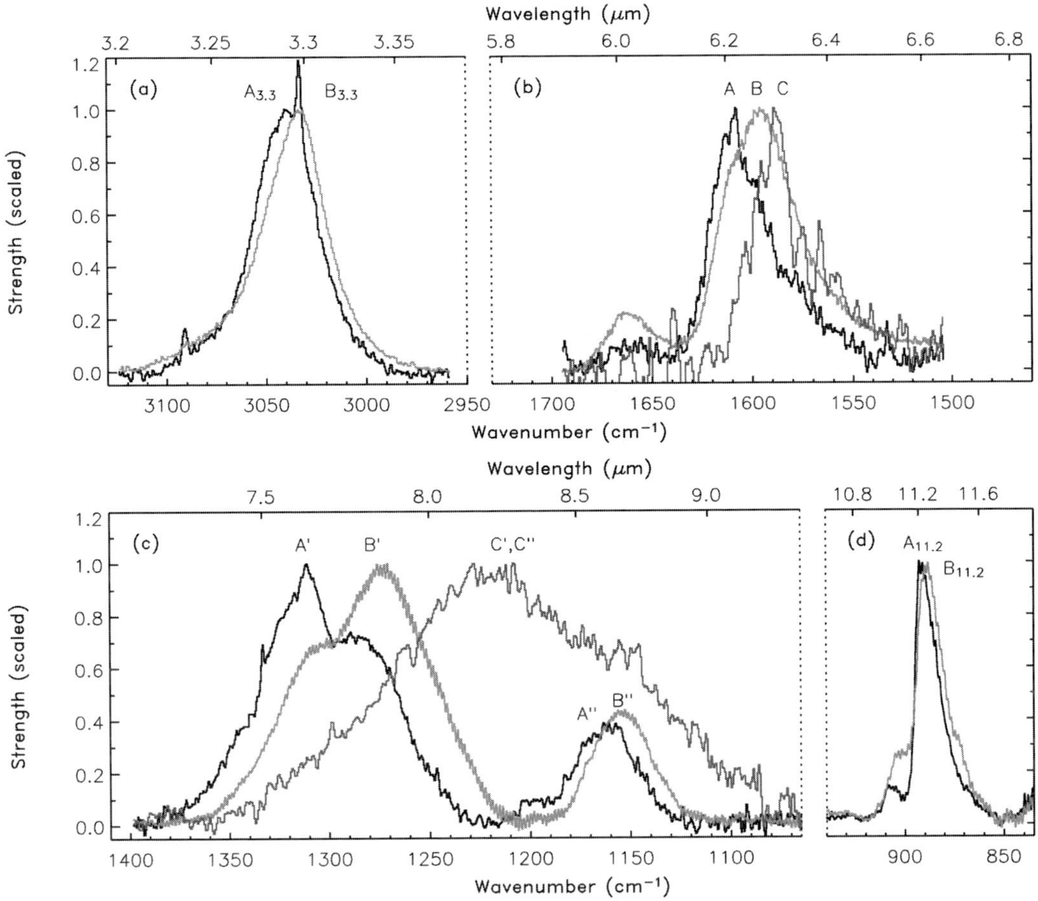

Figure 4. The source to source variations in position and profile of the main PAH bands. In particular large variations are evident in the 6 to 9 μm. Class A peaks at the shortest, 'nominal' wavelengths, class B at longer wavelengths, and class C at even longer wavelengths (Peeters *et al.* 2002; van Diedenhoven *et al.* 2004). Figure from van Diedenhoven *et al.* 2004.

and T-Tauri stars. In other words, interstellar material exhibit class A profiles while circumstellar material (CSM) shows class B and C profiles.

The dependence on the local environment is further established as follows. The central wavelength of the 7.7 μm complex strongly correlates with the radiation field G_0 and the PAH ionization parameter G_0/n_e (the ratio of the UV intensity G_0 to the electron density n_e) within reflection nebulae (Bregman & Temi 2005). Moreover, the central wavelength of the main PAH bands shows a remarkable anti-correlation with the effective temperature of the exciting star for a sample of post-AGB stars and isolated HAeBe stars, all of class B or C (Fig. 5; Sloan *et al.* 2007; Keller *et al.* 2008; Acke *et al.* 2010). This does not seem to hold in general. First, some class A objects, such as reflection nebulae for example, have a central star with low effective temperatures. Secondly, some planetary nebulae have high effective temperatures and class B profiles (e.g. NGC7027). Third, the PAH bands of Herbig AeBe stars with the same effective temperature can belong to different classes (Van Kerckhoven 2002; Boersma *et al.* 2008). This suggests that other parameters such as e.g. history, environment (ISM vs CSM), spatial structure (disk, collapsing cloud; Boersma *et al.* 2008) may play a role.

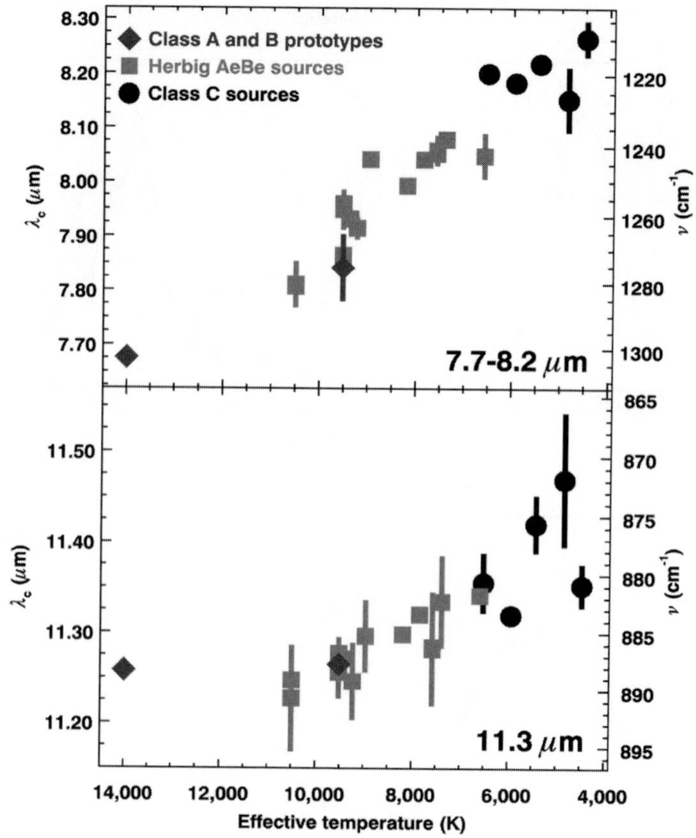

Figure 5. The central wavelengths of the 7.7–8.2 and 11.3 μm PAH features plotted versus the effective temperature of the host star. Figure from Keller *et al.* 2008.

4.2. *Implications*

These band profile variations primarily originates in a (chemical) modification of the PAHs. However, the specifics of this modification is still under debate. Recent studies focus on the CC stretching bands because i) the position of the class A 6.2 μm PAH band cannot be reproduced by the strongest pure CC stretching mode of pure PAHs (Peeters *et al.* 2002; Hudgins *et al.* 2005) using the current available laboratory and theoretical data, and ii) the 7.7 μm complex exhibit the most pronounced variability. Various PAH-related species and processes have been proposed:

- *Hetero-atom substituted PAHs* (e.g. Peeters *et al.* 2002; Hudgins *et al.* 2005; Bauschlicher *et al.* 2009). In particular, the substitution of a C atom by a N atom in a PAH, i.e. PANHs.
- *PAH-metal complexes* (e.g. Hudgins *et al.* 2005; Bauschlicher & Ricca 2009; Simon & Joblin 2010; Joalland *et al.* 2009). These are species in which a metal atom is located either below or above the carbon skeleton.
- *PAH clusters* such as dimers and trimers (e.g. Rapacioli *et al.* 2005; Simon & Joblin 2009).
- *Variations in the size distribution of the PAH family* (Bauschlicher *et al.* 2008, 2009; Cami 2011; Cami *et al.* 2011). Small PAHs ($N_C \leqslant 48$) emit at ~7.6 μm while large PAHs ($54 \leqslant N_C \leqslant 130$) emit at ~7.8 μm. The broad class C component, peaking at ~8.2 μm,

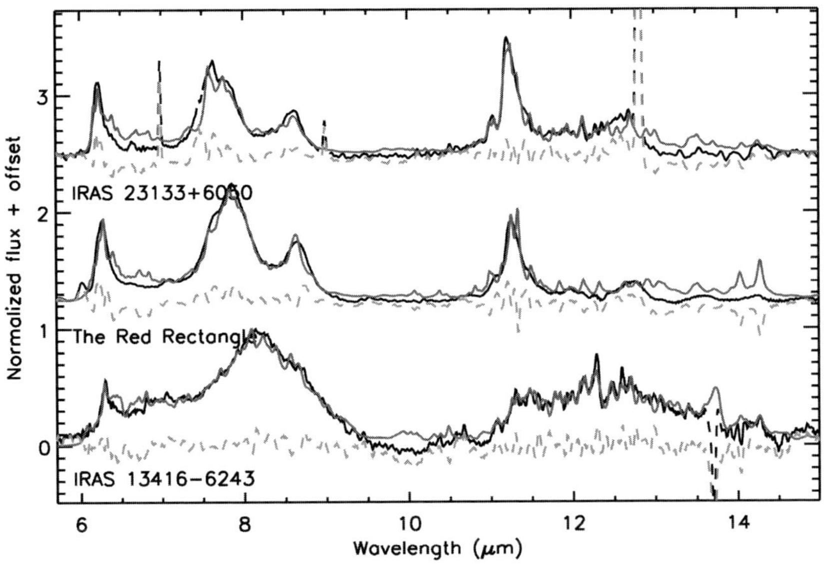

Figure 6. The astronomical PAH spectra (black) and the best fit PAH model spectra (ligther curves). Residuals are shown as dashed lines. The different classes are represented by the H II region IRAS 23133+6050 (class A, top), the Red Rectangle (class B, middle), and the proto-planetary nebula IRAS 13416-6243 (class C, bottom). Figure from Cami 2011.

can be modelled with a collection of small PAHs (Sect. 5). Hence, the PAH classes may then reflect different size distributions of the emitting PAHs.

• *Varying importance of aliphatics vs. aromatics* (e.g. Sloan *et al.* 2007; Boersma *et al.* 2008; Keller *et al.* 2008; Pino *et al.* 2008; Acke *et al.* 2010). Aliphatic carbonaceous species are proposed for the carrier of the class C 7.7 μm complex. UV processing of these species destroys the aliphatic bonds and hence increases their aromaticity. Different degrees of UV processing (e.g. due to an increase in effective temperature) is then invoked to explain the different classes. Since UV processing may also influence the PAH size distribution, both mechanisms may occur together. While increased UV processing can explain the evolutionary scenario from post-AGB stars to the ISM, the reversed process from aromatic material to more fragile aliphatic material is hard to image. The latter occurs when going from the ISM to proto-planetary environments. Hence, an active chemical equilibrium between aromatic and aliphatic species is proposed for all environments through hydrogenation, carbon reactions building (aliphatic) hydrocarbons and UV processing (Boersma *et al.* 2008).

5. Spectral decomposition

Over the years, a large amount of spectra from laboratory experiments and computational modeling of PAHs are obtained. Many of these are publicly available at the NASA Ames PAH spectral database (Bauschlicher *et al.* 2010) and the Cagliari/Toulouse database†. These databases can now be used to model the astronomical PAH spectra. Amazingly good fits are obtained to all three classes A, B and C with the full theoretically calculated NASA Ames PAH spectral database (Fig. 6; Cami 2011, Cami *et al.* 2011).

† At http://www.astrochem.org/pahdb/ and http://astrochemistry.ca.astro.it/database respectively.

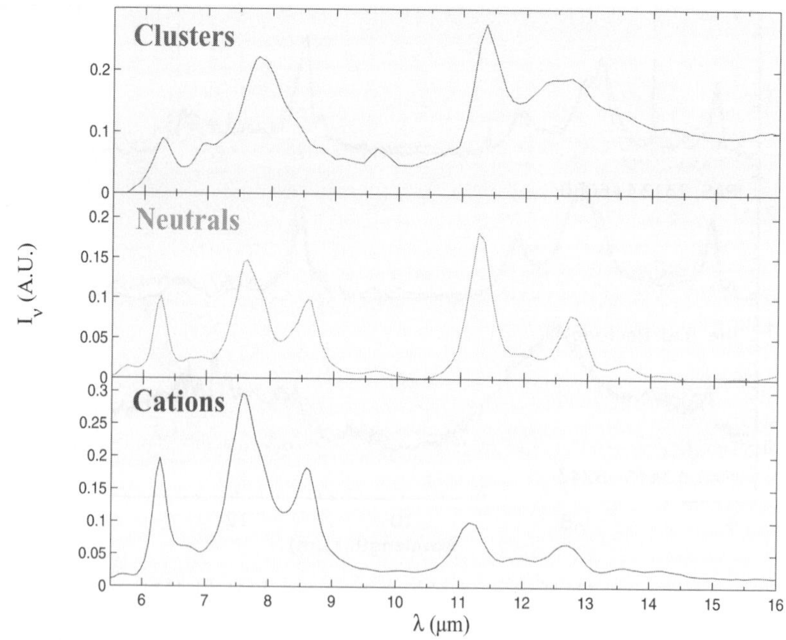

Figure 7. The three components in the singular value decomposition analysis of NGC 7023: PAH^+ (bottom), PAH^0 (middle) and VSG or PAH clusters (top). The components are normalized by their integrated intensity. Figure adapted from Rapacioli *et al.* 2005.

Although these fits are not unique and influenced by the bias present in the database, they can be broken down according to size, charge and composition and thus unveil the underlying origin of the spectral variations. For example, they show that the 6.2 μm band requires a significant contribution of nitrogen-substituted PAHs; the 7.7 μm band profile reflects the PAH size distribution; the 11.2 μm band is due to large, neutral and pure PAHs; and the 11.0 μm band is due to large PAH cations.

The mathematical method Blind Signal Separation is applied to spectral maps of PAH spectra in order to decompose the PAH spectra as a linear combination of a basis set of spatially different components (Boissel *et al.* 2001; Rapacioli *et al.* 2005; Berné *et al.* 2007, 2009). These studies found a basis set of three mathematically distinct components, referred to as PAH^0, PAH^+, and VSG†, which have different spectral characteristics (Fig. 7). Although this is a pure mathematical decomposition, the PAH^0 and PAH^+ components are attributed to respectively neutral and ionized PAHs and the VSG component to PAH clusters. Nevertheless, fitting of the PAH^0 and PAH^+ components with the NASA Ames PAH database confirms the attribution to neutral and ionized PAHs (Rosenberg *et al.* 2011). The PAH^0 and PAH^+ components are found closer to the edge of the PDR and the VSG component is located deeper into the PDR suggesting the destruction of the VSGs or PAH clusters by UV photons and subsequent formation of PAHs. These PAH clusters may be reformed in the denser and more shielded environments of molecular clouds. This method has only been applied to class A sources. Hence, in order to fit PAH spectra of class B and C, this basis set is expanded to include artificial templates (i.e. they do not result from a mathematical decomposition; Joblin *et al.* 2008). These four templates exhibit a very broad band at 8.2, 8.3 and 12.3 μm (thus mimicking

† This VSG component is *not* the same as the VSGs discussed in Sect. 3.1.

class C) and a third PAH template, PAH^x, which has peak positions consistent with class B.

6. The PAH hypothesis after 25 years

Although not a single PAH molecule has been identified to date†, the PAH hypothesis has not been refuted. One of its strongest criticisms has been the lack of a good match between a combination of experimental and/or theoretical PAH spectra and the astronomical PAH spectra. This criticism has been addressed and is now no longer in play (Sect. 5). However, weaknesses in the original PAH hypothesis have been revealed resulting in a refinement of the PAH hypothesis. Specifically, it now includes pure PAHs as well as PAH-related species such as PAHs containing impurities and (aliphatic) side-groups, hydrogenated PAHs, PAHs clusters, etc. Hence, a strict chemical definition of PAHs is no longer in use and it is arguable whether these species can really be dubbed PAHs. Nevertheless, this refined PAH hypothesis is alive and kicking.

To date, the wealth of astronomical PAH spectra and its large variability prompt more questions than can be answered with the current laboratory and theoretical data. It is clear that significant progress in our understanding of the astronomical PAH spectra can only be made by a joint effort of the observational, experimental and theoretical tools.

Acknowledgements

We thank O. Berné, J. Cami, D. Calzetti, F. Galliano, S. Hony, L. Keller and M. Rapacioli for providing figures for this article. Figures are reproduced by permission of the AAS, ESO and, EDP sciences.

References

Acke, B. & van den Ancker, M. E. 2004, *A&A* 426, 151
Acke, B., Bouwman, J., Juhász, A., *et al.*, 2010, *ApJ* 718, 558
Allamandola, L. J., Tielens, A. G. G. M., & Barker, J. R. 1985, *ApJL*, 290, L25
Allamandola, L. J., Tielens, A. G. G. M., & Barker, J. R. 1989, *ApJS*, 71, 733
Barker, J. R., Allamandola, L. J., & Tielens, A. G. G. M. 1987, *ApJ*, 315, L61
Bauschlicher, C. W. Jr., Peeters, E., & Allamandola, L. J. 2008, *ApJ*, 678, 316
Bauschlicher, C. W. Jr., Peeters, E., & Allamandola, L. J. 2009, *ApJ*, 697, 311
Bauschlicher, C. W. Jr. & Ricca, A. 2009, *ApJ*, 698, 275
Bauschlicher, C. W. Jr., Boersma, C., Ricca, A., *et al.*, 2010, *ApJS*, 189, 341
Bellamy, L. 1958, *The infra-red spectra of complex molecules*, 2nd ed. (New York: John Wiley & Sons, Inc.)
Bernard-Salas, J., Peeters, E., Sloan, G. C., *et al.*, 2009, *ApJ* 699, 1541
Berné, O., Joblin, C., Deville, Y., *et al.*, 2007 *A&A*, 469, 575
Berné, O., Fuente, A., Goicoechea, J. R., *et al.*, 2009, *A&A* 706, L160
Boersma, C., Bouwman, J., Lahuis, F., *et al.*, 2008, *A&A* 484, 241
Boersma, C., Bauschlicher, C. W. Jr., Allamandola, L. J., *et al.*, 2010 *A&A* 511, 32
Boissel, P., Joblin, C., & Pernot, P., 2001 *A&A* 373, L5
Borghesi, A., Bussoletti, E., & Colangeli, L. 1987, *ApJ* 314, 422
Boulanger, F., Boisssel, P., Cesarsky, D., & Ryter, C. 1998, *A&A*, 339, 194
Brandl, B. R., Bernard-Salas, J., Spoon, H. W. W., *et al.*, 2006, *ApJ* 653, 1129
Bregman, J. 1989, in IAU Symp. 135: *Interstellar Dust*, (eds.) Allamandola L.J. & Tielens A.G.G.M., 109

† Mainly because the mid-IR vibrational modes of PAHs show only a weak dependence on size and structure.

Bregman, J. & Temi, P., 2005, *A&A* 621, 831
Bregman, J. D., Bregman, J. N., & Temi, P., 2008 ASPC 381, 34
Calzetti, D., Kennicutt, R. C., Engelbracht, C. W., *et al.*, 2007, *ApJ* 666, 870
Calzetti, D. 2011, EAS Publications Series, Volume 46, 133 (http://www.eas-journal.org/)
Cami, J., Bernard-Salas, J., Peeters, E., & Malek, S. E. 2010 *Science* 329, 5996, 1180
Cami, J., 2011, EAS Publications Series, Volume 46, 117 (http://www.eas-journal.org/)
Churchwell, E., Babler, B. L., Meade, M. R., *et al.* 2009, Publications of the Astronomical Society of the Pacific, Vol. 121, issue 877, 213
Cohen, M., Tielens, A. G. G. M., Bregman, J., *et al.*, 1989, *ApJ*, 341, 246
Duley, W. W. & Williams, D. A. 1981, *MNRAS* 196, 269
Duley, W. W. & Williams, D. A. 1983, *MNRAS* 205, 67P
Engelbracht, C. W., Kundurthy, K. D., Gordon, K. D., *et al.*, 2006 *ApJ* 642, L127
Engelbracht, C. W., Gordon, K. D., Rieke, G. H., *et al.*, 2005, *ApJ* 628, 29
Galliano, F., Dwek, E., & Chanial, P. 2008a, *ApJ* 672, 214
Galliano, F., Madden, S., Tielens, A. G. G. M., *et al.*, 2008b, *ApJ* 679, 310
Geers, V. C., Augereau, J.-C., Pontoppidan, K. M., *et al.* 2006, *A&A* 459, 545
Genzel, R., Lutz, D., Sturm, E., *et al.*, 1998, *ApJ*, 498, 579
Gillett, F. C., Forrest, W. J., & Merril, K. M. 1973, *ApJ*, 183, 87
Gordon, K. D., Engelbracht, C. W., Rieke, G. H., *et al.*, 2008 *ApJ*, 682, 336
Holmlid, L. 2000, *A&A* 358, 276
Hony, S., Van Kerckhoven, C., Peeters, E., *et al.*, 2001, *A&A*, 370, 1030
Hudgins, D. M., Bauschlicher, C. W. Jr., & Allamandola, L. J. 2005, *ApJ* 632, 316
Hudgins, D. M. & Allamandola, L. J. 2004, Astrophysics of Dust, ASP Conference Series, Vol. 309, 665
Joalland, B., Simon, A., Marsden, C. J., & Joblin, C. 2009, *A&A* 494, 969
Joblin, C. & Tielens, A. G. G. M., eds. 2011, PAHs and the Universe: A Symposium to Celebrate the 25th Anniversary of the PAH Hypothesis , Vol. 46 (EAS Publication Series)
Joblin, C., Szczerba, R., Berné, O., & Szyszka, C. 2008, *A&A* 490, 189
Joblin, C., Tielens, A. G. G. M., Geballe, T. R., & Wooden, D. H. 1996, *ApJ*, 460, L119
Jones, A. P. & d'Hendecourt, L. 2000, *A&A* 355, 1191
Kaneda, H., Onaka, T., & Sakon, I. 2005 *ApJ* 632, L83
Keller, L. D., Sloan, G. C., Forrest, W. J., *et al.*, 2008, *ApJ* 684, 411
Kennicutt, R. C. Jr., Armus, L., Bendo, G., *et al.*, 2003, *PASP* 115, 810, 928
Lebouteiller, V., Brandl, B., Bernard-Salas, J., Devost, D., & Houck, J. R. 2007, *ApJ* 665, 390
Lutz, D., Spoon, H. W. W., Rigopoulou, D., Moorwood, A. F. M., & Genzel, R. 1998, *ApJ*, 505, L103
Madden, S., Galliano, F., Jones, A. P., & Sauvage, M. 2006, *A&A* 446, 877
Matsuura, M., Sloan, G. C., Bernard-Salas, J., *et al.*, 2010, Proceedings of the Asymmetrical Planetary Nebula V conference, Ed. A.A. Zijlstra *et al.*
Moutou, C., Sellgren, K., Verstraete, L., & Léger, A. 1999, *A&A*, 347, 949
Papoular, R., Conrad, J., Giuliano, M., Kister, J., & Mille, G. 1989, *A&A* 217, 204
Pech, C., Joblin, C., & Boissel, P. 2002, *A&A*, 388, 639
Peeters, E., Hony, S., van Kerckhoven, C., *et al.*, 2002, *A&A* 390, 1089.
Peeters, E., Allamandola, L. J., Bauschlicher, C. W. Jr. *et al.*, 2004a, *ApJ* 604, 252
Peeters, E., Spoon, H. W. W., & Tielens, A. G. G. M. 2004b, *ApJ* 613, 986
Peeters, E. 2011, EAS Publications Series, Volume 46, 13 (http://www.eas-journal.org/)
Peeters, E., Tielens, A. G. G. M., Allamandola, L. J., & Wolfire, M. 2011, *ApJ*, submitted
Petrie, S., Stranger, R., & Duley, W. W. 2003, *ApJ* 594, 869
Pino, T., Dartois, E., Cao, A.-T., *et al.*, 2008, *A&A* 490, 665
Puget, J. L., Léger, A., & Boulanger, F. 1985, *A&A* 142, L19
Puget, J. L. & Léger, A. 1989, ARAA, 27, 161
Rapacioli, M., Joblin, C., & Boissel, P. 2005, *A&A* 429, 193
Roche, P. F., Lucas, P. W., & Geballe, T. R. 1996, *MNRAS* 281, L25

Rosenberg, M. J. F., Berné, O., Boersma, C., Allamandola, L. J., & Tielens, A. G. G. M. 2011, *A&A*, in press
Russell, R. W., Soifer, B. T., & Forrest, W. J. 1975, *ApJL* 198, L41
Sakata, A., Wada, S., Tanabe, T., & Onaka, T. 1984, *ApJL*, 287, L51
Sellgren, K. 1984, *ApJ* 277, 623
Sellgren, K., Uchida, K. I., & Werner, M. W. 2007, *ApJ* 659, 1338
Sellgren, K., Werner, M. W., Ingals, J., *et al.*, 2010, *ApJ* 722, L54
Simon, A. & Joblin, C. 2009, J. Phys. Chem. A, 113, 4878
Simon, A. & Joblin, C. 2010, *ApJ*, 712, 69
Sloan, G. C., Jura, M., Duley, W. W., *et al.*, 2007, *ApJ* 664, 1144
Smith, J. D., Draine, B. T., Dale, D. A., *et al.*, 2007, *ApJ* 656, 770
Song, I.-O., McCombie, J., Kerr, T. H., & Sarre, P. J. 2007 *MNRAS*, 380, 979
Tielens, A. G. G. M. 2005, The Physics and Chemistry of the Interstellar Medium, ISBN 0521826349. Cambridge, UK: Cambridge University Press
Tielens, A. G. G. M. 2008, ARA*&A*, 46, 289
Tokunaga, A. T., Sellgren, K., Smith, R. G., *et al.*, 1991, *ApJ*, 380, 452
Uchida, K. I., Sellgren, K., Werner, M. W., & Houdashelt, M. L. 2000, *ApJ* 530, 817
van den Ancker, M. E., Tielens, A. G. G. M. & Wesselius, P. R. 2000 *A&A* 358, 1035
van Diedenhoven, B., Peeters, E., Van Kerckhoven, C., *et al.*, 2004, *ApJ* 611, 928.
Van Kerckhoven, C. 2002, Ph.D. thesis, Catholic University of Leuven
Vermeij, R., Peeters, E., Tielens, A. G. G. M., & van der Hulst, J. M. 2002, *A&A*, 382, 1042
Verstraete, L., Pech, C., Moutou, C., *et al.*, 2001, *A&A*, 372, 981
Yan, L., Sajina, A., Fadda, D., *et al.*, 2007, *ApJ* 658, 778

Discussion

KRELOWSKI: You have not mentioned that PAHs should have features in the visible and that the latter remain undetected.

PEETERS: That's correct. There is a talk later today by Nick Cox that will address this issue.

GEBALLE: You showed very accurate fits of an ensemble of PAHs to the 10 μm PAH spectrum. What is the situation regarding simultaneous fitting of the 3.3 and 3.4 μm features along with the 10 μm bands?

PEETERS: The intensity of the 3 μm bands in the theoretical spectra is off by a factor of $\sim$2. Hence, they currently cannot be fitted together with the bands at longer wavelengths.

MENTEN: For fitting the observed spectra with templates from the PAH database, do you have templates that were produced for different temperatures and do you use a single temperature for all the template spectra of different species?

PEETERS: No, we don't have temperature information for most PAHs in the database.

The Molecular Universe
Proceedings IAU Symposium No. 280, 2011
José Cernicharo & Rafael Bachiller, eds.

doi:10.1017/S1743921311024951

The Diffuse Interstellar Bands: an Elderly Astro-Puzzle Rejuvenated

Nick L. J. Cox[1]

[1]Instituut voor Sterrenkunde, K.U.Leuven,
B-3000, Leuven, Belgium
email: nick.cox@ster.kuleuven.be

Abstract. The interstellar medium constitutes a physically and chemically complex component of galaxies and is important in the cycle of matter and the evolution of stars. From various spectroscopic clues we now know that the interstellar medium is rich in organic compounds. However, identifying the exact nature of all these components remains a challenge. In particular the identification of the so-called diffuse band carriers has been alluding astronomers for almost a century.

In recent decades, observational, experimental and theoretical advances have rapidly lead to renewed interest in the diffuse interstellar bands (DIBs). This has been instigated partly by their perceived relation to the infrared aromatic emission bands, the UV extinction bump and far-UV rise, and the growing number of (small) organic molecules identified in space.

This chapter gives an overview of the observational properties and behaviour of the DIBs, and their presence throughout the Universe. I will highlight recent progress in identifying their carriers and discuss their potential as tracers and probes of (extra)-Galactic ISM conditions.

Keywords. astrochemistry, line: identification, line: profiles, polarization, ISM: abundances, ISM: atoms, ISM: clouds, (ISM:) dust, extinction, ISM: lines and bands, ISM: molecules

1. Introduction

In this contribution I will review and discuss the properties, behaviour, presence and absence of diffuse interstellar bands (DIBs) in space. To set the stage I will first discuss the diffuse interstellar medium and its chemical complexity (Section 2). Next I will discuss in detail the general properties of the DIBs (Section 3) and their behaviour (Section 4). Section 5 reviews recent work on the presence of DIBs in extra-galactic environments, while Section 6 focuses on observations of the carriers in other astrophysical environments. In the final part (Section 7) recent advances in laboratory astrophysics are highlighted and suggestions for further work and criteria for conclusive identification of carriers are proposed.

2. The diffuse ISM and its organic constituents

Diffuse to translucent interstellar clouds have typical temperatures in the range of 50 to 200 K, and particle volume densities between 100 to 1000 cm^{-3}. These colder clumps are immersed and dispersed throughout the low density, warm medium (e.g. Snow & McCall (2008), Ferrière (2001)). Visual extinctions for individual diffuse clouds are generally less than a few magnitudes. However, the total line-of-sight extinction can be much larger, in particular in the Galactic plane and towards the Galactic Centre. DIBs have been detected in sightlines with visual extinction of ∼10 mag (e.g. Cox *et al.* (2005)). For (individual) diffuse clouds the chemical networks are driven by photochemistry. The interstellar radiation field (ISRF) penetrates a significant fraction of the cloud's interior, even for translucent clouds (A_V > 4 mag; Snow & McCall (2008)). Nearby stars with strong UV flux can drastically increase the effective UV radiation. The (diffuse) ISM is rich in complex (organic) matter as evidenced from the various identified and unidentified spectral features (Kwok & Sandford (2008)).

Identified features

Already in the early 1900s a number of atoms, such as Na I, Ca II, Ti II were identified in reddened optical spectra of early type (OB) stars. The first molecules in space were detected through their UV-optical transitions, e.g. CH, CH^+, and CN.. Other molecules were later identified in the UV (e.g. C, CO, H_2) and the near-infrared (e.g. C_2, H_2, H_3^+). Polyatomic molecules, such as C_3 have been detected in the optical (Roueff *et al.* (2002), Krełowski *et al.* (2011)) as well as in the far-infrared/sub-mm (Cernicharo *et al.* (2000), Mookerjea *et al.* (2011)).

Unidentified features

Notwithstanding major progress in observational and laboratory astrophysics, several components of the (diffuse) ISM remain unknown. Their spectral signatures suggest primarily carbonaceous "organic" components. In Table 1 we give a brief overview of the unidentified spectral features detected in the ISM. Together the UIB, DIB, UV bump, and 3.4 μm feature constitute a significant part (> 30%) of the available cosmic carbon budget.

Table 1. Unidentified features in the ISM and a non-exhaustive summary of proposed carriers.

Feature	Wavelength	f_X	Proposed carriers	ref
UV bump	2175 Å	$\sim 5 \cdot 10^{-5}$	graphitic material, fullerenes, bucky onions, PAHs	1
Diffuse Interstellar Bands	4060 - 13175 Å	$\sim 10^{-7}$	PAHs, fullerenes, linear carbon chains	2
3.4μm absorption	3.4 μm		UV processed HACs	3
Infrared Emission Bands	3.3, 6.2, 7.7, 8.6, 11.3 μm	$\sim 10^{-6}$	PAHs, fullerenes	4

Notes: The fraction of abundance is defined as N_X/N_H. The uncertainty in $f_{\rm DIB}$ is as much as several orders of magnitude (Sect.3).
References: (1) Stecher (1965), Bless & Savage (1972), Menella *et al.* (1998), Iglesias-Groth (2004), Iglesias-Groth (2007), Mulas *et al.* (2011); (2) Heger (1922), Merrill & Wilson (1938), Herbig (1995), Sarre *et al.* (2006) (and references therein); (3) Butchart *et al.* (1986), McFadzean *et al.* (1989), Adamson *et al.* (1990), Pendleton *et al.* (1994), Sandford *et al.* (1995), Chiar & Pendleton, 2008; (4) Tielens (2011), Peeters, (2011), Cami *et al.* (2011), Draine (2011), Peeters, (2011).

3. The diffuse interstellar bands: from first discovery to extensive surveys

In the early 1900s several observers (e.g. Canon, Heger) noted broad "diffuse" absorption bands on photographic plates of spectra taken of bright stars. Heger (1922) first noted their stationary nature as observed towards a spectroscopic binary, indicating - like the known strong stationary sodium and calcium lines - their origin was not stellar, but rather interstellar. The seminal paper by Merrill & Wilson (1938) presented one of the first in depth systematic studies of the stronger diffuse bands and their relation to both the gas and dust in the ISM. From the broadness of the lines it was immediately evident that these species could not be atomic nor di-atomic in nature. Both atoms and diatomic molecules were detected in space by that time and their intrinsic line widths are much narrower (only a few km/s). Ever since Merrill's work, researchers have tried to answer the inevitable question raised in that paper:

"What are the carriers of the diffuse - interstellar - bands?
And are they due to grains or molecules?"

As of today, more than 400 DIBs have been confirmed towards a few individual lines-of-sight probing the diffuse ISM (Hobbs *et al.* (2008), Hobbs *et al.* (2009)). The strongest DIBs are detected in no more than a few hundred Galactic and extra-Galactic sightlines. All currently known DIBs lie between 4000 and 10000 Å, with two interstellar bands at 1.13 and 1.31 μm (Joblin *et al.* (1990), Rawlings *et al.* (2011)). A brief summary of the DIB observational properties is given in Table 2. The importance of the DIBs as a chemical component of the ISM is illustrated by the estimated fraction their carriers could constitute with respect the total abundance of PAHs in the ISM per unit reddening (Table 1): $N_{\rm DIB} \approx 10^{14}$ cm$^{-2} \approx 2{\cdot}10^{-8} N_H \approx 0.1 N_{\rm PAH}$ (with $N_{\rm PAH}/N_H \sim 3{\cdot}10^{-7}$ Tielens (2008) and N_H/E(B-V) $= 5 \cdot 10^{21}$ cm^{-2}/mag).

Table 2. Key properties of the diffuse interstellar bands

Property[1]	Typical value
FWHM	0.5 - 30 Å/ 2 - 80 cm^{-1} / 50 - 1000 km/s
Wavelength	4000-10000 Å
DIB strength - equivalent width	5 – 2000 mÅ per E(B-V)
DIB strength - central depth	up to 20% per E(B-V)
Peak of number density distribution	6500 Å
Peak of FWHM distribution	0.7 Å
Total equivalent widths[2]	$\sim$ 25 - 30 Å per E(B-V)
Total abundance N(DIB)[3]	$\sim 10^{14}$ cm^{-2}

Notes:
(1) Global properties gathered from surveys by Jenniskens & Desert (1994), Herbig (1995), Tuairisg *et al.* (2000), Hobbs *et al.* (2008); (2) This is an estimate since the DIB strength per unit reddening vary between lines-of-sight due to changing local environmental conditions; (3) This is a zeroth order estimate assuming a typical oscillator strength, f, of unity and transition wavelength of 6500 Å for all DIBs. Small PAHs generally have small f values between 0.01 and 0.1, thus requiring a higher abundance to give rise to bands of given strength (see Sect. 7).

4. Environmental behaviour of DIBs

Dust & gas

The equivalent width of diffuse interstellar bands shows a positive trend (albeit with only a weak to intermediate correlation) with the amount of dust (measured via reddening E(B-V)) in the line-of-sight† as well as with the amount of gas in the sightline (measured via absorption lines). In Figure 1 we show the relation of the 5780 Å DIB with dust and of the 6614 Å DIB with gas (N(K I)). Similar trends between DIB strength and reddening are observed for many other DIBs (see e.g. Friedman *et al.* (2011), Vos *et al.* (2011)). Also, most DIB strengths show a correlation with molecular tracers, such as CH and CH^+, which also correlate roughly with E(B-V) (Krełowski *et al.* (1999), Vos *et al.* (2011)). A few DIBs show a correlation with C_2 abundance (Thorburn *et al.* (2003)).

Although most DIBs are stronger in sightlines with more dust, their relative ratios can differ significantly, by as much as an order of magnitude for sightlines with similar E(B-V). A notable example, is the ratio between the 5780 and 5797 DIB (Krełowski & Westerlund (1988)). Krełowski & Sneden (1994) found that, for sightlines with similar reddening, the 5797 band is deeper than the 5780 band if the line-of-sight has as high molecular content (ζ-type), while it is shallower for sightlines with a low abundance of molecules

† This is partly a selection bias as many DIBs have been confirmed via their increase in strength with E(B-V).

(σ-type). Further observational studies and models show that the effective strength of the interstellar radiation field (ISRF) is very important in setting the balance between the formation/destruction of molecules as well as their hydrogenation and ionisation state (Le Page *et al.* (2003), Ruiterkamp *et al.* (2005), Cox & Spaans (2006)). Figure 2 shows a schematic drawing of the ISRF affecting diffuse clouds.

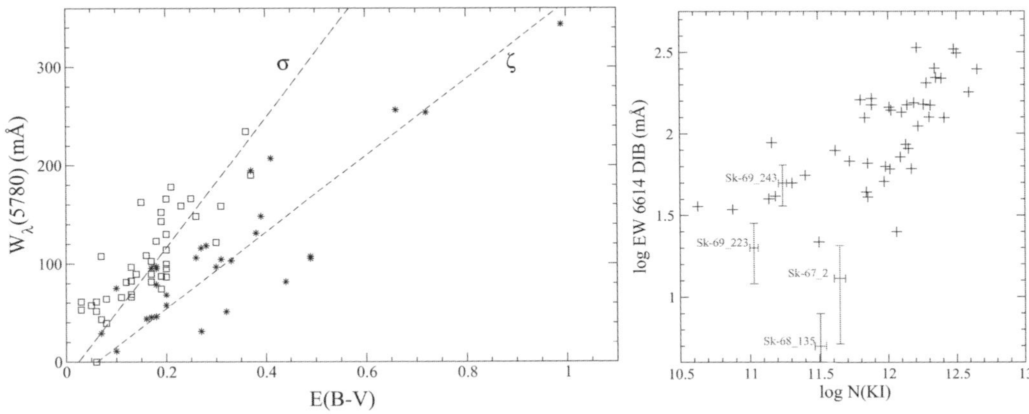

Figure 1. DIB strength versus dust and gas abundance. *Left*: The 5780 Å DIB equivalent width for Galactic sightlines is plotted as a function of reddening (adapted from Vos *et al.* 2011). Depending on the ratio between 5780 and 5797 DIBs there is a difference in the slopes of the linear relation for the two sets (e.g. σ versus ζ). *Right*: The 6613 Å DIB equivalent width is plotted as a function of N(K I) (adapted from Cox *et al.* 2006). Crosses (+) are for Galactic sightlines. Four LMC sources are indicated with error bars. The log-log plots shows a positive trend, although there is large scatter at the lower abundances.

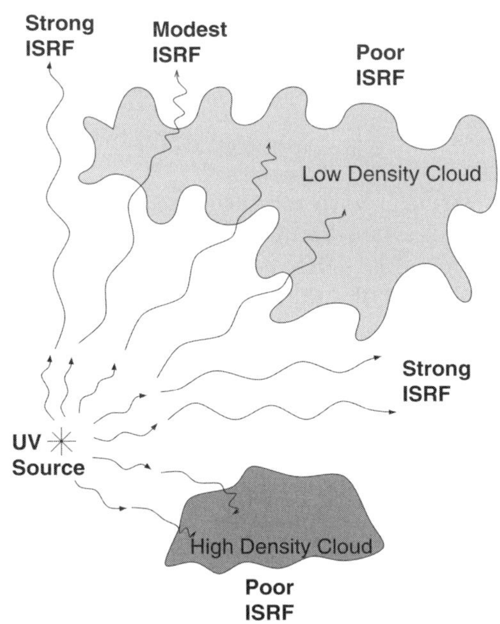

Figure 2. The relative strengths of several diffuse bands are affected by the effective interstellar radiation field (ISRF) impacting the respective interstellar clouds. The schematic drawing illustrates that different types of cloud are exposed to different levels of UV radiation. The low density - UV exposed - cloud is of σ-type, and the high-density - UV protected - cloud is of ζ-type. Figure courtesy of D. Vos.

Line polarisation

Interstellar grains are known to align due to the interstellar magnetic field. Thus carriers directly associated - or even attached to grains - could potentially align as well and thus give rise to a polarisation feature across the profile (Martin & Angel (1974)). Despite several studies of DIBs in sightlines with strong linear continuum polarisation no DIB polarisation features have been found. The most extensive surveys to date exclude the presence of line polarisation in 40 DIBs to a level of less than 0.01% (Cox *et al.* (2007), Cox *et al.* (2011)). Thus, DIB carriers (at least the ones studied) seem not to be related to large grains nor grain impurities.

The near-infrared DIB at 8621 Å

In particular, the 8621 Å DIB also does not reveal a polarisation signal. This DIB is unique in that it is reported to have a unusually tight correlation with reddening (Wallerstein *et al.* (2007), Munari *et al.* (2008)) although independent data show this is not so much the case (Krelowski - see Q&A section). Also, as it will be observed with the radial velocity spectrograph on the GAIA astrometry mission this DIB will be measured towards thousands of stars in the Milky Way, providing an independent reddening/extinction estimator to de-redden stellar SEDs and/or to construct a 3-dimensional Galactic dust extinction map. Also, this will confirm (hopefully) beyond doubt the ubiquitous presence of DIB carriers throughout the Galaxy (currently DIBs have been detected in only a few hundred lines-of-sight towards Galactic (field) stars).

Substructure in DIB profiles

High-resolution observations show that many of the narrow - both strong and weak - DIBs reveal sub-structure in their absorption profiles, whereas the broader DIBs have a smooth profile (Sarre *et al.* (1995), Foing & Ehrenfreund (1996), Kerr *et al.* (1996), Galazutdinov *et al.* (2003)). The observed double or triple peak sub-structures are typical of rotational contours of large (20 to 40 carbon atoms) gas-phase molecules (Danks & Lambert (1976)). The profiles do not vary in time, but do change with temperature of the ISM (Cami *et al.* (2004)).

DIB-DIB correlations

Since most DIBs grow stronger with increasing reddening or gas column density, it is not surprising there is a correlation between individual DIBs. However, it is surprising that almost none of the individual DIBs correlate perfectly with each other. This lack of correlation suggests that the DIBs do not share exactly the same carrier. This is the *1 DIB - 1 PAH* hypothesis. Recently, the 6196-6613 Å DIB pair has been reported to show a tight correlation, with a Pearson correlation coefficient of 0.986 (McCall *et al.* (2010)). This is indeed much better than the rather good correlation between the 5780 DIB and E(B-V) or N(H I) (r = 0.82 and r = 0.95, resp. Friedman *et al.* (2011)). This tight relation could hint at a common carrier. However, some problems remain. Firstly, the two DIBs have different line widths, shapes and strengths. Secondly, there remains significant scatter on the relation - in particular at low reddening. This puts, at best, strong constraints on the molecular constants of the carrier.

DIBs and the UV bump

The correlation between DIBs and the UV bump has also been of considerable interest as both features a believed to arise from the organic reservoir of interstellar matter. However, following up on earlier results by Megier *et al.* (2005)) who compared the far-UV rise with DIBs, a recent study by Xiang *et al.* (2011) shows no significant correlation between the UV bump or far-UV rise and DIB strengths.

Table 3. Global properties of the Local Group galaxies. See e.g. Cox *et al.* (2006), Cox *et al.* (2007), Cordiner *et al.* (2008a), Cordiner *et al.* (2008b), Cordiner *et al.* (2011) and references therein. Note that the gas-to-dust ratio is inversely proportional to the metalicity.

Galaxy	Metallicty	ISRF	UV extinction curve
Milky Way (MW)	~ 1	1 - 10	Nominal. Large scatter on far-UV rise
Large Magellanic Cloud (LMC)	~ 0.2 - 0.5	> 5	Weak far-UV rise, strong ISRF
Small Magellanic Cloud (SMC)	~ 0.1 - 0.2	> 30	No UV bump; strong far-UV rise
Andromeda (M31)	~ 1	< 1	Weak UV bump, anomalous extinction curve
Triangulum (M33)	~ 1	~ 1	

5. Ubiquitous presence of DIBs throughout Galactic and extra-Galactic ISM

In this section I will report on the presence of DIBs in extra-galactic environments. The study of DIBs throughout time and space is important to understand both the chemical evolution of Galaxies as well as to understand the properties of the carriers themselves. Observations of DIBs can potentially be used as proxies of local environmental conditions of the ISM in distant galaxies.

The Local Group

The Local Group galaxies offer a unique laboratory to study the behaviour of the DIBs in a wider range of physical conditions than possible within the Milky Way alone (Table 3). In particular the effects of variations in the gas-to-dust ratio, the metalicity, and the interstellar radiation field strength and hardness on the DIBs can be traced.

The LMC, SMC, M31 and M33 are sufficiently nearby (at about 50, 60, 780, and 880 kpc, respectively), to discern individual stars and to obtain spectra at high S/N. Nevertheless, 8+m class telescopes are required for high-resolution spectroscopy of stars in the SMC and LMC (Ehrenfreund *et al.* (2002). For M31 and M33 stars high-resolution spectra are still difficult to obtain, requiring several hours of integration time on the largest telescopes currently available. Low-resolution multi-object spectroscopy using Galactic templates has proven to be very successful (Cox & Cordiner (2008), Cordiner *et al.* (2011)). Studies of the Local Group thus offer additional insight into the formation and destruction of DIB carriers.

The LMC and SMC are the nearest Local Group members allowing for detailed studies of, for example, their stellar population and interstellar matter content. DIBs are under-abundant in the ISM of the SMC bar, but have normal Galactic strength in the wing (Cox *et al.* (2006), Welty *et al.* (2006), Cox & Cordiner (2008)). The extinction curve for the bar ISM reveals an absence of the UV bump carriers, while the wing ISM does show an UV bump.

M31 and M33 offer unique complete views of spiral galaxies, where we can easily probe different regions in the galaxies, unlike for the Milky Way due to our own vantage point in it. Due to this view from above most M31 and M33 sightlines probe only low to intermediate dust columns, unlike the high extinction lines-of-sight probing the Galactic plane. Although this ensures less confusion due to multiple clouds in a line-of-sight, it also yields intrinsically weaker and thus harder to detect diffuse bands (assuming the Galactic relation holds). A recent survey of DIBs towards 30 sightlines in M31 shows that their strengths show a trend with reddening similar to the one observed for the Milky Way (Cordiner *et al.* (2011)). The DIBs are on average a little stronger per unit reddening (thus more typical of σ-type clouds - see Fig. 1), but fall within the envelope of Galactic data points (Figure 3). This effect is likely an observational bias as stronger

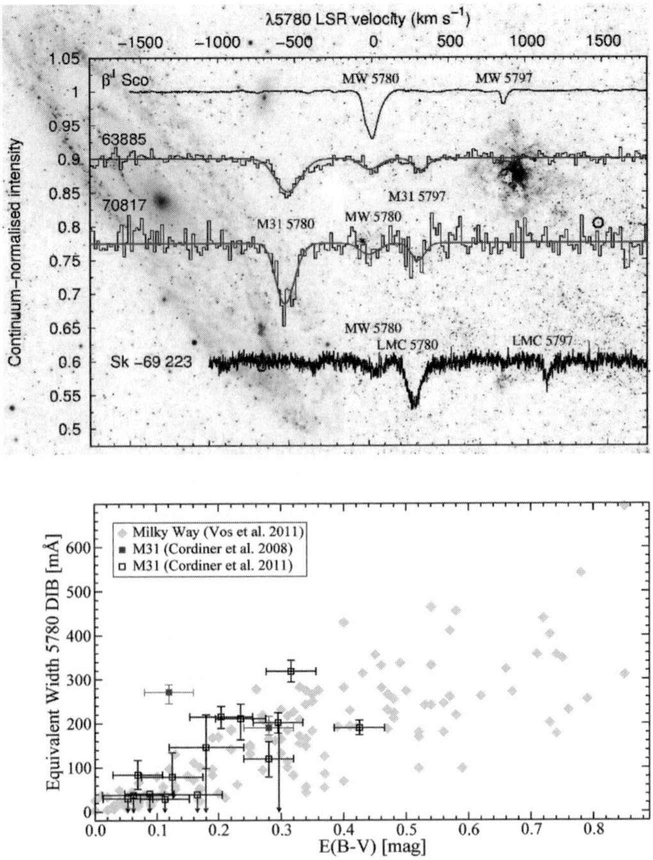

Figure 3. *Top:* The 5780 and 5797 Å DIBs observed towards the Milky Way, M31 and the LMC. *Bottom:* 5780 Å DIB equivalent width versus reddening for M31 and the Milky Way lines-of-sight (Adapted from Cordiner *et al.* 2008, 2011).

lines are more readily detected in low-resolution low-S/N spectra. It is also found that the DIB carrier abundance in M31 does not correlate with either the UV field strength nor the PAH emission. Further high-resolution spectra are needed to probe the physical conditions and radiation field via atomic and molecular lines.

Beyond the Local Group

Supernovae offer the unique opportunity to study the distant extra-galactic ISM. SNe can become bright enough to act - for a limited time - as background "light houses" suitable for high-resolution spectroscopy of the intervening ISM. Often the study of the ISM is obtained as an added bonus to studied of the temporal evolution of the expanding SNe photosphere. In this way DIBs have been detected in several galaxies such as NGC 1448 (d∼17 Mpc; Sollerman *et al.* (2005)), M100 (d∼16 Mpc; Cox & Patat (2008) and Fig. 4), and NGC 2770 (d∼26 Mpc; Thoöne *et al.* (2009)). The DIB strength ratios observed toward NGC 1448 suggest that the two supernovae, SN2001el and SN2003h, in this galaxy probe a σ and ζ environments, respectively. The line-of-sight towards SN2008D in NGC 2770 is highly extincted with an E(B-V) of 0.8 mag. The measured DIB strengths per unit reddening are similar to the Galactic trend indicating that multiple individual clouds in the ISM of this galaxy are probed. The ISM in M 100 towards SN 2006X is very dense with a high molecular content, and with DIBs per unit reddening more than 7 times weaker than for the Milky Way.

The 5780 and 6283 DIBs were detected in dusty starburst galaxies. The weaker DIBs at 5797 and 6613 were also detected in the galaxies M 82 and NGC 2146 at 4 and 14 Mpc, respectively (Heckman & Lehnert (2000)). Despite the harsh conditions expected for the dusty central regions of starburst galaxies the DIBs have similar strengths per unit reddening. Several broad and narrow DIBs have also been detected in the $z = 0.52$ Damped Lyα system toward AO 0235+164 DLA (Junkkarinen *et al.* (2004), York *et al.* (2006)). Thus, the formation and destruction of DIB carriers depends on the local physical and chemical conditions. And similar conditions exist among different galaxies.

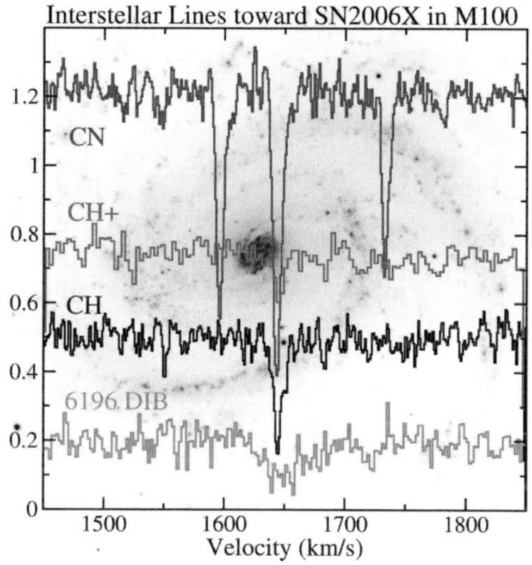

Figure 4. Detection of CN, CH^+, CH and 6196 Å DIB in M100 towards SN2006X. M100 is a spiral galaxy seen face-on at a distance of ~16 Mpc. Adapted from Cox & Patat (2008). Background image adapted from ESO VLT/FORS.

6. Diffuse (interstellar) band carriers in other astrophysical environments

Little is known about the presence of the carriers giving rise to the DIBs in other astrophysical environments. An important question to answer is if DIBs are exclusive to the ISM or that the carriers are present elsewhere? If the DIBs arise from large gas phase molecules such as PAHs and fullerenes it is suggested that they should also be present in other environments rich in organic material. However, diffuse *circumstellar* bands are absent in dusty circumstellar envelopes of AGB (Mauron & Kendall (2004)) and post-AGB stars (Luna *et al.* (2008)), as well as absent in atmospheres of cool stars (Destree *et al.* (2007)) and Herbig Ae/Be stars (Galazutdinov *et al.* (2003)). For all these cases the main difficulty is to distinguish the circumstellar diffuse bands from the interstellar diffuse bands. DIBs have been searched for - without success - in comets as well (O'Malia *et al.* (2010)). The search for diffuse band carriers in other environments is still very limited in scope mostly due to observational difficulties. For example, it is not known if diffuse band carriers could be present in HII regions, dense molecular clouds, planetary nebulae, supernova remnants, proto-planetary disks, or other environments in which PAHs are observed.

How can this lack be explained? The absence of diffuse band carriers in other environments could be due to the fact that the "parent" carriers get modified (such as ionised or deprotenated or hydrogenated) only sufficiently in the harsh environment of the diffuse

Table 4. Recent proposals for DIB carriers.

Candidate carrier	DIBs (Å)	Reference	Remark
C_{60}^{+}	9577, 9632	Foing & Ehrenfreund (1994), Foing & Ehrenfreund (1997)	No gas-phase spectra to verify peak positions
Carbon nano-tubes	–	Zhou *et al.* (2006)	Calculated transitions. Non-specific.
CH_2CN^-	8037	Cordiner & Sarre (2007)	Calculated. Match in position. Needs lab data to confirm.
Fullerenes	4428	Iglesias-Groth *et al.* (2008)	Theoretical calculation. Inaccurate peak position.
Large "floppy" molecule '	6863, 6844, 6203, 5545	Duley & Kuzmin (2010)	Origin bands of vibronic progression. Non-specific carrier.
Proflavine (neutral & ions)	4066, 4363, 4176, 4259	Bonaca & Bilalbegović (2010)	Calculated. Large offset in peak position.
HC_4H^+	5068.8	Krełowski *et al.* (2010)	Match in position, not of profile.
Acetylene plasma	5449	Linnartz *et al.* (2010)	Match in peak position and profile shape. Non-specific carrier.
l-C_3H_2	4881, 5449, 6150–6330	Maier *et al.* (2011), Oka & McCall, (2011)	Peak position and width match for 5449, less for 4887 band. No convincing match for 6150-6330 bands.

Notes: The recent tentative detection of naphthalene and anthracene Iglesias-Groth *et al.* (2008), Iglesias-Groth *et al.* (2008) refuted by Searles *et al.* (2011), Galazutdinov *et al.* (2011) is not included as it did not pertain to an existing DIB.

ISM such that the "new" species can give rise to optical absorption bands. Also, it is possible that the carriers are destroyed in PDRs/HII regions; carriers deplete from the gas phase into the ice-mantle of dust grains in cold, dense clouds.

7. Quest for the identity of the DIB carriers

As mentioned above the debate on the gas or dust origin of the diffuse band carriers was started already in the 1930s. Free flying molecules were slightly favoured in those early days. In the 40s and 50s dust grains were in vogue mainly due to many new developments in the study of dust particles and the idea that large molecules could not be formed / survive in space. With the detection of polyatomic molecules in space (via rotational line emission at radio frequencies) the question presented itself again if the DIBs could perhaps be gas phase molecules. The line widths indicated that these molecules had to be significantly larger than the known di-atomic species (Platt (1956), Danks & Lambert (1976)).

PAH-DIB hypothesis

With the discovery of infrared emission bands and their subsequent assignment to aromatics and fullerenes (Alamandola *et al.* (1985), Puget *et al.* (1985)), it became apparent that very large molecules are abundant in space. The link between PAHs and DIBs was quickly made (Crawford *et al.* (1985), Leger & d'Hendecourt (1985), Van der Zwet & Allamandola (1985)) as the optical properties (electronic absorption spectra) were known from laboratory studies (e.g. Birks (1970)). Apart from having (electronic) transitions in the UV, optical and near-infrared, PAHs are also very resistant to UV radiation. Thus the abundance of PAHs in space, their optical absorption spectrum and the presence of substructure in DIB profiles consistent with rotational profiles of intermediate-to-large PAHs all give strong support for the PAH-DIB hypothesis (Salama *et al.* (1999), Cox (2011)). A potential problem with the PAH-DIB (and PAH-UIR band) hypothesis is the lack of interstellar bands in both the UV (Snow & Destree (2011), Snow *et al.* (2011)) and near-infrared part of the astronomical spectrum.

Theoretical calculation of the main electronic transitions of small to intermediate PAHs (10 to 42 carbon atoms) show that the neutrals absorb in the UV, clustering between 2500-4000 Å. Cations and anions absorb in the visible/near-infrared (4000-14000 Å). For both the neutrals and ions the transitions are seen to shift redwards with increasing size of the PAH (Ruiterkamp *et al.* (2005)). Also, other carbon-based molecules, such linear chains and fullerenes, absorb in the optical. The latter have been detected in space via their infrared emission (Cami *et al.* (2010)). The addition of side groups to PAHs (such as a methyl group or a small carbon chain) or the nitrogen or isotopic substitution and (de)hydrogenation of PAHs all alter the specific electronic configuration and thus the corresponding absorption spectrum (see e.g. Hammonds *et al.* (2011), Hammonds *et al.* (2011) for the effect of (de)protonation). Not only PAHs, but also their derivatives, and related structures such as fullerenes, carbon chains and rings are now also to be considered inclusive in the term "PAH" when discussing the PAH-DIB hypothesis.

DIB carrier candidates

With the growing number of claims for DIB carriers independently verifiable criteria are probably warranted. In order of importance (high to low) the following criteria could be considered to be necessary for an identification of a DIB carrier (see e.g. also Galazutdinov *et al.* 2011):

(1) An exact match between the profile (position and shape) of the laboratory gas-phase low-temperature spectra and the DIB profile is required.

(2) All similarly strong transitions (preferably at least two) detected in the laboratory should also be confirmed in the astronomical spectra.

(3) Detection in at least two separate lines-of-sight

For the identification of a specific PAH (i.e. not related to a known DIB) the criteria for identification are equally strict but slightly different as the feature is potentially new. We add the following conditions, that ensure the new feature is real and of interstellar origin:

(4) Its interstellar nature needs to be confirmed; e.g. to be stationary in a spectroscopic binary and/or its strength shows a positive trend with reddening

(5) The features should be absent toward unreddened lines-of-sight.

(6) Independent detection with different instruments to avoid instrumental artefacts.

In order to solve the DIB problem, one needs to find a specific PAH or related structure (or class of molecules) that has a limited set of very strong absorption bands (and possibly a larger number of weaker bands) and is either more abundant in space or is a more efficient absorber than most other large carbonaceous molecules. However, none of the candidate PAHs, fullerenes, or chains, thus far measured in the laboratory has been found to match any of the DIBs (Salama *et al.* (2008)). This is partly due to the fact that most spectra have been obtained in matrix isolation and only very few species have been measured at cold temperatures in the gas phase. Recent studies have focused on obtaining laboratory spectroscopic data of low temperature gas-phase *neutral* PAHs to compare with astronomical spectra of diffuse and translucent clouds (e.g. Pino *et al.* (2011), Salama *et al.* (2011), Salama *et al.* (2011), Gredel *et al.* (2011), Dietsche *et al.* (2011)). Ten different PAHs (containing 2 to 13 benzene rings / 11 to 42 carbon atoms) were searched for but none were detected in the ISM by both Salama *et al.* (2011) and Gredel *et al.* (2011) independently. In some cases there is a decent match in shape and position for a single laboratory band with a DIB (e.g. Pentacene and 5362.8 Å DIB), however the shift and widths are still too different and other strong bands seen in the laboratory spectra are not observed in the ISM (Salama *et al.* (2011)). Typical column densities limits range from 10^{11} to 10^{14} cm^{-2} (Salama *et al.* (2011), Gredel *et al.* (2011)). Laboratory experiments on PAH cations and anions - whose spectra are expected to give more absorption bands in the red - are challenging with current experimental setups. Tentative detections of broad bands corresponding to naphthalene cation (Iglesias-Groth *et al.* (2008)) and anthracene cation (Iglesias-Groth *et al.* (2010)) were reported for one peculiar line-of-sight, but could not be confirmed in other sightlines (Searles *et al.* (2011), Galazutdinov *et al.* (2011)).

Table 4 gives a overview of recent DIB carrier identifications reported in literature, none of which is confirmed or without question. The tentative assignment of C_{60}^{+} to the 9577 and 9632 bands is still debated (Jenniskens *et al.* (1997), Galazutdinov *et al.* (2000), Ehrenfreund *et al.* (2006)). Unfortunately, the strong telluric absorption in this spectral range hampers detection of additional - weaker - bands from C_{60}^{+}. Results for calculated transitions or inferred vibronic progressions of large molecules are promising but await further study and confirmation from lab experiments. This is also the case for the good comparison between a hydrocarbon plasma band and the 5449 DIB (Linnartz *et al.* 2010). Mismatches, however small, for CH_2CN^- (Krelowski *et al.* 2010) and HC_4H^+ (Cordiner & Sarre 2007) do not allow for a positive carrier identification. The recent assignment of l-C_3H_2 to several DIBs (Maier *et al.* (2011), Raghunandan *et al.* (2011)) has been challenged by others, revealing inconsistencies between the observations and laboratory data (Oka & McCall, (2011), Krelowski (this volume)).

8. Conclusions

At present, the most promising candidate DIB carriers are large carbon molecules - in their various guises - that need to be stable, UV resistant, but sensitive to local cloud conditions (i.e. ISRF). However, alternative candidates cannot be dismissed a priori. Studying the presence and behaviour of DIBs in our Galaxy, Local Group and far beyond will hopefully lead to further insights into the formation and destruction mechanisms that govern the DIB carrier abundance. On the other hand, there exists now also an opportunity to use DIBs as a tracer of the physical and chemical conditions of the ISM in distant galaxies.

Spectroscopic signatures of PAHs and related structures are predicted to arise in *both* the UV (mainly neutrals) and near-infrared (predominantly large anions and cations). The observation of these transitions is crucial to both their assignment to the infrared emission bands as well as their link to the DIBs.

The apparent absence of DIB carriers - as opposed to PAHs - in various other environments such as circumstellar envelopes of evolved stars, comets, H II regions, etc., needs to be understood and can possible lead to new understandings leading to their identity.

Laboratory astrophysics is continuously developing new techniques to measure the optical spectra of complex molecules under interstellar space conditions. It is now increasingly feasible to make direct comparisons between laboratory data and astronomical observations. This leads to questions about the criteria that need to be met in order to firmly establish - beyond doubt - any identification. A possible set of criteria - open to discussion - is provided.

Acknowledgement

My thanks go to the SOC and LOC for organising an enjoyable and interesting meeting as well as for financial support. I thank Dennis Vos for preparing Figure 2. I also acknowledge the many fruitful and lively discussions I have had with my collaborators over the years and during this symposium in particular.

References

Adamson, A. J., Whittet, D. C. B., & Duley, W. W. 1990, *MNRAS*, 243, 400

Allamandola, L. J., Tielens, A. G. G. M., & Barker, J. R. 1985, *ApJ Suppl.*, 71, 733

Birks, J. B. 1970, *Photophysics of Aromatic Molecules*, John Wiley and Sons, Ltd., London

Bless, R. C. & Savage, B. D. 1972, *ApJ*, 171, 293

Bonaca, A. & Bilalbegović, G. 2010, *Chemical Physics (Lett.)*, 493, 33

Butchart, I., McFadzean, A., Whittet, D., Geballe, T., & Greenberg, J. 1986, *A&A*, 154, 5

Cami, J., Salama, F., Jiménez-Vicente, J., Galazutdinov, G., & Krełowski, J. 2004, *ApJ*, 611, 113

Cami, J., Bernard-Salas, J., Peeters, E., & Malek, S. E. 2010, *Science*, 329, 1180

Cami, J., Bernard-Salas, J., Peeters, E., & Malek, S. E. 2011, IAU Symp. 280, *this volume*

Cernicharo, J., Goicoechea, J. R., & Caux, E. 2000, *ApJ* (Lett.), 534, 199

Chiar, J. E. & Pendleton, Y. J. 2008, IAU Symposium 251, Eds. S. Kwok & S. Sandford, 35

Cordiner, M. A., Cox, N. L. J., Trundle, C., Evans, C. J., Hunter, I., Przybilla, N., Bresolin, F., & Salama, F. 2008, *A&A*, 480, 13

Cordiner, M. A., Smith, K. T., Cox, N. L. J., Evans, C. J., Hunter, I., Przybilla, N., Bresolin, F., & Sarre, P. J. 2008, *A&A*, 492, 5

Cordiner, M. A., Cox, N. L. J., Evans, C. J., Trundle, C., Smith, K. T., Sarre, P. J., & Gordon, K. D. 2011, *ApJ*, 726, 39

Cordiner, M., Cox, N., Smith, K., Evans, C., Trundle, C., Gordon, K., & Sarre, P. 2011, IAU Symp. 280, Poster 2.21

Cordiner, M. A. & Sarre, P. J. 2007, *A&A*, 472, 537

Cox, N. 2011, in: C Joblin & A. G. G. M. Tielens (eds.), *PAHs and the Universe*, EAS, 46, 349

Cox, N. L. J.., Boudin, N., Foing, B. H., Schnerr, R. S., Kaper, L., Neiner, C., Henrichs, H., Donati, J.-F., & Ehrenfreund, P. 2007, *A&A*, 464, 899

Cox, N. L. J. & Cordiner, M. A. 2011, IAU Symposium 251, 237

Cox, N. L. J., Kaper, L., Foing, B. H., & Ehrenfreund, P. 2005, *A&A*, 438, 187

Cox, N. L. J., Cordiner, M. A., Cami, J., Foing, B. H., Sarre, P. J., Kaper, L., & Ehrenfreund, P. 2006, *A&A*, 447, 991

Cox, N. L. J., Cordiner, M. A., Ehrenfreund, P., Kaper, L., Sarre, P. J., Foing, B. H., Spaans, M., Cami, J., Sofia, U. J., Clayton, G. C., Gordon, K. D., & Salama, F. 2007, *A&A*, 470, 941

Cox, N., Ehrenfreund, P., Foing, B., d'Hendecourt, L., Salama, F., & Sarre, P. 2011, *A&A*, 531, A25

Cox, N. L. J. & Patat, F. 2008, *A&A*, 485, 9

Cox, N. L. J. & Spaans, M. 2006, *A&A*, 451, 973

Crawford, M. K., Tielens, A. G. G. M., & Allamandola, L. J. 1985, *ApJ*, 293, 45

Danks. A. C., Lambert, D. L. 1976, *MNRAS*, 174, 571

Destree, J. D., Snow, T. P., & Eriksson, K. 2007, *ApJ*, 664, 909

Dietsche, R., Chakrabarty, S., Rice, C., Mazzotti, F., & Maier, J. 2011, IAU Symp. 280, Poster 3.25

Draine, B. 2011, in: C Joblin & A. G. G. M. Tielens (eds.), *PAHs and the Universe*, EAS, 46, 100

Duley, W., Kuzmin 2010, *ApJ* (Lett.), 712, 165

Ehrenfreund, P., Cami, J., Jiménez-Vicente, J., Foing, B. H., Kaper, L., Van der Meer, A., Cox, N., d'Hendecourt, L., Maier, J. P., Salama, F., Sarre, P. J., Snow, T. P., & Sonnentrucker, P., 2002, *ApJ* (Lett.), 576, 117

Ehrenfruend, P., Cox, N., & Foing, B. 2006, in "Natural Fullerenes and Related Structures of Elemental Carbon", Development in Fullerene Science, vol. 6, Rietmijer, F. (Ed.), 58

Ferrière, K. M. 2001, *Reviews of Modern Physics*, 73, 103

Foing, B. H. & Ehrenfreund, P. 1994, Nature, 369, 296

Ehrenfreund, P. & Foing, B. H. 1996, *A&A* (Lett.), 307, 25

Foing, B. H. & Ehrenfreund, P. 1997, *A&A* (Lett.), 317, 59

Friedman, S. D., York, D. G., McCall, B. J., Dahlstrom, J., Sonnentrucker, P., Welty, D. E., Drosback, M. M., Hobbs, L. M., Rachford, B. L., & Snow, T. P. 2011, *ApJ*, 727, 33

Galazutdinov, G., Krelowski, J., Musaev, F. A., Ehrenfreund, P., & Foing, B. H. 2000, *MNRAS*, 317, 750
Galazutdinov, G., Musaev, F. A., Bondar, A. V., & Krelowski, J. 2003, *MNRAS*, 345, 365
Galazutdinov, G., Lee, B-C., Song, I-O., Kazmierczak, M., & Krełowski, J. 2011, *MNRAS*, 412, 1259
Gredel, R., Carpentier, Y., Rouille, G., Steglich, M., Huisken, F., & Henning, Th. 2011, *A&A*, 530, A26
Hammonds, M., Candian, A., & Sarre, P. 2011, IAU Symp. 280, Poster 1.34
Hammonds, M., Pathak, A., Candian, A., & Sarre, P. 2011, in: C Joblin & A. G. G. M. Tielens (eds.), *PAHs and the Universe*, EAS, 46, 373
Heckman, T. & Lehnert, 2000, *ApJ*, 537, 690
Heger, M. 1922, Lick Observatory Bulletin, 10, 141
Herbig, G. H. 1995, *ARA&A*, 33, 19
Hobbs, L. M., York, D. G., Snow, T. P., Oka, T., Thorburn, J. A., Bishof, M., Friedman, S. D., McCall, B. J., Rachford, B., Sonnentrucker, P., & Welty, D. E. 2008, *ApJ*, 680, 1256
Hobbs, L. M., York, D. G., Thorburn, J. A., Snow, T. P., Bishof, M., Friedman, S. D., McCall, B. J., Oka, T., Rachford, B., Sonnentrucker, P., & Welty, D. E. 2009, *ApJ*, 705, 32
Iglesias-Groth, S. 2004, *ApJ* (Lett.), 608, 37
Iglesias-Groth, S. 2007, *ApJ* (Lett.), 661, 167
Iglesias-Groth, S., Manchado, A., García-Hernández, D. A., González Hernández, J. I., & Lambert, D. L. 2008, *ApJ*, 685, 55
Iglesias-Groth, S., Manchado, A., Rebolo, R., Gonzalez Hernandez, J. I., Garcia-Hernandez, D. A., & Lambert, D. L. 2010, *MNRAS*, 407, 2157
Jenniskens, P. & Désert, F.-X. 1994, *A&A* Suppl., 106, 39
Jenniskens, P., Mulas, G., Porceddu, I., & benvenuti, P. 1997, *A&A*, 327, 337
Joblin, C., D'Hendecourt, L., Leger, A., & Maillard, J. P. 1990, *Nature*, 346, 729
Junkkarinen, V. T., Cohen, R. D., Beaver, E. A., Burbidge, E. M., Lyons, R. W., & Madejski, G. 2004, *ApJ*, 614, 658
Kerr, T., Hibbins, R., Miles, J., Fossey, S., Somerville, W., & Sarre, P. 1996, *MNRAS*, 283, 105
Krełowski, J. & Sneden, C. 1994, *ASPC*, 58, 12
Krełowski, J. & Westerlund, B. E. 1988, *A&A*, 190, 339
Krełowski, J., Ehrenfreund, P., & Foing, B. H., *et al.* 1999, *A&A*, 347, 235
Krełowski, J., Beletsky, Y., Galazutdinov, G. A., Kołos, R., Gronowski, M., & LoCurto, G. 2010, *ApJ*, 714, 64
Krełowski, J. & Kazmierczak, M. 2011, IAU Symp. 280, Poster 2.50
Kwok, S. & Sandford, S. 2008, Eds. IAU Symposium 251
Leger, A. & D'Hendecourt, L. 1985, *A&A*, 146, 81
Leger, A., D'Hendecourt, L., Verstraete, L., & Schmidt, W. 1988, *A&A*, 203, 145
Le Page, V., Snow, T. P., & Bierbaum, V. M. 2003, *ApJ*, 584, 316
Linnartz, H., Wehres, N., van Winckel, H., Walker, G. A. H., Bohlender, D. A., Tielens, A. G. G. M., Motylewski, T., & Maier, J. P. 2010, *A&A*, 511, 3
Luna, R., Cox, N. L. J., Satorre, M. A., García Hernéz, D. A., Suárez, O., & García Lario, P. 2008, *A&A*, 480, 133
Maier, J. P., Lakin, N. M., Walker, G. A. H., & Bohlender, D. A. 2001, *ApJ*, 553, 267
Maier, J. P., Walker, G. A. H., Bohlender, D. A., Mazzotti, F. J., Raghunandan, R., Fulara, J., Garkusha, I., & Nagy, A. 2011, *ApJ*, 726, 41
Martin, P. G. & Angel, J. R. P. 1974, *ApJ*, 188, 517
Mauron, N. & Kendall, T. R., 2004, *A&A*, 428, 535
McCall, B. J., Drosback, M. M., Thorburn, J. A., York, D. G., Friedman, S. D., Hobbs, L. M., Rachford, B. L., Snow, T. P., Sonnentrucker, P., & Welty, D. E. 2010, *ApJ*, 708, 1628
McFadzean, A. D., Whittet, D. C. B., Bode, M. F., Adamson, A. J., & Longmore, A. J. 1989, *MNRAS*, 241, 873
Megier, A., Krełowski, J., & Weselak, T. 2005, *MNRAS*, 358, 563
Mennella, V., Colangeli, L., Bussoletti, E., Palumbo, P., & Rotundi, A. 1998, *ApJ* (Lett.), 507, 177

Merrill, P. W. & Wilson, O. C. 1938, *ApJ*, 87, 9

Mookerjea, B., Giesen, T., Stutzki, J., Cernicharo, J., Goicoechea, J. R., & Black, J. H. 2011, IAU Symp. 280, *this volume*

Mulas, G., Malloci, G., Joblin, C., & Cecchi-Pestellini, C. 2011, in: C Joblin & A. G. G. M. Tielens (eds.), *PAHs and the Universe*, EAS, 46, 327

Munari, U., Tomasella, L., Fiorucci, M., Bienaymé, O., Binney, J., Bland-Hawthorn, J., Boeche, C., Campbell, R., Freeman, K. C., Gibson, B., Gilmore, G., Grebel, E. K., Helmi, A., Navarro, J. F., Parker, Q. A., Seabroke, G. M., Siebert, A., Siviero, A., Steinmetz, M., Watson, F. G., Williams, M., Wyse, R. F. G., & Zwitter, T. 2008, *A&A*, 488, 969

Oka, T. & McCall, B. J. 2011, *Science*, 331, 293

O'Malia, K. K. J.; Thorburn, J. A.; Hammergren, M.; Snow, T. P.; Dembicky, J.; Hobbs, L. M.; York, D. G. 2010, *ApJ*, 708, 785

Peeters, E. 2011, in: C Joblin & A. G. G. M. Tielens (eds.), *PAHs and the Universe*, EAS, 46, 13

Peeters, E. 2011, IAU Symp. 280, *this volume*

Pendleton, Y. J., Sandford, S. A., Allamandola, L. J., Tielens, A. G. G. M., & Sellgren, K. 1994, *ApJ*, 437, 683

Pino, T., Carpentier Y., Feraud. G., Friha, H., Kokkin, D. L., Troy, T. P., Chalyavi, N., Brechignac, Ph., & Schmidt, T. W. 2011, in: C Joblin & A. G. G. M. Tielens (eds.), *PAHs and the Universe*, EAS, 46, 355

Platt, J. R., 1956, *ApJ*, 123, 486

Puget, J. L., Leger, A., & Boulanger, F. 1985, *A&A*, 142, 19

Raghunandan, R., Maier, J. P., Walker, G. A. H., Bohlender, D. A., Mazzotti, F. J., Fulara, J., Garkusha, I., & Nagy, A. 2011, IAU Symp. 280, *this volume*

Rawlings, M., Adamson, A., McCall, B., & Kerr, T. 2011, IAU Symp. 280, Poster 2.82

Roueff, E., Felenbok, P., Black, J. H., & Gry, C. 2002, *A&A*, 384, 629

Ruiterkamp, R., Cox, N.L.J., Spaans, M., Kaper, L., Foing, B.H., Salama, F., Ehrenfreund, P. *A&A*, 432, 515

Salama, F., Galazutdinov, G., Krełowski, J., Allamandola, L., & Musaev, F. 1999, *ApJ*, 526, 265

Salama, F. 2008, IAU Symposium 251, Eds. S. Kwok & S. Sandford, 357

Salama, F., Galazutdinov, G., Krelowski, J., Biennier, L., Beletsky, Y., & Song, I. O. 2011, *ApJ*, 728, 154

Salama, F., Galazutdinov, G., Krelowski, J., Biennier, L., Beletsky, Y., & Song, I. O. 2011, IAU Symp. 280, Poster 2.88

Sandford, S. A., Pendleton, Y. J., & Allamandola, L. J., 1995 *ApJ*, 440, 697

Sarre, P. J., Miles, J. R., Kerr, T. H., Hibbins, R. E., Fossey, S. J., & Somerville, W. B. 1995, *MNRAS*, 277, 41

Sarre, P. J. 2006, *Journal of Molecular Spectroscopy*, 238, 1

Searles, J., Destree, J., Snow, T., Salama, F., York, D., & Dahlstrom, J. 2011, *ApJ*, 732, 50

Snow, T. P. & Destree, J. D. 2011, in: C Joblin & A. G. G. M. Tielens (eds.), *PAHs and the Universe*, EAS, 46, 341

Snow, T. P. & McCall, B. J. 2008, *ARA&A*, 44, 367

Snow, T. P., Burgh, E. B., & Destree, J. P. 2011, IAU Symp. 280, Poster 1.83

Sollerman, J., Cox, N., Mattila, S., Ehrenfreund, P., Kaper, L., Leibundgut, B., & Lundqvist, P. 2005, *A&A*, 429, 559

Stecher, T. P. 1965, *ApJ*, 142, 1683

Thöne, C. C., Michałowski, M. J., Leloudas, G., Cox, N. L. J., Fynbo, J. P. U., Sollerman, J., Hjorth, J., & Vreeswijk, P. M. 2009, *ApJ*, 698, 1307

Thorburn, J. A., Hobbs, L. M., McCall, B. J., Oka, T., Welty, D. E., Friedman, S. D., Snow, T. P., Sonnentrucker, P., & York, D. G. 2003, *ApJ*, 484, 339

Tielens, A. 2008, *ARA&A*, 46, 289

Tielens, A. 2011, in: C Joblin & A. G. G. M. Tielens (eds.), *PAHs and the Universe*, EAS, 46, 3

Tuairisg, S.Ó., Cami, J., Foing, B. H., Sonnentrucker, P., & Ehrenfreund, P. 2000, *A&A* Suppl., 142, 225

Van der Zwet, G. P. & Allamandola, L. J. 1985, *A&A*, 146, 76

Vos, D. A. I., Cox., N. L. J., Kaper, L., Spaans, M., & Ehrenfreund, P. 2011, in preparation
Welty, D., Federman, S., Gredel, R., Thorburn, J., & Lambert, D. 2006, *ApJ* Suppl., 165, 138
Wallerstein, G., Sandstrom, K., & Gredel, R. 2007, *PASP*, 119, 1268
Xiang, F. Y., Li, A., & Zhong, J. X. 2011, *ApJ*, 733, 91
York, B. A., Ellison, S. L., Lawton, B., Churchill, C. W., Snow, T. P., Johnson, R. A., & Ryan, S. G., 2006, *2006*, 647, 29
Zhou, Z., Sfeir, M., Zhang, L., Hybertsen, M., Steigerwald, M., & Brus, L. 2006, *ApJ*, 638, 105

Discussion

Jacek Krelowski: According to our data the reported correlations between the 8621 DIB strength and reddening are too optimistic.

Nick Cox: This DIB is difficult to measure. It is relatively broad making the continuum placement more insecure. Also blending with stellar lines can be an issue, in particular in lower-resolution spectra of later spectral type stars used by Munari *et al.* In addition, it appears that also the uncertainties in E(B-V) are now becoming more important as measurements of DIB strengths are increasingly more accurate. New surveys with accurate measurements of this diffuse band and extinction are needed to confirm these results.

Gwendolyn Meeus: You said that DIBs are not present in circumstellar envelope of evolved stars. What is the situation for the circumstellar environment of young stars?

Nick Cox: Currently, DIB carriers are not yet - conclusively - detected in environments other than the ISM. Studies have been performed, for example, for evolved stars, Herbig Ae/Be stars, and comets (see Sect. 6 of this chapter).

Mark Rawlings: Have there been any attempts to measure solid state hydrocarbon features (e.g. 3.4 μm feature) towards the 8620.4 Å sightlines.

Nick Cox: Most sightlines for which the 8621 Å DIB has been measured are not heavily reddened (E(B-V) $<$ 2 mag). It would be interesting to explore the connection between the 3.4 μm feature (believed to be due to solid-phase aliphatic hydrocarbons) and the 8621 Å DIB (with an apparent strong relation to dust grains) in sufficiently reddened sightlines.

The Molecular Universe
Proceedings IAU Symposium No. 280, 2011
José Cernicharo & Rafael Bachiller, eds.

doi:10.1017/S1743921311024963

PDRs and XDRs

Rowin Meijerink[1]

[1]Leiden Observatory, Leiden University,
P.O. Box 9513, NL-2300 RA, Leiden, the Netherlands
email: meijerink@strw.leidenuniv.nl

Abstract. An overview of the important thermal and chemical processes in "photon-dominated regions" or "photo-dissociation regions" (PDRs) and "X-ray dominated regions" (XDRs) is presented. Applications of the models are shown to observations of the ultra-luminous infrared galaxy Mrk 231, and the starburst galaxy M 82.

Keywords. PDRs, XDRs, ISM, galaxies

1. Introduction

An overview of the theory and observations of "photon dominated Regions" or "photo-dissociation regions" (PDRs; far-ultraviolet; $6 < E < 13.6$ eV) and "X-ray dominated regions" (XDRs; $E > 1$ keV) is presented. In PDRs and XDRs, the chemical structure and thermal balance are completely determined by the radiation field. Therefore, PDRs and XDRs are direct manifestations of the energy balance of interstellar gas, and their study allows one to determine how the interstellar medium (ISM) survives the presence of stars and AGN (Tielens & Hollenbach 1985; Boland & de Jong 1982; van Dishoeck & Black 1988; Le Bourlot *et al.* 1993; Wolfire *et al.* 1993; Spaans *et al.* 1994; Sternberg & Dalgarno 1995; Stoerzer *et al.* 1998; Spaans 1996; Bertoldi & Draine 1996; Maloney *et al.* 1996; Lee *et al.* 1996; Kaufman *et al.* 1999; Le Petit *et al.* 2002; Meijerink & Spaans 2005; Meijerink *et al.* 2007, and references therein). First, the important processes in XDRs and PDRs are discussed. UV and X-ray irradiation give rise to different gas cloud thermal/chemical structures, and therefore also show different line emission spectra. X-rays penetrate through a larger column of gas (up to $N_{\rm H} \sim 10^{24}$ cm^{-2}) than UV photons ($N_{\rm H} \sim 10^{21} - 10^{22}$ cm^{-2}), and the heating efficiency of X-rays (10-50 percent) is much larger than that of UV irradiation (less than 1 percent). X-rays also give rise to significant ionization fractions (up to $x_{\rm e} \sim 10^{-2} - 10^{-1}$), driving an efficient ion-molecule chemistry, while UV photons are more efficient in dissociating molecules. Therefore, for comparable energy inputs, XDRs have much larger column densities of warm molecular gas than PDRs. Second, the models are applied to observations of the ISM in M 82 and Mrk 231, and the CO ladder in particular, as detected by the various Key Programs on the *Herschel* Space Observatory.

2. Physical processes

2.1. *PDRs*

The global properties of PDRs are determined by a number of physical processes: (i) heating through photo-electric emission (cf., Bakes & Tielens 1994; Weingartner & Draine 2001), (ii) heating by FUV pumping of H_2, followed by collisional de-excitation (cf., Hollenbach & McKee 1979), (iii) heating by cosmic rays (cf., Field *et al.* 1969), (iv)

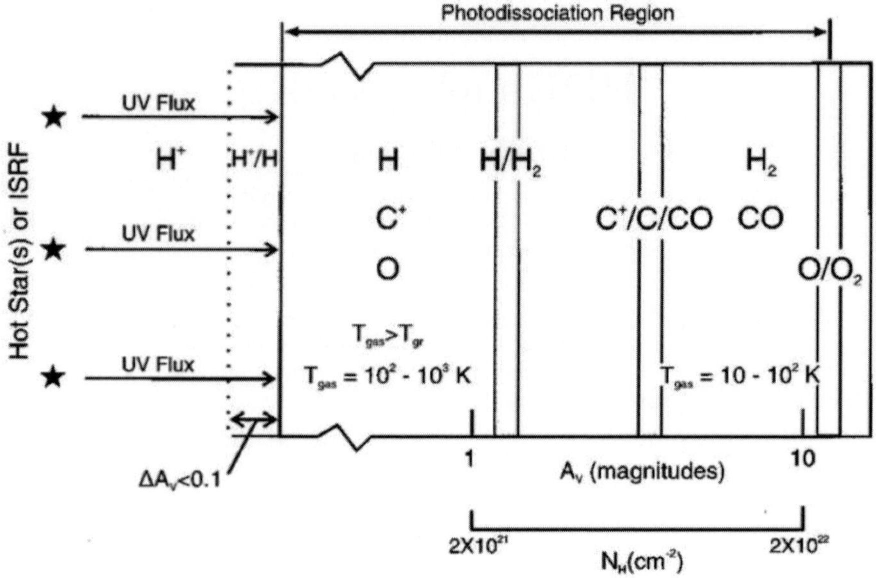

Figure 1. Schematic structure of PDR adapted from Tielens & Hollenbach (1985).

fine-structure line cooling of [CII] 158 μm, [OI] 63 μm, [CI] 609 μm (v) molecular line cooling by warm molecular gas containing CO, H_2, H_2O, OH, and CH (cf., Neufeld *et al.* 1995; Spaans & Silk 2000) (v) ion-molecule reactions driven by the ionization degree of $x_e \sim 10^{-4}$ maintained by the ionization of carbon in the FUV (cf., Black & Dalgarno 1977; van Dishoeck & Black 1986), and (vi) the ionization balance of atomic gas under the influence of photo-ionization driven by FUV photons and counteracting recombination and charge transfer reactions with metals and particularly PAHs (cf., Lepp & Dalgarno 1988; Bakes & Tielens 1994).

A schematic overview of a PDR is shown in Fig. 1. The thermal and chemical balance changes, as one moves into a PDR: The extinction along the line of side increases and the impinging radiation field is attenuated. There are two very fundamental chemical transitions, the H/H_2 and the $C^+/C/CO$ transition. The first one occurs because the Lyman and Werner electronic bands, that are able to dissociate H_2, become optically thick due to self-shielding. At slightly larger extinction, ionized carbon is converted into neutral carbon, as the FUV flux is attenuated by dust absorption. The transformation of neutral carbon to CO occurs when the FUV field is reduced by grain opacity, H_2 shielding and CO self-shielding.

The region, where most hydrogen is still in atomic form and the extinction still very small ($\sim$ few A_V), is usually referred to as the radical region since many carbon hydrides and their ions, e.g., CH, CH^+, CN, HCN, HCO^+, etc., reach their peak abundance here, caused by the presence of both C^+ and H_2 and high temperatures ($T \sim 10^2 - 10^3$ K). Many of the present atoms and molecules are collisionally excited at the abient densities and temperatures, and emit bright in the mid- and far-infrared, millimeterm and sub-millimeter.

There are four parameters that determine the global characteristerics of any PDR: (i) the strength of the impinging radiation field, in units of the Habing (1969) G_0 or Draine (1978) $I_{\rm UV}$ radiation field, (ii) the temperature and the ambient hydrogen, $n_{\rm H} = n({\rm H}) + n({\rm H_2})$, set the rates of the chemical reactions and excitation of the coolants, (iii)

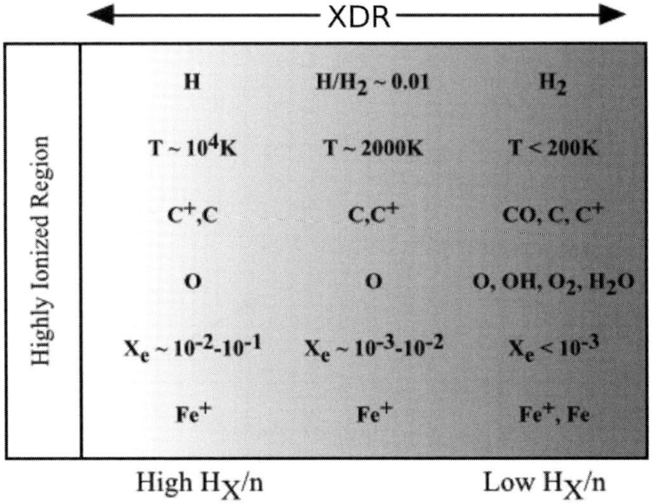

Figure 2. Schematic structure of XDR adapted from Maloney *et al.* (1996).

the metallicity, Z, in units of the solar value $Z_{\odot}$, determines the total abundances for carbon- and oxygen-bearing species, (iv) the spectral shape, characterized by the color temparature $T_{\rm eff}$, fixes the distribution of photon flux over energy.

2.2. *XDRs*

The following processes are important in XDRs: (i) photo-ionization heating (i.e., Coulomb interaction of keV electrons with thermal electrons), which is dominant by a large factory over photo-electric heating by dust grains (Maloney *et al.* 1996). (ii) cooling by metastable lines of [CI] 9823, 9850 Å and [OI] 6300 Å, fine-structure cooling by [OI] 63, 145 μm and [CII] 158 μm, as well as Lyman α emission (Maloney *et al.* 1996; Tielens & Hollenbach 1985; Spaans *et al.* 1994), (iii) molecular cooling by warm molecular gas, containing CO, H_2, H_2O and OH as well as gas-grain cooling where warm gas is cooled a the surfaces of lower temperature dust grains (cf., Neufeld *et al.* 1995; Spaans & Silk 2000) (iv) ion-molecule reactions driven by the ionization degree of 10^{-4} maintained by ionization of carbon (c.f., Black & Dalgarno 1977; van Dishoeck & Black 1986), and (v) the ionization balance of atomic gas under the influence of photo-ionization reactions driven by X-ray photons and charge transfer (c.f., Lepp & Dalgarno 1988; Maloney *et al.* 1996).

A schematic overview of an XDR is shown in Fig. 2. The global structure of the XDR is characterized by the density $n_{\rm H}$ and the energy deposition rate $H_{\rm X}$ per hydrogen atom. The heating in XDRs is driven by photo-ionization, and the heating efficiency is much larger in XDRs than in PDRs, where this is between 0.3 and 1.0 percent. The gas in XDRs is exposed to both X-ray and FUV, that is internally created by excitation of the Lyman-Werner and Lyman α bands. Attenuation of X-rays is done by atomic electronic absorptions. The lowest energy photons are attenuated most strongly.

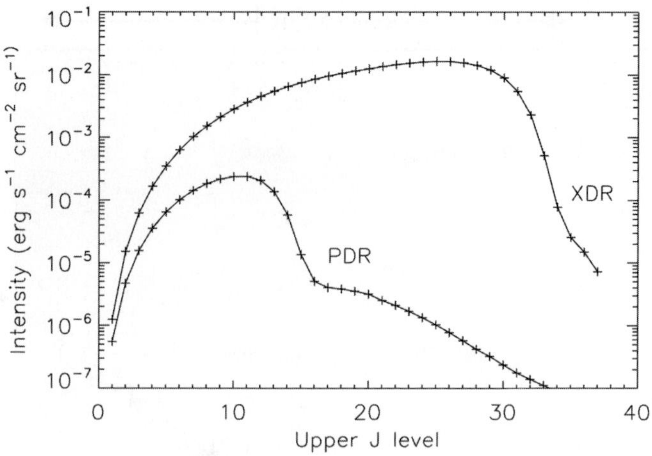

Figure 3. Comparison of PDR and XDR CO ladders (Meijerink 2006; Spaans & Meijerink 2008).

3. Differences and consequences for diagnostics

There are four very important differences between PDRs and XDRs: (i) X-rays penetrate much larger column densities than UV photons, (ii) the gas heating efficiencies in XDRs range from 10 to 50 percent, which is much larger than the $< 1\%$ in PDRs, (iii) dust heating is much more efficient in PDRs than in XDRs, and (iv) high ionization levels in XDRs drive an ion-molecule chemistry over a large column density.

As a result, XDRs produce larger column densities of warmer gas. Thus, identical incident energy densities give very different emergent continuum and line spectra. One example is given in Fig. 3 (Meijerink 2006; Spaans & Meijerink 2008), where the CO line spectral energy distribution is given for a model with density $n = 10^5$ cm^{-3} and $G_0 = 10^5$ ($F_X = 160$ erg cm^{-2} s^{-1}). The CO ladder for the PDR model turns over at the $J = 10-9$ transition, while CO fluxes of the XDR continue rising up to $J = 26-25$, for this particular model. It is therefore expected, that high-J CO lines are excellent XDR tracers, when a good sampling of the CO ladder is obtained.

4. Application of PDR and XDR models

The predictions derived from the PDR and XDR models, were in part the motivation for the Herschel Open Time Key Program *HerCULES* (PI: P.P. van der Werf), which uniformly and statistically measures the neutral gas cooling lines in a flux limited sample of 29 (U)LIRGs, selected from the Revised IRAS Bright Galaxy Sample (Sanders *et al.* 2003). The sample consists of 6 ULIRGs ($S_{60} > 12.19$ Jy) and 23 LIRGs ($S_{60} > 16.8$ Jy). Full high-resolution SPIRE FTS scan are made between wavelengths $\lambda \sim 200$ to 670 μm, covering CO(4-3) to CO(13-12), [CI], a number of H_2O lines, and several other bright lines. Also, PACS line scans of the [CII] 158 μm and [OI] 63,145 μm lines are obtained.

4.1. *Mrk 231*

One source in this sample is Mrk 231, the most luminous galaxy in the Revised IRAS Bright Galaxy Sample (Sanders *et al.* 2003). It has a redshift $z = 0.042170$ and a

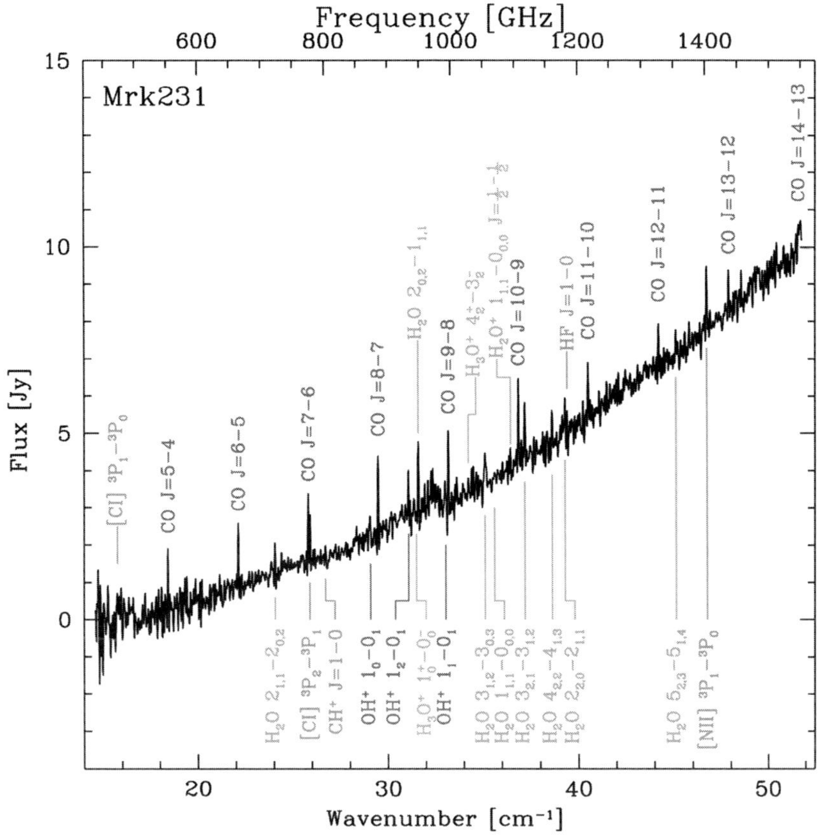

Figure 4. SPIRE FTS Spectrum Mrk 231 (van der Werf *et al.* 2010).

luminosity distance $D_L = 192$ Mpc, $L_{\rm IR}(8 - 1000\ \mu\rm m) = 4.0 \times 10^{12}\ L_\odot$. Mrk 231 contains a luminous, optically visible AGN, classified as s Seyfert 1 or a broad absorption line QSO (Boksenberg *et al.* 1977). A highly absorbed power-law X-ray spectrum was observed by Braito *et al.* (2004) with $L_{\rm X} = 6^{+0.6}_{-0.3} \times 10^{43}$ erg s^{-1} between 2 and 10 keV, so it is an excellent candidate to observe XDR signatures in the line emission. However, Mrk 231 also contains a kpc size disk harbouring intense star formation (Taylor *et al.* 1999), so a PDR contribution is also expected.

The obtained SPIRE FTS spectrum is shown in Fig. 4 (van der Werf *et al.* 2010). It contains at least the following lines: 10× CO (5-4 to 14-13), 2× [CI], [NII], 8× H_2O, 3× OH^+, H_2O^+, 2× H_3O^+, CH^+, and HF. There are still an number of unidentified lines at this point. Very surprising are the bright emission lines from OH^+ and H_2O^+, which are highly reactive, transient ionic species. Furthermore, the water lines have strenghts of 20-50 percent of the CO lines. This is not observed for the Orion Bar or the starburst galaxy M 82, where the strength of the water lines relative to CO is at least an order of magnitude lower.

A fit to the CO lines is shown in Fig. 5: a diffuse PDR component ($n = 10^{3.5}$ cm^{-3}, $G_0 = 10^2$) and a high density ($n = 10^5$, $G_0 = 10^{3.5}$) embedded PDR component are needed to fit the lower CO transition ($J < 10 - 9$), while for the higher-J CO lines an XDR component is fitted. This last component has a density $n = 10^{4.2}$ cm^{-3} and incident

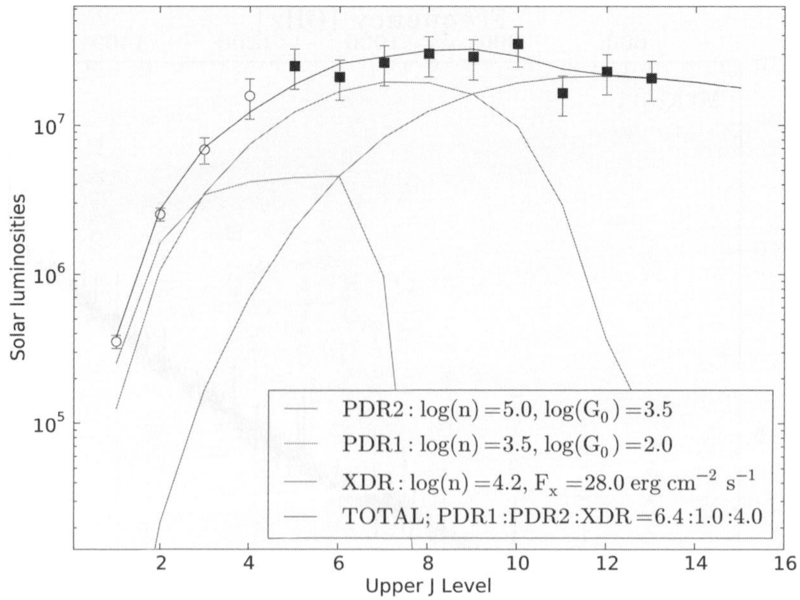

Figure 5. PDR/XDR fit to the CO ladder of Mrk 231 (van der Werf *et al.* 2010).

flux $F_x = 28$ erg s^{-1}cm^{-2}, a flux that can be produced out to a distance of 160 pc, with the observed AGN. The surface area ratios between the different component is PDR1: PDR2: XDR = 5.4:1.0:4.0.

It is also possible to fit a very dense, high illumination PDR to the very high-J CO lines, with $n = 10^{6.5}$ cm^{-3} and $G_0 = 10^5$. The ratio between the medium and high excitation component is 1.0:0.03. Here the smal surface are for the high excitation PDR indicates a number of small high density clumps in a very strong UV field. Since the radiating surface of the high excitation PDR is about 30× smaller than that of the medium excitation PDR, but its density about 30× larger, the H_2 masses in these two components must be comparable. In this scenario about half of the mass would be contained in 0.7% of the volume, and the dust in this component would be heated to $T_{\rm dust} = 170$ K, which is simply not observed for Mrk 231. Therefore, the possible presence of an extreme PDR component is not very likely.

4.2. *M 82*

It is very illustrative to compare the CO observations of Mrk 231 to those obtained for M 82 with *Herschel* HIFI (Loenen *et al.* 2010) and SPIRE (Panuzzo *et al.* 2010), and combined with ground based data (Ward *et al.* 2003). The two CO SLEDs, for ^{12}CO and ^{13}CO are shown in Fig. 6. The observations can be best reproduced with a combination of one low and two high-density components. The low-density component, which dominates in the lowest CO lines ($J \leqslant 3$), is found to have a density of $n = 10^{3.5}$ cm^3 and a UV flux of $G_0 = 10^2$. This component represents the extended molecular ISM. The mid-J CO lines ($4 < J < 7$) are mostly produced by clouds that have a density of $n = 10^5$ cm^3 and a radiation field of $G_0 = 10^{2.75}$. At transitions higher than 7 (and ^{13}CO $J > 5$), the line emission comes from clouds with even higher densities ($n = 10^6$ cm^3) and radiation fields ($G_0 = 10^{3.25}$). These two models represent the dense, star-forming molecular clouds.

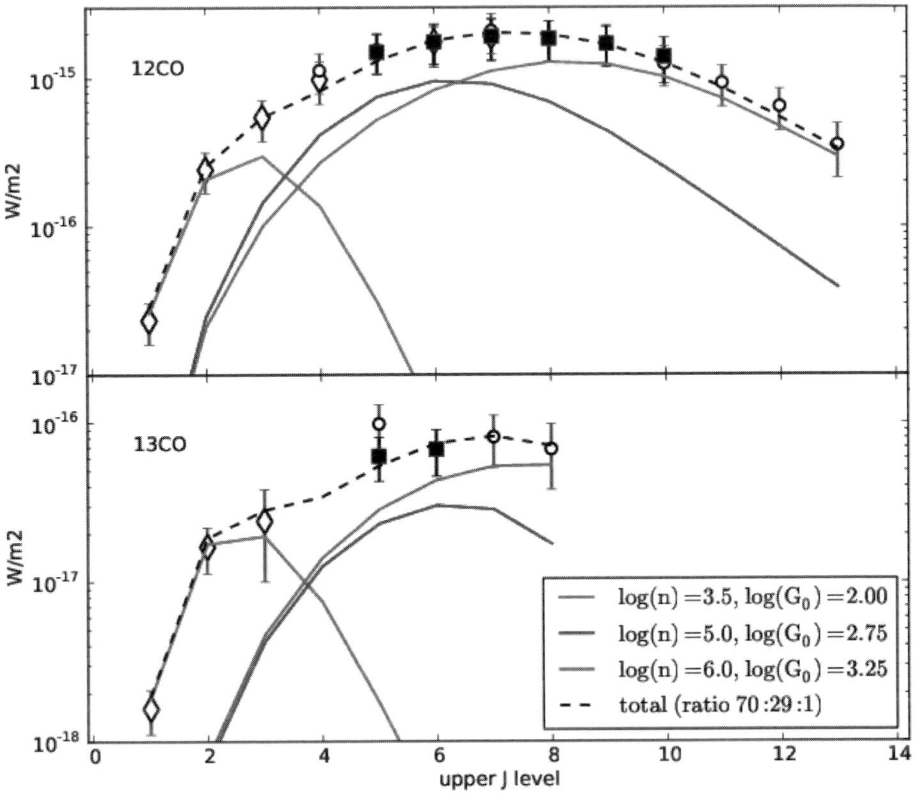

Figure 6. PDR fit to the ^{12}CO and ^{13}CO ladders of Mrk 82 (Loenen *et al.* 2010).

4.3. *Other diagnostics and caveats*

A major issue in the study of the interstellar medium in galaxies is that a single pointing usually contains a mix of components that is affected by different physical processes. It may contain (i) quiescent molecular and atomic gas (ii) star forming molecular gas (PDRs), (iii) AGN (X-ray) excited gas (XDRs), (iv) cosmic ray heated gas, (v) shocks (vi) mechanically (dissipation of turbulence) heated gas, and (vi) warm, very obscured gas (hot cores). As we saw for Mrk 231, it is possible to reproduce the high-J CO lines with an XDR component or a very extreme PDR component. Also, when a normal PDR component receives additional heating from, e.g., dissipating turbulence, it is possible to produce a flat CO ladder for Mrk 231 for the very high-J CO lines.

Additional information is given by other tracers. Bright OH^+ and H_2O^+ lines have been detected in Mrk 231. The estimated lower limits OH^+ and H_2O^+ abundances relative to H_2 are of order 2×10^{-10} in the central 160 pc. Such abundances require an efficient and penetrative source of ionization in the molecular gas, since the production of OH^+ is mainly driven by $H^+ + O \rightarrow O^+ + H$ followed by $O^+ + H_2 \rightarrow OH^+ + H$ and $H^+ + OH \rightarrow OH^+ + H$. The key species H^+ is 2 to 3 orders of magnitude more abundant in the XDR model than in the high excitation PDR, and the OH^+ abundance is larger by a comparable amount, see Fig. 7. A similar argument can be made for H_2O^+, which is formed by $OH^+ + H_2 \rightarrow H_2O^+ + H$. The extraordinary luminosity of the OH^+ and H_2O^+ (and to a lesser extent CH^+) lines in Mrk 231 is underlined by a comparison with the

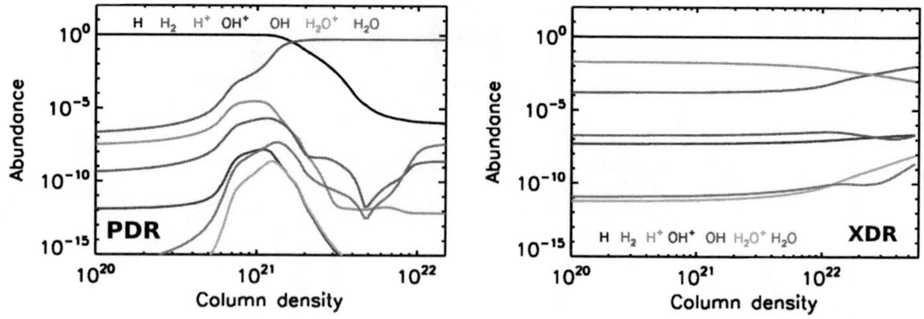

Figure 7. An extreme PDR and XDR used to fit the high-J CO lines (based on Meijerink *et al.* 2007). Note the overall much higher abundances of H^+, OH^+, and H_2O^+ in the XDR.

SPIRE spectrum of M82 and also the Orion bar PDR (Habart *et al.* 2010), which show no trace of OH^+ or H_2O^+, and only weak CH^+ emission in the Orion Bar, while in Mrk 231 the lines are only a factor 2 to 3 fainter than the CO lines.

Enhanced cosmic ray fluxes in a starburst environment will increase the degree of ionization and hence the production of OH^+ and H_2O^+ in a PDR (Meijerink *et al.* 2011), but cosmic ray fluxes have to be 2 to 3 orders of magnitude higher to obtain similar abundances as seen in the XDR. Also, they do not elevate the temperatures to the level required to produce the highest CO lines (Meijerink *et al.* 2006). It is thus the combination of strong high-J CO lines and high OH^+ and $H2O^+$ abundances that reveals X-ray driven excitation and chemistry in Mrk 231.

5. Summary & Conclusions

The *Herschel* Space Observatory opened a very new window in the study of the ISM in nearby galaxies. The SPIRE FTS instrument, covering the wavelength range between $\lambda = 200 - 670$ μm, contains a whealth of diagnostic lines, such as the CO $J = 5 - 4$ to $13 - 12$, water lines with excitation temperatures up to $T \sim 500$ K, and transient species OH^+ and H_2O^+.

This allowed us to study the excitation conditions of the gas in Mrk 231, and we found that both star formation as well as black hole accretion have to be considered simultaneously in order to model the obtained CO ladder. However, the PDR/XDR fit is not unique when considering only the CO lines, and information on dust temperature and other species has to be considered as well.

The spectra obtained in the HerCULES Open Time Key Program will be unique templates for the study of the ISM in high-redshift galaxies with ALMA. This will allow us, in combination with the PDR/XDR models, to characterize the star forming gas in the early universe.

References

Bakes, E. L. O. & Tielens, A. G. G. M. 1994, *ApJ*, 427, 822
Bertoldi, F. & Draine, B. T. 1996, *ApJ*, 458, 222
Black, J. H. & Dalgarno, A. 1977, *ApJS*, 34, 405
Boksenberg, A., Carswell, R. F., Allen, D. A., Fosbury, R. A. E., Penston, M. V., & Sargent, W. L. W. 1977, *MNRAS*, 178, 451
Boland, W. & de Jong, T. 1982, *ApJ*, 261, 110

Braito, V., *et al.* 2004, *A&A*, 420, 79
Draine, B. T. 1978, *ApJS*, 36, 595
Field, G. B., Goldsmith, D. W., & Habing, H. J. 1969, *ApJL*, 155, L149
Fischer, J., *et al.* 2010, *A&A*, 518, L41
Habart, E., *et al.* 2010, *A&A*, 518, L116
Habing, H. J. 1969, *Bull. Astr. Inst. Netherlands*, 20, 177
Hollenbach, D. & McKee, C. F. 1979, *ApJS*, 41, 555
Kaufman, M. J., Wolfire, M. G., Hollenbach, D. J., & Luhman, M. L. 1999, *ApJ*, 527, 795
Le Bourlot, J., Pineau Des Forets, G., Roueff, E., & Flower, D. R. 1993, *A&A*, 267, 233
Lee, H.-H., Herbst, E., Pineau des Forets, G., Roueff, E., & Le Bourlot, J. 1996, *A&A*, 311, 690
Le Petit, F., Roueff, E., & Le Bourlot, J. 2002, *A&A*, 390, 369
Lepp, S. & Dalgarno, A. 1988, *ApJ*, 324, 553
Loenen, A. F., *et al.* 2010, *A&A*, 521, L2
Maloney, P. R., Hollenbach, D. J., & Tielens, A. G. G. M. 1996, *ApJ*, 466, 561
Meijerink, R. 2006, Ph.D. Thesis, Leiden Observatory
Meijerink, R. & Spaans, M. 2005, *A&A*, 436, 397
Meijerink, R., Spaans, M., & Israel, F. P. 2006, *ApJL*, 650, L103
Meijerink, R., Spaans, M., & Israel, F. P. 2007, *A&A*, 461, 793
Meijerink, R., Spaans, M., Loenen, A. F., & van der Werf, P. P. 2011, *A&A*, 525, A119
Neufeld, D. A., Lepp, S., & Melnick, G. J. 1995, *ApJS*, 100, 132
Panuzzo, P., *et al.* 2010, *A&A*, 518, L37
Sanders, D. B., Mazzarella, J. M., Kim, D.-C., Surace, J. A., & Soifer, B. T. 2003, *AJ*, 126, 1607
Spaans, M. 1996, *A&A*, 307, 271
Spaans, M., Tielens, A. G. G. M., van Dishoeck, E. F., & Bakes, E. L. O. 1994, *ApJ*, 437, 270
Spaans, M. & Meijerink, R. 2008, *ApJL*, 678, L5
Spaans, M. & Silk, J. 2000, *ApJ*, 538, 115
Sternberg, A. & Dalgarno, A. 1995, *ApJS*, 99, 565
Stoerzer, H. & Hollenbach, D. 1998, *ApJ*, 495, 853
Taylor, G. B., Silver, C. S., Ulvestad, J. S., & Carilli, C. L. 1999, *ApJ*, 519, 185
Tielens, A. G. G. M. & Hollenbach, D. 1985, *ApJ*, 291, 722
van der Werf, P. P., *et al.* 2010, *A&A*, 518, L42
van Dishoeck, E. F. & Black, J. H. 1986, *ApJS*, 62, 109
van Dishoeck, E. F. & Black, J. H. 1988, *ApJ*, 334, 771
Ward, J. S., Zmuidzinas, J., Harris, A. I., & Isaak, K. G. 2003, *ApJ*, 587, 171
Weingartner, J. C. & Draine, B. T. 2001, *ApJS*, 134, 263
Wolfire, M. G., Hollenbach, D., & Tielens, A. G. G. M. 1993, *ApJ*, 402, 195

Discussion

D. Field: Could you use OH megamasers maps and assounded structural knowledge to create more detailed models

R. Meijerink: That is a very useful suggestion. I will take a look at the available observations to see whether it will better constrain the geometric, that is currently being constructed.

S. Muller: Did you see the broad wings due to the molecular outflow in your spectra of Mrk 231

R. Meijerink: No it is not observed in the SPIRE spectrum, as the spectral resolution is only of order $\sim$1500 km/s. They are however observed in the OH lines with PACS (Fischer *et al.* 2010).

D. Semenov: How sensitive are your results of the XDR chemical modeling to the shape & hardness of the X-ray spectrum.

R. Meijerink: We are using a power law spectrum. Making the spectrum shallower, will resul in a more extended XDR, and a temperature profile with a less strong gradient.

D. Lis: Mrk 231 is the only source that I know where HF is seen in emission. Is HF included in your models and, if so, can you reproduce the observed line intensity

R. Meijerink: It is currently not in the code, but it can be easily added. I have thus not tried to reproduce the HF line this far.

The Molecular Universe
Proceedings IAU Symposium No. 280, 2011
José Cernicharo & Rafael Bachiller, eds.

doi:10.1017/S1743921311024975

Turbulence in the Diffuse Interstellar Medium

Edith Falgarone[1], **Benjamin Godard**[2] **and Pierre Hily-Blant**[3]

[1]LERMA-LRA, Ecole Normale Supérieure & Observatoire de Paris, Paris, France
email: edith.falgarone@ens.fr

[2]CAB/CSIC, Madrid, Spain, email: bgodard@cab.inta-csic.es

[3]IPAG, Grenoble, France, email: pierre.hily-blant@obs.ujf-grenoble.fr

Abstract. The diffuse interstellar medium (ISM) hosts the first steps of interstellar chemistry and the seeds of dense structures. Since its turbulent pressure by far exceeds its thermal pressure, turbulence must play a prominent role in its evolution. Fed at galactic scales, turbulent energy cascades down to the dissipation scales, but as in both laboratory and atmospheric turbulence, it does so in an intermittent way : only a tiny fraction of the small-scales is fed by the turbulent cascade, so that dissipation occurs in bursts. In diffuse molecular clouds, where they can be observed, the signatures of intermittency are: (1) the non-Gaussian statistics of velocity increments, and (2) the existence of coherent structures of intense velocity-shear that appear to channel the large-scale turbulent energy down to milliparsec scales. Attempts at modelling the warm chemistry triggered in the diffuse ISM by bursts of turbulent dissipation are promising : in this framework, the so far unexplained molecular richness observed in this medium is naturally understood, in particular its CH^+, HCO^+ and CO abundances. Turbulent dissipation is also likely at the origin of the H_2 rotational line emission of the diffuse ISM and of a significant fraction of its [C II] emission.

Keywords. astrochemistry, turbulence, ISM: clouds, ISM: general, ISM : evolution, ISM: kinematics and dynamics, ISM: molecules, ISM: structure, submillimeter

1. Introduction: the diffuse neutral interstellar medium

The definition of the diffuse neutral interstellar medium (ISM) has evolved continuously since its fortuitous discovery by Hartmann in 1904, depending on the observational techniques used to probe it. It is now generally agreed that it is a low density medium ($n_H <$ a few 10^3 cm^{-3}) weakly shielded from the interstellar radiation field (ISRF) ($A_V <$ a few mag), therefore comprising the diffuse atomic, diffuse molecular and translucent media, following the terminology of Snow & McCall (2006). An even broader definition of the diffuse ISM is adopted in this paper. It comprises the above plus the edges of molecular clouds as traced by their $^{12}CO(J{=}1\text{-}0)$ emission, i.e. all the gas that is not in dense cores and/or star forming regions. Equivalently, the diffuse ISM can be considered to be all material with hydrogen column density less than a few $\sim 10^{21}$ cm^{-2} at the parsec scale. Such a broad definition is motivated by the shapes of the probability distribution functions (*pdf*) of the column density (inferred from dust extinction) of nearby molecular clouds : they have a log-normal part from $A_V \sim 0.5$ mag, up to $A_V \sim$ a few magnitude, and power-law tails at higher column densities, characteristics of dense cores and filaments, present only in clouds actively forming stars (Kainulainen *et al.*, 2009). The log-normal contribution comes from the diffuse ISM of these molecular clouds. The diffuse ISM, at large, may therefore be seen as all the gas that does not belong to small-scale self-gravitating structures such as dense cores, dense filaments and star-forming

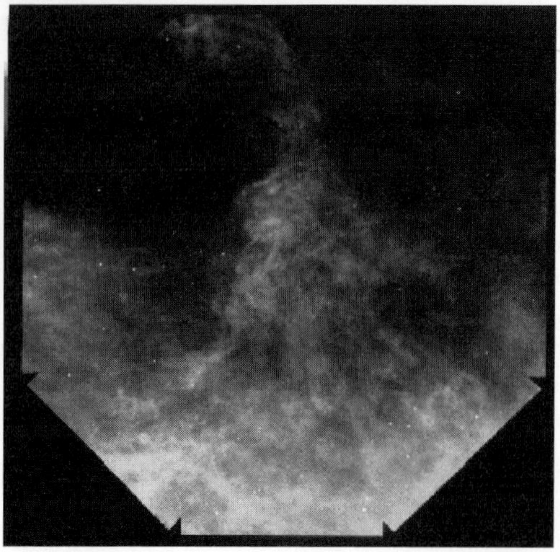

Figure 1. IRIS 100μm image of the dust thermal emission of the North Celestial Loop (Miville-Deschênes & Lagache 2005) rising $\sim$ 100 pc above the Galactic plane visible at the bottom. The size of the map is about 30°. The 40'×30' field of the Polaris Flare discussed in Sect. 4 is located close to the center of the image.

regions. Note that the diffuse medium, as defined above, comprises the bulk of the mass of molecular clouds, at the 30 pc scale (e.g. Goldsmith *et al.*, 2008).

1.1. *Thermal and non-thermal pressures*

The shape of the cooling function dominated by [C II] line emission makes the diffuse ISM thermally bistable for the pressure and metallicity conditions prevailing in the solar neighbourhood (Field *et al.*, 1969). The balance between heating and radiative cooling processes generates two stable phases at very different temperatures and densities, $T \sim 100$K and $n_{\rm H} \sim 30$ cm^{-3} for the cold neutral medium (CNM), $T \sim 8000$K and $n_{\rm H} \sim 0.3$ cm^{-3} for the warm neutral medium (WNM). However, a broad range of kinetic temperatures, pervading the thermally unstable range, is usually inferred from UV absorption lines sampling the diffuse ISM (e.g. Fitzpatrick & Spitzer, 1997), and recent sensitive HI observations also probe temperatures in this range (Begum *et al.*, 2010; Heiles & Troland, 2003). The latter authors infer that about half the WNM is at temperatures in the range 500 – 5000 K, thus evolving between the two stable phases.

The diffuse ISM is now modeled as a turbulent medium, evolving far out of thermal pressure equilibrium, as illustrated by the dust thermal emission of the North Celestial Loop (Fig. 1). Whether it is cooling (and condensing) WNM, destabilized in colliding flows, as in the numerical simulations of Audit & Hennebelle (2005), or CNM transiently heated by turbulent dissipation, in the framework discussed below, is presently an open issue.

Because it sits in the gravitational potential of the Galaxy, the diffuse ISM has to be supported by a pressure that is about 10 times its thermal pressure, $P_{tot} = 3\times10^{-12}$ dynes cm^{-2}, or $P_{tot}/k \sim 2\times10^4$ cm^{-3} K (Cox 2005). The non-thermal contributions to the total pressure are due to supersonic turbulence and magnetic fields, in rough equipartition, as shown by measurements of magnetic field intensity (Crutcher *et al.*, 2010) and HI linewidths (Haud & Kalberla, 2007). The distribution of thermal pressures in the solar neighbourhood, inferred from [CI] fine-structure absorption lines towards nearby stars (Jenkins & Tripp, 2011) peaks at about $P_{th}/k \sim 3\times10^3$ cm^{-3} K, with large fluctuations, up to a few 10^4 cm^{-3} K. It is interesting that the total non-thermal pressure in the Galactic plane is of the same order as the largest thermal pressure observed, suggesting that the non-thermal energy eventually and occasionally degrades into thermal energy.

These large fluctuations of thermal pressure are a sign of a complex dynamic evolution that may provide the missing important clues to the so-called "outstanding mysteries" raised by the observations of diffuse molecular gas (see the review of Snow & McCall, 2006): the ubiquitous small scale structure, down to AU-scales, and its remarkable molecular richness found in this hostile medium weakly protected from UV radiation (e.g. Liszt & Lucas, 1998).

1.2. *Two outstanding mysteries*

A resilient major puzzle is the existence in the diffuse ISM of molecular species that have a formation endothermicity far above the available thermal energy: their large observed abundances cannot be reproduced by chemistry models driven by UV photons and cosmic-rays. This has been known for 70 years in the case of CH^+ that not only forms through the highly endothermic reaction $C^+ + H_2$ ($\Delta E/k = 4940$K) but also is rapidly destroyed by collisions with H_2. Until recently, observation of its electronic transition in absorption against nearby stars was the only means of investigating this species in the diffuse gas (Crane *et al.*, 1995; Gredel *et al.*, 1997; Weselak *et al.*, 2008) because the frequency of its ground-state rotational transition (835 GHz) falls in an opaque domain of the atmosphere. Hence, only the solar neigbourhood had been sampled. The Herschel/HIFI observations of the CH^+(1-0) and $^{13}CH^+$(1-0) transitions in absorption against the dust continuum emission of remote sources (Falgarone *et al.*, 2010a,b) now extend the puzzle raised by the large abundances of CH^+ in the local diffuse medium to the whole Galactic disk (Fig. 2). Similar large CH^+ abundances are found over a broad range of total hydrogen column densities with a similar large scatter by a factor of 10 (or more) about the average value. The puzzle is even sharper now that SH^+ is also observed to be abundant in the diffuse ISM : this species has a larger formation endothermicity, via $S^+ + H_2$, than that of CH^+, $\Delta E/k = 9860$K (Menten *et al.*, 2011; Godard *et al.*, 2011).

Another outstanding and more recent puzzle is the origin of the H_2 pure rotational line emission of the diffuse ISM. Its discovery dates back to the ISO-SWS detection of the four lowest rotational transitions across a large pathlength of diffuse gas in the Galactic plane, $N_{\rm H} \sim 10^{22}$ cm^{-2}, equivalent to 16 mag (Falgarone *et al.*, 2005). Given the high energy of the upper level of these transitions, 510 K for the S(0) line, up to 2540 K for the S(3) line, and the absence of intense UV radiation field able to excite these levels by fluorescence, a non-thermal excitation of some kind is required. This Galactic emission was modelled by advocating heating of the gas by turbulent dissipation. The dissipation regions, distributed along the line of sight, were either hundreds of low velocity magneto-hydrodynamical (MHD) shocks (8 to 12 km s^{-1}) or thousands tiny regions of intense velocity-shears, heating the gas by ion-neutral friction and/or viscous dissipation. The interesting new result of these models was that a fraction as small as a few per cent of warm gas, heated by dissipation of mechanical energy, was sufficient to reproduce the observed line intensities, and their line ratios. New detections of these lines have been performed by Spitzer/IRS, in the diffuse medium sampled in the fields of supernova remnants (Hewitt *et al.*, 2009), in a translucent field of the Taurus molecular complex (Goldsmith *et al.*, 2010), in the Polaris Flare (Falgarone *et al.*, in prep.) and in diffuse clouds of the Chamaeleon complex and LDN 1780 (Ingalls *et al.*, in prep.): in these studies, UV excitation cannot explain the line intensities and non-thermal excitation, likely fed by turbulence, is required.

2. A few words on turbulence

The definition of turbulence is built on experiment. Turbulence is an instability of laminar flows that develops as soon as the inertial $\mathbf{v}\cdot\nabla\mathbf{v}$ forces greatly exceed the viscous

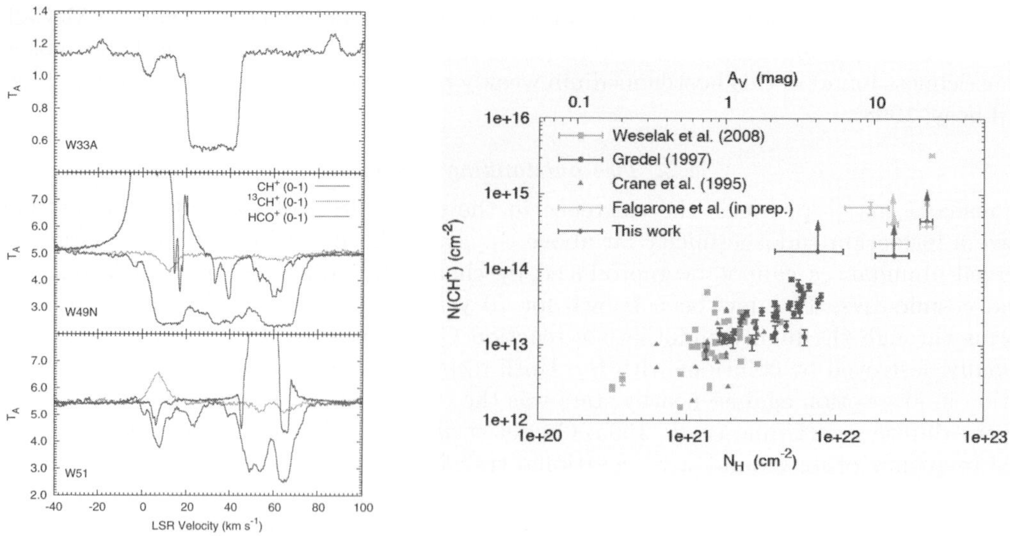

Figure 2. Left: CH^+(1-0), $^{13}CH^+$(1-0) and HCO^+(1-0) spectra obtained with the Herschel/HIFI instrument in the direction of three remote star forming regions. Note the saturated shape of the CH^+ lines over large velocity intervals and the difference between the lineprofiles of HCO^+ and CH^+. **Right:** CH^+ column density vs. total hydrogen column density on sight-lines towards nearby stars and far-IR star-forming regions. All data sets are consistent with a linear increase of CH^+ with the total hydrogen column density, with however a large scatter, providing abundances in the range $N(CH^+)/N_H = 10^{-9} - 2 \times 10^{-8}$.

$\nu\Delta\mathbf{v}$ forces (ν is the kinematic viscosity) i.e. when the Reynolds number $Re = lv_l/\nu$, at a scale l of characteristic velocity v_l, exceeds a few hundreds. This instability at scale l is at the origin of an energy transfer to smaller scales, that eventually become unstable too and transfer their kinetic energy to still smaller scales, etc... This is the turbulent cascade that develops between the integral scale, L, at which energy is injected, and the dissipation scale l_D, close to the particle mean-free-path, where energy is dissipated into heat due to the particle viscosity. The timescale for the growth of this instability is of the order of the turnover time $\tau_l = l/v_l$ at each length scale l. Kolmogorov (1941) predicted the self-similar behavior of the velocity field in incompressible turbulence by postulating a dissipationless cascade characterized by a transfer rate of kinetic energy independent of scale, $\epsilon \propto v_l^2/\tau_l = v_l^3/l$, hence the well-known scaling $v_l \propto l^{1/3}$. It is easy to demonstrate that this assumption leads to an energy spectrum $E(k) \propto k^{-5/3}$ known as the Kolmogorov spectrum. $E(k)$ has the dimension of a kinetic energy per unit mass and unit wavenumber because the average specific kinetic energy at scale $l = 2\pi/k$ is $\langle v_l^2 \rangle = \int_k^\infty E(k')dk'$. In Kolmogorov turbulence, the turnover timescale τ_l therefore decreases towards small-scales while the velocity gradient $v_l/l \propto l^{-2/3}$ slowly increases.

For half a century now, turbulence has been recognized to be intermittent i.e. the smaller the scale, the larger the spatio-temporal velocity fluctuations, relative to their average value (Landau & Lifchitz 1959, Kolmogorov 1962). Turbulent energy is not evenly distributed in space and time by the turbulent cascade : at each step of the cascade, the active sub-scales do not fill space so that the subset of space on which the active scales are distributed has a multifractal geometry (see the review of Anselmet *et al.*, 2001 and the book of Frisch, 1995). The statistical properties of the velocity fluctuations have been widely studied experimentally in laboratory and atmospheric flows : in all cases,

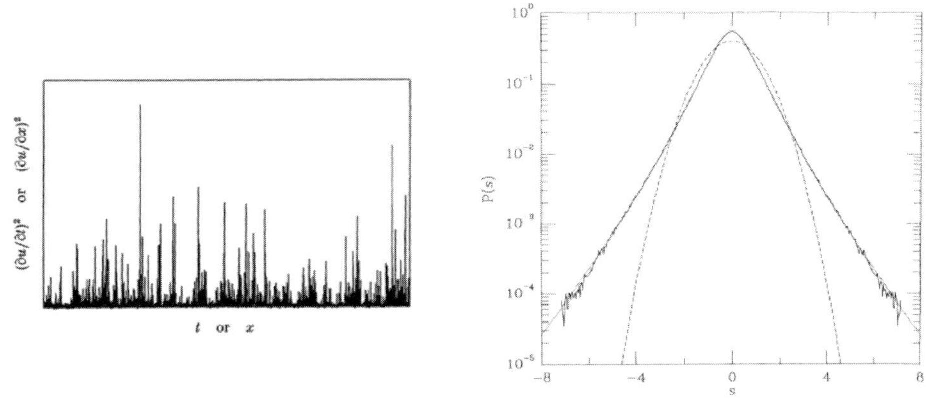

Figure 3. Left: A record of the square of the velocity derivative (with respect to time or space) measured in the turbulent wake of a cylinder (Sreenivasan *et al.*, 1989). This signal can be viewed as the rate of dissipation of turbulent kinetic energy. The rare largest excursions are more frequent than they would be if the signal were Gaussian. **Right:** Non-Gaussian probability distribution of the longitudinal velocity differences $\delta v_x(s) = v_x(x+s) - v_x(x)$ measured in such a flow over a lag s much smaller than the integral scale. The dashed line shows a Gaussian probability distribution with same dispersion.

the statistics of *velocity derivative and increments* signals are found to be non-Gaussian, with large departures from the average more frequent than for a Gaussian distribution (Fig. 3). The *pdf* of the turbulent velocity field, in turn, remains Gaussian. Moreover, the departure of the *velocity increments pdf*s from a Gaussian distribution increases as the lag over which the increments are measured decreases. All the functions of the velocity involving a spatial derivative have therefore non-Gaussian *pdf*s : the velocity gradients $(\partial_i v_i)$ and shears $(\partial_j v_i)$ and, accordingly, the rate-of-strain $S_{ij} = \partial_j v_i + \partial_i v_j$ and the dissipation rate $\epsilon_D = \frac{\nu}{2}\Sigma_{ij}(\partial_j v_i + \partial_i v_j)^2$, with non-Gaussian wings more pronounced at small scale.

The quantitative signature of intermittency appears in the behavior of the high-order structure functions of the longitudinal velocity field measured over a lag s, $\langle[\delta v_x(s)]^p\rangle \propto s^{\zeta_p}$. This relation is statistical, not deterministic, and the brackets hold for an average over an "appropriate" volume with respect to s^3. The anomalous scaling of the exponents $\zeta_p \neq p/3$ characterizes the degree of intermittency and provides the multifractal dimension of the most singular structures, i.e. that of the subset of space where the smallest active regions, and turbulent dissipation, are distributed (She & Levêque, 1994; Anselmet *et al.*, 2001). In incompressible turbulence, the so-called *active small scales* are those of largest vorticity or velocity shear.

Another essential facet of turbulent intermittency is that these active small scales are not randomly distributed in space but are organized into coherent structures – *the sinews of turbulence*, as qualified by Moffatt *et al.* (1994). In incompressible turbulence, these structures tend to be filamentary as shown in Fig. 4, while those of highest rate-of-strain and dissipation are rather in the form of sheets or ribbons, following the analysis of numerical simulations by Moisy & Jiménez (2004). These structures are remarkable in the sense that they are both large scales i.e. their length is comparable to the integral scale of turbulence, and small scale structures i.e. they have substructure down to the dissipation scale. The coupling between large scales and small scales operated by turbulence is repeatedly found in simulations. Mininni *et al.* (2006) find that small-scale intermittency is more pronounced in turbulent fields where the large-scale shear is larger, and

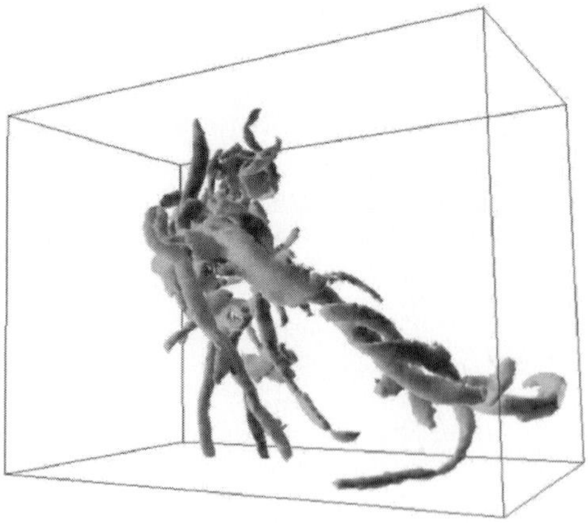

Figure 4. Structure of most intense vorticity found in numerical simulations of 3-dimensional incompressible hydrodynamical turbulence. The pattern shown, consisting of a large tube surrounded by smaller ones wrapped around it, is frequently found (from Moisy & Jiménez 2004). In magneto-hydrodynamical (MHD) turbulence, structures in which turbulent dissipation is concentrated appear to be more sheet-like (Politano & Pouquet, 1995).

Federrath *et al.* (2010) show that solenoidal (incompressible) forcing is able to provide a compressible turbulent flow with the statistical signatures of incompressible turbulence.

3. Questions raised by diffuse ISM turbulence

The nature and properties of turbulence in the diffuse ISM are still a highly debated and controversial issue in spite of dedicated observational and numerical efforts. This is due in part to the huge range of scales separating those of the energy injection (at the Galaxy scale and even beyond when infall is taken into account, see de Avillez & Breitschtweirdt, 2005) from those where it is dissipated, presumably below the milliparsec scale. It is also due to the fact that the turbulence is compressible, magnetized and multiphase. Unraveling the properties of interstellar turbulence is essential though because turbulence and magnetic fields are the main contributions to the pressure of the diffuse ISM (see Sect. 1) and the main support of molecular clouds against their self-gravity. Turbulent dissipation is therefore a key process among those leading to the formation of molecular clouds, star formation, and therefore galaxy evolution. (see the reviews of Elmegreen & Scalo, 2004 and Scalo & Elmegreen, 2004).

In the ISM, turbulence is observed to be highly supersonic with respect to the cold gas. Yet, one of the most impressive energy spectrum of turbulence in the diffuse ISM covering more than two orders of magnitude in scales (Fig 5, from Miville-Deschênes *et al.*, 2003) is consistent with the Kolmogorov spectrum of incompressible turbulence, which does not necessarily imply that ISM turbulence is incompressible (see Sect. 2). Interestingly, the energy spectrum of the HI emission has the same slope, $\beta = 11/3$, over the broad length-scale domain (0.1 – 20 pc), while that of molecular clouds traced in CO(1-0) or thermal dust emission is clearly smaller, with a steady increase from $\beta \sim 2$ when measured at the 10 – 20 pc scale to $\beta \sim 3.3$ at scales $<$ 1pc (see references in the review of Falgarone *et al.*, 2004). Another quantity, the turbulent energy transfer rate, $\epsilon_l = \frac{1}{2}\rho v_l^3/l$, seems to link turbulence in the diffuse atomic medium and that of diffuse molecular gas traced by ^{12}CO(1-0) emission because it has the same value $\epsilon \sim 2 \times 10^{-25}$ $\mathrm{erg\,cm^{-3}s^{-1}}$ in these two components of the local ISM (see Table II, Falgarone *et al.*, 2004).

Supersonic turbulence is expected to rapidly dissipate in shocks within a cloud crossing time, $\approx$ 1 Myr, for diffuse clouds of a few pc and internal velocity dispersion of a few km $\mathrm{s^{-1}}$. Numerical experiments show that magnetic fields do not significantly slow the dissipation (MacLow *et al.*, 1998). Whether turbulent dissipation occurs primarily

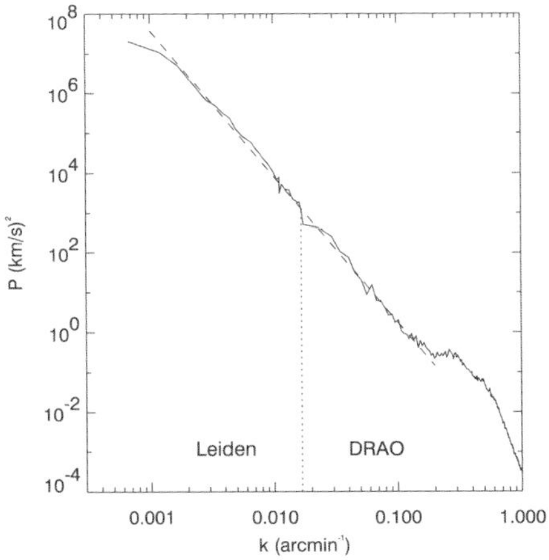

Figure 5. Power spectrum of the velocity field measured with HI emission line profiles in a nearby field at high galactic latitude (Miville-Deschênes *et al.*, 2003). Data obtained with the DRAO interferometer and the Leiden/Dwingeloo survey are merged to provide an unprecedented range of two orders of magnitude in the scale size sampled. The dashed line shows the Kolmogorov spectrum of incompressible turbulence

in compressible (curl-free) or in solenoidal (divergence-free) modes is still an open issue that has important bearings on the subsequent gas evolution and condensation. In their hydrodynamical simulations of mildly compressible turbulence, Porter *et al.* (2002) show that the compressible component of the velocity field is weaker than its solenoidal counterpart by a factor of ~ 6, independent of the nature of the driving process (compressible or solenoidal). Vestuto *et al.* (2003) find that the energy fraction in the solenoidal modes is dominant and increases with the magnetic field intensity in compressible MHD turbulence. These numerical experiments are still far from approaching conditions in the ISM, but they suggest that, even in the compressible (supersonic) case, turbulent dissipation may occur primarily in solenoidal modes, i.e. primarily velocity-shears, without direct gas compression.

4. Intermittency of turbulence in the diffuse ISM

4.1. *Parsec-scale coherent structures of intense velocity-shear*

Identifying regions of intermittency in interstellar turbulence is challenging for several independent reasons. Firstly, these regions are non-space filling and correspond to rare events in time and space (see Fig. 3): their finding requires the analysis of large homogeneous statistical samples of the velocity field. This means observations at high spectral and spatial resolution of large regions of diffuse material. Secondly, observations do not provide the full velocity field but only its line-of-sight (*los*) projection provided by the Doppler-shift of a molecular line, so statistical analysis similar to those performed in laboratory flows or in the solar wind are not possible. However, measurements of the spectrum of molecular lines with a frequency resolution finer than the sound speed in cold H_2 provide information on the velocity field. The ^{12}CO(1-0) line has turned out to be a most useful tool for this search, for reasons that will hopefully become clear to the reader by the end of this paper. Thirdly, only spatial variations of the *los* velocity in the plane-of-the-sky (*pos*) are provided by observations. Therefore, the velocity variations are by essence cross-variations $\partial_j v_i$. Finally, the line emission being integrated along a *los*, the velocity information at a given position is the full line profile and its moments, the first moment being the line centroid velocity. The statistics of the velocity increments are built through those of the line centroid velocity increments (CVI) measured between two positions separated by a lag l in the *pos*. The CVIs have been shown by Lis *et al.*

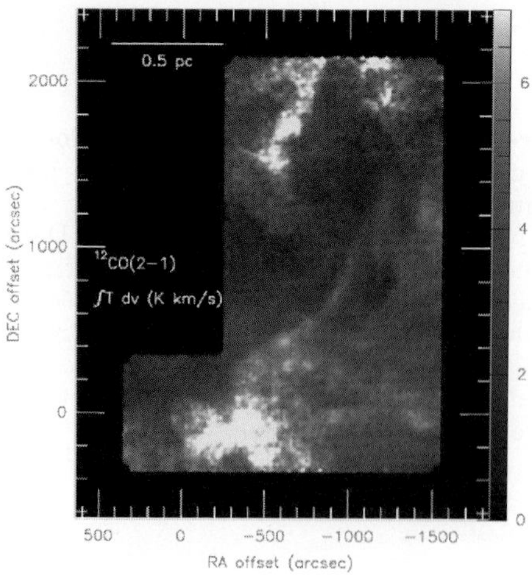

Figure 6. Map of the ^{12}CO(2-1) integrated intensity (in $\mathrm{K\,km\,s^{-1}}$) over a 40' × 30' area located close to the center of the Polaris Flare field shown in Fig 1. The number of independent spectra is $\sim 10^5$. The data have been obtained with the IRAM-30m/HERA mosaic of receivers, operating in the On-The-Fly and frequency-switching mode. The spatial resolution is 11 arcsec or 8mpc at the distance of the field, 150 pc.

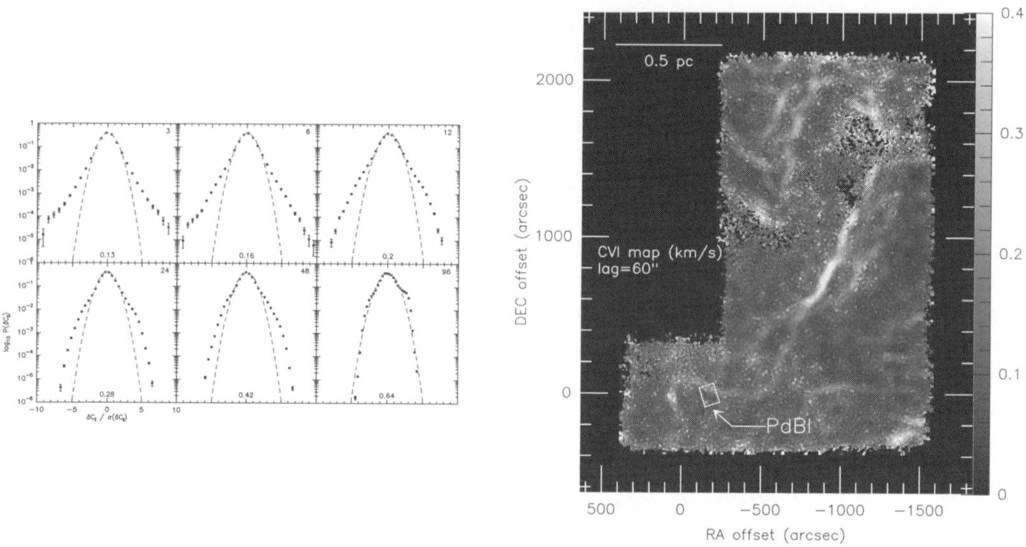

Figure 7. Left: Normalized *pdfs* of line centroid velocity increments (CVI) measured over variable lags, expressed in units of 15 arcsec (upper right corners), and computed within the field of Fig. 6. The Gaussians of same dispersion $\sigma(\delta C_l)$ (given in km s^{-1} at the bottom of each panel) are also drawn. The non-Gaussian wings of the *pdfs* increase as the lag decreases. Note that a probability of 10^{-5} is reached in the most extreme bins. **Right:** In the same field, locus of the positions populating the non-Gaussian wings of the *pdf* for a lag of 60 arcsec. The wedge is in km s^{-1}. The rectangle is the area observed with the IRAM-PdBI. (From Hily-Blant *et al.*, in prep.).

(1996) to be a proxi of the vorticity, being the *los* average of the *pos* projection of the vorticity.

The most recent maps carried out at the IRAM-30m telescope in the ^{12}CO(2-1) line, produce up to $\sim 10^5$ independent spectra sampling homogeneous turbulence in diffuse molecular gas. An example is shown in Fig. 6. (Hily-Blant & Falgarone, 2009, hereafter HF09, Hily-Blant *et al.*, in prep.). The CVI-*pdfs* in that field have the anticipated non-Gaussian wings that increase as the lag over which the increment is measured decreases (Fig. 7, left). The locus of the extreme CVIs (the E-CVIs) that build up the non-Gaussian wings of the *pdfs* is an elongated narrow structure ($\sim$ 0.03 pc thick), almost straight over more than one parsec and surrounded by an ensemble of weaker, shorter structures (Fig. 7, right). As expected, the lane of largest E-CVIs coincides with the region where the velocity-shears are the largest, 40 km s^{-1} pc^{-1} (see HF09). Most interestingly, it also coincides with a lane of weak ^{12}CO(2-1) emission and one of the weakest filaments of dust thermal emission detected at 250μm in that field with Herschel/SPIRE (Miville-Deschênes *et al.*, 2010, Men'shchikov *et al.*, 2010). Last, two low-mass dense cores in the field lie at the South-East tip of the E-CVIs locus (Heithausen *et al.*, 2002), suggesting a *causal link* between bursts of turbulence dissipation, the formation of CO molecule, the formation of tenuous dense filaments and that of low-mass dense cores. The first of these steps is detailed in Section 5.

A similar analysis has been performed in a cloud edge of the Perseus-Taurus-Auriga giant molecular complex. The field has the same total hydrogen column density as that in the Polaris Flare, but is less turbulent (half the pc-scale rms velocity dispersion). The CVI-*pdfs* show departures from a Gaussian distribution with an amplitude 2.5 times smaller than in the Polaris Flare (Hily-Blant *et al.*, 2008), in agreement with the theoretical predictions of Mininni *et al.* (2006).

This ensemble of properties, *(i)* the increasing departure of CVI-*pdfs* from a Gaussian distribution as the scale decreases, *(ii)* the spatial coherent structures of E-CVIs and *(iii)* the link between the large-scale properties of turbulence and the magnitude of the small-scale E-CVIs, suggests that the ^{12}CO(2-1) E-CVIs trace the intermittency of turbulence in diffuse molecular clouds. The multifractal geometry of the E-CVIs structures in the Polaris Flare probed by the non-linear dependence of ζ_p with p up to $p = 6$ discussed in Hily-Blant *et al.* (2008) is now confirmed, up to higher orders, by the analysis of this larger data set. Unexpectedly, the ζ_p departure from a linear dependence agrees, within the error bars, with the She & Levêque (1994) predictions for incompressible turbulence.

4.2. *Milliparsec-scale observations : approaching the dissipation scales*

A step further towards small-scales is provided by IRAM Plateau de Bure Interferometer (IRAM-PdBI) ^{12}CO(1-0) line observations of the field shown in Fig. 7 (right). A mosaic of 13 fields covers the 1'×2' area with a resolution of $\sim$ 4 arcsec or 3mpc (Falgarone *et al.*, 2009). These observations are unique so far : the lines detected are very weak, but the spatial dynamic range provided by the mosaic is large enough to allow the detection of 8 elongated structures with thickness as small as $\approx$ 3 mpc (600 AU) and length up to 70 mpc, the size of the mosaic. Three are shown in Fig. 8 (left panels). They are *not filaments* because once merged with short-spacing data, the PdBI-structures appear to be the sharp edges of extended CO emission (Fig. 8, right panels). Moreover, six out of eight form pairs of quasi-parallel structures at different velocities. This cannot be due to chance alignment. Upper limits of the velocity-shears estimated for the three pairs (i.e. not corrected for projection effects) include the largest values ever measured in non-star-forming regions, up to 780 km s^{-1} pc^{-1}: two velocity components separated by 3.5 km s^{-1}, a value of the order of the rms velocity dispersion of the CNM (Haud & Kalberla 2007), share an interface as thin as 4.5 mpc (in projection) with no shock signature nor density enhancement. Finally, the PdBI-structures are almost straight and their different position angles, $PA = 60^\circ$, 100° and 165°, cover the same broad range of values as the projection of the magnetic field inferred from absorption towards field stars over much larger scales (Hily-Blant *et al.*, in prep.). This suggests an interesting alignment of the *pos* projections of magnetic field and vorticity in these structures.

Falgarone *et al.* (2009) argue that, because the extended velocity components have sharp edges and because the small overlap of the pairs of velocity components cannot

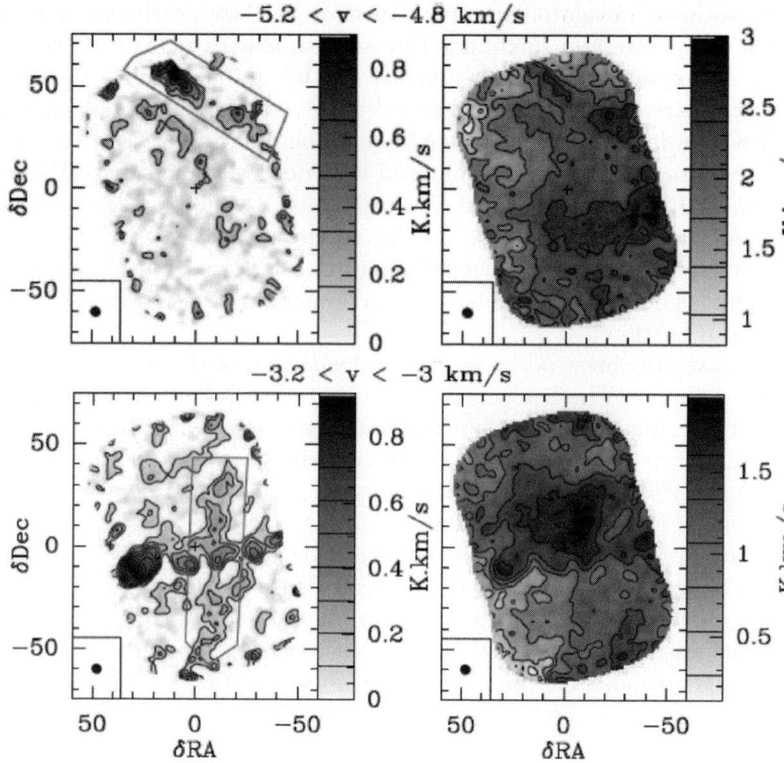

Figure 8. Three of the eight filamentary structures detected in the ^{12}CO(1-0) line with the IRAM-PdBI in the 2'×1' small field of the Polaris Flare drawn in Fig. 7. The synthesized beam ~ 4 arcsec (or 3 mpc) is shown. **Left panels:** ^{12}CO(1-0) emission (integrated over the velocity range indicated above the panels) detected by the interferometer. **Right panels:** ^{12}CO(1-0) emission integrated over the same velocity intervals and including the short-spacings provided by the IRAM-30m telescope. The elongated structures seen by the interferometer appear as the sharp edges of weak extended CO structures. The structure in the top panel is coinciding with the ~ 0.4 pc long E-CVI structure that crosses the PdBI mosaic frame (Fig. 7).

be due to a chance orientation and has to occur for any viewing angle, these extended components have to be *thin sheets of CO emission*, not volumes. A similar reasoning applies to the geometry of the large velocity-shear detected in the single-dish data (HF09). The edges detected by the PdBI mark a discontinuity, at the milliparsec scale, between a detected CO-rich and an undetected CO-poor component (presumably the CNM), coinciding with a large velocity-shear. A low density (~50 cm^{-3}) CO-rich component would have been detected, although sub-thermally excited, because the sensitivity of the ^{12}CO(1-0) observations allows the detection of column densities as low N(CO) ~ 10^{14} cm^{-2} in velocity bins of 0.1 km s^{-1}. These may be the first directly detected manifestations of the intermittency of interstellar turbulent dissipation. The large velocity-shears reveal an intense straining field, generating a local dissipation rate much larger than average, acting as a possible driver of CO formation (see Section 5).

Many surfaces in turbulent flows, in particular those of iso-dissipation, have been shown to be fractal with a dimension $D = 2 + \xi$ where ξ is the scaling exponent of the velocity structure function ($\xi = 1/3$ in Kolmogorov turbulence) (Sreenivasan *et al.*, 1989). If, as it seems to be the case, CO molecules form in the regions of intense turbulent dissipation in diffuse gas, it is not surprising that their spatial distribution be a fractal structure of dimension 1.3 in projection at the parsec scale. This result supports the conjecture linking the fractal distribution of CO emission in cloud edges to the power spectrum of turbulence made by Falgarone *et al.* (1991). Since, as expected from the behavior of ζ_p with p (Hily-Blant *et al.*, 2008), intermittency in the diffuse ISM is multi-fractal, the fractal dimension increases as the range of scales over which it is computed decreases, in agreement with the observed dependence of β with length-scale (Falgarone *et al.*, 2004).

These findings broaden the perspective on small-scale CO structures in diffuse molecular clouds. Isolated CO emission has been found in the high latitude sky (Heithausen, 2006) with bright sub-structures seen in PdBI observations (Heithausen, 2004). The sub-structures, although analysed as clumps by the author, are not randomly scattered in the field but form two elongated patterns, reminiscent for their thickness and length of what is found in FPH09, providing a velocity shear of 180 km s^{-1} pc^{-1}. Sakamoto & Sunada(2003) have discovered a number of CO small-scale structures at the edges of the Taurus molecular cloud. Their main characteristics are their large line-width and their sudden appearance and disappearance within 0.03 to 0.1 pc. Interestingly, the authors propose that their small-scale CO structures pinpoint CO molecule-forming regions, driven by the thermal instability in the turbulent diffuse ISM. Finally, sensitive ^{12}CO(1-0) observations of the high-latitude sky by Barriault *et al.* (2010) show a good coincidence between the emergence of ^{12}CO(1-0) emission and HI velocity-shears.

Models of non-equilibrium chemistry in diffuse gas have attempted to capture the essence of the coupling between the physics of turbulent dissipation and the warm chemistry it triggers (Falgarone *et al.*, 2005, Joulain *et al.*, 1998, Godard *et al.*, 2009). In the next Section, we summarize the most recent developments (Godard *et al.*, in prep.).

5. The predictions of the turbulent dissipation model

5.1. *Outline of the model*

The TDR model, (for Turbulent Dissipation Regions, Godard *et al.*, 2009, 2010) is based on the fact that turbulent dissipation that involves $\nabla \cdot \mathbf{v}$ and $\nabla \times \mathbf{v}$ is an intermittent quantity. The model relies on an analytical solution of the Helmholtz equation for the vorticity, the Burgers vortex. In addition to the traditional parameters of chemistry models, the density $n_{\rm H}$ and UV shielding characterized by the visual extinction A_V, the free parameters of the TDR model are constrained by the known large-scale properties of turbulence in the diffuse medium (see Joulain *et al.*, 1998): the rate-of-strain a – a quantity homogeneous to s^{-1} – and the maximum orthoradial gas velocity in the vortex. The rate-of-strain and the equilibrium radius of the vortex are related by $r_0^2 = 4\nu/a$. A hydrodynamical steady-state of a magnetized Burgers vortex of finite length is computed. The density is low enough that the neutral component decouples from the ions and magnetic field in the layers where the orthoradial velocity is the largest. The thermal and chemical evolution of a fluid cell crossing the steady-state vortex is computed at each time step, in a Lagrangian frame, with 127 species interacting via 2230 reactions. In this model, the dissipation is due to (1) viscous dissipation in the layers of intense velocity-shear at the vortex outer boundary and (2) ion-neutral friction induced by the decoupling of the ionized and neutral flows in the central regions.

Since the diffuse medium has a low density, its chemical and thermal inertia are large. The TDR model takes into account the long-lasting relaxation period that follows the end of any dissipative burst (the vortex lifetime i.e. the duration of the burst of dissipation, is set by energetic considerations, see Godard *et al.*, 2009). A random line of sight through the medium therefore samples three kinds of gas : active vortices, relaxation phases and the ambient medium. The importance of the relaxation phase in the final column

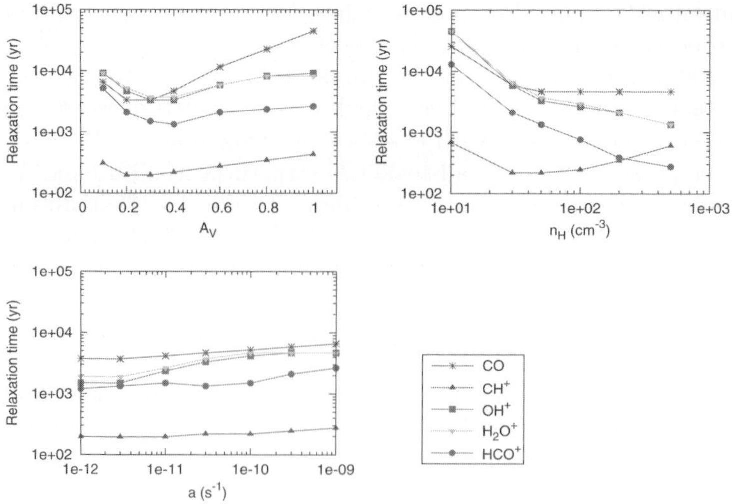

Figure 9. Relaxation times of different species as a function of the shielding from the ambient UV radiation field, A_V (top left), the density, n_H (top right) and the turbulent rate-of-strain, a (bottom left). Note that these quantities are not model dependent.

density of a given species is proportional to its duration. The relaxation times of a subset of species are displayed in Fig. 9 : they cover a broad range, from 200 yr for CH^+ up to 5×10^4 yr for CO. Note that these large differences are model independent. This introduces a potentially untractable complexity in the comparison of the abundances of different species (see below). For instance, it is only at relatively high density that CH can be considered as a tracer of H_2 in the ambient phase (Gerin *et al.*, 2010). In all other cases, it is, as is CH^+, a result of the warm chemistry.

5.2. *Scaling of* CH^+ *abundances with dissipation rate*

A large number of models have been run to explore the parameter space varying the density ($30 < n_H < 500$ cm^{-3}), the visual extinction ($0.1 < A_V < 2$ mag), the rate-of-strain ($10^{-12} < a < 10^{-10}$ s^{-1}) and the average turbulent energy transfer rate, ϵ, up to 10 times its solar neighbourhood value (Falgarone *et al.*, 2004). This quantity is assumed to be equal to the dissipation rate $\overline{\epsilon}_D$ averaged over time or space. The CH^+ abundance is found to be proportional to $\overline{\epsilon}_D$, which makes this radical a specific tracer of the dissipation of suprathermal energy in the ISM. The scaling of CH^+ with n_H and A_V is inferred from the model results:

$$N(CH^+)/N_H \sim 2\times10^{-8}(\epsilon/10^{-24}\,\mathrm{erg\,cm^{-3}s^{-1}})(n_H/50\,\mathrm{cm^{-3}})^{-2.3}(A_V/0.2)^{-1} \quad (5.1)$$

Note the interesting increase of the CH^+ abundance as the UV-shielding decreases. This unique behavior follows from the destruction route of CH^+ due primarily to collisions with H and H_2 rather than to photodissociation, with CH_2^+ and CH_3^+ being produced. Unlike CH^+, these species are destroyed by UV photons, reforming CH^+ as a photodissociation product. This was found in MHD-shock models of DR21 (Falgarone *et al.*, 2010a) where the behavior of H_2O and CH^+ as the UV radiation field increases are found to be anti-correlated.

5.3. CH^+, HCO^+ *and* CO, *tracers of turbulent dissipation in the diffuse ISM*

In the warm chemistry network of the TDR model, HCO^+ is a daughter molecule of CH^+, via the reaction $CH_3^+ + O \rightarrow HCO^+ + H_2$, while in the traditional UV-driven chemistry it is a daughter of CO^+ and water. CO in turn is produced by dissociative recombination of HCO^+. The predicted column densities of CH^+ and CO are displayed in Fig. 10 as a function of that of HCO^+ for a broad range of densities, UV-shieldings

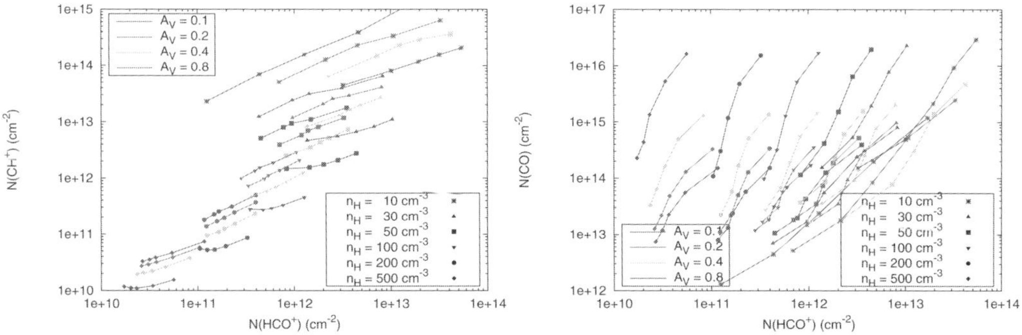

Figure 10. CH^+, HCO^+ and CO column densities obtained in the TDR model per magnitude of gas sampled, for a range of proton densities and screenings from the UV field. The free parameter along each curve is the rate-of-strain ranging from $a = 10^{-12}$ s^{-1} (upper right) to $a = 10^{-10}$ s^{-1} (bottom left), where the model reaches the limits of its validity. The column densities decrease as a increases because larger a imply smaller vortex radii, therefore more dissipative vortices, but less numerous because the transfer rate of turbulent energy in the cascade is kept the same.

and rates-of-strain, and for a total gas column density of $N_{\rm H} = 1.8 \times 10^{21}$ cm^{-2} or 1 mag.

While CH^+ and HCO^+ follow each other smoothly, the increase of CO with HCO^+ is much steeper as a increases. This is because CH^+ (and HCO^+ to a lesser extent) does not survive beyond the end of the active phase of the vortex, while its long relaxation time (Fig. 9) makes CO observable for a much longer time after the vortex blast. It is remarkable that the column densities of HCO^+ observed in the diffuse medium, in the range of a few 10^{11} to 8×10^{12} cm^{-2}, are naturally reproduced by most models as soon as CH^+ is in the observed range of a few 10^{12} to $\sim 8 \times 10^{13}$ cm^{-2}. Unfortunately, these two species have never been observed along the same *los*, but the above column densities all refer to low extinction *los* (i.e. total H column densities corresponding to $A_V = 1$ mag, within a factor of a few) since HCO^+ is observed in the direction of QSOs. The CO and HCO^+ column densities observed by Liszt & Lucas (1998) on the same *los* in the diffuse medium are bracketed by the model predictions for densities in the range 50–200 cm^{-3} and $A_V = 0.4$ to 0.8 mag. The lowest N(CO) observed in the diffuse ISM are those measured in absorption spectroscopy against nearby stars by Sonnentrucker *et al.* (2007) down to N(CO) $\sim$ a few 10^{12} cm^{-2} for $N(H_2)$ only a few 10^{19} cm^{-2}.

The observed values tend to be in better agreement with low rates-of-strain i.e. models in which dissipation is dominated by ion-neutral friction. Remarkably, the pc-scale velocity shear of 40 km s^{-1} pc^{-1} observed in the Polaris Flare would correspond to a straining field from the large scales $a = 1.5 \times 10^{-12}$ s^{-1}, while the intense small-scale shear of 780 km s^{-1} pc^{-1} detected at the mpc-scale is 20 times larger ($a = 3 \times 10^{-11}$ s^{-1}), comparable to the velocity-shear at the edge of the Burgers vortex for that rate-of-strain. It is therefore possible that, alike CH^+ and HCO^+, the CO emission detected in the diffuse molecular medium be a direct tracer of turbulent dissipation.

5.4. H_2 *and* C^+ *radiative cooling rates*

The last interesting prediction of the TDR model is the intensity of the radiative cooling during the thermal relaxation phase (see also Falgarone *et al.*, 2006). Fig. 11 displays the gas radiative cooling in the H_2 pure rotational lines, the fine-structure [C II] and [O I] lines and the CO rotational lines as a function of the gas temperature (left) and time (right). While the H_2 line emission is more intense than that of C^+, it lasts only for the

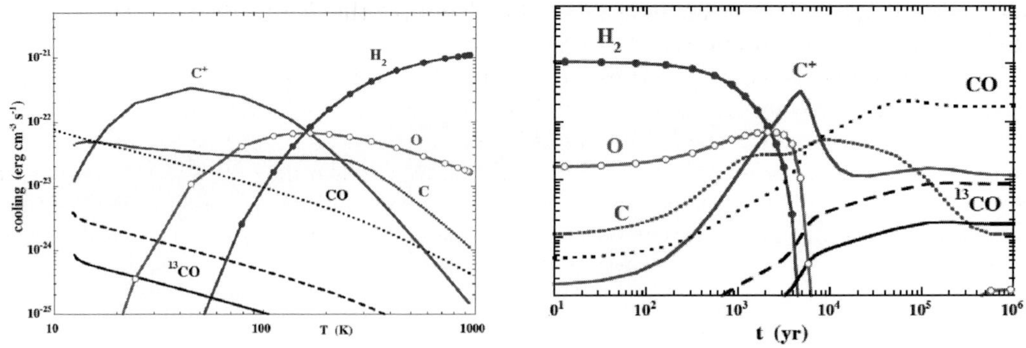

Figure 11. Radiative cooling rates of the pure rotational lines of H_2, the fine structure line of C^+ and OI and the ^{12}CO and ^{13}CO rotational lines during the isobaric thermal relaxation of a burst of dissipation, **left:** as a function of gas temperature and **right:** as a function of time. The latter display shows that the total energy loss in the H_2 and the C^+ lines are comparable.

short interval of time when the gas is warmer than 100 K. Then, as the gas cools down, the [C II] line emission takes over. It is remarkable that the total energy radiated by these two species during the relaxation phase is about the same, the latter being about ten times weaker on average over a duration ten times larger. The H_2 emission in the TDR model is comparable to that observed (Falgarone *et al.*, 2005, Ingalls *et al.*, in prep.), and the [C II] emission from a gas component heated by turbulent dissipation rather than by UV photons is predicted to be a possible significant fraction of the total [C II] emission of the diffuse medium.

6. Summary and perspectives

The signatures of space-time intermittency of turbulence in diffuse molecular gas are clearly present. $^{12}CO(2\text{-}1)$ maps of large fields carried out at high angular resolution provide an unprecedented view of diffuse molecular gas and allow a powerful statistical analysis of the velocity field. In particular, the non-Gaussian statistics of the velocity field reveal parsec-scale coherent structures of intense velocity-shear with sub-structures observed down to 3 mpc. The large velocity-shears associated with the mpc-scale structures and their topology suggest that these are the possible birth sites of CO molecules in the diffuse medium.

Models of turbulent dissipation regions (TDR) in which non-equilibrium chemistry is triggered by intermittent bursts of dissipation in the diffuse ISM are encouraging. They involve slightly helical magnetic fields aligned with the local intense vorticity and partial decoupling of the ions and neutrals in short-lived dissipative structures filling, on average, only a few per cent of the diffuse medium. The abundances of almost all molecular species so far unexplained in the diffuse ISM, and its pure rotational H_2 emission, are reasonably reproduced when a small fraction of active dissipative structures, and relaxation phases are considered in the *los* modelling . These models are the first attempt to bridge the chemistry to the huge range of scales coupled by the turbulent cascade, and do so in a theoretical framework. They provide relevant boundary conditions for the chemistry models, by providing a way to take into account, in a self-consistent way, the amount of supra-thermal energy that is injected at the mpc-scale, and the associated timescales. They have their limitations, though: the analytical solution of the Burgers vortex itself, the lack of retro-action of the gas chemistry (and therefore cooling, and ionization degree) on the dynamical (hydro and magnetic) steady-state, the *los* modelling ignoring any distribution of vortex properties, to name a few.

On the observational side, the nature of the dissipative structures (shocks or pure velocity-shears), the actual size of the dissipation scale, the fractal geometry of the smallest scales, and the topology of the magnetic field in these structures are open fields for future observations. ALMA should allow observations of the diffuse ISM between 1000 AU and 10 AU. On the theoretical side, the pictured evolution from coherent structures of intense velocity-shears to molecule formation, enhanced radiative cooling, gas condensation and formation of a network of dense gas filaments, is still to be computed.

References

de Avillez, M. A. & Breitschweirdt D. 2005 *A&A* 436 585
Anselmet, F., Antonia, R. A., & Danaila, L. 2001, *Planetary & Space Sciences*, 49, 1177
Audit, E. & Hennebelle, P. 2005, *A&A*, 433, 1
Barriault, L., Joncas, G., Falgarone, E. *et al.* 2010 *MNRAS* 406 2713
Begum, A., Stanimirovic, S., Goss, M., Heiles, C., Pavkovich, A., & Hennebelle, P. 2010 *ApJ* 725 1779
Cox, D. 2005, *ARAA* 43, 337
Crane, P., Lambert, D. L., & Sheffer, Y. 1995 *ApJS* 99, 107
Crutcher, R., Wandelt, B., Heiles, C., Falgarone, E., & Troland, T. 2010, *ApJ*, 725, 466
Elmegreen, B. G. & Scalo, J. 2004, *ARAA*, 42, 211
Falgarone, E., Phillips, T. G., & Walker, C. K. 1991, *ApJ*, 378, 186
Falgarone, E., Hily-Blant, P., & Levrier, F., 2004, *Ap&SS*, 292, 89
Falgarone, E., Verstraete, L., Pineau des Forêts, G., & Hily-Blant, P. 2005, *A&A*, 433, 997
Falgarone, E., Pineau des Forêts, G., Hily-Blant, P., & Schilke, P. 2006, *A&A*, 452, 511
Falgarone, E., Pety, J., & Hily-Blant, P. 2009, *A&A*, 507, 355
Falgarone, E., Ossenkopf, V., Gerin, M., *et al.*, 2010a, *A&A 518*, L118
Falgarone, E., Godard, B., Cernicharo, J. *et al.*, 2010b, *A&A* 521, L15
Federrath, C., Duval, J., Klessen, R. S., *et al.*, 2010, *A&A* 512, 81
Field, G., Goldsmith, D., & Habing, H. 1969 *ApJ* 155, L149
Fitzpatrick, E. L. & Spitzer, L. 1997, *ApJ*, 745, 623
Frisch, U. 1995, *Turbulence : the legacy of A.N. Kolmogorov*, Cambridge University Press, Cambridge
Gerin, M., de Luca, M., Goicoechea, J. *et al.*, 2010 *A&A* 521, L16
Godard, B., Falgarone, E., & Pineau des Forêts, G. 2009, *A&A*, 495, 847
Godard, B., Falgarone, E., Gerin, M. Hily-Blant, P., & de Luca, M., 2010, *A&A*, 520, A20
Godard, B., Falgarone, E., Gerin, M., *et al.*, 2011, submitted to *A&A*
Goldsmith, P. F., Heyer, M., Narayanan, G., *et al.*, 2008, *ApJ*, 680, 428
Goldsmith, P. F., Velusamy, T., Li, D., & Langer W. D. 2010, *ApJ*, 715, 1370
Gredel, R., 1997 *A&A* 320, 929
Hartmann J. 1904 *ApJ* 19, 268
Haud, U. & Kalberla, P. M. W. 2007, *A&A*, 466, 555
Heiles, C. & Troland, T. 2003 *ApJ* 586 1067
Heithausen, A. 2004, *ApJ*, 606, L13
Heithausen, A. 2006, *A&A*, 450, 193
Heithausen, A., Bertoldi, F., & Bensch, F. 2002, *A&A*, 383, 591
Hewitt J. W., Rho J., Andersen, M., & Reach W. T. 2009 *ApJ* 694, 1266
Hily-Blant, P. & Falgarone, E. 2007, *A&A*, 469, 173
Hily-Blant, P., Falgarone, E., & Pety, J. 2008, *A&A*, 481, 367
Hily-Blant, P. & Falgarone, E., 2009, *A&A*, 500, L29 (HF09)
Jenkins, E. B. & Tripp, T. M. 2011, *ApJ*, 734, 65
Joulain, K., Falgarone, E., Pineau des Forêts, G. P., & Flower, D. 1998, *A&A*, 340, 241

Kainulainen, J., Beuther, H., Henning, T., & Plume, R. 2009 *A&A* 508,L35
Kolmogorov, A. N. 1941, *Dokl. Akad. Nauk SSSR*, 26, 115
Kolmogorov, A. N. 1962, *J. Fluid Mech.*, 13, 82
Landau, L. E. & Lifchitz, E. M. 1959, *Fluid Mechanics*, Pergamon Press, Oxford
Lis, D. C., Pety, J., Phillips, T. G., & Falgarone, E. 1996, *ApJ*, 463, 623
Liszt, H. S. & Lucas, R. 1998, *A&A*, 339, 561
Mac Low, M.-M., Klessen, R. S., Burkert, A., & Smith, M. D. 1998, *Phys. Rev. Lett.*, 80, 2754
Menten, K., Wyrowski, F., Belloche, A., *et al.*, 2011 *A&A* 525, 77
Men'shchikov, A., André, P., Didelon, P., *et al.*, 2010 *A&A* 518 L103
Mininni, P. D., Alexakis, A., & Pouquet, A. 2006, *Phys. Rev. E*, 74, 6303
Miville-Deschênes, M.-A., Joncas G., Falgarone, E., & Boulanger, F., 2003, *A&A*, 411, 109
Miville-Deschênes, M.-A., Martin P., Abergel A., *et al.*, 2010, *A&A*, 518 L104
Miville-Deschênes, M.-A. & Lagache, 2005, *ApJS* 157, 302
Moffatt, H. K., Kida, S., & Ohkitani, K. 1994, *J. Fluid Mech.*, 259, 241
Moisy, F. & Jiménez, J. 2004, *J. Fluid Mech.*, 513, 111
Pety, J. & Falgarone, E. 2003, *A&A*, 412, 417
Politano, H. & Pouquet, A. 1995 *Phys. Rev. E*, 52, 636
Porter, D. H., Pouquet, A., & Woodward, P. R. 2002, *Phys. Rev. E*, 66, 026301
Sakamoto, S. & Sunada, K. 2003, *ApJ*, 594, 340
Scalo, J. & Elmegreen, B. G. 2004, *ARAA*, 42, 275
She, Z.-S. & Levêque, E. 1994, *Phys. Rev. Lett.* 72, 336
Snow, T. P. & McCall, B. J. 2006, *ARAA*, 44, 367
Sonnentrucker, P., Welty, D. E., Thorburn, J. A. & York, D. G. 2007 *ApJS* 168 58
Sreenivasan, K. R., Ramshankar, R., & Méneveau, C. 1989, *Proc. R. Soc. Lond. A*, 421, 79
Vestuto, J. G., Ostriker, E. C., & Stone, J. M. 2003, *ApJ*, 590, 858
Weselak, T., Galazutdinov, G., Musaev, F. & Krelowski, J., 2008 *A&A* 479 149

Discussion

AIKAWA: What is the initial conditions of the chemistry model? Is the criterion of $A_V \sim 0.2$ set by PDR models?

FALGARONE: The initial conditions of the chemistry are set by PDR models. The H_2 fraction at the low shielding value $A_V \sim 0.2$ is therefore very low.

OSSENKOPF: In a number of sources, we see CH^+ with very narrow linewidths of a few km s^{-1}. Isn't this a contradiction with the large-shear production?

FALGARONE: Such linewidths are not narrow, strictly speaking. They are comparable to those found in visible absorption lines sampling local diffuse gas and to the width of individual gas velocity components in submillimeter lineprofiles across the galactic diffuse medium. However, there is not one single process involved in the formation of CH^+. High velocity shocks produce CH^+ broad wings of several 10 km s^{-1} (e.g. Falgarone *et al.* 2010). In dense highly FUV illuminated PDRs (I do not know if these are the sources you refer to), the production of CH^+ is driven by reaction of C^+ with vibrationally excited H_2 excited by fluorescence (Agundez *et al.*, 2010). The supra-thermal energy being due to FUV photons, in that case, the CH^+ linewidths should not be any broader than those of other molecules. This mechanism is not efficient enough in diffuse clouds illuminated by the ambient interstellar radiation field (ISRF). The role of turbulent dissipation (and large velocity-shears) is dominant there. My talk was restricted to the diffuse medium irradiated by ISRF.

The Molecular Universe
Proceedings IAU Symposium No. 280, 2011
José Cernicharo & Rafael Bachiller, eds.

doi:10.1017/S1743921311024987

Molecular Evolution from AGB Stars to Planetary Nebulae

Sun Kwok[1]

[1]Faculty of Science, The University of Hong Kong, Hong Kong, China
email: sunkwok@hku.hk

Abstract. The late stages of stellar evolution from the Asymptotic Giant Branch (AGB) to planetary nebulae represent the most active phase of molecular synthesis in a star's life. Over 60 molecular species, including inorganics, organics, radicals, chains, rings, and molecular ions have been detected in the circumstellar envelopes of evolved stars. Most interestingly, complex organic compounds of aromatic and aliphatic structures are synthesized over very short time intervals after the end of the AGB. Also appeared during the post-AGB evolution are the unidentified 21 and 30 μm emission features, which are believed to originate from carbonaceous compounds.

The circumstellar environment is an ideal laboratory for the testing of theories of chemical synthesis. The distinct spectral behavior among AGB stars, proto-planetary nebulae (PPN), and planetary nebulae (PN) and the short evolutionary time scales that separate these stages pose severe constraints on models. In this paper, we will present an observational summary of the chemical synthesis in the late stages of stellar evolution, discuss chemical and physical processes at work, and speculate on the possible effects these chemical products have on the Galaxy and the Solar System.

1. Introduction

An intermediate mass (1-8 $M_\odot$) star undergoes H and He shell burning after having developed a C–O electron-degenerate core. The luminosity and size of the star increase continuously as the star climbs the asymptotic giant branch (AGB) on the Herztsprung-Russell diagram. At the same time, a stellar wind commences and gradually reduces the mass of H envelope. Over the last one million years of AGB evolution, up to several solar masses of circumstellar envelope can be built up by this mass loss process. This large-scale mass loss terminates when the entire H envelope is depleted. Continued H-shell burning will further deplete the remaining H envelope, giving the appearance of the star evolving to higher effective temperatures while maintaining a constant luminosity. This is the beginning of the post-AGB phase of evolution. When the stellar temperature reaches 30,000 K, the stellar Lyman continuum output is high enough to photoionize the circumstellar envelope, creating a planetary nebula. A detailed description on the origin and evolution of planetary nebulae can be found in Kwok (2000).

Manifestations of the mass loss process include the observation of rotational and vibrational transitions of molecules, and the vibrational modes of solid particles. The stellar wind in the AGB phase is expanding at $\sim$10 km s^{-1}, giving a dynamical age of the circumstellar envelope of $\sim 10^4$ yr. From stellar evolution models, the transition from the end of AGB to the beginning of PN is about several thousand years and objects in this phase of evolution are called proto-planetary nebulae (PPNs, Kwok 1993). Accelerated by a later-developed fast wind, PNs expand at higher speeds ($\sim$20-30 km s^{-1}), giving them a life time of 20,000-40,000 yr before being dispersed into the interstellar medium.

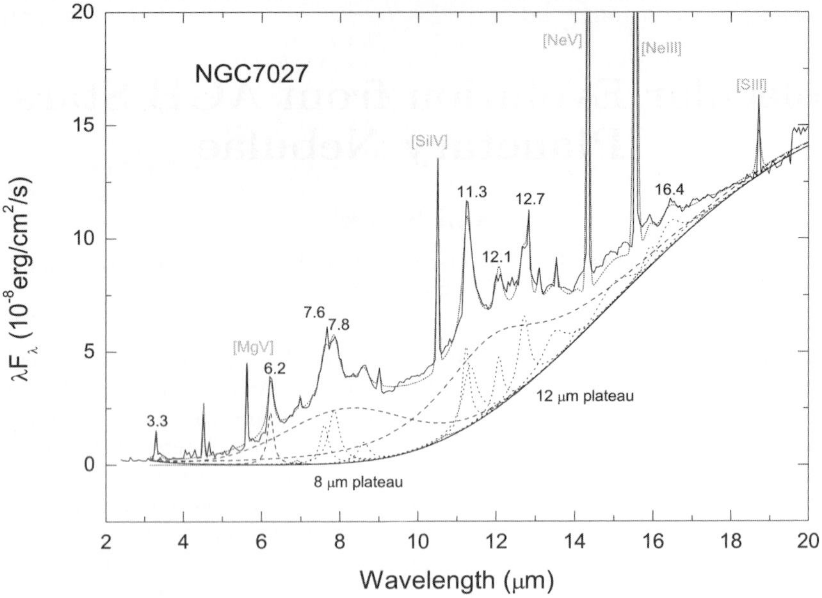

Figure 1. ISO spectrum of the planetary nebula NGC 7027. The thin solid lines are fits to discrete features and the dashed lines are fits to the plateau features.

The particle densities of the envelope are in the range of $10^2 - 10^6$ cm^{-3}. It is in this low-density, rapidly changing environment that molecular synthesis is taking place.

2. Synthesis of molecules and dust in AGB and post-AGB evolution

The circumstellar envelopes of stars in the late stages of stellar evolution represent an ideal laboratory to study the formation of organic compounds. The process begins with the synthesis of the element carbon by the triple-α nuclear reaction in the core of AGB stars. The dredge up of this element to the surface allows the formation of simple carbon molecules such as C_2, C_3, CN, etc., in the photosphere. The initiation of a stellar wind during the late AGB stage results in the formation of gas-phase molecules in the outflow. Over 60 molecular species have been detected, and they include inorganics (e.g., CO, SiO, SiS, NH_3, AlCl, etc.), organics (C_2H_2, CH_4, H_2CO, CH_3CN, etc.), radicals (CN, C_2H, C_3, HCO^+, etc.), chains (e.g., HCN, HC_3N, HC_5N, etc.), and rings (C_3H_2). Since the dynamical lifetime of the envelope is $\sim 10^4$ yr, the chemical reactions that lead to the formation of these species must be shorter than this time scale. Interferometric observations also allow the mapping of the molecular emission regions and therefore put further constraints on the reaction zone (see Guélin, these proceedings).

Accompanying this molecular outflow is the condensation of solid-state species, the most common of which are amorphous silicates and silicon carbide, both have strong spectroscopic features in the mid-infrared (Kwok, Volk & Bidelman 1997). Near the end of the AGB, the mass loss rate of the star is so high that the star itself can be totally obscured by its circumstellar dust envelope. In carbon-rich AGB stars, the continuum is often featureless and amorphous carbon has been suggested as the carrier of the continuum. It is also in this stage that the 13.7 μm ν_5 vibrational line of C_2H_2 can be seen in absorption against the dust continuum (Volk, Xiong & Kwok 2000). In the post-AGB phase of evolution, lines of C_4H_2, C_6H_2 and C_6H_6 appear (Cernicharo *et al.* 2001). The

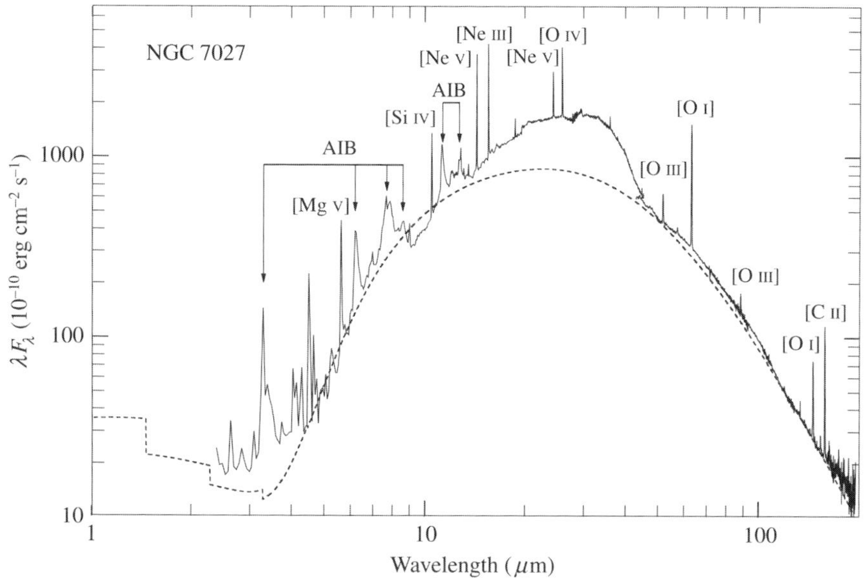

Figure 2. The *ISO SWS/LWS* spectrum of NGC 7027 between 1-200 μm. Some of the stronger atomic lines and AIB features are marked. The dashed line is a model fit composed of a sum of dust emission, *f-b* and *f-f* gas emission. The excess in 2-5 μm above the dashed line suggests that there is a warm dust component. The strong 30 μm feature can be clearly seen above the dust continuum.

first sign of aromatic features appear in PPN. The 3.3, 6.2, 7.7, 8.6, and 11.3 μm C–H and C–C stretching and bending modes of aromatic compounds can be seen in many C-rich PPN (Kwok, Volk & Hrivnak 1999, Hrivnak, Geballe & Kwok 2007). These aromatic infrared bands (AIB) are commonly found in C-rich PN and an example of PN spectrum is shown in Fig. 1.

3. Spectral features of PN and PPN

The infrared spectra of PPNs and PNs are characterized by the following characteristics:

- A strong continuum between 3 and 200 μm
- Aromatic features at 3.3, 6.2, 7.7, 8.6 and 11.3 μm.
- Aliphatic features at 3.4 and 6.9 μm.
- Other emission features at 15.8, 16.4, 17.4, 17.8 and 18.9 μm.
- Plateau emission features at 8, 12, and 17 μm.
- Unidentified broad emission features at 21 and 30 μm.

These spectral properties are discussed in detail below.

The infrared continuum. All PPN and PN (O- or C-rich) possess a strong infrared continuum stretching from the near infrared to millimeter wavelengths (Fig. 2). The color temperature of this continuum is ~50-100 K for PNs and higher for PPNs (Zhang & Kwok 1991). This continuum is generally assumed to be due to the remnant of the dust envelope from the AGB phase. Quite often a separate higher temperature dust continuum is needed to fit the underlying broad emission in the near infrared region (Zhang, Kwok, & Hrivnak 2010). This "warm" component is likely to be created during the post-AGB evolution. Such a broad emission continuum can only be produced by micron-size

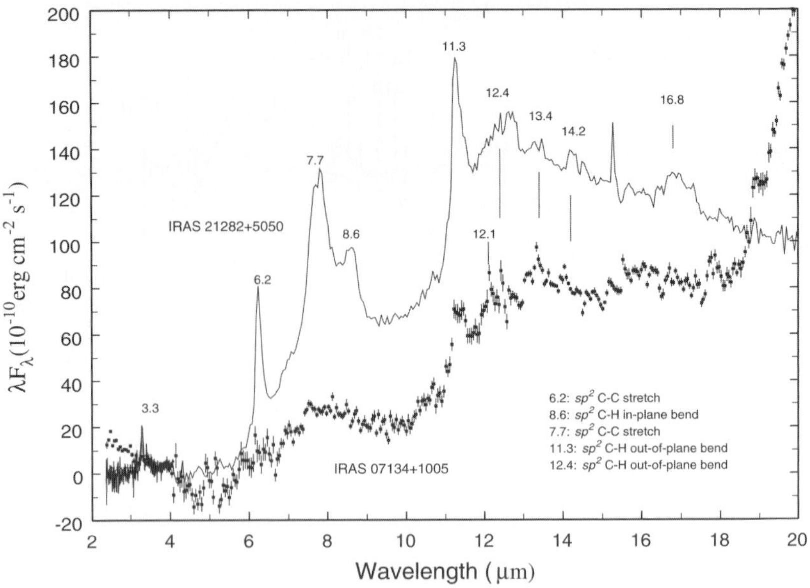

Figure 3. *ISO SWS01* spectra of the young PN IRAS 21282+5050 and the PPN IRAS 07134+1005, showing various aromatic C−H and C−C stretching and bending modes at 3.3, 6.2, 7.7, 8.6, and 11.3 μm. Beyond the 11.3 μm feature are the 12.1, 12.4, 13.3 μm out-of-plane bending mode features from small aromatic units.

solid-state particles, although the particles sizes responsible for the near-infrared continuum are likely to be smaller.

The aromatic infrared bands. The infrared emission features at 3.3, 6.2, 7.7, 8.6, and 11.3 μm are now firmly identified as arising from the stretching and bending modes of aromatic compounds (Knacke 1977, Duley & Williams 1981, Allamandola, Tielens & Barker 1989, Puget & Léger 1989) and they are collectively called the aromatic infrared bands (AIB).

The 11.3, 12.1, 12.4, 13.3 μm features can be seen in the spectra of PPN (Fig. 3) and can be identified as C−H out-of-plane bending modes of aromatic units with different number of exposed corners. The 11.3, 12.1, 12.4, 13.3 μm features correspond to a lone C−H group, two, three, or four adjacent C−H groups, respectively (Hudgins & Allamandola 1999).

The 3.4 and 6.9 μm features. The infrared spectra of PPNs differ from PNs in that the 3.4 μm feature can be as strong as the 3.3 μm feature (Geballe *et al.* 1992). The 3.4 μm feature has been identified as due to symmetric and antisymmetric stretching modes of methyl and methylene aliphatic groups, and these separate modes can be clearly identified (Hrivnak, Geballe & Kwok 2007). The 6.9 μm feature is strong in PPN and has been attributed to aliphatic C–H bending modes (Kwok, Volk & Hrivnak 1999, Kwok, Volk & Bernath 2001).

The 15.8 μm feature. The 15.8 μm feature ($\Delta\lambda \sim 1.3\mu$m) is strong in 21 μm sources (Fig. 4) (Hrivnak, Volk & Kwok 2009, Zhang, Kwok, & Hrivnak 2010). Figure 5 shows a comparison of the profiles of the 15.8 and 21 μm features.

The 8 and 12 μm plateau features. In addition to hydrogen, methyl ($-CH_3$) or methylene ($-CH_2$) side groups, it is also possible that other side groups such as carbonyl (C=O), aldehydic (−HCO), phenolic (OH), and amine (NH_2) be attached to the

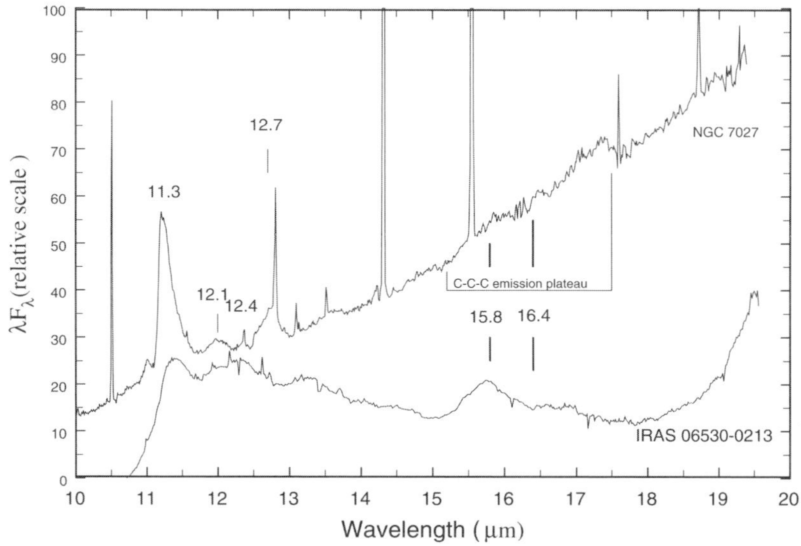

Figure 4. The *Spitzer IRS* spectra of young PN NGC 7027 and the PPN IRAS 06530−0213 showing 11.3, 12.1 μm AIB features and the unidentified 15.8 μm feature, and the 15-17.5 μm emission plateau in NGC 7027. The narrow feature in the NGC 7027 are atomic lines.

aromatic units. The broad emission plateaus at 8 and 12 μm are identified as superposition of in-plane and out-of-plane C–H bending modes of various aliphatic sidegroups attached to the aromatic rings (Fig. 6) (Kwok, Volk & Bernath 2001). Although these plateaus can be seen in the spectra of PN (Fig. 1), they are relatively more prominent in PPNs (Fig. 7).

The 17 μm plateau. A very broad emission plateau extending from 15 to 20 μm can be seen in the spectra of PPNs and PNs (Fig. 8). This feature has been suggested to be due to aromatic C-C-C in-plane and out-of-plane distortion modes (van Kerckhoven *et al.* 2000).

The 21 and 30 μm features. The AIB features are not the only infrared emission features that emerge in the post-AGB evolution. A strong emission feature at 21 μm was first discovered in four PPNs (Kwok, Volk & Hrivnak 1989) and is now seen in ~15 post-AGB stars in the Galaxy and ~10 in the Magellanic Clouds (Volk *et al.* 2011). High resolution *ISO* observations have shown that the 21 μm feature to have no detectable subfeatures. The intrinsic profiles of the feature is found to have a consistent shape among different sources, peaking at ~20.1±0.1 μm (Fig. 9) (Hrivnak, Volk & Kwok 2009).

The unidentified emission feature around 30 μm was first seen in carbon-rich AGB stars (e.g. IRC+10216 and AFGL 2688) and PN (Fig. 2). The 30 μm feature is now found to be common among carbon-rich PPN, especially those showing the 21 μm emission feature. This feature is very prominent in PNs and PPNs and is responsible for ~20% of the total infrared emission (Hrivnak, Volk & Kwok 2000). The possibility that thiourea ($SC(NH_2)_2$) and OH groups may contribute to the 30 μm feature has recently been suggested (Papoular 2011).

4. How do they form?

Although theoretically the condensation process is difficult to understand, observationally we see organics of aliphatic and aromatic structures form in PPNs on time scales

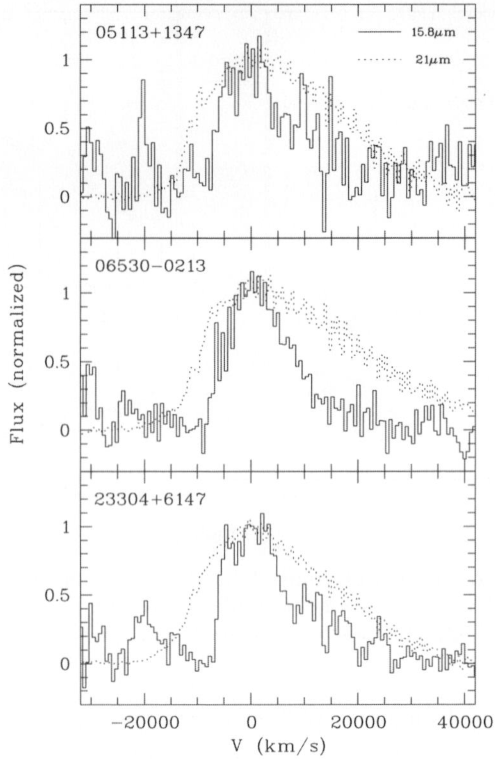

Figure 5. Comparison of the profiles of the 21 and 15.8 μm features in 3 PPNs. Figure adapted from Zhang, Kwok & Hrivnak (2010).

as short as a few hundred years. The best evidence that for their ease of formation can be found in novae, where the aliphatic features appear on a time scale of days. Figure 10 shows the infrared spectrum of the nova V2362 Cygni. Its spectrum changed from a pure gas spectrum of one showing strong dust continuum on time scale of months or

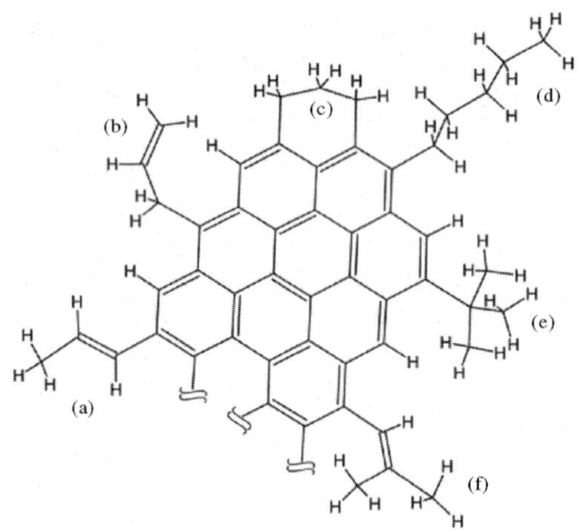

Figure 6. Schematic chemical structure illustrating the various possible side groups attached to aromatic rings that contribute to the plateau emissions (from Kwok, Volk & Bernath 2001).

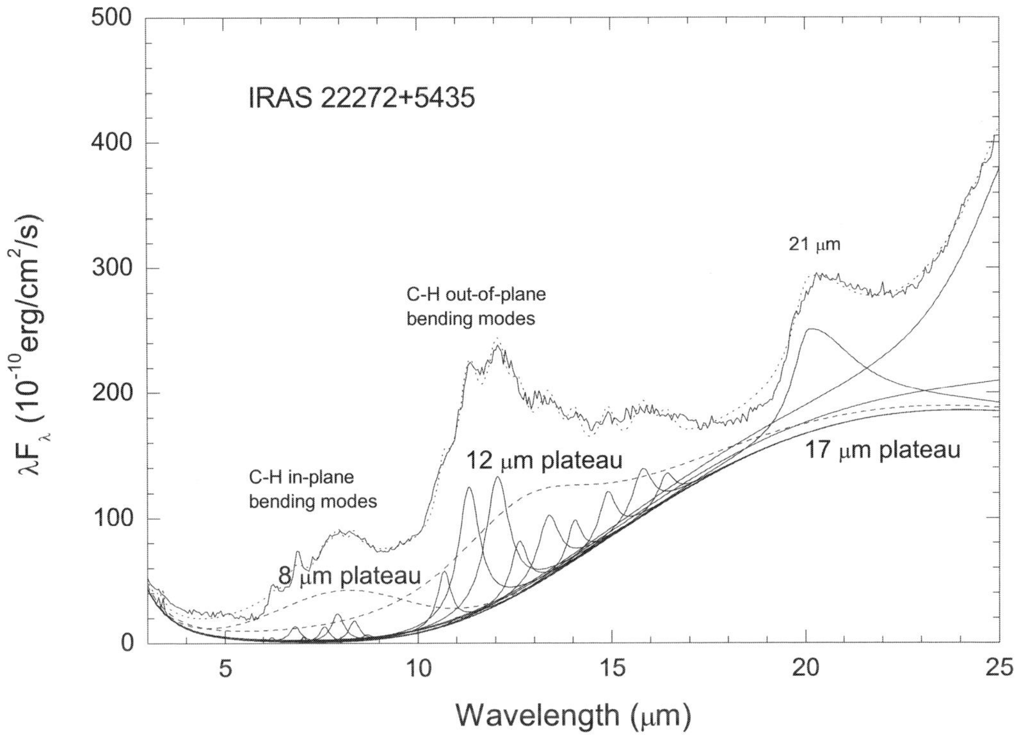

Figure 7. Spitzer IRS spectrum of the PPN IRAS 22272+5435. The thin solid lines are fits to discrete features and the dashed lines are fits to the plateau features. The rise at the long wavelength end is due to the 30 μm feature. Although the AIBs can be seen, most of the fluxes are emitted in the 8 and 12 μm plateau features and the 21 and 30 μm features.

even days. A strong 8 μm plateau feature is clearly evident in this spectrum. In the case of Nova V705 Cas, the 3.3, 3.4, 8.2, and 11.4 μm can be seen to emerge in the nova spectrum less than a year after outburst (Fig. 11) (Evans *et al.* 2005).

A similar example can be found in the optical transient source in the galaxy NGC 300. The 8 and 12 μm plateaus emissions can be seen about 3 months after outburst (Fig. 12).

5. Chemical structure of the carrier

Although the hypothesis that the AIB features originate from polycyclic aromatic hydrocarbons (PAH) molecules is popular in the astronomical community, there are a number of difficulties with this hypothesis. PAHs are fused ring molecules made up of C and H† and their vibrational bands are sharp and the peak wavelengths well defined. However, the observed UIE features are broad. In order to fit the observed astronomical spectra, it is necessary to use a complex mixture of PAHs of different sizes, structures, and charge states (Allamandola, Hudgins & Sandford 1999, Peeters *et al.* 2002), adopt empirical emission profiles (Draine & Li 2007), or a mixture of PAHs with each feature convolved with artificial profiles (Cami 2011). PAH molecules are primarily excited by

† In this conference, the term PAH (sometimes in quotation marks) is used to refer to various carbonaceous compounds containing both aromatic and aliphatic components. Since PAH is a precisely defined chemical term, it is the best if we end the confusion by confining the use of this term to fused aromatic rings.

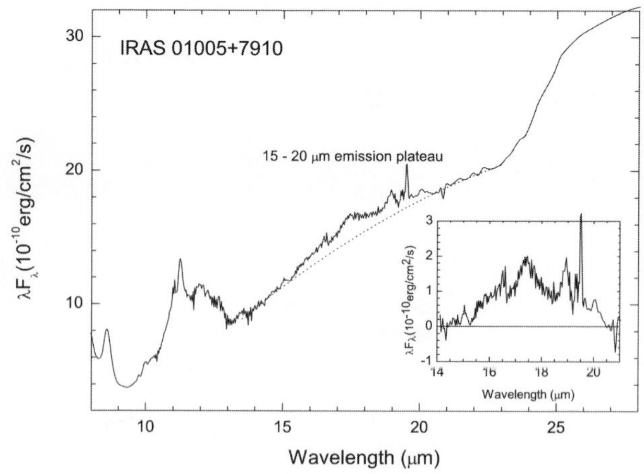

Figure 8. Spitzer IRS spectrum of the PPN IRAS 01105+7910 showing the 17 μm plateau emission. Figure adapted from Zhang, Kwok & Hrivnak (2010).

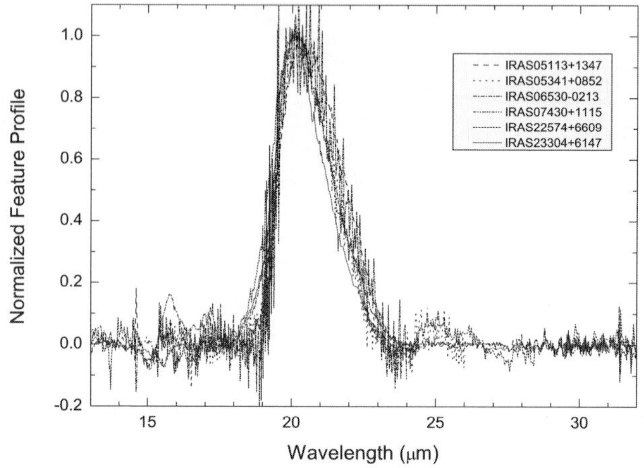

Figure 9. Comparison of the normalized 21 μm feature profile from the *Spitzer* spectra of 6 PPN. The features all have the same intrinsic profile and peak wavelengths. Figure adapted from Hrivnak, Volk & Kwok (2009).

UV radiation but AIBs are seen in PPNs and reflection nebulae with low temperature central stars. The observed shapes and peak wavelengths of the AIBs are independent of stellar temperature (Uchida *et al.* 2000). Furthermore, in spite of the well known vibrational and rotational spectra of PAH molecules, no individual PAH molecule has ever been detected either in absorption or emission (Clayton *et al.* 2003, Pilleri *et al.* 2009).

These issues are particularly relevant in evolved objects as the 3.4 and 6.9 μm aliphatic features and the 8 and 12 μm plateau features are strong. These spectral properties are similar to those observed in natural substances such as coal (Papoular *et al.* 1989, Guillois *et al.* 1996), kerogen, and petroleum fractions (Cataldo, Keheyan & Heymann 2002, Cataldo & Keheyan 2003, Cataldo, Keheyan & Heymann 2004). A number of artificial substances synthesized in the laboratory can also be candidates of the carrier. Hydrogenated amorphous carbon (HAC, Jones *et al.* 1990) are produced by laser ablation of graphite in a hydrogen atmosphere (Scott & Duley 1996). Quenched carbonaceous composites (QCC) are produced by hydrcarbon plasma deposition (Sakata *et al.* 1987).

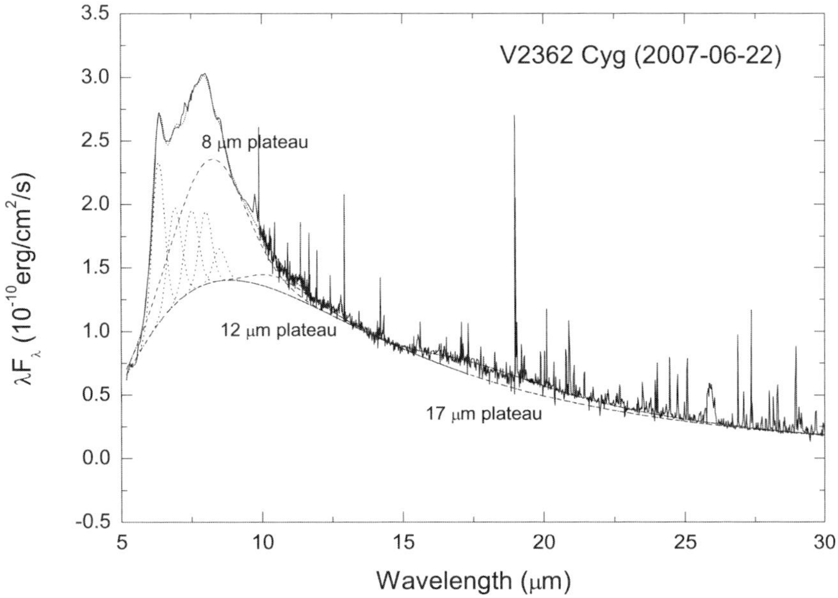

Figure 10. Spitzer IRS spectrum of nova V2362 Cygni at 446 days after outburst. A strong 8 μm plateau feature can clearly be seen above the dust continuum. The dotted lines are fits to discrete features at 6.2, 6.9, 7.6, 7.8, and 8.6 μm.

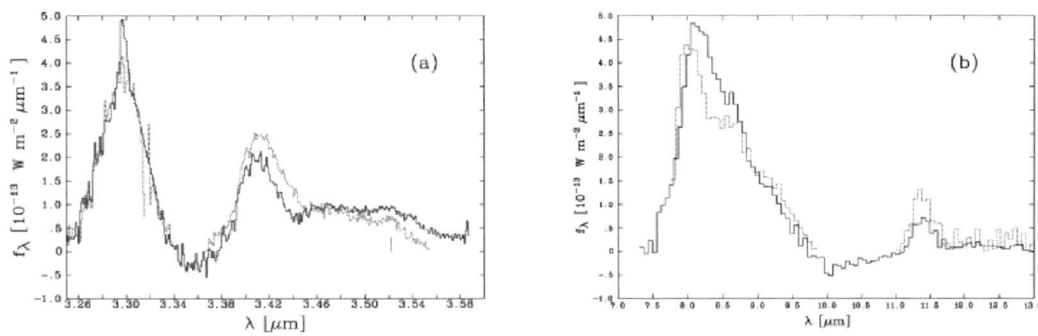

Figure 11. Profiles of the 3.3 μm and 3.4 μm features (left) and the 8.2 μm and 11.4 μm features (right) of Nova V705 Cas at two different epochs (253 and 320 days after outburst, shown as solid and broken lines respectively). Figure adapted from Evans *et al.* (2005).

Carbon nanoparticles of sizes 1-10 nm made up of primarily sp^2 carbon rings connected by a network of aliphatic chains can be made under different conditions and different initial gaseous compositions (Colangeli *et al.* 1995, Duley & Hu 2009, Jäger *et al.* 2009). These are all amorphous solids made up of pure H and C. When N is introduced, other organic solids are possible. Tholins is a refractory organic material formed by UV photolysis of reduced gas mixtures (N_2, NH_3, CH_4). HCN polymers are amorphous hydrogenated carbon nitride formed spontaneously from HCN.

We therefore suggest that the carrier of the UIE features in evolved stars are organic nanoparticles with mixed aromatic/aliphatic structures. The amorphous nature of the compound implies that there is a large variation of the bond lengths and bond angles, leading naturally to broad emission features. The small sizes of the particles will allow them to radiate in the near infrared region, and the mixed sp^2/sp^3 structure can naturally explain the observed spectral features. An example of such structure is shown in Fig. 13.

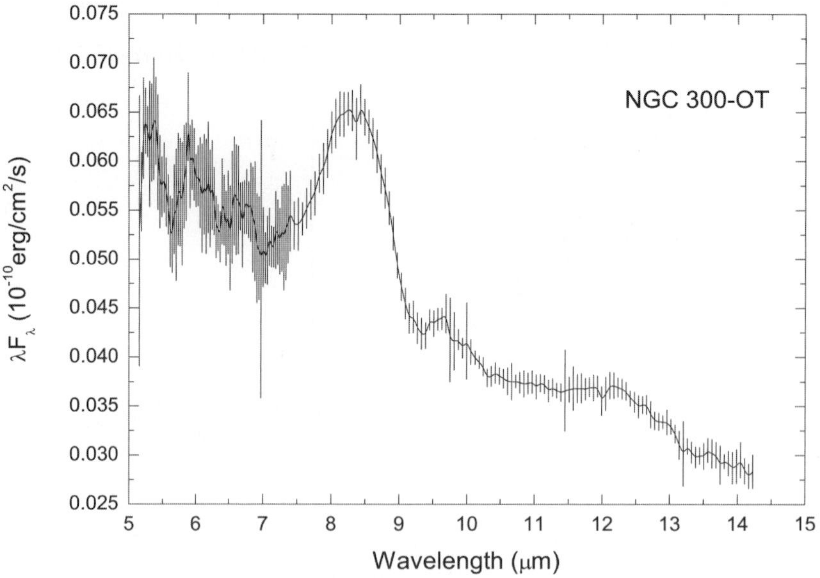

Figure 12. *Spitzer* IRS spectrum of the optical transient source NGC 300-OT 93 days after discovery. Broad emission features around 8 and 12 μm can be seen. Figure adapted from Prieto *et al.* 2009.

6. The stellar-solar system connection

In spite of the common perception that Solar System bodies are made up of minerals, metals, and ices, organics are in fact abundantly present in meteorites, comets, asteroids, and interplanetary dust particles (IDPs). Over 70% of the organics in carbonaceous chondrite meteorites is in the form of insoluble organic matter (IOM, Alexander *et al.* 2007). IOM is a macromolecular material with structures similar to that of kerogen (Kerridge 1999) and is possibly of interstellar origin due to excess of D, ^{13}C, and ^{15}N (see Alexander, these proceedings). Samples returned from the Comet Wild 2 also contain a significant organic component (Sandford *et al.* 2006). Spectra of IDPs show the 3.4 μm aliphatic feature and sometimes the C=O group (Flynn *et al.* 2003, Keller *et al.* 2004).

Using a variety of laboratory techniques such as Fourier transform infrared spectroscopy, nuclear magnetic resonance, electron paramagnetic resonance, X-ray absorption near-edge spectroscopy, high resolution electron microscopy, and thermal and chemical degradations, Derenne & Robert (2010) were able to determine the elemental and molecular contents of the Murchison IOM. A schematic of their model of the IOM is shown in Fig. 13. This structure is characterized by small units of aromatic islands and extensive aliphatic branches. Significant impurities such as O, N, and S atoms are also present.

7. Implications

We now know that PPNs and PNs are able to synthesize organic compounds of aromatic and aliphatic structures in the very low density circumstellar environment. The chemical structure of these organics undergo changes over time scales of hundreds of years. It is possible that photochemistry plays a part in this chemical evolution. Two unidentified emission features at 21 and 30 μm are observed. It is likely that these are carbonaceous compounds of unusual chemical compositions.

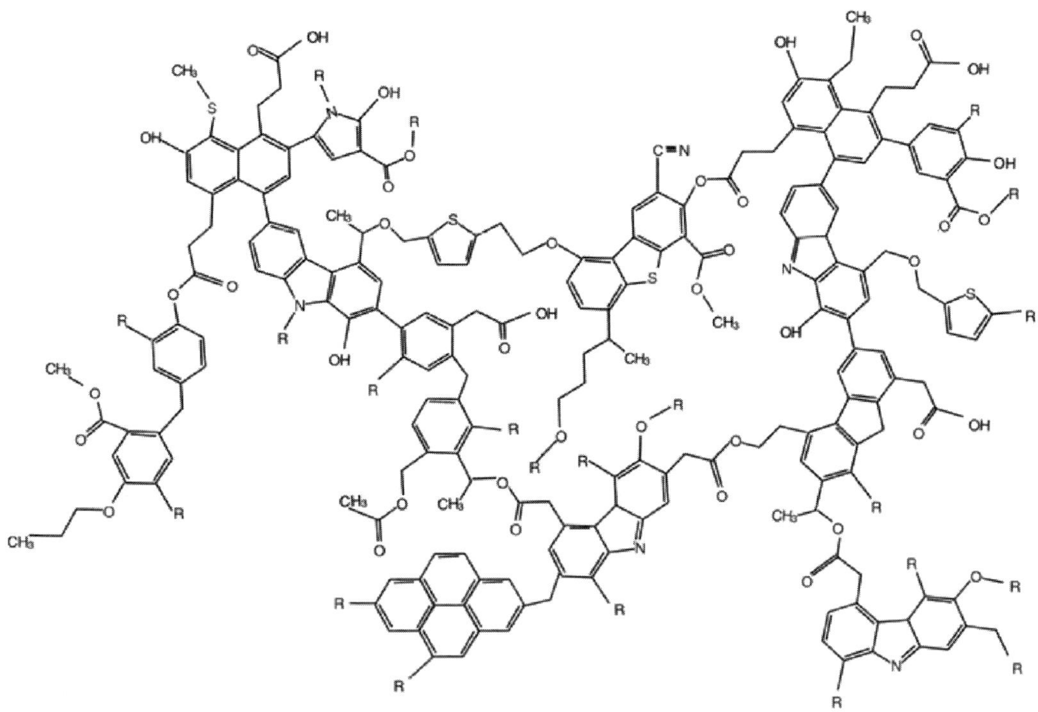

Figure 13. Model molecular structure of the IOM of Murchison meteorite. Figure adapted from Derenne & Robert (2010). This model is also applicable to the organic compound in PPN.

The strengths of the emission bands suggest that large quantities of organics are being synthesized and ejected into the interstellar medium. We now know that complex organics are commonly present in galaxies and they were produced in abundance in the early history of the Universe. Did evolved stars play an important role in the supply of organics in galaxies?

On the other side of the scale, is there a connection between organics produced by stars and those found in the Solar System? For this connection to exist, the organic grains have to survive in their journey through the ISM without extensive chemical processing either in the ISM or in the early Solar System. However, the discovery of pre-solar grains have confirmed that solid particles produced by AGB stars have reached the Earth (Zinner 1998). To what extent that the early Earth was enriched by organics from evolved stars (Kwok 2004, Kwok 2011) is an interesting topic for future studies.

Acknowledgements

The work was supported by a grant from the Research Grants Council of the Hong Kong Special Administrative Region, China (Project No. HKU 7027/11P).

References

Alexander, C. M. O. D. 2011, these proceedings

Alexander, C. M. O. D., Fogel, M., Yabuta, H., & Cody, G. D. 2007, Geochimica et Cosmochimica Acta, 71, 4380

Allamandola, L. J., Tielens, A. G. G. M., & Barker, J. R. 1989, *ApJS*, 71, 733

Allamandola, L. J., Hudgins, D. M., & Sandford, S. A. 1999, *ApJ*, 511, L115
Cami, J. 2011, in *PAHs and the Universe*, eds. C. Joblin & A. G. G. M. Tielens, EAS Publications Series, 46, 117
Cataldo, F., Keheyan, Y., & Heymann, D. 2002, *Int. J. Astrobiology*, 1, 79
Cataldo, F. & Keheyan, Y. 2003, *Int. J. Astrobiology*, 2, 41
Cataldo, F., Keheyan, Y., & Heymann, D. 2004, *Origins of Life and Evolution of the Biosphere*, 34, 13
Cernicharo, J., Heras, A. M., Tielens, A. G. G. M., Pardo, J. R., Herpin, F., Gulin, M., & Waters, L. B. F. M. 2001, *ApJ*, 546, L123
Clayton, G. C., *et al.* 2003, *ApJ*, 592, 947
Colangeli, L., Mennella, V., Palumbo, P., Rotundi, A., & Bussoletti, E. 1995, *A&AS*, 113, 561
Derenne, S. & Robert, F. 2010, *Meteoritics and Planetary Science*, 45, 1461
Draine, B. T. & Li, A. 2007, *ApJ*, 657, 810
Duley, W. W. & Williams, D. A. 1981, *MNRAS*, 196, 269
Duley, W. W. & Hu, A. 2009, *ApJ*, 698, 808
Evans, A., Tyne, V. H., Smith, O., Geballe, T. R., Rawlings, J. M. C., & Eyres, S. P. S. 2005, *MNRAS*, 360, 1483
Flynn, G. J., Keller, L. P., Feser, M., Wirick, S., & Jacobsen, C. 2003, *Geochimica et Cosmochimica Acta*, 67, 4791
Geballe, T. R., Tielens, A. G. G. M., Kwok, S., & Hrivnak, B. J. 1992, *ApJ*, 387, L89
Guélin, M. 2011, these proceedings
Guillois, O., Nenner, I., Papoular, R., & Reynaud, C. 1996, *ApJ*, 464, 810
Hrivnak, B. J., Volk, K., & Kwok, S. 2000, *ApJ*, 535, 275
Hrivnak, B. J., Geballe, T. R., & Kwok, S. 2007, *ApJ*, 662, 1059
Hrivnak, B. J., Volk, K., & Kwok, S. 2009, *ApJ*, 694, 1147
Hudgins, D. M. & Allamandola, L. J., 1999, *ApJ*, 516, L41
Jäger, C., Huisken, F., Mutschke, H., Jansa, I. L., & Henning, T. H. 2009, *ApJ*, 696, 706
Jones, A. P., Duley, W. W., & Williams, D. A. 1990, *Quart. J. Roy. Astr. Soc.*, 31, 567
Keller, L. P., Messenger, S., Flynn, G. J., Clemett, S., Wirick, S., & Jacobsen, C. 2004, *Geochimica et Cosmochimica Acta*, 68, 2577
Kerridge, J. F. 1999, *Sp. Sci. Rev.*, 90, 275
Knacke, R. F. 1977, *Nature*, 269, 132
Kwok, S. 1993, *Ann. Rev. Astr. Ap.*, 31, 63
Kwok, S. 2000, *The Origin and Evolution of Planetary Nebulae*, CUP
Kwok, S. 2004, *Nature*, 430, 985
Kwok, S. 2011, *Organic Matter in the Universe*, Wiley
Kwok, S., Volk, K., & Bidelman, W. P. 1997, *ApJS*, 112, 557
Kwok, S., Volk, K. M., & Hrivnak, B. J. 1989, *ApJ*, 345, L51
Kwok, S., Volk, K., & Hrivnak, B. J. 1999, *A&A*, 350, L35
Kwok, S., Volk, K., & Bernath, P. 2001, *ApJ*, 554, L87
Papoular, R. 2011, *MNRAS*, 415, 494
Papoular, R., Conrad, J., Giuliano, M., Kister, J., & Mille, G. 1989, *A&A*, 217, 204
Peeters, E., Hony, S., Van Kerckhoven, C., *et al.*, 2002, *A&A* 390, 1089
Pilleri, P. *et al.* 2009, *MNRAS*, 397, 1053
Prieto, J. L., Sellgren, K., Thompson, T. A., & Kochanek, C. S. 2009, *ApJ*, 705, 1425
Puget, J. L. & Léger, A. 1989, *Ann. Rev. Astr. Ap.*, 27, 161
Sakata, A., Wada, S., Onaka, T., & Tokunaga, A. T. 1987, *ApJ*, 320, L63
Sandford, S. *et al.* 2006, Science, 314, 1720
Scott, A. & Duley, W. W. 1996, *ApJ*, 472, L123
Uchida, K. I., Sellgren, K., Werner, M. W., & Houdashelt, M. L. 2000, *ApJ*, 530, 817
Van Kerckhoven, C., *et al.* 2000, *A&A*, 357, 1013
Volk, K., Xiong, G.-Z., & Kwok, S. 2000, *ApJ*, 530, 408
Volk, K. *et al.* 2011, *ApJ*, 735,127
Zhang, C. Y. & Kwok, S. 1991, *A&A*, 250, 179

Zhang, Y., Kwok, S., & Hrivnak, B. J. 2010, *ApJ*, 725, 990
Zinner, E. 1998, *Ann. Rev. Earth Planet. Sci.*, 26, 147

Discussion

DECIN: Prof. Kwok stated that PAH features were only detected in PPNe and PNe. Recently, K. Smolders *et al.* (2010) published the detection of PAH features in S type AGB stars, with a rich aliphatic component. This detection might help in understanding the excitation/formation of PAH.

KWOK: The AIB features in BZ CMa could be the result of a binary evolution. Thousands of AGB stars have been observed spectroscopically without showing AIB features.

HORNEKAER: Can you estimate the degree of inpurities (N, O, S) in the carbon particles?

KWOK: O, N, S impurities are found in the IOM of meteorites but spectroscopic identification of heterocyclic aromatic compounds in astronomical sources will need much higher spectral resolution than is available at present.

STERNBERG: Do the solid-state features disappear before the nebula disperses? Does the appearnce of the white dwarf affect the survival?

KWOK: As planetary nebulae expand, the flux of the infrared continuum also decreases as the result of geometric dilution of light from the central star. The strengths of the AIB features decreases correspondingly. As the central stars of PNe evolve to become WDs, the amount of UV light increases but there is no observed correlation between this and the strengths of the features.

MAUERSBERGER: Changes in proto-planetary nebulae spectra might be observed on a time scale of decades. It is therefore worthwhile to monitor such objects. This requires a thorough calibration and archiving of data.

KWOK: Since proto-planetary nebulae are evolving on time scales of 10^2-10^3 yr, it would certainly be worthwhile to monitor changes in the infrared spectra at decade intervals.

The Molecular Universe
Proceedings IAU Symposium No. 280, 2011
José Cernicharo & Rafael Bachiller, eds.

doi:10.1017/S1743921311024999

Fullerenes in Circumstellar and Interstellar Environments

Jan Cami[1,2], Jeronimo Bernard-Salas[3,4], Els Peeters[1,2] and Sarah E. Malek[1]

[1]Department of Physics & Astronomy, The University of Western Ontario, London, ON N6A 3K7, Canada
email: jcami@uwo.ca

[2]SETI Institute, 189 Bernardo Ave, Mountain View, CA 94043, USA

[3]Institut d'Astrophysique Spatiale, CNRS/Université Paris-Sud 11, 91405 Orsay, France

[4]Cornell University, 222 Space Sciences Bld., Ithaca, NY 14853, USA

Abstract. We recently identified several emission bands in the Spitzer-IRS spectrum of the unusual planetary nebula Tc 1 with the infrared active vibrational modes of the neutral fullerene species C_{60} and C_{70}. Since then, the fullerene bands have been detected in a variety of sources representing circumstellar and interstellar environments. Abundance estimates suggest that C_{60} represents ~0.1%–1.5% of the available carbon in those sources. The observed relative band intensities in various sources are not fully compatible with single-photon heating and fluorescent cooling, and are better reproduced by a thermal distribution at least in some sources. The observational data suggests that fullerenes form in the circumstellar environments of evolved stars, and survive in the interstellar medium. Precisely how they form is still a matter of debate.

Keywords. astrochemistry, circumstellar matter, ISM:molecules, infrared:ISM

1. Introduction

Fullerenes are large molecules made of carbon hexagons and pentagons that are organized in the shape of a hollow sphere or ellipsoid. The best-known member of the class is the archetypal so-called "buckminsterfullerene" C_{60} whose structure is the same as a soccer ball (see Fig. 1). Fullerenes were discovered in a series of laboratory experiments that simulated the circumstellar environment of carbon stars and that were aimed at understanding the formation of long carbon chains in such outflows (Kroto *et al.* 1985). Their unique properties has made them the topic of much research in very diverse fields on Earth – from superconductivity over nanotechnology to targeted drug delivery.

As soon as they were discovered on earth, it was suggested that fullerenes could also form in carbon star outflows and be injected into the interstellar medium (ISM). Since they are amongst the most stable carbon species known, they would then be ideally suited to survive the harsh conditions in the ISM.

Dedicated searches for the electronic bands of neutral C_{60} in the ISM and in circumstellar shells were negative (Snow & Seab 1989; Somerville & Bellis 1989; Herbig 2000). Foing & Ehrenfreund (1994) on the other hand reported the detection of two diffuse interstellar bands (DIBs) whose wavelengths are consistent with transitions of the C_{60}^+ cation as measured in solid matrices (Fulara *et al.* 1993); more recently, several more DIBs have been found that could be due to the same cation (Misawa *et al.* 2009). However, since the technique of matrix isolation spectroscopy introduces unpredictable shifts in the frequencies of the absorption bands, this promising case awaits confirmation from a comparison to a cold gas phase spectrum which is not yet available.

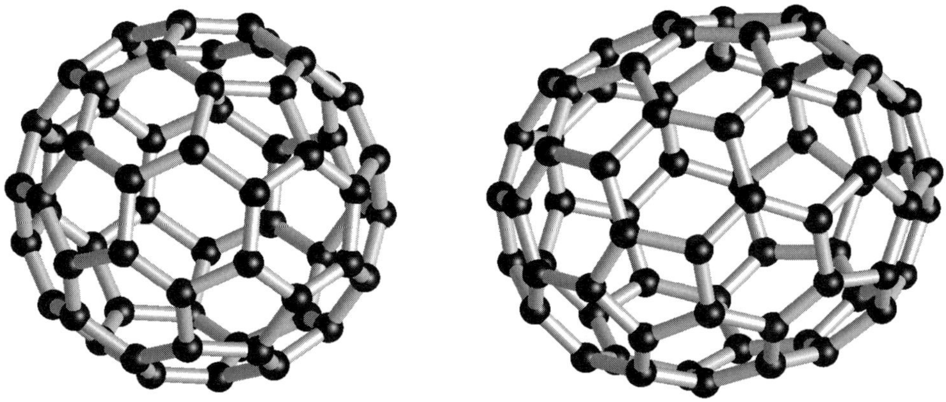

Figure 1. A representation of the molecular structure of C_{60} (left) and C_{70} (right).

Fullerenes can also be detected through their vibrational modes at infrared (IR) wavelengths (see Sect. 2), and several efforts have focused on these spectral signatures (Moutou *et al.* 1999; Clayton *et al.* 1995). Again, no conclusive evidence was found for the presence of C_{60} in space. More recently, Sellgren *et al.* (2007) analyzed the 15–20 μm spectrum of the reflection nebula NGC 7023 obtained with the Infrared Spectrograph (IRS, Houck *et al.* 2004) on board the Spitzer Space Telescope (Werner *et al.* 2004). They suggested that C_{60} could be one of the possible carriers for the spectral features in this source at 17.4 and 18.9 μm.

Cami *et al.* (2010) presented the unusual Spitzer-IRS spectrum of the planetary nebula (PN) Tc 1, and identified several emission features with the vibrational modes of the fullerene species C_{60} and C_{70}. Since then, fullerenes have been confirmed in many more sources of different nature, indicating that these species may be quite widespread in the Universe. A lively discussion has ensued especially over their excitation properties and formation mechanism.

Here, we briefly review some of the relevant properties of C_{60} and C_{70} in Sect. 2. We present the case for fullerenes in Tc 1 in detail, and summarize the other detections in Sect. 3. Some aspects of the excitation mechanism are discussed in Sect. 4, and the formation of fullerenes in space in Sect. 5.

2. Spectroscopic properties of C_{60} and C_{70}

A representation of a C_{60} molecule is shown in Fig. 1. The 60 carbon atoms are arranged in the shape of a truncated icosahedron (I_h symmetry): a hollow sphere that is made of 20 hexagons and 12 pentagons. C_{60} has 174 vibrational modes, but owing to its symmetry, many modes are degenerate and only 4 distinct modes are IR active: the $T_{1u}(1)$ mode at 18.9 μm; the $T_{1u}(2)$ mode at 17.4 μm; the $T_{1u}(3)$ mode at 8.5 μm and the $T_{1u}(4)$ mode at 7.0 μm. Laboratory experiments have measured the frequencies of these transitions in the pure, solid state (Krätschmer *et al.* 1990; von Czarnowski & Meiwes-Broer 1995) as well as in the gas-phase (Frum *et al.* 1991; Nemes *et al.* 1994). In those experiments, the measured bands are fairly broad and symmetric, and thus well characterized by a central wavelength and a width. The differences between the measured frequencies primarily reflect the effect of temperature: a higher temperature shifts the bands to lower frequencies and causes broadening (Nemes *et al.* 1994).

The intrinsic strength of the C_{60} bands is less well known, and the literature offers quite different values for the absorption coefficients (even relative to one another), obtained either from laboratory experiments or theoretical calculations. Cami *et al.* (2010) used data from Martin *et al.* (1993); Fabian (1996); Sogoshi *et al.* (2000); for the analysis of NGC 7023 by Sellgren *et al.* (2010), strengths from Choi *et al.* (2000) were used. Also Iglesias-Groth *et al.* (2011) presented molar absorptivities derived from laboratory experiments.

Whereas the model for the C_{60} geometry has traditionally been a soccer ball, the C_{70} molecule is a prolate top more like a rugby ball. The lower symmetry (D_{5h}) results in more IR active modes: of the 204 fundamental vibrational modes, 31 are IR active, and laboratory studies as well as theoretical DFT calculations are available (von Czarnowski & Meiwes-Broer 1995; Stratmann *et al.* 1998; Schettino *et al.* 2002). Most of the C_{70} IR active bands are weak, and furthermore blend with the C_{60} bands; but several isolated bands of medium strength are at 14.8, 21.8, 12.6 and 15.6 μm.

3. Fullerenes in astrophysical environments

3.1. *The planetary nebula Tc 1*

Cami *et al.* (2010) reported the identification of the IR C_{60} bands in the unusual Spitzer-IRS spectrum of the planetary nebula (PN) Tc 1. The high quality of the spectrum and the lack of contaminating features make this source the clearest case for the presence of fullerenes in space.

Fig. 2 shows the continuum-subtracted Spitzer-IRS observations of the planetary nebula Tc 1. Apart from the strong (typically forbidden) emission lines that are characteristic for PNe, the spectrum is very unusual, and exhibits clear emission bands at the wavelengths of all four fundamental modes of C_{60}†. Weaker and narrower features are present at the wavelengths of the strongest (unblended) C_{70} bands. In addition to the fullerene bands, a broad underlying plateau is evident between 6–9 μm, as is a broad emission feature at 11.5 μm that is generally attributed to SiC (Speck *et al.* 2009). There is not much else left in the spectrum though, and thus little contamination confuses the analysis of the fullerene spectral features.

As mentioned in Sect. 2, the experimentally obtained central wavelengths and band widths for the C_{60} bands depend on the temperature, and thus a comparison with laboratory data requires some estimate of a typical temperature for the fullerenes in Tc 1. Surprisingly, the observed band strengths for both C_{60} and C_{70} are consistent with a thermal population distribution over the vibrational levels (see Sect. 4). For C_{60}, we can thus derive an excitation temperature of 332 K; for C_{70} the derived temperature is 179 K. All other measurable quantities are consistent with such a temperature. Indeed, the central wavelengths of all C_{60} emission bands in the spectrum of Tc 1 correspond especially well to experimental values measured at ~300 K (rather than values measured at e.g. 1000 K). Further support comes from the band widths: laboratory spectra show temperature-dependent widths (FWHM) from about 8 cm^{-1} at 300 K to 13 cm^{-1} at 1000 K (Nemes *et al.* 1994) for both the 17.4 and 18.9 μm C_{60} bands. For Tc 1, the measured band widths are 9.9 and 12.6 cm^{-1} for the bands at 18.9 and 17.4 μm respectively, but since both bands suffer from contamination by the C_{70} bands, their true band widths are necessarily smaller, and thus only compatible with the lower temperature regime. The C_{60} bands at 7.0 and 8.5 μm are not resolved and thus offer no such information. Also the

† Note however that a significant fraction of the strong emission at 7.0 μm is due to an [ArII] line.

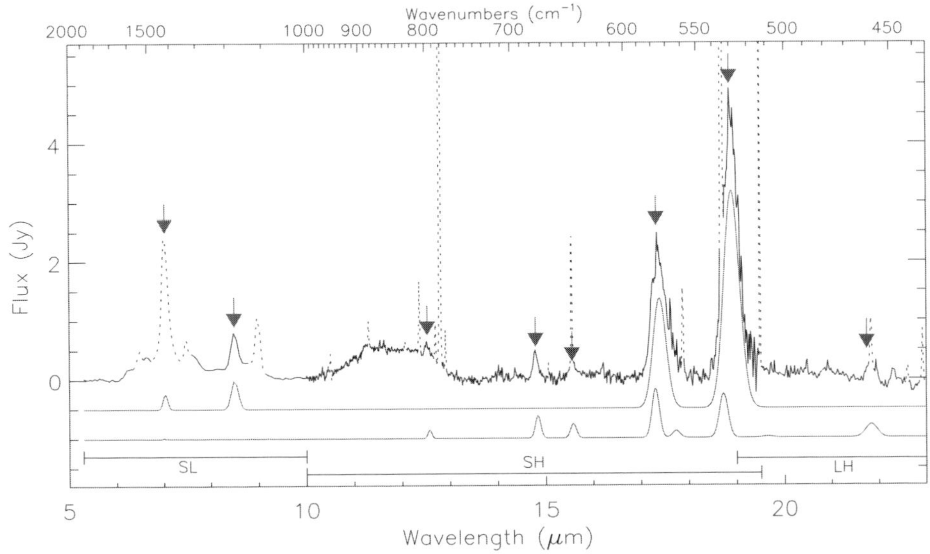

Figure 2. A composite IR spectrum of the planetary nebula Tc 1 as observed by Spitzer-IRS is shown on top. The spectrum is the combination of observations with the Short-Low module (SL, $\lambda/\Delta\lambda \sim 60-120$) and Short- and Long-High (SH and LH, $\lambda/\Delta\lambda \sim 600$) modules. A featureless dust continuum was subtracted from the spectrum. Known strong emission lines are masked and shown here as dotted lines. The curve below it is a thermal emission model for C_{60}; the bottom curve a similar model for C_{70}. Figure taken from Cami *et al.* (2010).

central wavelengths of the four isolated C_{70} features compare better to the frequencies measured at low temperatures.

Summarizing, we can conclude that the spectrum of Tc 1 shows the clear spectral fingerprint of the fullerene species C_{60} and C_{70}, and that all measurable quantities (wavelengths, widths and band ratios) are consistent with laboratory experiments at temperatures of a few hundred K.

3.2. *Other Evolved Objects*

With the clean fullerene spectrum of Tc 1 as a template, the presence of C_{60} has been confirmed in the circumstellar environments of a variety of evolved objects. García-Hernández *et al.* (2010) found the C_{60} bands in several other carbon-rich PNe. In most cases, the spectra are contaminated by PAH features, thus complicating the analysis; a notable exception is the PN SMP SMC 16 in the Small Magellanic Cloud that looks quite similar to Tc 1. No conclusive confirmation of the weaker C_{70} features has been reported in these objects though. At first, these detections suggested fullerenes form in the PN phase. However, they are already present in at least one proto-PN object (Zhang & Kwok 2011), and the C_{60} bands are also seen in the Spitzer-IRS spectra of a few post-AGB stars (Gielen *et al.* 2011).

In the search for fullerenes in space, special attention was given to the R Coronae Borealis (RCB) stars, a class of hydrogen-poor and helium-rich supergiants. Under typical conditions in circumstellar environments, the presence of hydrogen inhibits the formation of fullerenes (see Sect. 5), and thus the hydrogen-poor nature of the circumstellar environments of RCB stars was considered ideal to form C_{60}. The first search for C_{60} in such environments did not turn up any evidence for fullerenes (Clayton *et al.* 1995).

More recently, C_{60} has been detected in two RCB stars that were studied with the Spitzer Space Telescope (García-Hernández *et al.* 2011). Surprisingly, those two objects represent the most H-rich objects in a larger sample of RCB stars. In another study, an additional detection was made in one more RCB star (Clayton *et al.* 2011).

3.3. *Interstellar environments*

The spectral features of C_{60} have also been detected in interstellar environments – in the two reflection nebulae NGC 7023 and NGC 2023 (Sellgren *et al.* 2010), and also in the Orion Bar (Rubin *et al.* 2011). These sources are spatially resolved, and Sellgren *et al.* (2010) showed that in NGC 7023, the peak emission of PAHs originates from a different location as the peak emission from C_{60} (see Fig. 5). This has some implications for the formation and excitation of fullerenes (see Sect. 4 and 5).

3.4. *Abundances*

With the presence of fullerenes in space firmly established, an important question to address is what fraction of the cosmic carbon they represent, and different methods have been applied to estimate the fullerene abundances. In the evolved stars, of the order of 0.06–1.5% of the available carbon is estimated to be in the form of C_{60} (Cami *et al.* 2010; García-Hernández *et al.* 2010; Zhang & Kwok 2011). For NGC 7023 on the other hand, Sellgren *et al.* (2010) find that 0.1–0.6% of the interstellar carbon is in C_{60}. This latter is compatible with earlier estimates of 0.3–0.9% of the cosmic carbon that was estimated to be in the form of C_{60}^{+} (Foing & Ehrenfreund 1994).

It is interesting to see that the abundance estimates in the evolved objects are comparable to the interstellar fullerene abundances. If only a fraction of the carbon-rich evolved stars would produce fullerenes, one would expect the interstellar abundances to be much lower than the circumstellar abundances in individual objects. If the derived abundances are realistic, this then implies that the formation of fullerenes must be fairly common. In any case, the derived abundances represent a significant fraction of the cosmic carbon for a single molecular species.

The presence of C_{60} in diverse evolved objects suggests that fullerenes form in circumstellar outflows whenever conditions are right, and the detection of fullerenes in interstellar environments is then a testimony to the stability of these species that allows them to survive the harsh conditions in the ISM. In very general terms, the available observational data on the fullerenes thus confirm the early suggestion by Kroto *et al.* (1985): fullerenes form in carbon-rich circumstellar environments, and survive and thrive in the interstellar medium.

4. Excitation mechanism

Having identified several emission bands with a specific molecular carrier offers the prospect of using these spectral features as a diagnostic tool to extract quantitative information about the physical conditions that reign in the environment in which they reside. Precisely what information can be obtained depends somewhat on the molecules excitation mechanism at play.

Large, free molecules such as fullerenes and PAHs are generally thought to be excited by the absorption of a single photon (typically in the UV range). Rapid internal conversion then leaves the molecule in the electronic ground state, but in highly excited vibrational states; and as the molecule relaxes, it emits IR photons. For all relevant

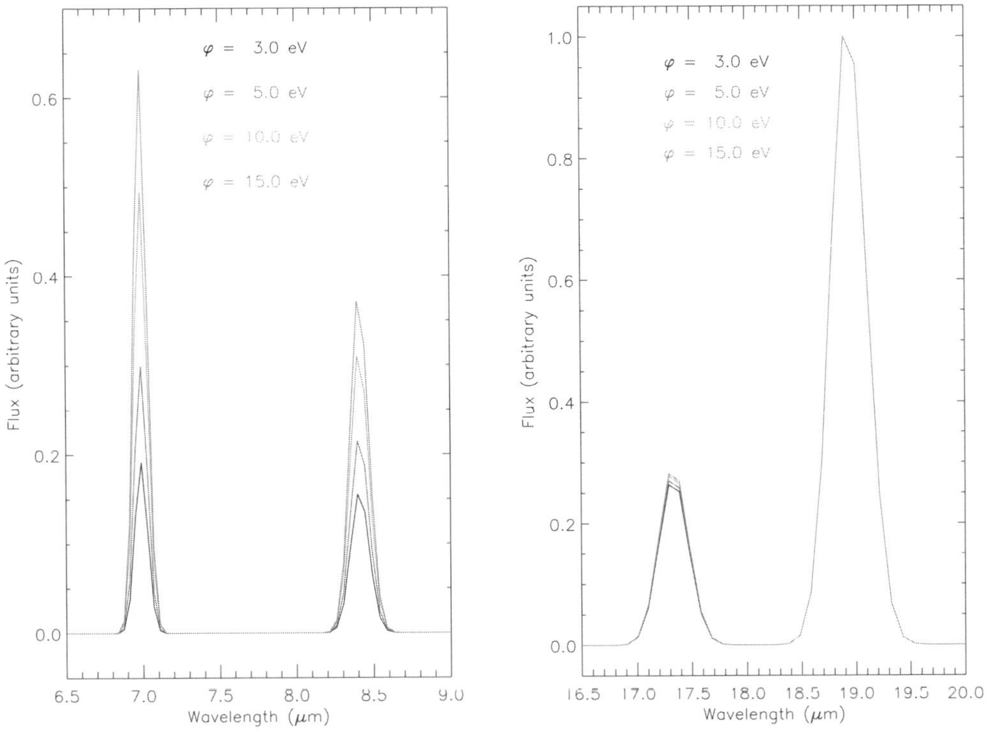

Figure 3. A few illustrative examples of how the C_{60} fluorescent emission spectrum changes for average photon energies of 3 eV, 5 eV, 10 eV and 15 eV (the lowest photon energies corresponding to the weakest emission). All spectra are normalized to the peak intensity of the 18.9 μm band. Widths were taken to be 10 cm^{-1} for the 17.4 and 18.9 μm bands and 20 cm^{-1} for the bands at 7.0 and 8.5 μm – consistent with the widths measured in the spectrum of Tc 1.

processes pertaining to the excitation & cooling mechanism, fullerenes have indeed very similar properties to PAHs (see e.g. Tielens 2005). Sellgren *et al.* (2010) analyzed the C_{60} in NGC 7023 using such an excitation model.

The resulting emission spectrum of this fluorescent cooling process can be calculated in various ways (e.g. ?Joblin *et al.* 2002). The only free parameter in these calculations is the average photon energy that is absorbed by the molecules – which is a function of the ambient radiation field and the absorption cross section of the molecule in question. Fig. 3 shows a few examples of how the resulting IR spectra change following absorption of photons with different energies. Note how there is little change in the ratio between the 17.4 μm band and the 18.9 μm band for the energies considered here. For typical PNe conditions ($\varphi \gtrsim 7$ eV) however, the bands at 7.0 and 8.5 μm should be fairly strong, and the band ratios $I_{7.0}/I_{18.9}$ and $I_{8.5}/I_{18.9}$ could in principle be used as a sensitive measure of the average photon energy.

However, the numerical values for the band ratios depend on the intrinsic band strengths used in the calculations; as mentioned in Sect. 2, these are not well known, and different values are available in the literature. Using the intensities as given by Choi *et al.* (2000) (as was done for the analysis of the C_{60} bands in the reflection nebula NGC 7023 presented by Sellgren *et al.* (2010)), the $I_{17.4}/I_{18.9}$ band ratio should be roughly constant at about 0.28 for photon energies relevant to PNe environments. The measured band ratio in Tc 1 however is 0.59; when accounting for the contamination by C_{70} the

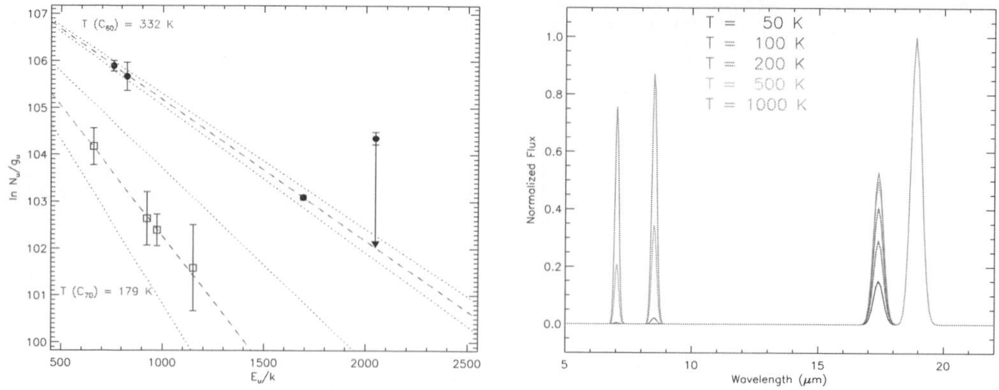

Figure 4. *(left)* Excitation diagram for the C_{60} and C_{70} bands in the Spitzer-IRS spectrum of Tc 1 (Fig. from Cami *et al.* 2010). *(right)* Illustrative examples of the spectral variations in the C_{60} bands. The bands at 7.0 and 8.5 μm are virtually absent for excitation temperatures below $\sim$200 K, whereas the 17.4/18.9 μm band ratio shows large variations for temperatures up to $\sim$500 K.

ratio is somewhat lower, but can never be brought in agreement with the expected ratio for fluorescent emission when adopting these intrinsic strengths. The 8.5/18.9 μm band ratio on the other hand is too low to be consistent with the same fluorescent cooling model.

One could argue that the adopted intensities for the bands are not correct, but also the other observations of the C_{60} bands pose difficulties in the framework of fluorescence. For all detected sources, the band ratios (including $I_{17.4}/I_{18.9}$) show significant variations from source to source (see e.g. García-Hernández *et al.* 2010). This is also evident from the different excitation temperatures ($\sim$200–700 K) that are derived assuming a thermal distribution over the vibrational bands. In many cases, the bands are contaminated by PAH features, but it is not clear if that alone can explain all the observed spectral variations. If true, these variations (and especially the $I_{17.4}/I_{18.9}$ band ratio) are not compatible with fluorescent cooling, irrespective of the intrinsic strength values used.

The fairly clean fullerene spectrum of Tc 1 offers the possibility of directly calculating the population distribution over the excited vibrational states from the total emitted power in the bands; and this can be done for both C_{60} and C_{70}. Surprisingly, the population distribution is fully consistent with a thermal distribution (see Fig. 4) over the vibrational states corresponding to an excitation temperature of 332 K for C_{60}, and 179 K for C_{70}.

Thermal models show quite different spectral variations as a function of temperature (Fig. 4) compared to the fluorescent cooling models and can more easily accommodate some of the observed changes in the $I_{17.4}/I_{18.9}$ band ratios. The cases where the 7.0 and 8.5 μm band are missing would then correspond to excitation temperatures $\lesssim 200\ K$, whereas an unacceptably low photon energy would be required in the fluorescence model.

It is not easy to understand how free C_{60} molecules in circumstellar or interstellar environments could produce an emission spectrum that is consistent with thermal excitation rather than with stochastic heating and fluorescent cooling. Since gas densities are typically orders of magnitude lower than the critical density, it is unlikely that gas collisions could keep the fullerenes thermalized. It was thus proposed that the fullerene molecules are not free, but in direct contact with the surface of dust grains, or in the solid state themselves Cami *et al.* (2010). In such a case, the fullerenes would be in equilibrium with the dust grains, which themselves are in equilibrium with the ambient radiation field. However, the derived excitation temperatures are hard to understand in this scenario. Further observations and analyses will be required to settle the excitation mechanism conclusively.

5. The formation of astrophysical fullerenes

From the fairly large number of evolved objects in which the C_{60} bands have been detected, it seems clear that these species do indeed form in the circumstellar environments of evolved stars. Precisely how this happens is far less clear though, and there has been some discussion about the precise pathways to fullerene formation†.

In this context, it is worthwhile to consider what we know from laboratory experiments. In the original experiments by Kroto *et al.* (1985), fullerenes self-assemble from much smaller carbon fragments by collisions. At very low gas densities, a broad distribution of cluster sizes is observed; but as the gas density is increased, collisions become more numerous, and only the most stable species (C_{60}) survives. If in carbon-rich circumstellar environments, collisional processes (occurring under the similar conditions) are driving the chemistry, fullerenes could form in the same way. The efficiency of the fullerene formation process is then a function of the density. However, an important condition in this formation process is the absence of hydrogen. It has long been realized that the presence of H will inhibit the formation of fullerenes (see e.g. Kroto & McKay 1988). Indeed, laboratory experiments show that the mere presence of hydrogen (as little as 0.4%) in such conditions already greatly reduces fullerene formation. At a hydrogen fraction of 10%, no fullerenes are formed anymore, and the experiments result in large amounts of PAHs instead (de Vries *et al.* 1993; Wang *et al.* 1995).

Based on the Spitzer-IRS spectrum of Tc 1 alone, it would seem that its circumstellar environment is fairly analogous to this laboratory setup. Indeed, the available material is clearly carbon-rich, while at the same time, the fullerene-rich IR spectrum does not show much evidence for the presence of PAHs nor any other H-containing species. It is thus conceivable that the region where the fullerenes are located is hydrogen-poor. However, it is clear that H is present at large in the nebula (see e.g. Milanova & Kholtygin 2009; Williams *et al.* 2008), and even in the photosphere of the central star (Mendez *et al.* 1988). Thus, if the formation of fullerenes in Tc 1 involves collisional growth of carbon clusters under similar conditions as in the laboratory experiments, there must have been H-poor regions in the nebula to allow the fullerenes to form.

García-Hernández *et al.* (2010) argue against a formation scenario in which a H-poor environment is required. They note that in their PNe data, the Spitzer-IRS spectra show the simultaneous presence of the C_{60} features and PAHs. From this, they conclude that both species co-exist, and that the fullerenes form at the same time and place as PAHs, in a H-rich environment. They propose photo-processing of solid Hydrogenated Amorphous Carbon (HAC) as the most plausible fullerene formation route in PNe environments. Support for that assertion is found in experimental results that show how such processing produces PAHs and fullerenes at the same time (Scott *et al.* 1997). Moreover, all C_{60} PNe furthermore show the so-called 30 μm feature that is generally attributed to MgS (see e.g. Hony *et al.* 2002), but that also has been linked to HACs (Grishko *et al.* 2001). They then interpret the PAH-poor and fullerene-dominated sources Tc 1 and SMP SMC 16

† Bettens & Herbst (1995) additionally discuss a possible formation route for fullerenes under interstellar conditions.

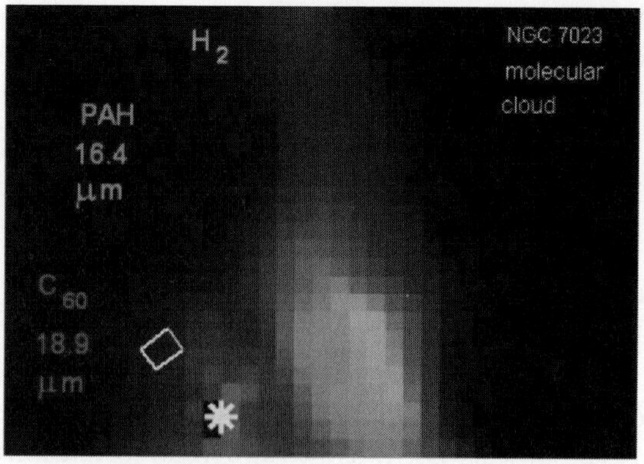

Figure 5. The spatial distribution of molecular hydrogen, PAHs and C_{60} in the reflection nebula NGC 7023 (Fig. from Sellgren *et al.* 2010)

as environments where the more hardy C_{60} molecules have survived whereas the PAHs have been destroyed.

It is clear that – in the context of astrophysical fullerenes – this scenario presents several difficulties as well. First, if HAC photo-processing is the general mechanism to produce PAHs and fullerenes in PNe, and if the 30 μm feature is indeed related to HACs, one would expect all PNe that exhibit the 30 μm feature to also exhibit PAH and fullerene bands. Second, it seems unlikely that UV photo-processing is a viable mechanism to explain the formation of C_{60} in evolved stars with much lower UV irradiation than is the case in the PNe – such as the proto-PN IRAS 01005+7910 (Zhang & Kwok 2011) but especially the two post-AGB stars presented by Gielen *et al.* (2011) with effective temperatures of the order of only 6,000 K. Finally, if Tc 1 and SMP SMC 16 represent environments where the fullerenes have survived and PAHs have not, it is hard to explain the numerous PNe that show clear PAH features but no C_{60} bands.

It is important to realize that the mixture of carbonaceous species that results from HAC photo-processing in the laboratory experiments is quite different from what is observed in the C_{60} PNe. The mass spectra reported by Scott *et al.* (1997) do indeed show the presence of fullerenes, but fullerenes certainly do not dominate the mass distribution. In all experiments, the photo-processing products include many more (dehydrogenated) PAHs, and low-mass molecules (and molecular fragments) containing H. Thus, to explain objects such as Tc 1 and SMP SMC 16 where abundant C_{60} is seen but no PAHs nor small species are present, significant *additional* processing of these sputtered species would be required to destroy the smaller molecules and the PAHs. UV radiation could conceivably dissociate some of the smaller molecules, but it is not clear how PAHs could be destroyed by radiation in the low-excitation environments represented by the C_{60} PNe. If, on the other hand, collisions are involved, the carbon clusters are processed under similar conditions as described above – and either result in fullerenes dominating the distribution (in the absence of H), or in the formation of PAHs.

Finally, it should be noted that the simultaneous presence of the C_{60} bands and the PAH features in the spectra of some PNe does not necessarily imply that they are co-located – merely that they are both located in the Spitzer beam. In fact, there is clear observational evidence in another object that PAHs and fullerenes are *not* co-located. In Fig. 5, we have reproduced the spatial distribution of molecular hydrogen, PAHs and fullerenes in the reflection nebula NGC 7023 (Sellgren *et al.* 2010, their Fig. 3). The C_{60} emission peaks on the central star, whereas the PAHs are clearly located further away from the star. Thus, although the spectra of the entire nebula show both PAH and fullerene features, both components show clearly different spatial distributions – no PAH emission is found where the C_{60} peaks. Such a distribution could well be the case for (some of) the PNe as well.

The two processes discussed above are not the only ways to form circumstellar fullerenes. An important third route involves high-temperature processes, and allows the formation of fullerenes in H-rich environments. Laboratory experiments carried out by Jäger *et al.* (2009) demonstrate that the temperature in the condensation zone determines the formation pathways of carbonaceous particles with a clear distinction between the resulting condensation products. At temperatures higher than 3500 K, the resulting particles are fullerene compounds and fullerene-like carbon grains (even in the presence of hydrogen); at temperatures lower than 1700 K, soot is formed with a large mixture of PAHs as by-products. This high-temperature mechanism may be relevant to explain the fullerenes in the RCBs (García-Hernández *et al.* 2011; Clayton *et al.* 2011), but a value of 3500 K is certainly not a typical temperature in the circumstellar environment of carbon stars. It remains to be seen whether such a formation route is also important for PNe.

Finally, a promising route in the astronomical settings presented here, is that fullerenes form following the destruction of PAHs. Micelotta *et al.* (2010a,b) studied the effects of shocks on interstellar PAHs, and found that the PAH structure is at least significantly altered (if not destroyed) for shocks with velocities between 75–150 km s^{-1} and in the hot post-shock gas. PAHs first lose their peripheral H-atoms, and then the resulting carbon clusters could assemble into fullerenes on fairly short timescales. This would certainly also be a viable route in PNe. Assembling fullerenes from PAH destruction could also explain the different spatial distribution of PAHs and fullerenes in NGC 7023 (see Fig. 5).

6. Conclusions & outlook

The fullerene species C_{60} and C_{70} have been identified in a variety of astronomical objects. It seems clear that fullerenes form in the circumstellar environments of evolved stars, and that they survive and thrive in the interstellar medium as well. It is less clear precisely how they form, and what their excitation mechanism is, and those topics remain a matter of debate and further research. The derived abundances indicate that $\sim$0.5% of the cosmic carbon could be in the form of C_{60}, and thus fullerenes represent a significant component of interstellar matter.

Fullerenes and fullerene compounds have often been suggested as possible carriers for some of the diffuse interstellar bands (DIBs). Neutral C_{60} is probably not a DIB carrier, but C_{60}^+ has been suggested as a carrier for some DIBs (Foing & Ehrenfreund 1994, 1997; Misawa *et al.* 2009).

Fullerene compounds offer many more possibilities. Fullerenes are readily protonated (see e.g. the review on C_{60} cation chemistry by Bohme 2009, and references therein), and it is thus quite possible that fulleranes (hydrogenated fullerenes) might be present in the ISM or in circumstellar environments as well. It has been suggested that fulleranes could play a role in the formation of H_2 (Cataldo & Iglesias-Groth 2009) and in interstellar extinction. C_{60}^+ was found to be generally quite unreactive at room temperature; nonetheless, many derivatization reactions have been studied in laboratory experiments, including reactions with species that have been detected in the ISM (again, see Bohme 2009) and could thus also be relevant for interstellar chemistry. Fullerenes are also unique in their ability to lock up atoms and small molecules inside their carbon cage. Such endohedral species have also been proposed as DIB carrier candidates, but have only been poorly studied in detail spectroscopically. Given the clear presence of fullerenes in space, we encourage further laboratory work and theoretical calculations on these species.

References

Bakes, E. L. O., Tielens, A. G. G. M., Bauschlicher, Jr. C. W., Hudgins, D. M., & Allamandola, L. J., 2001, *ApJ* 560, 261

Bettens, R. P. A. & Herbst, E., 1995, *International Journal of Mass Spectrometry and Ion Processes* 149, 321

Bohme, D. K., 2009, *Mass Spectrometry Reviews* 28, 672

Cami, J., Bernard-Salas, J., Peeters, E., & Malek, S. E., 2010, *Science* 329, 1180

Cataldo, F. & Iglesias-Groth, S., 2009 400, 291

Choi, C. H., Kertesz, M., & Mihaly, L., 2000, *The Journal of Physical Chemistry A* 104, 102

Clayton, G. C., De Marco, O., Whitney, B. A., *et al.*, 2011, *ArXiv e-prints*

Clayton, G. C., Kelly, D. M., Lacy, J. H., *et al.*, 1995, *AJ* 109, 2096

de Vries, M. S., Reihs, K., Wendt, H. R., *et al.*, 1993, *Geochim. Cosmochim. Acta* 57, 933

Fabian, J., 1996, *Phys. Rev. B* 53, 13864

Foing, B. H. & Ehrenfreund P., 1994, *Nat* 369, 296

Foing, B. H. & Ehrenfreund P., 1997, *A&A* 317, L59

Frum, C. I., Engleman, R., Hedderich, H. G., *et al.*, 1991, *Chem. Phys. Lett.* 176, 504

Fulara, J., Jakobi, M., & Maier, J. P., 1993, *Chem. Phys. Lett.* 211, 227

García-Hernández, D. A., Kameswara Rao, N., & Lambert, D. L., 2011, *ApJ* 729, 126

García-Hernández, D. A., Manchado, A., García-Lario, P., *et al.*, 2010, *ApJ Lett.* 724, L39

Gielen, C. A., Cami, J., Bouwman, J., & Min, M., 2011, *submitted to A&A*

Grishko, V. I., Tereszchuk, K., Duley, W. W., & Bernath, P., 2001, *ApJ Lett.* 558, L129

Herbig, G. H., 2000, *ApJ* 542, 334

Hony, S., Waters, L. B. F. M., & Tielens, A. G. G. M., 2002, *A&A* 390, 533

Houck, J. R., Roellig, T. L., van Cleve, J., *et al.*, 2004, *ApJS* 154, 18

Iglesias-Groth, S., Cataldo, F., & Manchado, A., 2011 413, 213

Jäger, C., Huisken, F., Mutschke, H., Jansa, I. L., & Henning, T., 2009, *ApJ* 696, 706

Joblin, C., Toublanc, D., Boissel, P., & Tielens, A. G. G. M., 2002, *Molecular Physics* 100, 3595

Krätschmer, W., Lamb, L. D., Fostiropoulos, K., & Huffman, D. R., 1990, *Nat* 347, 354

Kroto, H. W., Heath, J. R., Obrien, S. C., Curl, R. F., & Smalley, R. E., 1985, *Nat* 318, 162

Kroto, H. W. & McKay, K., 1988, *Nat* 331, 328

Martin, M. C., Koller, D., & Mihaly, L., 1993, *Phys. Rev. B* 47, 14607

Mendez, R. H., Kudritzki, R. P., Herrero, A., Husfeld, D., & Groth, H. G., 1988, *A&A* 190, 113

Micelotta, E. R., Jones, A. P., & Tielens, A. G. G. M., 2010a, *A&A* 510, A37+

Micelotta, E. R., Jones, A. P., & Tielens, A. G. G. M., 2010b, *A&A* 510, A36+

Milanova, Y. V. & Kholtygin, A. F., 2009, *Astronomy Letters* 35, 518

Misawa, T., Gandhi, P., Hida, A., Tamagawa, T., & Yamaguchi, T., 2009, *ApJ* 700, 1988

Moutou, C., Sellgren, K., Verstraete, L., & Léger, A., 1999, *A&A* 347, 949

Nemes, L., Ram, R., Bernath, P., *et al.*, 1994, *Chem. Phys. Lett.* 218, 295

Rubin, R. H., Simpson, J. P., O'Dell, C. R., *et al.*, 2011, *MNRAS* 410, 1320

Schettino, V., Pagliai, M., & Cardini, G., 2002, *JOURNAL OF PHYSICAL CHEMISTRY A* 106, 1815

Scott, A., Duley, W. W., & Pinho, G. P., 1997, *ApJ Lett.* 489, L193+

Sellgren, K., Uchida, K. I., & Werner, M. W., 2007, *ApJ* 659, 1338

Sellgren, K., Werner, M. W., Ingalls, J. G., *et al.*, 2010, *ApJ Lett.* 722, L54

Snow, T. P. & Seab, C. G., 1989, *A&A* 213, 291

Sogoshi, N., Kato, Y., Wakabayashi, T., *et al.*, 2000, *Journal of Physical Chemistry A* 104, 3733

Somerville, W.B. & Bellis J.G., 1989 240, 41P

Speck, A. K., Corman, A. B., Wakeman, K., Wheeler, C. H., & Thompson, G., 2009, *ApJ* 691, 1202

Stratmann, R., Scuseria, G., & Frisch, M., 1998, *Journal of Raman Spectroscopy* 29, 483
Tielens, A. G. G. M., 2005, The Physics and Chemistry of the Interstellar Medium
von Czarnowski, A. & Meiwes-Broer, K. H., 1995, *Chemical Physics Letters* 246, 321
Wang, X. K., Lin, X. W., Mesleh, M., *et al.*, 1995, *Journal of Materials Research* 10, 1977
Werner, M. W., Roellig, T. L., Low, F. J., *et al.*, 2004, *ApJS* 154, 1
Williams, R., Jenkins, E. B., Baldwin, J. A., *et al.*, 2008, *ApJ* 677, 1100
Zhang, Y. & Kwok S., 2011, *ApJ* 730, 126

Discussion

GUÉLIN: You were able to resolve spatially the PAH source from the C_{60} source. Do you see any shift in position between the lines attributed to fullerenes?

CAMI: The spatial distribution shown is for the reflection nebula NGC 7023 as presented by Sellgren *et al.* (2010). In that source, most C_{60} bands show contamination by neighbouring PAH features which certainly complicates matters; no shift in wavelength has been reported in that source.

STERNBERG: Does the preferential formation of PAHs versus C_{60} in the hydrogen-rich environments depend on whether the hydrogen is ionized or not?

CAMI: The laboratory experiments I referred to used an atmosphere of Ar and H_2; I am not aware of experiments with protons.

RAWLINGS: Would you care to comment on the expected degree of hydrogenation of the fullerenes in space?

CAMI: C_{60} is quite easily protonated, so I would certainly expect to see hydrogenated fullerenes in space.

The Molecular Universe
Proceedings IAU Symposium No. 280, 2011
José Cernicharo & Rafael Bachiller, eds.

doi:10.1017/S1743921311025002

Molecules in Supernova Ejecta

Isabelle Cherchneff & Arkaprabha Sarangi

Departement Physik, Universität Basel
Klingelbergstrasse 82, CH-4056, Basel, Switzerland
email: isabelle.cherchneff@unibas.ch

Abstract. The first molecules detected at infrared wavelengths in the ejecta of a Type II supernova, namely SN1987A, consisted of CO and SiO. Since then, confirmation of the formation of these two species in several other supernovae a few hundred days after explosion has been obtained. However, supernova environments appear to hamper the synthesis of large, complex species due to the lack of microscopically-mixed hydrogen deep in supernova cores. Because these environments also form carbon and silicate dust, it is of importance to understand the role played by molecules in the depletion of elements and how chemical species get incorporated into dust grains. In the present paper, we review our current knowledge of the molecular component of supernova ejecta, and present new trends and results on the synthesis of molecules in these harsh, explosive events.

Keywords. supernovae: general, molecular processes, astrochemistry

---

1. Introduction

Type II supernovae (SNe) are the end stage of the evolution of massive stars with initial masses on the Zero Age Main Sequence greater than 8 $M_{\odot}$. These explosive events deposit a tremendous energy to the surroundings with typical explosion energy of 10^{51} ergs per event. The explosion releases a burst of neutrons and synthesised elements heavier than iron through rapid neutron capture (i.e, r-process). The explosion shock wave crosses the helium core, induces a reverse shock at the interface between the heavy elements core and the hydrogen envelope of the stellar progenitor, and reprocesses the ejecta a few hours after explosion. This reverse shock triggers the onset of Raleigh-Taylor instabilities and results in macroscopic mixing between the various core layers. The nebular phase of the SN is reached after $\sim$ 100 days when the luminosity is sustained essentially by the radioactive decay of ^{56}Co. Assuming γ-rays full trapping, the bolometric light curve declines according to the mean lifetime of ^{56}Co decay into ^{56}Fe (t_{56} = 111.3 days). In view of this very harsh environment, the detection of molecules in the nebular phase a few months following the explosion of a SN came as a surprise. The locus of the first detection of molecules and dust in a SN event shortly after outburst was the Type II supernova SN1987A in the Large Magellanic Cloud. Indeed, the fundamental band $\Delta v = 1$ of carbon monoxide, CO, at 4.65 μm was observed between day 135 and day 260 (Catchpole *et al.* 1988), while the CO first overtone transition $\Delta v = 2$ in the region of 2.3 μm was detected at day 100 after explosion (Danziger *et al.* 1988). Excess emission around $\sim$ 9 μm was attributed to the fundamental band $\Delta v = 1$ of silicon monoxide, SiO, and was observed as early as day 160 (Aitken *et al.* 1988). Other molecular candidates were proposed as carriers of some unidentified bands in the IR spectrum and included CS (3.88 μm band), $H3^+$ (3.4 μm and 3.53 μm bands) and CO^+ (2.26 μm band). However, claims for their presence were not confirmed by theoretical models.

Since then, carbon and silicon monoxide ro-vibrational emission bands have been observed in other Type II SNe as well and appear to be the only molecules detected in

Table 1. A summary of current CO and SiO observations in Type II supernovae. The – sign refers to a non-detection of the molecular species.

Name	CO 1st overtone [2.3 μm]	CO fundamental [4.65 μm]	SiO fundamental [8.1 μm]	Reference
SN1987A	X			Spyromilio *et al.* (1988)
		X		Catchpole *et al.* (1988)
			X	Aitken *et al.* (1988)
	X	X		Danziger *et al.* (1989)
			X	Roche *et al.* (1991)
SN1995ad	X			Spyromilio & Leibundgut (1996)
SN1998S	X			Gerardy *et al.* (2000)
	X			Fassia *et al.* (2001)
SN1998dl	X			Spyromilio *et al.* (2001)
SN1999em	X			Spyromilio *et al.* (2001)
SN2002hh	X			Pozzo *et al.* (2006)
SN2004fj		X		Kotak *et al.* (2005)
		X	–	Szalai *et al.* (2011)
SN2005af		X	X	Kotak *et al.* (2006)
SN2004et		X	X	Kotak *et al.* (2009)

SN ejecta to date. This restricted number of observed species may point to the specific environment of SN ejecta characterised by harsh conditions, the presence of helium ions, and the lack of microscopically-mixed hydrogen. This hydrogen-deprived medium fosters a 'poor' chemistry. Indeed, because of the absence of radicals like hydrides (OH, CH etc.), a fast, hot chemistry that builds up larger, more complex species (e.g., hydrocarbons, water or carbon dioxide) cannot proceed, as it is the case in other circumstellar environments [e.g., AGB stellar winds (Cherchneff 2006)]. The chemical species thus formed are essentially simple diatomic molecules like CO and SiO.

2. Observations of molecules and existing models

The detection in the near IR of the first overtone band of CO has been positive in several Type II SNe at epochs ranging from 90 to 130 days after explosion. Along with the first overtone, the red wing of the fundamental band $\Delta v = 1$ at 4.65 μm is also detected in the mid-IR (e.g., Kotak *et al.* 2005). Models of the CO bands indicate conditions in the ejecta close to those provided by explosion models $\sim$ 130 days post-outburst. i.e., gas temperature close to 5000 K. The CO molecule is always present in SN ejecta some 100 days after outburst, pointing to an efficient chemical pathway to its synthesis at high temperatures. In SN1987A, the CO first overtone band flux peaks around 180 days before declining by several orders of magnitude up to day 600 where it became unobservable (Danziger *et al.* 1989). A similar behaviour is observed in other SNe, for example, SN2005af (Kotak *et al.* 2005). This decrease can be ascribed either to a change in CO mass with time or to different excitation conditions owing to the change in gas temperature, or both. In SN1987A, the fundamental band of CO is also observed to decrease in absolute strength with time but at a much smaller rate than the first overtone band (Danziger *et al.* 1989). The different behaviour of these two bands may indicate that they arise from different zones in the ejecta.

Silicon monoxide, SiO, has been detected in the mid-IR in three SNe through its fundamental band around 8.1 μm. Its first detection in SN1987A showed that emission was apparent from day 160 to day 520 (Aitken *et al.* 1988, Roche *et al.* 1991). No emission was recorded at later times and this decline coincided with the onset of dust formation at $\sim$ 530 days (Lucy *et al.* 1989). The depletion of SiO into dust was therefore proposed as a scenario for SiO emission decline. A similar behaviour was observed in SN2004et by Kotak *et al.* (2009).

Other molecular species have been proposed to explain unidentified bands in the IR spectrum of SN1987A. For example, the broad emission feature at 3.88 μm appearing at late time was proposed to come from carbon monosulphide, CS, by Meikle *et al.* (1993). These assignments were mainly made on wavelength coincidence and confirmation by chemical models was difficult (e.g., Liu (1998) for CS, Gearhart *et al.* (1999) for CO^+). As to the unidentified emission bands in the near-IR at 3.4 μm and 3.53 μm, Miller *et al.* (1992) ascribed the emission to H_3^+, while Yan & Dalgarno (1998) modelled the chemistry of a cool molecular clump where H_3^+ could form in large enough amount to satisfactorily reproduce the emission band intensity. However, their model assumed that the composition of the molecular clump comprised both hydrogen, H_3^+, CO and OH. As mentioned in Section 1, the scarcity of molecules observed in SN ejecta entails the unlikely presence of microscopic mixed hydrogen with atomic carbon and oxygen in the ejecta, thereby invalidating H_3^+ as a potential carrier for these bands. Therefore, the only two molecular species with unambiguous identification in SN ejecta remain CO and SiO, and their detection status is summarised in Table 1.

Chemical models have been attempted to explain the formation of CO and SiO in the ejecta of SN1987A. Petuchowski *et al.* (1989) and Lepp *et al.* (1990) proposed the first chemical models of the O- and C-rich zones including ion-molecule reactions, radiative association processes, and reactions with the fast Compton electrons created from the collisional degrading of γ-rays. They found that helium could not be microscopically mixed in the different layers because the reactive ion He^+ destroyed any molecular species. The prevalent formation processes for molecules were radiative association reactions whilst destruction was provided by collisions with Compton electrons. Liu *et al.* (1992) and Gearhart *et al.* (1999) modelled the emission bands of CO while Liu & Dalgarno (1994, 1996) modelled the emission bands of SiO. All models assumed similar chemical processes as Lepp *et al.* (1990) to derive molecular abundances and masses. Finally, Liu & Dalgarno (1995) explored the impact of molecular cooling in the ejecta as CO and SiO are strong coolants through their ro-vibrational lines. Their study of the O-rich core showed that the kinetic temperature could indeed drop from 5000 K to 1500 K when CO cooling was introduced in their model.

Central to these chemical models was the assumption that steady-state was quickly established for the chemistry at early times. However, this assumption was later disproved by Cherchneff & Lilly (2008) and Cherchneff & Dwek (2009) in their study of the chemistry of primitive SNe in the early universe. Based on a more extended chemical network (see Section 3), they showed that the chemical times of key formation and destruction processes were often shorter than the dynamical time, implying that an active chemistry could take place at time greater than 100 days post-outburst. For SN ejecta other than SN1987A, no chemical models have been attempted so far. We therefore present in the next section preliminary results for the non-equilibrium chemical kinetic model of a SN ejecta with a 15 $M_\odot$ progenitor mass of solar metallicity.

3. New model for stratified, homogeneous ejecta

We study the chemistry of the ejecta associated with a 15 $M_\odot$ progenitor of solar metallicity. A schematic view of conditions in the ejecta is given in Fig. 1. A pre-supernova

Table 2. Mass zones of the helium core for the 15 $M_\odot$ progenitor of solar metallicity. The mass cut is 1.78 $M_\odot$ and the ejected mass is $\sim$ 3.79 $M_\odot$ for a total He-core mass of 5.57 $M_\odot$.

Zone	Position ($M_\odot$)	Mass ($M_\odot$)	Major constituants
1A	1.79 - 1.89	0.1	Si-S-Fe
1B	1.89 - 1.98	0.09	Si-O
2	1.98 - 2.27	0.29	O-Si-Mg
3	2.27 - 2.62	0.35	O-Mg-Si
4A	2.62 - 2.81	0.19	O-C
4B	2.81 - 3.04	0.23	O-C-He
5	3.04 - 3.79	0.75	He-C-N

core explodes and experiences the passage of a blast wave which triggers a reverse shock at the base of the hydrogen envelope. This shock propagates inward and leaves a reversed pressure-gradient in its wake that produces Rayleigh-Taylor instabilities and macroscopic mixing in the helium core. The mixing has ceased after a few days (Joggerst *et al.* 2010) and the partial fragmentation of the helium core proceeds with time. Radioactive ^{56}Ni decays into ^{56}Co on short time scales, and the later decay of ^{56}Co in ^{56}Fe creates γ-rays that propagate in the ejecta. These γ-rays degrade to X-rays and ultraviolet (UV) photons by Compton scattering and create a population of fast Compton electrons in the ejecta. These Compton electrons ionise the elements, including helium. Typical temperatures and densities are also indicated as a function of time in the ejecta.

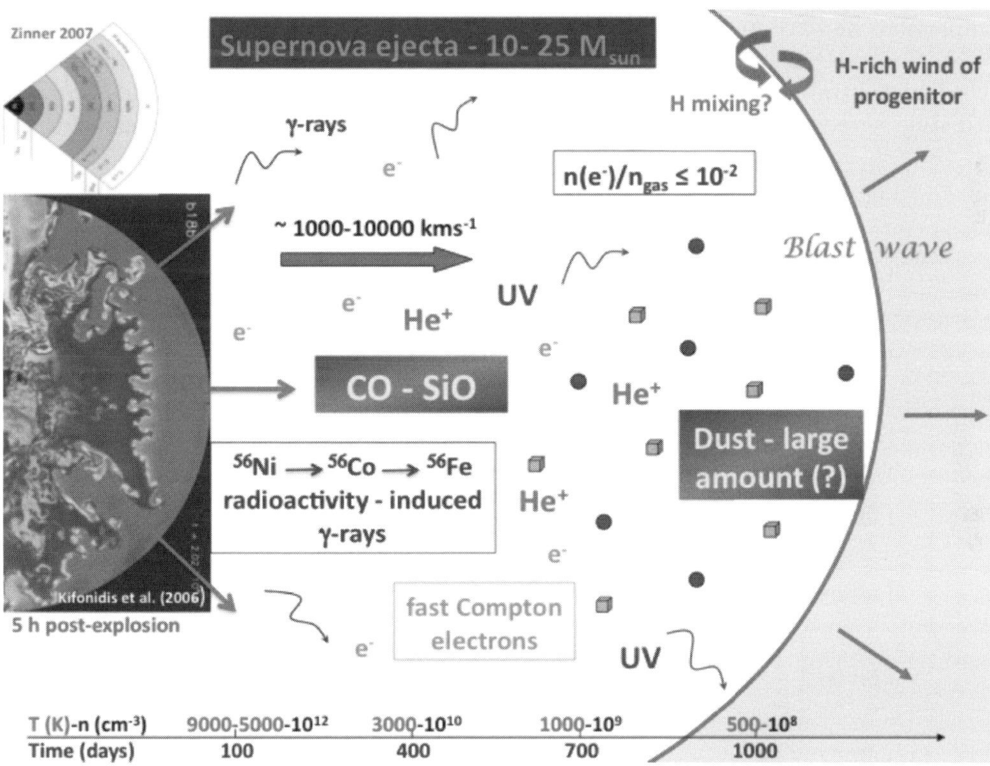

Figure 1. A schematic view of the ejecta of a Type II supernova. On the upper left-hand corner, the pre-supernova core reflects the various stages of nuclear burning and the stratified lay-out of the heavy-elements in shells (Zinner 2007). A few hours after the explosion, the helium core experiences Rayleigh-Taylor instabilities, mixing and fragmentation (Kifonidis *et al.* 2006). Other quantities are explained in the text.

Following Cherchneff & Dwek (2009), the ejecta is assumed homogeneous and consists of several mass zones characterised by an initial chemical composition taken from Rauscher *et al.* (2002) for their study of a 15 $M_{\odot}$ progenitor of solar metallicity. The elements in each zone are assumed to be microscopically mixed but no elemental mixing between zones is considered. We assume that the helium core is hydrogen-free as no microscopic mixing of hydrogen takes place inward in the core. Hydrogen thus stays confined to the most external layer that corresponds to the progenitor envelope. The zoning considered in this study is briefly summarised in Table 2. We model the formation of several key molecules, including CO, SiO, O_2, SO and SiS, as well as small molecular clusters of various solids (e.g., SiO_2, FeS, pure metals, carbon chains and rings) as in Cherchneff & Dwek (2010). We study the chemical processes active after outburst from 100 days to 1000 days, a time span relevant for the study of molecule and dust synthesis in the ejecta. They include the formation of molecules by termolcular reactions and by bimolecular processes like neutral-neutral reactions, radiative association reactions and ion-molecule reactions. The destruction processes include ionisation and dissociation processes by Compton electron and the UV radiation field, ion-molecule reactions, neutral-neutral processes and thermal fragmentation.

4. Results on molecules

The mass of SiO formed in the ejecta as a function of time is illustrated in Figure 2. For comparison, we have included the SiO masses derived from observations of various Type II SNe. Zones 1B, 2 and 3 are the prevalent providers of SiO through formation processes including neutral-neutral and radiative association reactions. The SiO mass variation with time agrees well with the decline of the observed SiO mass, in particular that of

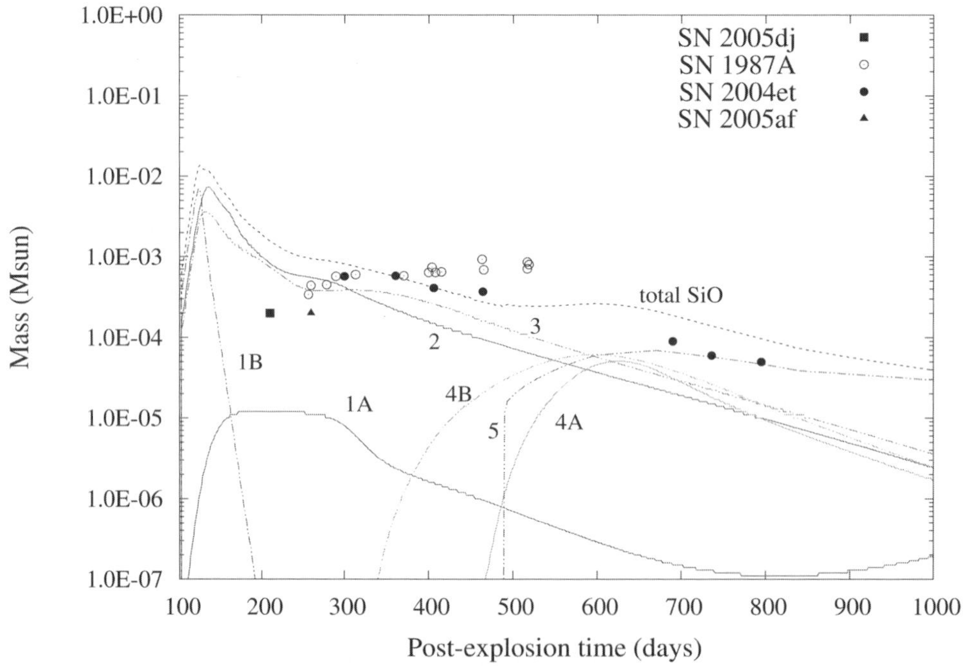

Figure 2. The mass of SiO as a function of post-explosion time for the various helium core mass zones (see text). SiO masses derived from observational data for several SNe are also shown.

SN2004et (Kotak *et al.* 2009). However, this agreement implies that the effective and steady conversion of gas-phase SiO into small molecular clusters of SiO_2 and Si proceeds at early time ($\sim$ 150 days post-outburst) for the three zones where SiO efficiently forms. In this regard, the SiO molecule is a direct tracer of the formation of silicate dust in SN ejecta. Observations of SiO at several post-explosion times thus probe the effectiveness of dust synthesis in these environments.

Results for CO as a function of mass zones and time are shown in Figure 3. CO masses derived for SN1987A from models by Liu *et al.* (1992) and Liu & Dalgarno (1995) are also plotted for comparison. However, a 15 $M_{\odot}$ progenitor model may not be appropriate for SN1987A, for which a more massive progenitor has been proposed (Woosley 1988). The metallicity of SN1987A progenitor is also sub-solar and similar to that of the LMC. The trends of the CO masses derived for SN1987A, i.e., an effective formation of CO in large amounts at very early times, are reproduced by our results although we produce more CO in the 15 $M_{\odot}$ ejecta. CO is prevalently formed in the two outer zones 4A and 4B where oxygen is still overabundant with respect to carbon and where helium is absent (zone 4A) or exists in small amounts (zone 4B). From our results, there exits no direct link between the presence of CO in the ejecta and carbon dust. The latter is traced by the formation of carbon chains and rings in our model, and forms in the outer, He/C-rich zone 5, when the He^+ abundance has diminished, either due to its effective recombination with thermal electrons or to non-effective ionisation of helium by Compton electrons. Small amounts of CO are formed in zone 5 but most of the carbon is either in atomic form or in carbon dust precursors. Therefore, CO formation is not linked to the condensation of carbon solids and CO is not a tracer of carbon dust in SN ejecta. CO is

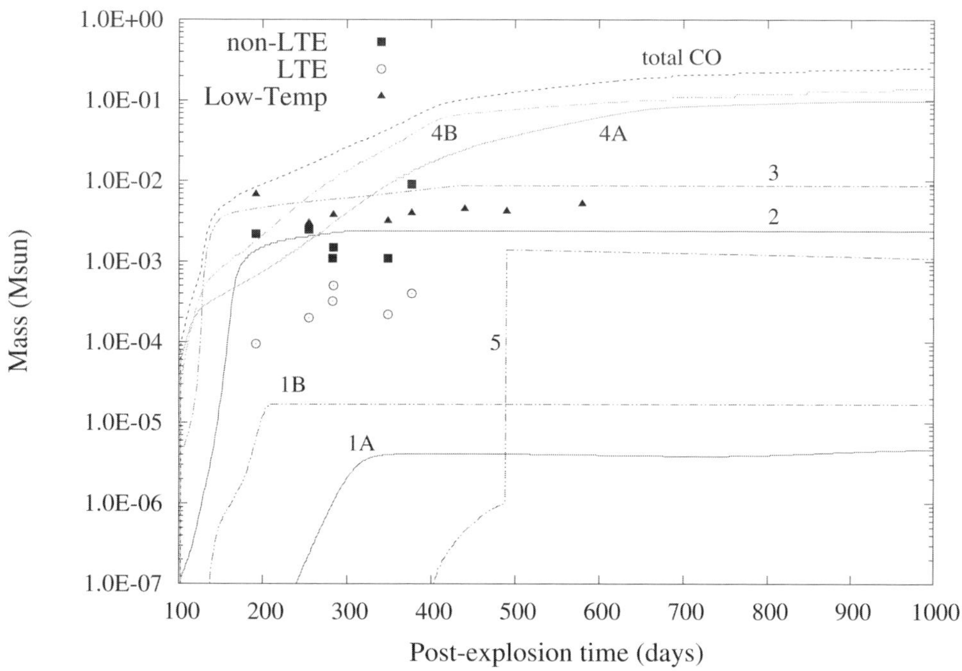

Figure 3. The mass of CO as a function of post-explosion time for the various helium core mass zones (see text). CO masses derived by Liu *et al.* (1992) and Liu & Dalgarno (1995) from observational data of SN1987A are shown.

just the prevalent molecule that forms in zones rich in oxygen and carbon, and poor in helium, whatever the initial C/O ratio of the zones.

Other molecular species are produced in large amounts in the ejecta and trace the specific chemical composition of their formation zones. In the inner most mass zone 1A, silicon sulphide, SiS, forms efficiently from radiative association and neutral-neutral reactions, pumping about 34% of the atomic Si present in the zone. The SiS mass at day 1000 is $\sim 3 \times 10^{-2}$ $M_{\odot}$. Because zones 2 and 3 are very rich in free atomic oxygen, the formation of dioxygen, O_2 proceeds as early as 300 days. In zone 1B and 2, sulphur is still abundant and its reaction with O_2 induces the formation of sulphur monoxide, SO, as soon as O_2 forms. The final masses of O_2 and SO at 1000 days post-explosion are $\sim 3 \times 10^{-1}$ $M_{\odot}$ and $\sim 1.5 \times 10^{-2}$ $M_{\odot}$, respectively.

5. Discussion

The ejecta of SNe form two types of molecular species, those directly involved in the formation processes of dust (e.g., SiO) and those only depleting gas-phase elements (e.g., CO, O_2). Members of the first category are rapidly depleted in the synthesis of dust and are thus direct tracers of the dust formation processes. Therefore, dedicated observational surveys of these specific species at various times after outburst would be extremely useful to directly assess the efficiency with which the SN ejecta produces dust. Molecules of the second kind are of great interest as well. For the present 15 $M_{\odot}$ progenitor model with solar metallicity, they represent $\sim$ 17% of the ejected mass at day 1000 and include CO, O_2, SiS and SO. Somewhat larger mass fractions of ejected molecules were derived for zero-metallicity SNe in the early universe (Cherchneff & Dwek 2009). These large molecular fractions impact the gas in two major ways. Firstly, the depletion by molecules of elements entering the processes of dust synthesis takes place and needs to be considered when assessing the amount of dust formed in SNe. This is well exemplified by the formation of gas-phase S_2 and SiS in zone 1A, which depletes atomic sulphur, thus impacting on the formation of FeS clusters. Secondly, molecules like CO are strong coolants through their ro-vibrational transitions and their presence will affect the thermal history of the environment where they reside. The cooling provided by CO molecules in O/C-rich clumps applied to the ejecta of SN1987A was addressed by Liu & Dalgarno (1995). They considered clumps with equal masses of atomic O and C and where 1% of the mass was in CO. The cooling provided by CO could decrease the gas kinetic temperature down by a factor of three compared to CO-free gas where cooling was essentially provided by the 6300 Å and 6363 Å forbidden lines of [OI]. In the present model, the CO mass corresponds to $\sim$ 2% of the O-rich core mass (i.e., zones 2 and 3) at day 400, where the mass fraction reaches $\sim$ 24% of the CO formation zone (i.e., zones 4A and 4B) at a similar time. The cooling provided by CO molecules must thus be strong as derived by Liu & Dalgarno (1995). Such a cooling will favour more efficient molecule and dust synthesis in the ejecta dense clumps.

Species of the second category are also crucial to the study of the SN ejecta evolution at very late time, i.e., in young SN remnants like Cas A. Molecules like CO or SiS are expected to survive in the dense, cool clumps where they formed and young remnants that have not yet been shocked by the reverse shock are expected to bear such a molecular signature. These species should thus be detectable at sub-millimetre wavelengths in young SN remnants and would not only provide clues on the local kinetics and the chemistry active in clumps or filaments but also on their chemical synthesis in the prior SN stage.

Acknowledgement

This research is supported by the Swiss National Science Foundation grant No 20GN21-128950 through the European Science Foundation Collaborative Research Project CoDustMas.

References

Aitken, D. K., Smith, C. H., James, S. D., *et al.* 1988, *M.N.R.A.S.*, 235, 19
Catchpole, R. M., Whitelock, P. A., Feast M. W., *et al.* 1988, *M.N.R.A.S.*, 231, 75
Cherchneff, I. 2006, *A&A* 456, 1001
Cherchneff, I. & Lilly, S. 2008, *ApJ* 683, L123
Cherchneff, I. & Dwek, E. 2009, *ApJ* 703, 642
Cherchneff, I. & Dwek, E. 2010, *ApJ* 713, 1
Danziger, I. J., Bouchet, P., Fosbury, R. A. E., *et al.* 1988, in: *Supernova 1987A in the Large Magellanic Cloud* (Cambridge and New York: Cambridge University Press), p. 37
Danziger, I. J., Lucy, L. B., Bouchet, P., *et al.* 1989, in: *Supernova 1987A in the Large Magellanic Cloud* (Cambridge and New York: Cambridge University Press), p. 37
Fassia, A., Meikle, P., Chugai, N., *et al.* 2001, *M.N.R.A.S*, 325, 907
Gearhart, R. A., Wheeler, J. C. & Swartz, D. A. 1999, *ApJ*, 510, 944
Gerardy, C. L., Fesen, R. A., Hoeflich, P., & Wheeler, J. C. 2000, *AJ* 119, 2968
Joggerst, C. C., Almgren, A., & Woosley, S. E. 2010, *ApJ* 723, 353
Kifonidis, K. Plewa, T., Scheck, L., *et al.* 2006, *A&A* 463, 661
Kotak, R., Meikle, P., van Dyck, S., *et al.* 2005, *ApJ* 628, L123
Kotak, R., Meikle, P., Pozzo, M., *et al.* 2006, *ApJ* 651, L117
Kotak, R., Meikle, P., Farrah, D., *et al.* 2009, *ApJ* 704, 306
Lepp, S., Dalgarno, A., & McCray, R. 1990, *ApJ*, 358, 262
liu, W., Dalgarno, A., & Lepp, S. 1992, *ApJ*, 396, 679
Liu, W. & Dalgarno, A. 1994, *ApJ*, 438, 789
Liu, W. & Dalgarno, A. 1995, *ApJ*, 454, 472
Liu, W. & Dalgarno, A. 1996, *ApJ*, 471, 480
Liu, W. 1998, *ApJ*, 496, 967
Lucy, L. B., Danziger, I. J., Gouiffes, C., & Bouchet, P. 1989, in: *Supernova 1987A in the Large Magellanic Cloud* (Cambridge and New York: Cambridge University Press), p. 37
Meikle, P., Spyromilio, J., Allen, D. A., *et al.* 1993, *M.N.R.A.S.* 261, 235
Miller, S., Tennyson, J., Lepp, S., & Dalgarno, A. 1992, *Nature*, 355, 420
Petuchowski, S. J., Dwek, E., Allen, J. E. & Nuth III, J. A. 1989, *ApJ*, 342, 406
Pozzo, M., Meikle, Rayner, T. J., *et al.* 2006, *M.N.R.A.S*, 368, 1169
Rauscher, T., Heger, A., Hoffman, R. D., & Woosley, S. E. 2002, *ApJ*, 576, 323
Roche, P. F., Aitken, D. K., & Smith, C. H. 1991, *M.N.R.A.S*, 252, 39
Spyromilio, J., Meikle, W. P. S., Learner, R. C. M., & Allen, D. A. 1988, *Nature*, 334, 327
Spyromilio, J. & Leibundgut, B. 1996, *M.N.R.A.S*, 283, L89
Spyromilio, J., Leibundgut, B., & Gilmozzi, R. 2001, *A&A*, 376, 188
Szalai. T, Vinkó, Balog, Z., *et al.* 2011, *A&A*, 527, 61
Woosley, S. E. 1988, *ApJ* 218, 253
Yan, Min. & Dalgarno, A. 1998, *ApJ*, 500, 1049
Zinner, E. 2007, in: K. K. Turekian, H. D. Holland & A. M. Davis (eds.), *Treatise in Geochemistry 1* (Oxford and San Diego: Elsevier), p. 1

Discussion

Posch: How efficient do you expect dust formation by supernovae to be in the end, considering dust destruction in supernova remnants?

Cherchneff: My impression is that they are efficient dust makers in the end. But an answer to your question requires accurate models of dust formation in supernovae and the impact of the reverse shock on dust in the remnant. Those models do not yet exist. Existing studies consider the classical nucleation theory to model dust formation which is not appropriate for the case of a supernova ejecta. The fragmentation and clumpiness of the medium must also be taken into account. In the remnant, the impact of the reverse shock on ejecta dust has also been studied, but assuming a homogenous ejecta. Destruction efficiencies vary between 70% and 99% of the initial dust mass depending on the pre-shock gas density. So a very quick assessment of the dust mass surviving in the remnant would be between 0.001 $M_\odot$ and 0.03 $M_\odot$ assuming an average dust mass of 0.1 $M_\odot$ synthesised by a 20 $M_\odot$ progenitor. These values are comparable to the amount of dust produced by a carbon star over its lifetime. Furthermore, recent observations with Spitzer and Herschel of Cas A have detected warm and cool dust in the remnant. And supernova dust is extracted from meteorites. So somehow, supernovae form dust and some of it survives in the remnant and continues its journey to the interstellar medium.

The Molecular Universe
Proceedings IAU Symposium No. 280, 2011
José Cernicharo & Rafael Bachiller, eds.

doi:10.1017/S1743921311025014

Spectral Line Surveys of Evolved Stars

J. Cernicharo[1], M. Agúndez[2] and M. Guélin[3]

[1]CAB. INTA-CSIC. Crta Torrejón km 4. 28850 Torrejón de Ardoz. Madrid. Spain
email: jcernicharo@cab.inta-csic.es

[2]LUTH, Observatoire de Paris Meudon. Paris. France
email: marcelino.agundez@obspm.fr

[3]IRAM. 300 Av. de la Piscine. F-38406 St. Martin d'Hères, France
email: guelin@iram.fr

Abstract. The observation of evolved stars in selected evolutionary stages allow us to track the evolution of the physical properties and chemical composition of the matter that is being returned to the interstellar medium during these last stages of the life of stars. While the dust component can be characterized through the observation of the spectral energy distribution in the infrared part of the spectrum, spectral line surveys carried out in a wide spectral band provide the best probe of the physical properties and chemical composition of the gas phase. In this lecture we review the different line surveys carried out toward these objects and their impact in our understanding of the chemical complexity evolution in the circumstellar envelopes around evolved stars.

Keywords. astrochemistry, line: identification, surveys, stars: AGB and post-AGB

---

1. Introduction and history

Stars evolving out of the main sequence experience some of the most drastic changes of all its life, from the huge expansion that convert them into giant or supergiant stars and the mass loss process that creates a circumstellar envelope around them, to the transformation into a white dwarf or supernova at the very end of its life which injects huge amounts of energy into the circumstellar and nearby interstellar medium.

Among the various types of objects embraced by the generic term "evolved stars", a particularly interesting one are Asymptotic Giant Branch (AGB) stars, which are red giant stars resulting from low mass progenitors ($\lesssim 8$ $M_\odot$) that undergo mass loss processes producing extended circumstellar envelopes composed of gas phase molecules and tiny dust grains. Dredge-up processes during the AGB phase selectively bring carbon from the interior of the star to the stellar surface so that the amount of carbon can eventually exceed that of oxygen. Thus, AGB stars can be M stars ($[C]/[O] < 1$) or C stars ($[C]/[O] > 1$) depending on the carbon-to-oxygen abundance ratio in the stellar surface. The physical conditions in the atmosphere of AGB stars (with temperatures of 2000-3000 K and densities higher than 10^{13} cm^{-3}) make the material to be mainly molecular with the composition being determined by thermochemical equilibrium. In this region stable molecules dominate, with CO reaching a large abundance and the $[C]/[O]$ ratio determining which other abundant molecules will be present: stable oxygen bearing molecules in M stars and carbon-bearing molecules in C stars. These stable molecules incorporate into the stellar wind and in the outer layers exposed to the interstellar ultraviolet (UV) field start to be photodissociated releasing radicals which then react forming new species. The evolution of the star from a cool red giant to a hot white dwarf controls the transition from the AGB to the planetary nebula phase, in which the star now emits a large flux of

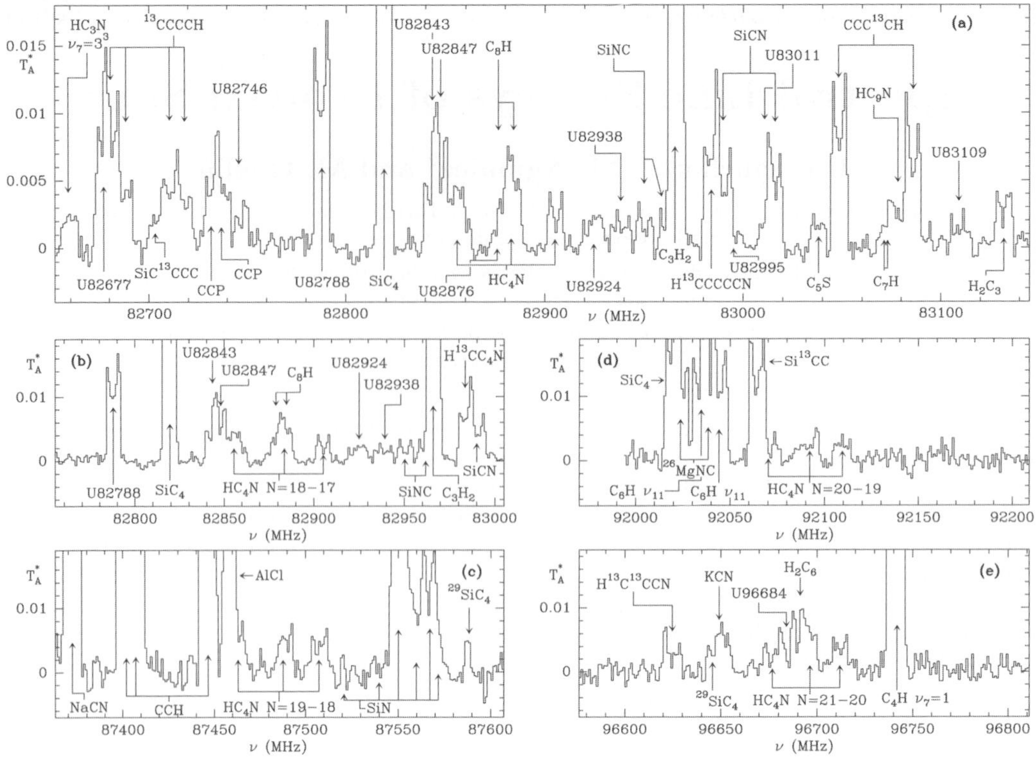

Figure 1. Spectra of IRC +10216 observed with the IRAM 30-m telescope as a part of the 3mm line survey. This zoom in frequency to the line survey shows a forest of lines coming from several species, their isotopes, and even their vibrationally excited states. Figure updated from the detection of HCCCCN by Cernicharo *et al.* (2004).

UV photons that irradiate the expanding circumstellar gas inducing important changes in the chemical composition.

Circumstellar envelopes (CSEs) of evolved stars foster a remarkably complex chemistry (see Fig. 1,2). The mechanisms behind the chemical synthesis, how it varies from the inner to the outer layers of the expanding envelope, its relationship to dust grain formation, and its evolution from the asymptotic giant branch (AGB) to the post AGB-phases, are all important issues that have yet to be understood (Cernicharo *et al.* 2004, Herpin & Cernicharo 2001, Herpin *et al.* 2002). With this purpose, observations at different wavelength ranges are clearly required. The atmosphere and the inner dust condensation regions are best probed by infrared (IR) ro-vibrational lines or by high-J lines in the ground and vibrationally excited states of abundant species such as CO, HCN and SiS (Fonfría *et al.* 2008, Cernicharo *et al.* 1996a, Decin *et al.* 2010). The outer and colder layers of the CSE are best probed by molecular rotational lines at millimeter wavelengths (Cernicharo *et al.* 2000, Kemper *et al.* 2003, Agúndez & Cernicharo 2006, He *et al.* 2008). Based on a large set of high spectral resolution data, the physical conditions of the different regions of the CSE can be well constrained providing a complete picture of the object (see, e.g., Agúndez 2009, Cernicharo *et al.* 2000, Pardo *et al.* 2004, Pardo *et al.* 2005, Pardo *et al.* 2007a,b).

The observation of evolved stars in selected evolutionary stages allow us to track the evolution of the physical properties and chemical composition of the matter that is being returned to the interstellar medium during these last stages of the life of stars. While the dust component can be characterized through the observation of the spectral energy distribution in the infrared part of the spectrum, spectral line surveys carried out in a wide spectral band provide the best probe of the physical properties and chemical composition of the gas phase.

Carbon-rich evolved stars.

The by far most studied C star is IRC +10216, which has been the target of different spectral line surveys. In the early 80's Johansson *et al.* (1984) carried out the first such survey with the 20-m Onsala telescope at $\lambda \sim 3.8$ mm, covering the frequency range 72.2-91.1 GHz. Some years after IRC +10216 was surveyed in the $\lambda \sim 0.9$ mm band with the JCMT (339.6-364.6 GHz; Avery *et al.* 1992) and the CSO (330.2-358.1 GHz; Groesbeck *et al.* 1994) telescopes, in the $\lambda \sim 8$ mm band (28-50 GHz) with the 45-m Nobeyama telescope (Kawaguchi *et al.* 1995), and in the $\lambda \sim 2$ mm spectral region (130-160 GHz) with the IRAM 30-m telescope (Cernicharo *et al.* 2000). These line surveys showed us that the spectrum of IRC +10216 at millimeter wavelengths is plenty of emission lines corresponding to rotational transitions of different types of molecules present in the circumstellar envelope. An important outcome of these line surveys (see Fig. 1,2) has been the identification of new molecules in space, such as the metal halides NaCl, AlCl, KCl, and AlF (Cernicharo & Guélin 1987b), the metal-bearing molecules MgNC, MgCN, AlNC, AlCN, SiCN, and NaCN (Guélin *et al.* 1986; Kawaguchi *et al.* 1993; Ziurys *et al.* 1993,2002; Guélin *et al.* 2000,2004; Turner *et al.* 1994). Due to the carbon-rich nature of the circumstellar gas, most of the lines arise from carbon-bearing molecules such as the carbon chain radicals C_2H, C_4H, C_5H, C_6H, C_7H, and C_8H (see Cernicharo & Guélin 1996b, and references therein), the carbenes H_2C_3 and H_2C_4 (Cernicharo *et al.* 1991a,b), the cyclic hydrocarbons c-C_3H_2 and c-C_3H, HCN and the cyanopolyynes HC_3N, HC_5N, HC_7N, and HC_9N (Cernicharo *et al.* 2000), the related radicals CN, C_3N and C_5N, the sulfur carbon chains CS, C_2S, C_3S (Cernicharo *et al.* 1987a), and C_5S (Bell *et al.* 1993; Cernicharo *et al.*, in preparation), and the silicon carbon-bearing molecules SiC (Cernicharo *et al.* 1989), SiC_2 (Thaddeus *et al.* 1984), c-SiC_3 (Apponi *et al.* 1999), and SiC_4 (Ohishi *et al.* 1989), and many diatomic and triatomic molecules among others (Cernicharo *et al.* 2000). These molecules are either concentrated around the central star or distributed in a hollow shell of radius ~ 15" (Guélin *et al.* 1993, Guélin *et al.* 1997), and have rotational temperatures in the range 15-90 K (Agúndez 2009). Depending on the spectral region covered, each line survey sample different types of molecules and excitation conditions. For example, at centimeter wavelengths observed lines either belong to heavy species or are low excitation transitions of molecules of medium weight, whereas at wavelengths shorter than 1 mm most of the lines either arise from relatively light molecules or correspond to high excitation transitions of medium weight molecules.

The new generation of low-noise and wide-band heterodyne receivers installed on radiotelescopes in recent years has made possible to carry out sensitive line surveys of IRC +10216 such as those performed by He *et al.* (2008) at $\lambda \sim 2$ mm (130-160 GHz) with the 12-m ARO telescope and at $\lambda \sim 1.3$ mm (219.5-267.5 GHz) with the 10-m SMT telescope, that carried out with IRAM 30-m telescope at $\lambda \sim 3$ mm in the frequency range 80-115.8 GHz (Cernicharo *et al.* in preparation), the one made by Tenenbaum *et al.* (2010) with the 10-m SMT telescope at $\lambda \sim 1.3$ mm (214.5-285.5 GHz), and the recent line survey carried out with the SMA interferometer by Patel *et al.* (2011) in the frequency range 293.9-354.8 GHz. These sensitive line surveys have detected many weak lines arising from rare isotopologues and vibrationally excited states of abundant molecules (see Fig. 1,2), even reaching the confusion limit in some regions of the spectrum. They have also permitted the detection of new molecules with relatively low abundances, such as the radicals

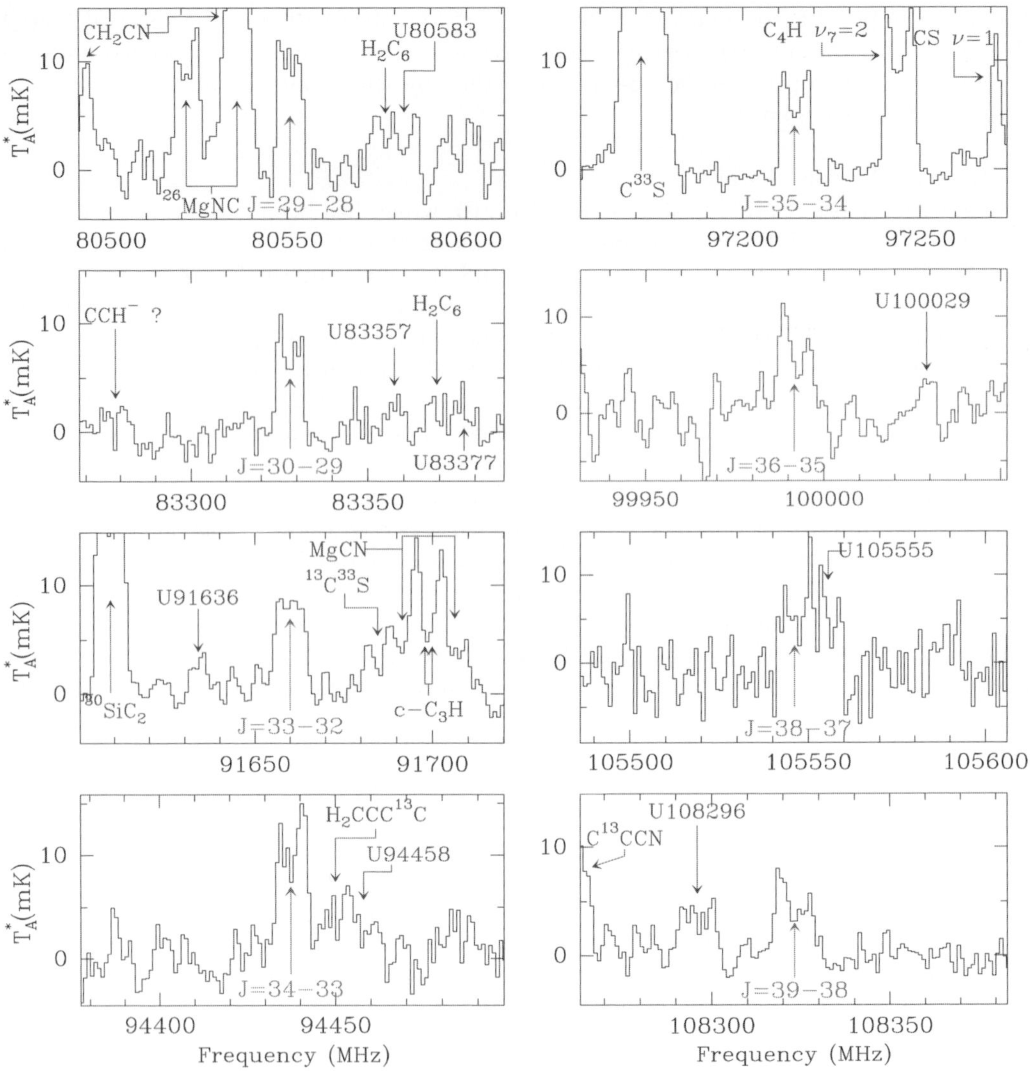

Figure 2. Spectra of IRC +10216 observed with the IRAM 30-m telescope, showing a series of lines from C_5N^-. The marginal weak line U83278 is worth noting, because it is within 0.1 MHz of the J=1-0 line of CCH^- (see Cernicharo *et al.* 2008).

SiNC (Guélin *et al.* 2004) and HC_4N (Cernicharo *et al.* 2004), the cyanides KCN and FeCN (Pulliam *et al.* 2010; Zack *et al.* 2011), as well as the recently discovered negatively charged molecules C_6H^-, C_4H^-, C_8H^-, C_3N^-, C_5N^- (see Fig. 2), and CN^- (McCarthy *et al.* 2006; Cernicharo *et al.* 2007; Remijan *et al.* 2007; Thaddeus *et al.* 2008; Cernicharo *et al.* 2008; Agúndez *et al.* 2010a).

Of great interest is also the extension of spectral line surveys from the radio region to the infrared part of the spectrum, as it allows to probe warm regions of the envelope closer to the star. Low spectral resolution spectra in the far-infrared region, λ = 45-197 μm, have been obtained with the Long Wavelength Spectrometer (LWS) on board the Infrared Space Observatory (ISO) toward several carbon-rich objects such as IRC +10216, CRL 618, and NGC 7027 (Cernicharo *et al.* 1996a; Herpin *et al.* 2002). These spectra show emission lines corresponding to highly excited rotational transitions

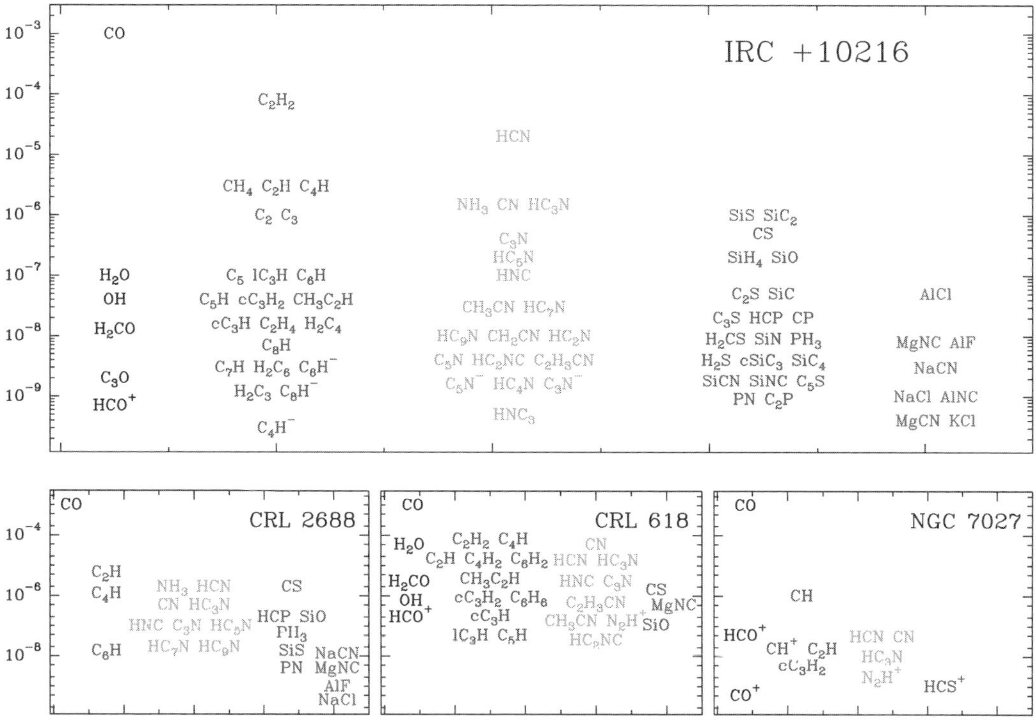

Figure 3. Molecular abundances relative to H_2 in the carbon-rich evolved stars IRC +10216, CRL 2688, CRL 618, and NGC 7027 (from Agúndez 2009).

of molecules which are abundant in the warm circumstellar regions, such as CO and HCN. Another nice example is the line survey of IRC +10216 carried out by Fonfría *et al.* (2008) with the IRTF telescope in the wavelength range 11-14 μm, where a forest of lines corresponding to rovibrational transitions of C_2H_2 and HCN is observed and used to derive their abundances in the 1-100 R_* zone, as well as the physical conditions in the inner circumstellar envelope, where the high-J lines of these species are formed.

The proximity of IRC +10216 to us ($\sim$ 130 pc) and the relatively large mass loss rate ($\sim 2 \times 10^{-5}$ $M_\odot$ yr^{-1}) makes it to be the prototype of AGB star and justifies the extensive observational studies carried out on it. Spectral line surveys have however been performed on other C stars such as CIT 6, observed with the 12-m ARO and SMT telescopes by Zhang *et al.* (2009a) in the frequency ranges 131-160 GHz, 219-244 GHz, and 252-268 GHz, and CRL 3068, which has been surveyed with the same telescopes by Zhang *et al.* (2009b) in the frequency ranges 130-162 GHz and 219.5-267.5 GHz. These two objects, CIT 6 is located at $\sim$ 400 pc and has a mass loss rate of $\sim 3 \times 10^{-6}$ $M_\odot$ yr^{-1} while CRL 3068 is at $\sim$ 1200 pc and has a mass loss rate of $\sim 1.2 \times 10^{-4}$ $M_\odot$ yr^{-1}, have less intense lines than IRC +10216, so that only the lines of the most abundant molecules can be detected. At this level of sensitivity both sources show a chemical composition very similar to that of IRC +10216.

More evolved carbon-rich stars have also been studied through line surveys. The pre-planetary nebula CRL 2688, which is in a slightly more evolved stage than IRC +10216, has been surveyed at λ = 3 mm (85-116 GHz) by Park *et al.* (2008), who have found a molecular composition quite similar to that of IRC +10216. Another pre-planetary nebula in a more evolved stage is CRL 618, which has been observed with the IRAM

30-m telescope in the frequency range 80-276 GHz (Pardo *et al.* 2004; Pardo *et al.* 2005; Pardo *et al.* 2007a,b), allowing for an accurate description of the molecular envelope. In this object the central star emits an important UV flux which alter the chemistry of the molecular envelope ejected during the previous AGB phase (Cernicharo 2004b). A more extreme object is the planetary nebula NGC 7027, whose extremely hot central star ($T_* \sim 220{,}000$ K) irradiates with a strongly ionizing radiation field the circumstellar gas. Zhang *et al.* (2008) have carried out a line survey of this object at millimeter wavelengths (71-111 GHz, 157-161 GHz, and 218-267 GHz) with some of the detected lines arising from rotational transitions of molecules and a large fraction of them from recombination transitions of H^+ and He^+. The evolution of the molecular abundances with the evolutionary phase along the sequence of carbon-rich objects IRC +10216 $\rightarrow$ CRL 2688 $\rightarrow$ CRL 618 $\rightarrow$ NGC 7027 is shown schematically in Fig. 3. We see that the pre-planetary nebula CRL 2688, which is in a more evolved stage than IRC +10216, has a chemical composition that is still quite similar to that of the latter object, with high abundances of organic molecules with a large unsaturated character, such as C_2H, C_4H, CN, HCN, or HC_3N. CRL 2688 is about ten times further than IRC +10216, although its mass loss rate is about ten times larger, so that it is still possible to detect species with moderately low abundances such as HCP, PH_3, as well as several metal-containing molecules. The pre-planetary nebula CRL 618 is clearly in a more evolved stage than CRL 2688, so that it starts to show important differences in the chemical composition, as is the relatively large abundance of oxygen-bearing molecules such as H_2O and H_2CO. It is also worth to note that the abundance ratios [HNC]/[HCN] or [HC_3N]/HCN are noticeably enhanced and the appearance of aromatic hydrocarbons such as benzene. Most of these difference are due to the strong enhancement of the UV radiation field which emanates from the central white dwarf, which makes the circumstellar material to be severely photoprocessed. Finally, the planetary nebula NGC 7027 shows a chemical composition which is clearly different from the previous less evolved objects. The presence of positive ions is much more important, with species such as CO^+, CH^+, and HCS^+, and in general terms the chemical complexity of the molecules decreases as a consequence of the huge UV field emitted by the very hot central star (see e.g. Bachiller *et al.* 1997; Herpin *et al.* 2002).

Oxygen-rich evolved stars.

The chemical composition of the circumstellar gas around oxygen-rich evolved stars is generally thought to be noticeably less rich than that around carbon-rich objects. Therefore less efforts have been undertaken to carry out line surveys on oxygen-rich objects than on carbon-rich ones. Only recently, the new generation of low-noise and wide-band receivers has stimulated astronomers to perform sensitive line surveys of oxygen-rich evolved stars at millimeter wavelengths. The first such survey was carried out towards the supergiant VY Canis Majoris (Tenenbaum *et al.* 2010) with the 10-m SMT telescope at $\lambda \sim 1.3$ mm (214.5-285.5 GHz). Observations at millimeter wavelengths of this object have revealed a chemical complexity with unexpected molecules such as NaCl, PN, HNC, and HCO^+ being present in the different circumstellar outflows around the central star (Ziurys *et al.* 2007).

Sensitive line surveys of the oxygen-rich evolved stars OH 231.8+4.2 and IK Tau are also being carried out with the IRAM 30-m telescope covering all accessible spectral windows, from $\lambda = 3$ mm to $\lambda = 0.9$ mm, and preliminary results indicate the existence of a prominent chemical diversity with molecules such as SiO, NS, HNCO, CH_3OH among others (Sánchez Contreras *et al.* 2011). IK Tau has also been recently surveyed by Kim *et al.* (2010) at selected wavelengths in the submillimeter domain using the APEX telescope.

Observations of oxygen-rich evolved stars such as VY Canis Majoris, IK Tau, TX Cam, RX Boo, IRC +10011, and R Cas have also been carried out by Polehampton *et al.* (2010) in the far-infrared region of the spectrum using the LWS instrument on board ISO.

2. Results from the Herschel Space Observatory

IRC +10216 (CW Leo) is one of the brightest infrared sources in the sky, thus an ideal target to be observed with the *Herschel Space Observatory* (Pilbratt *et al.* 2010). The HIFI instrument (de Graauw *et al.* 2010) provides both a high resolution and a wide spectral coverage. The first is necessary for resolving the complex kinematics characteristic of IRC +10216, allowing us to distinguish between the contribution from the inner acceleration zone (Fonfría *et al.* 2008, Cernicharo *et al.* 2011) and from the expanding envelope where the gas reaches the terminal velocity (Cernicharo *et al.* 2000, Agúndez 2009). The wide spectral coverage is mandatory to obtain a complete inventory of lines to study in detail the chemical content and molecular excitation. A complete line survey of this source has been carried out with HIFI. Preliminary results are shown by Cernicharo *et al.* (2010a,b) (see Fig. 4). In addition to HCN masers and lasers from all vibrationally excited states of this molecule below 7000 K (see also Cernicharo *et al.* 2011; Cernicharo *et al.* 1996), the HIFI line survey of IRC +10216 shows a forest of lines of SiC_2 arising from levels with energies up to 1000 K and the presence of some hydrides such as HCl (Cernicharo *et al.* 2010b) and HF (Agúndez *et al.* 2011). Complete line surveys of the O-rich evolved stars VY CMa and OH231.8 will be performed in the next months with HIFI.

In addition to these sources a large number of evolved stars have been observed with low spectral resolution using SPIRE and PACS (Griffin *et al.* 2010; Poglitsch *et al.* 2010) within the Key Program (KP) MESS (Groenewegen *et al.*). The main results from this KP have been shown in the special issue of Astronomy and Astrophysics devoted to these instruments on board Herschel (vol. 518). Among them it is worth to the the detection of warm water in IRC +10216 (Decin *et al.* 2010) and its interpretation in terms of a photochemistry in the inner region due to the clumpy structure of the envelope (see also Agúndez *et al.* 2010b). While HCN dominates the far-IR spectrum of C-rich objects, H_2O is the main contributor to the SPIRE and PACS spectra of O-rich evolved stars.

Finally, the HIFISTARS KP is devoted to the observation with HIFI of selected lines of CO, H_2O, HCN, SiO, SiS, NH_3 and many other species in a large sample of evolved stars (see e.g., Bujarrabal *et al.* 2010; see also the different papers from this KP in the special issue of A&A devoted to HIFI -vol 521).

3. Chemistry in evolved stars

The basic chemical processes taking place in the envelopes around AGB stars were already summarized by Lafont *et al.* (1982), focusing on IRC +10216, although they are extensible to any other AGB star. They identified four different formation mechanisms of molecules: thermochemistry in the regions near the stellar photosphere and freeze out of the abundances in the immediately outer layers, gas phase chemical reactions dominated by kinetics, grain processes, and photochemistry.

It is now well established since the pioneering work of Tsuji (1973) that the [C]/[O] abundance ratio in the stellar photosphere is the most important parameter which determines the type of molecules found in circumstellar envelopes. The high stability of CO causes this molecule to have a large abundance, locking most of the limiting reactant and allowing for the reactant in excess to form either carbon-bearing molecules when [C]/[O]

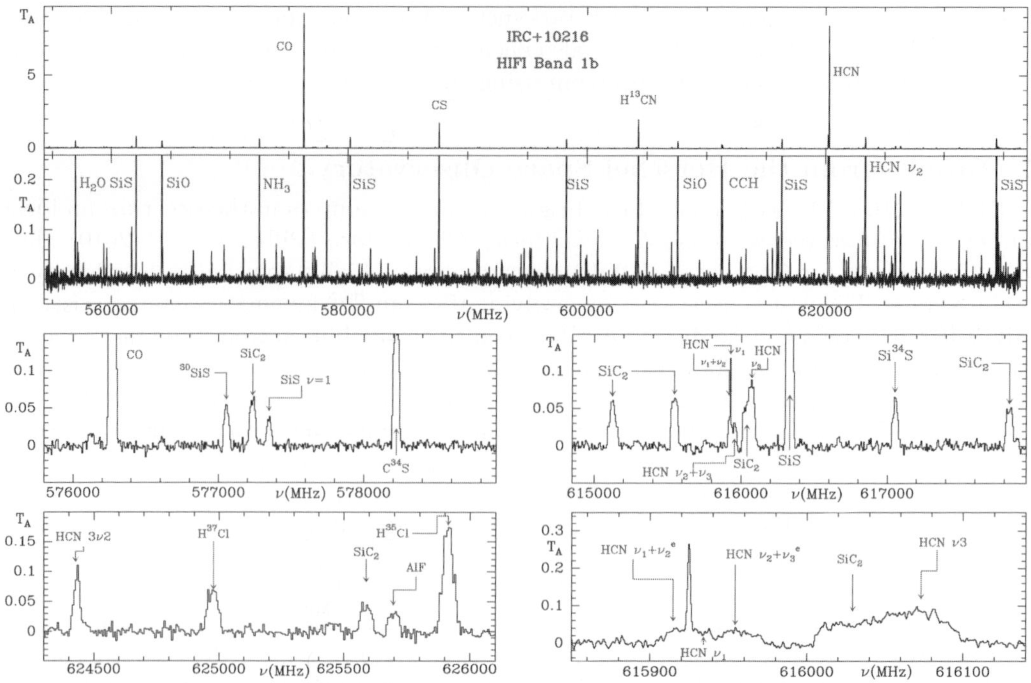

Figure 4. Spectra of IRC +10216 observed with HIFI band 1b. The two upper panels present the complete spectrum at two different intensity scales. The panels below show different 3 GHz wide ranges of the survey. All data have been smoothed to a spectral resolution of 2.8 km s^{-1} except for the right bottom panel which shows the spectrum around several vibrational lines of HCN with the nominal WBS resolution (1.1 MHz, $\simeq$0.5 km s^{-1}). Note the narrow feature of HCN shown by the $\nu_1 + \nu_2^{1e}/\nu_1$ J=6-5 line at $\sim$615.92 GHz (see Cernicharo *et al.* 2010a).

$>$ 1, and oxygen-bearing molecules when [C]/[O] $<$ 1. This simple fact provides a good explanation of the overall chemical composition in the inner envelopes of AGB stars, although there are notable exceptions that thermochemical equilibrium calculations cannot explain, such as the presence, with moderately high abundances, of HCN, CS, and NH_3 in oxygen-rich objects and that of water vapor and ammonia in carbon-rich objects. This points towards the existence of non-equilibrium chemical processes being at work in the inner regions of circumstellar envelopes, that could be either induced by shocks (Cherchneff 2006) or perhaps by UV photons penetrating from the interstellar medium to the inner regions through a clumpy envelope (Agúndez *et al.* 2010b).

In the intermediate layers of circumstellar envelopes, when the temperature has decreased below $\sim$ 1000 K, refractory elements start to condense and form solid dust grains. An important effect is that some molecules containing refractory elements such as Si decrease their abundances in the gas phase as they incorporate into dust grains. Another effect, yet to be studied in detail, is that some molecules may form mantles on the surface of dust grains and may participate in dust surface reactions forming new molecules that can return to the gas phase by several desorption mechanisms, such as thermal desorption or desorption induced by interstellar UV photons or cosmic rays.

As the gas expands density decreases and the column density of material located towards the interstellar medium do so. Thus, in the outer circumstellar layers molecules start to be exposed to the ambient interstellar UV field and are then photoprocessed,

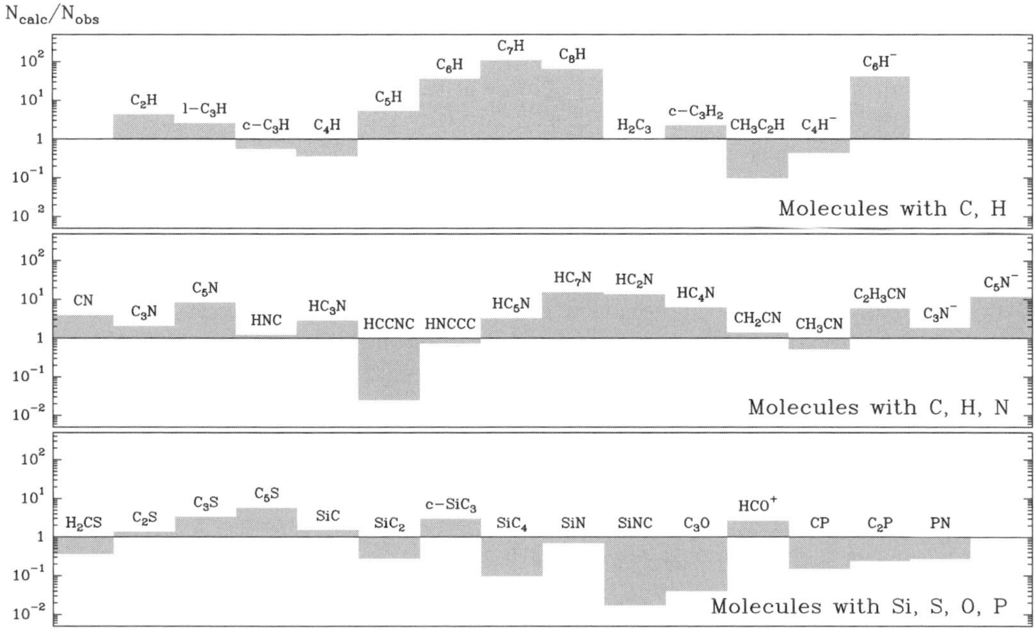

Figure 5. Comparison between column densities calculated (N_{calc}) with a chemical model and values derived from observations (N_{obs}) of IRC +10216 (from Agúndez 2009).

they are mostly dissociated but also ionized, producing radicals and ions which can participate in rapid gas phase reactions to form new molecules. This photochemistry has been extensively studied through chemical models, most of which have focused on the carbon star IRC +10216 as this object harbors the largest variety of molecules among all known evolved stars (Cherchneff *et al.* 1993; Millar & Herbst 1994; Cordiner & Millar 2009). Such models have successfully explained the formation of a large fraction of the molecules observed in the outer layers of IRC +10216 and of other carbon-rich AGB stars. To illustrate this we show in Fig. 5 a comparison between the column densities calculated by a chemical model of IRC +10216 and the values derived from observations carried out at millimeter wavelengths with the IRAM 30-m (see Agúndez 2009). We see that for most of the molecules the differences between calculated and observed column densities are within a order of magnitude, which we may consider as a reasonably good agreement taking into account the uncertainties of both the abundances derived from observations and calculated with the chemical model. There are however important discrepancies for some molecules, such as the long carbon-chain radicals C_6H, C_7H, and C_8H whose abundances are overestimated by the chemical model, probably due to an underestimation of the photodestruction rates, or species such as HCCNC, SiNC, or C_3O which are underestimated due to the lack of efficient formation routes in the chemical model.

The chemistry of oxygen-rich AGB stars has been less well studied through chemical models. A few ones have been carried out in the past (e.g. Willacy & Millar 1997), although the recent discovery of an enhanced chemical complexity in some oxygen-rich evolved stars will certainly need of a revisit of chemical models aiming to explain the formation of the newly detected molecules.

4. The future

ALMA will provide an angular resolution permiting to resolve the molecular emission in the dust formation zone of AGB stars. Toward these objects in will be possible to reach the maximum angular resolution provided by ALMA as the brightness temperature of the CO, HCN, SiO, SiS lines will be above 1000 K. The available data in the submillimeter domain with the SMA towards IRC+10216 (Patel *et al.* 2011) shows the presence of a large number of unknown narrow lines certainly formed in the innermost zone of the envelope. These future high angular resolution and extremely sensitive observations will reveal new families of complex molecular species that participate in the formation and growth of dust grains. The physical conditions of the dust formation zone have been traced so far only through the most abundant molecules (CO, HCN, SiS, SiO). The chemistry of the most complex molecules in the innermost zone remains still a mystery that ALMA will help to solve.

References

Agúndez, M. & Cernicharo, J., 2006, *ApJ*, 650, 374
Agúndez, M. 2009, PhD Thesis, Universidad Autónoma de Madrid (Spain)
Agúndez, M., Cernicharo, J., Guélin, M., *et al.* 2010a, *A&A*, 517, L2
Agúndez, M., Cernicharo, J., & Guélin, M. 2010b, *ApJ*, 724, L133
Agúndez, M., Cernicharo, Waters, L. B. F.. M., *et al.*, 2011, *A&A*, 533, L6, *A&A*, 521, L3
Apponi, A. J., McCarthy, M. C., Gottlieb, C. A., & Thaddeus, P., 1999, *ApJ*, 516, L103
Avery, L. W., Amano, T., Bell, M. B., *et al.* 1992, *ApJS*, 83, 363
Bachiller, R., Forveille, T., Huggins, P. J., & Cox, P. 1997, *A&A*, 324, 1123
Bell, M. B., Avery, L. W., & Feldamn, P. A., 1993, *ApJ*, 417, L37
Cernicharo, J. & Guélin, M. 1987a, *A&A*, 181, L9
Cernicharo, J. & Guélin, M. 1987b, *A&A*, 183, L10
Cernicharo, J., Gottlieb, C. A., Guélin, M., *et al.* 1989, *ApJ*, 341, L25
Cernicharo, J., Gottlieb, C. A., Guélin, M., *et al.* 1991a, *ApJ*, 368, L39
Cernicharo, J., Gottlieb, C. A., Guélin, M., *et al.* 1991b, *ApJ*, 368, L43
Cernicharo, J., Barlow, M. J., Gonzlez-Alfonso, E., *et al.* 1996a, *A&A*, 315, L201
Cernicharo, J. & Guélin, 1996b, *A&A*, 309, l27
Cernicharo, J., Guélin, M., & Kahane, C. 2000, *A&AS*, 142, 181
Cernicharo, J., Guélin, M., & Pardo, J. R. 2004, *ApJ*, 615, L145
Cernicharo, J., *ApJ*, 608, L41
Cernicharo, J., Guélin, M., Agúndez, M., *et al.* 2007, *A&A*, 467, L37
Cernicharo, J., Guélin, M., Agúndez, M., *et al.* 2008, *ApJ*, 688, L83
Cernicharo, J., Guélin, M., Agúndez, M., *et al.* 2010a, *A&A*, 521, L8
Cernicharo, J., Guélin, M., Agúndez, M., *et al.* 2010b, *A&A*, 518, L136
Cernicharo, J., Guélin, M., Agúndez, M., *et al.* 2011, *A&A*, 529, L3
Cherchneff, I., Glassgold, A. E., & Mamon, G. A. 1993, *ApJ*, 410, 188
Cherchneff, I. 2006, *A&AS* 456, 1001
Cordiner, M. A. & Millar, T. J. 2009, *ApJ*, 697, 68
Decin *et al.*, 2010, *A&A*, 518, L143
de Graauw T. *et al.* 2010, *A&A*, 518, L6
Fonfría, J. P., Cernicharo, J., Richter, M. J., & Lacy, J. H. 2008, *ApJ*, 673, 445
Griffin *et al.* 2010, *A&A*, 518, L3
Groenewegen, M. A. T.., Waelkens, C., Barlow, M. J., *et al.*, 2011, *A&A*, 526, 162
Groesbeck, T. D., Phillips, T. G., & Blake, G. A. 1994, *ApJS*, 94, 147
Guélin, M., Gómez González, J., Cernicharo, J., & Kahane, C. 1986, *A&A*, 157, L17

Guélin, M., Lucas, R., & Neri, R. 1993, *A&A*, 280, L19

Guélin, M., Lucas, R., & Neri, R. 1997, *in CO: twenty-five years of millimeter-wave spectroscopy, Proc. IAU*, 170, 359

Guélin, M., Muller, S., Cernicharo, J., *et al.*, 2000, *A&A*, 363, L9

Guélin, M., Muller, S., Cernicharo, J., *et al.* 2004, *A&A*, 426, L49

He, J. H., Dinh-V-Trung, Kwok, S., Müller, H. S. P., *et al.* 2008, *ApJS*, 177, 275

Herpin, F. & Cernicharo, J., 2001, *ApJ*, 530, L129

Herpin, F., Goicoechea, J. R., Pardo, J. R., & Cernicharo, J. 2002, *ApJ*, 577, 961

Johansson, L. E. B., Andersson, C., Elldér, J., *et al.* 1984, *A&A*, 130, 227

Kawaguchi, K., Kagi, E., Hirano, T., *et al.* 1993, *ApJ*, 406, L39

Kawaguchi, K., Kasai, Y., Ishikawa, S., & Kaifu, N. 1995, *PASJ*, 47, 853

Kemper, F., Stark, R., Justtanont, K., *et al.*, 2003, *A&A*, 407, 609

Kim, H., Wyrowski, F., Menten, K. M., & Decin, L. 2010, *A&A*, 516, A68

Lafont, S., Lucas, R., & Omont, A. 1982, *A&A*, 106, 201

McCarthy, M. C., Gottlieb, C. A., Gupta, H., & Thaddeus, P. 2006, *ApJ*, 652, L141

Millar, T., J. & Herbst, E. 1994, *A&AS*, 288, 561

Ohishi, M., Kaifu, N., Kawaguchi, K., *et al.*, 1989, *ApJ*, 345, L83

Pardo, J. R., Cernicharo, J., Goicoechea, J. R., & Phillips, T. G., 2004, *ApJ*, 615, 495

Pardo, J. R., Cernicharo, J., & Goicoechea, J. R., 2005, *ApJ*, 628, 275

Pardo, J. R., Cernicharo, J., Goicoechea, J. R., *et al.* 2007a, *ApJ*, 661, 250

Pardo, J. R., Cernicharo, J., Goicoechea, J. R., *et al.*, 2007b, *ApJ*, 661, 250

Park, J. A., Cho, S.-H., Lee, C. W., & Yang, J. 2008, *AJ*, 136, 2350

Patel, N. A., Young, K. H., Gottlieb, C. A., *et al.* 2011, *ApJS*, 193, 17

Pilbratt *et al.*, 2010, *A&A*, L1

Poglitsch *et al.*, 2010, *A&A*, 518, L2

Polehampton, E. T., Menten, K. M., van der Tak, F. F. S., & White, G. J. 2010, *A&A*, 510, A80

Pulliam, R. L., Savage, C., Agúndez, M., *et al.* 2010, *ApJ*, 725, L181

Remijan, A. J., Hollis, J. M., Lovas, F. J., *et al.* 2007, *ApJ*, 664, L47

Sánchez Contreras, C., Velilla Prieto, L., Cernicharo, J. *et al.* 2011, *in the Molecular Universe, Proc. IAU*, 180, poster 1.075

Tenenbaum, E. D., Dodd, J. L., Milam, S. N., *et al.* 2010, *ApJS*, 190, 348

Turner, B. E., Steimle, T. C., & Meerts, L., 1994, *ApJ*, 426, L97

Thaddeus, P., Cummins, S. E., & Linke, R. A., 1984, *ApJ*, 283, L45

Thaddeus, P., Gottlieb, C. A., Gupta, H., *et al.* 2008, *ApJ*, 677, 1132

Tsuji, T. 1973, *A&AS*, 23, 411

Willacy, K. & Millar, T. J. 1997, *A&A*, 324, 237

Zack, L. N., Halfen, D. T., & Ziurys, L. M. 2011, *ApJ*, 733, L36

Zhang, Y. & Kwok, S., Dinh-V-Trung 2008, *ApJ*, 678, 328

Zhang, Y. & Kwok, S., Dinh-V-Trung 2009a, *ApJ*, 691, 1660

Zhang, Y., Kwok, S., & Nakashima, J. 2009b, *ApJ*, 700, 1262

Ziurys, L. M., Guélin, M., Apponi, A. J., & Cernicharo, 1993, *ApJ*, 445, L47

Ziurys, L. M., C. Savage, J. L. Highberger, *et al.*, 2002, *ApJ*, 564, L45

Ziurys, L. M., Milam, S. N., Apponi, A. J., & Woolf, N. J. 2007, *Nature*, 447, 1094

Discussion

Paola Caselli: Why did you not expect to see so many SiC_2 lines and which fraction of Si is locked in SiC_2?

José Cernicharo: From millimeter data alone the SiC_2 emission seems to arise from the external shell where the kinetic temperature is low. The fact that with Herschel/HIFI we have detected lines involving levels with energies as high as 1000 K means that SiC_2 is produced in the inner envelope with an abundance of 1-2 10^{-7}. This abundance is perhaps enhanced by a factor of 2 when the gas reaches the external envelope (Cernicharo *et al.* 2010a).

The Molecular Universe
Proceedings IAU Symposium No. 280, 2011
José Cernicharo & Rafael Bachiller, eds.

doi:10.1017/S1743921311025026

Chemistry of the Solar System Revealed in the Interiors of the Giant Planets

Jonathan I. Lunine[1]

[1]Department of Astronomy, Cornell University,
402 Space Sciences Building, Ithaca, NY 14853 USA
email: `jlunine@astro.cornell.edu`

Abstract. The giant planets of our solar system contain a record of elemental and isotopic ratios of keen interest for what they tell us about the origin of the planets and in particular the volatile compositions of the solid phases. In situ measurements of the Jovian atmosphere performed by the Galileo Probe during its descent in 1995 demonstrate the unique value of such a record, but limited currently by the unknown abundance of oxygen in the interior of Jupiter–a gap planned to be filled by the Juno mission set to arrive at Jupiter in July of 2016. Our lack of knowledge of the oxygen abundance allows for a number of models for the Jovian interior with a range of C/O ratios. The implications for the origin of terrestrial water are briefly discussed. The complementary data sets for Saturn may be obtained by a series of very close, nearly polar orbits, at the end of the Cassini-Huygens mission in 2016-2017, and the proposed Saturn Probe. This set can only obtain what we have for Jupiter if the Saturn Probe mission carries a microwave radiometer.

Keywords. giant planets, noble gases, water, oxygen

1. Introduction

The giant planets of our solar system were traditionally considered to be cosmic oddballs, interesting specimens of the solar nebula (protoplanetary disk) from which the Sun and planets formed some 4.55 billion years ago. Indeed, speculation based on pioneering models of planetary accretion suggested that giant planet formation might be relatively rare Wetherill (1992). However, the current tally of discoveries of giant planets indicates that roughy 10% of Sun-like stars possess planets within an order of magnitude the mass of Jupiter Marcy and X (2010). Ground-based microlensing survey have found hundreds of unbound (free-floating) or loosely bound planets with masses of order that of Jupiter; given the efficiency of detection, this implied that two giant planets have formed for every star in the galaxy OGLE and M (2011).

Since giant planets appear to be a common product of star formation, albeit with only a minority remaining in tightly bound orbits around their parent stars, understanding the composition and structure of such bodies should be an important goal of astrophysics. Study of Jupiter, Saturn, Uranus and Neptune no longer can be regarded as a planetological specialty, but rather as an important branch of astrophysics that informs the study of the countless numbers of giant planets likely to be present in the cosmos.

Of particular interest in this regard are in situ measurements of composition, by entry probe, combined with remote sensing of deep atmospheric species and mapping of gravitational and magnetic fields–remote sensing that requires close range (though see dePater *et al.* (2001) for examples of microwave radiometry of the giant planets done from Earth). The next decade promises an unprecedented opportunity to build on the Galileo entry probe mission of 1995 (Fig. 1), through close orbital sensing of Jupiter and Saturn from Juno and Cassini. The possibility of a Saturn entry probe akin to the Galileo

Jupiter probe, in the decade following, would complete the basic set of measurements needed to characterize in detail both of the solar system's largest planets.

Here I will focus on the chemical and isotopic information contained in Jupiter and Saturn that informs the details of solar system formation, and the observations that will deliver this information in the near future.

2. An overview of the chemical and isotopic signatures of Jupiter and Saturn

A summary of the key chemical and isotopic signatures in Jupiter and Saturn is given in Table 1. Jupiter's composition was measured *in situ* down to 22 bars; Saturn's was measured remotely by a variety of ground- and space-based instruments.

Helium. The helium abundance is thought to have been uniform throughout the protosolar nebula during the formation of the planets and, therefore, any variation from the generally accepted protosolar value of 0.28 (expressed as a mass fraction; Profitt 1994) should reflect processes internal to the giant planets themselves. The Jovian value is close to but slightly below the protosolar one, with error bars based on the disagreement between the mass spectrometer and the helium detector (refractometer) instruments. The Saturnian value is well below the solar value but with larger error bars because it is derived from remote sensing data. The values given here are from preliminary analyses of

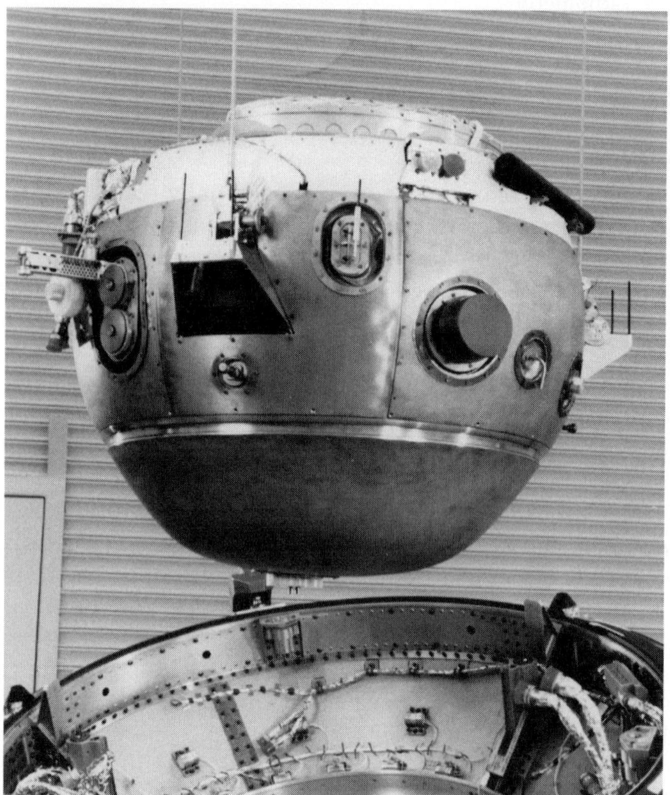

Figure 1. Descent module of the Galileo probe, suspended above the deceleration module/aeroshell. Diameter of the descent module is about 0.9 meters (NASA).

Table 1. Some chemical and isotopic signatures in Jupiter and Saturn[1]

Element	Abundance in Jupiter (solar = 1)	Isotopes in Jupiter	Abundance in Saturn (solar = 1)	Isotopes in Saturn[3]
hydrogen	*defined* = 1	$D/H = 2.25^+_-0.35x10^{-5}$	defined = 1	$D/H = 1.70^{+0.75}_{-0.45}x10^{-5}$
helium	$0.9^+_-0.5$	$^3He/^4He = 1.66^+_-0.05x10^{-4}$	$\approx 0.6 - 0.86$	?
neon	$0.059^+_-0.004$	?	?	?
argon	$3.0^+_-1.0$	?	?	?
krypton	$2.03^+_-0.38$	?	?	?
xenon	$2.11^+_-0.40$	?	?	?
carbon	$3.82^+_-0.66$	$^{12}C/^{13}C = 92.6^+_-0.4$	$10.4^+_-0.4$	$91.8^{+5.5}_{-5.3}$
nitrogen	$4.90^+_-1.87$	$^{14}N/^{15}N = 440^+_-60$	4.5^+_-1	?
oxygen	$0.29^+_-0.09$	?	?	?
sulfur	$2.88^+_-0.69$	?	≈ 17	?

Notes:
[1] Jovian abundances from various sources as tabulated and cited in Marty *et al.* (2008) except for helium which is from von Zahn *et al.* (1998) and Niemann *et al.* (1998), argon from Mahaffy *et al.* (2000), D/H as reported in Fouchet *et al.* (2009), and oxygen, which is from Wong *et al.* (2008), using the solar abundances of Grevesse *et al.* (2005) (but see also Asplund *et al.* 2009). Solar He/H_2 is from Profitt 1994. Saturnian abundances taken from Fouchet *et al.* (2009) based on various published and unpublished sources.

Cassini infrared data, which relies on the sensitivity to the helium fraction of the shape of the continuum infrared absorption. However, this analysis does not yield the same result as a different analysis which combines data from radio occultations with infrared data, and further there is disagreement with the same type of analysis from the earlier Voyager flybys. This complex situation is well described in Fouchet *et al.* (2009).

Despite the uncertainties the interpretation seems straightforward, which is that helium is not completely miscible with hydrogen at the temperatures relevant to the interiors of Jupiter and Saturn, and the zone of immiscibility is larger in Saturn than it is in Jupiter–consistent with the lower temperature at a given pressure in Saturn relative to Jupiter. Immiscibility of helium means that droplets of helium form in Saturn's deep interior; the sinking of these results in enrichment of helium in the deep interior while the outer layers are depleted (Stevenson and Salpeter, 1977)). The sinking of the droplets releases gravitational energy which should be measurable as an added heat source, and indeed the thermal emission from Saturn (Hanel *et al.* 1983)) is larger than expected for a planet which is the age of the Sun. Jupiter, on the other hand, has an infrared emission consistent with its assumed age of 4.55 billion years, assuming that the emissions sources are only thermalized solar energy and the original virtualized energy of formation of the self-gravitating body. While the coarse outlines of this story seem complete, there remain disagreements on the details (such as whether helium differentiation is occurring in Jupiter as well) because of uncertainties in the hydrogen-helium equation of state at high pressure (Hubbard *et al.* 2009, Wilson and Militzer 2010).

The ratio of the light isotope of helium ^{3}He relative to hydrogen is lower in Jupiter than the generally accepted value for both the solar wind (a sample of gases from the interior of the Sun today) and the local interstellar medium, while the D/H value is larger. The sum of the ^{3}He and deuterium abundances, normalized to the hydrogen abundance, is within the (somewhat large) error bars the same as the solar and ISM values; see (Fig. 2) During stellar nucleosynthesis deuterium is destroyed and ^{3}He produced. Since the Sun's deep interior has been undergoing fusion, and the local interstellar medium is contaminated with material from massive stars that similarly reflects fusion processes within Population 1 stars, a sample of gas unaltered by fusion from the time of the solar system's formation should have lower ^{3}He and higher deuterium than that found in the solar wind (a sample of the solar interior) or local ISM, but the sum should be the same. As figure 2 shows, this is indeed the case for Jupiter. (Owen and Encrenaz 2003). A similar test for Saturn awaits measurement of the ^{3}He abundance, which can only be done *in situ* by an entry probe.

Noble gases and C,N,S,P bearing species. Besides that of helium, the abundances of all of the other noble gases (neon, argon, krypton, and xenon) have been measured in Jupiter by the Galileo mass spectrometer. Remote sensing measurements of these species in the well-mixed part of the atmosphere is extremely difficult, and hence there are no values reported for Saturn. Initial work analyzing the data used solar abundances that suggested a uniform enrichment of argon, krypton and xenon relative to solar (Mahaffy *et al.* (2000)), but more recent updates to the solar elemental abundances (Asplund 2009) lead to a slightly larger enrichment of argon relative to solar compared to that of krypton and xenon. The neon abundance Ne/H_2 is strongly depleted relative to solar, and unlike water (see below) this is not a meteorological effect in the observable atmosphere. It is also unlikely to be a depletion in the nebula, since the He/H_2 ratio is not similarly depleted. Most likely it is a miscibility effect in the deep interior (Wilson and Militzer 2010). Unlike neon, which is simply too volatile to incorporate in icy planetesimals, argon, krypton and xenon directly condensed or were trapped or adsorbed in water ice.

There has been a lot of debate on the question of the origin of the heavy noble gas enrichments, as well as that of carbon, nitrogen, sulfur and phosphorous. The uniformity is most easily explained if all of the species condensed out as ices, and given their differing

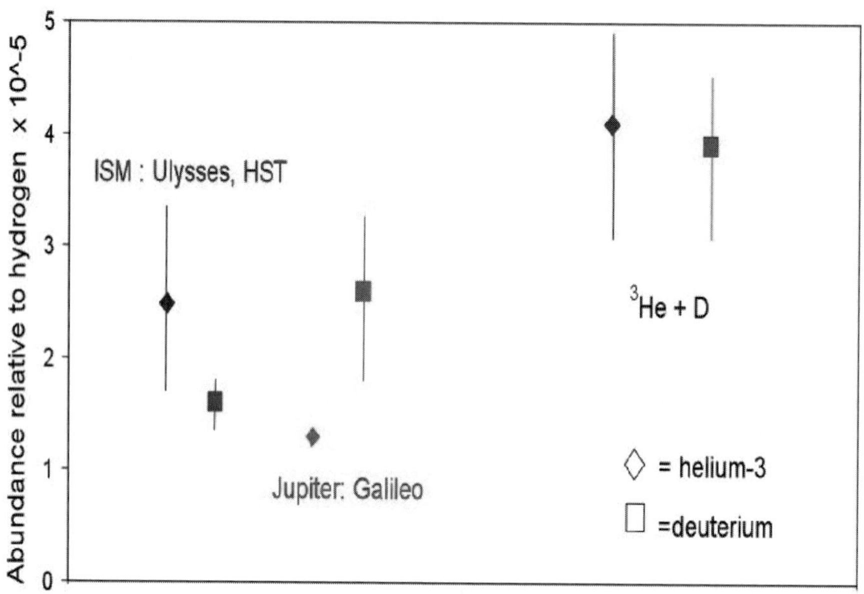

Figure 2. ^{3}He and deuterium (D) abundances in Jupiter, and their sum, plotted relative to the hydrogen abundance in Jupiter. The red points and error bars are the Galileo probe data; the blue come from Ulysses solar wind measurements and Hubble Space Telescope observations of the local interstellar medium. Figure based on an argument in Owen and Encrenaz (2003).

volatilities, this required very low temperatures in the protoplanetary disk, 30 K or perhaps lower (Owen *et al.* 1999). An alternative model invoked trapping in clathrate hydrate (Hersant *et al.* 2004). Revisions to solar abundances led to enrichments that were no longer quite so uniform (Grevesse *et al.* (2005)), which allowed the clathrate model to more easily fit the data (Mousis *et al.* 2009). The clathrate model requires nebular temperatures of between 30-40K, not quite as cold as direct condensation, but also implies a larger fraction of water ice accompanying the carbon-, nitrogen- and sulfur-bearing species than does the direct condensation model.

The carbon isotopic ratio $^{12}C/^{13}C$ in Jupiter is similar to the terrestrial value in those species in Jupiter not affected by photochemical processes, while the $^{14}N/^{15}N$ is enriched by a factor of two relative to terrestrial (Owen and Encrenaz, 2003). This suggests that while the carriers of carbon to the Earth and to Jupiter were perhaps similar, those of nitrogen were not.

Oxygen. The deep oxygen abundance in Jupiter is not known because the abundance of water–the primary carrier of oxygen in the Jovian atmosphere–is in principle strongly affected by condensation and rainout associated with meteorological processes (Lunine and Hunten 1987), by advection (Showman and Dowling 2000), or both. The Galileo probe fell into a relative dry region known as a "hot spot" (for the excess brightness observed in such regions at 5 μm wavelengths), and the water abundance was observed to increase with depth, or equivalently, with pressure. The sparseness of the measurements made it impossible to know whether the water had "leveled out" at a value corresponding to 1/3 solar or would have increased further had the probe returned data below the 21 bar level where communication ceased. Predictions for either the direct condensation model or the clathrate model are for oxygen abundances several to many times solar in the Jovian deep interior.

3. A carbon-rich Jupiter

Although atmospheric explanations for the water abundance in Jupiter are attractive, one must not rule out the possibility that water truly is depleted in the Jovian interior–that is, the oxygen-to-hydrogen ratio in Jupiter is subsolar. One justification for making such a case is that a planet with a C/O > 1–a "carbon-rich planet"–has been discovered (Madhusudhan *et al.* 2011a), companion to the star WASP12a with a C/O ratio of 0.44, roughly solar. One explanation is that the portion of this system's protoplanetary disk was somehow depleted in water at the time the planet formed and acquired its heavy element inventory (Madhusudhan *et al.* 2011). Prior to this discovery, the possibility of a carbon-rich Jupiter was considered on the basis of the Galileo results alone by Lodders (2004), who proposed that the point in our protoplanetary disk at which water ice condensed–the snowline– might have been further from the Sun than the point at which Jupiter formed, and volatiles adhering to solid organics rather than water ice were carried into Jupiter. Indeed, for an increasingly oxygen-depleted (or carbon-enriched) protoplanetary disk, the point at which water ice forms from water vapor moves progressively outward (Fig. 3). One difficulty with this hypothesis is the presence of water in the asteroid belt in the form of the main belt comets (Hsieh and Jewitt 2006), although the relatively high rock-to-ice ratio in the Jovian vs Saturnian satellite systems might be an argument in favor.

Mousis *et al.* (2011) looked at the possibility that Jupiter might be a carbon-rich planet and examined element-by-element the fit to the Galileo probe data of two contrasting models: one in which the planetesimal building blocks of Jupiter derived from a disk with

C/O= 1/2 (roughly, the solar value), and the other in which C/O=1. To achieve the latter they assumed that the oxygen abundance in the disk had been depleted by a factor of two.. They then invoked planetesimal trapping in clathrate hydrates that provides a good fit to the elemental abundances in the Galileo probe data for the C/O=solar case (Mousis *et al.* 2009, and reexamined the fit in the case for which C/O = 1. Within the error bars of the probe data, there is little to distinguish the two bulk elemental abundances the one from the other. Both fit carbon, nitrogen, sulfur and argon equally well. Neither fits phosphorous, krypton or xenon, overestimating all of them. The solar C/O case is closer to P and Xe than the carbon-rich case, but both are well outside the probe error bars. Mousis *et al.* (2011) point out that the measured phosphorous abundance is likely a lower limit to the bulk abundance of phosphorous in Jupiter, because highly non-volatile phases may be carrying much of this element. Krypton and xenon cannot be so explained, but Mousis *et al.* (2009) had already invoked complexing of these noble gases with H_3^+ as a way of limiting their trapping in clathrate hydrates in the protoplanetary disk.

Regardless of the details, the real test is in the oxygen abundance, which must be seven times solar in Jupiter in the C/O=solar case to account for the trapping of the volatiles, and roughly solar in the C/O=1 case. In both cases, the O/H_2 abundance in the Jovian envelope is larger than that in the protoplanetary disk overall because of the inefficiency of trapping of the volatiles in the ice. For direct condensation of all volatiles with no clathrate, the water enhancement is about 4 times solar in the C/O=solar case and below solar in the C/O= 1 case (remember that Mousis *et al.* (2011) label these cases by the C/O values in the source material of the planetesimals in the original disk). The hypothesis of Lodders (2004), that the volatiles are actually trapped within organic solids, would lead to an even smaller water abundance in Jupiter. Thus, any determination of the deep water abundance in Jupiter yielding a value relative to hydrogen of solar or less would be a strong argument in favor of a carbon-rich Jupiter.

All of this begs the question of the source of the depletion. A wide range of oxidation states existed in the protoplanetary disk at different times and locations. For example,

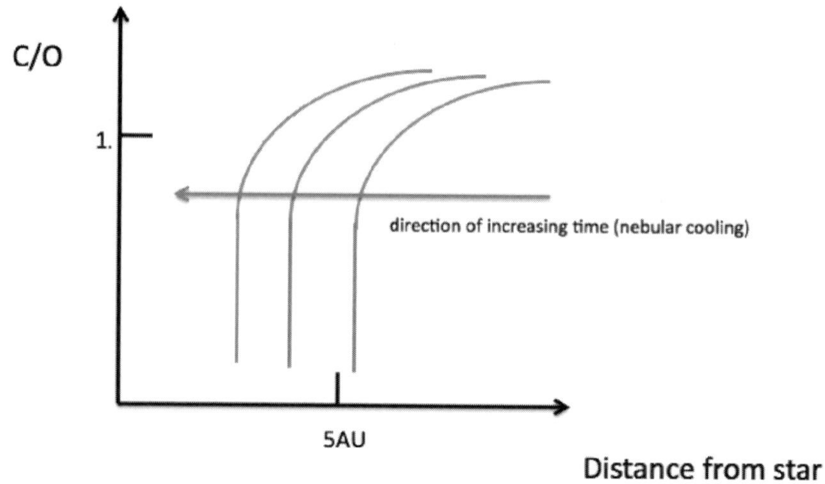

Figure 3. Qualitative behavior of the snowline– the radial position at which water ice condenses–in a protoplanetary disk, for varying elemental ratios of carbon to oxygen. Because silicates and metals sequester some of the oxygen at high temperatures, the amount of water drops precipitously and the snow line moves rapidly outward for C/O values approaching, but still below, unity.

the enstatite chondrites are thought to have come from parent bodies formed inward of all the other parent bodies, and their mineralogy suggests reducing conditions–consistent with a depleted water vapor abundance–in the region of the nebula where they formed (Krot *et al.* 2000). That region was almost certainly inward of the nebula snowline, where radial zones of enrichment and depletion of water vapor associated with the cold finger effect of condensation and inward radial drift of the icy particles existed (Ciesla and Charnley 2006. Recent models of the early evolution of Jupiter and Saturn hypothesize that these giant planets moved inward significantly during the late stages of their formation, reaching 1.5 AU in the case of Jupiter (Walsh *et al.* 2011), almost certainly inward of the snowline (Fig. 4). If planetesimals in this region accreted volatiles on refractory organic and silicate surfaces, as in the Lodders (2004) idea, which were then incorporated into Jupiter, the latter would appear carbon-rich and oxygen-depleted. However, temperatures in that region may not have been low enough to provide sufficient amounts of the more volatile phases. Further, this seems inconsistent with the nitrogen isotopic measurements for Jupiter, which suggest different molecular carriers for nitrogen in Jupiter versus the terrestrial planet region.

A second possibility is that Jupiter's migration scattered these water poor planetesimals to the colder outer solar system, where they trapped noble gases and carbon-and nitrogen-bearing species at lower temperature–but water ice, already frozen out, was not available. Again, these models are of interest only should it turn out that the oxygen depletion measured by Galileo reflects a bulk oxygen depletion, rather than being atmospheric. Essential is the measurement of this quantity, in Jupiter and if possible, in Saturn.

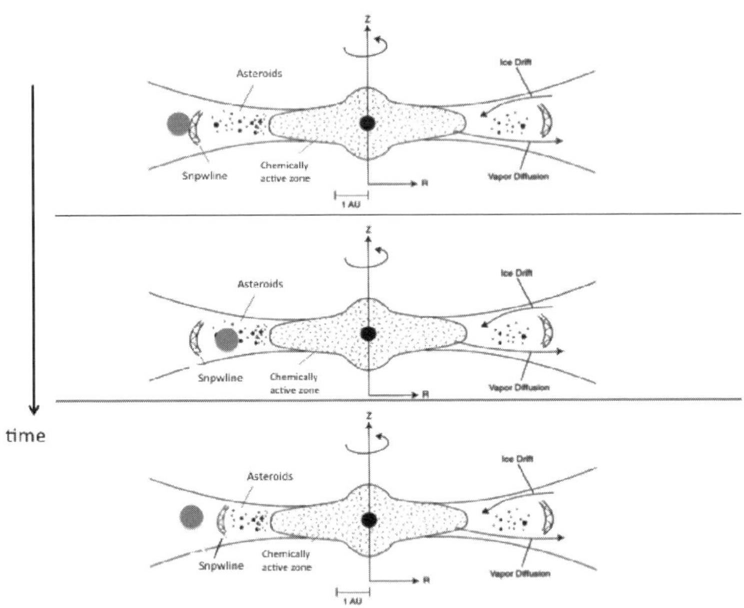

Figure 4. Schematic of the movement of Jupiter and the nebular snow line. In the top panel, Jupiter represented by the grey disk lies beyond the snowline, represented by the crescent. In the middle panel, later, Jupiter has moved inward of the snowline thanks to the influence of Saturn (Walsh *et al.* 2011). In the bottom panel, Jupiter has moved out again, while the snowline is moving inward at much slower pace (longer timescale). Base figure from Cyr *et al.* 1998

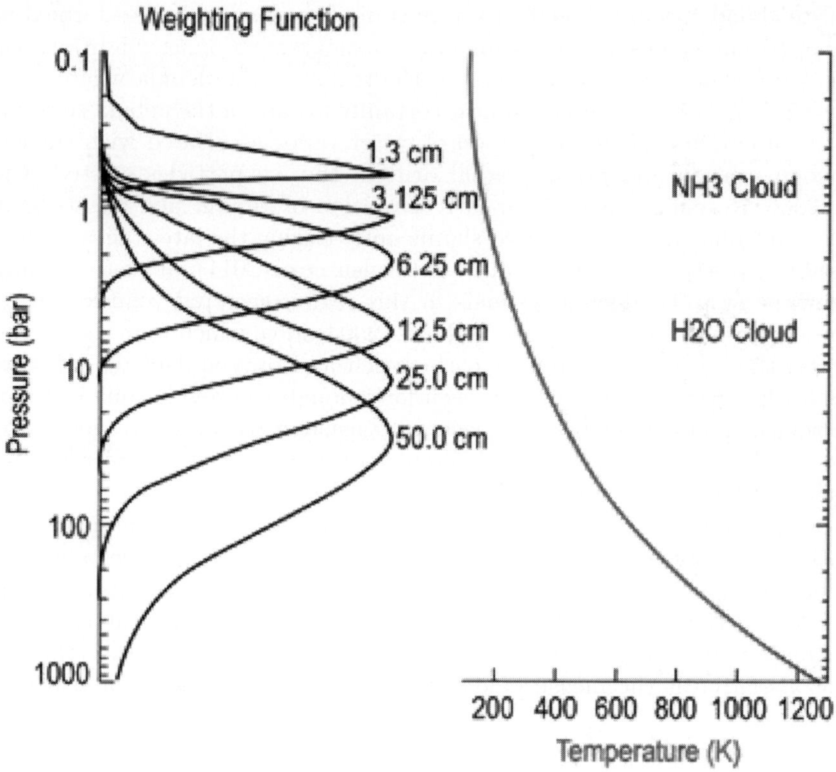

Figure 5. Weighting functions for the microwave radiometer on Juno, showing that the instrument is sensitive to water abundances at tens of bars bar pressure levels on Jupiter. Notional cloud decks are shown on the right based on the drawn temperature profile and the vapor pressures of the cloud formers. From Janssen *et al.* (2005).

4. The future: measuring oxygen in Jupiter and noble gases in Saturn

August 5, 2011 marked the launch of NASA's Juno mission to Jupiter, carrying eight instruments including two from the Italian Space Agency (ASI). Among the instrument suite is a microwave radiometer (MWR) built at NASA's Jet Propulsion Laboratory (Janssen *et al.*, 2005) with the capability of measuring the abundance of water down to many tens of bars below the Galileo limit (Fig. 5). Complementary to the MWR is a near-infrared spectrometer JIRAM (Jovian Infrared Auroral Mapper), built by ASI, that will obtain the water abundance in the meteorological layer (Adriani *et al.* 2008), thereby providing a cross-check for the water abundance measured at the shortest wavelengths by the MWR. The two instruments together will be able to provide a definitive answer for whether the water abundance is below or above solar, and in the latter case, by how much. Measurements will begin when Juno enters Jupiter orbit in summer 2016.

Further in the future is the possibility of sending a probe into the Saturn atmosphere akin to Galileo. Such a mission would complement the detailed remote sensing being done by the Cassini Saturn Orbiter, which beginning in 2016 will make very close flybys of Saturn similar to what Juno will do at Jupiter, to map the gravitational and magnetic fields, as well as examine the atmospheric composition at close range in the infrared. Unfortunately, a microwave radiometer akin to that on Juno is not present on

Figure 6. The author in front of a duplicate of the Galileo Probe mass spectrometer (on table) built by Hasso Niemann of the NASA Goddard Spaceflight Center. The detector is on the left; electronics in the center, and the upper housing on the right. Photo by M. Amato.

the Cassini Orbiter. Current strategic plans for the US planetary program (Space Studies Board 2011) limit Saturn Probe missions to the medium "New Frontiers" class, which means that a very deep probe (100 bars), will not be practical. Nonetheless, even flying a current version of the venerable Galileo Neutral Mass Spectrometer (Fig. 6) will provide a wealth of data to compare with that from Jupiter, and the possibility of deep probing of the Saturn atmosphere by microwave radiometry from a carrier, to complete the compositional diagnosis by obtaining the oxygen abundance, remains under study.

The fascinating possibility that Jupiter and Saturn might have distinct oxygen abundances because they sampled at different times and to differing extents regions of the nebula heterogeneous in oxygen (i.e., water) abundance is testable if a Saturn probe can measure the bulk oxygen abundance. In particular, should Saturn have a significantly higher C/O abundance than Jupiter (even if both were above solar), it would support the idea that the region inward of the nebular snowline was significantly depleted in water, giving us a unique insight into nebular condensation and transport processes. Complementary to this information are the "hot Jupiters" like WASP12-b which are easier targets for measuring C/O ratios, because their elevated temperature profiles allow both water and carbon-bearing species to be measured without interference from condensation processes (although cloud-forming species more refractory than water ice can reduce spectral contrast for some temperature ranges).

5. Implications for the origin of water on terrestrial planets

In this chapter I have pointed out that we know less about the bulk elemental abundance of the solar system's largest reservoir of planetary material than one might have otherwise assumed. Another area of deep ignorance is the origin of Earth's water, which recently was succinctly summarized (Cowen 2011). Dynamical simulations argue against a cometary origin for Earth's water (Lunine 2011), as does the high cometary D/H ratio (see this conference)–but with the caveat of Brown *et al.* (2011). This leaves local sources such as may be derived from adsorption onto grains (de Leeuw *et al.* 2010), or the asteroid belt. However, while meteorite water contents suggest there is plenty of water in main belt carbonaceous chondrite parent bodies, matching the geochemical constraints is difficult (see the discussion and references in Lunine *et al.* 2011). Others have suggested the possibility that the main belt comets might be examples of bodies from which the water was derived (Hsieh and Jewitt, 2006; see also the discussion in Lunine *et al.* 2011). However, we know little about these bodies.

The notion that Jupiter and Saturn might have moved significantly in terms of radial distance in the final stages of their formation makes it more likely that icy bodies from beyond the snowline were introduced into the terrestrial planet region, making the acquisition of water on the Earth easier. However, we do not have any information on the D/H values for water in this material, unless we think that these bodies are similar to comets, in which case D/H is not consistent with the terrestrial value. And the water adsorption hypothesis invokes material for which we have no geochemical information at all. It is a daunting problem.

Ultimately, there may be a connection between the problem of the elemental abundances of our giant planets and the sources of water for the Earth. The material that Jupiter and Saturn swept up in the late stages of their formation should be related to that which they threw outwards, to make the comets, and inward to seed the terrestrial planets with water and organics. And so determining the full set of elemental abundances in Jupiter and Saturn is a part of the process of testing hypotheses for how the Earth acquired its water. Whether there is enough information to make headway–and whether studies of the elemental abundances in extrasolar giant planets relative to their parent stars will reveal important systematics–will not be known for some time to come.

Acknowledgement

The talk was presented while the author was a Visiting Professor at the University of Rome, Tor Vergata, Rome Italy. Some of the work was financed within the scope of the program "Incentivazione alla mobilita' di studiosi straineri e italiani residenti all'estero." This work was also supported by NASA through the Juno mission to Jupiter.

References

Adriani, A., Coradini, A., Filacchione, G., Lunine, J. I., Bini, A., Pasqui, C., Colamai, L., Colosimo, F., Dinelli, B. M., Grassi, D., Magni, G., Moriconi, M. L., & Orosei, R. 2008 *Astrobiology*, 8, 613

Asplund, M., Grevesse, N., Jacques Sauval, A., & Scott, P. 2009 *ARAA*, 47, 481.

Brown, R. H. B., Lauretta, D. S., Schmidt, B., & Moores, J. 2011, *PSS*, in press.

Ciesla, F. J. & Carnley, S. B. 2006, in: D. S. Lauretta and H. Y.McSween (eds.), *Meteorites and the Early Solar System II* (Tucson: University of Arizona Press), p. 209.

Cowen, R. 2011 *Science News* January 15 , 2011, 26

Cyr, K. E., Sears, W. D., & Lunine, J. I. 1998 *Icarus*, 135, 537

de Leeuw, N. H., Catlow, R. A., King, H. E., Putnis, A., Muralidharan, K., Deymier, P., Stimpfl, M., & Drake, M. J. 2010 *Chem. Comm.*, 46, 8923.

dePater, I., Dunn, D., Romani, P., & Zahnle, K. 2001, *Icarus*, 149, 66
Fouchet, T., Moses, J. I., & Conrath, B. J. 2009, in: M. K. Dougherty, L. W. Esposito & S. M. Krimigis (eds.), *Saturn from Cassini-Huygens* (Dordrecht: Springer), p. 83
Grevesse, N., Asplund, M., & Sauval, A. J. 2005, in G. Alecian, O. Richard, S. Vauclair (eds.), *Element stratification of stars, 40 years of atomic diffusion* (Les Ulis: EDP Sciences), p. 21
Hanel, R. A., Conrath, B. J., Kunde, V. G., Pearl, J. C., & Pirraglia, J. A. 1983 *Icarus*, 53, 262
Hersant, F., Gautier, D., & Lunine, J. I. 2004, *Planetary and Space Science*, 52, 623
Hsieh, H. & Jewitt, D. C. 2006, *Science*, 312, 561
Hubbard, W. B., Dougherty, M. K., Gautier, D., & Jacobson, R. 2009, in: M. K. Dougherty, L. W. Esposito & S. M. Krimigis (eds.), *Saturn from Cassini-Huygens* (Dordrecht: Springer), p. 75
Janssen, M. A., Hofstader, M. D., Gulkis, S., Ingersoll, A. P., Allison, M. A., Bolton, S. J., Levin, S. M., & Kamp, L. W. 2005, *Icarus*, 173, 447
Krot, A. N., Fegley, B., Jr., Lodders, K., & Palme, H. 2000, in: V. Mannings, A. P. Boss & S. S. Ressell (eds.), *Protostars and Planets IV* (Tucson: University of Arizona Press), p. 1019.
Lodders, K. 2004, *ApJ*, 611, 587
Lodders, K. & Fegley, B. 1998, *Meteorit. Planet. Sci.*, 33, 871
Lunine, J. I., O'Brien, D. P., Raymond, S. N., Morbidelli, A., Quinn, T. & Graps. A. L. 2011 *Advanced Science Letters*4, 325.
Lunine, J. & Hunten, D. M. 1987, *Icarus*, 69, 566
Madhusudhan, N., Harrington, J., Stevenson, K., Nymeyer, S., Campo, C., Wheatley, P., Deming, D., Blecic, J., Hardy, R., Lust, N., Anderson, D., Collier-Cameron, A., Britt, C., Bowman, W., Hebb, L., Hellier, C., Maxted, P., Pollacco, D., & West, R. 2011, *Nature*, 64, 469
Madhusudhan, N., Mousis, O., Johnson, T. V., & Lunine, J. I.2011, *ApJ* in press
Mahaffy, P. R., Niemann, H. B., Alpert, A., Atreya, S. K., Demick, J., Donahue, T. M., Harpold, D. N., & Owen, T. C. 2000, *J. Geophys. Res.*, 105, 15,061.
Marty, B., Guillot, T., Coustenis, A., & 55 others 2008 *Experimental Astronomy* DOI 10.1007/s10686-008-9094-9
Mousis, O., Madhusudhan, N., Lunine, J. I., & Johnson, T. V. 2011, *ApJ*, submitted.
Mousis, O., Marboeuf, U., Lunine, J. I., Alibert, Y., Fletcher, L. N., Orton, G. S., Pauzat, F., and Ellinger, Y. 2009, *ApJ*, 696, 1348.
Niemann, H. B., Atreya, S. K., Carignan, G. R., Donahue, T. M., Haberman, J. A., Harpold, D. N., Hartle, R. E., Hunten, D. M., Kasprzak, W. T., Mahaffy, P. R., Owen, T. C., & Way, S. H. 1998, *J. Geophys. Res.*, 103, 22,831
Owen, T., Mahaffy, P. R., Niemann, H. B., Atreya, S., Donahue, T., Bar-Nun, A., & de Pater, I. 1999, *Nature*, 402, 269
Owen, T. & Encrenaz, T. 2003, *Space Science Rev.*, 106, 121
Proffitt, C. R. 1994, *ApJ*, 425, 849
Showman, A. P. & Dowling, T. E. 2000, *Science*, 289, 1737
Space Studies Board 2011 *Vision and Voyages for Planetary Science in the Decade 2013-2022* (Washington: National Academies Press)
Stevenson, D. J. & Salpeter, E. E.1977 *ApJ Suppl*, 35, 221
Wong, Von Zahn, U., Hunten, D. M., & Lehmacher, G. 1998 *JGR* 103, 22815
Walsh, K. J., Morbidelli, A., Raymond, S. N., O'Brien, D. P., & Mandell, A. M. 2011, *Nature*, 475, 206
Wilson, H. F. & Militzer, B. 2010, *Physical Review Letters*, 104, 147
Wong, M. H., Lunine, J. I., Atreya, S. K., Johnson, T., Mahaffy, P. R., Owen, T. C., & Encrenaz, T. 2008 *Reviews in Mineralogy & Geochemistry*. 68, 219.

Discussion

VAN DISHOECK: Can you elaborate on the similarities and differences between Main Belt comets and normal comets, in terms of composition, especially water/organics ratio?

LUNINE: To my knowledge we have no information on the composition of the Main Belt comets, certainly not the level of detail needed for the ratio of water to organics.

IRVINE: Do we know enough about cometary D/H to rule out traditional comets as the source of Earth's water?

LUNINE: Yes, one can rule out comets, if one assumes that the D/H value observed in the cometary comae are representative of the value in the bulk nucleus. Although it seems to be a reasonable assumption, there are several papers that argue D/H values may be significantly altered by vaporization processes (for example, Brown *et al.* 2011).

The Molecular Universe
Proceedings IAU Symposium No. 280, 2011
José Cernicharo & Rafael Bachiller, eds.

doi:10.1017/S1743921311025038

An Overview of Comet Composition

Dominique Bockelée-Morvan

LESIA, Observatoire de Paris, F92195, Meudon, France
email: dominique.bockelee@obspm.fr

Abstract. Comets are made of ices, organics and minerals that record the chemistry of the outer regions of the primitive solar nebula where they agglomerated 4.6 Gyr ago. Compositional analyses of comets can provide important clues on the chemical and physical processes that occurred in the early phases of Solar System formation, and possibly in the natal molecular cloud that predated the formation of the solar nebula. This paper presents a short review of our present knowledge of the composition of comets. Implications for the origin of cometary materials are discussed.

Keywords. comets : general, astrochemistry, Solar System: formation

1. Introduction

Comets are remnants of the planetesimals that formed the outer planets. Formed at large distances from the primordial Sun, they have remained for most of their lifetime outside of the orbit of Pluto, either in the trans-Neptunian scattered disk (associated to the Kuiper Belt) or in the Oort cloud. These two dynamically instable comet reservoirs are supplying the ecliptic short-period (also called Jupiter family comets, JFC) and long-period dynamical classes of comets observed in the inner Solar System. Though a variety of subtle evolutionary mechanisms operated for cometary nuclei during their long storage in the Oort cloud and Kuiper belt (Stern 2003) and, for short-period comets, during their many passages close to the Sun, the bulk composition of cometary nuclei is still regarded to be in large part pristine, except possibly for the most volatile ices. Thus, observing comets today provides a window through which we can view an earlier time when the planets were forming. Their composition provides a record of the chemistry of the primitive solar nebula, in the region and at the time of their formation.

Comets have been studied by a number of in situ investigations from spacecraft encounters (Keller *et al.* 2005), the most recent ones being the Deep Impact and EPOXI missions towards 9P/Tempel 1 and 103P/Hartley 2, respectively (A'Hearn *et al.* 2005, 2011) and the Stardust mission which collected and returned dust samples of comet 81P/Wild 2 to Earth for laboratory analyses (Brownlee *et al.* 2006). Although the in situ measurements have provided important new findings concerning comet composition, the majority of results on the icy component of cometary nuclei have been derived from remote spectroscopic observations of gas phase species in the atmosphere. This paper presents a short review of our present knowledge of the molecular and isotopic composition of comets, focussing on volatiles. Detailed reviews have been published by Bockelée-Morvan *et al.* (2005), Mumma & Charnley (2011). Implications on the origin of cometary materials are discussed in Section 6. More detailed discussions are presented in Rodgers & Charnley (2008), Ehrenfreund *et al.* (2005) and Irvine *et al.* (2000).

2. Molecular composition of Comets

2.1. *Molecular inventory of cometary volatiles*

Our present knowledge of the composition of cometary nuclear ices is essentially based on investigations of the coma that develops when comets are approaching the Sun and ices sublimate. Indeed, direct investigations of the nucleus surface (only possible from a spacecraft) only showed the near-infrared spectral signature of water ice on the nucleus of 9P/Tempel 1 (Sunshine *et al.* 2007), and a still unidentified feature at 2.39 μm on that of 19P/Borrelly observed from the Deep Space 1 spacecraft (Soderblom *et al.* 2002).

About two dozens molecules (not including isotopologues, molecular ions, atoms and radicals) have now been identified in cometary atmospheres (Fig. 1). These identifications were obtained through millimetre, submillimetre and infrared spectroscopy, mass spectrometry aboard the Giotto and Vega spacecraft, and, at a lesser extent, using ultraviolet spectroscopy (see the review of spectroscopic identifications of Bockelée-Morvan *et al.* 2005). Most identifications were obtained in the exceptionally bright comets C/1995 O1 (Hale-Bopp) and C/1996 B2 (Hyakutake). Observations in the visible wavelength range sample signatures of radicals, atoms and ions that are photolysis and chemical products of parent molecules released from the nuclear ices.

The main components of cometary nuclear ices are water (about 80 % by number) followed by CO, CO_2, CH_3OH, CH_4, H_2S and NH_3 (Fig. 1). A recent investigation of the CO_2 abundance in 17 comets using the AKARI space telescope shows that CO_2 dominates over CO in most comets (Ootsubo *et al.* 2010; Hamada *et al.* 2010). Abundances of detected cometary parent molecules range from less than 0.01 % to 20 % relative to water, and generally decrease with increasing complexity, except for hydrocarbons. Complex species, such as methyl formate and ethylene glycol, are detected. The cumulative histogram of the number of detected species, N, as a function of their abundance, X, suggests that observational biases affect the detection of species with abundances $< 10^{-3}$ and that a large number of molecules remains to be identified in cometary atmospheres (Crovisier *et al.* 2004).

2.2. *Chemical diversity*

According to current theories, comet formation in the solar nebula extended over a wide range of heliocentric distances for both Oort cloud comets, and Jupiter family comets issued from the trans-neptunian scattered disk. The current vue is that the Uranus-Neptune zone is the main source region of Oort cloud comets, with some contribution from the Jupiter-Saturn region, while Jupiter family comets formed beyond Neptune. If comets originated from distinct regions of the solar nebula, they could display diversity in their chemical composition depending on the local temperature and nebular composition where they formed. If there was significant mixing of nebular material across large ranges in heliocentric distance, even individual cometary nuclei could exhibit chemical inhomogeneity.

Chemical diversity among comets is indeed observed for both parent volatiles and daughter species. From a study of radicals (OH, CN, C_2, C_3, NH) in 85 comets, A'Hearn *et al.* (1995) proposed the existence of two classes of comets, depending on their C_2 and C_3 abundances: "typical" comets and "carbon-depleted" comets. They found that about one-half of the Jupiter-family comets are C_2 and C_3-depleted, but the fraction of carbon-depleted Oort cloud comets is much smaller. An update of this study confirming these two compositional classes is presented by Schleicher (2008). Note that the parent of C_2 radical is likely C_2H_2 (via C_2H), but the origin of C_3 remains unclear (Helbert *et al.* 2005).

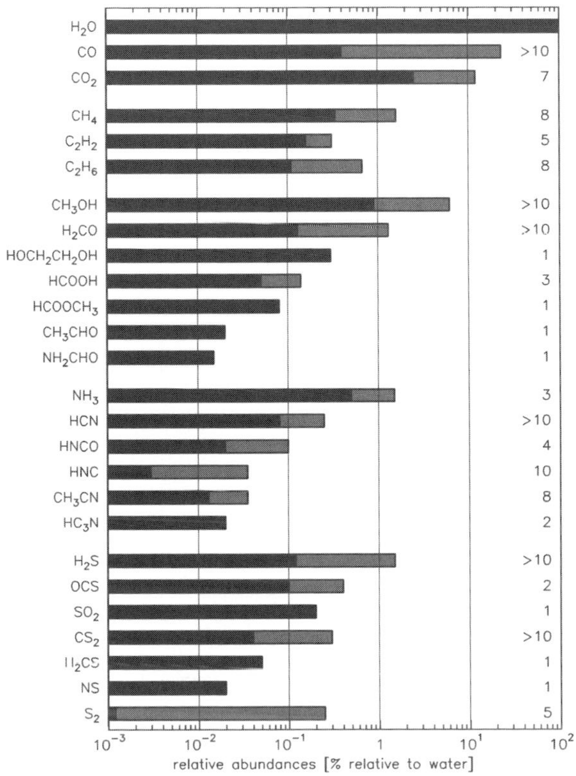

Figure 1. Abundances relative to water of cometary parent molecules. The range of measured values is shown in the gray (red in the color version) portion. The number of comets in which the molecule has been detected is given in the right (adapted from Bockelée-Morvan *et al.* (2005), courtesy J. Crovisier).

Surveys of parent volatile abundances show strong evidence for chemical diversity among comets. Depending on the molecule, abundances vary by a factor of a few to several tens (for CO) among comets (Fig. 1,2). Diversity is observed both within the population of Oort cloud comets and Jupiter family comets (Fig. 2). No obvious correlation is observed between the abundances of the hypervolatile species and the dynamical class. No trend between volatility and comet-to-comet variability is observed either (Biver *et al.* 2002; Gibb *et al.* 2003; Disanti & Mumma 2008). Three compositional classes, based on the organic composition (HCN, C_2H_2, C_2H_6 and CH_3OH), are possibly emerging (Mumma *et al.* 2008). Enriched and severely organic-depleted comets are present in both reservoirs (Mumma *et al.* 2008).

So far, Jupiter family comets have been still poorly investigated, because of their relatively low level of activity. A number of molecules have only been detected in comet C/1995 O1 (Hale-Bopp) or in a small number of comets.

Questions arise whether the molecular abundances measured in the coma are representative of those inside the nucleus, and to which extent the nucleus composition has been altered since comet formation. The chemistry could have evolved both during the long storage within the Oort cloud or Kuiper belt as well as during close passages to the Sun after removal from these reservoirs. Evolutionary effects are most important in the outer layers of a comet while the interior of the nucleus should be in large part

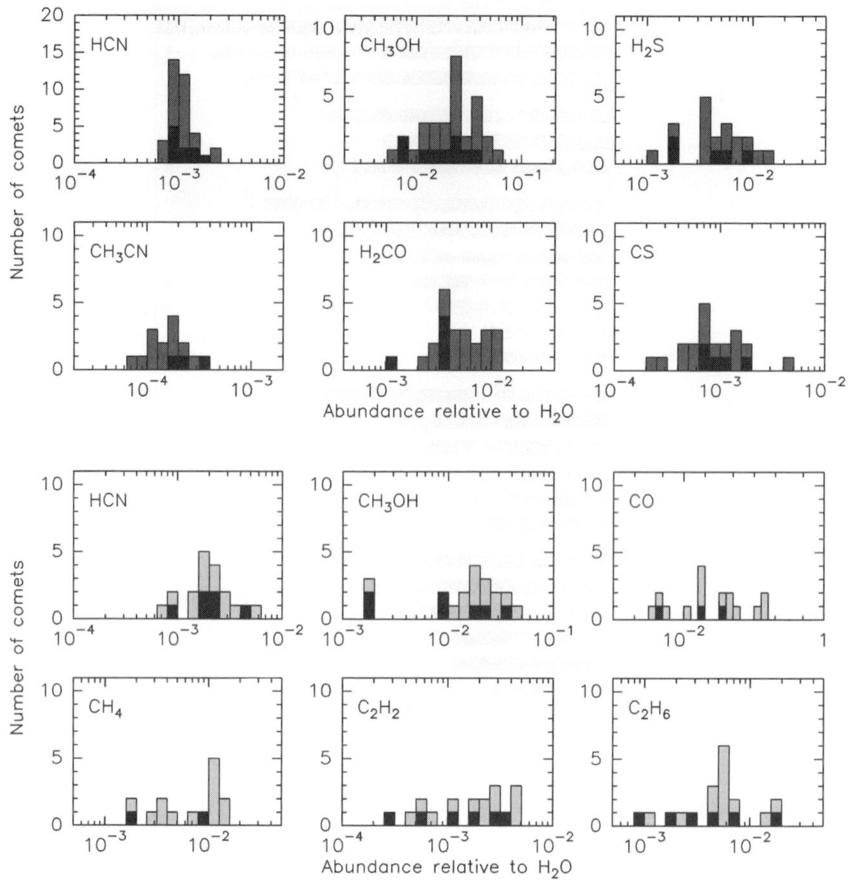

Figure 2. Histogram of molecular abundances relative to water from radio observations (top, Crovisier *et al.* 2009) and IR observations (bottom). The histogram for Jupiter family comets is in black.

pristine except for perhaps the most volatile species. The consistency of the volatile composition of 73P/Schwassmann-Wachmann 3 as a function of date and in its individual fragments suggests that evolutionary effects are small (Dello Russo *et al.* 2007). This might be related to the very low thermal conductivity of surface layers (Groussin *et al.* 2007; Gortsas *et al.* 2011). The fact that short-period and long-period comets display similar composition diversity, even in highly volatile species (the exception being CO, which is found abundant only in long-period comets) argues for the composition diversity to be largely pristine.

3. Isotopic composition of volatiles

Isotopic ratios are important diagnostics of how cometary materials formed because isotopic fractionation is very sensitive to chemical and physical conditions. So far, only a few isotopic ratios have been measured in cometary gases as summarized in Table 1 (see Jehin *et al.* 2009, for an extended review).

HDO has been observed from submillimetre spectroscopy in the bright comets Hale-Bopp and C/1996 B2 (Hyakutake), from which a D/H ratio of 3×10^{-4} consistent with

Table 1. Isotopic abundances in cometary volatiles

Isotopic ratio	Species	Value	Comet	References
D/H	[H_2O]	$(3.06\pm0.34)\times 10^{-4}$	1P/Halley	[1] [2]
		$(2.9\pm1.0)\times10^{-4}$	C/1996 B2 (Hyakutake)	[3]
		$(3.3\pm0.8)\times10^{-4}$	C/1995 O1 (Hale-Bopp)	[4]
		$(2.5\pm0.7)\times10^{-4}$	C/2002 T7 (LINEAR)	[5]
		$(4.1\pm1.5)\times10^{-4}$	8P/Tuttle	[6]
		$(4.6\pm1.4)\times10^{-4}$	C/2001 Q4 (NEAT)	[7]
		$< 2.5\times10^{-4}$	153P/Ikeya-Zhang	[8]
		$(1.61\pm0.24)\times10^{-4}$	103P/Hartley 2	[9]
	[HCN]	$(2.3\pm0.4)\times10^{-3}$	C/1995 O1 (Hale-Bopp)	[10]
$^{12}C/^{13}C$	[C_2]	93 ± 10	4 comets	[11]
	[CN]	91 ± 4	21 comets	[12]
	[HCN]	111 ± 12	C/1995 O1 (Hale-Bopp)	[13]
		114 ± 26	17P/Holmes	[14]
$^{14}N/^{15}N$	[HCN]	205 ± 70	C/1995 O1 (Hale-Bopp)	[14]
		139 ± 26	17P/Holmes	[14]
	[CN]	141 ± 29	21 comets	[12]
		220 ± 40	73P (fragment C)	[12]
$^{16}O/^{18}O$	[H_2O]	518 ± 45	1P/Halley	[2]
		470 ± 40	1P/Halley	[1]
		530 ± 60	4 comets	[15]
		425 ± 55	C/2002 T7 (LINEAR)	[5]
$^{32}S/^{34}S$	[CS]	27 ± 3	C/1995 O1 (Hale-Bopp)	[13]
	[CS]	16 ± 3	17P/Holmes	[16]
	[S^+]	23 ± 6	1P/Halley	[17]
	[H_2S]	16 ± 3	C/1995 O1 (Hale-Bopp)	[18]

[1] Eberhardt *et al.* (1995); [2] Balsiger *et al.* (1995); [3] Bockelée-Morvan *et al.* (1998); [4] Meier *et al.* (1998a); [5] Weaver *et al.* (2011); [6] Villanueva *et al.* (2009); [7] Hutsemékers *et al.* (2008); [8] Biver *et al.* (2006); [9] Hartogh *et al.* (2011); [10] Meier *et al.* (1998b); [11] Wyckoff *et al.* (2000); [12] Manfroid *et al.* (2009); [13] Jewitt *et al.* (1997); [14] Bockelée-Morvan *et al.* (2008); for Hale-Bopp, revised value from the data set of Jewitt *et al.* (1997) and Ziurys *et al.* (1999); [15] Biver *et al.* (2007); [16] Biver *et al.* (2008); [17] Altwegg (1996); [18] Crovisier *et al.* (2004); .

Giotto measurements in comet 1P/Halley has been derived (Table 1). More recently, similar values were obtained in other Oort cloud comets from observations of D, OD, and HDO in the UV, near UV, and near infrared, respectively. This is about twice higher than the D/H value in terrestrial water, and approximately one order of magnitude larger than the protosolar value in H_2 ($\sim 2\times10^{-5}$) (Fig. 3). The upper limit of 2.5×10^{-4} measured in comet 153P/Ikeya-Zhang might suggest that the D/H ratio is variable among comets (Biver *et al.* 2006). DCN has been detected in comet Hale-Bopp, yielding a D/H ratio in HCN seven times higher than that measured in cometary water (Table 1). Measurements of deuteration in other molecules have been attempted (Crovisier *et al.* 2004). An upper limit of 5×10^{-3} was derived for the D/H in CH_4 (Bonev *et al.* 2009; Kawakita and Kobayashi 2009). Deuterium enrichments are characteristic of interstellar or protosolar chemistry, and result from ion-molecule or grain-surface reactions at low temperature (Ehrenfreund *et al.* 2005).

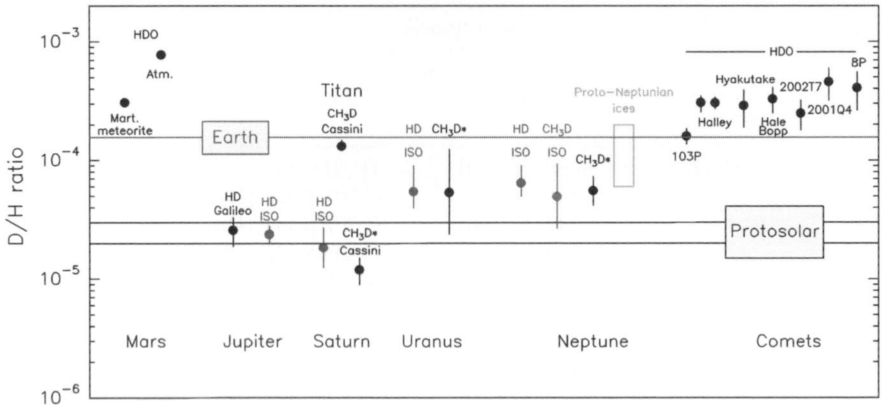

Figure 3. D/H in the Solar System.

The detection of deuterated molecules in the atmospheres of Jupiter family comets is difficult because of their faintness. HDO was detected in the JFC 103P/Hartley 2 with Herschel, from which a D/H value equal to the Earth value was derived (Hartogh *et al.* 2011). Hence, all comets do not share the same D/H.

Measurements have been obtained for the $^{12}C/^{13}C$ ratio in C_2, CN and HCN, for the $^{32}S/^{34}S$ ratio in CS, S^+, H_2S, and for the $^{16}O/^{18}O$ ratio in water: in first approximation they are consistent with terrestrial values ($^{12}C/^{13}C = 89$; $^{32}S/^{34}S = 23$; $^{16}O/^{18}O = 499$). Some measurements suggest an excess in ^{18}O and ^{34}S (Table 1).

The $^{14}N/^{15}N$ ratio has been measured in the CN radical in ~20 comets (Arpigny *et al.* 2003; Hutsemékers *et al.* 2005; Manfroid *et al.* 2005; Jehin *et al.* 2004; Manfroid *et al.* 2009). The average value is $^{14}N/^{15}N = 148 \pm 6$, with no significant differences emerging between Oort cloud and Jupiter family comets. Only comet 73P/Schwassmann-Wachmann 3 presents a value which is out of the typical range (210±50 and 220±40, for fragment B and C, respectively, Manfroid *et al.* 2009). The average value corresponds to a factor of two enrichment in ^{15}N with respect to the Earth atmospheric value $^{14}N/^{15}N = 272$. The $^{14}N/^{15}N$ ratio was measured in HCN in comets C/1995 O1 (Hale-Bopp) and 17P/Holmes, belonging to the Oort cloud and JFC classes, respectively. For comet Hale-Bopp, an $HC^{14}N/HC^{15}N$ ratio marginally higher than the terrestrial value was reported (Jewitt *et al.* 1997; Ziurys *et al.* 1999), but a reanalysis of the data suggests instead a value consistent with the value in CN (Bockelée-Morvan *et al.* 2008). The $HC^{14}N/HC^{15}N$ ratio measured in comet 17P/Holmes is also consistent with the value measured in CN (Table 1)(Bockelée-Morvan *et al.* 2008). There is some observational evidence that HCN photolysis is not the sole production mechanism of CN radicals. Indeed, for some comets, the CN production rate exceeds the HCN production rate (Fray *et al.* 2005). CN radicals are also produced by HC_3N and CH_3CN but those species are minor constituents of cometary atmospheres (Fig. 1). This led to the hypothesis that significant amounts of CN radicals could be produced by the thermal degradation of refractory organics present in dust grains. The similar isotopic properties of CN and HCN point to HCN being the prime parent of CN in cometary atmospheres, or imply that refractory organics producing CN radicals are equally enriched in ^{15}N.

^{15}N excesses observed in CN and HCN are similar to values measured in carbonaceous meteorites and Interplanetary Dust Particles (IDPs) (Floss *et al.* 2006; Busemann *et al.*

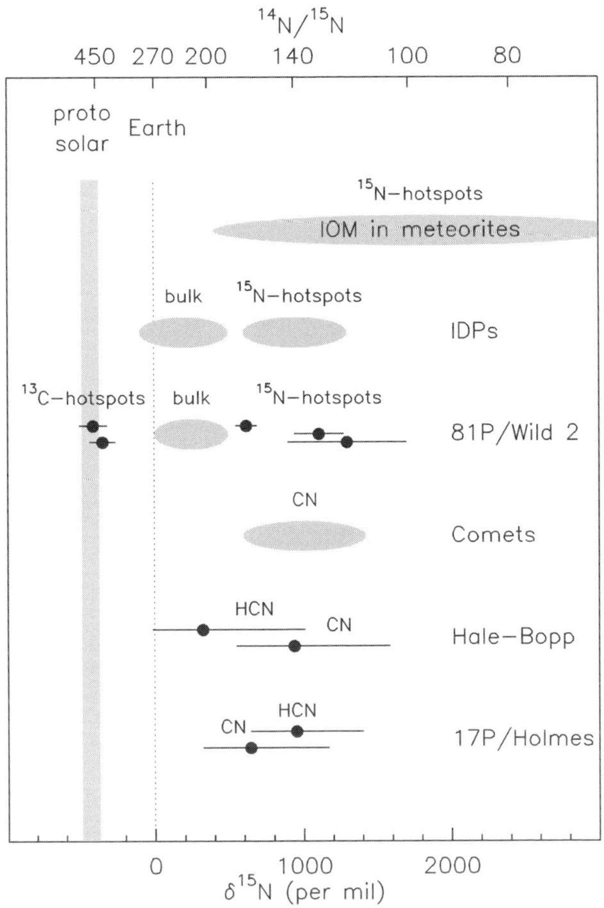

Figure 4. N isotope measurements in primitive matter of the Solar System, given either as $^{14}N/^{15}N$ ratios or $\delta^{15}N$ values ($\delta^{15}N$ = ($^{14}N/^{15}N$) /($^{14}N/^{15}N)_{Earth}$-1). Shown are the protosolar value measured by the Genesis mission (Marty *et al.* 2011), values for comet 17P/Holmes (Bockelée-Morvan *et al.* 2008), comet Hale-Bopp (Arpigny *et al.* 2003; Bockelée-Morvan *et al.* 2008) and other comets (Jehin *et al.* 2009), dust particles of comet 81P/Wild 2 collected by the Stardust mission (McKeegan *et al.* 2006), interplanetary dust particles IDPs (Floss *et al.* 2006), and insoluble organic matter (IOM) in carbonaceous meteorites(Busemann *et al.* 2006). So–called hotspots are regions that present strong isotopic enrichments relative to the surrounding material.^{13}C hotspots refer to 81P/Wild 2 submicrometre grains enriched in ^{13}C with respect to the terrestrial $^{12}C/^{13}C$ ratio and depleted in ^{15}N. Points with error bars correspond to single measurements. When many measurements are available, ellipses encompassing the range of measured values are drawn. Adapted from Bockelée-Morvan *et al.* (2010).

2006)(Fig. 4). ^{15}N excesses are also present at the submicrometre scale in the dust particles collected by the Stardust mission in comet 81P/Wild 2 (McKeegan *et al.* 2006). Altogether, it indicates that N-bearing volatiles in the solar nebula underwent important N isotopic fractionation at some stage of Solar System formation.

4. Ortho-to-para ratios

Molecules with H atoms at symmetrical positions may exist in different nuclear-spin species, e.g. ortho and para species for molecules with two H atoms. The ortho/para

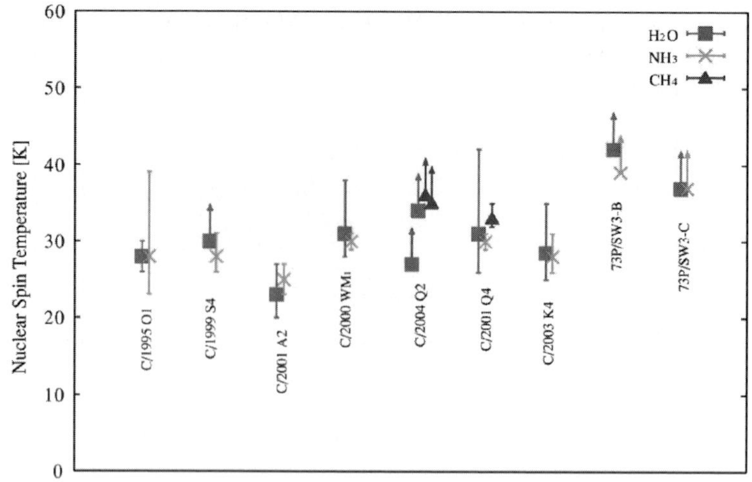

Figure 5. Comparison of nuclear spin temperatures of H_2O, NH_3 and CH_4 observed in comets. From Shinnaka *et al.* (2011).

ratio (OPR) in water and ammonia has been measured in a number of comets (see e.g., Kawakita *et al.* 2004, Bonev *et al.* 2007, Shinnaka *et al.* 2011). Measurements are also available for CH_4 (Kawakita *et al.* 2004) and CH_3OH (Pardanaud *et al.* 2007). Most measurements indicate a spin temperature within 25–35 K, though some comets exhibit higher spin temperatures, corresponding to equilibrated OPR. Interestingly, water, ammonia and methane shares the same nuclear spin temperature (Fig. 5). No relationship was found between the spin temperature and the comet dynamical origin, or with the heliocentric distance at which the comets were observed, and their volatile composition. However, there may be some link between the $^{14}N/^{15}N$ in CN and the spin temperature of NH_3 to be confirmed in the future: fragments B and C of comet 73P/Schwassmann-Wachmann 3 displays a $^{14}N/^{15}N$ ratio distinct from the average value (Sect. 3) and an equilibrated OPR in NH_3.

It is often argued in the literature that un-equilibrated spin temperatures have a meaning related to the temperature environment of the species in the nucleus or even before, i.e., they could indicate that molecules formed at ~30 K or were last processed at about 30 K (e.g., Shinnaka *et al.* 2011). However the real signification of the spin temperature is debated (Crovisier 2007). Buntkowsky *et al.* (2008) show, by quantum mechanics calculations, that the inter-conversion between para and ortho H_2O states is very fast inside the ice, and argue that the actual spin temperatures measured in cometary atmospheres depend both on the ice temperature and of the evaporation history of the ice. Thermal desorption is more efficient for para-water than for ortho-water: Dulieu *et al.* (2011) have estimated to $\sim$ 2.4 ($T_{spin} \sim$ 28 K) the OPR in water set by this process. Further work (both experiments and models) is needed to validate those interpretations with the constraint that the similarity in the spin temperatures of H_2O, NH_3 and CH_4 should be explained.

5. Dust composition

First insights on the composition of the refractory component of cometary nuclei were obtained from in situ measurements in comet 1P/Halley. Mass spectrometry analysis of cometary dust revealed the presence of both rocky (Mg, Si, Fe, Ca) and organic refractory particles, the so-called CHON organic grains (Jessberger *et al.* 1988). Amorphous carbon grains were also identified. Altogether, the dust grains of 1P/Halley were found to have a chondritic CI-like elemental composition for the most refractory elements, whereas enrichments are observed for the light elements C, H, N (Jessberger *et al.* 1988). The concentrations of C, H, O and N could not be determined in the dust grains of comet 81P/Wild 2 collected by Stardust but most other elements were also found to be in chondritic abundance (Flynn *et al.* 2006). A notable exception is sulfur, which was found to be depleted. Whether this depletion is real or reflects capture heating is debated (Zolensky *et al.* 2006).

The mid and far-IR spectra of comets show that silicates in comets are dominated by Mg-rich olivines and pyroxenes and are in two forms : amorphous and crystalline (Hanner *et al.* 2005). Crystalline silicates are grains that condense at high temperatures or can be formed from the annealing of amorphous silicates at somewhat lower temperature. The scarcity of the crystalline silicates in the interstellar medium indicates that these compounds formed in the inner hot regions of the solar nebula. Other high-temperature minerals are present in comet Wild 2 Stardust samples, such as Calcium Aluminium-rich inclusions (CAIs) or iron sulfides (Zolensky *et al.* 2006). Amorphous silicates are presumably relics of the presolar cloud and analogous to the glassy silicates grains (GEMS) found in IDPs. GEMS-like material is found in Stardust but was possibly created during capture (Ishii *et al.* 2008).

Our knowledge of the nature of the refractory organic matter is rather limited. Based on the mass spectra recorded by the Giotto and Vega spacecraft, a number of compounds have been hypothesized (see the review of Cottin *et al.* 1999). The presence of formaldehyde polymers is proposed to explain the extended distribution of formaldehyde in cometary atmospheres (Cottin *et al.* 2004). Spectroscopic studies of the 3.4 μm C-H signature of Stardust samples indicate that the aliphatic chains of the organics are longer than in carbonaceous meteorites, and in the diffuse interstellar medium. They are also significantly longer than in IDPs (Keller *et al.* 2006). The aromatic fraction of the total organic matter is small, but is rich in both oxygen and nitrogen compared with meteoritic insoluble organic matter (Sandford *et al.* 2006; Clemett *et al.* 2010). Glycine and PAHs have been identified in Stardust samples (Elsila, Glavin, & Dworkin 2009; Clemett *et al.* 2010).

6. Implications for the origin of cometary material

The identification of interstellar-like organic and other exotic gases sublimated from cometary nuclei gives the definite impression that comets preserve a record of the interstellar composition characteristic of the presolar cloud, or that cometary and interstellar molecules formed by similar processes (Irvine *et al.* 2000; Bockelée-Morvan *et al.* 2000; Ehrenfreund *et al.* 2005). The gas phase composition in the interstellar medium depends on many variables, but almost every parent species observed in comets has also been detected in the interstellar medium, either in cold dark molecular clouds or in dense cloud cores where star formation is taking place. The molecular composition of comets, interstellar ices, hot cores and hot corinos are all characterized by the presence of large quantities of saturated/hydrogenated species (e.g. H_2O, CH_4, CH_3OH, NH_3, H_2S), coexisting with oxidized molecules (as CO, CO_2 and SO_2). For interstellar ices, it is believed that this composition traces primarily a combination of ion-molecule and grain-surface processes at low temperature. Complex organic species can then form through energetic processing by UV photons and cosmic rays. Complex species are not abundant enough

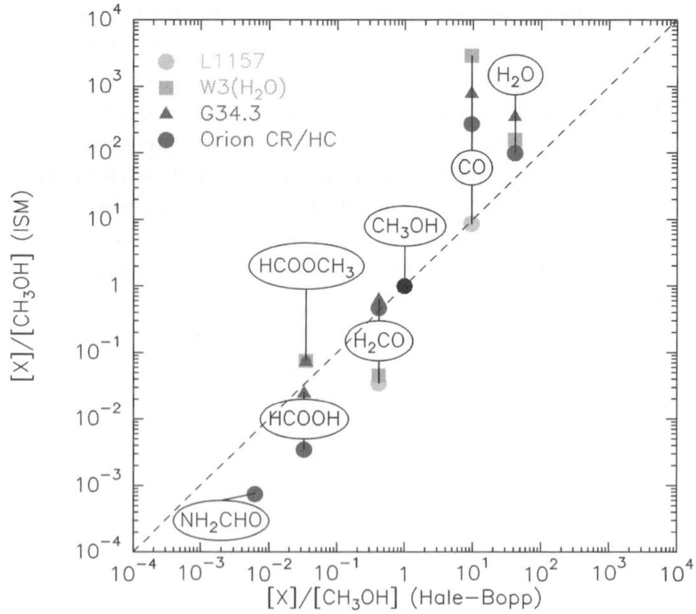

Figure 6. Molecular abundances in comet Hale-Bopp (CHO-bearing species) compared to those in the bipolar flow L1157 and several hot molecular core. From Bockelée-Morvan *et al.* (2000).

to be detected in interstellar ices, but are found in the hot cores containing massive protostars, or in the 'hot corinos' around low-mass protostars, where the hot environment allows the icy grain mantles to be evaporated. Figure 6 illustrates a quantitative similarity between interstellar and cometary volatile material. Discrepancies do exist. For example, the cometary abundance of ethylene glycol relative to methanol is much larger than measured in the hot core Sgr (B2) (Crovisier *et al.* 2004). The cometary abundance of methanol is generally much smaller when compared to interstellar ices (Ehrenfreund *et al.* 2005).

Another indicator of interstellar-like chemistry is the D/H ratio, as already discussed in Sect.3. However there is evidence that cometary D/H values are lower than in most interstellar sources, in particular around protostars where extreme deuteration is observed (Ceccarelli *et al.* 2007; Bergin *et al.* 2010). This low deuterium fractionation in comets is interpreted by ion-molecule chemistry in gas phase at relatively high temperature (25–35 K, Meier *et al.* 1998a, 1998b) or, alternatively, by mixing of D-enriched material with material reprocessed in the inner solar nebula (Hersant, Gautier, & Huré 2001). Ion-molecule reactions in the outer solar nebula could also act to lower the interstellar fractionation (Aikawa and Herbst 1999). Extreme D-enrichments are observed in primitive meteorites and IDPs. It was previously thought that these enrichments were reflecting those acquired in the presolar cloud. The negative correlation between the D/H ratios of the CH bonds and their binding energy in the Insoluble Organic Matter of Orgueil carbonaceous meteorite suggests that deuterium enrichments were acquired at the surface of the protosolar disk through intense UV irradiation from the proto-Sun (Remusat *et al.* 2006). Whether this mechanism affected the D-fractionation observed in cometary HCN and H_2O has still to be investigated. It is interesting to note that the cometary dust particles collected by Stardust are also less D-enriched than carbonaceous meteorites and IDPs (McKeegan *et al.* 2006).

The lower D/H value found in the water of the Jupiter family comet 103P/Hartley 2 with respect to Oort cloud comets was not anticipated by models which consider isotopic exchanges in the solar nebula (Hersant, Gautier, & Huré 2001; Kavelaars *et al.* 2011). Indeed, these models predict that comets formed at higher distances from the Sun should exhibit higher D/H ratios as they incorporated less processed material from the inner solar nebula. Further measurements of the D/H ratio in the two dynamical families are required to investigate whether a dichotomy is present, or whether the two families display an isotopic diversity. The current understanding of deuteration in different solar system reservoirs is clearly incomplete.

The isotopic composition of the main carrier of nitrogen in the solar nebula (N_2) has been recently evaluated to $^{14}N/^{15}N = 441 \pm 6$ from the study of solar wind ions sampled by the Genesis mission (Marty *et al.* 2011), a value close to measurements in the atmosphere of Jupiter (Fouchet *et al.* 2004). The interpretation of the ^{15}N enrichment in cometary HCN and CN (a factor of 3 with respect to the protosolar value) is not as compelling as for deuterium. There is indeed still little evidence for N isotopic fractionation in the interstellar medium (Ikeda *et al.* 2002; Gerin *et al.* 2009; Lis *et al.* 2010), though recent assessments of the N isotopic fractionation for starless and prestellar cores at various stages of evolution are promising (Milam & Charnley al. 2011). Rodgers & Charnley (2008) showed that interstellar chemistry can potentially produce ^{15}N enrichments in both nitrile and amine groups. On the other hand predicted ^{15}N enhancements in HCN for exchange reactions in the solar nebula involving the main nitrogen reservoir N_2 are modest (Terzieva & Herbst 2000). Alternative mechanisms include a nucleosynthetic origin and photochemical self-shielding in the solar nebula, as proposed for explaining the oxygen isotope anomalies in meteorites. Irrespective of the mechanism, protosolar HCN never isotopically equilibrated with nebular gas at later phases of Solar System evolution.

The molecular and isotopic compositions of cometary volatiles suggest an origin in chemistry at low temperature. Whether this occurred in the cold molecular cloud that predated the Solar System or in the outer regions of the solar nebula cannot be decided based on the available data. However, the composition of cometary dust shows that cometary nuclei also incorporated nebular products formed in the hot inner regions of the solar nebula, revisiting previous views according to which comets could be pure interstellar condensates (Greenberg 1982). By some process (e.g., turbulence, see Bockelée-Morvan *et al.* 2002, Ciesla 2007) high-T temperature materials were transported out to the region of formation of comets before their agglomeration, indicating that the solar nebula underwent large-scale radial mixing at early times. Surprisingly, the most recognizable characteristics of presolar material were found to be rare in Stardust samples compared to their occurrence in chondritic porous IDPs, believed to be of cometary origin (Ishii *et al.* 2008). Only a very few numbers of presolar grains identified by non-solar isotopic compositions have been found (McKeegan *et al.* 2006). Altogether, the mineralogical and isotopic composition of comet Wild 2 grains suggests that its refractory material more closely resembles an inner Solar System asteroid than an outer solar system comet with primitive unaltered dust. Likely, comet Wild 2 is not representative of the whole population of comets. Cometary nuclei may present a large diversity in their refractory composition, as observed for their icy composition, relevant to the place and time of their formation in the solar nebula.

Progresses in our knowledge of comet composition will benefit from future space missions to comets (e.g., the Rosetta mission of ESA) and from telescopic observations with yet available or new instrumentation (e.g., the Herschel space observatory, ALMA). Our understanding of comet composition in the context of the origin and evolution of our Solar System will greatly benefit from theoretical models of chemical and physical processes in the solar nebula, and from upcoming new results on the composition of protostellar sources and protoplanetary disks at different stages of evolution.

References

A'Hearn, M. F., Millis, R. L., Schleicher D. G., Osip D. J., & Birch P. V. 1995, *Icarus* 118, 223
A'Hearn, M. F., Belton, M. J. S., Delamere, W. A., *et al.* 2005, *Science* 310, 258
A'Hearn, M. F., Belton, M. J. S., Delamere, W. A., *et al.* 2011, *Science* 332, 1396
Aikawa, Y. & Herbst, E. 1999, *ApJ* 526, 314
Altwegg, K. 1996, *Habilitationsschrift*, University of Bern
Altwegg, K. & Bockelée-Morvan, D. 2003, *SSRv* 106, 139
Arpigny, C., Jehin, E., Manfroid, J., *et al.* 2003, *Science* 301, 1522
Balsiger, H., Altwegg, K., & Geiss, J. 1995, *JGR* 100, 5827
Bergin E. A., Phillips, T. G. & Comito, C., *et al.* 2010, *A&A* 521, L20
Biver, N., Bockelée-Morvan, D., Crovisier, J., *et al.* 2002, *Earth, Moon, and Planets* 90, 323
Biver, N., Bockelée-Morvan, D., Crovisier, J., *et al.* 2006, *A&A* 449, 1255
Biver, N., Bockelée-Morvan, D., Crovisier, J., *et al.* 2007, *P&SS* 55, 1058
Biver, N., Bockelée-Morvan, D., Crovisier, J., *et al.* 2008, *LPI Contrib.* 1405, 8149
Bockelée-Morvan, D. 2010, *EAS* 41, 313
Bockelée-Morvan, D., Lis, D. C., Wink, J. E., *et al.* 2000, *A&A* 353, 1101
Bockelée-Morvan, D., Gautier, D., Hersant, F., Huré, J.-M., & Robert, F. 2002, *A&A* 384, 1107
Bockelée-Morvan, D., Crovisier, J., Mumma, M. J., & Weaver, H. A. 2005, in: M.C. Festou, H.U. Keller, & H.A. Weaver (eds), *Comets II* (Univ. of Arizona Press: Tucson), p. 391
Bockelée-Morvan, D., Biver, N., Jehin, E., *et al.* 2008, *ApJ* 679, L49
Bonev, B. P., Mumma, M. J., Villanueva, G. L., *et al.* 2007, *ApJ* 661, L97
Bonev, B. P., Mumma, M. J., Gibb, E. L., *et al.* 2009, *ApJ* 699, 1563
Brownlee, D., Tsou, P., Aléon, J., *et al.* 2006, *Science* 314, 1711
Buntkowsky, G., Limbach, H. H., Walaszek, B., *et al.* 2008, *Z. Phys. Chem.* 222, 1049
Busemann, H., Young, A. F., O'D. Alexander, C. M., *et al.* 2006, *Science* 312, 727
Ceccarelli, C., Caselli P., Herbst E., Tielens A. G. G. M., Caux E. 2007, in: B. Reipurth, D. Jewitt, & K. Keil (eds), *Protostars and Planets V* (University of Arizona Press: Tucson), p. 47
Ciesla, F. J. 2007, *Science* 318, 613
Clemett, S. J., Sandford S. A., Nakamura-Messenger K., Hörz F., McKay D. S. 2010, *M&PS* 45, 701
Cottin, H., Bénilan, Y., Gazeau, M.-C., & Raulin, F. 2004, *Icarus* 167, 397
Cottin, H., Gazeau, M. C., & Raulin, F. 1999, *P&SS* 47, 1141
Crovisier, J. 2007, in: Proceedings of the XVIIIemes Rencontres de Blois: Planetary Science: Challenges and Discoveries, astro-ph/0703785
Crovisier, J., Bockelée-Morvan D., Colom P., *et al.* 2004, *A&A* 418, 1141
Crovisier J., Biver N., Bockelée-Morvan D., *et al.* 2009, *EM&P* 105, 267
Dello Russo N., Vervack R. J., Weaver H. A., *et al.* 2007, *Nature* 448, 172
Disanti M. A., Mumma M. J. 2008, *SSRv* 138, 127
Dulieu, F. 2011, IAU Symposium 280, this book
Eberhardt, P., Meier, R., Krankowsky, D., & Hodges, R. R. 1994, *A&A* 288, 315
Ehrenfreund, P., Charnley, S. B., & Wooden, D. 2005, in: M. C. Festou, H. U. Keller, & H. A. Weaver (eds), *Comets II* (Univ. of Arizona Press: Tucson), p. 115
Elsila J. E., Glavin D. P., & Dworkin J. P. 2009, *M&PS* 44, 1323
Floss, C., Stadermann, F. J., Bradley, J. P., *et al.* 2006, *Geochim. Cosmochim. Acta* 70, 2371
Flynn, G. J., Bleuet, P., Borg, J., *et al.* 2006, *Science* 314, 1731

Fouchet, T., Irwin, P. G. J., Parrish, P., *et al.* 2004, *Icarus* 172, 50
Fray, N., Bénilan, Y., Cottin, H., Gazeau, M.-C., & Crovisier, J. 2005, *P&SS* 53, 1243
Gerin M., Marcelino N., Biver N., *et al.* 2009, *A&A* 498, L9
Gibb, E. L., Mumma M. J., Dello Russo N., DiSanti M. A., & Magee-Sauer K. 2003, *Icarus* 165, 391
Gortsas N., Kührt E., Motschmann U., & Keller H. U. 2011, *Icarus* 212, 858
Greenberg, J. M. 1982, in: L.L. Wilkening (ed.), *Comets* (Univ. of Arizona press: Tucson), p. 1131
Groussin O., A'Hearn, M. F., Li, J.-Y., *et al.* 2007, *Icarus* 191, 63
Hamada S., Yamaguchi M., Ootsubo T., *et al.* 2010, DPS, 42, 962
Hanner, M. S. & Bradley, J. P. 2005, in: M.C. Festou, H.U. Keller, & H.A. Weaver (eds), *Comets II* (Univ. of Arizona Press: Tucson), p. 555
Hartogh, P., Lis, D. C., Bockelée-Morvan, *et al.* 2011, submitted
Helbert J., Rauer H., Boice D. C., & Huebner W. F. 2005, *A&A* 442, 1107
Hersant F., Gautier D., & Huré J.-M. 2001, ApJ 554, 391
Hutsemékers, D., Manfroid, J., Jehin, *et al.* 2005, *A&A* 440, L21
Hutsemékers, D., Manfroid J., Jehin E., Zucconi J.-M., & Arpigny C. 2008, *A&A* 490, L31
Ikeda, M., Hirota, T., & Yamamoto, S. 2002, *ApJ* 575, 250
Irvine, W. M., Schloerb, F. P., Crovisier, J., Fegley, B., & Mumma, M. J. 2000, in: V. Mannings, A.P. Boss, & S.S. Russel (eds), *Protostars and Planets IV* (Univ. of Arizona press: Tucson), p. 1159
Ishii, H. A., Bradley, J. P., Dai, Z. R., *et al.* 2008, *Science* 319, 447
Jehin, E., Manfroid, J., Cochran, A. L., *et al.* 2004, *ApJ* 613, L161
Jehin, E., Manfroid, J., Hutsemékers, D., Arpigny, C., & Zucconi, J.-M. 2009, *EM&P* 105, 167
Jessberger, E. K., Christoforidis, A., Kissel, J., *et al.* 1988, *Nature* 332, 691
Jewitt, D., Matthews, H. E., Owen, T., & Meier, R. 1997, *Science* 278, 90
Kavelaars, J. J., Mousis, O., Petit, J.-M., & Weaver, H. A. 2011, textitApJ 734, L30
Kawakita, H. & Kobayashi, H. 2009, ApJ 693, 388
Kawakita, H., Watanabe, J., Furusho, R., *et al.* 2004, *ApJ* 601, 1152
Kawakita H., Dello Russo, N., & Furusho, R., *et al.* 2006, *ApJ* 643, 1337
Keller, H. U., Britt, D., & Buratti, B. J., Thomas N. 2005, in: M.C. Festou, H.U. Keller, & H.A. Weaver (eds), *Comets II* (Univ. of Arizona Press: Tucson), p. 211
Keller, L. P., Bajt, S., Baratta, G. A., *et al.* 2006, *Science,* 314, 1728
Lis D. C., Wootten A., Gerin M., & Roueff, E. 2010, *ApJ* 710, L49
Manfroid, J., Jehin, E., Hutsemékers, D., *et al.* 2005, *A&A* 432, L5
Manfroid J., Jehin, E., Hutsemékers, D., *et al.* 2009, *A&A* 503, 613
Marty B., Chaussidon, M., Wiens, R. C., *et al.* 2011, *Science*, 332, 1533
McKeegan, K. D., Aléon, J. Bradley, J., *et al.* 2006, *Science* 314, 1724
Meier, R., Owen, T. C., Matthews, H. E., *et al.* 1998a, *Science* 279, 842
Meier, R., Owen, T. C., Jewitt, D. C., *et al.* 1998b, *Science* 279, 1707
Milam, S. & Charnley, S. B. 2011, IAU Symposium 280, 259P
Mumma, M. J., Disanti M. A., Bonev B. P., *et al.* 2008, *LPI Contrib.* 1405, 8282
Mumma, M. J. & Charnley, S. B. 2011, *ARAA* in press
Ootsubo, T., Usui, F., & Kawakita, H., *et al.* 2010, *ApJ* 717, L66
Pardanaud, C., Crovisier, J., Bockelée-Morvan, D., & Biver, N. 2007, in: J.L. Lemaire, & F. Combes (eds), *Molecules in Space and Laboratory*, p. 135 (ISBN: 9782901057581).
Remusat, L., Palhol, F., Robert, F., Derenne, S., & France-Lanord, C. 2006, *E&PSL* 243, 15
Rodgers, S. D. & Charnley, S. B. 2008, *MNRAS* 385, L48
Sandford, S. A., Aléon, J., Alexander, C. M. O.'D., *et al.* 2006, *Science* 314, 1720
Schleicher, D. G. 2008, *AJ* 136, 2204
Shinnaka, Y., Kawakita H., Kobayashi H., *et al.* 2011, *ApJ* 729, 81
Soderblom, L. A., Becker, T. L., Bennett, G., *et al.* 2002, *Science* 296, 1087
Stern, S. A. 2003, *Nature* 424, 639
Sunshine, J., Groussin, O., Schultz, P. H., *et al.* 2007, *Icarus* 191, 73

Terzieva, R. & Herbst, E. 2000, *MNRAS* 317, 563
Villanueva, G. L., Mumma, M. J., Bonev, B. P., *et al.* 2009, *ApJ* 690, L5
Weaver, H.A., Feldman, P.D., Combi, M.R., *et al.* *Icarus*, preprint
Wyckoff, S., Kleine, M., Peterson, B. A., Wehinger, P. A., & Ziurys, L. M. 2000, *ApJ* 535, 991
Ziurys, L. M., Savage, C., Brewster, M. A., *et al.* 1999, *ApJ* 527, L67
Zolensky, M. E., Zega, T. J., Yano, H., *et al.* 2006, *Science* 314, 1735

The Molecular Universe
Proceedings IAU Symposium No. 280, 2011
José Cernicharo & Rafael Bachiller, eds.

doi:10.1017/S174392131102504X

The Power of Sample Return Missions - Stardust and Hayabusa

Scott A. Sandford[1]

[1]Astrophysics Branch, NASA-Ames Research Center,
Mail Stop 245-6, Moffett Field, CA 94035 USA
email: Scott.A.Sandford@nasa.gov

Abstract. Sample return missions offer opportunities to learn things about other objects in our Solar System (and beyond) that cannot be determined by observations using *in situ* spacecraft. This is largely because the returned samples can be studied in terrestrial laboratories where the analyses are not limited by the constraints - power, mass, time, precision, etc. - imposed by normal spacecraft operations. In addition, the returned samples serve as a scientific resource that is available far into the future; the study of the samples can continue long after the original spacecraft mission is finished. This means the samples can be continually revisited as both our scientific understanding and analytical techniques improve with time.

These advantages come with some additional difficulties, however. In particular, sample return missions must deal with the additional difficulties of proximity operations near the objects they are to sample, and they must be capable of successfully making a round trip between the Earth and the sampled object. Such missions therefore need to take special precautions against unique hazards and be designed to successfully complete relatively extended mission durations.

Despite these difficulties, several recent missions have managed to successfully complete sample returns from a number of Solar System objects. These include the Stardust mission (samples from Comet 81P/Wild 2), the Hayabusa mission (samples from asteroid 25143 Itokawa), and the Genesis mission (samples of solar wind). This paper will review the advantages and difficulties of sample return missions in general and will summarize some key findings of the recent Stardust and Hayabusa missions.

Keywords. Stardust Mission, Hayabusa Mission, comets: general, comets: individual (Wild 2), asteroids, sample return

1. Introduction

Spacecraft are now 'routinely' used to study a host of astrophysical objects within our Solar System and beyond. The majority of these missions are designed to study their subjects remotely (space telescopes) or *in situ* (spacecraft sent to various destinations within the Solar System), and their use has revolutionized our understanding of space. These missions do, however, suffer from a number of limitations, particularly when it comes to determination of the detailed compositional nature of the objects studied. Sample return missions offer a means of augmenting the study of the composition of extraterrestrial objects in unprecedented detail.

The planetary sciences have long benefited from the availability of free sample returns that arrive at the Earth in the form of meteorites and cosmic dust and these materials have provided the context of much of what we know about the formation and evolution of our Solar System. However, these samples generally arrive at the Earth as 'orphans' whose original parent bodies are unknown. Under these conditions we lack the context needed to fully take advantage of the samples. Sample return missions can provide the material for study along with contextual information gathered from the *known* parent body.

Sample return began with samples returned from the Moon (Heiken *et al.* 1991) and these samples have amply demonstrated the advantages of sample return. These samples have now been available for over 40 years and continue to be the focus of active research. Subsequent to the lunar sample returns there was little in the way of sample return from space for many years, and what sample return there was did little to provide the parent body context of a focused sample return mission. For example, samples of extraterrestrial materials have been returned from space on exposed impact surfaces (i.e. Zolensky & Kinard 1993), but these samples were similar to collected meteorites and stratospheric dust in the sense that the samples' parent bodies were unknown.

However, in recent years there have been a handful of missions that have flown and successfully returned to Earth with samples of targeted objects. These include the Genesis mission, which returned samples of solar wind, the Stardust mission, which returned dust samples from the coma of Comet 81P/Wild 2, and the Hayabusa mission, which retuned dust particles from the surface of asteroid 25143 Itokawa. All three of these missions have demonstrated the scientific value of sample return missions. In the discussions that follow, a general review of the advantages and difficulties associated with sample return missions will be presented. In addition, a number of specific examples of how sample return missions have proven their worth will be presented. These examples will focus on sample return from small, primitive Solar System objects, i.e. on the Stardust and Hayabusa missions. For further information about the Genesis mission, the reader is invited to see Burnett *et al.* (2003). Finally, the discussion will provide some summary comments concerning important lessons learned from these missions that are specific to sample return before ending with a brief look to what is on the horizon for future sample return missions for small Solar System bodies.

2. Advantages of Sample Return

There are numerous advantages to sample return for the study and understanding of our Solar System, particularly when it comes to the determination of compositions. Having actual extraterrestrial samples on hand in terrestrial laboratories allows for the use of state-of-the-art analytical techniques and equipment, providing for the ultimate in current precision, sensitivity, resolution, and reliability. In contrast, spacecraft instruments making *in situ* measurements are, of necessity, not state-of-the-art since they fly instrument designs that must be locked into place well before launch and frequently have extended flights to their targets. In fact, sample return missions have the advantage that they are a resource for both current *and* future studies. As laboratory analytical techniques improve, samples can always be reexamined to take advantage of these improvements.

Sample return missions also avoid limitations associated with cost, power, mass, and reliability that are imposed on spacecraft instruments. Spacecraft are limited on what they can carry and this constrains the measurements they can make. In contrast, samples returned to the Earth's surface can be studied using any analytical technique one cares to bring to bear, vastly increasing the information that can be gleaned from the parent body. It is interesting to note that some of the terrestrial analytical devices used to study Stardust and Hayabusa samples are not only more massive than the spacecraft that returned them, they are more massive than the launch pads from which they left the Earth's surface!

The study of returned samples also enjoys the benefit that the samples can be studied by a broad international community that can use many different analytical techniques in an iterative and fully adaptive fashion. The work of these scientists need not be limited by instrument designs or ideas current at the time of the launch of the original spacecraft. Also, since the samples reside in terrestrial laboratories and can be made available to multiple researchers, analyses can be made using multiple, fully calibrated techniques and instruments, and verified by multiple researchers. Finally, because the actual analyses are done on Earth, sample return takes advantage of a tremendous resource not fully available to non-return missions, namely the expertise of the world's analytical chemists, physicists, and meteoriticists.

3. Special Difficulties of Sample Return

Of course, sample return missions also must deal with some special difficulties that do not exist, or are less problematic, for other missions. One principal additional difficulty that must be accommodated by sample return missions is that they must make round trips. Most *in situ* space missions arrive at their destination, do their work, and are then more-or-less abandoned when their tasks are completed. In contrast, sample return mission must not only launch, travel to their collection site, and collect a sample, but must also return to the Earth and deliver the sample safely to the surface. These issues add complications to the spacecraft's design that are associated with the sample reentry system and can make for longer mission lifetimes that place greater demands on the reliability of spacecraft systems.

Of course, one unique need that all sample return missions will share is the availability of a suitable sample collection system. The details and associated difficulties of such collection systems will vary from mission to mission and are strongly dependent on the nature of the body from which the sample is being collected, the type of sample desired, and so on. However, all such systems will share a number of common requirements. Chief among these is the need to collect, preserve, and deliver the sample with a minimum amount of alteration or contamination. Most spacecraft missions go to some pains to ensure they launch a 'clean' spacecraft to minimize risk to spacecraft operations and data taken by onboard instrumentation. Sample return missions generally share these same concerns, but have the added need to ensure the collected samples remain as pristine as is practically possible from the time of their collection to their ultimate analysis in a terrestrial laboratory. This requires that special care be taken in the design, assembly, testing, and flight of the spacecraft to mitigate contamination of the collected samples. In reality, it is generally not possible to guarantee a perfectly contamination-free environments. Thus, it is necessary for sample return missions to also make a special effort to not only carry out contamination control processes, but to also do contamination assessment. In this manner, contamination that can be eliminated is removed, and contamination that cannot is at least identified and characterized so that it will not be confused with parent body materials during analysis after sample return. Such assessments need to be made at all stages of the mission from manufacture of the spacecraft to ultimate removal of the sample from the sample return system.

While not always the case (Genesis is an example of an exceptions), sample return missions usually require the spacecraft operate in near proximity to the parent body from which the sample is to be collected. This can greatly complicate spacecraft operations and generally results in risks that are not associated with missions that have larger 'stand-off' distances.

Finally, sample return missions are, of course, most powerful when they are equipped with many of the same sorts of *in situ* analytical instruments associated with one-way, non-sample return missions. Such instrumentation (imagers, spectrometer, etc.) benefit sample return missions in both practical and scientific ways. Scientifically, such measurements can greatly increase the value of the returned samples by providing context from their parent body, as well as yielding results in their own right. On a more practical note, such instrumentation can also provide a means of establishing what are the best possible sampling sites, how representative these sampling sites are of the entire body, and so on. While the additional instruments add complexity to the overall spacecraft, they generally do not raise concerns beyond those normal for non-return missions (with the exception that one needs to ensure that these instruments will not be a source of sample contamination and will work properly in the near-body sampling environment).

4. The *Stardust* Comet Sample Return Mission

4.1. *Overview of the Stardust Mission*

Stardust, a NASA Discovery class mission, was the first spacecraft in history to return solid samples from an astronomical body beyond the Earth-Moon system (Brownlee *et al.* 2003; Brownlee *et al.* 2004; Brownlee *et al.* 2006). The mission retrieved samples from Comet 81P/Wild 2, an ~4.5 km-diameter body that was presumably formed and spent most of the past 4.5 or so billion years in the Kuiper Belt. This comet was somehow perturbed from its original orbit and on 10 September 1974 it had a close encounter with Jupiter that placed it in its current orbit, which lies between Mars and Jupiter. In its current orbit, the comet has an expected dynamical lifetime of $\sim 10^4$ years before it hits a larger object or is ejected from the Solar System (Levison & Duncan 1997). Comet 81P/Wild 2 now approaches the Sun close enough that solar heating causes classic cometary activity near perihelion.

During the Stardust flyby on 2 January 2004, the comet was active, and images showed the presence of at least 20 dust jets coming from the nucleus (Brownlee *et al.* 2004; Sekanina *et al.* 2004) (Fig. 1). Stardust approached to within 234 km of the surface of Wild 2, when the comet was at a solar distance of 1.86 AU. Particles were collected from the comet's coma when they impacted at the 6.12 km/sec encounter velocity into low-density silica aerogel (Tsou *et al.* 2003; Brownlee *et al.* 2003; Brownlee *et al.* 2006). The collecting aerogel consisted of a porous glass comprised of nanometer-sized silica filaments with bulk density that varied from <0.01 g/cm^3 at the impact surface to 0.05 g/cm^3 at 3-cm depth (Tsou *et al.* 2003). The aerogel's composition was predominantly SiO_2, but contaminants, including some C-bearing components (largely -CH_3 and -CH_2- groups) were also present (Sandford *et al.* 2010). Stardust aerogel tiles collected over a thousand 5-300 μm (and many more smaller) comet particles. Onboard impact sensors indicate that most of the collected particles were associated with just a few specific dust jets (Tuzzolino *et al.* 2004). Particles ejected from the comet were exposed to space for only a few hours before collection, but solar heating was probably sufficient to vaporize most ices during transit from Wild 2 to the Stardust spacecraft. The collected materials were successfully delivered to Earth's surface by the Stardust Sample Return Capsule on 15 January 2006 (Brownlee *et al.* 2006), which was subsequently opened in a specially designed cleanroom at the NASA Curatorial Facility at Johnson Space Center.

Particle impacts into the aerogel produced a variety of differently shaped tracks. Non-fragmenting particles produced carrot-shaped tracks with length/diameter ratios of >25. However, many tracks show bulbous upper regions and sometimes multiple 'roots'

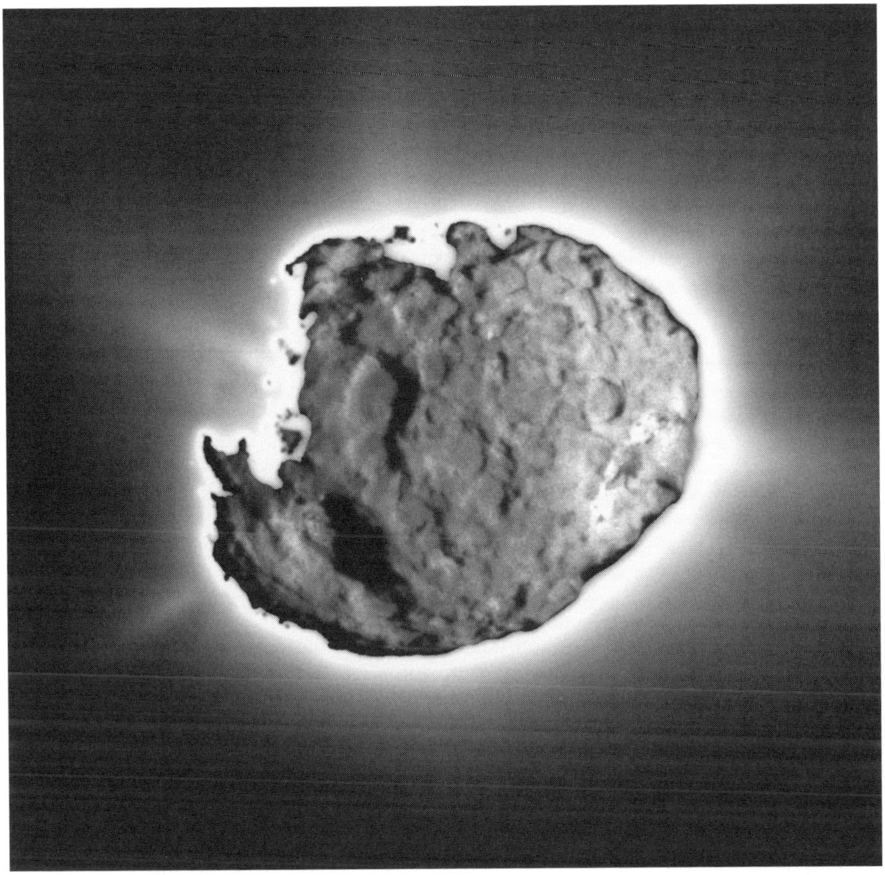

Figure 1. A composite of long and short exposure images of Comet Wild 2 showing both the complex nature of the surface of the surface of the nucleus and the many jets of dust and gas leaving its surface (see Brownlee *et al.* 2006).

(Fig. 2). These tracks were produced by weakly bound aggregate particles that broke apart on impact with the aerogel (Brownlee *et al.* 2006; Hörz *et al.* 2006). Typically the upper portions of tracks are lined with melted aerogel containing dissolved projectile material, the mid-track regions contain less melt and more preserved projectile material and compressed aerogel, and the ends of the tracks contain largely unmelted materials (terminal particles). The deepest penetrating particles are generally solid mineral grains or aggregates comprised of micrometer-size or larger grains. Particle impacts were also collected on the aluminum foils and frames that held the aerogel tiles in place in the collector grid. Impacts on these produced bowl-shaped craters lined with melted, and in some cases unmelted, projectile residues (Hörz *et al.* 2006).

For a period of six months afterwards a small fraction of these returned samples were examined by a large Preliminary Examination Team (PET) that consisted of over 175 scientists from around the world. PET members studied the samples using a host of different analytical techniques and their results have been reported in papers in special issues of *Science* (Brownlee *et al.* 2006; Flynn *et al.* 2006; Keller *et al.* 2006; McKeegan *et al.* 2006; Sandford *et al.* 2006; Zolensky *et al.* 2006) and *Meteoritics and Planetary Science* (see *MAPS*, vol 43, No. 1/2 2008). At the conclusion of the preliminary examination period the samples were formally turned over to the NASA Curatorial Facility to

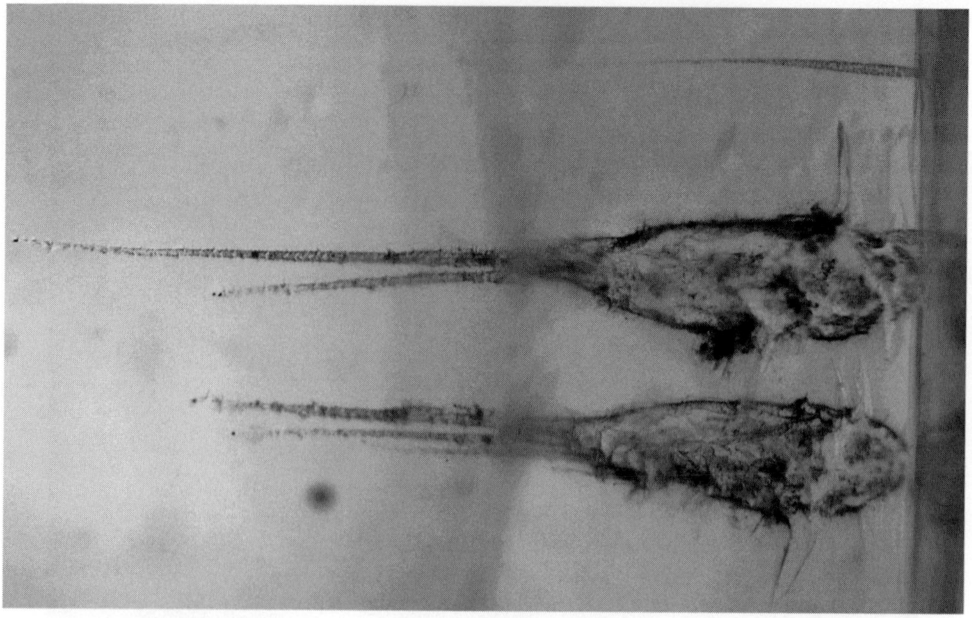

Figure 2. Many Wild 2 particles that impacted the Stardust aerogel collector were composites of many smaller grains. These particles broke up upon impact with the aerogel surface resulting in complex track forms with cometary materials deposited all along their lengths. In this image the surface of the aerogel collector is to the right and particles entered the aerogel moving from right to left. A much simpler, narrower 'carrot' track due to the impact of a single mineral grain can be seen near the top of the image. Tracks lengths depend on the size and structure of the impacting particle but are typically less than 1 cm in length.

be made available for allocation to the world's scientists in the same manner as Apollo samples, stratospheric cosmic dust samples, and Antarctic meteorites. Numerous samples have been distributed since that time and papers resulting from study of Stardust samples now appear regularly in the peer-reviewed literature. Some of the highlights of these studies are summarized in the section that follows.

4.2. *Highlights from Studies of the Returned Comet Wild 2 Samples*

Even in low-density aerogel, the hypervelocity capture of dust particles from Wild 2 resulted in the destruction and alteration of a significant fraction of the impacting material. This requires that care be taken to identify 'pristine' cometary materials from altered cometary materials and secondary impact products - a process made more difficult by the small nature of the samples. Despite these difficulties, it has been possible to learn an impressive number of things from the returned samples. A few points of interest and importance include:

(1) *Comet 81P/Wild 2 is a repository of largely unprocessed protosolar nebular materials.* The distribution of minerals and organics (abundance and composition) is heterogeneous both within and between particles (Flynn *et al.* 2006; Sandford *et al.* 2006; Zolensky *et al.* 2006; Lanzirotti *et al.* 2008). These materials represent a *highly unequilibrated* reservoir of materials, suggesting they experienced very little parent body processing after incorporation into the comet.

(2) *The protosolar nebula experienced mixing that spanned essentially its entire radial extent prior to the formation of the comet.* Wild 2 samples contain everything from very high temperature minerals (Zolensky *et al.* 2006; Zolensky *et al.* 2008) to highly volatile

organics (Sandford *et al.* 2006; Cody *et al.* 2008). These various components could not have been made at the same times in the same locations. Some must have formed very near the protosun while others could never have been inside the orbit of Jupiter and survived. And yet somehow they all got together in the comet formation zone.

(3) *Comet 81P/Wild 2 is not simply an assemblage of circumstellar and interstellar materials.* Most of the minerals in the samples are crystalline and look very 'meteoritic' in composition and isotopic composition (McKeegan *et al.* 2006; Zolensky *et al.* 2006), implying they are of *protosolar* origin, not *presolar* origin. Isotopically anomalous *circumstellar* grains are seen but are relatively uncommon (Stadermann *et al.* 2008). *Interstellar* materials are more common but still represent a minor fraction of the total material (see item 5 below; McKeegan *et al.* 2006).

(4) *Organics are present and show a large and unprecedented compositional diversity.* The population of organics found in the samples are distinct from those seen in primitive meteorites and IDPs, but some show similarities to both (Sandford *et al.* 2006; Cody *et al.* 2008; Matrajt *et al.* 2008; Rotundi *et al.* 2008; Clemett *et al.* 2010). The organics are very rich in both oxygen and nitrogen and contain a diverse set of both aromatic and non-aromatic compounds. New materials not seen before in other extraterrestrial materials include a volatile, largely non-aromatic population that has a relatively simple chemistry (Sandford *et al.* 2006; Cody *et al.* 2008). Included in the organics is the amino acid glycine (Glavin *et al.* 2008) whose cometary origin has been confirmed by isotopic studies (Elsila *et al.* 2009).

(5) *Many of the organics have an interstellar/protonebular heritage.* While isotopically anomalous *circumstellar* grains are very uncommon, organics with excesses of D and ^{15}N are less so, implying these materials or their immediate precursors have a pre-solar heritage (McKeegan *et al.* 2006; Stadermann *et al.* 2008).

In summary, the samples from Comet Wild 2 show no clear affinities with any *specific* meteorite type, although most of its major compositional components, particularly the minerals, are similar to those found in primitive meteorites. While some 'exotic' circumstellar and interstellar materials are present, the samples are clearly dominated by protosolar nebular materials that were apparently sampled from across the majority of the protosolar disk, perhaps over an extended period. However, the highly unequilibrated nature of the samples demonstrates that, once these materials were gathered into the comet, they experienced very little parent body processing. Thus, comets (at least as represented by Comet Wild 2) would seem to be very 'primitive' in the sense that they have preserved considerable material that was largely unaltered by parent body processes, but not particularly 'primitive' in terms of containing higher proportions of presolar materials than other available meteoritic samples. It is probably worth stressing one additional point - the *enormous heterogeneity* seen in the Stardust samples highlights the dangers of inferring cometary compositions on the basis of measurements, remote or otherwise, of individual species. For example, the D/H ratio measured telescopically in an individual molecular species in a cometary coma is *highly* unlikely to represent a true measure of the comet's D content as a whole.

Of course, one of the advantages of sample return missions is that the items listed above represent a hard lower limit to what will be learned from these samples. The samples are still being be measured in laboratories around the world and will continue to be studied for decades into the future. Further discoveries are sure to come.

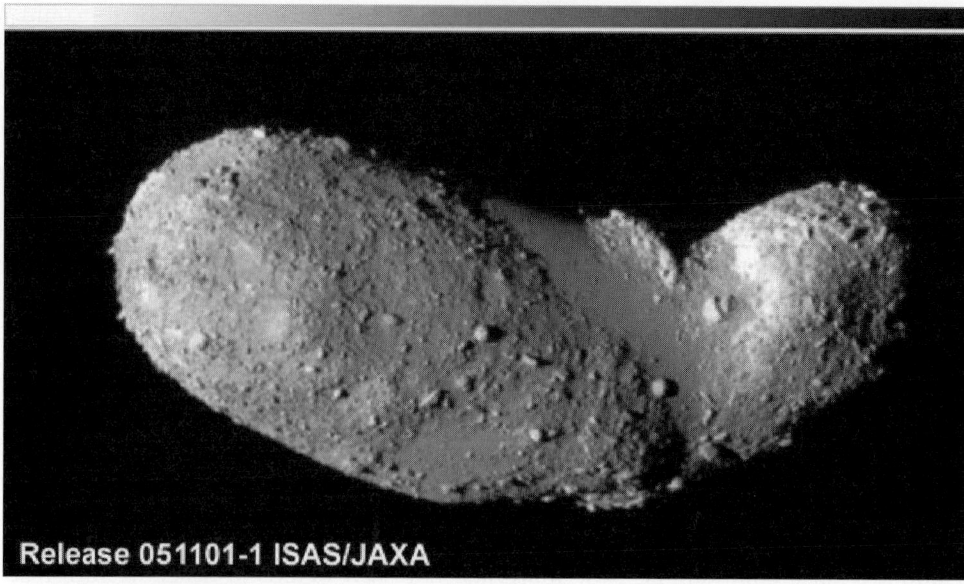

Figure 3. Asteroid Itokawa as seen by the Hayabusa spacecraft (see Yano *et al.* 2006). Itokawa is a small (length ∼540 m) 'rubble pile' asteroid that shows evidence of having been collisonally disrupted in the past.

5. The *Hayabusa* Asteroid Sample Return Mission

5.1. *Overview of the Hayabusa Mission*

The Japanese Hayabusa spacecraft was launched from Earth by JAXA on an MV-5 rocket in May 2003. The spacecraft rendezvoused with its target, the small S(IV)-type asteroid 25143 Itokawa (Fig. 3), in September 2005 (Fujiwara *et al.* 2006), after which it carried out an extensive series of remote sensing measurements. These included measurements by an onboard near-infrared spectrometer (Abe *et al.* 2006) and an X-ray fluorescence detector (Okada *et al.* 2006). These were used to estimate the approximate mineral and chemical composition of the surface of the asteroid. The combined results of these measurements suggested that asteroid Itokawa is a 'rubble pile asteroid' consisting of materials similar to primitive LL5-6 Ordinary Chondrites. The spacecraft twice descended to the surface of Itokawa to collect samples, once on 20 November 2005 and again on 26 November 2005 in the vicinity of the MUSES-C Regio, an area on the asteroid dominated by cm-sized gravel (Yano *et al.* 2006). Although the spacecraft's sampling system failed to operate properly during both descents, the inlet sampling horn of this system undoubtedly impacted with the surface of the asteroid both times (Yano *et al.* 2006). The escape velocity of the asteroid is sufficiently small (on the order of ∼0.2 m/sec) that these touchdown impacts were sufficient to mobilize small dust grains from the asteroid's surface to the spacecraft's sample container.

Despite a number of difficulties with the spacecraft, it was ultimately able to return to Earth and successfully deliver its Sample Return Capsule (SRC) to Australia in 2010 (Abe *et al.* 2011). Subsequent opening of the sample canister in a specially prepared cleanroom in Sagamihara, Japan demonstrated that, while the sampling system had not recovered the full amount of material desired, it did contain thousands of particles in the 1-200 micron diameter range. Analysis of these grains has since demonstrated that these particles contain a mixture of contaminants associated with the spacecraft and dust grains from the asteroid's surface, with at least 1500 of the dust particles being asteroidal

(Nakamura *et al.* 2011; Tsuchiyama *et al.* 2011). These samples constitute the second extraterrestrial regolith to have been sampled, the only other being the Moon, which was sampled by collections made by the Apollo and Luna missions (Heiken *et al.* 1991).

5.2. *Highlights from Studies of the Returned Itokawa Samples*

The Itokawa samples returned by Hayabusa have only been available for study for a relatively short time, and as a result our understanding of them is still far from complete. Nonetheless, the samples have been examined using a variety of analytical techniques and a number of important findings have already been made. A few points of interest and importance include:

(1) *Itokawa materials are a match to LL Ordinary Chondrites.* Particles found in the collector range from monomineralic to polymineralic and contain both poorly- and highly-equilibrated phases (~20% and 80%, respectively. The types, compositions, relative abundances and densities of minerals in the returned samples provide a good match to LL Ordinary Chondrites with the closest match being to petrographic grade LL5-6 (Nakamura *et al.* 2011; Tsuchiyama *et al.* 2011). Preliminary oxygen isotopic analyses are consistent with this identification and oxygen isotopic ratios clearly demonstrate the samples are extraterrestrial (Yurimoto *et al.* 2011).

(2) *The returned dust particles show evidence of relatively short surface exposures.* Noble gases extracted from individual grains show the presence of non-terrestrial solar wind gases whose compositions and concentrations imply relatively short surface exposures (<35 My) (Nagao *et al.* 2011). Additionally, approximately half of the examined particles show some evidence of space weathering at their surfaces, although the weathered layers are not as 'mature' as those seen in lunar soils, implying lower total exposures (Noguchi *et al.* 2011). Together, these results suggest that Itokawa has a relatively immature regolith.

The examination of samples returned by the Hayabusa spacecraft is still in its earliest stages at the time of this writing and only a small fraction of the available material has been given more than a cursory examination. It is certain that additional discovers remain to be made with these new and interesting samples.

6. Lessons Learned from Recent Missions

The needs and critical issues of sample return missions will vary with the nature of the samples desired and the source bodies from which they are taken. However, all sample return missions share a number of common requirements, and valuable lessons have been learned from the recent Stardust and Hayabusa missions that will be of value for future sample missions. Chief among these are issues associated with contamination control and assessment. A discussion of many of these issues in context of the Stardust mission can be found in Sandford *et al.* (2010). Summarized below are some lessons learned that should pertain to most sample return missions:

(1) Sample return missions need to address both *contamination control*, i.e., minimization of contamination, and *contamination assessment*, and these activities should be carried out throughout the mission from mission design to sample recovery. Contaminants should be minimized where possible, but it should also be recognized that even with contamination controls in place, there will likely still be some contamination. This is particularly problematic for collections of organic materials. Given that some contamination is likely to be inevitable, it is critical to characterize the nature of known or potential contaminants so they can be distinguished from returned samples. Ideally,

both contamination control assessment activities should be fully integrated into a sample return mission's budget and schedule.

(2) Contamination control and assessment requires cooperative efforts be made that involve the spacecraft manufacturers, the spacecraft operators, the mission's Science Team, and whatever organization will ultimately be responsible for the distribution and long-term curation of the returned samples. Each of these groups should identify an individual or individuals responsible for organizing contamination control and assessment activities, and these individuals need to work closely together across organizational boundaries.

(3) Oddly enough, one important issue that missions need to address early is to define in a fairly rigorous way what is meant by the word 'clean' and how this definition will translate into operational activities. It is surprising how this simple word can mean very different things to different people - a spacecraft engineer may see 'clean' as an issue of eliminating surface particulates of a certain size and clean something by wiping it with a solvent soaked wipe, while an organic chemist may flinch in horror at the uniform organic residue left behind when the solvent evaporates. This is one of many areas where there are enormous benefits to clear and direct communications between mission scientists and engineers. Of course, the agreed-on definition of 'clean' may well differ from mission to mission, depending on the nature of the expected sample and how it will be analyzed.

(4) During construction of the spacecraft and sampling system, it is critical to document what components/materials are used. Where possible, samples of these materials should be collected and archived for ultimate storage *at the time of manufacture.* These materials can then serve as controls and standards to help assess possible contamination of the returned samples by spacecraft materials.

(5) It is important to use "witness coupons" to track the introduction of contaminates during the manufacture, flight, and recovery of the spacecraft, and during the subsequent removal of the samples from the sampling system. These coupons need to be removed and examined quickly so that problems associated with unexpected or problematic contaminants can be dealt with rapidly (again, these activities should be explicitly funded and scheduled within the mission).

(6) Witness coupons need to be designed so that they can easily be divided and distributed to multiple analysts in parallel. It is often not practical to do contamination assessment measurements using multiple analytical techniques on the same coupon in serial fashion. Sample return spacecraft should also carry a significant number of relevant witness coupons. These material represents the all-important 'blank' or 'control' sample that is critical to the interpretation of many studies of the returned samples and there will be considerable demand for them. Also, since it is not always clear what materials will work best for contamination control and assessment and different materials can result in different analytical constraints, it is generally desirable to use more than one type of witness coupon.

(7) Plans must be made in advance so the organization responsible for the ultimate curation of the returned samples is prepared to store and distribute not only the returned samples, but also the associated contamination control and assessment materials (witness coupons, samples of spacecraft materials, etc.).

7. Future Missions

Previous programs and missions like Apollo, Genesis, Stardust, and Hayabusa have now amply demonstrated the enormous value of returning samples to Earth for study. Given the value of these samples, and the wide variety of Solar System objects that remain unsampled, it is therefore not surprising that a number of future sample return missions are being considered. In terms of missions to small, primitive Solar System bodies, future sample return missions to both comets and asteroids are currently being considered and, in several cases, being built.

In the case of future missions to asteroids, there are currently two missions preparing for flight, the Japanese Hayabusa 2 mission and the NASA OSIRIS-REx New Frontiers mission. Hayabusa 2 is essentially a follow-on mission to the original Hayabusa mission and is currently targeted to visit asteroid (162173) 1999 JU3. The current schedule has a proposed launch date of July 2014 with an expected arrival at the target in 2018. The asteroid would be surveyed for one and a half years and sampled before the spacecraft departed for Earth in December 2019, with a return to Earth in December 2020.

OSIRIS-REx was selected for flight by NASA in May 2011 and is scheduled to launch in 2016. It arrives at its target, asteroid 1999 RQ36, in late 2019 Once there, the spacecraft uses a number of instruments, including several different cameras, a laser altimeter, and two spectrometers, to take *in situ* observations to characterize the asteroid and establish potential sampling sites. Once a final sampling site is identified and a sample is obtained, the spacecraft will leave the asteroid in early 2021 and return its sample to the Earth on 24 September 2023.

Additional asteroid sample return missions are also in the wings, including the Marco Polo mission under consideration by ESA.

There are currently no selected comet sample return missions in flight or in preparation for flight, but a number of such missions are under consideration, and a comet sample return mission is one of the key mission types identified as important in the most recent Planetary Sciences Decadal Survey. It is therefore highly likely that one or more comet sample return missions will be in the works in the not too distant future.

8. Conclusion

Sample return missions provide a powerful means of learning about the nature and history of non-terrestrial bodies in our Solar System. Past sample return missions have successfully returned samples of the Moon, the solar wind, Comet 81P/Wild 2, and the asteroid 25143 Itokawa. Each of these sample returns has resulted in scientific advances that could not have been made in any other way. Combined, they amply demonstrated the value of sample return missions and strongly suggest that future such missions will yield many additional new and important insights.

Acknowledgements

The author is grateful to the organizers of this meeting for being invited to present this review and is grateful for past and current support from NASA grants from the Origins of Solar System, Exobiology, Cosmochemistry, and Discovery Mission Programs, which all have played key roles in the progression of sample return science.

References

Abe, M., *et al.* 2006, *Science*, 312, 1334
Abe, M., *et al.* 2011, *LPSC*, 42, 1638
Brownlee, D. E., *et al.* 2003, *J. Geophys. Res.*, 108, 8111
Brownlee, D. E., *et al.* 2004, *Science*, 304, 764
Brownlee, D. E., *et al.* 2006, *Science*, 314, 1711
Burnett, D. S., Barraclough, B. L., Bennett, R., Neugebauer, M., Oldham, L. P., Sasaki, C. N., Sevilla, D., Smith, N., Stansbery, E., Sweetnam, D., & Wiens, R. C. 2003, *Space Sci. Rev.*, 105(3-4), 509

Clemett, S. J., Sandford, S. A., Nakamura-Messenger, K., Hörz, F., & McKay, D. S. 2010, *Meteoritics and Planetary Science*, 45, 701
Cody, G. D., *et al.* 2008, *Meteoritics and Planetary Science*, 43, 353
Elsila, J. E., Glavin, D. P., & Dworkin, J. P. 2009, *Meteoritics and Planetary Science*, 44, 1323
Flynn, G. J., *et al.* 2006, *Science*, 314, 1731
Fujiwara, A., *et al.* 2006, *Science*, 312, 1330
Glavin, D. P., Dworkin, J. P., & , Sandford, S. A. 2008, *Meteoritics and Planetary Science*, 43, 399
Heiken, G. H. *et al.* (eds.) 1991, *Lunar Source Book*, (Cambridge: Cambridge University Press), p. 736
Hörz, F., *et al.* 2006, *Science*, 314, 1716
Keller, L. P., *et al.* 2006, *Science*, 314, 1728
Lanzirotti, A., Sutton, S. R., Flynn, G. J., Newville, M., & Rao, W. 2008, *Meteoritics and Planetary Science*, 43, 187
Levison, H. F. & Duncan, M. J. 1997, *Icarus*, 127, 13
Matrajt, G., Ito, M., Wirick, S., Messenger, S., Brownlee, D. E., Joswiak, D., Flynn, G., Sandford, S., Snead, C., & Westphal, A. 2008, *Meteoritics and Planetary Science*, 43, 315
McKeegan, K. D., *et al.* 2006, *Science*, 314, 1724
Nagao, K., *et al.* 2011, *LPSC*, 42, 2119
Nakamura, T., *et al.* 2011, *Science*, in press
Noguchi, T., *et al.* 2011, *LPSC*, 42, 1596
Okada, T., *et al.* 2006, *Science*, 312, 1338
Rotundi, A., *et al.* 2008, *Meteoritics and Planetary Science*, 43, 367
Sandford, S. A., *et al.* 2006, *Science*, 314, 1720
Sandford, S. A., *et al.* 2010, *Meteoritics and Planetary Science* , 45, 406
Sekanina, Z., Brownlee, D. E., Economou, T. E., Tuzzolino, A. J., & Green, S. F. 2004, *Science*, 304, 769
Stadermann, F. J., Hoppe, P., Floss, C., Heck, P. R., Hörz, F., Huth, J., Kearsley, A. T., Leitner, J., Marhas, K. K., McKeegan, K. D., & Stephan, T. 2008, *Meteoritics and Planetary Science*, 43, 299
Tsou, P., Brownlee, D. E., Sandford, S. A., Hörz, F., & Zolensky, M. E. 2003, *J. Geophys. Res.*, 108, 8113
Tsuchiyama, A., *et al.* 2011, *Science*, in press
Tuzzolino, A. J., *et al.* 2006, *Science*, 304, 1776
Yano, H., *et al.* 2004, *Science*, 312, 1350
Yurimoto, H., *et al.* 2011, *LPSC*, 42, 1755
Zolensky, M. E. & Kinard, W. H. 1993, *Adv. Space Res.*, 13, (8)75
Zolensky, M. E., *et al.* 2006, *Science*, 314, 1735
Zolensky, M. E., *et al.* 2008, *Meteoritics and Planetary Science*, 43, 261

Discussion

CHIAR: Can you elaborate on the size range of individual grains as well as the size(s) of the aggregates collected by the Stardust mission?

SANDFORD: Typical particles that hit the Stardust collector were in the 5-50 micron diameter range. Some of these particles were single mineral grains, but many were aggregate particles that consisted of micron and sub-micron grains. The Dust Flux Monitor detected impacts of particles up to nearly 1 cm in size, but these were rare and none struck the aerogel collector itself so little is known about their structure. Smaller grains (<5 micron diameter) also struck the collector but these are difficult to count because they create small, short tracks in the aerogel and are harder to find.

CUYLLE: You mentioned the detection of amino acids in Stardust samples. Is anything known about their chirality?

SANDFORD: Unfortunately, the only amino acid that has been identified in Stardust samples so far is glycine, which is not a chiral molecule, so nothing can currently be said about the chirality of cometary amino acids.

DE VRIES: Can you say anything about the Fe/Mg ratios of the olivines returned by the Hayabusa mission?

SANDFORD: The suite of minerals seen in the returned Hayabusa samples, their relative abundances, and their chemical compositions are all good matches to what is seen in LL Ordinary Chondrites. Olivines, which make up approximately a third of the samples examined so far, typically show [Fe/(Fe+Mg)] values that fall in the 0.25-0.35 range, which is also consistent with LL Ordinary Chondrites.

The Molecular Universe
Proceedings IAU Symposium No. 280, 2011
José Cernicharo & Rafael Bachiller, eds.

doi:10.1017/S1743921311025051

A Common Origin for Organics in Meteorites and Comets: Was It Interstellar?

Conel M. O'D. Alexander

DTM, Carnegie Institution of Washington,
5241 Broad Branch, Washington DC 20015, USA
email: alexander@dtm.ciw.edu

Abstract. The insoluble organic material preserved in primitive chondritic meteorites shares many similarities with the refractory organic material in interplanetary dust particles and comets, suggesting that there is a genetic link between the organic matter in objects that formed between ~3 AU and ~30 AU from the Sun. These similarities include large D and ^{15}N enrichments in bulk and even more extreme enrichments in isotopic hotspots. The enrichments attest to formation in very cold environments, either in the outer Solar System or the protosolar molecular cloud. There are many properties of this organic material that are consistent with an interstellar origin, but a Solar System origin cannot be ruled out. Similar organic material is presumably an important component of most protoplanetary disks, and heating or sputtering of this material would be a source of PAHs in disks. The soluble organic matter was more heavily effected by processes on the chondritic parent bodies than the insoluble material. Amino acids, for instance, probably formed by reaction of ketones and aldehydes with NH_3 and HCN. The accretion of the relatively volatile NH_3 and HCN, presumably in ices, strengthens the chondrite-comet connection. However, unlike most comets the water in chondrites, when it was accreted, had D/H ratios that were similar to or depleted relative to Earth.

Keywords. Comets: general, meteorites, dust, asteroids, Solar System: formation, molecular processes, astrochemistry, astrobiology

1. Introduction

The three basic types of extraterrestrial material that are accreted by the Earth and survive atmospheric entry are classified according to size into: interplanetary dust particles (IDPs), micrometeorites (and the related fully melted cosmic spherules) and meteorites. IDPs and micrometeorites almost certainly have both cometary and asteroidal sources, although the proportions of cometary and asteroidal particles remain a matter of debate (Dermott *et al.* 2002; Nesvorny *et al.* 2010). The most recent dynamical arguments suggest that most come from Jupiter family comets (Nesvorny *et al.* 2010). With a few exceptions, those that come from the Moon and Mars, meteorites are fragments of main-belt asteroids (2-4 AU), with a strong bias towards inner belt asteroidal sources (Morbidelli *et al.* 2002). While there is an overall gradient of spectral classes in the asteroid belt (Gradie, Chapman & Tedesco 1989), there has been considerable mixing and it is likely that the meteorite collection has sampled many of these classes (Burbine *et al.* 2002).

Entry velocity, size, density and entry angle together determining the degree of frictional heating that a particle experiences during atmospheric entry and, therefore, the extent to which organics and other temperature sensitive materials are preserved. The IDPs that are collected in the upper atmosphere are small (typically <50 μm) and low density, so that they decelerate relatively gently. Nevertheless, even the least heated of them probably experience temperatures of at least several degrees Celsius for several seconds. Denser particles from the same bodies as the IDPs would probably be destroyed during atmospheric entry, so it is unclear how representative of their parent bodies IDPs are. Micrometeorites (typically 100-1000 μm) are generally severely heated on atmospheric entry, and so most are unlikely to retain unmodified organic material. However, organics do survive in a rare group of C-rich particles that may have links to comets (Duprat *et al.* 2010). Evaporation, ablation and poor thermal conduction confine heating to the outer few millimeters of recovered meteorites.

The most primitive meteorites, the chondrites, are composed of materials that formed in the solar nebula or that even predate it. Chondrites can contain up to several weight percent of organic material. The complex suite of compounds that constitute this organic matter can have large D and ^{15}N enrichments that are usually interpreted to indicate that they or their precursors formed at very low temperatures. Similar enrichments are seen in IDPs (Messenger 2000; Aléon *et al.* 2003; Floss *et al.* 2006; Busemann *et al.* 2009) and Stardust particles returned from comet Wild 2 (McKeegan *et al.* 2006; Matrajt *et al.* 2008). Thus, it seems likely that there is a genetic link between the organics in chondrites and comets. It is still a matter of contention whether the organics formed in the presolar molecular cloud and/or in the early Solar System. A detailed understanding of the compositions of the various components of the chondritic organic matter may ultimately help resolve this debate.

Most chondrites also show evidence of having accreted water, presumably as ice, when they formed. It seems likely that the organics and water ice would have been intimately associated when they accreted, as they are in comets. In which case, the origins of the organics and water may be linked. Understanding the origin of the water in chondrites may also elucidate the origin of the water in the inner Solar System and transport processes in the solar protoplanetary disk (or solar nebula).

2. Meteorite Classification

Meteorites are broadly divided into two, the chondrites and achondrites (Krot *et al.* 2003). Only the most primitive types of meteorite, the chondrites, are generally thought to contain indigenous organic material. This is because achondrite meteorites come from parent bodies that experienced extensive melting, and even wholesale differentiation into silicate mantles and iron cores. Chondrites are comprised of three main components: refractory inclusions, chondrules and fine-grained matrix. Refractory inclusions and chondrules are high temperature (1400-1800°C) objects that formed in the solar nebula. Organics would not have survived the high formation temperatures of refractory inclusion or chondrule. Consequently, it is only in the matrix that cements the meteorites together that primary organic matter should be found. It is in the matrix that one also finds presolar circumstellar grains that formed around highly evolved stars, such as supernovae and asymptotic giant branch (AGB) stars (Nittler 2003; Alexander 2009), and were inherited from the protosolar molecular cloud.

Historically, the chondrites have been divided into three classes based on their compositions and mineralogies (ordinary, carbonaceous and enstatite). These in turn have been subdivided (Scott & Krot 2003; Weisberg, McCoy & Krot 2006) into a number of groups: ordinary chondrites into H, L and LL, carbonaceous chondrites into CI, CM, CR, CV, CO, CB, CH and CK, and enstatite chondrites into EH and EL. The name carbonaceous chondrite is a historical one and is a bit misleading since some ordinary and enstatite chondrites contain more C than some carbonaceous chondrites. The chondrite classification scheme is still evolving as more meteorites are found - two new classes (R and K chondrites) have been identified, and a number of individual meteorites do not belong to any recognized group.

The CIs have bulk compositions that are essentially identical to that of the rock-forming component (excluding H, He, etc.) of the Sun (Lodders 2003). Because solar (or CI) is the starting composition from which all Solar System materials evolved, the compositional variations of chondrites and their components are generally expressed as deviations relative to CI. The deviations of the compositions of chondrites (and their components) from CI are presumed to be the results of their formation processes in the solar nebula.

After formation, the chondrites experienced secondary modification - protracted episodes of thermal metamorphism and aqueous alteration, and in some cases brief shock heating events - on their parent bodies. A petrographic classification scheme for secondary processes divides the chondrites into 6 types - types 3 to 6 reflect increasing extent of thermal metamorphism, and types 3 to 1 reflect increasing degrees of aqueous alteration. By convention, the chemical classification is followed by the petrologic one (e.g., CI1, CM2, CV3).

3. Organics in Chondrites

There have been a number of recent reviews of the organic matter in meteorites and IDPs (Botta & Bada 2002; Sephton 2002; Gilmour 2003; Sephton 2005; Pizzarello, Cooper & Flynn 2006). However, with the renewed interest in the subject, the introduction of new techniques, and new samples from Antarctica and the Stardust sample return mission to Comet 81P/Wild 2, the field is evolving rapidly. Here we focus on the most recent work. The organic matter is normally divided into solvent soluble (SOM) and insoluble (IOM) fractions, and consequently they are discussed separately. The abundances of IOM and the known components of the SOM are summarized in Table 1. Also included in Table 1 are the matrix volume percents because the organic matter would have been accreted by the chondrites in the matrix material.

3.1. *Insoluble Organic Matter*

The IOM shows tremendous variation in its elemental and isotopic compositions both within and between chondrite groups (Alexander *et al.* 2007b, 2010), but these variations are due, at least in large part, to processes that occurred on the asteroidal parent bodies of the chondrites (Alexander *et al.* 2007b; Alexander *et al.* 2010; Herd *et al.* 2011). In the meteorites (C1-2) that appear to have seen the most benign parent body conditions, the bulk composition of the IOM, normalized to 100 Cs, is $\sim C_{100}H_{70-79}N_{3-4}O_{11-21}S_{1-5}$ (Alexander *et al.* 2007b). This is remarkably similar to the average composition of the comet Halley CHON particles - $C_{100}H_{80}N_4O_{20}S_2$ - detected by the Vega 1 spacecraft (Kissel & Krueger 1987).

Even amongst the C1-2 chondrites there is considerable elemental and isotopic variation in their IOM. The chondrites that harbor the apparently most primitive IOM are the CRs. Not only are the bulk H and N isotopic compositions of the IOM in CRs more anomalous than IOM from almost any other chondrite, but their IOM also contains small ($\leqslant 1\ \mu m$) isotopic hotspots that are more numerous and more extreme than in other chondrites (Busemann *et al.* 2006). The D/H ratios of the hotspots can be up to ~40 times the terrestrial ratio, and $^{15}N/^{14}N$ ratios can be up to ~4 times the terrestrial ratio. The hotspot D and ^{15}N enrichments overlap with the range of gas phase molecules in the ISM, but the two isotope systems are not correlated in any simple way. The bulk CR IOM

Table 1. The abundances of insoluble and soluble organic compounds (in μg/g or ppm) found in carbonaceous chondrites. For the CMs, all data are from the Murchison CM2 meteorite, unless otherwise noted (updated from Botta & Bada 2002).

	CI	**CM**	**CR**	**Tag. Lake**
Matrix (vol%)	100	~50	~35	~80
IOM	~20,000	~10,000	~5,000	~18,000
Amino acids	~5[a]	14-71[d]	1-250[b]	0.04-5.6[c]
Aromatic hydrocarbons		3[d]	16[e]	
Aliphatic hydrocarbons		>35		
Monocarboxylic acids		>300	96[e]	165-448[c]
Hydroxy- and dicarboxylic acids		14-15	212[e]	
Purines and pyrimidines		1.3		
Basic N heterocycles		7		
Amines		8	103[e]	
Alcohols		11		
Aldehydes and Ketones		27		
Sulphonic acids		68		
Phosphonic acids		2		
Polyols		>8[f]		

Notes:
[a] Average for Orgueil and Ivuna (Ehrenfreund *et al.* 2001).
[b] Range from Martins *et al.* (2007).
[c] Herd *et al.* (2011).
[d] For Y-791198 (Naraoka *et al.* 1998).
[e] For GRA 95229 (Pizzarello *et al.* 2008).
[f] Lower limit of glyceric acid (Cooper *et al.* 2001).

isotopic compositions, and the range of hotspot compositions and sizes are very similar to those seen in the most primitive anhydrous chondritic IDPs (Messenger 2000; Aléon *et al.* 2003; Floss *et al.* 2006; Busemann *et al.* 2009), whose H and N isotopic compositions are dominated by those of their organics. The CR IOM H and N isotopic compositions are generally more anomalous than the compositions of organic particles identified in the Stardust mission samples (McKeegan *et al.* 2006; Matrajt *et al.* 2008). However, this could be because of modification during capture and, possibly, contamination.

The H/C ratios of the primitive IOM are not unlike those of a fairly mature coal. Thus, it is not surprising that the IOM is fairly aromatic. Nuclear Magnetic Resonance (NMR) spectroscopy suggests that ~50% of the C in CR IOM is in small PAHs (one to a few benzene rings), that these PAHs are heavily substituted and are cross-linked by short, highly branched aliphatic chains and O functionality (Cody & Alexander 2005). IOM from CI and CM chondrites tend to be slightly more aromatic, but otherwise similar in structure to IOM in CRs (Gardinier *et al.* 2000; Cody, Alexander & Tera 2002; Cody & Alexander 2005; Cody *et al.* 2011). This confirms earlier work based on degradative chemistry (Hayatsu *et al.* 1977) and various pyrolysis techniques (Sephton, Pillinger & Gilmour 1999; Sephton & Gilmour 2001; Remusat *et al.* 2005; Sephton *et al.* 2005; Wang *et al.* 2005). The variations in aromatic C content are probably the result of parent body processes, and largely involved the conversion of aliphatic C to aromatic C and associated H loss (Herd *et al.* 2011). There also seems to be variable amounts of H isotope exchange and/or preferential D loss during this process. This suggests that the D is preferentially concentrated in the aliphatic fraction, an inference that appears to be confirmed by recent pyrolysis (Remusat *et al.* 2006) and NMR measurements (Wang *et al.* 2011). Simple heating of IOM above a few hundred degrees Celsius in an O-free atmosphere (pyrolysis) releases a lot of small PAHs and leaves a char. Heating of IOM-like material would be one mechanism for producing the PAHs that have been seen in protoplanetary disks. Sputtering of IOM-like material by energetic particles would also release PAHs.

The isotopic hotspots in IOM are not abundant enough to account for the bulk D and ^{15}N enrichments. However, these hotspots are often associated with so-called globules or nanoglobules (Nakamura-Messenger *et al.* 2006). These globules are spherical to irregular in shape and often hollow (Nakamura *et al.* 2002; Garvie & Buseck 2004; Garvie, Baumgardener & Buseck 2008). Similar globules have now been found in IDPs and Stardust particles (Busemann *et al.* 2009; De Gregorio *et al.* 2010). Raman spectroscopy also suggests that the structure of the bulk of the organic material in the most primitive IDPs is similar to that of IOM from primitive chondrites, although on average it may be more disordered (Busemann *et al.* 2009). There is also a suggestion that the D in IDPs is largely associated with aliphatic material (Keller *et al.* 2004). However, based on IR spectra in the 3.4 μm C-H stretch region, higher CH_2/CH_3 ratios in IDPs than in chondritic IOM suggests that the aliphatic material in IDPs tends to be shorter and/or more branched (Flynn *et al.* 2003; Keller *et al.* 2004; Muñoz Caro *et al.* 2006; Kebukawa, Alexander & Cody 2011). Whether the greater disorder and higher CH_2/CH_3 ratios in IDPs is indigenous or was induced by irradiation in interplanetary space and atmospheric entry heating remains to be seen.

From the above, it is likely that there is a genetic relationship between the IOM and the organic matter in IDPs/comets, although the abundance of organic matter in IDPs and comets is much higher. The organic C contents are perhaps most usefully expressed in terms of the fractions of the bulk solar C. Based on Mg/C ratios, the percentage of solar C in CI chondrite IOM is $\sim$7%, it is $\sim$30% on average in IDPs and $\geqslant$30% in comet Halley CHON particles (Alexander *et al.* 2007a). Thus, if there is a genetic relationship between IOM in chondrites and IDPs/comets, IOM-like material seems to have been widely distributed in the solar nebula but it formed or was better preserved in the outer Solar System. This along with the structural and isotopic properties of the IOM places considerable constraints on the various formation scenarios that have been proposed for the IOM.

One often invoked formation mechanism is Fischer-Tropsch-type (FTT) synthesis in the solar nebula (Hayatsu & Anders 1981; Ferrante *et al.* 2000; Nuth, Hill & Kletetschka 2000). Typically, FTT synthesis takes place on catalytic mineral surfaces, but there is no systematic association of the IOM with any mineral in meteorites (Remusat *et al.* 2010). Also, FTT synthesis would be most efficient in a narrow zone in the warm inner Solar System (Kress & Tielens 2001). Hence it is unclear why IOM-like organics would be more abundant in the outer Solar System. Finally, FTT synthesis cannot explain the D and ^{15}N enrichments in IOM. It has been pointed out that the conditions in the outer Solar System would have been quite similar to those in the protosolar molecular cloud. Therefore, the D enrichments in the IOM could have formed by reaction of IOM-like material with H_2D^+ in the Solar System (Gourier *et al.* 2008; Remusat *et al.* 2009). However, if the conditions in the outer Solar System and the protosolar molecular cloud were so similar, this begs the question - what process destroyed the condensed organic C that appears to be so abundant in the diffuse and dense ISM, and then allowed organic C to reform so efficiently as IOM elsewhere in the Solar System? FTT synthesis also does not explain why there is a fairly constant ratio of circumstellar grains to IOM in chondrites (Alexander *et al.* 2007a,b).

Based largely, but not solely, on the isotopic composition of the IOM, an ISM origin for the IOM has been invoked (e.g., Robert & Epstein 1982; Yang & Epstein 1984; Alexander *et al.* 2007b). An ISM origin is strengthened by the similarity of the IR 3.4 μm C-H stretch adsorption feature of IOM and of carbonaceous dust in the diffuse ISM (Ehrenfreund *et al.* 1991; Pendleton *et al.* 1994). Some proposed structures for the diffuse ISM carbonaceous dust invoke small, highly substituted PAHs and short, highly branched aliphatic chains (Jones, Duley & Williams 1990; Sandford *et al.* 1991; Dartois *et al.* 2005) and are remarkably similar to that of the IOM. On the other hand, Pendleton & Allamandola (2002) concluded that the PAHs in the diffuse ISM dust were large and that the dust contains few N or O heteroatoms. Oxygen isotopes suggest that much of the O functionality in chondritic IOM is probably the result of parent body processes (Alexander *et al.* 2007b), and it is not clear whether the N content of IOM is inconsistent with the spectroscopic constraints for diffuse ISM dust. It is also important to point out that the spectra Pendleton & Allamandola (2002) used is in fact a 600^oC sublimate from IOM and is, therefore, not representative of the bulk IOM. Nevertheless, this does not explain the absence of large PAHs in IOM. Pendleton & Allamandola (2002) concluded that the diffuse ISM dust probably formed in the outflows of C-stars. This too would be inconsistent with the IOM because the C-star outflows will be depleted in D and ^{15}N, and can have C isotopic compositions that are far from solar. Equally problematic to an interstellar origin for the IOM is that the 3.4 μm C-H stretch feature seen in the diffuse ISM dust and IOM is not seen in molecular clouds.

To overcome some of these problems, Alexander *et al.* (2008) suggested that in very cold molecular clouds cosmic ray irradiation of ice-coated diffuse ISM dust led to dehydrodgenation of aliphatic material and radiation damage of any large PAHs. Upon warming up, either as the dust enters a forming system or re-enters the diffuse ISM, the radicals in the damaged dust may react with radicals in the ice (e.g., H, OH, NH_2, CH_3, etc.) rather than reform their original bonds. Since the ice is likely to be D- and ^{15}N-rich, this process could explain both the heteroatom content, and D and ^{15}N enrichments in the IOM. Dust re-entering the diffuse ISM would have to loose its O functionality and remake any large PAHs, processes that would presumably have to be driven by the intense UV radiation field that exists there. This model should be testable experimentally. However, it does not explain the solar C isotopic composition of the IOM or the formation of globules. The C isotopes might be explained if in fact the typical lifetime of carbonaceous dust in the ISM is short compared to the production rate by C-stars (Serra Díaz-Cano & Jones 2008). In this case, there must be an efficient mechanism for making the dust in the ISM, and the bulk C isotopic composition of the dust that forms in the ISM is likely to have been close to the bulk isotopic composition of the ISM (i.e., solar) at that time. Nakamura-Messenger *et al.* (2006) suggested that globules could have formed by irradiation of ISM ices, but in their mechanism the wall thicknesses of the globules would reflect the penetration depth of the radiation (10s to a few hundred nanometers), which is too little for cosmic rays and probably requires unreasonably high C contents for the ice.

A quite different scenario for IOM formation is envisaged by Cody *et al.* (2008; 2011). They suggest that the IOM formed by polymerization of formaldehyde when accreted ice melted as meteorite parent bodies warmed due to the decay of short-lived radionuclides. This process naturally produces globule-like objects and can produce material with NMR spectra that resembles IOM on heating. If the ice formed in the ISM or outer Solar System, it and the formaldehyde it contained would probably be D-rich and contain ^{15}N-rich

N-bearing species that could be incorporated into the IOM. This model faces two hurdles. The first is whether there would have been enough formaldehyde in the ices to explain the abundances of IOM or IOM-like material in chondrites and, particularly, in IDPs and comets. This is especially problematic given that polymerization is an inherently inefficient process. The second is whether liquid water was present in cometary bodies, and whether the kinetics of the transformation of the initial polymer to more IOM-like material could take place at reasonable temperatures and over reasonable timescales for either comets or chondrites. The latter is experimentally testable, and the former has gained at least some recent support from the Stardust mission (Berger *et al.* 2011).

Ultimately, whether the organics in chondrites, IDPs and comets are solar or interstellar in origin may be determined not by the properties of the organics themselves, but rather by the abundance and origin of amorphous silicate material in IDPs and comets (Alexander *et al.* 2007a). Silicates in the ISM are essentially entirely amorphous. In IDPs and comets, on the other hand, crystalline silicates are a major component and must have formed at temperatures (>1000 K) that would have destroyed any interstellar organics. There is amorphous silicate material in primitive chondrites and IDPs, but there is an ongoing debate about whether the bulk of it is interstellar or not. Isotope analyses show that a few percent of the amorphous material in IDPs is circumstellar in origin. Given the short lifetime of circumstellar dust in the ISM, it seems likely that a significant fraction of the remaining amorphous material must be interstellar. Unfortunately, it is difficult to prove an interstellar origin and energetic Solar System processes (e.g., irradiation, impacts and condensation from a hot gas) can produce material with similar properties. Stardust samples do contain amorphous material, but it is hard to distinguish between indigenous material and that produced during capture. Comets are also known to contain amorphous silicates, but the results of the Deep Impact mission show that the grains being emitted by comets may not be representative of the bulk (Lisse *et al.* 2007). Nevertheless, one can roughly explain the abundances of organic C in IDPs and comet Halley if: (1) one assumes that all amorphous silicates in IDPs are interstellar and that ~50% of comet Halley silicates are amorphous and interstellar, and (2) adopt the range of abundances for carbonaceous dust and PAHs in the ISM inferred by Zubko, Dwek & Arendt (2004). Obviously, if a large fraction of the amorphous silicates proves to be solar in origin, then a proportional fraction of the organics must also be solar.

3.2. *Soluble Organic Matter*

The SOM in primitive chondrites is a complex suite of soluble compounds. The best characterized of these are summarized in Table 1. However, it seems likely that the bulk of the SOM has yet to identified (Schmitt-Kopplin *et al.* 2010). This poorly characterized material is composed of >46,000 compounds with masses of up to at least 1000 daltons and an approximate average composition of $C_{100}H_{155}N_3O_{20}S_3$ that resembles an IOM composition but with roughly twice the H content. Thus, it seems likely that the IOM and the uncharacterized SOM are related, but the uncharacterized SOM is presumably more aliphatic. Even within the classes of well-characterized SOM compounds there is tremendous structural diversity, although abundances tend to decrease significantly with increasing C number and extent of branching. The abundances of volatile compounds could also have been modified during the residence of the meteorites on Earth. Until we have a better understanding of the nature of the uncharacterized material, the work on the known soluble components probably presents a rather distorted picture of the origins of the SOM. Nevertheless, important features of the well-characterized SOM are briefly discussed below.

Even more than the IOM, the soluble organics are susceptible to modification (isotopic exchange, synthesis and destruction) during parent body processing. For instance, amino acid abundances decline dramatically within a chondrite group with increasing degree of alteration (Martins *et al.* 2007; Martins & Sephton 2009; Glavin *et al.* 2010; Herd *et al.* 2011). In the least altered chondrites, the amino acids are racemic within error. Surprisingly, the development of L enantiomeric excesses in amino acids like isovaline that do not easily racemize accompanies this decrease in abundance with increasing alteration (Glavin *et al.* 2010). Apparently, aqueous alteration either produced the enantiomeric excesses or amplified small excesses that already existed. The mechanism that produced or amplified these enantiomeric excesses is unknown, but one intriguing possibility is that it involves the IOM, which is known to possess chiral centers (Kawasaki *et al.* 2006).

The amino acids themselves are thought to be products of Strecker synthesis, the reaction of ketones and aldehydes with NH_3 and HCN. Presumably amino acid synthesis generally took place very early on in the alteration of the meteorites, but in the case of the Tagish Lake meteorite there may have been a second episodes of synthesis after alteration was fairly advanced (Herd *et al.* 2011). The fact that the relatively volatile NH_3 and HCN were in the ices that were accreted by the chondrites again suggests a link with comets. The HCN in comets is enriched in ^{15}N by roughly a factor of two over terrestrial (Shinnaka *et al.* 2011). The N isotopic composition in cometary NH_3 has not been measured, but in the ISM it is depleted in ^{15}N relative to terrestrial (Lis *et al.* 2010). The amino acids and NH_3 in two relatively unaltered CRs (GRA 95229 and LAP 02342) are only enriched in ^{15}N by about 10% over terrestrial (Pizzarello & Holmes 2009). In the Strecker synthesis, it is the N from the NH_3, not the HCN, that is inherited by the amino acid. The N from the HCN is converted into NH_3, which may eventually be incorporated into amino acids but only after significant dilution by pre-existing NH_3, which may explain the modest ^{15}N enrichments in NH_3 and amino acids.

The H isotopic compositions of amino acids have also been measured in one relatively unaltered CR (GRA 95229), and two more altered CMs (Murchison and Murray) (Pizzarello & Huang 2005; Pizzarello, Huang & Alexandre 2008). In a given meteorite, the D enrichment in straight chain amino acids tends to increase with increasing C number, and branched amino acids are more enriched than both the straight chain ones and the bulk IOM from the same meteorites. The amino acids in the more altered CMs are systematically less D-rich than in the less altered CR. The isotopic compositions of the putative precursor aldehydes and ketones are unknown, so it cannot be ruled out that some of the isotopic variations in the amino acids were inherited. Nevertheless, the patterns of isotopic variations are suggestive of isotopic exchange with the aqueous fluid during or after synthesis. Whatever the cause of the variations, the D enrichments (up to 8 times terrestrial) indicate that, like the IOM, the amino acids or their precursors formed in cold environments, either in the outer Solar System or the ISM. However, the C isotopic compositions of the amino acids are distinctly (>3-5%) more ^{13}C-rich than the IOM and are closer to those of the carbonate minerals that are present in aqueously altered chondrites (Engel, Macko & Silfer 1990; Pizzarello, Huang & Fuller 2004; Martins *et al.* 2007).

In the more altered CM Murchison, carboxylic acids rather than amino acids are the most abundant of the well-characterized SOM. In the Tagish Lake meteorite, in which a range of alteration is recorded, carboxylic acid abundances do seem to be a function of the extent of aqueous alteration (Herd *et al.* 2011). As with the amino acids, the H isotopic compositions of the straight chain carboxylic acids vary with C number, and the branched carboxylic acids are more D-rich (Huang *et al.* 2005; Herd *et al.* 2011). Again, the D enrichments of the carboxylic acids indicate that they or their precursors have a cold chemistry heritage. It is tempting to invoke a simple conversion of amino acids to carboxylic acids, but the C isotopic compositions of the carboxylic acids are inconsistent with this. The ^{13}C abundances of the straight chain carboxylic acids in Murchison decrease with increasing C number down to isotopic compositions that resemble those of the IOM (Huang *et al.* 2005). Based on both the H and C isotopic compositions, it is possible that the carboxylic acids in Murchison and Murray are related to or were derived from the aliphatic component of the IOM (Huang, Alexandre & Wang 2007; Herd *et al.* 2011). The variation in C isotopic composition with chain length seen in Murchison and one Tagish Lake sample could reflect isotopic exchange with dissolved CO_2 with a composition like that of the carbonate minerals (Herd *et al.* 2011).

Comets contain simple alkanes and alkenes (Mumma & Charnley 2011). The same is true at least of the CM Murchison (Yuen *et al.* 1984). However, Murchison also contains CO and CO_2, several well-known mineral catalysts of FTT synthesis and would have generated considerable amounts of H_2 during alteration. Thus, at this stage it should not be ruled out that the simple hydrocarbons, as well as other components of the SOM are the products of FTT synthesis during alteration rather than being primordial.

4. Water in Chondrites

The O isotopic compositions of inner Solar System objects all appear to be more ^{16}O-poor than the composition of the Sun. UV self-shielding during the photodissociation of CO in the outer Solar System or near the surface of a molecular cloud can produce ^{16}O-poor water ice. Hence, the favored explanation for the O isotopic composition of the terrestrial planet region is that there was a massive influx of outer Solar System and/or interstellar water ice into the inner Solar System (Yurimoto & Kuramoto 2004; Lyons & Young 2005). This ice would have been accompanied by organic material and other primitive materials. Chondrites, except probably the enstatite chondrites, seem to have accreted water ice, indicating that they formed beyond the snowline. When they formed, all chondrites also contained circumstellar grains and, as argued above, organics that are genetically linked to organics in IDPs/comets. In comets, NH_3 and HCN are major volatile components. The fact that these relatively volatile compounds are needed for making the amino acids in chondrites strengthens the comet-chondrite connection.

The D/H ratio in water ice is expected to increase with radial distance from the Sun as temperatures and pressures decrease, and conditions approached those in molecular clouds. In most, but not all, comets the H isotopic compositions of the water are enriched in D by roughly a factor of 2 relative to terrestrial values (Mumma & Charnley 2011). Given the apparent relationship between chondritic and IDP/comet organics, one might expect that, as with comets, chondritic water would be similarly enriched in D.

The H isotopic composition of chondritic water is difficult to measure directly. This is because it is now primarily incorporated into hydrous silicates as OH^-. Heating of the meteorites does release the water, but also releases H from organics that will exchange/mix with the H from the hydrous silicates. The H isotopes of hydrated silicates can be measured at small scales (<30 μm) in materials that are free or organics or using techniques that preferentially measure H in hydrous minerals (Deloule, Robert & Doukhan 1998; Grossman *et al.* 2002; Bonal *et al.* 2011). The average H isotopic compositions of the water in some chondrites have also been estimated by subtracting the known organic materials that have been isolated from them (Robert 2002; Alexander *et al.* 2010).

From the range of D enrichments recorded in these kinds of measurements and estimates, it has been argued that the water in chondrites is composed of two components, one that is much more D-rich than comets and may be interstellar, and another that formed at relatively high temperatures in the inner Solar System so that it has a H isotopic composition that is only slightly more D-rich than the solar composition (Deloule *et al.* 1998; Robert 2001; Robert 2002). More recently it has been suggested that the H isotopic composition of water in chondrites at all spatial scales has been enriched in D to varying degrees by two parent body processes (Alexander *et al.* 2010; Alexander *et al.* 2011; Bonal *et al.* 2011): (1) the oxidation of Fe by water and the consequent loss of H_2, and (2) by isotopic exchange with D-rich organics. If correct, none of the chondrites accreted water ice that was as D-rich as in most comets. The exact H isotopic compositions at the time of accretion are still uncertain, but in the CM and CI chondrites their D/H ratios were probably significantly below the terrestrial composition. Presumably the CM and CI water formed or re-equilibrated at smaller radial distances than typical cometary ices. How this can be reconciled with the presence of IDP/comet-like organics in chondrites, as well as NH_3 and HCN in the chondritic ices, remains to be seen.

While it is still not clear when the Earth and the other terrestrial planets accreted their volatiles, if it was at a late stage it would have been after parent body processes on the chondrites' parental asteroids had ceased. Thus, if the chondrites were the sources of the volatiles, their present bulk compositions (water and organics) should reflect what would have been accreted. Not only the H isotopes constrain the likely sources of the volatiles, but also the relative abundances and isotopic compositions of N and the noble gases. At present, no single chondrite group can explain all of these constraints.

5. Summary and Implications

Chondritic meteorites preserve a record of materials that were present in the first few million years of solar nebula evolution. These materials include demonstrably presolar circumstellar grains that were inherited from the protosolar molecular cloud. Most, if not all, chondrites also accreted organic matter when they formed, but generally this material has been heavily modified in the asteroidal parent bodies. In three chondrite groups (CI, CM, and CR) the organic matter has been reasonably well preserved, although it has been modification to varying degrees by hydrothermal alteration. The organic matter in the CI-CM-CR chondrites is broadly divided into solvent soluble and insoluble fractions.

The insoluble fraction (IOM) is a structurally complex, macromolecular material composed of small aromatic moieties that are cross-linked by short, highly branched aliphatic chains. It has a bulk elemental composition that is similar to the average of comet Halley CHON particles, and is dispersed in chondrite matrices as particles that are typically <1 μm across. In bulk, the insoluble material is significantly enriched in D and ^{15}N. Even more extreme isotopic enrichments can be found in localized (usually <1 μm across) hotspots, some of which are associated with so-called globules - spherical to irregular organic particles that are often hollow.

Similar globules, isotopic hotspots and/or bulk isotopic enrichments are found in the organic matter of chondritic interplanetary dust particles and comet 81P/Wild 2 samples. Thus, there is a genetic link between objects that formed between $\sim$3 AU and $\sim$30 AU from the Sun. The enrichments attest to formation in very cold environments, either in the outer Solar System or the protosolar molecular cloud. There are many properties of this organic material that are consistent with an interstellar origin, e.g., hotspot isotopic compositions overlap with those of gas-phase molecules in molecular clouds, a 3.4 μm IR adsorption feature resembles that of diffuse ISM dust and abundances relative to circumstellar grains are similar in all chondrites. However, at present a Solar System origin cannot be ruled out. Since the early Solar System does not appear to have been particularly unusual and this organic material was so widespread in the solar nebula, presumably similar material is an important component of most protoplanetary disks. Heating or sputtering of this material would be a source of PAHs in disks.

The solvent-extractable organic material (SOM) is a complex suite of compounds. Many of these compounds show almost complete structural diversity for a given carbon number, and also have large enrichments in D and, if they contain N, in ^{15}N. The amino acids are thought to have formed by reaction of aldehydes and ketones with NH_3 and HCN very early in the aqueous alteration of the CI-CM-CRs. The fact that volatile NH_3 and HCN were accreted, presumably in ice, strengthens the chondrite-comet connection. Some amino acids in CI-CM-CR chondrites do not readily racemize exhibit 0-20% L-enantiomeric excesses. The L-enantiomeric excesses seem to increase with increasing hydrothermal alteration and decreasing amino acid abundances. Thus, the L-enantiomeric excesses appear to be the product of the alteration or the alteration has amplified small excesses that were inherited at the time of accretion.

Despite the links between chondrites and comets suggested by the organic matter, unlike most comets the water in chondrites, when it was accreted, had D/H ratios that were similar to or depleted relative to Earth. The water in chondrites seems to have formed or isotopically re-equilibrated with H_2 at the higher temperatures of the inner Solar System. How this can be reconciled with the preservation of primitive organics and accretion of volatiles like NH_3 and HCN remains to be seen.

References

Aléon, J., Robert, F., Chaussidon, M., & Marty, B. 2003, *Geochim. Cosmochim. Acta*, 67, 3773

Alexander, C. M. O'D. 2009, in: F. Boulanger, C. Joblin, A. Jones, & S. Madden (eds.) *Interstellar dust from astronomical observations to fundamental studies*, (European Astronom. Soc. Publ. Series), p. 75

Alexander, C. M. O'D., Boss, A. P., Keller, L. D., Nuth, I., J. A., & Weinberger, A. 2007a, in: B. Reipurth, D. Jewitt, & K. Keil (eds.) *Protostars and Planets V*, (Tucson: University of Arizona Press), p. 801

Alexander, C. M. O'D., Cody, G. D., Fogel, M., & Yabuta, H. 2008, in: S. Kwok, & S. A. Sandford (eds.) *Organic Matter in Space* (Hong Kong: Cambridge University Press), p. 293

Alexander, C. M. O'D., Fogel, M., Yabuta, H., & Cody, G. D. 2007b, *Geochim. Cosmochim. Acta*, 71, 4380

Alexander, C. M. O'D., Howard, K. T., Bowden, R., Fogel, M., & Bonal, L. 2011, *Lunar Planet. Sci.*, 42, #1869

Alexander, C. M. O'D., Newsome, S. N., Fogel, M. L., Nittler, L. R., Busemann, H., & Cody, G. D. 2010, *Geochim. Cosmochim. Acta*, 74, 4417

Berger, E. L., Zega, T. J., Keller, L. P., & Lauretta, D. S. 2011, *Geochim. Cosmochim. Acta*, 75, 3501

Bonal, L., Alexander, C. M. O'D., Huss, G. R., & Nagashima, K. 2011, *Lunar Planet. Sci.*, 42, #1287

Botta, O. & Bada, J. L. 2002, *Surveys in Geophysics*, 23, 411

Burbine, T. H., McCoy, T. J., Meibom, A., Gladman, B., & Kiel, K. 2002, in Asteroids III, eds. W. F. Bottke, Jr., A. Cellino, P. Paolicchi, & R. P. Binzel (Tuscon: University of Arizona Press), 653

Busemann, H., *et al.* 2009, *Earth Planet. Sci. Lett.*, 288, 44

Busemann, H., Young, A. F., Alexander, C. M. O'D., Hoppe, P., Mukhopadhyay, S., & Nittler, L. R. 2006, *Science*, 314, 727

Cody, G. D. & Alexander, C. M. O'D. 2005, *Geochim. Cosmochim. Acta*, 69, 1085

Cody, G. D., Alexander, C. M. O'D., Kilcoyne, A. L. D., & Yabuta, H. 2008, in: S. Kwok, & S. A. Sandford (eds.) *Organic matter in space*, (Hong Kong: Cambridge University Press), p. 277

Cody, G. D., Alexander, C. M. O'D., & Tera, F. 2002, *Geochim. Cosmochim. Acta*, 66, 1851

Cody, G. D., Heying, E., Alexander, C. M. O'D., Nittler, L. R., Kilcoyne, A. L. D., Sandford, S. A., & Stroud, R. M. 2011, *Proc. Nat. Acad. Sci.*, 108,

Cooper, G., Kimmich, N., Belisle, W., Sarinana, J., Brabham, K., & Garrel, L., 2001, *Nature*, 414, 879

Dartois, E., Muñoz Caro, G. M., Deboffle, D., Montagnac, G., & d'Hendecourt, L. 2005, *Astron. Astrophys.*, 432, 895

De Gregorio, B. T., Stroud, R. M., Nittler, L. R., Alexander, C. M. O'D., Kilcoyne, A. L. D., & Zega, T. J. 2010, *Geochim. Cosmochim. Acta*, 74, 4454

Deloule, E., Robert, F., & Doukhan, J. C. 1998, *Geochim. Cosmochim. Acta*, 62, 3367

Dermott, S. F., Durda, D. D., Grogan, K., & Kehoe, T. J. J. 2002, in: W. F. Bottke, Jr., A. Cellino, P. Paolicchi, & R. P. Binzel (eds.) *Asteroids III*, (Tucson: University of Arizona Press), p. 423

Duprat, J., *et al.* 2010, *Science*, 328, 742

Ehrenfreund, P., Glavin, D. P., Botta, O., Cooper, G., & Bada, J. L. 2001, *Proc. Nat. Acad. Sci.*, 98, 2138

Ehrenfreund, P., Robert, F., d'Hendencourt, L., & Behar, F. 1991, *Astron. Astrophys.*, 252, 712

Engel, M. H., Macko, S. A., & Silfer, J. A. 1990, *Nature*, 348, 47

Ferrante, R. F., Moore, M. H., Nuth, J. A., & Smith, T. 2000, *Icarus*, 145, 297

Floss, C., Stadermann, F. J., Bradley, J. P., Dai, Z. R., Bajt, S., Graham, G. A., & Lea, A. S. 2006, *Geochim. Cosmochim. Acta*, 70, 2371

Flynn, G. J., Keller, L. P., Feser, M., Wirick, S., & Jacobsen, C. 2003, *Geochim. Cosmochim. Acta*, 67, 4791

Gardinier, A., Derenne, S., Robert, F., Behar, F., Largeau, C., & Maquet, J. 2000, *Earth Planet. Sci. Lett.*, 184, 9

Garvie, L. A. J., Baumgardener, G., & Buseck, P. R. 2008, *Meteor. Planet. Sci.*, 43, 899

Garvie, L. A. J. & Buseck, P. R. 2004, *Earth Planet. Sci. Lett.*, 224, 431

Gilmour, I. 2003, in: A. M. Davis (ed.) *Meteorites, Comets and Planets*, (Oxford: Elsevier-Pergamon), p. 269

Glavin, D. P., Callahan, M. P., Dworkin, J. P., & Elsila, J. E. 2010, *Meteor. Planet. Sci.*, 45, 1948

Gourier, D., Robert, F., Delpoux, O., Binet, L., Vezin, H., Moissette, A., & Derenne, S. 2008, *Geochim. Cosmochim. Acta*, 72, 1914

Gradie, J. C., Chapman, C. R., & Tedesco, E. F. 1989, in: R. P. Binzel, T. Gehrels, & M. S. Matthews (eds.) *Asteroids II*, (Tucson: Univ. Arizona Press), p. 316

Grossman, J. N., Alexander, C. M. O'D., Wang, J., & Brearley, A. J. 2002, *Meteor. Planet. Sci.*, 37, 49

Hayatsu, R. & Anders, E. 1981, in: F. L. Boschke (ed.) *Cosmo- and Geochemistry* (Berlin: Springer-Verlag), p. 1

Hayatsu, R., Matsuoka, S., Scott, R. G., Studier, M. H., & Anders, E. 1977, *Geochim. Cosmochim. Acta*, 41, 1325

Herd, C. D. K., *et al.* 2011, *Science*, 332, 1304

Huang, Y., Alexandre, M. R., & Wang, Y. 2007, *Earth Planet. Sci. Lett.*, 259, 517

Huang, Y., Wang, Y., Alexandre, M. R., Lee, T., Rose-Petruck, C., Fuller, M., & Pizzarello, S. 2005, *Geochim. Cosmochim. Acta*, 69, 1073

Jones, A. P., Duley, W. W., & Williams, D. A. 1990, *QJRAS*, 31, 567

Kawasaki, T., Hatase, K., Fujii, Y., Jo, K., Soai, K., & Pizzarello, S. 2006, *Geochim. Cosmochim. Acta*, 70, 5395

Kebukawa, Y., Alexander, C. M. O'D., & Cody, G. D. 2011, *Geochim. Cosmochim. Acta*, 75, 3530

Keller, L. P., Messenger, S., Flynn, G. J., Clemett, S., Wirick, S., & Jacobsen, C. 2004, *Geochim. Cosmochim. Acta*, 68, 2577

Kissel, J. & Krueger, F. R. 1987, *Nature*, 326, 755

Kress, M. E. & Tielens, A. G. G. M. 2001, *Meteor. Planet. Sci.*, 36, 75

Krot, A. N., Keil, K., Goodrich, C. A., Scott, E. R. D., & Weisberg, M. K. 2003, in: A. M. Davis (ed.) *Meteorites, comets and planets*, (Oxford: Elsevier-Pergamon), p. 83

Lis, D. C., Wootten, A., Gerin, M., & Roueff, E. 2010, *ApJ*, 710, L49

Lisse, C. M., Kraemer, K. E., Nuth, J. A., Li, A., & Joswiak, D. 2007, *Icarus*, 191, 223

Lodders, K. 2003, *ApJ*, 591, 1220

Lyons, J. R. & Young, E. D. 2005, *Nature*, 435, 317

Martins, Z., Alexander, C. M. O'D., Orzechowska, G. E., Fogel, M. L., & Ehrenfreund, P. 2007, *Meteor. Planet. Sci.*, 42, 2125

Martins, Z. & Sephton, M. 2009, in: A. B. Hughes (ed.) *Origins and synthesis of amino acids*, (Weinheim: Wiley-VCH), p. 3

Matrajt, G., *et al.* 2008, *Meteor. Planet. Sci.*, 43, 315

McKeegan, K. D., *et al.* 2006, *Science*, 314, 1724

Messenger, S. 2000, *Nature*, 404, 968

Morbidelli, A., Bottke, W. F., Jr., Froeschle, C., & Michel, P. 2002, in: W. F. Bottke, Jr., A. Cellino, P. Paolicchi, & R. P. Binzel (eds.) *Asteroids III*, (Tucson: University of Arizona Press), p. 409

Mumma, M. J. & Charnley, S. B. 2011, *ARAA*, 49, 471

Muñoz Caro, G. M., *et al.* 2006, *Astron. Astrophys.*, 459, 147

Nakamura, K., Zolensky, M. E., Tomita, S., Nakashima, S., & Tomeoka, K. 2002, *Int. J. Astrobiol.*, 1, 179

Nakamura-Messenger, K., Messenger, S., Keller, L. P., Clemett, S. J., & Zolensky, M. E. 2006, *Science*, 314, 1439

Naraoka, H., Shimoyama, A., Komiya, M., Yamamoto, H., & Harada, K. 1988, *Chem. Lett.*, 17, 831

Nesvorny, D., Jenniskens, P., Levison, H. F., Bottke, W. F., Vokrouhlicky, D., & Gounelle, M. 2010, *ApJ*, 713, 816

Nittler, L. R. 2003, *Earth Planet. Sci. Lett.*, 209, 259

Nuth, J. A., Hill, H. G. M., & Kletetschka, G. 2000, *Nature*, 406, 275

Pendleton, Y. J. & Allamandola, L. J. 2002, *ApJS*, 138, 75

Pendleton, Y. J., Sandford, S. A., Allamandola, L. J., Tielens, A. G. G. M., & Sellgren, K. 1994, *ApJ*, 437, 683

Pizzarello, S., Cooper, G. W., & Flynn, G. J. 2006, in: D. S. Lauretta, & H. Y. McSween Jr. (eds.) *Meteorites and the Early Solar System II*, (Tucson: University of Arizona Press), p. 625

Pizzarello, S. & Holmes, W. 2009, *Geochim. Cosmochim. Acta*, 73, 2150

Pizzarello, S., Huang, Y., & and Alexandre, M. R. 2008, *Proc. Nat. Acad. Sci.* 105, 3700

Pizzarello, S. & Huang, Y. 2005, *Geochim. Cosmochim. Acta*, 69, 599

Pizzarello, S., Huang, Y., & Alexandre, M. R. 2008, *Proc. Nat. Acad. Sci.*, 105, 3700

Pizzarello, S., Huang, Y., & Fuller, M. 2004, *Geochim. Cosmochim. Acta*, 68, 4963

Remusat, L., Derenne, S., Robert, F., & Knicker, H. 2005, *Geochim. Cosmochim. Acta*, 69, 3919

Remusat, L., Guan, Y., Wang, Y., & Eiler, J. M. 2010, *ApJ*, 713, 1048

Remusat, L., Palhol, F., Robert, F., Derenne, S., & France-Lanord, C. 2006, *Earth Planet. Sci. Lett.*, 243, 15

Remusat, L., *et al.* 2009, *ApJ*, 698, 2087

Robert, F. 2001, *Science*, 293, 1056

Robert, F. 2002, *Planet. Space Sci.*, 50, 1227

Robert, F. & Epstein, S. 1982, *Geochim. Cosmochim. Acta*, 46, 81

Sandford, S. A., Allamandola, L. J., Tielens, A. G. G. M., Sellgren, K., Tapia, M., & Pendleton, Y. J. 1991, *ApJ*, 371, 607

Schmitt-Kopplin, P., *et al.* 2010, *Proc. Nat. Acad. Sci.*, 107, 2763

Scott, E. R. D. & Krot, A. N. 2003, in: A. M. Davis (ed.) *Meteorites, Comets and Planets*, (Oxford: Elsevier-Pergamon), p. 143

Sephton, M. A. 2002, *Nat. Prod. Rep.*, 19, 292

Sephton, M. A. 2005, *Phil. Trans. Roy. Soc. A*, 363, 2729

Sephton, M. A. & Gilmour, I. 2001, *Planet. Space Sci.*, 49, 465

Sephton, M. A., Love, G. D., Meredith, W., Snape, C. E., Sun, C.-G., & Watson, J. S. 2005, *Planet. Space Sci.*, 53, 1280

Sephton, M. A., Pillinger, C. T., & Gilmour, I. 1999, *Planet. Space Sci.*, 47, 181
Serra Díaz-Cano, L. & Jones, A. P. 2008, *Astron. Astrophys.*, 492, 127
Shinnaka, Y., Kawakita, H., Kobayashi, H., Jehin, E. l., Manfroid, J., Hutsemékers, D., & Arpigny, C. 2011, *ApJ*, 729, 81
Wang, Y., Huang, Y., Alexander, C. M. O'D., Fogel, M., & Cody, G. 2005, *Geochim. Cosmochim. Acta*, 69, 3711
Wang, Y., Kebukawa, Y., Cody, G. D., & Alexander, C. M. O'D. 2011, *Lunar Planet. Sci.*, 42, #2380
Weisberg, M. K., McCoy, T. J., & Krot, A. N. 2006, in: D. S. Lauretta, & H. Y. McSween, Jr. (ed.) *Meteorites and the Early Solar System II*, (Tucson: University of Arizona Press), 19
Yang, J. & Epstein, S. 1984, *Nature*, 311, 544
Yuen, G., Blair, N., Des Marais, D. J., & Chang, S. 1984, *Nature*, 307, 252
Yurimoto, H. & Kuramoto, K. 2004, *Science*, 305, 1763
Zubko, V., Dwek, E., & Arendt, R. G. 2004, *ApJS*, 152, 211

Discussion

IRVING: How are the enantiomeric excesses in some SOM compounds produced?

ALEXANDER: At this point, it is a complete mystery, but the answer obviously has tremendous astrobiological implications.

SARRE: It somewhat surprises me that PAHs are not seen in comets. Could it be that they are tied up in IOM-like material?

ALEXANDER: In meteorites there are small amounts of free PAHs, but the vast majority are tied up in IOM. So if, as seems likely, the organics in chondrites are related to those in comets, tying up most of the PAHs in IOM-like material seems a likely explanation for why PAHs are not seen in comets.

ELLINGER: Do you have structural information about the position of the D in the IOM and SOM?

ALEXANDER: In the IOM several lines of evidence suggest that the D is primarily associated with the aliphatic material. For the individual SOM compounds, there is not information about whether specific functional groups are more deuterated than others.

The Molecular Universe
Proceedings IAU Symposium No. 280, 2011
José Cernicharo & Rafael Bachiller, eds.

doi:10.1017/S1743921311025063

Biomarkers of Habitable Worlds - Super-Earths and Earths

L. Kaltenegger[1,2]

[1]MPIA, Koenigstuhl 17, 69117 Heidelberg, Germany
email: kaltenegger@mpia.de

[2]CfA, 60 Garden street, Cambridge, 02138 MA, USA

Abstract. A decade of exoplanet search has led to surprising discoveries, from giant planets close to their star, to planets orbiting two stars, all the way to the first extremely hot, rocky worlds with potentially permanent lava on their surfaces due to the star's proximity. Observation techniques have reached the sensitivity to explore the chemical composition of the atmospheres as well as physical structure of some detected planets. Recent advances in detection techniques find planets of less than 10 M_{Earth} (so called Super-Earths), among them some that may potentially be habitable. Two confirmed non-transiting planets and several transiting Kepler planetary candidates orbit in the Habitable Zone of their host star. The detection and characterization of rocky and potentially Earth-like planets is approaching rapidly with future ground- and space-missions, that can explore the planetary environments by analyzing their atmosphere remotely. The results of a first generation space mission will most likely be an amazing scope of diverse planets that will set planet formation, evolution as well as our planet in an overall context.

Keywords. Earth, atmospheric effects, astrobiology

1. Introduction

The current status of exoplanet characterization shows a surprisingly diverse set of giant planets. For a subset of these, some properties have been measured or inferred using radial velocity (RV), micro-lensing, transits, and astrometry. These observations have yielded measurements of planetary mass, orbital elements and (for transits) the planetary radius and during the last few years, physical and chemical characteristics of the upper atmosphere of some of the transiting planets (see e.g. a review by (Seager & Deming 2010) and references therein). Specifically, observations of transits, combined with RV information, have provided estimates of the mass, radius and density of a subset of these planets, ranging from Giant planets to rocky planets like Corot 7b (Léger *et al.* 2009) and Kepler 10b (Batalha *et al.* 2011). Most detected planets orbit extemely close to their host star due to a detection bias and therefore receive high amounts of stellar irradiation and have subsequent high surface temperature. Recent investigations of high precision radial velocity data samples have shown that between 20% and 50% of all sample stars exhibit RV variations indicating the presence of super-Earths or ice giants (Lovis *et al.* 2009), (Howard *et al.* 2011). Among the hundreds of confirmed planets, already two close-by, low mass RV planets, Gl 581 d (Udry *et al.* 2007), (Mayor *et al.* 2009) and HD 85512 b (Pepe *et al.* 2011), with minimum masses of 7 and 3.6 Earth masses, respectively, orbit in the Habitable Zone of their parent star on the outer and inner edge respectively. These close-by planets provide excellent targets for future atmospheric exploration. Several Kepler transit planetary candidates from the February 2011 data release (Borucki *et al.* 2011), that are consistent with rocky models, orbit their host stars also in the Habitable Zone, providing first statistics of the number of planets and eta_{Earth} (see e.g. (Traub 2011)) and a more complete sample is expected in the second Kepler data release in 2012. The discovery of transiting planets with masses below 10 Earth masses and radii consistent with rocky planetary models, answered the important question if planets more massive than Earth could potentially be rocky. 10 Earth masses are used from formation theories as the upper limit for rocky planet formation, for comparison, Uranus has about 15 Earth masses. Above that mass the planet is thought to accumulate a substantial amount of gas that makes it akin to a gas giant, not a rocky planet with an outgassed atmosphere. Where exactly such a cut-off mass is - if it exists at all - is an open question. Recent discoveries by ground based, as well as the Corot and Kepler mission, found planets with masses below 10 Earth masses and densities akin to Neptune as well as Earth, suggesting that there is not one cut-off mass above which a planet is like Neptune and below which it is rocky like Earth or Venus. The first planets below 10 M_{Earth}, with both mass estimates and radius measurements, have provided a wide range of observed radii and densities. Especially in the mass range below $5M_{Earth}$, two planets in the multiple planet system, Kepler 11b and Kepler 11f (Lissauer *et al.* 2011), with 4.3 and $2.3M_{Earth}$ have radii of 1.97 and 2.61 Earth radii and mean densities of 3.1 and 0.7 g/cm^3, respectively. These derived densities also allow substantial envelopes of light gases for this mass range. Recent atmosphere observations for a 6.55 $\pm$ 0.98 Earth mass planet, GJ 1214 b, with a mean density of 1.8 g/cm^3 (Charbonneau *et al.* 2009) (for atmosphere observations see e.g. (Bean *et al.* 2010), (Désert *et al.* 2011)), indicates either hazes or high cloud cover in an expanded atmosphere.

Observing mass and radius alone can not break the degeneracy of a planet's nature due to the effect of an extended atmosphere that can also block the stellar light and increase the observed planetary radius significantly (see e.g. (Adams et al. 2008)). Even if a unique solution would exist, planets with similar density, like Earth and Venus, present very different planetary environments in terms of habitable conditions. Therefore the question refocusses on atmospheric features to characterize a planetary environment. The atmosphere is the only accessible quantity that allows exploring a planetary environment remotely.

2. The Habitable Zone

Different aspects of what determines the boundaries of the HZ have been discussed broadly in the literature. The main differences among these studies are the climatic constraints imposed. Here we focus on the circumstellar HZ defined by J. Kasting in 1993 (Kasting *et al.* 1993) as an annulus around a star where a planet with a $CO_2/H_2O/N_2$ (Earth-like composition) atmosphere, surface pressures between 1 and 10 bar and a sufficiently large water content, like Earth, can host liquid water permanently on a solid surface. In this definition, the two edges of the HZ (see Fig. 1) as well as the equilibrium temperature of the planet, T_{eq}, depend on the Bond albedo of the planet A, the luminosity of the star L_{star}, the planet's semi major axis D, as well as the eccentricity e, of the orbit, and in turn the average stellar irradiation at the planet's location (see also (Selsis *et al.* 2007)). Note that this definition of the HZ implies surface habitability and in turn allows remote detectability of life as we know it. Subsurface life that could exist on planets with very different surface temperatures is not considered here, because of the lack of remotely detectable atmospheric to assert habitability.

A more eccentric orbit increases the annually averaged irradiation proportional to $(1-e^2)^{-1/2}$ (see (Williams & Pollard 2002)). The inner edge of the HZ denotes the location

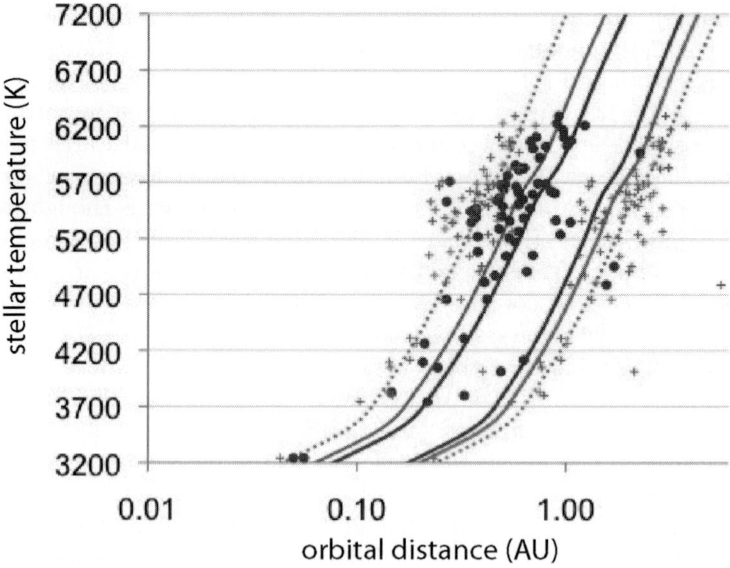

Figure 1. Extent of the HZ for water loss limit for 0 % and 50 % cloud coverage (inner limits) and 100 % cloud coverage (outer limit dashed line) and position of potentially habitable Kepler planetary candidates in the HZ, individual HZ limits are indicated with crosses (Kaltenegger & Sasselov 2011).

where the entire water reservoir can be vaporized by runaway greenhouse conditions, followed by the photo-dissociation of water vapor and subsequent escape of free hydrogen into space. The outer boundary denotes the distance from the star where the maximum greenhouse effect fails to keep CO_2 from condensing permanently, leading to runaway glaciation. Note that at the limits of the HZ, the Bond albedo of a habitable planet is fully determined by its atmospheric composition and de- pends on the spectral distribution of the stellar irradiation. To simply estimate if an Earth-like planet as defined above is potentially habitable ($175K < T_{eq} < 270K$) (see e.g. (Kaltenegger *et al.* 2011a)), one can approximate T_{eq} for Earth-like planets around different stars using the maximum albedo for rocky planet atmospheres, that varies depending on the planet's cloud cover as well as host star. Applying this definition to the Kepler Feb 2011 data release, assuming circular orbits and albedo corresponding to 50% cloud coverage (that is also consistent with the empirical "Venus"-limit of the HZ), leads to 27 Kepler planetary candidates with $175K < T_{eq} < 270K$ (see Fig. 1). Among those are 3 planetary candidates that have radii smaller than 2 Earth radii (Kaltenegger & Sasselov 2011). The potentially rocky planet candidates in multiple systems (see e.g. (Lissauer *et al.* 2011)) are extremely interesting objects because their mass could be determined using transit time variations to calculate a mean density and potentially confirm high density and rocky characteristics.

The HARPS Radial Velocity program recently published the RV data and the orbital parameters of a new low-mass planet of 3.6 ± 0.5 M_{Earth} min. mass around the star HD 85512 (Pepe *et al.* 2011). HD 85512 is a K5V star with 0.126 ± 0.008 L_{Sun}, 0.69 M_{Sun} and an effective temperature, T_{eff}, of 4715 ± 102. HD 85512 b orbits its star in 58.43 ± 0.13 days at 0.26 ± 0.005 AU with an eccentricity of 0.11 ± 0.1, which places it inside the and near to the inner edge of the Habitable Zone (HZ). The planet is one of the least massive planets detected to date. It is the least massive planet confirmed in the HZ. Low mass Main Sequence M dwarfs are the most abundant stars in the

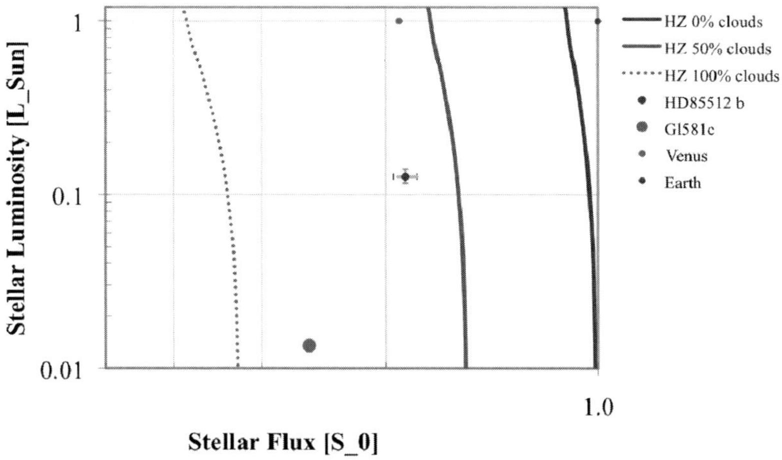

Figure 2. L_{star} in L_{Sun}, versus stellar flux on top of the planetary atmosphere in S_0. The extent of the inner edge of the HZ for the water loss limit for 0%, 50% and 100% (dashed line) cloud coverage (from right to left) and Earth, Venus and Gl 581c are added for comparison (Kaltenegger *et al.* 2011b) . Note that the x-axis decreases towards the right.

galaxy, representing about 75% of the total stellar population. Many planets including potentially rocky planets within the Habitable Zone (HZ) like Gl581d (Selsis *et al.* 2007), (von Bloh *et al.* 2007), (von Paris *et al.* 2011),(Wordsworth *et al.* 2011), (Kaltenegger *et al.* 2011a), (Hu & Ding 2011)) are thus likely to be found in the near future, providing excellent targets ((Scalo *et al.* 2007), (Tarter *et al.* 2007), (Kaltenegger *et al.* 2010a)), that can be probed for atmospheric components, especially for hot planets with extended atmospheres.

Fig. 2 shows the position and borders of the HZ. For comparison the values for Earth and Venus as well as the exoplanet Gl 581c are shown.

Several designs for future space missions that have the explicit purpose of detecting other Earth-like worlds as well as analyzing their characteristics, determining the composition of their atmospheres, and searching for signs of life are under study at ESA as well as NASA. They also have the capacity to investigate the physical properties and composition of a broader diversity of planets, to understand the formation of planets and some even have the capability to interpret potential biosignatures.

3. A SPECTRAL FINGERPRINT OF AN ATMOSPHERE

A planet is a very faint, small object close to a very bright and large object, its parent star. In the visible part of the spectrum we observe the starlight, reflected off the planet, in the IR we detect the planets own emitted flux. The Earth-Sun intensity ratio is about 10^{-7} in the thermal infrared (10 μm), and about 10^{-10} in the visible (0.5 μm) (see Fig. 3). Nevertheless the contrast ratio of hot extrasolar Giant planets (EGP) to their parent stars flux as well as the contrast ratio of a planet to a smaller parent star is much more favorable, making Earth-like planets around small stars very interesting targets (see e.g. (Scalo *et al.* 2007), (Tarter *et al.* 2007)).

Different strategies exist to characterize a planet's atmosphere: direct detection that resolves the planet and star individually and transmission as well as secondary eclipse measurements that subtract the stellar light from a combined star-planet detection. The first imaged exoplanet candidates around young stars show the improvement in direct detection techniques that are designed to resolve the planet and collect its photons. This can currently be achieved for widely separated young objects and has already detected exoplanets candidates (see e.g. (Kalas *et al.* 2008), (Lagrange *et al.* 2010), (Marois *et al.* 2008)). Future telescopes will allow for direct detection in even closer orbits for close-by stars. In the next years ground-based as well as space missions will give us statistics on the number, size, period and orbital distance of planets, extending to terrestrial planets on the lower mass range end as a first step, while future space missions are designed to characterize their atmospheres.

The atmosphere of a planet contains the detectable information to explore the planetary environment remotely. The spectrum of a planet can contain signatures of atmospheric species, what creates its spectral fingerprint. On Earth some atmospheric species exhibiting noticeable spectral features in the planet's spectrum result directly or indirectly from biological activity: the main ones are O_2, O_3, CH_4, and N_2O (see Fig. 3 and (Kaltenegger & Traub 2009), (Rauer *et al.* 2011) (Pallé *et al.* 2009)). CO_2 and H_2O are in addition important as greenhouse gases in a planet's atmosphere and potential sources for high O_2 concentration from photosynthesis (see e.g. (Des Marais *et al.* 2002), (Kaltenegger *et al.* 2010c). Note that similar features could be detectable on Earth-like moons (Williams *et al.* 1997) (Kaltenegger 2010).

Sagan *et al.* (Sagan *et al.* 1993) analyzed a spectrum of the Earth taken by the Galileo probe in 1993, searching for signatures of life and concluded that the large amount of O_2 and the simultaneous presence of CH_4 traces are strongly suggestive of biology for a planet around a Sun-like star. Fig. 3 shows the detectable features in the planet's reflection, emission and transmission spectrum using the Earth itself as a proxy for observations and model fits to data of spectra of the Earth.

Our search for signs of life is based on the assumption that extraterrestrial life shares fundamental characteristics with life on Earth, in that it requires liquid water as a solvent and has a carbon-based chemistry (see e.g. (Brack 1993), (Owen 1980). Life on the

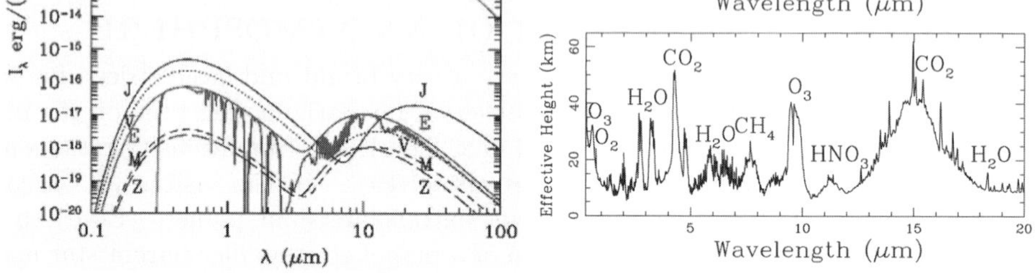

Figure 3. Model of our Solar System (left) (assumed here to be Black Bodies with Earth spectrum shown). Synthetic transmission spectra (right) of the Earth from UV to IR shown. The intensity is given as a fraction of solar intensity as well as the relative height in the atmosphere. The atmospheric features are indicated (figure adapted from (Kaltenegger *et al.* 2010c) and (Kaltenegger & Traub 2009))

base of a different chemistry is not considered here because the vast possible life-forms produce signatures in their atmosphere that are so far unknown. Therefore we assume that extraterrestrial life is similar to life on Earth in its use of the same input and output gases, that it exists out of thermodynamic equilibrium (see e.g. (Lovelock 1975) (Rosenqvist & Chassefiere 1995)). Biomarkers is used here to mean detectable species, or set of species, whose presence at significant abundance strongly suggests a biological origin (e.g. couple CH_4+O_2, or CH_4+O_3). Bio-indicators are indicative of biological processes but can also be produced abiotically. It is their quantities, and detection along with other atmospheric species, in a certain context (for instance the properties of the star and the planet) that points toward a biological origin.

It is relatively straightforward to remotely ascertain that Earth is a habitable planet, replete with oceans, a greenhouse atmosphere, global geochemical cycles, and life - if one has data with arbitrarily high signal-to-noise and spatial and spectral resolution. The interpretation of observations of other planets with limited signal-to-noise ratio and spectral resolution as well as absolutely no spatial resolution, as envisioned for the first generation instruments, will be far more challenging and implies that we need to gather information on the planet environment to understand what we will see (see e.g. (Kaltenegger *et al.* 2010c)): After detection, we will focus on main properties of the planetary system, its orbital elements as well as the presence of an atmosphere using the light curve of the planet or/and a crude estimate of the planetary nature using very low-resolution information (3 or 4 channels).

Presently, radius measurements can only be performed when the planet transits in front of its parent star, by an accurate photometric technique. If the secondary eclipse of the transiting planet can be observed (when the planet passes behind the star), then the thermal emission of the planet can be measured, allowing the retrieval of mean brightness temperature thanks to the knowledge of the radius from the primary transit. If a non-transiting target is observed in both visible and IR ranges, the albedo can be estimated in the visible once the radius is inferred from the IR spectrum, and compared with one derived from the thermal emission only.

Then a higher resolution spectrum will be used to identify the compounds of the planetary atmosphere, constrain the temperature and radius of the observed exoplanet. In that context, we can test if we have an abiotic explanation of all compounds seen in the atmosphere of such a planet. If we do not, we can work with the exciting biotic hypothesis. O_2, O_3, CH_4 are good biomarker candidates that can be detected by a low-resolution (Resolution < 80) spectrograph. Note that the presence of biogenic gases such as O_2/O_3 + CH_4 may imply the presence of a massive and active biosphere, but their absence does not imply the absence of life. Life existed on Earth before the interplay between oxygenic photosynthesis and carbon cycling produced an oxygen-rich atmosphere (see Fig. 4).

The coronagraph and occulter concepts detect the reflected light of a planet and operate in the visible and near infrared (0.5-1μm) with a minimum resolution of 80. The interferometric systems suggested for future direct imaging missions operate in the mid-IR (6-20 μm) and observe the thermal emission emanating from the planet with a minimum resolution of 25. The viewing geometry results in different flux contribution of the overall detected signal from the bright and dark side, for the reflected light, and the planet's hot and cold regions for the emitted flux. Both spectral regions contain the signature of atmospheric gases that can be observed with low resolution and can indicate habitable conditions and, possibly, the presence of a biosphere: CO_2, H_2O, O_3, CH_4, and N_2O in the thermal infrared, and H_2O, O_3, O_2, CH_4 and CO_2 in the visible to near-infrared. The presence or absence of these spectral features (detected collectively) will indicate similarities or differences with the atmospheres of terrestrial planets, and its astrobiological potential.

4. Evolution of biomarkers over geological times on Earth

One crucial factor in interpreting planetary spectra is the point in the evolution of the atmosphere when its biomarkers and its habitability become detectable. The spectrum of the Earth has not been static throughout the past 4.5 billion years ago (Ga). This is due to the variations in the molecular abundances, the temperature structure, and the surface morphology over time. At about 2.3 Ga oxygen and ozone became abundant, affecting the atmospheric absorption component of the spectrum. At about 0.44 Ga, an extensive land plant cover followed, generating the red chlorophyll edge in the reflection spectrum. The composition of the surface (especially in the visible), the atmospheric composition, and temperature-pressure profile can all have a significant influence on the detectabilty of a signal.

Fig. 4 shows theoretical visible and mid-infrared spectra of the Earth over the main six epochs during its geological evolution. The epochs are chosen to represent major developmental stages of the Earth, and life on Earth. If an extrasolar planet is found with a corresponding spectrum, we can use the stages of evolution of our planet to characterizing it, in terms of habitability and the degree to which it shows signs of life, assuming similar evolution time scales. Furthermore we can learn about the evolution of our own planet's atmosphere and possible the emergence of life by observing exoplanets in different stages of their evolution. Earth's atmosphere has experienced dramatic evolution over 4.5 billion years, and other planets may exhibit similar or greater evolution, and at different rates. It shows epochs that reflect significant changes in the chemical composition of the atmosphere. The ozone/oxygen absorption features could have been used to indicate the presence of biological activity on Earth in combination with a reducing gas like CH_4 anytime during the past 50% of the age of the solar system. Different signatures in the atmosphere are clearly detectable over Earth's evolution and observable with low resolution.

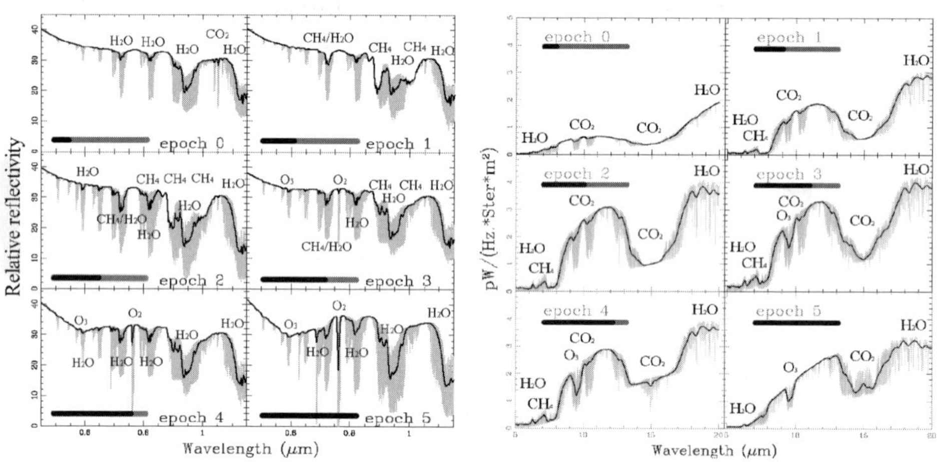

Figure 4. The visible to near-IR (left) and mid IR (right) spectral features on an Earth-like planet change considerably over its evolution from a CO_2 rich (epoch 0) to a CO_2/CH_4-rich atmosphere (epoch 3) to a present-day atmosphere (epoch 5). The bold lines show spectral resolution of 80 and 25 comparable to the proposed visible TPF-C/New World Observer and Darwin/TPF-I mission concept, respectively.

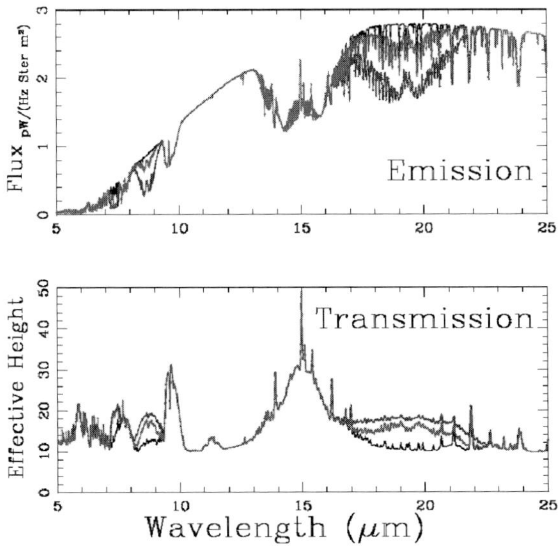

Figure 5. Current Earth spectra model: emergent (top) and transmission (bottom) spectra 5 μm to 25 μm of an Earth atmosphere including water clouds at 1km, 5km and 12km for three volcanic SO_2 concentrations (Black: No eruption. Red: 10x Pinatubo eruption (baseline). Blue: 100x baseline). The calculations are performed at very high resolution (0.1 wavenumbers) and subsequently smoothed for display (R = 150) (adapted from (Kaltenegger *et al.* 2010b)).

Such spectra will be used as part of a big grid to characterize any exoplanets found and influences the design requirements for a spectrometer to detect habitable planets, a grid different international groups are working on (see e.g. (Rauer *et al.* 2011), (Segura *et al.* 2003), (Spiegel *et al.* 2010), (Forget & Pierrehumbert 1997), (Cowan *et al.* 2009), (Goldblatt & Zahnle 2011), (Edson *et al.* 2011), (Kitzmann *et al.* 2010), (Domagal-Goldman *et al.* 2011),(Kaltenegger *et al.* 2010c) and references therein). Other geochemical cycles as well as volcanic events can also leave observable features in a planet's atmosphere that can potentially be detectable (see Fig. 5) (see also e.g. (Kaltenegger *et al.* 2010b)).

5. Summary

Spectroscopy of the atmosphere of extrasolar planets allows to remotely explore a planet's environment, distinguishing Mini-Neptunes from rocky Super-Earths, exploring atmospheric compositions as well as searching for indications for habitability. Any information we collect on habitability is only important in a context that allows us to interpret, what we find. To search for signs of life we need to understand how the observed atmosphere physically and chemically works. Knowledge of the temperature and planetary radius is crucial for the general understanding of the physical and chemical processes occurring on the planet. These parameters as well as an indication of habitability can be determined with low resolution spectroscopy and low photon flux, as assumed for first generation space missions. The combination of spectral information in the visible (starlight reflected off the planet) as well as in the mid-IR (planet's thermal emission) allows a confirmation of atmospheric species, a more detailed characterization of individual planets but also to explore a wide domain of planet diversity. Being able to measure the outgoing shortwave and longwave radiation as well as their variations along the orbit, to determine the albedo and identify greenhouse gases, would in combination allow us to explore the climate system at work on the observed worlds, as well as probe planets similar to our own for habitable conditions.

Acknowledgement

L.K. acknowledges support from NAI and DFG funding ENP Ka 3142/1-1.

References

Adams, E. R., Seager, S., & Elkins-Tanton, L. 2008, *ApJ*, 673, 1160
Batalha, N. M., Borucki, W. J., Bryson, S. T., *et al.* 2011, *ApJ*, 729, 27
Bean, J. L., Miller-Ricci Kempton, E., & Homeier, D. 2010, *Nature*, 468, 669
Borucki, W. J., Koch, D. G., Basri, G., *et al.* 2011, *ApJ*, 736, 19
Brack, A. 1993, Origins of Life and Evolution of the Biosphere, 23, 3
Charbonneau, D., Berta, Z. K., Irwin, J., *et al.* 2009, *Nature*, 462, 891
Cowan, N. B., Agol, E., Meadows, V. S., *et al.* 2009, *ApJ*, 700, 915
Des Marais, D. J., Harwit, M. O., Jucks, K. W., *et al.* 2002, Astrobiology, 2, 153
Désert, J.-M., Bean, J., Miller-Ricci Kempton, E., *et al.* 2011, *ApJ*, 731, L40
Domagal-Goldman, S. D., Meadows, V. S., Claire, M. W., & Kasting, J. F. 2011, Astrobiology, 11, 419
Edson, A., Lee, S., Bannon, P., Kasting, J. F., & Pollard, D. 2011, *Icarus*, 212, 1
Forget, F. & Pierrehumbert, R. T. 1997, Science, 278, 1273
Goldblatt, C. & Zahnle, K. J. 2011, *Nature*, 474
Howard, A. W., Marcy, G. W., Bryson, S. T., *et al.* 2011, ArXiv e-prints
Hu, Y. & Ding, F. 2011, *A&A*, 526, A135
Kalas, P., Graham, J. R., Chiang, E., *et al.* 2008, Science, 322, 1345
Kaltenegger, L. 2010, *ApJ*, 712, L125
Kaltenegger, L., Eiroa, C., Ribas, I., *et al.* 2010a, Astrobiology, 10, 103
Kaltenegger, L., Henning, W. G., & Sasselov, D. D. 2010b, *AJ*, 140, 1370
Kaltenegger, L. & Sasselov, D. 2011, *ApJ*, 736, L25
Kaltenegger, L., Segura, A., & Mohanty, S. 2011a, *ApJ*, 733, 35
Kaltenegger, L., Selsis, F., Fridlund, M., *et al.* 2010c, Astrobiology, 10, 89
Kaltenegger, L. & Traub, W. A. 2009, *ApJ*, 698, 519
Kaltenegger, L., Udry, S., & Pepe, F. 2011b, ArXiv e-prints
Kasting, J. F., Whitmire, D. P., & Reynolds, R. T. 1993, *Icarus*, 101, 108
Kitzmann, D., Patzer, A. B. C., von Paris, P., *et al.* 2010, *A&A*, 511, A66
Lagrange, A.-M., Bonnefoy, M., Chauvin, G., *et al.* 2010, Science, 329, 57
Léger, A., Rouan, D., Schneider, J., *et al.* 2009, *A&A*, 506, 287
Lissauer, J. J., Fabrycky, D. C., Ford, E. B., *et al.* 2011, *Nature*, 470, 53
Lovelock, J. E. 1975, Royal Society of London Proceedings Series B, 189, 167
Lovis, C., Mayor, M., Bouchy, F., *et al.* 2009, in IAU Symposium, Vol. 253, IAU Symposium, 502–505
Marois, C., Macintosh, B., Barman, T., *et al.* 2008, Science, 322, 1348
Mayor, M., Bonfils, X., Forveille, T., *et al.* 2009, *A&A*, 507, 487
Owen, T. 1980, in Astrophysics and Space Science Library, Vol. 83, Strategies for the Search for Life in the Universe, ed. M. D. Papagiannis, 177
Pallé, E., Zapatero Osorio, M. R., Barrena, R., Montañés-Rodríguez, P., & Martín, E. L. 2009, *Nature*, 459, 814
Pepe, F., Lovis, C., Ségransan, D., *et al.* 2011, ArXiv e-prints
Rauer, H., Gebauer, S., Paris, P. V., *et al.* 2011, *A&A*, 529, A8
Rosenqvist, J. & Chassefiere, E. 1995, 43, 3

Sagan, C., Thompson, W. R., Carlson, R., Gurnett, D., & Hord, C. 1993, 365, 715
Scalo, J., Kaltenegger, L., Segura, A. G., *et al.* 2007, Astrobiology, 7, 85
Seager, S. & Deming, D. 2010, *ARA&A*, 48, 631
Segura, A., Krelove, K., Kasting, J. F., *et al.* 2003, Astrobiology, 3, 689
Selsis, F., Kasting, J. F., Levrard, B., *et al.* 2007, *A&A*, 476, 1373
Spiegel, D. S., Raymond, S. N., Dressing, C. D., Scharf, C. A., & Mitchell, J. L. 2010, *ApJ*, 721, 1308
Tarter, J. C., Backus, P. R., Mancinelli, R. L., *et al.* 2007, Astrobiology, 7, 30
Traub, W. A. 2011, ArXiv e-prints
Udry, S., Bonfils, X., Delfosse, X., *et al.* 2007, *A&A*, 469, L43
von Bloh, W., Bounama, C., Cuntz, M., & Franck, S. 2007, *A&A*, 476, 1365
von Paris, P., Gebauer, S., Godolt, M., Rauer, H., & Stracke, B. 2011, *A&A*, 532, A58
Williams, D. M., Kasting, J. F., & Wade, R. A. 1997, *Nature*, 385, 234
Williams, D. M. & Pollard, D. 2002, International Journal of Astrobiology, 1, 61
Wordsworth, R. D., Forget, F., Selsis, F., *et al.* 2011, *ApJ*, 733, L48

Discussion

MELNICK: How doe modest changes in eccentricity affect the atmosphere of Super-Earth and Earth-like planets?

KALTENEGGER: The effect on the spectrum of an Earth-like atmosphere would be small if the eccentricity is moderate, see e.g. the work by (Williams & Pollard 2002). If the eccentricity is high, that can have an influence on the planet's climate see e.g. (Spiegel *et al.* 2010) and therefore also show detectable features in its spectrum. It depends on how dense the atmosphere is and what the heating & cooling times of the atmosphere are in comparison to the time the planet spends close or far away from its star due to eccentricity.

ENCRENAZ: On two occasions the Earth almost froze over.Volcanoes warmed up the crust, probably due to greenhouse effects generated by SO_2. We should also investigate this molecule in exoplanets.

KALTENEGGER: There are different opinions what warmed the atmosphere and got us out of Earth's *snowball* phase(s). The main contenders now are the greenhouse effect generated by CO_2 and/or as you mentioned by SO_2. We explored (Kaltenegger *et al.* 2010b) if you could potentially detect SO_2 and in turn volcanic activity on exoplanets, or if you want to turn the question around, what the minimum amount of SO_2 in Earth's current and Archean atmosphere is, to be detectable with low resolution (see Fig. 4). We found that 10 to 100 times a Pinatubo eruption would leave a remotely detectable SO_2 feature in the infrared spectrum in both emission and transmission. That is very interesting because it would allow us to test whether volcanism is a common occurrence on rocky exoplanets, and if it - and underlying concepts like tectonic or a planetary liquid interior - are common and if they depend on the mass of the planet.

Regarding the snowball phase: One interesting point is that the climate on Earth - and on a geological active rocky planet - is buffered by geological cycles, in our case the carbonate-silicate cycle, that regulates the concentration of CO_2 in Earth's atmosphere. If the surface temperature gets colder, CO_2 does not get rained out as efficiently of our atmosphere (and subsequently subducted), builds up and generates a stronger greenhouse effect that in turn warms the planet, generating liquid rainout and removal of excess

CO_2 from the atmosphere again. That is the underlying concept of why we think that the CO_2 greenhouse effect helped de-freeze snowball Earth. Btw the carbonate-silicate cycle is also the reason why we have a Habitable *Zone*, not just a habitable *Distance*, because the feedback between surface temperature and CO_2 concentration keeps the surface temperature of a rocky planet above freezing for a range of distances from it's star - as well as counters the initial low stellar flux Earth received.

The Molecular Universe
Proceedings IAU Symposium No. 280, 2011
José Cernicharo & Rafael Bachiller, eds.

doi:10.1017/S1743921311025075

The Chemistry of the Early Universe

S. C. O. Glover

Institut für Theoretische Astrophysik, Zentrum für Astronomie der Universität Heidelberg, Albert-Ueberle-Str. 2, 69120 Heidelberg, Germany

Abstract. The chemistry of the early Universe is a fascinating field of study. Even in the absence of any elements heavier than lithium, a surprising degree of chemical complexity proves to be possible, giving the topic considerable interest in its own right. In addition, the fact that molecular hydrogen plays a key role in the formation of the first stars and galaxies means that if we want to understand the formation of these objects, we must first develop a good understanding of the chemical evolution of the gas. In this review, I first give a brief introduction to the chemistry occurring in the gas prior to the formation of the first stars and galaxies, and then go on to discuss in more detail the main chemical processes occurring during the gravitational collapse of gas from intergalactic to protostellar densities, and how these processes influence the final outcome of the collapse.

Keywords. stars: formation – galaxies: formation – cosmology: theory

1. Introduction

Chemistry in the early Universe has proved to be an interesting topic of study for several reasons. Before the first stars and galaxies formed, the elemental composition of the gas was very simple – it consisted only of hydrogen and helium (and their isotopes) plus a tiny trace of lithium – and yet this is sufficient to give rise to a surprising degree of chemical complexity. In addition, the interaction between the gas and the cosmic microwave background (CMB) plays an important role in regulating the chemistry of the gas, and moreover one which may lead to observable changes in the CMB (see e.g. Maoli, Melchiorri & Tosti 1994; Dubrovich, Bajkova & Khaikin 2008; Schleicher *et al.* 2008). Furthermore, understanding the chemical evolution of the gas at these early epochs is vital if one wants to understand the formation of the first stars and galaxies, given the crucial role played by molecular hydrogen (H_2) in this process.

There is a wide range of topics that one could discuss in a review of the chemistry of the early universe, ranging from the physics and chemistry of the recombination epoch at redshift $z > 800$ to the reionization of the Universe at $z < 15$. However, to keep this review to a manageable length, I will focus on only a few key issues, and will not discuss either the recombination epoch or the process of cosmological reionization. Both of these topics are treated in detail elsewhere (see e.g. Fendt *et al.* (2009) and references therein for a lengthy discussion of the physics of the recombination epoch, or the recent review by Morales & Wyithe (2010) for a discussion of the epoch of reionization).

In the next Section, I review what we know about the chemical evolution of the IGM from the end of the recombination epoch until the time at which the earliest protogalaxies form. I follow this up in Section 3 by discussing the chemical evolution of the gas within these early protogalaxies. I show how an understanding of the gas chemistry can allow us to determine the mass scale of the first star-forming minihalos, outline the main chemical changes that occur in the gas during its collapse to protostellar densities, and highlight the important role played by molecular hydrogen throughout this process.

2. Pregalactic chemistry

2.1. *The chemical composition of the gas at the end of the recombination epoch*

At the end of the recombination epoch, at a redshift $z \sim 800$, the chemical makeup of the gas in the Universe – what we might refer to as pregalactic gas – was very simple. The main constituent of the gas was neutral atomic hydrogen, accounting for roughly 75% of the total mass. A small amount of hydrogen remained ionized, with the ratio by number of ionized to neutral hydrogen being roughly $H^+/H \simeq 2.5 \times 10^{-3}$ (Wong, Moss & Scott 2008; note that this value continues to decline at lower redshift, decreasing by about an order of magnitude by the time we reach $z = 100$). Neutral atomic helium (primarily ^{4}He, but also including a small fraction of ^{3}He) accounted for most of the remaining mass, but in this case the abundance of singly or doubly ionized helium was very close to zero, as helium recombination comes to an end at a significantly higher redshift than hydrogen recombination (Wong, Moss & Scott 2008). Deuterium was also present, with a fractional abundance relative to hydrogen $D/H \simeq 2.5 \times 10^{-5}$ (Cyburt, Fields & Olive 2008), and a ratio of D^+ to D that was essentially the same as the ratio of ionized to neutral hydrogen. Finally, there was a trace amount of lithium ($Li/H \simeq 5 \times 10^{-10}$; Cyburt, Fields & Olive 2008), the majority of which was in the form of Li^+. The fractional abundances of all other ionized or molecular species were smaller than 10^{-12} at this point (Alizadeh & Hirata 2010).

2.2. *The chemical evolution of the IGM*

Starting from this simple beginning, many other chemical species were produced as the Universe evolved. Detailed chemical models for the evolution of pregalactic gas have been constructed by a number of authors (see e.g. Stancil, Lepp & Dalgarno 1996, 1998; Galli & Palla 1998; Lepp, Stancil, & Dalgarno 2002; Puy & Signore 2002; Schleicher *et al.* 2008), but in this discussion I will focus on a few general principles that can help us to understand the chemical evolution of the gas at these early times.

As a starting point, it is useful to draw a distinction between what I will call "primary" species, i.e. species that are formed directly from chemical reactions amongst the main chemical constituents of the gas (H, H^+, D, D^+, He and Li^+), and "secondary" species that are formed mainly from reactions involving one or more of the primary species. Good examples of primary species are H^-, formed via the reaction

$$H + e^- \rightarrow H^- + \gamma, \tag{2.1}$$

or H_2^+, formed via the reaction

$$H + H^+ \rightarrow H_2^+ + \gamma. \tag{2.2}$$

The prime example of a secondary species is H_2, as direct formation of H_2 via the reaction

$$H + H \rightarrow H_2 + \gamma \tag{2.3}$$

occurs at a negligible rate (Gould & Salpeter 1963), and most H_2 in metal-free gas forms via the reactions

$$H^- + H \rightarrow H_2 + e^- \tag{2.4}$$

and

$$H_2^+ + H \rightarrow H_2 + H^+. \tag{2.5}$$

The evolution of the fractional abundance of a primary species X is largely governed by the relative sizes of three important timescales: the formation time, defined as

$$t_{\mathrm{form,X}} = \frac{n_{\mathrm{X}}}{R_{\mathrm{form,X}}} \tag{2.6}$$

where n_{X} is the number density of species X and $R_{\mathrm{form,X}}$ is the formation rate per unit volume of X; the photodissociation time,

$$t_{\mathrm{pd,X}} = \frac{1}{R_{\mathrm{pd,X}}}, \tag{2.7}$$

where $R_{\mathrm{pd,X}}$ is the photodissociation rate of species X; and the Hubble time

$$t_{\mathrm{H}} = \frac{1}{H(z)}, \tag{2.8}$$

where $H(z)$ is the Hubble constant.

At high redshift, photodissociation of most molecules and molecular ions by the CMB occurs rapidly, and the photodissociation time is small. Similarly, the high gas density means that the formation timescale of most species is also small. In this regime, both timescales are typically much shorter than the Hubble time, and hence the abundance of species X evolves until it reaches an equilibrium value set by the balance between formation and photodissociation (in which case $t_{\mathrm{form,X}} = t_{\mathrm{pd,X}}$). Since photodissociation is very effective at high redshift, this equilibrium value is generally very small.

As we move to lower redshifts, the cosmological background density decreases, since $\rho(z) \propto (1+z)^3$, and hence the formation time of any given primary species tends to increase. However, the energy density of the CMB decreases even more rapidly, $\rho_{\mathrm{rad}} \propto (1+z)^4$, and the CMB temperature decreases as $T_{\mathrm{CMB}} \propto (1+z)$. What this means is that we eventually reach a redshift at which the number density of photons capable of photodissociating species X starts to fall off exponentially, as the threshold energy for photodissociation starts to fall within the Wien tail of the CMB spectrum.† Once we reach this point, t_{pd} rapidly increases, and we are left with a situation in which t_{form} is by far the shortest of the three timescales. The abundance of species X therefore increases rapidly until $t_{\mathrm{form,X}} \simeq t_{\mathrm{H}}$. Beyond this point, the abundance of X does not increase significantly, as the time required would be longer than the age of the Universe, and we speak of the abundance of the species "freezing out" at some asymptotic value.

The evolution with redshift of secondary species such as H_2 is harder to generalize in this way, as it often depends on the behaviour of more than one primary species, each of which passes through the three stages noted above at a different redshift. For example, H_2 formation via the H_2^+ ion becomes possible at a redshift $z \sim 400$, but H_2 formation via the H^- ion is suppressed until $z \sim 100$, owing to the fact that H^- has a much lower binding energy than H_2^+ (0.755 eV for the former, 2.65 eV for the latter), and hence can be destroyed by photons with significantly smaller energies. The H_2 abundance in the pregalactic gas therefore passes through two periods of relatively rapid growth, one occurring around $z \sim 400$ and the second around $z \sim 100$. Following this second period of growth, the fractional abundance of H_2 is of the order of 10^{-6}, and H_2 is the most abundant molecule in the gas. However, this is still far smaller than the typical amount

† This simple picture is complicated somewhat by the presence of non-thermal photons produced during hydrogen recombination that slightly distort the spectral shape of the CMB (Switzer & Hirata 2005; Hirata & Padmanabhan 2006), but even in this case the general principle is the same.

of H_2 produced within the earliest star-forming protogalaxies, and hence although the chemical evolution of the pregalactic gas is interesting, it is of limited influence: almost all of the chemistry affecting the formation of the first stars and galaxies occurs within the protogalaxies themselves, as I explore in more detail in the next section.

3. Protogalactic chemistry

As we move to lower redshifts, the distribution of matter in the Universe becomes increasingly inhomogeneous. The small perturbations in the dark matter distribution that were present during the recombination epoch grow with time, evolving initially as $\delta \propto t^{2/3}$, but eventually entering the regime in which their evolution becomes non-linear. These perturbations undergo runaway gravitational collapse and virialization, forming dense, gravitationally bound structures that are often referred to as dark matter halos. In the ΛCDM model, the earliest halos to form do so on the smallest scales, with larger, more massive halos forming at later times.

Very little gas accumulates within the earliest halos, as it is prevented from undergoing gravitational collapse by its own thermal pressure, and also by the significant relative velocity that exists between the gas and the dark matter (Tseliakhovich & Hirata 2010). However, once the dark matter halos become sufficiently massive, their gravitational attraction dominates over these effects and substantial quantities of gas flow into them, being heated by adiabatic compression and shocks in the process. In regions where the relative velocity between gas and dark matter is very small, this occurs once the mass of the dark matter halo exceeds $M_{\rm min} \sim 2 \times 10^4 \, {\rm M_\odot}$ (Naoz & Barkana 2007), but in more typical regions where there is a relative velocity between gas and dark matter of several kilometers per second, the minimum halo mass is higher, $M_{\rm min} \sim 10^5 \, {\rm M_\odot}$ (Tseliakhovich, Barkana & Hirata 2010). Low-mass halos with masses close to $M_{\rm min}$ – often referred to as "minihalos" to distinguish them from their considerably more massive, low-redshift brethren – are therefore the first objects that have the potential to form stars. Whether or not the gas within them can actually form stars depends on how rapidly it can cool. This in turn depends on the chemical evolution of the gas within the minihalos, a subject that I address below.

3.1. *H_2 formation and the onset of cooling*

As gas falls into a typical minihalo, it is heated up to a temperature close to the virial temperature of the minihalo

$$T_{\rm vir} \simeq 1200 \left(\frac{M}{10^6 h^{-1} \, {\rm M_\odot}} \right)^{2/3} \left(\frac{1+z}{10} \right) \, {\rm K}, \tag{3.1}$$

(where this expression assumes that the gas is predominantly atomic hydrogen, and that $z \geqslant 10$; see Barkana & Loeb 2001 for a more general expression). In the post-shock gas, the electron fraction decreases due to radiative recombination, but at the same time H_2 forms, primarily via reactions 2.1 and 2.4. The amount of H_2 that forms in the gas can be estimated using a simple toy model for the chemistry first introduced by Tegmark *et al.* (1997). We start by assuming that radiative recombination is the only process affecting the electron abundance, and writing the rate of change of the electron number density as

$$\frac{{\rm d}n_{\rm e}}{{\rm d}t} = -k_{\rm rec} n_{\rm e} n_{\rm H^+}, \tag{3.2}$$

where $n_{\rm e}$ is the number density of electrons, $n_{\rm H^+}$ is the number density of protons, and $k_{\rm rec}$ is the recombination coefficient. If we assume that ionized hydrogen is the only source of free electrons, implying that $n_e = n_{H^+}$, and that the temperature remains roughly constant during the evolution of the gas, then we can solve for the time evolution of the electron fraction:

$$x = \frac{x_0}{1 + k_{rec} n t x_0}, \tag{3.3}$$

where $x \equiv n_e/n$, n is the number density of hydrogen nuclei, and x_0 is the initial value of x. If we now assume that all of the H_2 forms via the pair of reactions given above, and that the only process competing with reaction 2.4 for the H^- is mutual neutralization

$$H^+ + H^- \rightarrow H + H, \tag{3.4}$$

then we can write the time evolution of the H_2 fraction, $x_{H_2} \equiv n_{H_2}/n$, as

$$\frac{dx_{H_2}}{dt} = k_{ra} x n_H \, p_{ad}, \tag{3.5}$$

where k_{ra} is the rate coefficient of reaction 2.1, and p_{ad} is the probability that any given H^- ion will be destroyed by associative detachment (reaction 2.4) rather than by mutual neutralization. Given our assumptions above, this can be written as

$$p_{ad} = \frac{k_{ad} n_H}{k_{ad} n_H + k_{mn} n_{H^+}}, \tag{3.6}$$

where k_{ad} is the rate coefficient for reaction 2.4 and k_{mn} is the rate coefficient for reaction 3.4. If we again assume that $n_e = n_{H^+}$, and in addition assume that the H_2 fraction and the fractional ionization are small, so that $n_H \simeq n$, then the expression for p_{ad} can be simplified to

$$p_{ad} = \left(1 + \frac{k_{mn}}{k_{ad}} x\right)^{-1}. \tag{3.7}$$

Substituting this into Equation 3.5, we obtain

$$\frac{dx_{H_2}}{dt} = k_{ra} x n_H \left(1 + \frac{k_{mn}}{k_{ad}} x\right)^{-1}. \tag{3.8}$$

We can solve for the time evolution of the H_2 fraction

$$x_{H_2} = \frac{k_{ra}}{k_{rec}} \ln\left(\frac{1 + x_0 k_{mn}/k_{ad} + t/t_{rec}}{1 + x_0 k_{mn}/k_{ad}}\right), \tag{3.9}$$

where $t_{rec} = (k_{rec} n x_0)^{-1}$ is the recombination time, and where we have once again assumed that $n_H \sim n$. If the initial fraction ionization is small, so that $x_0 \ll k_{ad}/k_{mn}$, then we can write this expression in the simpler form

$$x_{H_2} = \frac{k_{ra}}{k_{rec}} \ln\left(1 + t/t_{rec}\right). \tag{3.10}$$

This analysis tells us two important things. First, it shows us that the H_2 fraction evolves only logarithmically with time: most of the H_2 forms within the first few recombination times, with H_2 formation at later times being strongly suppressed owing to the loss of the free electrons from the gas due to recombination. Second, it shows us that the amount of H_2 that forms depends on the ratio of the rate coefficients for H^- formation (reaction 2.1) and H^+ recombination, k_{ra}/k_{rec}. The value of this ratio is strongly temperature dependent, and can be written approximately as

$$\frac{k_{ra}}{k_{rec}} \simeq 10^{-8} T^{1.5}. \tag{3.11}$$

The amount of H_2 that forms in the gas therefore increases significantly with increasing temperature. As the H_2 cooling rate is also a strongly increasing function of temperature (see e.g. Flower *et al.* 2000), one finds that there is a reasonably sharp divide occurring at a critical virial temperature $T_{\rm crit} \sim 1000$ K between minihalos that contain gas that can cool rapidly and form stars, and minihalos that contain gas that barely cools over a Hubble time (see e.g. Tegmark *et al.* 1997; Yoshida *et al.* 2003). If we convert this critical virial temperature into a corresponding critical minihalo mass using Equation 3.1, we find that

$$M_{\rm crit} \simeq 7.6 \times 10^5 h^{-1} \left(\frac{1+z}{10}\right)^{-3/2} {\rm M}_\odot. \tag{3.12}$$

At $z > 40$, this value is smaller than the minimum mass $M_{\rm min} \sim 10^5$ M$_\odot$ required to overcome the effects of the streaming of the gas relative to the dark matter, which therefore sets the minimum mass scale for a star-forming minihalo. At $z < 40$, however, $M_{\rm crit} > M_{\rm min}$, and it is the amount of H_2 that can form within the gas that is responsible for setting the minimum mass scale. One implication of this result is that there will be a population of small minihalos that forming at $z < 40$ that contain significant gas fractions, but that do not form stars, because their gas is unable to cool in less than a Hubble time. These small starless minihalos may be important sinks for ionizing photons during the epoch of reionization (Haiman, Abel & Madau 2001).

3.2. *The initial collapse phase*

In minihalos where H_2 cooling is efficient, the gas cools and collapses under its own self-gravity, collecting at the centre of the minihalo. The evolution of the gas at this stage depends on the amount of H_2 it is able to form, which in turn depends upon the initial ionization state of the gas.

During the formation of the very first Population III stars, the initial fractional ionization of the gas is the same as the residual ionization in the intergalactic medium, i.e. $x_0 \sim 2 \times 10^{-4}$ at $z \sim 30$. In this case, the amount of H_2 that forms in the gas is typically enough to cool it to a temperature of $T \sim 200$ K but not below. At this temperature, deuterated hydrogen, HD, forms easily from H_2 via

$$H_2 + D^+ \rightarrow HD + H^+, \tag{3.13}$$

but its destruction via the inverse reaction

$$HD + H^+ \rightarrow H_2 + D^+ \tag{3.14}$$

is starting to become ineffective, owing to the fact that this reaction is endothermic by around 0.04 eV. At $T = 200$ K, chemical fractionation therefore boosts the HD/H_2 ratio by a factor of 20 compared to the cosmic deuterium-to-hydrogen ratio. This boost to the HD abundance, together with the fact that HD is a much more effective coolant than H_2 at low temperatures (see e.g. Flower *et al.* 2000), allows HD to become an important coolant at 200 K, but detailed studies have shown that the extra cooling provided by HD is not enough to reduce the temperature significantly below 200 K (Bromm, Coppi & Larson 2002). Therefore, H_2 continues to dominate the cooling and control the further evolution of the gas. In this scenario, the collapse of the gas is greatly slowed once its density reaches a value of around 10^4 cm^{-3}, corresponding to the critical density $n_{\rm crit}$, at which the rotational and vibrational level populations of H_2 approach their local thermodynamic equilibrium (LTE) values. At densities higher than this critical density, the H_2 cooling rate per unit volume scales only linearly with n (compared to a quadratic dependence, $\Lambda_{H_2} \propto n^2$ at lower densities), while processes such as compressional heating continue to increase more rapidly with n. As a result, the gas temperature begins to increase once the density exceeds $n_{\rm crit}$.

Gas reaching this point in the collapse enters what Bromm, Coppi & Larson (2002) term a "loitering" phase, during which cold gas accumulates in the centre of the halo but only slowly increases its density. This loitering phase ends once the mass of cold gas that has accumulated exceeds the local value of the Bonnor-Ebert mass (Bonnor 1956; Ebert 1955), given in this case by (Abel, Bryan, & Norman 2002)

$$M_{\rm BE} \simeq 40 T^{3/2} n^{-1/2} \, {\rm M}_{\odot}, \tag{3.15}$$

which for $n \sim 10^4$ cm^{-3} and $T \sim 200$ K yields $M_{\rm BE} \sim 1000 \, {\rm M}_{\odot}$. Once the mass of cold gas exceeds $M_{\rm BE}$, its collapse speeds up again, and becomes largely decoupled from the larger-scale behaviour of the gas. The next notable event to occur in the gas is the onset of three-body H_2 formation, which is discussed in Section 3.3 below.

If the initial fractional ionization of the gas is significantly higher than the residual fraction in the IGM, then a slightly different chain of events occurs. A larger initial fractional ionization implies that $t_{\rm rec}$ is shorter and hence leads to a logarithmic increase in the amount of H_2 formed after a given physical time. The higher H_2 fraction produced in this way allows the gas to cool to a slightly lower temperature, thereby boosting the HD abundance to a significantly higher level than is reached in the low ionization case. This allows HD to take over as the dominant coolant, driving the temperature down further. A number of studies have examined this HD-dominated regime (see e.g. Nakamura & Umemura 2002; Nagakura & Omukai 2005; Johnson & Bromm 2006; Yoshida *et al.* 2007; McGreer & Bryan 2008; Kreckel *et al.* 2010), and have shown that the minimum temperature reached in this case can be as low as the CMB temperature, and that the higher critical density of HD, $n_{\rm crit,HD} \sim 10^6$ cm^{-3}, means that the gas does not reach the loitering phase until much later in its collapse. Once the gas does reach this phase, however, its subsequent evolution is very similar to that in the low-ionization case discussed above. Cold gas accumulates at $n \sim n_{\rm crit}$ until its mass exceeds the Bonnor-Ebert mass, which in this case is $M_{\rm BE} \sim 40 \, {\rm M}_{\odot}$ if $T = 100$ K and $n = 10^6$ cm^{-3}. Once the gas mass exceeds $M_{\rm BE}$, the collapse speeds up again, and the gas begins to heat up. Aside from the substantial difference in the size of $M_{\rm BE}$, the main difference between the evolution of the gas in this case and in the low ionization case lies in the fact that in the high ionization case, the gas reheats from $T \sim 200$ K or below to $T \sim 1000$ K much more rapidly than in the low ionization case. As we shall see later, this period of rapid heating may exert an important influence on the ability of the gas to fragment.

Several different scenarios have been identified that lead to an enhanced fractional ionization in the gas, and that potentially allow the gas to reach the HD-dominated regime. Gas within minihalos with $T_{\rm vir} > 9000$ K will become hot enough for collisional ionization of hydrogen to supply the necessary electrons, but halos of this size will typically have at least one star-forming progenitor (Johnson, Greif & Bromm 2008; Greif *et al.* 2010), and hence typically will have already been enriched with metals. Other possibilities include the formation of protogalaxies in "fossil" HII regions, i.e. regions that are recombining after having been ionized by a previous Population III protostar (Oh & Haiman 2003; Nagakura & Omukai 2005; Yoshida *et al.* 2007), or that are irradiated with a significant flux of X-rays (e.g. Glover & Brand 2003) or cosmic rays (Stacy & Bromm 2007; Jasche, Ciardi & Ensslin 2007). However, recent work by Wolcott-Green & Haiman (2011) has shown that the HD forming within a primordial minihalo is very susceptible to photodissociation by even a weak extragalactic UV background (see also Yoshida, Omukai

& Hernquist 2007), and so it is possible that gas will reach the HD-dominated regime in only a few minihalos.

3.3. *Three-body H_2 formation*

Once the collapsing gas reaches a density of around 10^8–10^9 cm^{-3}, its chemical makeup starts to change significantly. The reason for this is that at these densities, the formation of H_2 via the three-body reactions (Palla, Salpeter, & Stahler 1983)

$$\mathrm{H + H + H \rightarrow H_2 + H,} \tag{3.16}$$

$$\mathrm{H + H + H_2 \rightarrow H_2 + H_2,} \tag{3.17}$$

$$\mathrm{H + H + He \rightarrow H_2 + He,} \tag{3.18}$$

starts to become significant. These reactions quickly convert most of the hydrogen in the gas into H_2. At the same time, however, they generate a substantial amount of thermal energy: every time an H_2 molecule forms via one of these three-body reactions, its binding energy of 4.48 eV is converted into heat, and so at these densities, three-body H_2 formation heating can become the main process responsible for heating the gas. For this reason, even though the abundance of H_2, the dominant coolant during this phase of the collapse, increases by more than two orders of magnitude, the gas typically does not cool significantly. Indeed, the temperature often increases.

One major uncertainty in current treatments of the gas chemistry at this point in the collapse is exactly how quickly the gas becomes molecular. Although reaction 3.16 is the dominant source of H_2 at these densities, the rate coefficient for this reaction is poorly known, with published values differing by almost two orders of magnitude at 1000 K, and by an even larger factor at lower temperatures (Glover 2008; Turk *et al.* 2011). The effects of this uncertainty have recently been studied by Turk *et al.* (2011). They show that it has little effect on the density profile of the gas, and only a limited effect on the temperature profile. However, it has far more significant effects on the morphology of the gas and on its velocity structure. Simulations in which a high value was used for the three-body rate coefficient find that gravitational collapse occurs more rapidly, and show that the molecular gas develops a much more flattened, filamentary structure. Significant differences are also apparent in the infall velocities and the degree of rotational support. The effect that this has on the ability of the gas to fragment (see Section 3.5 below) is not yet understood.

3.4. *The final stages of collapse*

As the gas collapses to even higher densities, several important events occur which affect its ability to cool. The first of these occurs at a density of around 10^{10} cm^{-3}, where the main rotational and vibrational lines of H_2 start to become optically thick. This reduces the effectiveness of H_2 cooling, leading to an ongoing rise in the gas temperature. In one-dimensional simulations (e.g. Omukai & Nishi 1998; Omukai *et al.* 1998; Ripamonti *et al.* 2002; Ripamonti & Abel 2004), it is possible to treat optically thick H_2 cooling accurately by solving the full radiative transfer problem. These models show that although the optical depth of the gas becomes large at frequencies corresponding to the centers of the main H_2 emission lines, the low continuum opacity of the gas allows photons to continue to escape through the wings of the lines, with the result that the H_2 cooling rate is suppressed far less rapidly as the collapse proceeds than one might at first expect (see Omukai *et al.* 1998 for a detailed discussion of this point). In three-dimensional simulations, more approximate treatments of the cooling are necessary (see e.g. Yoshida *et al.* 2006; Turk, Abel & O'Shea 2009; Clark *et al.* 2011a), but the general findings are the same – H_2 cooling is suppressed to some extent, but still provides enough dissipation of energy to allow the collapse to continue without a large increase in the temperature.

A second important event occurs once the number density reaches $n \sim 10^{14}$ cm^{-3}. At this density, collision-induced emission from H_2 (the inverse of the more familiar collision-induced absorption; see e.g. Frommhold 1993) becomes important. An isolated H_2 molecule has no dipole moment, but the activated complex formed during the collision of two H_2 molecules can have a dipole moment, and hence can emit radiation via dipole transitions. In principle, this can occur in gas of any density, but the probability of a photon being emitted in any given collision is very small, owing to the short lifetime of the collision state ($\Delta t < 10^{-12}$ s at the temperatures relevant for Pop. III star formation; see Ripamonti & Abel 2004). For this reason, collision-induced emission (CIE) becomes an important process only at very high gas densities. Another consequence of the short lifetime of the collision state is that the individual lines associated with the dipole transitions become so broadened that they actually merge into a continuum. This is important, as it means that the high opacity of the gas in the rovibrational lines of H_2 does not significantly reduce the amount of energy that can be radiated away by CIE. Therefore, once the gas reaches a sufficiently high density, CIE becomes the dominant form of cooling, as pointed out by several authors (Omukai & Nishi 1998; Ripamonti *et al.* 2002; Ripamonti & Abel 2004).

The most detailed study of the effects of CIE cooling on the collapse of primordial gas was carried out by Ripamonti & Abel (2004). They showed that CIE cooling could actually become strong enough to trigger a thermal instability, However, the growth rate of this instability is longer than the gravitational free-fall time, meaning that it is unlikely that this process can drive fragmentation during the initial collapse of the gas.

The phase of the collapse dominated by CIE cooling lasts for only a relatively short period of time. The gas becomes optically thick in the continuum once it reaches a density $n \sim 10^{16}$ cm^{-3} (Omukai & Nishi 1998; Ripamonti & Abel 2004), which strongly suppresses any further radiative cooling. Once this occurs, the gas temperature rises until it reaches a point at which the H_2 begins to dissociate. At these densities, this occurs at a temperature $T \sim 3000$ K. The dissociation of H_2 slows the temperature rise for a while, as most of the energy released during the collapse goes into dissociating the H_2 rather than raising the temperature. However, once the H_2 is exhausted, the temperature once again begins to climb steeply. The thermal pressure in the interior of the collapsing core rises rapidly and eventually becomes strong enough to halt the collapse. At the point at which this occurs, the size of the dense core is around 0.1 AU, its mass is around $0.01\,M_\odot$ and its mean density is of order 10^{20} cm^{-3} (Yoshida, Omukai & Hernquist 2008). It is bounded by a strong accretion shock. This pressure-supported, shock-bounded core is the structure that we identify as a Population III protostar.

3.5. *Formation and fragmentation of the protostellar accretion disk*

Most numerical simulations of the formation of Population III stars stop at or before the point at which the protostar forms, owing to the increasingly small timesteps that need to be taken to follow the evolution of the gas. As significant fragmentation of the gas during the initial collapse is uncommon (see e.g. Turk, Abel & O'Shea 2009, who find that fragmentation into a binary system occurs in only one out of five cases), most studies have assumed that the gas surrounding the protostar does not fragment at later times either, but instead is simply accreted by the existing protostar, either directly or via a protostellar accretion disk. A good overview of the results of these studies can be found in Bromm & Yoshida (2011).

However, in the past few years, a number of simulations have been performed that attempt to directly model the evolution of the gas after the formation of the first Population III protostar using a technique developed for studies of contemporary star formation (see e.g. Clark *et al.* 2008; Stacy, Greif & Bromm 2010; Clark *et al.* 2011a,b; Greif *et al.* 2011; Smith *et al.* 2011). Gravitationally bound regions of gas that become smaller than some pre-selected size scale are replaced by what are usually termed sink particles (in an SPH code; see e.g. Bate, Bonnell & Price 1995) or sink cells (in a grid code; see e.g. Federrath *et al.* 2010). These sinks can accrete gas from their surroundings and continue to interact gravitationally with the surrounding gas, but allow one to neglect the very small-scale hydrodynamical flows that would otherwise force one to take very small numerical timesteps owing to the Courant condition.

One of the main results of these simulations is the demonstration that the first protostar to form does not simply accrete all of the available gas. The infalling gas forms an accretion disk around the protostar, but this disk is unable to transfer gas inwards, onto the protostar, rapidly enough to balance the rate at which new material is falling onto the disk. The disk therefore grows in mass until it becomes gravitationally unstable, at which point it begins to fragment, forming multiple protostars (Clark *et al.* 2011b). The gravitational interaction between these protostars and the surrounding gas can trigger further fragmentation, and close encounters between protostars may lead to mergers, or in some cases to protostars being ejected from the center of the minihalo (Greif *et al.* 2011).

Molecular hydrogen plays a crucial role in this process. The midplane temperature of the protostellar accretion disk is typically 1500–2000 K, and the disk is maintained at this temperature primarily by a combination of H_2 line cooling and CIE cooling (Clark *et al.* 2011b). If the molecular hydrogen were not present, the main disk coolant would be bound-free emission from H^-, which becomes effective only at a temperature $T \sim 6000$ K. The disk would therefore be considerably hotter, allowing it to transfer gas more rapidly onto the central protostar, and making it less susceptible to gravitational instability. In this case, previous studies have shown that the disk would probably remain stable and would not fragment (Tan & McKee 2004; Tan & Blackman 2004; Mayer & Duschl 2005a).

As well as regulating the disk temperature, molecular hydrogen also provides the cooling necessary to allow the fragments forming in the disk to survive as distinct, gravitationally-bound objects. Studies of the fragmentation of self-gravitating accretion disks have shown that in order for fragments to survive, they must be able to cool within roughly half of an orbital period (Gammie 2001). Fragments that cannot cool this quickly undergo a thermal bounce and are then sheared apart by disk rotation, rather than continuing to collapse. In the case of the fragments forming in a Pop. III accretion disk, two key processes provide the necessary cooling: CIE cooling at $n < 10^{16}$ cm^{-3} and the collisional dissociation of H_2 at $n > 10^{16}$ cm^{-3} (Clark *et al.* 2011b).

Finally, there are some indications that the chemical state of the gas at early times during its collapse exerts a substantial influence on the degree of fragmentation that occurs at much later times. As we have already seen, if HD cooling becomes efficient during the collapse of the gas, then the gas can cool to a significantly lower temperature than in the standard H_2-dominated case. However, a consequence of this is that the gas undergoes a more rapid period of reheating once the HD molecules reach LTE, during which it has a very stiff effective equation of state. This leads to a pronounced loss of small-scale structure from the gas, which appears to reduce the amount of fragmentation that occurs at later times in the evolution of the minihalo (Clark *et al.* 2011a; Greif *et al.* 2011).

Acknowledgement

Financial support for this work was provided by the Baden-Württemberg-Stiftung via their program International Collaboration II (grant P-LS-SPII/18), from the German Bundesministerium für Bildung und Forschung via the ASTRONET project STAR FORMAT (grant 05A09VHA), and by a Frontier grant of Heidelberg University sponsored by the German Excellence Initiative.

References

Abel, T., Bryan, G. L., & Norman, M. L. 2002, *Science*, 295, 93
Alizadeh, E. & Hirata, C. M. 2010, *Phys. Rev. D.*, submitted; arXiv:1012.2378
Barkana, R. & Loeb, A., 2001, *Phys. Rep.*, 349, 125
Bate, M. R., Bonnell, I. A., & Price, N. M., 1995, *MNRAS*, 277, 362
Bonnor, W. B., 1956, *MNRAS*, 116, 351
Bromm, V., Coppi, P. S., & Larson, R. B., 2002, *ApJ*, 564, 23
Bromm, V. & Yoshida, N., 2011, *ARA&A*, in press; arXiv:1102.4638
Clark, P. C., Glover, S. C. O., & Klessen, R. S., 2008, *ApJ*, 672, 757
Clark, P. C., Glover, S. C. O., Klessen, R. S., & Bromm, V. 2011a, *ApJ*, 727, 110
Clark, P. C., Glover, S. C. O., Smith, R. J., Greif, T. H., Klessen, R. S., & Bromm, V. 2011b, *Science*, 331, 1040
Cyburt, R. H., Fields, B. D., & Olive, K. A., 2008, *JCAP*, 11, 012
Dubrovich, V., Bajkova, A., & Khaikin, V. B., 2008, New Astron., 13, 28
Ebert, R., 1955, Z. Astrophys., 37, 217
Federrath, C., Banerjee, R., Clark, P. C., & Klessen, R. S., 2010, *ApJ*, 713, 269
Fendt, W. A., Chluba, J., Rubiño-Martín, J. A., & Wandelt, B. D., 2009, *ApJS*, 181, 627
Flower, D. R., Le Bourlot, J., Pineau des Forêts, G., & Roueff, E., 2000, *MNRAS*, 314, 753
Frommhold, L., 1993, Collision-Induced Absorption in Gases. (Cambridge Univ. Press: Cambridge)
Galli, D. & Palla, F. 1998, *A&A*, 335, 403
Gammie, C. F., 2011, *ApJ*, 553, 174
Glover, S. C. O., 2008, in First Stars III, eds. O'Shea, B. W., Heger, A., Abel, T., AIP Press, pp. 25-29
Glover, S. C. O. & Brand, P. W. J. L., 2003, *MNRAS*, 340, 210
Gould, R. J. & Salpeter, E. E. 1963, *ApJ*, 138, 393
Greif, T. H., Glover, S. C. O., Bromm, V., & Klessen, R. S., 2010, *ApJ*, 716, 510
Greif, T., Springel, V., White, S., Glover, S., Clark, P., Smith, R., Klessen, R., & Bromm, V., 2011, *ApJ*, in press; arXiv:1101.5491
Haiman, Z., Abel, T., & Madau, P., 2001, *ApJ*, 551, 599
Hirata, C. M. & Padmanabhan, N. 2006, *MNRAS*, 372, 1175
Jasche, J., Ciardi, B., & Ensslin, T. A., 2007, *MNRAS*, 380, 417
Johnson, J. L. & Bromm, V., 2006, *MNRAS*, 366, 247
Johnson, J. L., Greif, T. H., & Bromm, V., 2008, *MNRAS*, 388, 26
Kreckel, H., Bruhns, H., Čížek, M., Glover, S. C. O., Miller, K. A., Urbain, X., & Savin, D. W., 2010, *Science*, 329, 69
Lepp, S. H., Stancil, P. C., & Dalgarno, A. 2002, *J. Phys. B*, 35, 57
Maoli, R., Melchiorri, F., & Tosti, D., 1994, *ApJ*, 425, 372
Mayer, M. & Duschl, W. J., 2005, *MNRAS*, 356, 1
McGreer, I. D. & Bryan, G. L., 2008, *ApJ*, 685, 8
Morales, M. F. & Wyithe, J. S. B. 2010, *ARA&A*, 48, 127
Nagakura, T. & Omukai, K., 2005, *MNRAS*, 364, 1378
Nakamura, F. & Umemura, M., 2002, *ApJ*, 569, 549
Naoz, S. & Barkana, R. 2007, *MNRAS*, 377, 667
Oh, S. P. & Haiman, Z., 2003, *MNRAS*, 346, 456

Omukai, K. & Nishi, R., 1998, *ApJ*, 508, 141
Omukai, K., Nishi, R., Uehara, H., & Susa, H., 1998, *Prog. Theor. Phys.*, 99, 747
Palla, F., Salpeter, E. E., & Stahler, S. W., 1983, *ApJ*, 271, 632
Puy, D. & Signore, M., 2002, New Astron. Reviews, 46, 709
Ripamonti, E., Haardt, F., Ferrara, A., & Colpi, M., 2002, *MNRAS*, 334, 401
Ripamonti, E. & Abel, T., 2004, *MNRAS*, 348, 1019
Schleicher, D. R. G., Galli, D., Palla, F., Camenzind, M., Klessen, R. S., Bartelmann, M., & Glover, S. C. O. 2008, *A&A*, 490, 521
Smith, R. J., Glover, S. C. O., Clark, P. C., Greif, T., & Klessen, R. S., 2011, *MNRAS*, 414, 3633
Stacy, A. & Bromm, V., 2007, *MNRAS*, 382, 229
Stacy, A., Greif, T. H., & Bromm, V., 2010, *MNRAS*, 403, 45
Stancil, P. C., Lepp, S., & Dalgarno, A. 1996, *ApJ*, 458, 401
Stancil, P. C., Lepp, S., & Dalgarno, A. 1998, *ApJ*, 509, 1
Switzer, E. R. & Hirata, C. M. 2005, Phys. Rev. D, 72, 083002
Tan, J. C. & Blackman, E. G., 2004, *ApJ*, 603, 401
Tan, J. C. & McKee, C. F., 2004, *ApJ*, 603, 383
Tegmark, M., Silk, J., Rees, M. J., Blanchard, A., Abel, T., & Palla, F., 1997, *ApJ*, 474, 1
Tseliakhovich, D., Barkana, R., & Hirata, C. 2010, arXiv:1012.2574
Tseliakhovich, D. & Hirata, C. 2010, Phys. Rev. D, 82, 083520
Turk, M. J., Abel, T., & O'Shea, B. 2009, *Science*, 325, 601
Turk, M. J., Clark, P. C., Glover, S. C. O., Greif, T. H., Abel, T., Klessen, R. S., & Bromm, V., 2011, *ApJ*, 726, 55
Wolcott-Green, J. & Haiman, Z., 2011, *MNRAS*, 412, 2603
Wong, W. Y., Moss, A., & Scott, D. 2008, *MNRAS*, 386, 1023
Yoshida, N., Abel, T., Hernquist, L., & Sugiyama, N. 2003, *ApJ*, 592, 645
Yoshida, N., Omukai, K., Hernquist, L., & Abel, T., 2006, *ApJ*, 652, 6
Yoshida, N., Omukai, K., & Hernquist, L., 2007, *ApJ*, 667, L117
Yoshida, N., Oh, S. P., Kitayama, T., & Hernquist, L., 2007, *ApJ*, 663, 687
Yoshida, N., Omukai, K., & Hernquist, L., 2008, *Science*, 321, 669

The Molecular Universe
Proceedings IAU Symposium No. 280, 2011
José Cernicharo & Rafael Bachiller, eds.

doi:10.1017/S1743921311025087

Observations of Molecules in High Redshift Galaxies

Kirsten Kraiberg Knudsen[1]

[1]Department of Earth and Space Sciences, Chalmers University of Technology
Onsala Space Observatory, SE-439 92 Onsala, Sweden
email: kraiberg@chalmers.se

Abstract. I present an overview of the molecular gas observations in high redshift galaxies. This field has seen tremendous progress in the past few years, with an increased number of detections of other molecules than CO. The molecular line observations are done towards different classes of massive starbursts, including submillimeter galaxies, quasars, and massive gas-rich disks. I will highlight results of detections of HCN, HCO+, and other small molecules, as well as the Spitzer detections of PAHs. Additionally, I will discuss about the excitation of CO and other species in the high-z galaxies and put this in the context of new telescopes such as ALMA.

Keywords. ISM: molecules, galaxies: evolution, galaxies: ISM, galaxies: high-redshift, galaxies: starburst, quasars: emission lines, infrared: ISM, radio lines: galaxies, submillimeter

1. Introduction

The number of observations of molecular gas in high redshift galaxies has increased tremendously over the past decade. The first detection of CO in a high redshift, $z > 1$(†), galaxy was towards IRAS F10214+4724 ($z = 2.2867$; Brown & Vanden Bout, 1991; Solomon, Radford & Downes, 1992), which is an IR-luminous galaxy. For a long time it was also the best studied high redshift galaxy as it is one of the brightest IR galaxies at cosmological distance due to gravitational lensing. Observations of molecular gas in high redshift galaxies an important part of studying galaxy formation and evolution as the molecular gas traces the gas that is available for star formation and the gas that is present in the star forming regions.

In this contribution I will give an overview of the numerous molecular line studies in high redshift galaxies. This field has expanded rapidly in the past few years with many new and exciting results, and thus I will not attempt to give a complete collection of all high redshift molecular line detections, but rather give a broad overview highlighting how these studies provide valuable insight in the nature of high redshift star forming galaxies and how it places the galaxies in the context of galaxy formation and evolution. Even though the combination of molecular and atomic lines allows for a detailed analysis of the underlying physics, I focus here on the molecular line results. Observations of atomic fine-structure lines towards high-z galaxies is another major challenge.

About half of the stars that we see in present galaxies has been formed during the first few billion years after Big Bang. Through observations of high-z starbursts, we are able to study the early phases of massive galaxies, that is so to say the childhood of the massive galaxies. Massive starbursts are rare phenomena in the local Universe, typically seen in ultra-luminous IR galaxies (ULIRGs). However, at high redshift $z > 1$ starbursts prevail and produce about 30% of the total energy output, hence constituting a significant

† Assuming $H_0 = 71$, $\Omega_M = 0.27$, and $\Omega_\Lambda = 0.73$, $z = 1$ corresponds to a Universe age of 5.9 Gyr, similarly, $z = 3$ and 6 correspond to 2.1 and 0.95 Gyr, respectively.

population at early times. Most high-z starbursts with star formation rates (SFR) larger $100\,M_\odot/yr$, and often $> 1000\,M_\odot/yr$, have been discovered through submm continuum observations (so-called submillimeter galaxies, SMGs; e.g. Smail *et al.* 1997). The submm light is the redshift far-infrared dust emission that arise from the dust-enshrouded massive star forming regions.

Molecular line studies of high redshift galaxies are used to address many different scientific questions, including:

– What is the total molecular gas mass and gas fraction? How does the gas fraction evolve?
– What is the evolution the molecular gas density on cosmological time scales?
– What is the dynamics of the gas?
– What are the excitation conditions in high-z galaxies?
– What fuels the gas reservoir, mergers vs. accretion?
– What is the relative importance of star formation and AGN, both in quasars and in starburst galaxies?

I will show results for observations of small molecules and the excitation conditions that are probed using their rotational transitions. I will also highlight results for large molecules, namely the PAHs.

2. Small molecules

Despite H_2 being the most prevalent molecule in space, it is difficult to observe because it has no permanent dipole moment. Hence many other molecules are used to probe the conditions in different environments of the ISM, star forming regions, and AGN.

2.1. *Molecular hydrogen*

UV-lines of the rotational-vibrational state of molecular hydrogen are redshifted into bands that are observable using ground-based telescopes. These lines would be very difficult to detect in emission, however, a number of projects have successfully detected H_2 in absorption towards damped Lyα absorption systems (DLAs).

Ledoux *et al.* (2006) detected molecular hydrogen in absorption in the very high redshift damped Lyα absorption system (DLA) at $z = 4.224$ towards the quasar PSS J1443+2724. The detected absorption lines are rotational levels of the vibrational ground state at rest wavelengths in the far-UV part of the spectrum. They derive column densities of $\log N(H_2) \sim 17.5$. Ledoux *et al.* (2006) finds that the density of the gas is $n < 50\,cm^{-3}$ and suggests that the gas seen in the absorbing galaxy has been left over from previous star formation activity.

Observations of H_2 emission lines is very challenging and only few lines have been successfully detected. In IRS spectroscopy of a multiply imaged galaxy behind the Bullet Cluster, Gonzalez *et al.* (2010) find tentative detections of two pure rotational lines, H_2 0–0 $S(4)$ and H_2 $S(5)$. This is the first detection of these lines in high redshift galaxies. For a single temperature gas, the observed flux ratio gives a temperature $T \sim 375\,K$. With an estimated warm molecular gas mass of $M_{H_2} \sim 2.2 \times 10^8\,M_\odot$ (assuming a lensing magnification of 100), this suggest a relatively cool dusty starburst in a low-mass dwarf galaxy ($M_\star \sim 4 \times 10^9\,M_\odot$; Gonzalez *et al.* 2010).

2.2. *Carbon monoxide*

With a low dipole moment of 0.1 Debye, hence easily excited, and being the second most abundant molecule in interstellar space, CO is commonly used as a tracer of molecular gas. With an excitation temperature of 5.52 K for the rotational $J = 1 - 0$ transition, CO(1-0) traces essentially all molecular gas, predominantly excited by H_2 collisions.

The majority of our knowledge about molecular gas comes from observations of CO – in particular for high redshift galaxies. Depending on the exact redshift, the CO(3-2) or CO(4-3) are redshifted into the mm bands. These lines are generally brighter than the ground transition, and there are many atmospheric windows at millimeter wavelengths, hence making those transitions a prime choice for molecular gas searches.

In the 1990's the number of CO observations in high-z galaxies was limited to a small number of extremely luminous galaxies, mostly quasars (e.g. Ohta et al., 1996; Barvainis *et al.*, 1997; Downes *et al.*, 1999; Guilloteau *et al.*, 1999). In the past decade the number CO observations has sky-rocketed thanks to a number of improvements: (i) Large samples of high-redshift galaxies have been established through observations across different wavelengths. For example today more than 35000 quasars are known at $z > 2$. Through systematic searches with the instruments SCUBA, MAMBO, LABOCA, Spitzer MIPS, and very recently with Herschel PACS / SPIRE, the number of high-z massive starburst galaxies, such as SMGs, continues to rise. (ii) The advances in receiver and correlator technology, which have provided more sensitive instruments. Moreover, the increased bandwidth coverage that is now reaching up to several GHz makes it possible to observe several thousand km/s velocity ranges in one go. As a result, it is possible to search for CO based on UV redshifts that might be offset from the systemic redshift of the targeted galaxy — this is in particular important for quasars.

Despite these important improvements, molecular gas observations at high-z are challenging. So far obtaining a CO detection typically requires several hours of observations per individual object. This means that this research field is only slowly moving from studying a few individual sources to systematic surveys of large samples – to contrast the number of known quasars at $z > 2$, the number of $z > 2$ quasars with a CO detection is currently around 30.

I here summarize some of the main results in high-z CO observations:

Quasars: It has been found that about 1/3 of optically selected quasars have thermal far-infrared emission with dust temperatures comparable to what is found in the starburst galaxies, $T_{\rm dust} \sim 30 - 50\,\rm K$, with implied far-infrared luminosities of $> 10^{12-13}\,\rm L_\odot$ (e.g. Omont et al., 2001; Carilli *et al.*, 2001). It is expected that a significant fraction of the FIR luminosity is indeed powered by star formation. One of the arguments for this is the detection of molecular through CO, showing the existence of gas reservoirs with masses of the order $M_{H_2} \sim 10^{10}\,\rm M_\odot$. Qualitatively quasars are found to follow the $L'_{\rm CO} - L_{\rm FIR}$ relation known for starbursts (illustrated in Fig. 1), however, there is a large scatter around the relation, which can be assigned to the fact that the far-infrared luminosity might be powered by both the star formation and AGN, as well as to the fact that many quasars have only been observed in CO(3-2), (4-3), or (5-4) transitions which might be subthermally excited. In some case CO is not detected towards a quasar. For example in the case of PSS J1057, which has a very high far-infrared luminosity, the CO non-detection implies an upper limit significantly away from the $L'_{\rm CO} - L_{\rm FIR}$ relation (Knudsen *et al.*, in preparation). This indicates that the far-infrared emission is powered by the AGN. Among the numerous results that have been derived from CO studies of quasars include the location of $z = 6$ quasars the $M_{\rm BH} - M_{\rm bulge}$ correlation from nearby massive galaxies (Häring & Rix, 2004). This is the relation between the mass of the central super-massive black hole and the mass of the bulge and this relation suggests that the evolution of the SMBH and the bulge is coeval. The results from Wang *et al.* (2010)

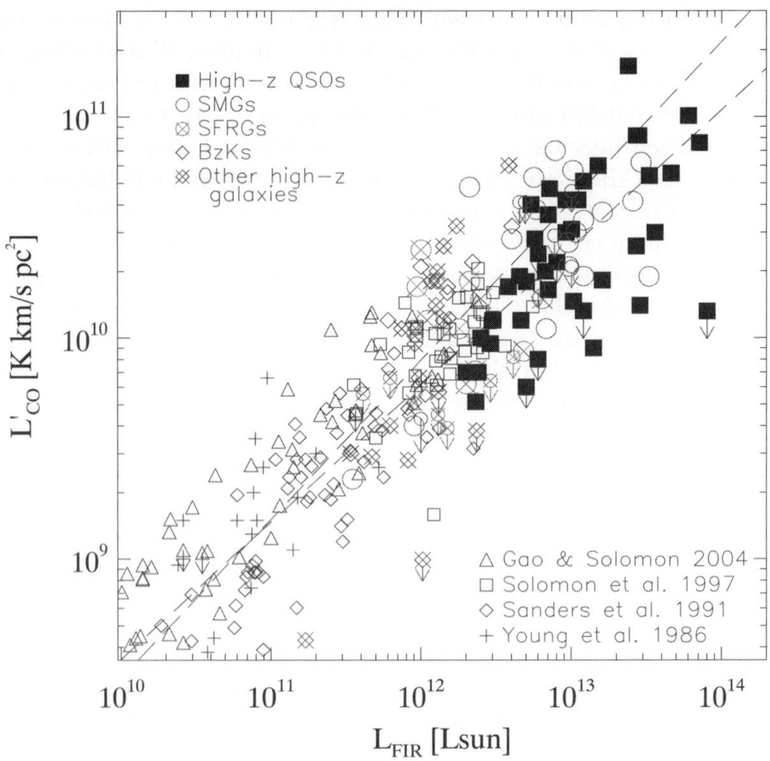

Figure 1. $L_{\rm FIR}$ and $L'_{\rm CO}$ plotted for nearby spiral and starburst galaxies (Young *et al.* 1986; Sanders *et al.* 1991; Solomon *et al.* 1997; Gao & Solomon, 2004a), high redshift starburst galaxies (e.g. Neri *et al.*, 2003; Baker *et al.* 2004; Kneib *et al.* 2005; Greve et al., 2005; Tacconi *et al.*, 2006, 2010; Knudsen *et al.*, 2009; Daddi *et al.*, 2010; Casey *et al.* 2011), and high redshift quasars (e.g. Downes *et al.*, 1999; Guilloteau *et al.*, 1999; Walter *et al.*, 2003; Hainline *et al.*, 2004; Riechers *et al.*, 2006; Maiolino *et al.*, 2007; Coppin *et al.*, 2008; Wang *et al.*, 2010; Knudsen *et al.*, in prep). The dashed lines show the best-fit to the $\log L_{\rm FIR} - \log L'_{\rm CO}$ from Greve *et al.* (2005) and Riechers *et al.* (2006a).

shows that the $z = 6$ quasars have a much large $M_{\rm BH}$ than nearby massive galaxies, when using sensitive CO observations as one of the only means to derive the $M_{\rm bulge}$.

Submillimeter galaxies: Obtaining CO observations for SMGs has been very challenging with the most important reason for this being the difficulties of identifying the correct counterpart galaxy and obtaining a spectroscopic redshift. A CO detection helps confirm the validity of the counterpart and redshift determination. Intensive efforts using in particular the IRAM PdBI have so far produced a similar number of detections as for the quasars. Among the largest samples so far are Greve *et al.* (2005) and Tacconi *et al.* (2006, 2008). The majority of the CO detected SMGs have been detected in blank-field studies meaning that they are more luminous than local ULIRGs and thus representing starburst environments that are found in the local Universe. SMGs found in galaxy cluster fields have been gravitationally lensed and their properties are more like those found in local LIRGs and ULIRGs — however, the number of studied lensed SMGs is small (e.g. Kneib *et al.* 2005; Knudsen *et al.* 2009).

The molecular gas masses found in SMGs are similar to those found in quasars, namely on the order of 10^{10} $M_\odot$. The gas depletion time is estimated to be around 10-40 Myr (e.g. Tacconi *et al.*, 2006), and implied gas fractions of $20-50\%$ e.g. Tacconi *et al.* (2008).

High angular resolution studies of the molecular gas are used for dynamical studies of the SMGs, and e.g. Engel *et al.* (2010) has shown that most SMGs appear to be major merger events. Other studies also suggest that some SMGs are not major mergers, but could be disks at early stages that are growing through accretion of "cold" intergalactic gas where some of the support for this is derived from the CO detections, namely through mapping combined with multiple J transition (e.g. Carilli *et al.* 2010)

Other starburst galaxies: Also other starburst galaxies have had successful CO detections, this include Lyman Break Galaxies (e.g. Baker *et al.*, 2004; Coppin *et al.* 2007) and submillimeter-faint star-forming radio galaxies (Casey *et al.* 2011). Here I highlight the results for BzK selected galaxies. Daddi *et al.* (2008, 2010) have observed CO towards six BzK-selected galaxies, which have been observed from mm to UV wavelengths. They have UV morphologies that are consistent with clumpy unstable disks – contrary to the majority of SMGs, which appear to be interacting or merging galaxies (e.g. Engel *et al.* (2010). Other properties where BzKs appear to be fundamentally different from SMGs include the size of the gas reservoirs and gas depletion time scale. The size of the molecular gas is estimated to be $6 - 11$ kpc in BzKs, while $< 2 - 4$ kpc in SMGs (e.g. Daddi *et al.* 2010; Tacconi *et al.* 2008). The gas depletion time for BzKs are ~ 0.5 Gyr, which is about an order of magnitude longer than for SMGs that have $\tau_{\rm dep} \sim 10 - 50$ Myr.

2.3. *Water*

Detection of water emission lines in high redshift galaxies is difficult. Several searches for different H_2O lines in both quasars and starburst galaxies, have yielded non-detections (e.g. Riechers *et al.* 2006, Wagg & Momjian 2009, Edmonds *et al.* 2009, McKean 2011). A few tentative detections (Bradford *et al.* 2009, Lupu *et al.* 2011, McKean *et al.* 2011) and a three detections have been published. In the Cloverleaf quasar, Casoli *et al.* (1994) detected the H_2O $2_{02} \rightarrow 2_{11}$ line using the IRAM 30m.

Impellizzeri *et al.* (2008) detected the 22 GHz maser line in the $z = 2.64$ quasar MG J0414+0534. The maser is likely induced by a relativistic jet interacting a gas cloud or associated with the circumnuclear accretion disk.

For the H-ATLAS source SDP.17b (J090302.9-014127B), which is a strongly lensed galaxy at $z = 2.3$, Omont *et al.* (2011) report a detection of the H_2O $2_{02} \rightarrow 2_{11}$ line. The results suggest that SDP.17b could be the host of a luminous AGN similar to the nearby Mrk 231 galaxy.

2.4. *HCN, HCO$^+$, and CN – tracers of dense gas*

While, due partly to its abundance, the rotational emission lines from CO are several time brighter than those of other small molecules, other molecules such as HCN, HCO^+, and CN, provide a valuable insight into the denser gas. The dipole moment is substantially larger for HCN ($\mu_{\rm HCN} = 2.98$D), HCO^+ ($\mu_{\rm HCO^+} > 3$D), and CN ($\mu_{\rm CN} = 1.45$D), than for CO ($\mu_{\rm CO} = 0.12$D), and hence the critical density for collisional thermal excitation is around 10^{4-5} cm^{-3} in the ground transition, which is significantly larger than for CO. Therefore these molecules are expected to probe the dense gas regions of the actual star formation as is also the case in the Galaxy. Studies of local starbursts and AGN show that the ratio between HCN and HCO^+ can be used in analysis of distinguishing between starburst and AGN due to the difference in their physical and chemical properties (e.g. Imanishi *et al.* 2009).

The HCN and HCO^+ lines are many times fainter than the CO lines, and so far the successful detections towards high redshift galaxies have only been made towards very bright objects, mostly quasars. A number of searches have been carried out, especially for

HCN (e.g. Vanden Bout *et al.* 2004; Carilli *et al.* 2005). The Cloverleaf quasar ($z = 2.5$) was the first high-z galaxy for which an HCN-detection was confirmed (Solomon *et al.* 2003), where the HCN line luminosity imply that the mass of the very dense gas is $\sim 10^{10}$ $M_{\odot}$. The detection of HCO^+(4-3) and (1-0) in the Cloverleaf shows that HCO^+(4-3) is subthermally excited similarly to local starburst galaxies (Riechers *et al.*, 2006b, 2011a). Same conclusions are not reached for extreme $z = 3.9$ quasar APM08279+5525 (Wagg *et al.*, 2005; Garcia-Burillo et al. 2006; Riechers *et al.* 2010).

In nearby galaxies, there is a stronger correlation between the HCN line luminosity and the far-infrared luminosity (used as a proxy for the SFR) than is found for CO (Gao & Solomon 2004b). Gao *et al.* (2007) combined high-z HCN observations and showed that the high-z galaxies systematically lie above the local $L_{\rm FIR} - L'_{\rm HCN}$ relation — this included upper limits derived from HCN-non-detections. This suggests that the SFR per unit dense molecular gas is higher than that of ULIRGs and nearby star forming galaxies. Moreover, when comparing the HCN with CO, there is a strong indication of a higher fraction dense molecular gas in the high-z galaxies.

A detection of HNC(5-4) has been published by Guélin *et al.* (2006) in APM08279. HNC(5-4) has a critical density around $\sim$ $10^7\,{\rm cm}^{-3}$, and the abundance ratio between HNC and HCN is known to have large variations in different environments. For APM08279, the abundance ratio [HNC]/[HCN] is found to be close to one, which is much larger than what is found Galactic hot, dense molecular clouds (Guélin *et al.* 2006).

Finally, the cyanide radical, CN, which is typically found in regions that are exposed to UV radiation field from stars, has also been detected at high-z. Tracing the dense interfaces between a molecular clouds and ionizing fronts, CN is likely a good tracer of the photochemistry in those layers. The ratio between HCN and CN might provide diagnostics for the relative importance of PDRs in a source, hence also trace the UV field. The first high-z detection of CN(3-2) was reported by Riechers *et al.* (2007), who detected this line in the Cloverleaf quasar. The CN line luminosity is slightly brighter than that of HCN and HCO^+. Riechers *et al.* (2007) also find that the Cloverleaf CN-detection follows the $L'_{\rm CN} - L_{\rm FIR}$ relation known from local galaxies, and thus shows promise as a valuable line for high-z galaxy studies.

3. Excitation of molecular gas

With a series of different J-transition observations for a specific molecule it is possible to estimate or determine the excitation conditions of the molecular gas. Typically the derived physical properties are temperature and density, however, depending on the model employed this could also include intensity of the radiation field. The most commonly used molecule is CO because of the relatively brighter lines. The low-J transitions will trace the cold and less dense gas, while the higher-J transitions will trace the dense gas. SMGs, quasars, and massive BzK-selected galaxies have proven to have different excitation conditions and different dominant gas components.

3.1. *Excitation in starburst galaxies*

Through multi-J ^{12}CO observations from $J = 10 - 9$ to $J = 3 - 2$ and $J = 1 - 0$ in the SMG SMMJ2135-0102, $z = 2.3$, a.k.a. the Cosmic Eyelash, the phases of the gas components have been studied and modeled in detail (Danielson *et al.* 2011). The CO observations have been combined with detections of [CI], [CII], and HCN. With 14 detections and 20 upper limits from molecular lines and atomic fine-structure lines, this is one of the best studied high-z starburst galaxies. Using both LVG models (large velocity gradient) and PDR models (photodissociation region), Danielson *et al.* (2011) found that the CO line SED would be best fit with two-phase gas components, one component being a warm, dense luminous component with $T \sim 60\,\text{K}$ and densities of $n_{H_2} \sim 10^{3.6}\,\text{cm}^{-3}$, and the other component being a low-excitation, cool extended massive component with $T \sim 25\,\text{K}$ and $n_{H_2} \sim 10^{2.7}\,\text{cm}^{-3}$. Furthermore, the UV radiation field is about 1000 times more intense that in the Galaxy.

The analysis of the CO line SED of the Cosmic Eyelash, demonstrates very well the amount detail that can be derived about high-z starbursts, however, it is also a demanding study requiring molecular observations from several different mm and submm telescopes. Some other studies of SMGs have obtained multi-J CO observations (e.g. Weiss *et al.*, 2005, 2007; Riechers *et al.*, 2011d; Carilli *et al.* 2010). These have not had the same number of CO transitions as in the case of the Eyelash, however, they provide results for a larger number of sources. In Weiss *et al.* (2005, 2007) presented high-J ($J > 3$) observations of both SMGs and quasars, with the finding that SMGs typically have a turn-over in CO SED around $J = 5 - 4$, which is a bit lower than in quasar starbursts. Furthermore, the CO line SEDs can be fitted with a single gas component (using LVG models). It is however important to keep in mind that the Weiss *et al.* (2007) study does not include CO(1-0) and is therefore not sensitive to an extended, low-dense gas component. Ivison *et al.* (2011) and Riechers *et al.* (2011b) combined $J = 1 - 0$ observations with $3-2$, or $4-3$ and $6-5$, observations of $z = 2$ and $z = 3.4$ SMGs and demonstrated that CO SED fitting require more than one gas component, similarly as described above for the Eyelash. Additional to the dense, warm gas present in the star forming region, evidence for an extended, low density is found. The extended component have typical sizes $> 10\,\text{kpc}$ with no evidence for large-scale flows of cold gas (Ivison *et al.*, 2011). These extended, low dense gas component likely contributes at least 30-50% to the total molecular gas reservoir in SMGs.

3.2. *Excitation in quasars*

Contrary to the SMGs, quasars appear not to have an extended molecular gas reservoir. This has been found in two studies of both $z \sim 4$ and $z \sim 2$ quasars using the CO(1-0) line compared to high-J transitions (Riechers *et al.*, 2006a, 2011e).

In the study from Weiss *et al.* (2007), quasars have also been observed. The CO line SEDs can be fitted with a single temperature / density model (using LVG models), in good agreement with the CO(1-0) results from Riechers *et al.* (2006, 2011e). A notable difference between the CO line SEDs of quasars and SMGs is the fact that the line SED has a turn-over at higher J, in most cases around $J = 7-6$ or $J = 8-7$. The molecular gas is possibly concentrated in a more compact region in the quasars, and the presence of the quasar is expected to influence the excitation. In the case of APM08279, the CO line SED differs from all other known cases. The SED turns over only around $J = 10-9$, and the fitted LVG model suggests a kinetic temperature of $\sim 200\,\text{K}$ close to the dust temperature and a density $\sim 2.5 \times 10^4\,\text{cm}^{-3}$ (Weiss *et al.*, 2007).

For the two quasars, APM08279 and the Cloverleaf, there exists observations of several HCN and HCO^+ lines. These two quasars show different excitation properties. For the Cloverleaf, the HCO^+ line SED indicate a subthermal excitation of the $J = 4 - 3$ transition with $(4-3)/(1-0)$ ratio similar to the local starburst galaxy NGC253 (Riechers *et al.* 2011a; Knudsen *et al.*, 2007). The excitation is consistent with being purely collisionally excited. The results look remarkably different for APM08279, where strong HCN(6-5), HNC(6-5), and HCO^+(6-5), emission has been detected (Riechers *et al.*, 2010). In this case, modeling of the HCN-excitation indicate that collisional excitation cannot alone account of the high strength of the line emission, but radiative excitation through strong IR-radiation field, which efficiently pumps the high-J lines, likely provide sufficient excitation. APM08279 is an extreme and unique case, for which so far there has been no other high-z galaxy found with similar excitation properties.

3.3. *Excitation in massive disk galaxies*

While SMGs and quasars show compact and high-excitation molecular gas, recent studies of BzK-selected massive star forming disk galaxies ($z \sim 1.5$) yield results that show different properties.

Observations of CO in $J = 3-2$, $2-1$, and $1-0$ in these massive disk galaxies have revealed low gas excitation. The CO(3-2) appears to be subthermally excited (Dannerbauer *et al.* (2009). A subthermal excitation might also be the case for the CO(2-1) line, however, the data are not sufficiently sensitive to confirm this (Aravena *et al.*, 2010). The galaxy BzK-21000 has the best sampled CO SED, and the results of LVG model fitting imply a temperature of $T_{\rm kin} = 20 - 150\,{\rm K}$ and density of $n_{H_2} = 400 - 2500\,{\rm cm}^{-3}$ (Dannerbauer *et al.* 2009; Aravena et al., 2010), which are very different from the temperature and density estimated for quasars and SMGs, as discussed above. While the CO SED is different from both nearby and high-z starbursts, it very similar to what is seen in our own Milky Way. Low excitation of the molecular gas as found in the BzK star forming galaxies could provide a possible explanation why CO(4-3) was not detected in the lensed $z_{\rm spec} = 4.04$ SMG SMMJ163555+661300 (Knudsen *et al.*, 2010). In general, while it is thought that SMGs are progenitor of nearby massive elliptical galaxies, the BzK selected massive disk galaxies are related to the nearby massive disks galaxies (which later might become part of a major merger event).

4. Observations of PAHs

Polycyclic aromatic hydrocarbons (PAHs) are large molecules (with sizes up to a few nm) found in the star forming regions. The vibrational modes of PAHs produce emission lines that are seen in the mid-infrared part of the spectrum. The main spectral features are observed at rest frame wavelengths 6.2, 7.7, 8.6, 11.3, 12.7, and 17 μm coming from stretching and bending of the C$-$C or C$-$H bonds (e.g. Draine 2003). PAHs are excited by the UV photons from star formation. Calzetti *et al.* (2005) find in a study of M51, that the PAHs are heated by photons coming from current star formation (< 10 Myr) as well as recent star formation (< 100 Myr). However, in intense UV-radiation fields, the PAHs could be ionized or destroyed (e.g. Tacconi-Garman *et al.*, 2005, and references therein). A correlation between the PAH emission and the star formation rate has been found, though there is also a dependence on, e.g., metallicity (Calzetti (2011) and references therein).

In the local Universe, both normal and starburst galaxies show strong PAH emission and with little variation in strength and shape (e.g. Rigopoulou *et al.*, 1999). However, towards AGNs the PAH emission is weak, potentially caused by destruction of the PAHs or by a dilution of the PAHs emission feature by the warm continuum from the AGN. Hence, in AGN and quasar studies the mid-infrared continuum from hot dust is expected to dominate the mid-infrared spectra.

With the launch of the Spitzer Space Telescope, the infrared spectrograph IRS enabled observations of PAHs towards high redshift galaxies. A large series of projects have been carried out for galaxies ranging from lyman break galaxies to quasars.

4.1. *Ly-α blobs*

While high-z PAHs studies has been observations of individual galaxies, Colbert *et al.* (2011) obtained Spitzer spectroscopy of Ly-α blobs (LABs). The LABs are Lyα

nebulae with Lyα fluxes $\sim 10^{43-44}$ erg/s extended over $\sim 50 - 150$ kpc and typically found within high-z galaxy over-densities. The flux is comparable to that of high-z radio galaxies, however, there is no evidence that it should arise from interaction with jets. The actual source of energy exciting the hydrogen gas is not known. Models and observations point towards different types of scenarios, which include superwind outflows and cooling flows (e.g. Matsuda *et al.*, 2004). In the study from Colbert *et al.* (2011), two out of the six sample sources that are infrared-bright sources associated with LABs show only continuum emission suggesting that they are dominated by an AGN. The other four sources show PAH emission line, implying signification star formation. Despite that in two of the PAH detected sources have estimated SFRs that would imply sufficient energy from supernova outflows to power the LABs, there is no conclusive evidence that these infrared-bright galaxies could be the single source responsible for the LABs.

4.2. *Starburst galaxies*

Massive starbursts like those seen in SMGs are probably the most obvious targets for observations of PAHs at high redshifts. With star formation rates on the order of $1000\,M_\odot$/yr the PAH emission lines are expected to be bright, and therefore also provide a means to, for example, determine the redshift for sources with very faint optical counterparts. A large number of other PAH studies of high-z starburst galaxies that have been selected through different selection criteria have been published (e.g. Yan *et al.* 2005; Rigby *et al.*, 2008; Dasyra *et al.* 2009; Desai *et al.*, 2009; Siana *et al.*, 2009; Fadely *et al.* 2010). Here I focus on the SMGs.

Both Pope *et al.* (2008) and Menéndez-Delmestre *et al.* (2007, 2009) have obtained IRS spectra of more than 30 SMGs in two different studies. The SMGs follow the relation between infrared luminosity and PAH line luminosity that is established for local starburst galaxies (e.g. Peeters *et al.* 2004). An important implication of this is that the PAH luminosity can be used for estimating the SFR in SMGs. Comparing the extinction properties derived from the mid-infrared spectroscopy of SMGs with those of local ULIRGs, Menéndez-Delmestre *et al.* (2009) argues that the PAH emission arise from a more extended region ($> 1 - 2$ kpc) than is seen in the compact star forming regions of ULIRGs.

The PAH results for the SMGs suggest that the mid-infrared spectral properties are somewhat different from those of $z \sim 2$ 24μ-selected ULIRGs (e.g. Sajina *et al.* 2007; Dasyra *et al.* 2009), as the latter displays mid-infrared spectra of both star formation dominated and AGN dominated emission (e.g. Pope *et al.* 2008). An important point about PAH studies of SMGs is that while X-ray studies can show the presence of an AGN, the mid-infrared spectroscopy can be used for determining the relative contribution of the AGN and starburst to the generated energy output (e.g. Pope *et al.* 2008).

4.3. *Quasars and AGN*

Even though quasar host galaxies in many cases harbor a massive starburst, the mid-infrared spectra appear different because of the AGN hot dust emission, which contribute significantly at these wavelengths. Lutz *et al.* (2008) obtained IRS spectra for 12 $z = 1.8 - 2.8$ radio-quiet quasars, which all show signs of intense star formation activity, however, the far-infrared emission alone does not provide conclusive evidence for the relative heating of the dust by star formation and AGN activity. Using the PAH emission lines, it is possible to model how much dust emission arise from the starburst and how much from the AGN. The reason for this is that the PAH emission from a starburst region is less likely to be dominated by emission from the AGN compared to a number of other star formation tracers. Mid-infrared spectroscopy was also used for local PG quasars to estimate the star formation contribution (e.g. Schweitzer *et al.*, 2006). In the spectra of the 12 quasars, Lutz *et al.* (2008) show PAH detections in nine cases. For the other three the non-detection possibly reflects structure in the continuum emission and the silicate feature rather than an absence of PAH emission. Based on the PAH detections, derived SFRs estimates are up to $\sim 3000\,M_\odot/yr$, strongly supporting that the far-infrared emission is dominated the intense star formation.

In a study by Coppin *et al.* (2010), mid-infrared spectroscopy of PAH emission was used to investigate the relative importance of AGN in SMGs that show signs of the presence of powerful AGN. The near-infrared and mid-infrared photometric colors suggest that these are AGN dominated SMGs; majority of SMGs are known not to have a dominating AGN (e.g. Pope *et al.* 2008). In all eight SMGs PAHs are seen in the spectra, however, there is also a strong mid-infrared continuum. Combined with previous results for other SMGs (also see discussion above), Coppin *et al.* (2010) conclude that about 15% of SMGs have mid-infrared spectra dominated by AGN emission. Though with only two of the eight observed galaxies having bolometrically important AGN, this implies that the total infrared luminosity, L_{IR}, is dominated by AGN emission in about 5% of all blank-field SMGs. These AGN-dominated SMGs could represent a later stage in an evolutionary sequence, where the SMGs are followed by an AGN or quasar phase.

5. Outlook

As ALMA is gradually becoming available with cycle-0 in the near future and the complete array only a couple of years away, doing detailed molecular line spectroscopy will be possible even for high redshift galaxies. Once complete, ALMA provides a very desirable combination of high sensitivity and bandwidth of up to 8 GHz, and thereby making it possible to obtain multiple molecules lines within the same observations as well as obtaining observations of other lines than CO for a large number of galaxies. For example, as highlighted above, CN could provide a useful tracer of the dense gas. The CN(3-2) is only about 5.5GHz away from the CO(3-2) line in rest-frame frequency, meaning that for a $z \sim 2-3$ galaxy both lines can observed simultaneously. Within the same spectra other fainter molecular lines are likely to emerge, providing a new challenge for high-z studies, namely astrochemistry.

One of the recent, interesting results is the detection of extended, low density, cool massive molecular gas reservoirs in the high-z star forming galaxies. It is possible that some of these reservoirs are related to secular galaxy evolution, which might be fed through "cold" streams of gas from the intergalactic medium (e.g. Dekel *et al.* 2009). One of the challenges for studying this at increasing redshift is the fact that the CMB temperature rises with redshift. Simple modeling shows that extended molecular disks will be subject to dimming due to the CMB (Weiss *et al.*, in preparation), impacting in particular on the low-J CO transitions.

Finally, deriving the cosmic evolution of the molecular gas space density, Ω_{H_2}, will provide important input for models of galaxy formation and evolution. Ω_{H_2} is predicted to rise rapidly and peak around $z \sim 2-3$ before declining to the known level of the present-day Universe (Obreschkow & Rawlings, 2009), similar to the cosmic star formation rate density. To address, large scale blind searches of molecular gas are needed, requiring a large sensitivity, large velocity/redshift range coverage, and a large field of view. A challenge for the EVLA and the SKA precursor MeerKAT, as these will be able to survey the CO(1-0) emission line from redshift $z \sim 1.5$ to $z > 10$.

6. Summary

• *Small molecules*: Carbon monoxide is the most commonly observed molecule in high redshift galaxies. It provides a confirmation that a molecular gas reservoir is present and that this could be the gas fueling the starburst in the high redshift galaxies so far selected for studies of molecular emission lines. The CO line observations are used for studying many different aspects of the individual galaxies as well as galaxy evolution. For example high angular resolution imaging provide a wealth of information on the dynamics of the starburst region.

Despite that H_2 is the most abundant molecule, the number of successful detections of H_2 at high-z is extremely limited. However, aside from CO, a few other small molecules have been detected in high-z galaxies and used as probes for different conditions. HCN and HCO^+ have been detected towards a small number of bright quasars. Additionally, there is a detection of CN and HNC. The number of detections of these four molecules along with others is expected to increase within the next couple of years as ALMA will be completed.

Water has turned out to be difficult to detect with so far less than a handful of successful detections — all towards galaxies where the AGN might play an important role.

• *Gas excitation*: Quasars appear to have higher excitation conditions than starbursts. Furthermore, quasars appear not to have a significant extended molecular gas component with low density. Starbursts on the other hand do appear to have two important phases in the molecular gas - a dense molecular gas reservoir and the a low-dense gas possibly extended over a larger region. The extended gas reservoirs contain a significant fraction of the total molecular gas. Extreme starbursts like SMGs have higher excitation of the molecular gas than massive gas-rich galaxies found through the BzK selection technique.

• *PAHs at high-z*: Numerous studies of mid-infrared spectroscopy of high redshift galaxies, have shown that PAH observations provide a strong tool for differentiating between star formation dominated and AGN dominated mid-infrared emission. Furthermore, the PAH spectral lines provide a means for determination of the redshift for SMGs that have very faint optical counterparts. The SFR can be estimated from the PAH luminosity. Using this it is possible to separate the contribution to the infrared luminosity from the starburst and the AGN in, e.g., quasars.

• *Future*: With the new and significantly improved capabilities of ALMA, and other radio / mm telescopes, the molecular line observations of high redshift galaxies will move away from single molecule observations into an era where emission lines from several molecules might show up in the same spectrum. This means that it will very soon become possible to do astrochemical studies in galaxies that represent early and extreme phases in the evolution of galaxies.

References

Aravena, M., Carilli, C., Daddi, E., *et al.* 2010, *ApJ*, 718, 177

Baker, A. J., Tacconi, L. J., Genzel, R., Lehnert, M. D., & Lutz, D. 2004, *2004*, 604, 125

Barvainis, R., Maloney, P., Antonucci, R., & Alloin, D. 1997, *ApJ*, 484, 695

Bradford, C. M., Aguirre, J. E., Aikin, R., *et al.* 2009, *ApJ*, 705, 112

Brown, R. L. & Vanden Bout, P. A. 1991, *AJ*, 102, 1956

Calzetti, D., *et al.* 2005, *ApJ*, 633, 871

Calzetti, D. 2011, *EAS Publications Series* (Proceedings eds. C. Joblin and A.G.G.M. Tielens, vol 46, 133.

Casey, C. M., Chapman, S. C., Neri, R., *et al.* 2011, *MNRAS*, accepted (arXiv:0910.5756)

Carilli, C. L., Bertoldi, F., Omont, A., Cox, P., McMahon, R. G., & Isaak, K. G. 2001, *AJ*, 122, 1679

Carilli, C. L., Solomon, P., Vanden Bout, P., *et al.* 2005, *ApJ*, 618, 586
Carilli, C. L., Daddi, E., Riechers, D., *et al.* 2010, *ApJ*, 714. 1407
Casoli, F., Gerin, M., Encrenaz, P. J., & Combes, F. 1994, *A&A*, 287, 716
Colbert, J. W., Scarlata, C., Teplitz, H., Francis, P., Palunas, P., Williger, G. M., & Woodgate, B. 2011, *ApJ*, 728, 59
Coppin, K. E. K., Swinbank, A. M., Neri, R., *et al.* 2007, *ApJ*, 665, 936
Coppin, K. E. K., *et al.* 2008, *MNRAS*, 389, 45
Coppin, K. E. K., Pope, A., Menéndez-Delmestre, K., *et al.* 2010, *ApJ*, 713, 503
Daddi, E., Dannerbauer, H., Elbaz, D., Dickinson, M., Morrison, G., Stern, D., & Ravindranath, S. 2008, *ApJL*, 673, 21
Daddi, E., *et al.* 2010, *ApJ*, 713, 686
Danielson, A. L. R., Swinbank, A. M., Smail, I., *et al.* 2011, *MNRAS*, 410, 1687
Dannerbauer, H., Daddi, E., Riechers, D. A., Walter, F., Carilli, C. L., Dickinson, M., Elbaz, D., & Morrison, G. E. 2009, *ApJL*, 698, 178
Dasyra, K. M., Yan, L., Helou, G., *et al.* 2009, *ApJ*, 701, 1123
Dekel, A., *et al.* 2009, *Nature*, 457, 451
Desai, V., Soifer, B. T., & Dey, A. 2009, *ApJ*, 700, 1190
Downes, D., Neri, R., Wiklind, T., Wilner. D. J., & Shaver, P. A. 1999, *ApJL*, 513, L1
Draine, B. T. 2003, *ARA&A*, 41, 241
Edmonds, R., Wagg, J., Momjian, E., Carilli, C. L., Wilner, D. J., Humphreys, E. M. L., Menten, K. M., & Hughes, D. H. 2009, *AJ*, 137, 3293
Engel, H., Tacconi, L. J., Davies, R. I., Neri, R., Smail, I., *et al.* 2010, *ApJ*, 724, 233
Fadely, R., Allam, S. S., Baker, A. J., Lin, H., Lutz, D., Shapley, A. E., Shin, M.-S., Allyn Smith, J., Strauss, M. A., & Tucker, D. L. 2010, *ApJ*, 723, 729
Gao, Y. & Solomon, P. M. 2004a, *ApJS*, 152, 63
Gao, Y. & Solomon, P. M. 2004b, *ApJ*, 606, 271
Gao, Y., Carilli, C. L., Solomon, P. M., & Vanden Bout, P. A. 2007, *ApJ*, 660, L93
García-Burillo, S., Graciá-Carpio, J., Guélin, M., *et al.* 2006, *ApJL*, 645, L17
Gonzalez, A. H., Papovich, C., Bradač, M., & Jones, C. 2010, *ApJ*, 720, 245
Greve, T. R., Bertoldi, F., Smail, I., *et al.* 2005, *MNRAS*, 359, 1165
Guélin, M., Salomé, P., Neri, R., *et al.* 2006, *A&A*, 462, L45
Guilloteau, S., Omont, A., Cox, P., McMahon, R. G., & Petitjean, P. 1999, *A&A*, 349, 363
Hainline, L. J., Scoville, N. Z., Yun, M. S., Hawkins, D. W., Frayer, D. T., & Isaak, K. G. 2004, *ApJ*, 609, 61
Häring, N. & Rix, H.-W. 2004, *ApJL*, 604, L89
Imanishi, M., Nakanishi, K., Tamura, Y., & Peng, C.-H. 2009, *AJ*, 137, 3581
Impellizzeri, C. M. V., McKean, J. P., Castangia, P., Roy, A. L., Henkel, C., Brunthaler, A., & Wucknitz, O. 2008, *Nature*, 456, 927
Ivison, R. J., Papadopoulos, P. P., Smail, I., Greve, T. R., Thomson, A. P., Xilouris, E. M., & Chapman, S. C. 2011, *MNRAS*, 412, 1913
Kneib, J.-P., Neri, R., Smail, I., Blain, A., Sheth, K., van der Werf, P., & Knudsen, K. K. 2005, *A&A*, 434, 819
Knudsen, K. K., Walter, F., Weiss, A., Bolatto, A., Riechers, D. A., & Menten, K. 2007, *ApJ*, 666, 156
Knudsen, K. K., Neri, R., Kneib, J.-P., & van der Werf, P. P. 2009, *A&A*, 496, 45
Knudsen, K. K., Kneib, J.-P., Richard, J., Petitpas, G., & Egami, E. 2010, *ApJ*, 709, 210
Ledoux, C., Petitjean, P., & Srianand, R. 2006, *ApJL*, 640, L25
Lupu, R. E., Scott, K. S., Aguirre, J. E. *et al.* 2011, *ApJ*, submitted, (arXiv:1009.5983)
Lutz, D., Sturm, E., Tacconi, L. J., Valiante, E., Schweitzer, M., Netzer, H., Maiolino, R., Andreani, P., & Shemmer, O. 2008, *ApJ*, 684, 853
Matsuda, Y., Yamada, T., Hayashino, T., *et al.* 2004, *AJ*, 128, 569
Maiolino, R., *et al.* 2007, *A&A*, 472, L33

McKean, J. P., Impellizzeri, C. M. V., Roy, A. L., Castangia, P., Samuel, F., Brunthaler, A., Henkel, C., & Wucknitz, O. 2011, *MNRAS*, 410, 2506

Menéndez-Delmestre, K., *et al.* 2007, *ApJL*, 655, L65

Menéndez-Delmestre, K., Blain, A. W., Smail, I., Alexander, D. M., Chapman, S. C., Armus, L, Frayer, D., Ivison, R. J., & Teplitz, H. 2009, *ApJ*, 699, 667

Neri, R., *et al.* 2003, *ApJL*, 597, L113

Obreschkow, D. & Rawlings, S. 2009, *ApJL*, 696, L129

Ohta, K., Yamada, T., Nakanishi, K., Kohno, K., Akiyama, M., & Kawabe, R. 1996, *Nature*, 382, 426

Omont, A., Cox, P., Bertoldi, F., McMahon, R. G., Carilli, C., & Isaak, K. G. 2001, *A&A*, 374, 371

Omont, A., Neri, R., Cox, P., Lupu, R., Guélin, M., *et al.* 2011, *A&A*, 530, L3

Peeters, E., Spoon, H. W. W., & Tielens, A. G. G. M. 2004, *ApJ*, 613, 986

Pope, A., Chary, R.-R., Alexander, D. M., Armus, L., Dickinson, M., Elbaz, D., Frayer, D., Scott, D., & Teplitz, H. 2008, *ApJ*, 675, 1171

Riechers, D. A., Weiss, A., Walter, F., Carilli, C. L., & Knudsen, K. K. 2006a, *ApJ*, 649, 635

Riechers, D. A., Walter, F., Carilli, C. L., Weiss, A., Bertoldi, F., Menten, K. M., Knudsen, K. K., & Cox, P. 2006b, *ApJL*, 645, L13

Riechers, D. A., Walter, F., Cox, P., Carilli, C. L., Weiss, A., Bertoldi, F., & Neri, R. 2007, *ApJ*, 666, 778

Riechers, D. A., Weiss, A., Walter, F., & Wagg, J. 2010, *ApJ*, 725, 1032

Riechers, D. A., Walter, F., Carilli, C. L., Cox, P., Weiss, A., Bertoldi, F., & Menten, K. M. 2011a, *ApJ*, 726, 50

Riechers, D. A., Hodge, J., Walter, F., Carilli, C. L., & Bertoldi, F. 2011b, *ApJL*, accepted (arXiv:1105.4177)

Riechers, D. A., Carilli, C. L., Maddalena, R. J., *et al.* 2011c, *ApJL*, accepted (arXiv:1106.2553)

Rigby, J. R., Marcillac, D., Egami, E., *et al.* 2008, *ApJ*, 675, 262

Rigopoulou, D., Spoon, H. W. W., Genzel, R., Lutz, D., Moorwood, A. F. M., & Tran, Q. D. 1999, *AJ*, 118, 2625

Sajina, A., Yan, L., Armus, L., Choi, P., Fadda, D., Helou, G., & Spoon, H. 2007, *ApJ*, 664, 713

Sanders, D. B., Scoville, N. Z., & Soifter, B. T. 1991, *ApJ*, 370, 158

Schweitzer, M., Lutz, D., Sturm, E., *et al.* 2006, *ApJ*, 649, 79

Siana, B., Smail, I., Swinbank, A. M., Richard, J., Teplitz, H. I., Coppin, K. E. K., Ellis, R. S., Stark, D. P., Kneib, J.-P., & Edge, A. C. 2009, *ApJ*, 698, 1273

Smail, I., Ivison, R. J., & Blain, A. W. 1997, *ApJL*, 490, L5

Solomon, P. M., Radford, S. J. E., & Downes, D. 1992, *Nature*, 356, 318

Solomon, P. M., Downes, D., Radford, S. J. E., & Barrett, J. W. 1997, *ApJ*, 478, 144

Solomon, P., Vanden Bout, P., Carilli, C., & Guelin, M. 2003, *Nature*, 426, 636

Tacconi, L. J., Neri, R., Chapman, S. C., *et al.* 2006, *ApJ*, 640, 228

Tacconi, L., *et al.* 2008, *ApJ*, 680, 246

Tacconi, L. J., Genzel, R., Neri, R., *et al.* 2010, *Nature*, 463, 781

Tacconi-Garman, L. E., *et al.* 2005, *A&A*, 432, 91

Vanden Bout, P. A., Solomon, P. A., & Maddalena, R. J. 2004, *ApJL*, 614, 97

Wagg, J., Wilner, D. J., Neri, R., Downes, D., & Wiklind, T. 2005, *ApJL*, 634, L13

Wagg, J. & Momjian, E. 2009, *AJ*, 138, 895

Walter, F., *et al.* 2003, *Nature*, 424, 406

Wang, R., Carilli, C. L., Neri, R. *et al.* 2010, *ApJ*, 714, 699

Weiss, A., Downes, D., Walter, F., & Henkel, C. 2005, *A&A*, 440, L45

Weiss, A., Downes, D., Walter, F., & Henkel, C. 2007, *ASP Conference Series* (eds. A.J. Baker, J. Glenn, A.I. Harris, J.G. Mangum, M.S. Yun), 375, 25

Yan, L., Chary, R., Armus, L., Teplitz, H., Helou, G., Frayer, D., Fadda, D., Surace, J., & Choi, P. 2005, *ApJ*, 628, 604

Young, J. S., Schloerb, F. P., Kenney, J. D., & Lord, S. D. 1986, *ApJ*, 304, 443

Discussion

R. MAUERSBERGER: The SEDs of CO lines in distant galaxies look very different from those of Galactic clouds. IS there still something like an X-factor to convert CO intensity into an H_2 column density?

KNUDSEN: An expected question! The different types of high-z galaxies represent a variety of physical conditions, and therefore different CO/H_2 conversion factors have to be chosen. As an example, Daddi *et al.* (2008) made careful considerations about this in their study of massive disk galaxies (BzK selected) and assumed a different conversion factor than is normally used in high redshift submm galaxies. Tacconi *et al.* (2008) made a detailed analysis to estimate an X-factor for high-z starburst galaxies.

GUELIN: For the type of temperatures (T_K) and densities (n_{H_2}) that prevail in high-z circumnuclear disks, CO lines alone cannot determine independently T_K and n_{H_2}, as these parameters are degenerated. You have to turn to high density tracers like HCN, but those may not be coeval with CO. With all its power, ALMA may not be able to help on this matter as there is little hope to resolve the clouds inside the very distant disks.

KNUDSEN: Thank you for raising this point. To elaborate a bit more on your comment: A combination of different models (e.g. LVG and PDR) could provide a better insight. Furthermore, including observations of atomic fine-structure lines could add extra information, assuming the emission is from the same region.

M. GERIN: Have you considered looking for absorption lines from hydrides such as ground state water lines?

KNUDSEN: I have not looked into this option. I expect it would be difficult.

The Molecular Universe
Proceedings IAU Symposium No. 280, 2011
José Cernicharo & Rafael Bachiller, eds.

doi:10.1017/S1743921311025099

Absorption Line Surveys at Intermediate Redshift

Sébastien Muller
Onsala Space Observatory, Sweden
email: mullers@chalmers.se

Abstract. The z=0.89 spiral galaxy located on the line of sight to the quasar PKS1830-211 is right now the extragalactic object with the largest number of detected molecular species (34, plus isotopic variants). This remarkable molecular inventory was recently built after an unbiased spectral survey of molecular absorption lines with the Australia Telescope Compact Array in the 7 mm band. We present the results of this survey, as a showcase of the various interests of molecular absorption studies at intermediate redshift: chemical inventory and gas properties in distant galaxies, chemical enrichment history of the Universe, measurements of the CMB temperature as a function of z, and constraints on the constancy of fundamental constants.

Keywords. quasars: absorption lines, quasars: individual (PKS1830-211), radio lines: galaxies, galaxies: ISM, ISM: molecules, surveys, cosmic microwave background.

1. Introduction

Interstellar molecules were first detected in absorption along the line of sight toward bright nearby stars. Now, we know that the interstellar medium (ISM) is home of more than 150 molecular species and astrochemistry has been a rapidly growing field. Moreover, recent surveys have open the door of extragalactic astrochemistry (e.g., Wang *et al.* (2004), Martín *et al.* (2006), Martín *et al.* (2011), Costagliola *et al.* (2011)), although only the central region of the bright and nearby sources reveal their molecular richness.

After the discovery of molecular absorption in five objects located at intermediate redshifts (0.24<z<0.89, cf a review by Combes *et al.*2008), it is now possible to explore the chemical content in the disk of galaxies up to significant redshifts. The interests of absorption studies for distant objects are manifold:

• There is no distance dilution, absorption can be detected as easily at high redshift as in the Local Universe. The only limitation is the need for background continuum.

• It offers outstanding spatial resolution, at the apparent angular size of the continuum source. For example at mm wavelengths, the continuum beam pencil through the z=0.89 galaxy located in front of PKS1830-211 is of the order of the parsec scale, projected in the plane of this galaxy.

• Excitation is simple in low density (~100-1000 cm^{-3}) gas, where collisions are inefficient and the excitation is coupled with the radiative field. In the absence of other local heating source, the excitation is dominated by CMB photons, so that rotation temperatures nearly equal the CMB temperature.

• Optical depths can be obtained directly from the line-to-continuum flux density ratio, and column densities can be derived straightforwardly, as a consequence of the two previous points. This is not the case in extragalactic emission studies, limited by the beam resolution.

• Deep inventory of the molecular content can be obtained, including species that would be nearly impossible to detect in emission due to distance dilution.

• High-z molecular absorbers are valuable cosmological probes (see § 5 hereafter).

• Transitions difficult to observe from the ground, due to poor atmospheric transitions or challenging techniques, may be redshifted to a more accessible frequency window. For example, submillimeter lines of light hydrides (cf. the wealth of Herschel results presented in this symposium) could be studied in absorption, at redshifted frequency, with ALMA.

2. Extragalactic radio-mm molecular absorbers

The number of extragalactic radio-mm molecular absorbers known is very limited, and only five have been discovered to date at intermediate redshifts ($0.24<z<0.89$, cf a review by Combes (2008)), despite active searches (e.g., Wiklind & Combes (1996a), Drinkwater *et al.* (1996), Murphy *et al.* (2003), Curran *et al.*(2004, 2006, 2008)). They are listed in Table 1, together with some other nearby extragalactic absorbers. We do not include here the H_2/CO absorbers detected from optical/UV spectroscopy, such as those discussed by Noterdaeme *et al.*(2011), since a deep exploration of their molecular content remains currently out of reach.

For the majority of absorbers, the absorption occurs in the host-galaxy of the continuum source. In the case of the last three entries in Table 1, the absorbers correspond to intervening galaxies physically unrelated to the quasar. As molecular gas is mostly concentrated within the inner disk of spiral galaxies, it is not surprising that their background quasars are strongly gravitationally lensed, as a result of the small impact parameter.

The most notable mm molecular absorber at intermediate redshift is the z=0.89 spiral galaxy toward the quasar PKS1830-211. It has the highest redshift, the brightest millimeter continuum and the largest amount of absorbing material. The intervening galaxy appears as a face-on spiral (Winn *et al.* (2002); Fig.1 *left*) and acts as a gravitational lens, splitting the background quasar into two main compact cores (NE and SW), separated by 1" and embedded in a fainter Einstein ring, seen at cm wavelengths (Jauncey *et al.* (1991)). Molecular absorption is seen toward both compact images (Wiklind & Combes (1998), Muller *et al.* (2006)), separated in velocity by $\sim$150 km s^{-1} (Fig.1 *right*). The SW and NE images of the quasar fall on either side of the z=0.89 galaxy bulge at galactocentric radii of $\sim$2 and $\sim$4 kpc, respectively.

The SW line of sight is the most molecular rich, with a H_2 column density of $\sim 2\times10^{22}$ cm^{-2}, while the NE line of sight is only detected in a handful of molecular species ($N(H_2)\sim10^{21}$ cm^{-2}). These H_2 column density estimates are consistent with X-ray absorption measurements (Mathur & Nair (1997)), the differential A_V extinction between the NE and SW quasar images (Falco *et al.* (1999)) and the scaling from C_2H column densities (Muller *et al.* (2011)) using a Galactic conversion factor (Lucas & Liszt (2000), Gérin *et al.* (2010)). The density is of the order of 2000 cm^{-3} toward the SW absorption (Henkel *et al.* (2009)).

3. The ATCA 7 mm spectral survey toward PKS1830-211

With the Australian Telescope Compact Array (ATCA), we have recently conducted the first unbiased spectral survey toward PKS1830-211 (cf. Muller *et al.* (2011)), covering the frequency range 30–50 GHz, which corresponds to the rest-frame interval 57–94 GHz at z=0.89. Such a survey has been made possible by the new broadband backend system (CABB) on ATCA, providing an instantaneous bandwidth of 4 GHz and 1 MHz spectral

Table 1. Extragalactic radio-mm molecular absorbers.

Source	z_{abs}	Continuum flux density @3mm (Jy)[a]	$N(H_2)$ (10^{20} cm^{-2})	Molecules detected	Ref.
		Absorption in the host-galaxy of the continuum source			
Cen A	0.002	6	2	CO, OH, NH_3, CN, HCO^+, HCN, N_2H^+, CS, H_2CO, c-C_3H_2	(1)
NGC1052	0.005	0.6	25	OH, CO, HCO^+, HCN	(2)
3C293	0.045	0.19	0.2	CO, HCO^+, HCN	(3)
4C31.04	0.060	0.16	0.1	CO, HCO^+, HCN	(4)
PKS 1413+135	0.247	0.5	5	CO, CN, HCO^+, HCN, HNC	(5)
B 1504+377	0.673	0.4	5	CO, HCO^+, HCN, HNC	(6)
		Absorption in an intervening galaxy			
B 0218+357	0.685	0.5[b]	40	CO, NH_3, H_2O, HCO^+, HCN, HNC, H_2CO, CS, H_2S	(7)
PKS 0132-097	0.765	0.4	–	OH (only, so far)	(8)
PKS 1830-211	0.886	3[b]	200/10[c]	now 34 species (not including isotopic variants), see Table 2	(9)

Notes:
[a] most of the sources have time variable flux.
[b] gravitationally lensed. This is the total flux density.
[c] toward the SW and NE images, respectively.
References:
(1): Gardner & Whiteoak (1976), Seaquist & Bell (1986), Seaquist & Bell (1990), Israel *et al.* (1990), Eckart *et al.* (1990), Wiklind & Combes (1997a), Muller & Dinh-V-Trung (2009), Espada *et al.* (2010); (2): Omar *et al.* (2002), Liszt & Lucas (2004); (3): Evans *et al.* (1999), García-Burillo *et al.* (in prep.); (4): García-Burillo *et al.* (2007); (5): Wiklind & Combes (1997b); (6): Wiklind & Combes (1996a); (7): Wiklind & Combes (1995), Menten & Reid (1996), Henkel *et al.* (2005), Muller *et al.* (2007), Kanekar (2011), Muller *et al.* (in prep.); (8): Kanekar *et al.* (2005); (9): Wiklind & Combes (1996b), Gérin *et al.* (1997), Wiklind & Combes (1998), Chengalur *et al.* (1999), Menten *et al.* (1999), Muller *et al.* (2006), Henkel *et al.* (2008), Menten *et al.* (2008), Bottinelli *et al.* (2009), Henkel *et al.* (2009), Muller *et al.* (2011).

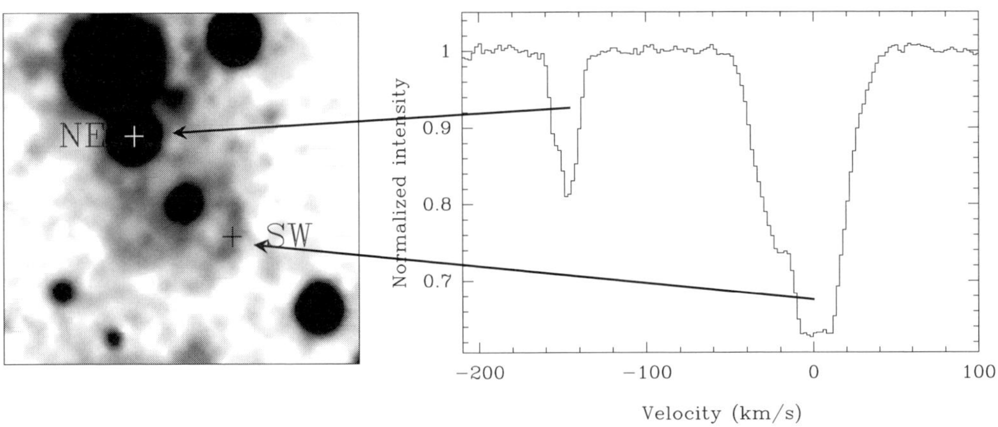

Figure 1. *Left:* Hubble Space Telescope image of the z=0.89 face-on spiral galaxy toward the quasar PKS1830-211. The positions of the NE and SW lensed images of the background quasar, separated by 1", are indicated by the crosses. Note the very high extinction along the SW line of sight, where the image of the quasar is invisible in this optical image. *Right:* Spectrum of the HCO^+ 2-1 line observed with the Plateau de Bure interferometer, showing the two velocity components toward the NE and SW images of the quasar. Adapted from Muller *et al.* (2006).

resolution. The whole survey was done in a total of 18 hours of telescope time, allowing us to achieve a resulting optical depth sensitivity of a few per mil.

There are only few lines with optical depth greater than 0.1 in the survey, as can be seen on Figure 2, where the cumulative number of detected lines is given as a function of the optical depth. The median opacity is ~0.01, given the noise level. In total, we have been able to identify 28 different molecular species (plus some isotopic variants). A family picture is shown in Figure 3. The line identification was rather easy (no U-lines), because the line density is low (~4 lines/GHz on average over the whole 7 mm band), and, as the excitation is coupled with the CMB, we could filter only low-energy transitions (E_{low}<20–30K) during the search in molecular databases, thereby reducing heavily the list of line candidates. Here, we especially acknowledge and emphasize the convenience of on-line molecular databases, such as the Cologne Database for Molecular Spectroscopy (Müller *et al.* (2001)) and the Jet Propulsion Laboratory Molecular Spectroscopy (Pickett *et al.* (1998)), which we used intensively for line identification.

The molecular inventory, listed in Table 2, includes common molecules detected in the ISM, formed of the H, C, N, O, S and Si elements. There are ions (e.g., HCO^+, HOC^+, N_2H^+, SO^+), highly reactive radicals (e.g., C_2H, C_4H), unsaturated carbon chains (e.g., l-C_3H, l-C_3H_2, C_4H), saturated species (CH_3OH, CH_3NH_2), all together demonstrating that a complex and active chemistry is at work in these molecular clouds located in the disk of a z=0.89 galaxy. Certainly, the molecular inventory in Table 2 is not complete, owing to the observational bias toward species with a high dipole moment, a low partition function and low energy transitions within the frequency coverage of the survey. Future observations will undoubtedly increase the list. Still, it is interesting to note that this z=0.89 galaxy toward PKS1830-211 is right now the extragalactic source with the largest number of detected molecular species !

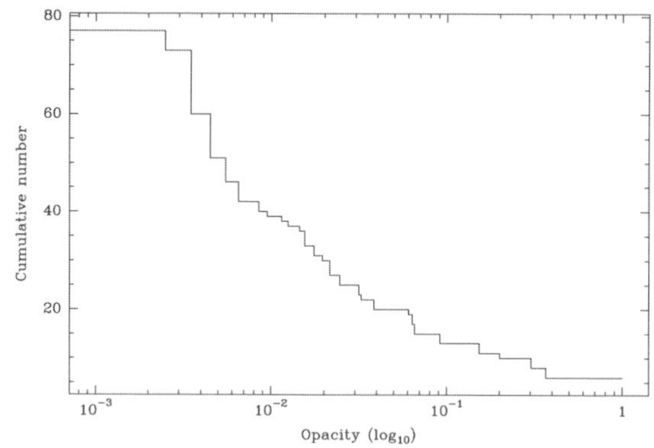

Figure 2. Cumulative number of lines detected in the ATCA 7 mm survey toward PKS1830-211, as a function of opacity. This illustrates the obvious needs for large instantaneous bandwidth and sensitivity for spectral survey.

Most of the lines are optically thin, so that the associated column densities can be derived directly and accurately (assuming that the rotation temperature $T_{rot}=T_{CMB}=5.14$ K at z=0.89). Taking the H_2 column densities given above, we can then obtain the fractional molecular abundances.

It is very tempting to compare these abundances with those observed in local sources, in order to characterize the chemistry. Such a comparison is shown in Figure 4, where we include the fractional abundances measured in diffuse clouds (Liszt *et al.* (2008) and references therein) in translucent clouds (Turner (2000)), in the dark cloud TMC 1 (Ohishi *et al.* (1992)), in the Galactic Center molecular cloud Sgr B2(N) (Nummelin *et al.* (2000)) in the the N113 region in the Large Magellanic Cloud (Wang *et al.* (2009)) and in the nuclear region of the nearby starburst galaxies NGC253 (Martín *et al.* (2006)) and NGC4945 (Wang *et al.* (2004)). There is a large scatter of up to nearly two orders of magnitude between the abundances in different sources, reflecting the different

Table 2. Census of species detected at z=0.89 toward PKS1830-211(SW).

1 atom	2 atoms	3 atoms	4 atoms	5 atoms	6 atoms	7 atoms
(H) [d]	(OH) [d]	(H_2O) [h]	(NH_3) [gh]	**CH_2NH**	**CH_3OH**	**CH_3NH_2**
(C) [i]	(CO) [bci]	C_2H [e]	H_2CO [ce]	c-C_3H_2 [e]	**CH_3CN**	**CH_3C_2H**
	(CS) [af]	HCN [aef] △	**l-C_3H**	**l-C_3H_2**		**CH_3CHO**
	SiO [j] †	HNC [aef] ◇	**HNCO**	**H_2CCN**		
	NS	N_2H^+ [a]	**H_2CS**	**H_2CCO**		
	SO	HCO^+ [aef] ○		**C_4H**		
	SO^+	**HCO**		HC_3N [ej]		
		HOC^+				
		(H_2S) [f]				
		C_2S				

Notes: In parenthesis, species detected in other studies; bold face, new detections; underlined: first extragalactic detections.
○ including $H^{13}CO^+$, $HC^{18}O^+$ and $HC^{17}O^+$; △ including $H^{13}CN$ and $HC^{15}N$; ◇ including $HN^{13}C$ and $H^{15}NC$; † including ^{29}SiO.
References: (a) ?; (b) Gérin *et al.* (1997); (c) Wiklind & Combes (1998); (d) Chengalur *et al.* (1999); (e) Menten *et al.* (1999); (f) Muller *et al.* (2006), including isotopologues; (g) Henkel *et al.* (2008); (h) Menten *et al.* (2008); (i) Bottinelli *et al.* (2009); (j) Henkel *et al.* (2009).

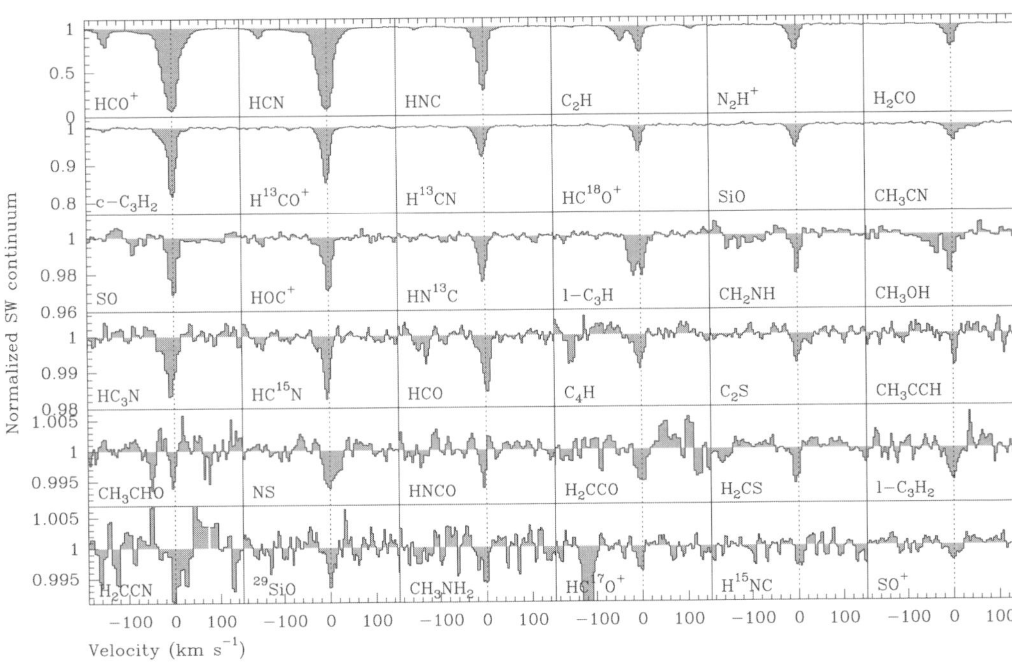

Figure 3. A family picture of molecular species detected in the ATCA 7 mm survey toward PKS1830-211. Note the change of absorption depth scale (*top-left to bottom-right corner*) from the deepest (100%) HCO^+/HCN lines down to the faintest (a few per mil) lines ($HC^{17}O^+$/$H^{15}NC$/SO^+). The absorption toward the NE image of the quasar is seen at a velocity of -150 km s^{-1} on the HCO^+ spectrum.

environment conditions and chemical setups. Here, the common sources of uncertainty have to be kept in mind: beam averaging (especially for emission lines in extragalactic sources), uncertainty on H_2 column densities, excitation and opacity effects, as well as on the background continuum covering factor for absorption lines. For the latter, we assume

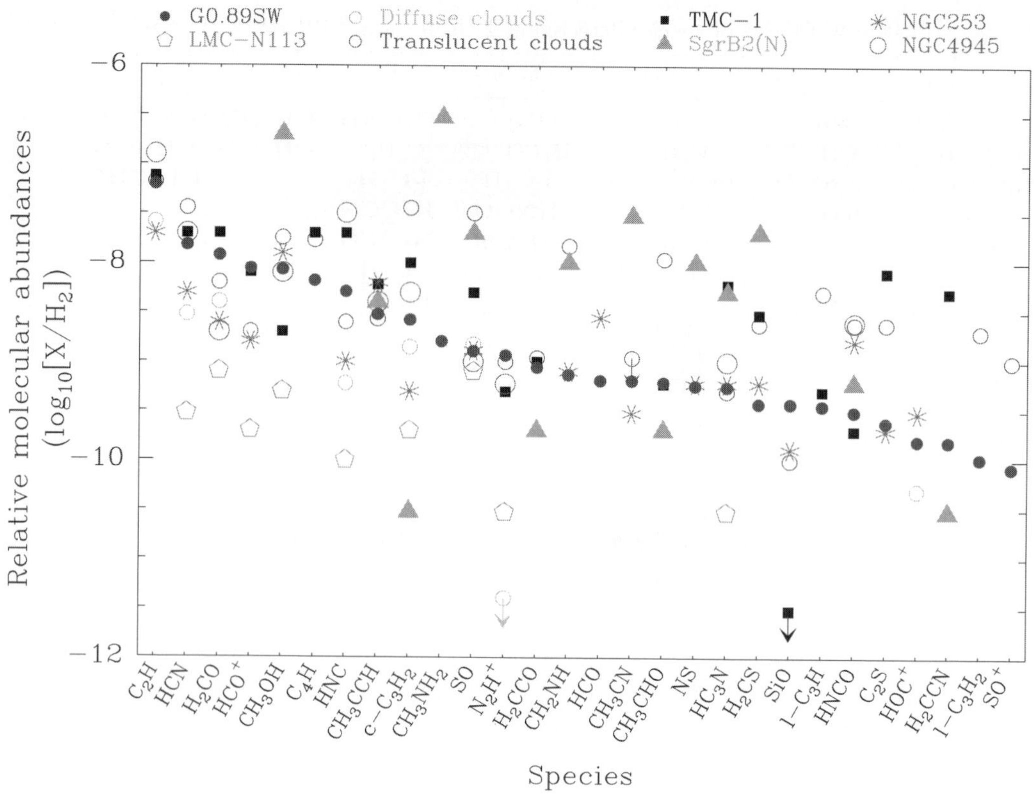

Figure 4. Comparison of fractional molecular abundances in the z=0.89 galaxy (SW image) and that in various other objects. Species are ordered from the most to less abundant toward PKS1830-211(SW).

a source covering factor of unity for all species, so that the apparent optical depths and column densities are strictly speaking lower limits.

Overall, we find that the molecular abundances toward the SW image are intermediate between those in diffuse and translucent clouds, and clearly deviate from those observed toward the Galactic Center molecular cloud Sgr B2N. Toward the NE image, molecular abundances are close to those observed in diffuse clouds. Despite the look-back time of ~7 Gyr, the chemistry does not appear significantly different from that in Galactic clouds. It will be interesting to obtain additional information on the metallicity, elemental abundances and PAHs. We note that the H_2CO/H_2CS ratio of 32±6 is consistent with the present O/S cosmic abundance ratio (33:1, Lodders (2003)). In contrast, isotopic ratios are markedly different from that in local sources (see §5.2).

4. Time variations of the absorption profiles

The monitoring of the HCO^+ 2–1 line over 13 years (1995-2008) has revealed the time variability of the absorption profile (Muller & Guélin (2008)). The changes in absorption depths are shown in Figure 5. For example, one can see that the line at ~−150 km s^{-1} (arising toward the NE image) almost completely vanished in 2006, and reappeared in 2008. Changes in the NE line and wings of the SW line are correlated, indicating that a change in the morphology of the quasar is in cause. Past VLBA observations by Jin *et al.* (2003) have revealed variations in the distance between the NE and SW cores, of up to 200 micro-arcseconds within 8 months, corresponding to 1.6 pc projected in the plane of the z=0.89 galaxy. Those changes were interpreted as due to the recurrent ejection of plasmons along a helicoidal jet by Nair *et al.* (2005). As a consequence, the continuuum illumination is scanning a slightly changing line of sight through the disk of the intervening galaxy.

This configuration offers the unique chance to investigate the molecular clouds structure in the disk of a z=0.89 galaxy. Very-long-baseline mm-interferometry could help monitoring the continuum morphology and location of molecular absorption at regular time interval, yielding a measurement of their size. Alternatively, a high-sensitivity spectral monitoring of the absorption profile of various molecular species could probe their spatial correlation at sub-parsec scale.

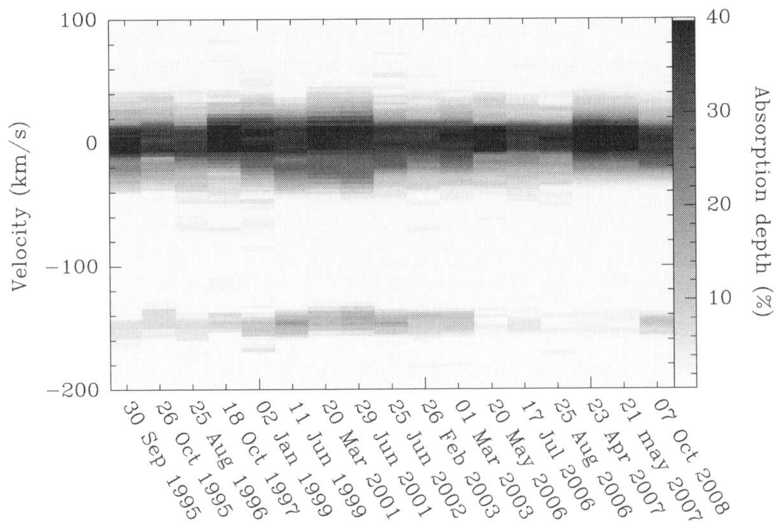

Figure 5. Evolution of the HCO^+ 2–1 absorption line profile observed toward PKS1830-211. Note the deepening of the -147 km s^{-1} feature between 1999 and 2003, its quasi-disappearance between 2006-2007, and re-appearance in the last 2008 spectrum.

5. Molecular absorbers as cosmological probes

Besides allowing detailed inventory of the molecular content in distant galaxies, radio-mm molecular absorbers are also valuable cosmological probes. The variety of molecular species provides various diagnostic tools for constraining physical properties (e.g., T_{CMB} as a function of redshift), tracing the chemical enrichment history of the Universe (e.g., via measurements of isotopic ratios) and testing fundamental physics (e.g., the constancy of fundamental constants).

5.1. *The Cosmic Microwave Background temperature as a function of z*

Measuring the CMB temperature at different redshifts is important to test cosmology models: the adiabatic expansion of the Universe implies $T_{CMB}\propto(1+z)$, but models of decaying dark energy predict a different dependency: $T_{CMB}\propto(1+z)^{(1-\alpha)}$ with $\alpha>0$, (see e.g., Lima (1996), Jetzer *et al.* (2011)).

Observations with the COBE satellite have demonstrated that the CMB at z=0 corresponds to a nearly-perfect black body and T_{CMB} at z=0 is measured with a very high accuracy: 2.72548 $\pm$ 0.00057 K (Fixsen (2009)). At z>0, two methods can be used to probe T_{CMB}:

- multifrequency Sunyaev-Zeldovich observations toward galaxy clusters (e.g., Horellou *et al.* (2005), Luzzi *et al.* (2009));
- spectroscopic absorption study toward quasars.

Most of the measurements done so far at high redshifts have used UV spectroscopy (e.g., of CI, CII and CO, see Noterdaeme *et al.*(2011)). Strictly speaking, only upper limits of T_{CMB} are derived, as the contributions from other local sources of excitation (collisions, pumping), yet largely uncertain, have to be subtracted. On the other hand, radio-mm molecular absorbers are particularly interesting, because they provide more constraints on the local excitation and physical conditions, when a variety of molecular species are detected. For example, from our ATCA 7 mm survey we derive rotation temperatures of 5.58$\pm$0.56 K and 6.23$\pm$0.41 K for c-C_3H_2 and HC_3N, respectively – two species with high dipole moments (3.43 and 3.73 Debye) – while T_{CMB} is expected to be 5.14 K from the standard model with adiabatic expansion. As already said, these values are upper limits of T_{CMB}, and accurate collisional rates are required for further interpretation of these rotation temperatures. We note that the two lines of sight toward the SW and NE images of PKS1830-211, presumably with different local excitation conditions, offer the possibility of a robust determination of T_{CMB} at z=0.89.

5.2. *Isotopic ratios*

Heavy elements (i.e. heavier than lithium) are produced in stellar interiors by the nucleosynthesis activity, and isotopic abundance ratios are good tracers of the nucleosynthesis history and thus of the chemical enrichment history of the Universe.

At z=0.89, the look-back time is about half the age of the Universe, and the spiral galaxy toward PKS1830-211 should have an age of less than 6 Gyr. Given this timescale, stars with mass of less than $\sim$1.5 $M_\odot$ barely had the time to contribute to the enrichment of the ISM. The elemental composition of the ISM should therefore be dominated by massive stars processing.

Isotopic ratios can be derived through abundance ratios of different isotopologues of a same species. However, the interpretation of these measurements can be complicated by several effects, such as opacity of the main isotopologue, fractionation or selective photodissociation. It is therefore primordial to check the systematics by measuring the isotopologues abundance ratios in many species.

Toward PKS1830-211, we have been able to estimate the isotopic ratios of C, N, O, Si and S elements (Muller *et al.*(2006, 2011)). The values are given in Table 3, where we also give for comparison the values of isotopic ratios in the Solar System (terrestrial) and in the carbon-rich circumstellar envelope (CSE) IRC+10216, reflecting the isotopic composition of material processed in the core of a $\sim$2 $M_\odot$ star. Apart from the $^{12}C/^{13}C$ ratio, there is a clear evolution sequence from the ratios at z=0.89 to their terrestrial values and that in the CSE of IRC+10216. No significant differences are found between the isotopic ratios of silicon and sulfur – both processed in massive stars – in the Solar System and in IRC+10216. In contrast, their values in the z=0.89 galaxy is about half lower. This may suggest an initial mass function different from that in the Milky Way.

We do not detect the deuterated variants of the HCO^+ or HCN molecules in our survey, placing an upper limit of $D/H<7.1\times10^{-4}$ and suggesting no strong deuterium fractionation enhancement.

Clearly, it would be interesting to extend these isotopic ratios measurements at various redshifts to trace the chemical enrichment history of the Universe.

Table 3. Comparison of isotopic ratios at different states of stellar processing.

Source	$^{12}C/^{13}C$	$^{14}N/^{15}N$	$^{16}O/^{18}O$	$^{18}O/^{17}O$	$^{28}Si/^{29}Si$	$^{32}S/^{34}S$
PKS1830-211 z=0.89(SW) [a]	$33 \pm 5^{\dagger}$	152 ± 27	69 ± 11	13^{+3}_{-2}	11^{+4}_{-2}	10.5 ± 0.6
Solar System (terrestrial) [b]	89	272	499	5.3	19.7	22.7
IRC+10216 [c]	45 ± 3	>4400	1260^{+315}_{-240}	0.7 ± 0.2	$15.4 \pm 1.1^{\dagger}$	21.8 ± 2.6

Notes: † likely underestimated due to opacity effects.
References: (a) Muller *et al.* (2011); (b) Lodders (2003); (c) Cernicharo *et al.* (2000).

5.3. *Constancy of fundamental constants*

A cosmological variation of the value of a fundamental constant such as the proton-to-electron mass ratio (μ), would induce a shift in frequency (different from the Doppler shift) between a spectral line observed at z=0 (e.g., in the lab) and at high-redshift. A recent and thorough review of the field is given by Uzan (2011). We focus here on μ and constraints on its variation set by molecular transitions. If all the lines had the same dependency on μ, it would be impossible to distinguish a variation of μ from a Doppler shift. However, certain molecular transitions have different sensitivity to changes in μ. This is the case for inversion lines of ammonia (Flambaum & Kozlov (2007)) and some lines of methanol (Jansen *et al.* (2011)) with respect to pure rotational lines. This has pushed several groups to use the only two redshifted ammonia absorbers currently known, those toward B0218+357 and PKS1830-211 (Henkel *et al.*(2005, 2008)), to constrain variations of μ by comparing the velocity offsets between inversion lines of ammonia and rotational lines of other species (see Table 4).

Among the systematics that could affect the measurement, the kinematical bias (which species to take as velocity reference ?) is a severe limitation, as is proven in Figure 6, where we compare the velocity centroids of all species detected in our survey toward PKS1830-211. Several species have significant velocity offsets with respect to the bulk of other species, in particular CH_3OH and CH_3NH_2 (as well as NH_3, Henkel *et al.* (2009)), which have to be further investigated at better velocity resolution/sensitivity.

Interestingly, different lines of methanol have different sensitivity to changes in μ, which could minimize the kinematical bias by removing the need for other species as velocity reference. The best case would be provided by methanol transitions with very different sensitivity to changes in μ, but at similar frequencies to limit continuum morphology changes (and potential different lines of sight), and similar excitation.

Alternatively, a statistical approach to address the systematics would have to wait for the discovery of a large number of new millimeter molecular absorbers.

Table 4. Constraints on the variation of the proton-to-electron mass ratio toward redshifted mm molecular absorbers.

Method	Target	$\Delta\mu/\mu$ (10^{-6})	Ref.
Inversion lines of NH_3 *vs* (HCO^+, HCN)	B0218+357	$0.74 \pm 0.47_{\rm stat} \pm 0.76_{\rm sys}$	(1)
Inversion lines of NH_3 *vs* HC_3N	PKS1830-211	0.08 ± 0.47	(2)
Inversion lines of NH_3 *vs* (CS, H_2CO)	B0218+357	-0.35 ± 0.12	(3)
Inversion lines of NH_3 *vs* (average of 22 species)	PKS1830-211	-2.04 ± 0.74	(4)
CH_3OH *vs* (average of 22 species)	PKS1830-211	-1.95 ± 0.47	(4)

References: (1) Murphy *et al.* (2008); (2) Henkel *et al.* (2009); (3) Kanekar (2011); (4) Muller *et al.* (2011).

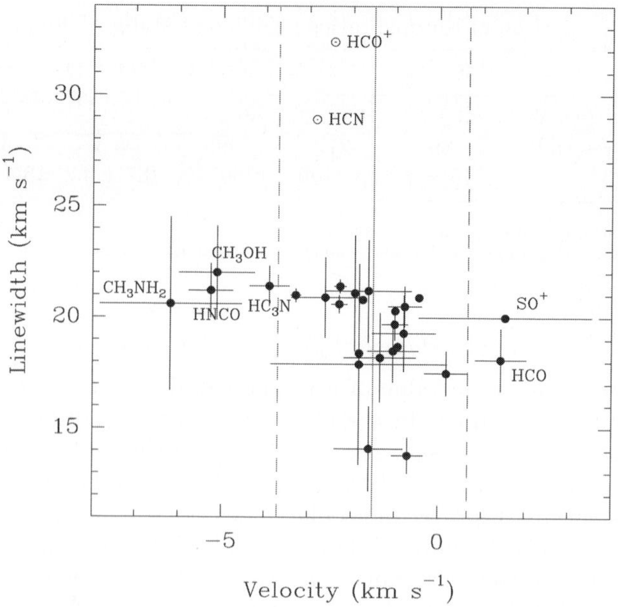

Figure 6. Diagram of the FWHM versus velocity centroids, derived from Gaussian fits of the lines toward PKS1830-211(SW). The average velocity (full line) and 3σ dispersion (dashed lines), calculated from 22 species, are indicated. Note that CH_3NH_2, HNCO, CH_3OH, HC_3N, HCO and SO^+ have velocity centroids significantly different from the bulk of other species. The lines of HCO^+ and HCN are saturated.

6. Summary and perspectives

Redsifted millimeter molecular absorbers are currently the only objects allowing us to study the chemistry evolution at intermediate redhifts. The first unbiased spectral survey toward such a molecular absorber, that at z=0.89 toward PKS1830-211, has yielded the (somehow surprising) detection of a large number of molecules, which can be used as powerful diagnostic tools to investigate the property of the gas in a galaxy at a look-back time of half the age of the Universe.

With the sensitivity of the ALMA and EVLA arrays, it is now possible to conduct large and deep radio/(sub)mm spectral scans: especially, redshifted molecular absorbers offer the possibility to search from the ground for molecules which could not be observed at z=0 due to atmospheric attenuation, e.g., such as light hydrides. The sensitivity improvement increases the number of targetable continuum sources, hence chance of discovering new molecular absorption systems.

Acknowledgement

I am grateful to the organizers for inviting me to present this work and acknowledge the help and support of my collaborators.

References

Bottinelli, S., Hughes, A. M., van Dishoeck, E. F., *et al.*, 2009, *ApJ*, 690, 130
Cernicharo, J., Guélin, M., & Kahane, C., 2000, *A&A Supp.*, 142, 181
Chengalur, J. N., de Bruyn, A. G., & Narasimha, D., 1999, *A&A*, 343, 79
Combes, F. & Wiklind, T., 1997, *ApJ*, 486, 79
Combes, F., 2008, *Ap&SS*, 313, 321

Costagliola, F., Aalto, S., Rodriguez, M. I., *et al.*, 2011, *A&A*, 528, 30
Curran, S. J., Murphy, M. T., Pihlström, Y. M., *et al.*, 2004, *MNRAS*, 352, 563
Curran, S. J., Whiting, M. T., Murphy, M. T., *et al.*, 2006, *MNRAS*, 371, 431
Curran, S. J., Whiting, M. T., Wiklind, T., *et al.*, 2008, *MNRAS*, 391, 765
Drinkwater, M. J., Combes, F., & Wiklind, T., 1996, *A&A*, 312, 771
Eckart, A., Cameron, M., Genzel, R., *et al.*, 1990, *ApJ*, 365, 522
Espada, D., Peck, A. B., Matsushita, S. *et al.*, 2010, *ApJ*, 720, 666
Evans, A. S., Sanders, D. B., Surace, J. A., & Mazzarella, J. M., 1999, *ApJ*, 511, 730
Falco, E. E., Impey, C. D., Kochanek, C. S., *et al.*, 1999, *ApJ*, 523, 617
Fixsen D. J., 2009, *ApJ*, 707, 916
Flambaum, V. V. & Kozlov, M. G., 2007, *Phys. Rev. Lett.*, 99, 801
García-Burillo, S., Combes, F., Neri, R., *et al.*, 2007, *A&A*, 468, L71
Gardner, F. F. & Whiteoak, J. B. ,*MNRAS*, 175, 9
Gérin, M., Phillips, T. G., Benford, D. J., *et al.*, 1997, *ApJ* 488, 31
Gérin, M., de Luca, M., Goicoechea, J. R., *et al.*, 2010, *A&A*, 521, L16
Henkel, C., Jacq, T., Mauersberger, R., Menten, K. M., & Steppe, H., 1987, *A&A*, 188, L1
Henkel, C., Mauersberger, R., Wiklind, T., *et al.*, 1993, *A&A*, 268, L17
Henkel, C.; Jethava, N.; Kraus, A., *et al.*, 2005, *A&A*, 440, 893
Henkel, C., Braatz, J. A., Menten, K. M., & Ott, J., 2008, *A&A*, 485, 451
Henkel, C., Menten, K. M., Murphy, M. T., *et al.*, 2009, *A&A*, 500, 725
Horellou C., Nord M., Johansson D., & Lévy A., 2005, *A&A*, 441, 435
Israel, F. P., van Dishoeck, E. F., Baas, F., *et al.*, 1990, *A&A*, 227, 342
Jauncey, D. L., Reynolds, J. E., Tzioumis, A. K., *et al.*, 1991, *Nature*, 352, 132
Jansen, P., Xu, L.-H., Kleiner, I., Ubachs, W., & Bethlem, H. L., 2011, *Phys. Rev. Lett.*, 106, 801
Jetzer P., Puy D., Signore M., & Tortora C., 2011, *General Relativity and Gravitation*, 43, 1083
Jin, C., Garrett, M. A., Nair, S., *et al.*, 2003, *MNRAS*, 340, 1309
Kanekar, N., Carilli, C. L., Langston, G. I., *et al.*, 2005, *Phys. Rev. Lett.*, 95, 1301
Kanekar, N., 2011, *ApJl*, 728, L12
Lima J. A. S., 1996, *Physical Review D*, 54, 2571
Liszt, H. & Lucas, R., 2004, *A&A*, 428, 445
Liszt, H. S., Pety, J., & Lucas, R., 2008, *A&A*, 486, 493
Lodders, K., 2003, *ApJ*, 591, 1220
Lucas, R. & Liszt, H., 1998 ,*A&A*, 337, 246
Lucas, R. & Liszt, H. S., 2000, *A&A*, 358, 1069
Luzzi G., Shimon M., Lamagna L., *et al.*, 2009, *ApJ*, 705, 1122
Martín, S., Mauersberger, R., Martín-Pintado, J., Henkel, C., & García-Burillo, S., 2006, *ApJS*, 164, 450
Martín, S., Krips, M., Martín-Pintado, J., *et al.*, 2011, *A&A*, 527, 36
Mathur, S. & Nair, S., 1997, *ApJ*, 484, 140
Menten, Karl M. & Reid, Mark J., 1996, *ApJ*, 465, L99
Menten, K. M., Carilli, C. L., & Reid, M. J., 1999, ASP Conf. Ser., 156, 218
Menten, K. M., Güsten, R., Leurini, S., *et al.*, 2008, *A&A*, 492, 725
Müller, H. S. P., Thorwirth, S., Roth, D. A., & Winnewisser, G., 2001, *A&A*, 370, 49
Muller, S., Guélin, M., Dumke, M., *et al.*, 2006, *A&A*, 458, 417
Muller, S., Guélin, M., Combes, F., & Wiklind, T., 2007, *A&A*, 468, L53
Muller, S. & Guélin, M., 2008, *A&A*, 491, 739
Muller, S. & Dinh-V-Trung, 2009, *ApJ*, 696, 176
Muller, S., Beelen, A., Guélin, M., *et al.*, 2011, *A&A*, in press, [arXiv:1104.3361]
Murphy, M. T., Curran, S. J., & Webb, J. K., 2003, *MNRAS*, 342, 830
Murphy, M. T., Flambaum, Victor V., Muller, S., & Henkel, C., 2008, *Science*, 320, 1611
Nair, S., Jin, C., & Garrett, M. A., 2005, *MNRAS*, 362, 1157
Noterdaeme, P., Petitjean, P., Srianand, R., *et al.*, 2011, *A&A*, 526, L7
Nummelin, A., Bergman, P., Hjalmarson, Å., *et al.*, 2000, *ApJ Supp.*, 128, 213

Ohishi, M., Irvine, W. M., & Kaifu, N., 1992, *IAU Symp.* 150, 171
Omar, A., Anantharamaiah, K. R., Rupen, M., & Rigby, J., 2002, *A&A*, 381, L290
Patnaik, A. R., Browne, I. W. A., King, L. J., *et al.*, 1993, *MNRAS*, 261, 435
Pickett, H. M., Poynter, R. L., Cohen, E. A., *et al.*, 1998, *J. Quant. Spectrosc. & Rad. Transfer*, 60, 883
Seaquist, E. R. & Bell, M. B., 1986, *ApJ*, 303, L67
Seaquist, E. R. & Bell, M. B., 1990, *ApJ*, 364, 94
Turner, B. E., 2000, *ApJ*, 542, 837
Uzan, J.-P., 2011, *Living Reviews in Relativity*, 14, 2
Wang, M., Henkel, C., Chin, Y.-N., *et al.*, 2004, *A&A*, 422, 883
Wang, M., Chin, Y.-N., Henkel, C., Whiteoak, J. B., & Cunningham, M., 2009, *ApJ*, 690, 580
Wiklind, T. & Combes F. 1995, *A&A*, 299, 382
Wiklind, T. & Combes F. 1996a, *A&A*, 315, 86
Wiklind T. & Combes F. 1996b, *Nature*, 379, 139
Wiklind T. & Combes F. 1997a, *A&A*, 324, 51
Wiklind T. & Combes F. 1997b, *A&A*, 328, 48
Wiklind T. & Combes F. 1998, *ApJ*, 500, 129
Wiklind, T. & Combes, F., 1999, arXiv:astro-ph/9909314
Winn, J. N., Kochanek, C. S., McLeod, B. A., *et al.*, 2002, *ApJ*, 575, 103

Discussion

KARL MENTEN: In your plot of derived T_{CMB} values, two species have about 3 K with apparently small error bars. What is the explanation of this ?

S. MULLER: Those two points correspond to the rotation temperatures derived for H_2CCO and H_2CCN. Both are detected with limited signal-to-noise ratios, and H_2CCN has a myriad of hyperfine components. We are investigating if those points and their error bars were not underestimated.

RAINER MAUERSBERGER: Two of the strongest local central sources of galaxies, Sgr A and Cen A, show absorption. Why does this encouraging statistics not extend to more distant sources ?

S. MULLER: This is likely due to the combination of small number of bright radio-mm sources at high-z and clumpiness of the molecular gas, resulting in a rare chance of alignment. In addition, there is a selection bias against background sources with high extinction, for which the redshift of an intervening object could not be known from optical spectroscopy. This might change soon with the increase in instantaneous bandwidth and sensitivity of the new generation of telescopes at mm/cm wavelengths.

MICHEL GUÉLIN: I would like to stress that molecular line absorption measurements yield more reliable molecular abundance ratios than emission line measurements, especially in medium and low density gas, as the excitation temperature is close to the cosmic background temperature. This is important for the derivation of isotopic abundance ratios.

The Molecular Universe
Proceedings IAU Symposium No. 280, 2011
José Cernicharo & Rafael Bachiller, eds.

doi:10.1017/S1743921311025105

Extragalactic Line Surveys

Sergio Martín[1]

[1]European Southern Observatory,
Alonso de Córdova 3107, Vitacura, Casilla 19001, Santiago 19, Chile
email: smartin@eso.org

Abstract. 40 years have passed since the first molecular detection outside our Galaxy. Since then, our knowledge on the distribution, kinematics and composition of the molecular material in the extragalactic ISM has built up significantly based not only on the carbon monoxide observations but also in the more than 50 molecular species detected. In particular, line surveys have been proven to be excellent tools to study the chemical composition in the nuclei of galaxies. Such studies have been favored by the increasing instantaneous bandwidth of current mm and sub-mm facilities. Here I will summarize the highlights of extragalactic molecular spectroscopy, mostly focusing in the results from molecular line surveys published in the last few years as well as the aims of still ongoing projects.

Keywords. Astronomical data bases: surveys, Galaxies: abundances, Galaxies: active, Galaxies: ISM, Galaxies: starburst

1. Introduction: Chemical variations in galactic nuclei

To date, all of the spectral line surveys towards extragalactic sources have been carried out at low spatial resolution. The large beam sizes in most of these studies just gather the averaged emission over the central few hundred parsecs in the central regions of nearby galaxies. Almost two decades ago, the review paper by Mauersberger & Henkel (1993) suggested that we could naively expect that the chemical variations in the different molecular cloud complexes, affected by different physical and chemical proccesses, would cancel out when observing the averaged emission over large regions. However, based on a number of previous studies and the more than two dozen of extragalactic molecular detections, they claimed that "surprisingly, the variations in the chemical composition do not cancel out" (Mauersberger & Henkel 1993). Their comparison of the molecular abundances of just seven species in the starburst galaxy prototypes, M 82 and NGC 253, where only three species were detected in both galaxies, showed that significant differences could be identified. Such abundance differences were attributed to the different heating mechanisms driving the chemistry in their central regions. It became then clear the potential of chemical composition comparative studies to probe the physical processes in the heavily obscured central regions of galaxies. Spectral line surveys, particularly at mm and submm wavelenghts, became key to understand and stablish the chemical classification and evolution of galactic nuclei.

2. Targeted vs. Unbiased surveys

Starburst galaxies have been the evident targets for the first molecular line surveys given the strong molecular emission. Moreover, these galaxies are indeed the most prominent molecular emitters outside the galaxy. The study by Wang *et al.* (2004) towards

Table 1. Instantaneous bandwidth in GHz of currently available mm and submm facilities

Single Dish		Interferometer	
MOPRA	... 8.2	SMA	... 8
IRAM 30 m	... 8	CARMA	... 8
Nobeyama 45 m	... 32	PdBI	... 4
APEX	... 8	ATCA	... 4
JCMT	... 1.8	ALMA	... 8

the central region of NGC 4945 is likely the first systematic detailed study of the chemical composition in an external galaxy. Using the Swedish-ESO Submillimeter Telescope (SEST), they observed 80 transitions from 19 different species over a frequecy range between 82 and 354 GHz. The derived molecular abundances and isotopic ratios measured in NGC 4945 were compared to those available in the literature from M 82 and NGC 253. Though quite complete, this comparison was limited by the selection of the observed molecular species. Line surveys over wide frequency ranges, on the other hand, provide a complete and unbiased description of the chemical composition in galactic nuclei. The first unbiased spectral line survey was carried out with the IRAM 30m telescope towards the central region of the starburst galaxy NGC 253 and covered the 2 mm atmospheric window between 129 and 175 GHz (Martín *et al.* 2006). This scan resulted in the detection of 111 spectral features from 25 molecular species as well as three hydrogen recombination lines. The follow up towards M 82, the brightest molecular emitter together with NGC 253, surveyed the same 2 mm window as well as the 1.2 mm band between 241 and 260 GHz (Aladro *et al.* Submitted). Although expected from previous observations, the detection rate in M 82 was significantly lower, with 72 spectral features from 18 molecular species. The combination of both line surveys resulted in the most complete comparison between the molecular composition of two extragalactic sources. Indeed, the observed chemical differences are claimed to be directly linked to the different state of evolution of their nuclear starbursts, and therefore to the different leading heating mechanisms affecting the ISM within their central few hundred parsecs. While a number of species, such as CH_3OH, HNCO, NH_3, SiO, *NS*, *HOCO*, and *CH_2NH*,claimed to be formed in dust grains and injected into gas phase via shocks, were found to be enhanced in NGC 253, other molecules, such as CO^+, HCO, HOC^+, $c-C_3H_2$, CH_3CCH, and *NH_2CN*, mostly formed in the gas phase and likely enhanced by intense UV fields, are significantly more abundant towards M 82. Even though some of these differences were previously known, a number of them (in *italics* in the list above) were pointed out from the observations covered by the line surveys. In particular, line surveys allow the identification of particularly interesting and contrasted species such as the case of methanimine (*NH_2CN*) which shows rotational temperatures similar to those measured towards the Galactic hot core Sgr B2(M) in both M 82 and NGC 253, but shows an abundance enhancement towards the former.

3. Receiver Bandwidth Upgrades: Ongoing surveys

At that time of the first line surveys described in previous section, the available instantaneous bandwidth in both SEST and the IRAM 30m telescopes was ~2 GHz. Even with such not inconsiderable bandwidths, spectral scans projects were still significantly expensive in terms of telescope time. To illustrate this, the surveys towards M 82 and NGC 253 each required more than one hundred hours of telescope time at the IRAM 30m telescope. However, in the last few years, major upgrades have been performed in all the mm and submm facilities around the globe. By the end of 2005, the MOPRA telescope commissioned the MOPS digital filterbank spectrometer which provided an $\sim$ 8 GHz instantaneous bandwidth. Though unique at the time, most single-dish telescopes and interferometers catched up with similarly wide band receivers as shown in Table 1. Just recently, the Nobeyama 45m telescope started commissioning the SAM45 spectrometer, consisting of a copy of the ALMA/ACA correlator that allows a maximum of a 32 GHz bandwidth.

These major telescope upgrades have open the possibility of extending the exploration of wider spectral bands towards larger sample of extragalactic objects given the significantly reduced time required for such spectral scans. Consequently, a number of spectral line surveys have been carried out or are currently ongoing in almost every mm and submm facility. Table 2 summarizes the observational details of some of the most relevant line surveys carried out at mm and submm wavelenghts. Here we describe the main aims of theses surveys:

IRAM 30m .– Following the success of the 2mm line surveys of NGC 253 and M 82, a series of deep line survey follow up at 3mm are currently ongoing and near completion. This project sample consist of 6 galaxies, including two young starbursts (M 83 and NGC 253), two evolved starbursts (M 82 and M 51), the active galactic nucleus (AGN) prototype NGC 1068 and the ULIRG Arp 220. A total of 36 molecular species are detected in the covered band. This study aims to set the basis of a chemical classification of the extragalactic ISM in the central few hundred parsecs of nearby bright galaxies. The new instrumentation at the IRAM 30m telescope reduced the time required for each spectral scan by almost an order of magnitude, down to $\simeq$ 10 hours per source.

Nobeyama 45m .– Three line surveys in the 3 mm atmospheric band had just been finished at the 45 m facility making use of the recently upgraded spectrometer (Poster 2.69). With more than a dozen molecular species detected, these surveys will be able to establish the differences between the chemistry in starburst galaxies (NGC 253 and IC 342) and the AGN dominated NGC 1068. Thus the main motivation of this work is to study the chemical fingerprints of the AGN influence on the surrounding molecular gas. The relatively small ($18''$) beam of the 45 m telescope is smaller than the starburst ring in NGC 1068, allowing the observation of the chemical composition towards the AGN without contamination from the molecular material in starburst.

APEX 12m .– The spectral scans towards NGC 4945, NGC 253 and Arp 220 with the APEX telescope (Poster 2.83) are exploiting the capabilities of the FLASH receiver at APEX which processes an 8 GHz instantaneous bandwidth. These surveys are exploring the higher energy transitions in the 0.9 mm atmospheric window. A total of 16 molecular species are detected at the highest frequencies ever surveyed outside the Galaxy. Additionally, NGC 253 has been surveyed in the 1.3 mm atmospheric window from 185 to 275 GHz (Poster 2.83). These APEX scans complete the spectral coverage towards NGC 253 over a total band of 274 GHz between 3.5 and 0.8 mm (see Table 2).

SMA .– The Submillimeter Array was used to carry out the first interferometric extragalactic unbiased survey (see Section 6). The large width of the emission lines in Arp 220 together with the strong line confusion in this object, made it necessary to use the stable baselines achieved with the interferometer. A total of 70 spectral features from 15 molecular species and 6 isotopologues were identified. The survey was partially confusion limited, with an average of 1.8 lines per GHz. This spectral scan was the first

Table 2. Extragalactic spectral scans at mm and submm wavelenghts

Source	Frequency Range (GHz)	Telescope	Resolution ($km\,s^{-1}$)	rms (mK)	Reference
NGC 253	86-116	IRAM 30m	12	2	Aladro *et al.* In prep.
	85-116	Nobeyama 45m	20	2-13	Nakajima *et al.* In prep.
	129-175	IRAM 30m	8	2-6	Martín *et al.* (2006)
	185-275	APEX 12m	12	2	Requena-Torres *et al.* In prep.
	280-360	APEX 12m	20	2	Requena-Torres *et al.* In prep.
M 82	86-116	IRAM 30m	12	2	Aladro *et al.* In prep.
	130-175	IRAM 30m	8	2-5	Aladro *et al.* Submitted
	241-260	IRAM 30m	5	4-6	Aladro *et al.* Submitted
NGC 1068	86-116	IRAM 30m	12	2	Aladro *et al.* In prep.
	85-116	Nobeyama 45m	20	1-4	Nakajima *et al.* In prep.
M 83	86-116	IRAM 30m	12	2	Aladro *et al.* In prep.
M 51	86-116	IRAM 30m	12	2	Aladro *et al.* In prep.
Arp 220	86-116	IRAM 30m	12	2	Aladro *et al.* In prep.
	202-242	SMA	250	4 [1]	Martín *et al.* (2011)
	280-360	APEX 12m	20	2	Requena-Torres *et al.* In prep.
IC 342	85-116	Nobeyama 45m	20	1-2	Nakajima *et al.* In prep.
NGC 4945	280-360	APEX 12m	20	2	Requena-Torres *et al.* In prep.
PKS 1830-211	30-50	ATCA	6-10	...	Muller *et al.* (2011)

Notes:
[1] rms in $mJy\,beam^{-1}$ units

carried out towards a ULIRG at a distance of 70 Mpc, an order of magnitude farther than nearby starburst such as NGC 253 and M 82. Some of the highlights of this study are summarized in Section 6.

4. Going wider: high redshift wide band receivers

In the last few years, a number of extremely wide band receivers have been developed to measure the redshift of distant heavily obscured objects. These receiver cover wide frequency bands at a low spectral resolution. Even if this spectral resolution is similar or even wider than the tipical linewidth of local galaxies, it is still possible to study the brightest molecular transitions in relatively large samples of galaxies at a relatively low observing time expense. In this section we summarize some of the latest publications making use of these wide band instruments.

Redshift Search Receiver (RSR) at FCRAO 14m .– The RSR, design as a facility instrument for the 50m LMT, is a dual polarization dual beam instrument with 4 broadband receivers that covers the 37 GHz accross the 3mm atmospheric window, between 74 and 111 GHz. The spectral resolution of $\sim 100\,km\,s^{-1}$ provided is barely enough to resolve the typical linewidth of $\sim 200\,km\,s^{-1}$ in the central region of local galaxies. This spectrometer was used to survey the 3 mm atmospheric band towards a sample of 10 galaxies with different types of nuclear activity (see Table 1 in Snell *et al.* 2011). A total of 33 transition of 13 molecular species were detected. The low number of spectral features as compared with similar scans with the IRAM 30m or Nobeyama 45m is a consequence of the line smoothing due to the low spectra resolution. Still line ratios between the brighter spectral features could be extracted from this survey. For example the HCO^+/HCN ratio was observed to be enhanced in AGN dominated galaxies when compared to starburst (SB) dominated galaxies (Snell *et al.* 2011). This result, in disagreement with the result from Krips *et al.* (2008), where the ratio HCO^+/HCN is observed to be enhanced in SB galaxies, shows the difficulty of establishing activity templates in extragalactic sources. It is indeed difficult to disantangle the contribution from AGN and SB at the low resolution provided by single dish observations. As such, the highest HCO^+/HCN ratios from the surveys from Snell *et al.* (2011) were derived from the LIRGs NGC 3690 and NGC 6240 where, even if known to host an active nucleus, the contribution from the star formation cannot be excluded.

Z-Spec at CSO 10.4m .– Z-Spec is a millimeter wave grating spectrometer dispersing the light to an array of 160 bolometers. Though covering an enormous bandwidth of almost 120 GHz from 190 to 307 GHz, the spectral resolution is very low. Velocity resolution ranges from $700\,\mathrm{km\,s^{-1}}$ at the lower frequencies to $1200\,\mathrm{km\,s^{-1}}$ at the higher end. This resolution is significantly coarser than the observed molecular linewidth even in distant galaxies. Still, this receiver has been used to survey large frequency bands in local galaxies such as the 3 positions observed towards the starburst galaxy M 82 (Naylor *et al.* 2010). With an integration of just 1 hour per position, 10 molecular species were detected within the band. Even though the resolution was heavily limited, it is possible to simultaneous observe three transitions of CS and therefore to constrain and compare the physical properties of the molecular gas in the three observed positions. Similarly, Kamenetzky *et al.* (2011) observed the central region of NGC 1068 with Z-spec, where they identified 12 molecular species.

Though not intended as a molecular line survey, the observations with Z-Spec towards teh Cloverleaf galaxy at z$\sim$2.56 covered the rest frame wavelenghts between 272 and 444 μm (Bradford *et al.* 2009). At these wavelenghts, four transition of CO (from $J = 6-5$ up to $9-8$) were clearly detected and [CI] at 370 μm blended to the CO $7-6$ transition. Additionally, tentative detection of H_2O in emission and CH^+ and LiH^+ in absorption were reported. These CO simultaneous measurements, together with previous lower-J detections with the IRAM PdBI, allow the detailed study of the CO spectral line energy distribution (SLED) in the cloverleaf (Figure 2 in Bradford *et al.* 2009). The studies at high-z with broad band spectrometers at ground based telescopes are similar to those carried out with Herschel space observatory toward nearby galaxies. For example, the observations with the SPIRE spectrometer covering the frequencies from 467 to 989 GHz measured the CO emission from $J = 5-4$ to $13-12$ towards the ULIRG Mrk 231 (van der Werf *et al.* 2010). The resulting CO SLED has been modelled with a combination of dense gas exposed to a strong UV radiation and a significant X-ray contribution from the supermassive black hole to explain the line luminosity of the higher transitions. Similar SLED were modelled towards M 82 based on both SPIRE and HIFI spectrometers on board Herschel (Panuzzo *et al.* 2010; Loenen *et al.* 2010). The SPIRE spectrum towards Mrk 231 (van der Werf *et al.* 2010) also resulted in the detection rotational lines of H_2O, OH^+, H_2O^+, CH^+, and HF, which shows the potential of molecular studies in the submillimeter to infrared wavelenghts.

5. Absorption systems

Among the surveys compiled in this review, those towards absorption systems deserve an special mention due to its enormous potential to carry out molecular studies at intermediate redshift. Such is the case of the spectral scan carried out with ATCA towards

Table 3. Census of extragalactic molecular species and isotopologues detected

2 atoms	3 atoms	4 atoms	5 atoms	6 atoms	7 atoms
OH	H_2O, $H_2^{18}O$	H_2CO	c-C_3H_2	CH_3OH, $^{13}CH_3OH$	CH_3C_2H
CO { ^{13}CO, $C^{18}O$, $C^{17}O$ }	HCN { $H^{13}CN$, $HC^{15}N$, DCN }	NH_3	HC_3N { $H^{13}CCCN$, $HC^{13}CCN$, $HCC^{13}CN$ }	CH_3CN	CH_3NH_2
H_2, HD	HCO^+ { $H^{13}CO^+$, $HC^{18}O^+$, DCO^+ }	HNCO	CH_2NH		CH_3CHO
CH	C_2H	H_2CS	NH_2CN		
CS { ^{13}CS, $C^{34}S$, $C^{33}S$ }	HNC { $HN^{13}C$, DNC }	$HOCO^+$	CH_2CO		
CH^+	N_2H^+, N_2D^+	C_3H	l-C_3H_2		
CN	OCS	H_3O^+	H_2CCN		
SO, ^{34}SO	HCO		H_2CCO		
SiO, ^{29}SiO	H_2S		C_4H		
CO^+	SO_2				
NO	HOC^+				
NS	C_2S				
LiH	H_3^+				
CH	H_2O^+				
NH	l-C_3H				
OH^+					
HF					
SO^+					

Notes: Updated Table from Martín *et al.* (2011) with the new detections in the absorption system PKS 1830-211 by Muller *et al.* (2011).

the absorption system PKS 1830-211 (Muller *et al.* 2011). The survey (see Invited talk in this volume by Sebastien Muller for an extensive description of the results) covered the frequency range from 30 to 50 GHz, equivalent to the $5-3$ mm rest frame between 57 and 111 GHz at the redshift of z=0.89. As a result of the spectral scan 28 molecular species and 8 isotopologues were identified. This study turns PKS 1830-211 into one of the extragalactic sources with the largest number of molecular detections. The accurate fractional abundances derived from these observations allows complete comparison with Galactic sources (see Fig. 8 in Muller *et al.* 2011) which showed the similar molecular composition between this absorption system and that found in typical Galactic diffuse and translucent clouds.

6. The chemistry in the Arp 220 ULIRG

At a redshift of $z=0.018$, Arp 220 is the nearest ULIRG. With a star formation rate of SFR$\sim 300\,M_{\odot}\,yr^{-1}$, this galaxy could be considered as a scaled down version of the galaxies observed at high redshift, with SFR up to an order of magnitude higher than that in Arp 220. The 1.3 mm line survey carried out with the SMA (Sect. 3) aimed to establish the differences and similarities between the molecular composition in ULIRGs and that found in local nearby starburst galaxies. Moreover, it aimed to study whether the nuclear power source, AGN and/or SB, had an imprint in the chemical composition. Previous to this spectral scan, a number of molecular studies had been carried out towards this source. The paper by Greve *et al.* (2009) presented a complete compilation of all single dish observations of CO, HCN, HCO^+, HNC and HNCO, as well as CO isotopologues in Arp 220. A detailed study of the molecular content was also carried out in absorption with the Infrared Space Observatory (ISO) between 25 to 1300 μm (González-Alfonso *et al.* 2004).

The overall chemical composition derived from the 1.3 mm band emission lines resemble that in starbursts such as NGC 253. However, the emission from vibrationally excited transitions of HC_3N and CH_3CN, never detected in local starbursts, was detected towards Arp 220 (Martín *et al.* 2011). Vibrational temperatures derived from these transitions are in the range of $T_{\rm vib} = 300 - 500\,$K. Within our Galaxy, vibrationally excited emission of HC_3N is observed to be tracing the hottest gas around star forming cores (de Vicente *et al.* 1997, 2000). In the extragalactic ISM, vibrational emission of HCN and HC_3N have also been recently detected towards the also heavily obscured nucleus of the LIRG NGC 4418 (Costagliola & Aalto 2010; Sakamoto *et al.* 2010), as well as $v_2 = 1$ l-type absorption lines of HCN towards Arp 220 (Salter *et al.* 2008) and mid-infrared vibration-rotation bands of C_2H_2, HCN, and CO_2 in a sample of (U)LIRGs (Lahuis *et al.* 2007). The high vibrational temperatures derived from these detections is suggested to be the consequence of the IR-pumping (Sakamoto *et al.* 2010). Though a hot component associated to a buried AGN cannot be excluded (Costagliola & Aalto 2010), the detected vibrationally excited emission has been claimed to be most likely tracing the warmest gas in the extremely compact star forming regions within these heavily obscured environments (Martín *et al.* 2011). This scenario is supported by the detection of the ^{18}O isotopologue of water in the Arp 220 1.3mm line survey. The water abundance estimated from the $H_2^{18}O/C^{18}O$ ratio of $\sim 2 \times 10^5$ is similar to that measured in Galactic hot cores (Gensheimer *et al.* 1996; Cernicharo *et al.* 2006). Such water detection should imply an enormous star formation concentration of the order of a few 10^6 Sgr B2(N)-like hot cores in a $\sim$ 700 pc region (Martín *et al.* 2011).

7. Molecular Census Evolution

To date, a total of 54 molecular species and 23 isotopologues have been claimed to be detected in the extragalactic interestellar medium. Table 3 shows a similar table to that presented in Martín *et al.* (2011) which became outdated by the recent detection of SO^+, l-C_3H, l-C_3H_2, H_2CCN, H_2CCO, C_4H, CH_3NH, and CH_3CHO in the spectral scan towards the absorption system PKS 1830-211 (see Sect. 5). It is definitely worth having a look on how the census of extragalactic molecular detections has evolved as a function of time. Fig. 1 shows the evolution with time of the cumulative number of molecular species detected in the extragalactic ISM. Since the first molecular detection 40 years ago (OH in absorption, Weliachew 1971) there has been a steady increase in the number of molecular detections. However, we can note two periods in which the rate of detections has been quicken. First, mostly in the period between 1985 and 1990, the increase was due to the availability of sensitive radio facilities, and in particular to the large collecting area of the IRAM 30m telescope. Only in the last few years, from 2003 up until now, spectral line scans have played the dominant role in terms of detection rate. The advent of ALMA, with a significant increase in the collecting area of about an order of magnitud compared to the largest current facilities, will be the next step in sensitivity. In the next decade such increase in sensitivity will allow us to match the number of

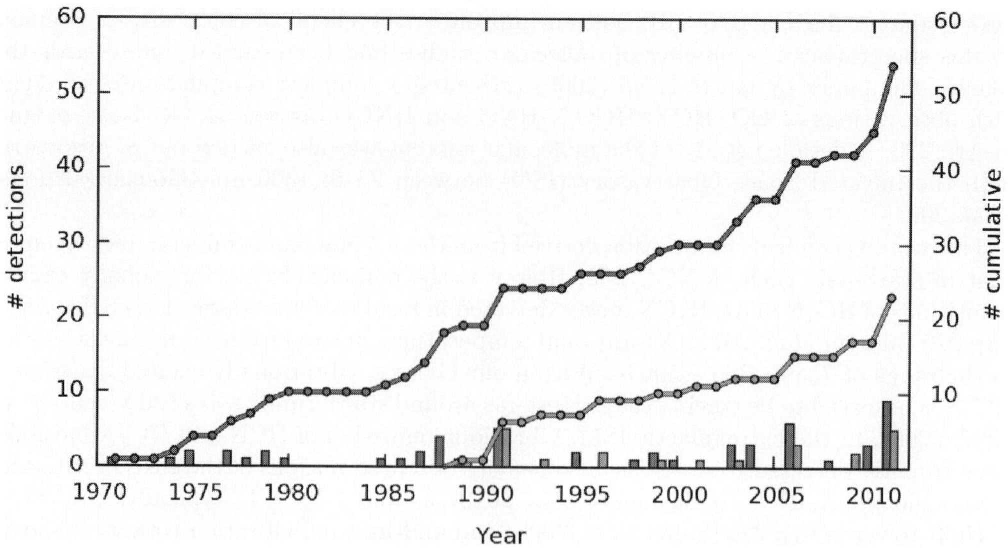

Figure 1. Histogram shows the number of extragalactic molecular detections per year for both the main (in red) and rarer (in green) isotopologues. The connected circles shows the cumulative number of detections.

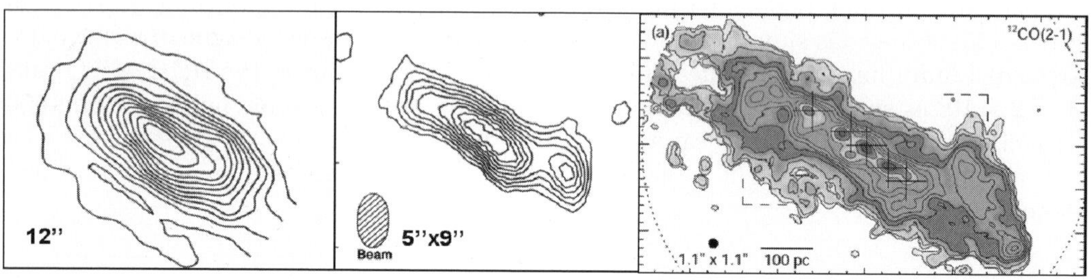

Figure 2. CO integrated emission maps of the starburst galaxy NGC 253 at different spatial resolutions. (left) CO $1-0$ map at $12''$ resolution by Mauersberger *et al.* (1996) with the IRAM 30m telescope. (center) CO $2-1$ map at $5'' \times 9''$ resolution by Canzian *et al.* (1988) with the OVRO interferometer. (right) CO $1-0$ map at $1.1''$ resolution by Sakamoto *et al.* (2011).

extragalactic molecular detections to the number of detected species in the Galactic ISM.

8. High Resolution Line Surveys

Even though sensitivity is an key factor for spectral scans, the next step in molecular line surveys does not only applies to sensitivity but also to resolution. Fig. 2 shows the increase in both richness and complexity of the molecular structure in the central few hundred parsecs of NGC 253 as the resolution improves from single dish resolution to high resolution interferometric observations. The high resolution maps by Sakamoto *et al.* (2011) reached a similar spatial resolution to that achieved in large scale surveys towards the Galactic center with small telescopes (⌀ =60 cm, Sawada *et al.* 2001). The imaging capabilities of instruments like ALMA, together with the large bandwidth of the receivers will allow to carry out deep 3D spectral scans. It will be possible to study the chemical composition variations accross the center of nearby galaxies at a similar resolution to that currently achieved by large single dish telescopes towards our very Galactic center. Thus, we expect line surveys will move a step further from studying the averaged molecular composition over the central few hundred parsecs of nearby galaxies to spatially resolve the chemical composition of individual molecular complexes.

References

Bradford, C. M., Aguirre, J. E., Aikin, R., *et al.* 2009, *ApJ*, 705, 112
Canzian, B., Mundy, L. G., & Scoville, N. Z. 1988, *ApJ*, 333, 157
Cernicharo, J., Goicoechea, J. R., Pardo, J. R., & Asensio-Ramos, A. 2006, *ApJ*, 642, 940
Costagliola, F. & Aalto, S. 2010, *A&A*, 515, A71
de Vicente, P., Martín-Pintado, J., Neri, R., & Colom, P. 2000, *A&A*, 361, 1058
de Vicente, P., Martin-Pintado, J., & Wilson, T. L. 1997, *A&A*, 320, 957
Gensheimer, P. D., Mauersberger, R., & Wilson, T. L. 1996, *A&A*, 314, 281
González-Alfonso, E., Smith, H. A., Fischer, J., & Cernicharo, J. 2004, *ApJ*, 613, 247
Greve, T. R., Papadopoulos, P. P., Gao, Y., & Radford, S. J. E. 2009, *ApJ*, 692, 1432
Kamenetzky, J., Glenn, J., Maloney, P. R., *et al.* 2011, *ApJ*, 731, 83
Krips, M., Neri, R., García-Burillo, S., *et al.* 2008, *ApJ*, 677, 262
Lahuis, F., Spoon, H. W. W., Tielens, A. G. G. M., *et al.* 2007, *ApJ*, 659, 296
Loenen, A. F., van der Werf, P. P., Güsten, R., *et al.* 2010, *A&A*, 521, L2
Martín, S., Krips, M., Martín-Pintado, J., *et al.* 2011, *A&A*, 527, A36
Martín, S., Mauersberger, R., Martín-Pintado, J., Henkel, C., & García-Burillo, S. 2006, *ApJS*, 164, 450
Mauersberger, R. & Henkel, C. 1993, in Reviews in Modern Astronomy, Vol. 6, Reviews in Modern Astronomy, ed. G. Klare, 69–102
Mauersberger, R., Henkel, C., Wielebinski, R., Wiklind, T., & Reuter, H. 1996, *A&A*, 305, 421
Muller, S., Beelen, A., Guélin, M., *et al.* 2011, ArXiv e-prints
Naylor, B. J., Bradford, C. M., Aguirre, J. E., *et al.* 2010, *ApJ*, 722, 668
Panuzzo, P., Rangwala, N., Rykala, A., *et al.* 2010, *A&A*, 518, L37
Sakamoto, K., Aalto, S., Evans, A. S., Wiedner, M. C., & Wilner, D. J. 2010, *ApJ*, 725, L228
Sakamoto, K., Mao, R.-Q., Matsushita, S., *et al.* 2011, ArXiv e-prints
Salter, C. J., Ghosh, T., Catinella, B., *et al.* 2008, *AJ*, 136, 389
Sawada, T., Hasegawa, T., Handa, T., *et al.* 2001, *ApJS*, 136, 189
Snell, R. L., Narayanan, G., Yun, M. S., *et al.* 2011, *AJ*, 141, 38
van der Werf, P. P., Isaak, K. G., Meijerink, R., *et al.* 2010, *A&A*, 518, L42
Wang, M., Henkel, C., Chin, Y., *et al.* 2004, *A&A*, 422, 883
Weliachew, L. 1971, *ApJ*, 167, L47

Discussion

CASELLI: You talk about "hot core" species. Did you try to compare abundance ratios in Arp 220 with typical "hot core" abundances, to test if we are indeed observing $\sim 10^6$ Sgr B2 hot cores?

MARTIN: Such comparison was performed with the abundances measured towards the starburst galaxy NGC 253, but not yet with Arp 220. However, Arp 220 and NGC 253 show significantly similar abundances but for the vibrational emission observed towards Arp 220. This difference points out towards a "hot core" origin but a closer comparison with Galactic templates is yet to be performed.

The Molecular Universe
Proceedings IAU Symposium No. 280, 2011
José Cernicharo & Rafael Bachiller, eds.

doi:10.1017/S1743921311025117

Gas-Phase Reactions in the ISM: Rate Coefficients, Temperature Dependences, and Reaction Products

Ian W. M. Smith

Department of Chemistry
University of Cambridge
Lensfeld Road
Cambridge CB2 1EW, UK

Abstract. Information about the rate coefficients and products of processes that occur in the interstellar medium are required as input to computer models that seek to reproduce the abundances of the rich variety of molecules that have been observed in different regions of the interstellar medium. In this brief review, I seek to identify the different kinds of gas-phase processes for which information is required and to consider the experimental, theoretical, and semi-empirical methods which are employed to measure or predict rate coefficients, $k(T)$, and how they depend on temperature (T) – and also how the products of reactions can, in favourable cases, be observed.

1. Introduction

This is the third time that I have been asked to review aspects of chemical reaction kinetics before an audience of astronomers and astrochemists. On both previous occasions (Smith 1988; Smith 1997), my brief was to consider reactions between electrically neutral species, with some special emphasis on these reactions at low temperatures. This time, I have been asked to cast my net wider: to consider kinetics between a wider range of types of species and over a wider range of temperature. Given the time and space available, I could only cover the complete ranges of species and temperature by being extremely superficial. To avoid that, I shall be selective, but I start by summarising in Table 1 the range of gas-phase processes which play a role in the chemistry of the interstellar medium (ISM), giving examples of each type. In the last column of this Table, I give the number of examples of each kind of process that is included in one of the main kinetic data bases for astrochemistry – that maintained by Eric Herbst's group (`http://physics.ohio-state.edu/~eric/research.html`)

In this review, I shall concentrate on (i) ion-neutral reactions, and (ii) neutral-neutral reactions. More detailed discussions, how kinetic information is obtained for these kinds of reactions, and information about the other processes listed in Table 1 can be found in two recent publications (Wakelam *et al.* 2010a; Smith 2011).

The primary information for any reaction that might play a role in the chemistry of the ISM is of two kinds. First, there is the rate coefficient, $k(T)$, and how it depends on temperature (T). The great majority of relevant processes are second-order: that is their rate is proportional to the product of the concentrations of two reactant species, say A and B, so that the change in concentration of each as a result of their reaction can be written as:

Table 1. Gas-phase reactions of different types included in the OSU database (version osu-09-2008) for astrochemistry.

Type of process	Example	Number in model
Gas-grain interactions	$H + H + grain \rightarrow H_2 + grain$	14
Direct cosmic ray processes	$H_2 + \zeta \rightarrow H_2^+ + e$	11
Cation-neutral reactions	$H_2^+ + H_2 \rightarrow H_3^+ + H$	2933
Anion-neutral reactions	$C^- + NO \rightarrow CN^- + O$	11
Radiative associations (ion)	$C^+ + H_2 \rightarrow CH_2^+ + h\nu$	81
Associative detachment	$C^- + H_2 \rightarrow CH_2 + e$	46
Chemi-ionization	$O + CH \rightarrow HCO^+ + e$	1
Neutral-neutral reactions	$C + C_2H_2 \rightarrow C_3H + H$	382
Radiative association (neutral)	$C + H_2 \rightarrow CH_2 + h\nu$	16
Dissociative recombination	$N_2H^+ + e \rightarrow N_2 + H$	539
Radiative recombination	$H_2CO^+ + e \rightarrow H_2CO + h\nu$	16
Anion-cation recombination	$HCO^+ + H^- \rightarrow H_2 + CO$	36
Electron attachment	$C_6H + e \rightarrow C_6H^- + h\nu$	4
External photo-processes[a]	$C_3N + h\nu \rightarrow C_2 + CN$	175
Internal photo-processes[a]	$CO + h\nu \rightarrow C + O$	192

Notes: [a] External photo-processes are those induced by the 'interstellar radiation field'; internal photo-processes are induced by radiation generated in dark clouds by the interaction of molecules, especially H_2, with cosmic rays.

$$-d[A]/dt = -d[B]/dt = k(T)[A][B] \tag{1.1}$$

In the laboratory, $k(T)$ is usually determined by observing changes in the concentration of one reactant, say A, in a large excess of the second reactant, B, so that

$$[A]_t = [A]_0\ exp(-k_{1st}t) \quad \text{where} \quad k_{1st} = k(T)[B] \tag{1.2}$$

In a series of experiments, $[B]$ is systematically varied and $k(T)$ is obtained from the variation of k_{1st} with $[B]$. In data bases for interstellar chemistry (and elsewhere), the temperature-dependence of rate constants is frequently expressed using a modified form of the celebrated Arrhenius equation:

$$k(T) = \alpha(T/300)^{\beta}\ exp(-\gamma/T) \tag{1.3}$$

However, for only very few reactions, is the kinetic information sufficiently accurate over a wide enough range of temperatures for the parameters, α, β and γ, to be cited with much confidence.

The other piece of information that is required for each reaction is the nature of the products. Sometimes they can be inferred from thermodynamic arguments. Where this is not the case and more than one set of products is energetically accessible, the experimental determination of the branching ratios into different channels can be demanding.

Table 2. Methods used to study the kinetics of ion-neutral reactions with the ranges of temeperatures over which experiments have been performed

Method	Range T/K	Reference
ICR – ion cyclotron resonance	300	Anicich (1993a), Anicich (1993b)
FA – flowing afterglow	82–600	Durkin *et al.* (1968), Ferguson *et al.* (1969)
SIFT – selected ion flow tube	80–480 18–295 up to 1400	Herbst *et al.* (1987), McIntosh, Adams, & Smith (1988) Bohringer, & Arnold (1986) Viggiano (2006)
SIFDT – selected ion flow/drift tube	kinectic energies equivalent to 1800 K	Viggiano (2006)
CRESU – reaction kinetics in uniform supersonic flows	8–300	Rowe *et al.* (1984), Rowe *et al.* (1985)

2. Ion-Neutral Reactions

As has been recognised for many years (e.g., Herbst & Klemperer 1973), reactions between ions, especially cations, and neutral atoms and molecules are of primary importance in the ISM, especially in the cold cores of dark, dense clouds. The data base of rate coefficients for such reactions is extensive (e.g., Anicich 1993a,b; Snow & Bierbaum 2008). The experimental techniques employed to obtain these data can be divided into two main categories (i) those in which the ionic reactant is trapped and then the rate of its loss observed as a neutral co-reactant is introduced; and (ii) those in which the ionic reactant is injected into a flowing gas and the loss of the ion is observed at a point downstream as a neutral reactant is added to the gas flow. In trapping experiments, it is necessary to work at low total gas density (*ca.* 10^{11} cm^{-3}), whereas flow experiments are conducted in excess helium at a density of *ca.* 10^{16} cm^{-3}. In Table 2 I list the variants of these methods and the temperature range over which they have been employed to provide rate coefficients for ion-neutral reactions.

Ion cyclotron resonance (ICR) experiments provided many of the early rate data on ion-neutral reactions but were limited to room temperature (Anicich 1993a,b). Later experiments using ion traps enabled measurements at temperatures as low as 10 K and are reviewed by Dunn (1995) and Gerlich (2008). These experiments have been especially useful in providing rate coefficients for important radiative association processes such as:

$$C^+ + H_2 \rightarrow CH_2^+ + h\nu \quad \text{and} \quad CH_3^+ + H_2 \rightarrow CH_5^+ + h\nu \qquad \text{(R1a), (R1b)}$$

Several important developments in the deployment of flow methods should be noted.

• First was the development of the selected ion flow tube (SIFT) by Adams & Smith (1976). In this technique, ions are created upstream of the main flow tube and then the desired ion is selected by a quadrupole mass filter before injection into the main gas flow. The SIFT method has been adapted to study (a) the kinetics of anion-neutral reactions (see Bierbaum elsewhere in these Proceedings), and (b) the reactions of ions with atomic radicals, such as H, N and O atoms (Snow & Bierbaum 2008). Early SIFT experiments covered the temperature range 80–480 K (Herbst *et al.* 1987; McIntosh *et al.* 1988). In one heroic experiment, a SIFT apparatus was cooled using liquid He to reach

18 K (Bohringer & Arnold 1986), and, at the Air Force Research Laboratories, SIFT experiments have been performed at temperatures as high as 1400 K (Viggiano 2006).

• Second was the addition of an electric field to the flow region of a SIFT apparatus allowing the ions to be accelerated and hence the relative translational energy of the collisions between the ion and the neutral co-reactant to be systematically increased. This method called the selected ion flow drift tube (SIFDT or SIDF) allows reaction rate coefficients to be studied as a function of the reactant collision energy up to energies equivalent to 1800 K (Viggiano 2006). Because the applied field does not affect the internal energies of the ions, comparison of the rate coefficients under equilibrated high temperature conditions and those measured in SIFDT experiments allows conclusions to be reached about the effect of internal energy on the rate coefficients for ion-neutral reactions.

• Third was the development of the CRESU (Cinetique de Reactions Ecoulement Supersonique Uniforme, or reaction kinetics in uniform supersonic flows) technique by Rowe and his colleagues. Expansion of gas through specially designed convergent-divergent Laval nozzles generated gas flows with temperatures as low as 20 K (Rowe *et al.* 1984), a limit which was further lowered to 8 K by pre-cooling the gas using liquid N_2 (Rowe *et al.* 1985).

For many – but not all – reactions between ions and neutral species, there is no activation barrier to inhibit reaction. Rather, the rate coefficient is controlled by the ability of the long-range attraction between the reactants to bring them into close contact. The simplest treatment of this capture process is appropriate when the interaction arises between the charge on the ion and the induced dipole on the neutral molecule. If the polarizability (α) of the molecule is independent of its orientation and the ion carries a single electronic charge (e), then a simple treatment, generally referred to as the Langevin model, leads to the following expression for the rate coefficient for reaction between two species with a collisional reduced mass of μ:

$$k_L = 2\pi\ e\ (\alpha/\mu)^{1/2} \tag{2.1}$$

If cgs units are used for the quantities on the right-hand-side of this equation, k_L is given in cm^3s^{-1} and has a typical value of *ca.* 10^{-9} cm^3s^{-1}.

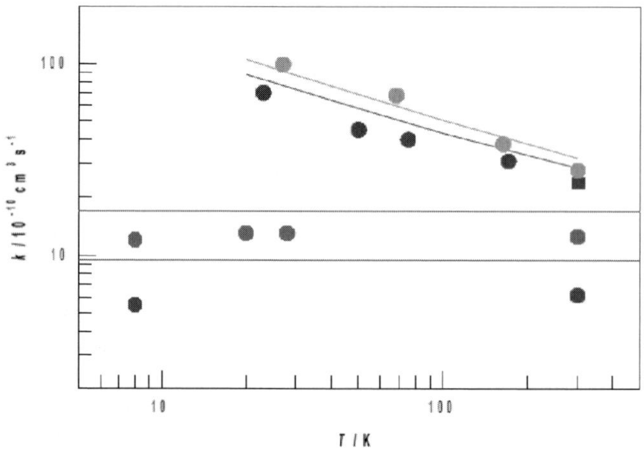

Figure 1. Example of rate coefficients for ion-neutral reactions. The symbols show experimental results for $N^+ + H_2O$ (green circles), $N^+ + NH_3$ (purple circles and squares), $He^+ + N_2$ (red circles) and $N^+ + O_2$ (blue circles). The lines are theoretical fits according to the equations in the text. The sources of the experimental data are given in Smith (2011).

Table 3. Methods used to study the kinetics of neutral-neutral reactions with the ranges of temeperatures over which experiments have been performed

Method	Range of T and p	Reference
DF – discharge-flow	typically: *ca.* 200–600 K up to *ca.* 1 bar in the turbulent flow reactor	`http://iupac-kinetic.ch.cam.ac.uk/`
HTFFR – high temperature fast flow reactor	up to 1320 K	Cosic & Fontijn (2000)
PLP – pulsed laser photolysis	typically: *ca.* 200–600 K down to 80 K up to *ca.* 700 bar	`http://iupac-kinetic.ch.cam.ac.uk/` Fulle *et al.* (1996)
HTPR – high temperature photolysis reactor	up to 1430 K	Fontijn *et al.* (2006)
PLP and **CRESU** apparatus	13–300	Sims *et al.* (1994), Smith (2006)

When the neutral reactant possesses an electric dipole moment (μ_D), in addition to the (assumed) isotropic attraction between the charge on the ion and dipole induced in the molecule, there is an anisotropic force between the ion and the molecular dipole. There have been a number of theoretical treatments of this situation, all of which are again based on the assumption that reaction will inevitably follow 'capture'. Wakelam *et al.* (2010a) recommend that the ratio of the rate coefficient for such a reaction (k_D) to the Langevin value can be estimated in terms of a parameter x, defined by:

$$x = \mu_D/(2\alpha k_B T)^{1/2} \tag{2.2}$$

In terms of this parameter, rate coefficients are given by:

$$k_D/k_L = 0.4767\,x + 0.620 \text{ if } x \geqslant 2 \text{ and } = [(x+0.5090)^2/10.526] + 0.9754 \text{ if } x < 2 \tag{2.3}$$

Fig.1 shows rate coefficients obtained at low temperatures (T$\leqslant$300 K) for reactions between atomic cations and two non-polar and two polar molecular reactants. They fit the formulae given above quite well – in particular, showing that the values of $k(T)$ are independent of temperature in the case of non-polar reactants but show a negative T-dependence in the reactions with polar molecules. It is interesting to note that the Langevin formula, Eq. 2.1, and its extension, represented by Eqs. 2.2 and 2.3, match the experimental results rather well, despite the fact that these reactions occur by charge exchange (that is, transfer of an electron) which does not obviously require a 'close collision'.

3. Neutral-Neutral Reactions

There is a large kinetic data base for neutral-neutral chemical reactions, mainly obtained for the purposes of modelling atmospheric chemistry (see `http://iupac-kinetic.ch.cam.ac.uk/` and `http://jpldataeval.jpl.nasa.gov/`). For higher temperatures, kinetic data relevant to combustion systems have been evaluated by Baulch *et al.* (2005). Most of the kinetic data for the lower range of temperatures have been obtained through the use of two experimental techniques: discharge-flow methods and pulsed laser photolysis (Pilling & Seakins 1995). In Table 1, I summarise the temperature ranges over which these methods have been employed to provide rate coefficients for neutral-neutral reactions.

Of the 159 species identified in the ISM by July 2010 (Wakelam *et al.* 2010a), 136 are electrically neutral and most of these are potentially reactive as they are either free radicals, which have unpaired electrons or empty, low-lying, molecular orbitals, or unsaturated molecules, which contain one or more double or triple bonds. Clearly, this implies that neutral-neutral reactions can be important – even at very low temperatures, if the value of γ in Eq.1.3 is close to zero.

The rate coefficients for reactions between radicals and saturated molecules typically, but not always, exhibit a positive dependence on temperature; that is, $k(T)$ increases with T due to the presence of a barrier on the potential energy path leading from reactants to products. Two good examples of this behavior are provided by the reactions of the radicals CN and C_2H with H_2:

$$CN + H_2 \rightarrow HCN + H \quad \text{and} \quad C_2H + H_2 \rightarrow C_2H_2 + H \qquad \text{(R2), (R3)}$$

The extensive experiments on these reactions, from 300 to 3500 K in the case of (R2) and from 180 to 3000 K in the case of (R3), have been evaluated by Baulch *et al.* (2005), They express their recommendations for $k(T)$) in the form of Eq.1.3, with $\alpha = 5.0\times10^{-13}$ cm^3s^{-1}, $\beta = 2.60$ and $\gamma = 960$ K for (R2), and $\alpha = 1.95\times10^{-12}$ cm^3s^{-1}, $\beta = 2.32$ and $\gamma = 444$ K for (R3). Despite the significant positive values of γ, the reactions (R2) and (R3) could play a significant role in warmer regions of the ISM not least because the co-reactant is the abundant molecule H_2. The rate coefficients for reactions of free radicals with unsaturated molecules and for reactions between pairs of radicals show milder, and even negative, dependences on temperature, and consequently these reactions can be significant in cooler regions of the ISM.

Experiments on neutral-neutral reactions utilize the CRESU method to attain low temperatures (Sims *et al.* 1994; Smith 2006) and these experiments employ pulsed laser photolysis to generate atomic or molecular free radicals (such as C, Al, Si, B, O, CN, OH, CH, C_2H, C_2, C_4H) in the presence of neutral co-reactants (especially, saturated and unsaturated hydrocarbons such as CH_4, C_2H_4, C_2H_2). Generally, the rate coefficients for those reactions that are fast enough to be measured in CRESU experiments increase as the temperature is lowered – but the form of this variation is different for different reactions. Smith *et al.* (2006) have proposed two semi-empirical criteria that can be used to predict whether a reaction between a radical and an unsaturated molecule will be rapid (i.e., $k(10\text{ K}) \geqslant 10^{-10}$ cm^{-3} s^{-1}) at the lowest temperatures found in interstellar clouds: (a) if the ($k(298\text{ K}) >$ *ca.*5×10^{-12} cm^{-3} s^{-1}), and (b) if the difference between the ionization energy (IE) of the molecule and electron affinity (EA) of the radical is $<$ *ca.* 9 eV. These criteria have been supported by measurements on the reaction of oxygen atoms with a range of alkenes (Sabbah *et al.* 2007). Reactions between pairs of free radicals are likely to proceed over potential energy surfaces that exhibit no barrier to reaction: hence γ in Eq.1.1 is likely to be zero. However, reactions between radicals are very difficult to examine in CRESU experiments and only the reaction $O(^3P) + OH \rightarrow O_2 + H$ has been studied up to the present time (Carty *et al.* 2006).

A second problem in respect of neutral-neutral reactions, not only at low temperatures, is the measurement of product yields or branching ratios (Seakins 2007). Progress in this area has been made using infrared absorption techniques to determine the product yields relative to reactant loss or by comparing the relative yields of different products (Taatjes & Hershberger 2001). Another technique, useful for a number of reactions of potential astrochemical importance, is to measure the relative H-atom yields from several related reactions, either in discharge-flow (Bergeat & Loison 2001) or pulsed laser photolysis

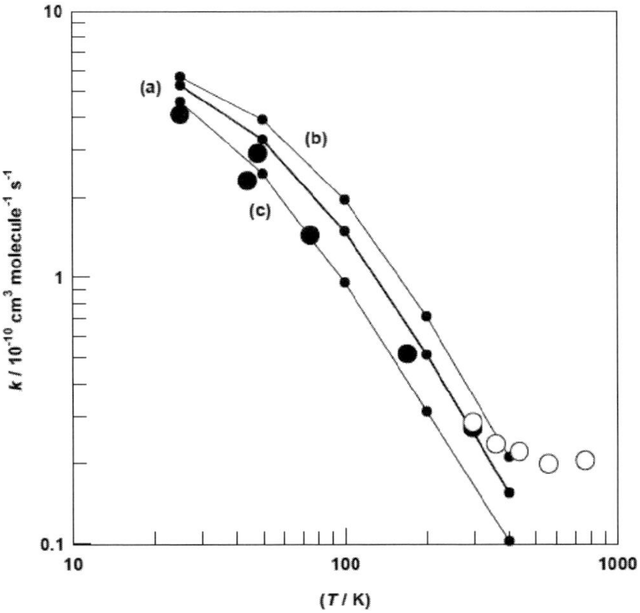

Figure 2. Comparison of experimental rate constants - filled circles (Sims *et al.* 1994) and open circles (Sims & Smith 1988) – with theoretical values of the rate constants (Talbi & Smith 2009): (a) with the zero-point level of the inner transition state 10.9 kJ mol^{-1} below the energy of the separated reactants; (b) with this energy difference increased to 13.4 kJ mol^{-1}; and (c) with this energy difference lowered to 8.4 kJ mol^{-1}.

experiments (Gannon *et al.* 2007). The relative yields can be converted to absolute yields if one of the reactions is known to generate H-atoms in unit yield.

Finally, it should be pointed out that *ab initio* quantum chemical calculations can be very useful. To illustrate this point, and to show how calculations on a chosen reaction can provide detailed information, I refer to the reactions between CN radicals and NH_3 molecules, which was one of the first two neutral-neutral reactions whose kinetics were studied in CRESU experiments (Sims *et al.* 1994). Between 295 and 25 K, its rate coefficient increased markedly, the results being fitted to the expression: $k(T) = (2.77{\pm}0.67){\times}10^{-11}(T/298)^{-1.14\pm0.15}$ $cm^{-3}s^{-1}$. In the experiments the rate of loss of CN radicals in the presence of known concentrations of NH_3 was measured, so no information was obtained about the products and two sets of products are energetically possible:

$$CN + NH_3 \rightarrow HCN + NH2 \text{ for which } \Delta H_{298} = -\ 78.9{\pm}0.8 \text{ kJ mol}^{-1} \qquad \text{(R4a)}$$

$$\text{and } CN + NH_3 \rightarrow H_2NCN + H \text{ for which } \Delta H_{298} = -\ 38{\pm}10 \text{ kJ mol}^{-1} \qquad \text{(R4b)}$$

Quantum chemical calculations (Talbi & Smith 2009) demonstrated that there is no low energy path leading to H_2NCN + H. These calculations further showed that the lowest energy path, leading to HCN + NH_2 passes initially through a potential well, 39.3 kJ mol^{-1} ($\equiv$ 4730 K) below the energy of the separated reactants, and subsequently over a 'submerged barrier', with a crest 10.9 kJ mol^{-1} (1310 K) below the energy of the reactants, which serves as an 'inner transition state' (Georgievskii & Klippenstein 2007; Sabbah *et al.* 2007). The dynamics of reaction (R4a) were investigated (Talbi and Smith 2009) using the two transition state model of Georgievskii & Klippenstein (2007). As Fig.2 shows the rate constants calculated for temperatures between 25 and

200 K are in excellent agreement with the experimental values. The conclusion that the reaction between CN and NH_3 occurs exclusively by channel (R4a) has been confirmed in experiments in which the H-atom yield from (R2) was compared with that from CN + C_2H_2 (Blitz *et al.* 2009). The latter reaction proceeds 100% to HC_2CN + H but no significant H-atom signal was obtained from CN + NH_3. The example just cited demonstrates what can be achieved from combined theoretical and experimental studies on specific reactions. More generally, long-range transition state theory (Georgievskii & Klippenstein, 2005) can be used to estimate upper limits to the rate coefficients for reactions where no barrier or inner transition state is likely to reduce the rate of formation of any intermediate complex. In applying this theory to radical-radical reactions, it may be necessary to multiply the result by a factor that allows for the fact that several potential energy surfaces correlate with the separated reactants (Wakelam *et al.* 2010a).

4. Rate coefficients in chemical models

Chemical modeling of regions of the ISM has a long history. Rate coefficients for the processes summarized in Table 1 are a major part of the input data for such models. The need for chemical databases became clear in the 1980s (Millar *et al.* 1991) and two such databases have now been available on the internet for some time (`http://www.ufda.net/` and `http://physics.ohio-state.edu/~eric/research.html`). They contain between 400 and 500 species and 4,000 and 5,000 reactions.

A third database, named KIDA (Kinetic Database for Astrochemistry) has recently been developed (`http://kida.obs.u-bordeaux1.fr/`). It incorporates two innovations: (i) outsiders can make recommendations that will be incorporated into the database after examination by a KIDA 'expert', and (ii) for each key reaction, the entry includes a data sheet which gives reasons for the recommended values of the rate coefficient and its uncertainty.

As well as comparing predicted molecular abundances with those observed, chemical models can be used to identify those reactions in the model which are key in predicting the abundances of particular species (Wakelam 2010a). This can encourage experimentalists and theoreticians to obtain new or improved data for these reactions. The inclusion of uncertainties in the recommended rate coefficients for reactions in the KIDA database is also important since they can be used to estimate the uncertainties in predicted abundances and hence make it possible to examine whether differences between observed and predicted molecular abundances are significant (Wakelam *et al.* 2010a, Wakelam *et al.* 2010b)

5. Concluding remarks

The limitations of time (in my talk at the Symposium) and space (for this article) mean that this short review has been highly selective. Amongst the topics that I have mentioned, it seems fair to conclude that the rate coefficients for most cation-neutral reactions are relatively well-known. The data base for anion-neutral reactions is smaller but rapidly expanding. Knowledge of the rate coefficients for neutral-neutral reactions has improved rapidly in recent years and the combination of quantum chemistry and rate theories is making the calculation of those for reactions where measurement is problematic much more certain. One experimental development, which should reduce the uncertainties in rate coefficients at the temperatures prevalent in the cold cores of dense interstellar clouds (*ca.* 10 K), is the extension of CRESU measurements to lower temperatures, which is anticipated through the deployment of a pulsed version of the CRESU apparatus (Canosa *et al.* 2008).

I have briefly mentioned radiative association in connection with reactions (R1a) and (R1b). The ability to measure rate coefficients for these processes rests upon the facts that (i) ions can be trapped for "many minutes or even hours" (Gerlich 1995), and (ii) the H_2 will remain in the gas-phase at sufficient concentration even when the trap is cooled to *ca.* 10 K. Rate coefficients for other ion-neutral radiative processes must be inferred from measurements on the three-body association process and estimates of the rate of radiative stabilization. No experiments on radical-neutral radiative associations have been reported. Model calculations (Wakelam *et al.* 2010a) suggest that the reaction

$$C + H_2 \rightarrow CH_2 + h\nu \qquad (R5)$$

is important in the cold cores of dense clouds. Determining its rate coefficient presents a formidable challenge to both experiment and theory.

Finally, I note that I have not mentioned several of the processes listed in Table1. Details of experimental measurements can be found in a longer review published elsewhere (Smith 2011). Dissociative recombination processes figure prominently in chemical models. Their rate coefficients have been determined in FALP (flowing afterglow Langmuir probe) experiments (Adams & Smith 1988) and in experiments employing storage rings. The latter experiments (Geppert and Larsson 2008) have two advantages: (i) they allow time for any internal excitation of the reactant ions to cool before the neutral is introduced; (ii) they provide information about the neutral products of the process. One important instance of the latter type of measurements is the establishment of the branching ratios in the dissociative recombination of H_3O^+ with electrons (Jensen *et al.* 2000).

References

Adams, N. G. & Smith, D. 1976, *Int. J. Mass spectrom. Ion Proc.*, 21, 349

Adams, N. G. & Smith, D. 1988, *in Rate Coefficients for Astrochemistry*, T.J. Millar and D.A. Williams (Ed.) (Kluer, Dordrecht) p.173

Anicich, V. G. 1993a, *ApJS*, 84, 215

Anicich, V. G. 1993b, *Phys. Chem. Ref. Data*, 22, 1469

Baulch, D. L., Bowman, C. T., Cobos, C. J., Cox, R. A., Just, Th., *et al.*, 2005, *J. Phys. Chem. Ref. Data*, 34, 757

Bergeat, A., & Loison, J. C., 2001, *Phys. Chem. Chem. Phys.*, 3, 2038

Blitz, M. A., Seakins, P. W., & Smith, I. W. M., 2009, *Phy. Chem. Chem. Phys.*, 11, 10824

Bohringer, H., & Arnold, F., 1986, *J. Chem. Phys.*, 48, 1459

Canosa, A., Goulay, F., Sims, I. R., & Rowe, B. R., 2008, *Low Temperatures and Cold Molecules*, I.W.M. Smith (Ed.) (Imperial College press, London) p.55

Carty, D., Goddard, A., Köhler, S. P. K., Sims, I. R., & Smith, I. W. M., 2006, *J. Phys. Chem. A*, 110, 3101

Cosic, B., & Fontjin, A., 2000, *J. Phys. Chem. A*, 104, 5517

Dunn, G. H., 1995, *Phys. Scr.*, T59, 249

Durkin, D. B., Fehsenfeld, F. C., Schmeltekoff, A. L., & Ferguson, E. E., 1968, *J. Chem. Phys.*, 49, 1365

Ferguson, E. E., Bohme, D. K., Fehsenfeld, F. C., & Durkin, D. B., 1969, *J. Chem. Phys.*, 50, 5039

Fontijn, A., Shamsuddin, S. M., Anderson, W. R., & Marshall, P., 2006, *Combust. Flame.*, 145, 543

Fulle, D., Hamann, H. F., Hippler, H., & Troe, J., 1996, *J. Chem. Phys.*, 105, 963

Gannon, K. L., Glowacki, D. R., Blitz, M. A., Hughes, K. J., Pilling, M. J., & Seakins, P. W., 2007, *J. Phys. Chem. A.*, 111, 6676

Georgievskii, Y., & Klippenstein, S. J., 2005, *J. Chem. Phys.*, 122, 194103

Georgievskii, Y., & Klippenstein, S. J., 2007, *J. Phys. Chem. A.*, 111, 3802

Gerlich, D., 1995, *Phys. Scr*, T59, 256

Gerlich, D., 2008, *Low Temperatures and Cold Molecules*, I.W.M. Smith (Ed.) (Imperial College press, London) p. 121

Herbst, E., & Klemperer, W., 1973, *ApJ*, 185, 505

Herbst, E., Smith, D., Adams, N. G., & DeFrees, D. J., 1987, *ApJ*, 312, 351

Jensen, M. J., Bilodeau, R. C., Safvan, C. P., Seirsen, K., Andersen, L. H., Pedersen, H. B., & Heber, O., 2000, *ApJ*, 543, 764

McIntosh, B. J., Adams, N. G., & Smith, D., 1988, *Chem. Phys. Lett.*, 148, 143

Millar, T. J., Rawlings, J. M. C., Bennett, A., & Brown, P. D., Charnley,S. B., 1991, *A&A Supplt.*, 87, 585

Pilling, M. J., & Seakins, P. W., 1995, *Reaction Kinetics* (New York, Oxford University Press)

Rowe, B. R., Dupeyrat G., Marquette, J. B., & Gaucherel, P., 1984, *J. Chem. Phys.*, 80, 4915

Rowe, B. R., Marquette, J. B., Dupeyrat G., & Ferguson, E. E., 1985, *Chem. Phys. Lett.*, 113, 403

Sabbah, H., Biennier, L., Sims, I. R., Georgievskii, Y., Klippenstein, S. J., & Smith, I. W. M., 2007, *Science*, 317, 102

Seakins, P. W., 2007, *Ann. Rep. Prog. Chem. Sect. C*, 103, 173

Sims, I. R., & Smith, I. W. M., 1988, *J. Chem. Soc. Farad.*, 84 527

Sims, I. R., Queffelec, J. L., Defrance, A., Rebrion-Rowe, C., Travers, D., Bocherel, P., Rowe, B. R., & Smith, I. W. M., 1994, *J. Chem. Phys.*, 100, 4229

Smith, I. W. M., 1988, in *Rate Coefficients for Astrochemistry*, T. J. Millar and D. A. Williams (Ed.) (Kluer, Dordrecht) p. 103

Smith, I. W. M., 1997, in *Molecules in Astrophysics: Probes and Processeses*, IAU Symposium 178, E.F. van Dishoeck (Ed.) p. 253

Smith, I. W. M., 2006, *Angew. Chem. Int. Ed.*, 45, 2842

Smith I. W. M., Sage, A. M., Donahue, N. M., Herbst, E., & Quan, D., 2006, *Faraday Discuss.*, 133, 137

Smith, I. W. M., 2011, *ARAA*, to be published in September 2011.

Snow, T. P., & Bierbaum, V. M., 2008, *Ann. Rev. Anal. Chem.*, 1, 229

Taatjes, C. A., & Hershberger, J. F., 2001, *Ann. Rev. Phys. Chem.*, 52, 41

Talbi, D., & Smith, I. W. M., 2009, *Phys. Chem. Chem. Phys.*, 11, 8477

Wakelam, V., Smith, I. W. M., Herbst, E., Troe, J., Geppert, W., Linnnartz, H., Őberg, K., Roueff, E., Agundez, M., Pernot, P., *et al.*, 2010a, *Space Science Reviews*, 156, 13

Wakelam, V., Herbst, E., Le Bourlot, J., Hersant, F., Selsis, F., & Guilloteau, S., 2010b, *A&A*, 156, 13

Viggiano, A. A., 2006, *Phys. Chem. Chem. Phys.*, 8, 2557

Discussion

L. Decin: You have shortly touched upon the field of ab-initio calculations/ quantum chemistry. Where do you expect to see the largest progress in next few years when combining laboratory experiments and quantum chemistry?

I. Smith: I believe that quantum chemical calculations will play an increasing role in identifying the reation products a) where multiple channels are energetically accesible, and b) where the minimum energy paths exhibit multiple transition states.

X. Li: The differences of rate constants between different modeling methods are apparent, especially at lower temperatures. Could you give some possible reasons of that? Thanks.

I. Smith: I believe that some of these differences may arise when some calculations do not correctly include the long-range interactions between the reactants. This is done in the transition state methods described later in Dr. Klippenstein's lecture.

E. Herbst: i) To what temperature do shock tubes provide rate constants? ii) Are internal degrees of freedom excited thermally

I. Smith: i) I am not sure of the high temperature record but I know that the reaction $CN + H_2$ has been studied to 3000 K and $C_2H + H_2$ has been studied to 3500 K. ii) Rotational state distributions will be rapidly thermalised. Vibrational degrees of freedom can be more problematic – indeed shock tube methods are frequently employed to study vibrational relaxation at high temperatures.

M. Agundez: Concerning state-to-state chemical kinetics, i.e. measuring the rate constant of a reaction with one of the reactants in a given rotational, vibrational or electronic state, how feasible and trustable would be such measurements?

I. Smith: There is information on reaction rates from vibrational and electronic states for a few systems. It is more difficult to obtain rotationally selected rate coefficients since rotational relaxation is rapid with essentially all collision partners.

The Molecular Universe
Proceedings IAU Symposium No. 280, 2011
José Cernicharo & Rafael Bachiller, eds.

doi:10.1017/S1743921311025129

Long-Range Interaction Potential of Open Shell Atoms with Neutral Molecules : Application to the Calculation of the Rate Constant for the $C_2H(^2\Sigma^+) + O(^3P)$ Reaction

Yuri Georgievskii and Stephen J. Klippenstein
Chemical Sciences and Engineering Division, Argonne National Laboratory, Argonne, IL 60439

Abstract. An accurate knowledge of the long-range interaction potential for the ground and first few excited electronic states is needed for quantitative prediction of the rate coefficients for astrochemical reactions at low temperatures. Some reactions important for astrochemical modeling include an open-shell atom as one of the fragments. Due to the interplay between the spin-orbit and quadrupole interactions such reactions require a special treatment. In this paper we derive the general expressions for the energy levels for such systems, apply them to the $C_2H(^2\Sigma^+) + O(^3P)$ reaction, and compare the results with *ab initio* calculations.

1. Introduction

Chemical modeling of the interstellar medium currently involves a large number of species (>400) and reactions between them (>4000) following the dramatic increase in the number of interstellar and circumstellar molecules detected via their rotational (sub)millimeter-wave spectra [Wakelam *et al.* (2010)]. Knowledge of the rate coefficients for these reactions is an important prerequisite of such modelling. Many of these reactions occur between highly reactive radical species under extremely low temperatures (i.e., <50 K and typically near 10 K). Direct experimental measurements of the rate coefficients for such reactions are very difficult to obtain.

An important feature of many of these reactions is that they proceed via formation of a metastable reactive complex that may subsequently dissociate into different bimolecular products or be stabilized via radiative transitions and/or molecular collisions. A key complication arises from the fact that, due to the highly unsaturated nature of many of the species involved, not only the ground, but several excited electronic states with different spin may allow for formation a bound complex. However, not all of these electronic states contribute to the total rate since at low temperatures only barrierless entrance channels are accessible.

The master equation approach provides a standard procedure for calculating the rate coefficient for a complex-forming reaction [Miller & Klippenstein (2006)]. In this approach all possible conformers of the reactive complex are properly characterized in terms of their densities of states and the transitions between them and to the reactants and products are characterized in terms of the corresponding numbers of states. Subsequent solution of the master equation provides the rate coefficients for reactions leading from reactants to all products. For the temperature range pertinent to astrochemical modelling, however, the situation is considerably simplified and the solution of the master equation is not required. Simple analysis shows that if there is a pathway from reactants to products with no barrier above the entrance channel the efficiency of the reaction is 100 % and the corresponding rate coefficient is simply the rate for complex formation (i.e., the capture rate). On the other hand, if such a barrierless path does not exist, the rate coefficient is very small and can be neglected.

Thus, the rate coefficient calculation involves first the determination of the presence or absence of any barriers above the entrance channel for each of the possible electronic states follows by the evaluation of the capture rate for each reactive electronic state. The total rate coefficient is then the sum of these partial rate coefficients weighted by the appropriate Boltzmann factors.

The capture rate coefficient may be evaluated with variational transition state theory (VTST). We have recently developed the long-range transition state theory (LRTST), which is a special version of VTST adopted for low temperature applications [Georgievskii & Klippenstein (2005)]. This theory is purely classical. Quantum mechanical treatments for non-reactive degrees of freedom are also available. Two essentially equivalent quantum approaches have been developed by Clary and coworkers [Clary (1990)], under the name of adiabatic capture centrifugal sudden approximation, and by Troe and coworkers [Troe (1997)], known as the statistical adiabatic channel model. The quantum treatment, however, is considerably more complicated and requires much more computational effort. Importantly, quantum effects are only important for temperatures of the order of and lower than the rotational constant of one of the fragments.

The long-range potential, which is needed for the capture rate coefficient calculation, is usually represented as a sum of a small number of terms representing interactions between major permanent multipole moments of the fragments and the C_6-term which represents induced dipole and dispersion interactions [Georgievskii & Klippenstein (2005)]. The situation in which one of the fragments is an atom with an open electronic shell, such as $O(^3P)$ or $C(^3P)$, is more complicated. The major term in the long-range part of the potential for such atoms is the interaction of the atomic quadrupole, which depends on electronic angular momentum, with the multipole moments of the second fragment. Its interplay with the spin-orbit interaction of the atom requires a special treatment.

In this paper we derive general expressions for the energy levels of the atom in the ^3P electronic state in an arbitrary electric field. These expressions generalize the results of Gentry & Giese (1977), who considered the interaction of an open-shell atom with an ion, and can be applied to the calculation of the interaction between an open-shell atom and a neutral molecule with the dipole, quadrupole, and higher multipole moments. We then apply the obtained expressions to calculate the rate coefficient for $C_2H(^2\Sigma^+)$ reacting with $O(^3P)$ and compare the result with corresponding results employing directly evaluated ab initio potentials. This reaction was found to be a key reaction in the recent review of interstellar chemistry by Wakelam *et al.* (2010).

2. Potential Energy Surface

The Hamiltonian of a quadrupole in an external electric field is given by [Landau & Lifshitz (1991)],

$$\hat{V} = \frac{1}{6}\frac{\partial^2 \phi}{\partial x_j \partial x_k}\hat{Q}_{jk}, \tag{2.1}$$

where ϕ is the electric field potential and $\hat{Q}_{jk}$ is the operator for the quadrupole moment. The quadrupole moment, when averaged over all degrees of freedom but the direction of the angular momentum, may be expressed in terms of the angular momentum components as,

$$\hat{Q}_{jk} = \frac{3Q}{2J(2J-1)}\left(\hat{J}_j\hat{J}_k + \hat{J}_k\hat{J}_j - \frac{2}{3}J(J+1)\delta_{jk}\right), \tag{2.2}$$

The Hamiltonian in Eq. 2.1 can be transformed to a more convenient form by appropriate rotation of the system of coordinates,

$$\hat{V} = p_x\hat{Q}_{xx} + p_y\hat{Q}_{yy} + p_z\hat{Q}_{zz}, \tag{2.3}$$

where p_i are the eigenvalues of the $\frac{1}{6}\frac{\partial^2\phi}{\partial x_j \partial x_k}$ matrix and the components of the quadrupole tensor are given in the new coordinate system. Note, that $p_x + p_y + p_z = 0$. The important feature of this Hamiltonian is that it is invariant under the $\mathbf{D_2}$ group, and the representations of that group can be used for classification of its energy levels.

For the dipole-quadrupole interaction the electric dipole potential is given by

$$\phi(\mathbf{r}) = \frac{\mathbf{d}\cdot\mathbf{r}}{r^3}, \tag{2.4}$$

where $\mathbf{d}$ is the dipole moment and $\mathbf{r}$ is the vector connecting the fragments. The second derivative of this potential is given by

$$\frac{\partial^2\phi}{\partial x_j \partial x_k} = 15\frac{\mathbf{d}\cdot\mathbf{r}}{r^7}x_j x_k - 3\frac{d_j x_k + d_k x_j}{r^5} - 3\frac{\mathbf{d}\cdot\mathbf{r}}{r^5}\delta_{kj}. \tag{2.5}$$

Without restriction of generality one can assume that $\mathbf{r}$ is directed along the z-axis and $\mathbf{d}$ belongs to the zx-plane. Then, the following expressions for the p_i parameters are obtained with rotation around the y-axis,

$$\begin{pmatrix} p_x \\ p_y \\ p_z \end{pmatrix} = \frac{d}{2r^4}\begin{pmatrix} 1/2\cos\theta - \sqrt{1 + 5/4\cos^2\theta} \\ -\cos\theta \\ 1/2\cos\theta + \sqrt{1 + 5/4\cos^2\theta} \end{pmatrix}, \tag{2.6}$$

where θ is the angle between the $\mathbf{d}$ and $\mathbf{r}$ directions. It is worth noting that $p_x < p_y < p_z$

We will consider first the strong and the weak spin-orbit coupling limits and then the general case.

2.1. *Strong spin-orbit coupling limit*

This case corresponds to low temperatures where the relevant separations between the fragments a large. In this limit the total angular momentum is a good quantum number and each spin-orbit manifold should be considered separately. The spin-orbit splitting at infinite separation provides the reference energy for each manifold. In this limit one should understand under $\hat{\mathbf{J}}$ in Eq. 2.2 the total electronic angular momentum $\hat{\mathbf{J}} = \hat{\mathbf{L}} + \hat{\mathbf{S}}$ and use the effective quadrupole moment Q_J instead of Q in Eq. 2.2. The value of Q_J depends on J.

J=2 quintet.

In this spin-orbit multiplet $Q_J = Q$. Under the $\mathbf{D_2}$ group it is split into two A representations, $|A'\rangle \equiv |0\rangle$ and $|A''\rangle \equiv \sqrt{1/2}(|2\rangle + |-2\rangle)$, and three B representations, with one each of the B_x, $\sqrt{1/2}(|1\rangle + |-1\rangle)$, B_y, $\sqrt{1/2}(|1\rangle - |-1\rangle)$, and B_z representations, $\sqrt{1/2}(|2\rangle - |-2\rangle)$. Here by B_i we denote the representation which is unchanged under C_2^i transformation, A is the totally symmetric representation, and $|M_J\rangle$ is a state with a given $\hat{J}_z$ value. In the following table we show the $\hat{J}_x^2$, $\hat{J}_y^2$, and $\hat{J}_z^2$ matrices and the

Hamiltonian eigenvalues in each of these subspaces:

R	$\vert\rangle$	$\hat{J}_x^2 - 2$	$\hat{J}_y^2 - 2$	$\hat{J}_z^2 - 2$	E
A	$\vert 0\rangle$ $\sqrt{\frac{1}{2}}(\vert 2\rangle + \vert -2\rangle)$	$\begin{pmatrix} 1 & \sqrt{3} \\ \sqrt{3} & -1 \end{pmatrix}$	$\begin{pmatrix} 1 & -\sqrt{3} \\ -\sqrt{3} & -1 \end{pmatrix}$	$\begin{pmatrix} -2 & 0 \\ 0 & 2 \end{pmatrix}$	$\pm\sqrt{\frac{3}{2}}Q\times$ $\sqrt{p_x^2 + p_y^2 + p_z^2}$
B_x	$\sqrt{\frac{1}{2}}(\vert 1\rangle + \vert -1\rangle)$	2	-1	-1	$\frac{3}{2}Qp_x$
B_y	$\sqrt{\frac{1}{2}}(\vert 1\rangle - \vert -1\rangle)$	-1	2	-1	$\frac{3}{2}Qp_y$
B_z	$\sqrt{\frac{1}{2}}(\vert 2\rangle - \vert -2\rangle)$	-1	-1	2	$\frac{3}{2}Qp_z$

(2.7)

For the dipole-quadrupole interaction it gives the following energies,

R	E
A	$\pm\frac{\sqrt{3}}{2}V_0\sqrt{2\cos^2\theta + 1}$
B_x	$\frac{3}{8}V_0(\cos\theta - \sqrt{5\cos^2\theta + 4})$
B_y	$-\frac{3}{4}V_0\cos\theta$
B_z	$\frac{3}{8}V_0(\cos\theta + \sqrt{5\cos^2\theta + 4})$

, (2.8)

where we have used the notation $V_0 = dQ/r^4$ for the strength of the dipole-quadrupole interaction.

J=1 triplet.

In this spin-orbit multiplet $Q_J = -Q/2$. Under the $\mathbf{D_2}$ group it is split into three, with one each of the B_x, B_y, and B_z representations. The corresponding states $|x\rangle$, $|y\rangle$, and $|z\rangle$ are defined by having zero projection of the angular momentum on the corresponding axis, $\hat{J}_i = 0$.

R	$\vert\rangle$	$\hat{J}_x^2 - \frac{2}{3}$	$\hat{J}_y^2 - \frac{2}{3}$	$\hat{J}_z^2 - \frac{2}{3}$	E
B_x	$\vert x\rangle$	$-2/3$	$1/3$	$1/3$	$\frac{3}{2}Qp_x$
B_y	$\vert y\rangle$	$1/3$	$-2/3$	$1/3$	$\frac{3}{2}Qp_y$
B_z	$\vert z\rangle$	$1/3$	$1/3$	$-2/3$	$\frac{3}{2}Qp_z$

(2.9)

2.2. *Strong quadrupole limit*

In this case one can neglect the spin-orbit interaction and consider only the orbital part of the wave function. The three-dimensional orbital space is split into three one-dimensional representations, $|i\rangle$, $i = x, y, z$, which denote the states with zero projection of the angular momentum on the i-th axis $L_i = 0$,

$$
\begin{array}{|c|c|c|c|c|c|}
\hline
R & |\rangle & \hat{L}_x^2 - \frac{2}{3} & \hat{L}_y^2 - \frac{2}{3} & \hat{L}_z^2 - \frac{2}{3} & E \\
\hline
B_x & |x\rangle & -2/3 & 1/3 & 1/3 & -3Qp_x \\
\hline
B_y & |y\rangle & 1/3 & -2/3 & 1/3 & -3Qp_y \\
\hline
B_z & |z\rangle & 1/3 & 1/3 & -2/3 & -3Qp_z \\
\hline
\end{array}
\tag{2.10}
$$

For the dipole-quadrupole interaction it gives the following energies,

$$
\begin{array}{|c|c|}
\hline
R & E \\
\hline
B_x & \frac{3}{4}V_0(-\cos\theta + \sqrt{5\cos^2\theta + 4}) \\
\hline
B_y & \frac{3}{2}V_0\cos\theta \\
\hline
B_z & \frac{3}{4}V_0(-\cos\theta - \sqrt{5\cos^2\theta + 4}) \\
\hline
\end{array}
\tag{2.11}
$$

The last expression shows that there are six attractive and three repulsive potential energy surfaces in the dipole-quadrupole approximation.

2.3. *General case*

The total electronic Hamiltonian with the spin-orbit coupling is

$$\hat{H} = \hat{V}_{dQ} - A\hat{\mathbf{S}} \cdot \hat{\mathbf{L}}. \tag{2.12}$$

We have deliberately chosen the minus sign in front of the spin-orbit interaction term to stress that $J = 2$ is the ground spin-orbit multiplet for oxygen atom. Under the $\mathbf{D_2}$ group the nine-dimensional spin-orbit space of this Hamiltonian is split into the three A representations and two of each B_x, B_y, and B_z representation. In the following table the diagonal components of the quadrupole operator and the full $\mathbf{S} \cdot \mathbf{L}$ matrix are shown in each of these subspaces in the appropriate basis.

$$
\begin{array}{|c|c|c|c|c|c|}
\hline
R & |\rangle & \hat{L}_x^2 - \frac{2}{3} & \hat{L}_y^2 - \frac{2}{3} & \hat{L}_z^2 - \frac{2}{3} & \mathbf{S} \cdot \mathbf{L} \\
\hline
A & \begin{array}{c} |x^{(L)}\rangle|x^{(S)}\rangle \\ |y^{(L)}\rangle|y^{(S)}\rangle \\ |z^{(L)}\rangle|z^{(S)}\rangle \end{array} & \begin{array}{c} -2/3 \\ 1/3 \\ 1/3 \end{array} & \begin{array}{c} 1/3 \\ -2/3 \\ 1/3 \end{array} & \begin{array}{c} 1/3 \\ 1/3 \\ -2/3 \end{array} & \begin{pmatrix} 0 & -1 & -1 \\ -1 & 0 & -1 \\ -1 & -1 & 0 \end{pmatrix} \\
\hline
B_x & \begin{array}{c} |y^{(L)}\rangle|z^{(S)}\rangle \\ |z^{(L)}\rangle|y^{(S)}\rangle \end{array} & \begin{array}{c} 1/3 \\ 1/3 \end{array} & \begin{array}{c} -2/3 \\ 1/3 \end{array} & \begin{array}{c} 1/3 \\ -2/3 \end{array} & \begin{pmatrix} 0 & 1 \\ 1 & 0 \end{pmatrix} \\
\hline
B_y & \begin{array}{c} |x^{(L)}\rangle|z^{(S)}\rangle \\ |z^{(L)}\rangle|x^{(S)}\rangle \end{array} & \begin{array}{c} -2/3 \\ 1/3 \end{array} & \begin{array}{c} 1/3 \\ 1/3 \end{array} & \begin{array}{c} 1/3 \\ -2/3 \end{array} & \begin{pmatrix} 0 & 1 \\ 1 & 0 \end{pmatrix} \\
\hline
B_z & \begin{array}{c} |x^{(L)}\rangle|y^{(S)}\rangle \\ |y^{(L)}\rangle|x^{(S)}\rangle \end{array} & \begin{array}{c} -2/3 \\ 1/3 \end{array} & \begin{array}{c} 1/3 \\ -2/3 \end{array} & \begin{array}{c} 1/3 \\ 1/3 \end{array} & \begin{pmatrix} 0 & 1 \\ 1 & 0 \end{pmatrix} \\
\hline
\end{array}
\tag{2.13}
$$

Using these expressions one can explicitly write down the expressions for the energies. For the B_x subspace one obtains

$$E_{1,2}^{(x)} = \frac{3}{2}Qp_x \pm \sqrt{A^2 + \frac{9}{4}Q^2(p_z - p_y)^2}. \tag{2.14}$$

The energies for the B_y and B_z subspaces are obtained from Eq. 2.14 by permutation of indices. In the case of the A representation the energies are the eigenvalues of the following Hamiltonian,

$$\begin{pmatrix} -3Qp_x & A & A \\ A & -3Qp_y & A \\ A & A & -3Qp_z \end{pmatrix}. \tag{2.15}$$

We explicitly consider the important specific case, $p_x = p_y = -p_z/2$. In this case Eq. 2.15 has a simple solution,

$$E_1^{(a)} = -A + \frac{3}{2}Qp, \tag{2.16}$$

$$E_{2,3}^{(a)} = -A + \frac{3}{2}\left(A - \frac{Qp}{2} \pm \sqrt{A^2 + AQp + \frac{9}{4}Q^2p^2}\right), \tag{2.17}$$

where $p \equiv p_z$.

In the general case, instead of writing the solution of Eq. 2.15 explicitly, which is rather cumbersome, we show how the energies correlate with the strong quadrupole and strong spin-orbit coupling limits. Assuming that $p_x < p_y < p_z$ one obtains the following correlation table,

R	dQ − limit	T	SL − limit	T
A	$-3Qp_x + \frac{A^2}{3Q}(\frac{1}{p_y - p_x} + \frac{1}{p_z - p_x})$	R	$2A$	?
	$-3Qp_y + \frac{A^2}{3Q}(\frac{1}{p_x - p_y} + \frac{1}{p_z - p_y})$	A	$-A + \sqrt{\frac{3}{2}}Q\sqrt{p_x^2 + p_y^2 + p_z^2}$	R
	$-3Qp_z + \frac{A^2}{3Q}(\frac{1}{p_x - p_z} + \frac{1}{p_y - p_z})$	A	$-A - \sqrt{\frac{3}{2}}Q\sqrt{p_x^2 + p_y^2 + p_z^2}$	A
B_x	$-3Qp_y + \frac{A^2}{3Q}\frac{1}{p_z - p_y}$	A	$A + \frac{3}{2}Qp_x$	A
	$-3Qp_z + \frac{A^2}{3Q}\frac{1}{p_y - p_z}$	A	$-A + \frac{3}{2}Qp_x$	A
B_y	$-3Qp_x + \frac{A^2}{3Q}\frac{1}{p_z - p_x}$	R	$A + \frac{3}{2}Qp_y$	A
	$-3Qp_z + \frac{A^2}{3Q}\frac{1}{p_x - p_z}$	A	$-A + \frac{3}{2}Qp_y$	A
B_z	$-3Qp_x + \frac{A^2}{3Q}\frac{1}{p_y - p_x}$	R	$A + \frac{3}{2}Qp_z$	R
	$-3Qp_y + \frac{A^2}{3Q}\frac{1}{p_x - p_y}$	A	$-A + \frac{3}{2}Qp_z$	R

(2.18)

From this table one can see that there are two surfaces, $A^{(2)}$ and $B_z^{(2)}$, which should have barriers (repulsive behavior at large separation changing to attraction at smaller distances) and one surface with a well, $B_y^{(1)}$ (long-range attraction changes to repulsion at shorter separations). In Fig. 1 a diagram is shown that correlates the energy levels for the dipole-quadrupole interaction in the strong and weak spin-orbit limits.

3. $C_2H(^2\Sigma^+) + O(^3P)$ Reaction

If one neglects for a moment the spin-orbit interaction, there are three triply-degenerate electronic states for $O(^3P)$ and one doubly-degenerate electronic state for $C_2H(^2\Sigma^+)$. Altogether it gives eighteen electronic surfaces for the reactive complex. When strongly coupled, it provides three doublet ($S = 1/2$) and three quartet ($S = 3/2$) electronic PESs.

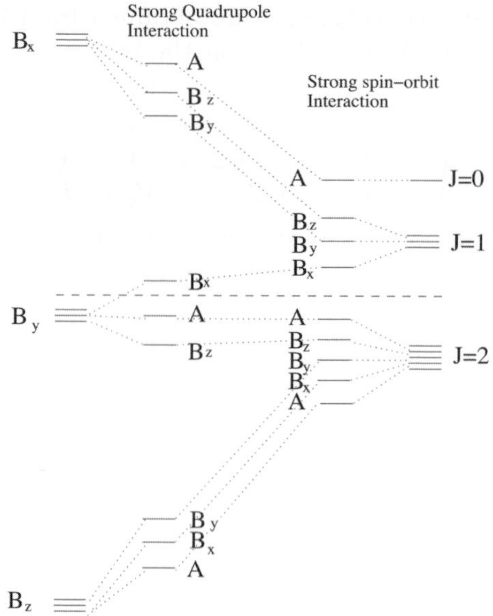

Figure 1. The correlation diagram for the energy levels of the quadrupole in an external electric field.

To figure out which of these surfaces are attractive we have performed minimum energy path (constrained optimization) calculations for all three doublet and three quartet states. We evaluate these interactions with the CASPT2 method employing nine-electron, eight-orbital (9e,8o) active space consisting of three p-orbitals for O (4e,3o) and two π, two π^* (full π, π^* space), and one radical σ orbitals for C_2H (5e,5o). These CASPT2 calculations employ Dunning's correlation-consistent, polarized valence, double-ζ basis set [Dunning (1989)] and were done using the formalism of Celani and Werner [Celani & Werner (2000)] as implemented in the MOLPRO quantum chemistry package [Molpro (2010)].

Our calculations show that there are two attractive doublet potential energy surfaces which correlate with the two CO + CH($^2\Pi$) product states and one attractive quartet surface which correlates with the CO+CH($^4\Sigma^-$) products. All other PESs have potential barriers higher than the reactants and can be neglected.

Thus, there are altogether eight spin-orbital states of the reactive complex which contribute to the total rate constant. In the considered analytical approximation for each spin-orbital state of the oxygen atom in the electric field of the C_2H fragment an additional degeneracy of two should be added because of the spin degeneracy of the $C_2H(^2\Sigma^+)$ fragment itself. Thus, the capture rate coefficients should be calculated only for four lowest spin-orbital states of the oxygen atom in the electric field of the C_2H fragment. These four states correlate with four $J = 2$ multiplet states of the oxygen atom at infinite separation.

The important feature of the $O(^3P) + C_2H(^2\Sigma^+)$ reaction is that the dipole moment d of the C_2H fragment is relatively small ($d \simeq 0.31$ a.u.) while the quadrupole moment $Q \equiv Q_{zz}$ is relatively large ($Q \simeq 7.1$ a.u.). Thus, to correctly reproduce *ab initio* data in addition to the dipole-quadrupole interaction one should at least include the quadrupole-quadrupole interaction term.

To compare the analytical and *ab initio* PESs we consider the linear configuration with the oxygen atom approaching the C_2H fragment from both the carbon and hydrogen sides. In this situation $p_x = p_y = -p_z/2$ and Eqs. 2.16 and 2.17 are applicable. The *ab initio* calculations have been performed at the CASPT2(5e,4o)/cc-pvqz level. In Fig. 2 both the four lowest analytical and *ab initio* PESs are shown for the oxygen atom on the carbon side of the C_2H fragment. The C_6 dispersion coefficient was approximated

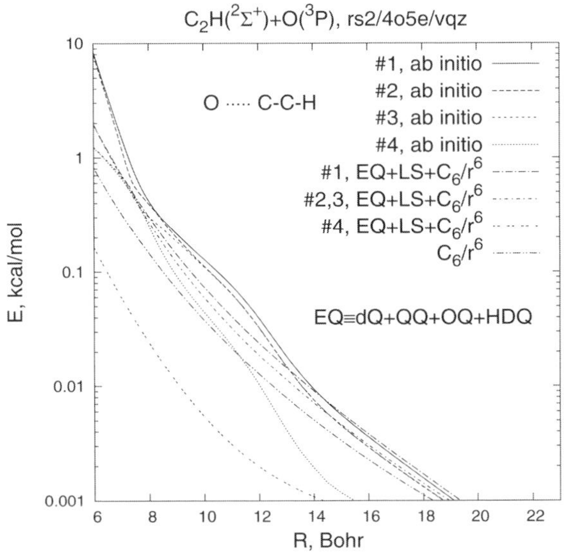

Figure 2. The four lowest potential energy surfaces for the $O(^3P) + C_2H(^2\Sigma^+)$ reaction with the oxygen approaching C_2H on the carbon side.

according to the standard formula [Hirschfelder, Curtiss, & Bird (1967)],

$$C_6 = 1.5\alpha_1\alpha_2 I_1 I_2/(I_1 + I_2)CF, \tag{3.1}$$

where α_i, I_i are the polarizability and the ionization energy of the i-th fragment, and the correction factor $CF = 2.3$ was chosen to better fit *ab initio* energies at large separation distances R.

In Fig. 3 both the four lowest analytical and *ab initio* PESs are shown for the oxygen atom on the hydrogen side of the C_2H fragment.

While at larger distances, $R > 14$ Bohr, the agreement is almost perfect, in the intermediate distance range, $8 < R < 12$, Bohr the analytical and *ab initio* energies may differ by up to a factor of two. We note that this difference strongly depends on the basis set. One of the reasons for this difference may be related to the basis set superposition error in the *ab initio* calculations. At smaller distances, $R < 8$ Bohr, chemical interaction becomes the dominant one and should be included in the energy estimates.

To calculate the $C_2H + O$ reaction rate coefficients for each electronic PES the variable reaction coordinate transition state theory code [Georgievskii & Klippenstein (2003)], which was developed to treat the radical-radical association reactions, has been used. *Ab initio* energies were calculated directly "on the fly" using cc-pvqz basis set. The non-diagonal spin-orbit matrix elements were calculated on the MCSCF(4e,3o) level and the diagonal energies on the CASPT2(4e,3o) level. The analytical calculations included dipole-quadrupole, quadrupole-quadrupole, dipole-induced dipole, and dispersion interaction terms. In Fig. 4 the partial rate coefficients calculated both with analytical and *ab initio* methods are shown for different electronic surfaces as functions of temperature as well as the total rate coefficient. One can see that the difference between the analytical

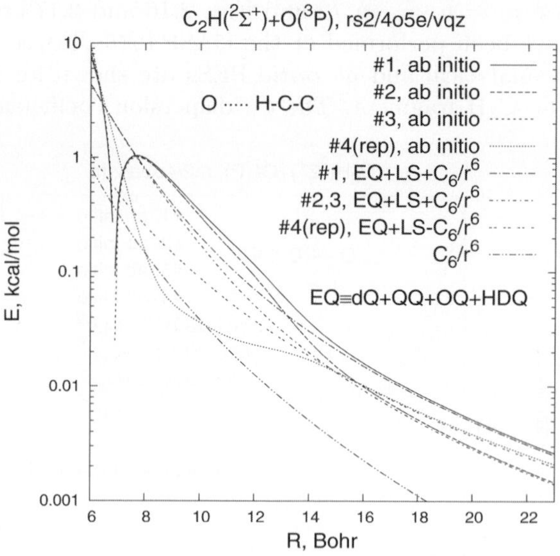

Figure 3. The four lowest potential energy surfaces for the $O(^3P) + C_2H(^2\Sigma^+)$ reaction with the oxygen approaching C_2H on the hydrogen side.

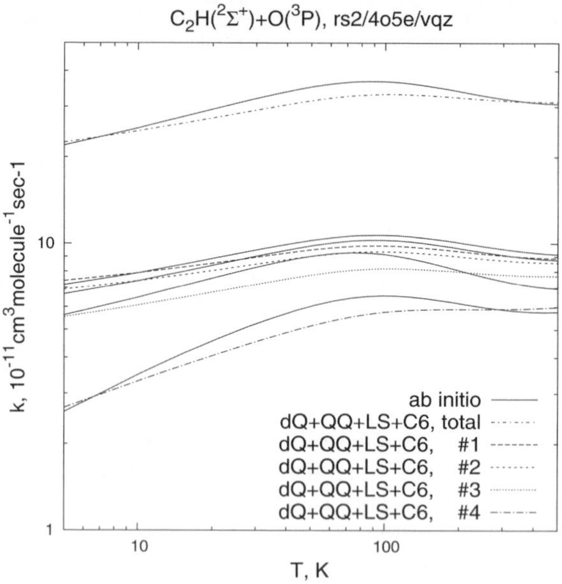

Figure 4. The capture rate coefficients for the $O(^3P) + C_2H(^2\Sigma^+)$ reaction as functions of temperature.

and *ab initio* potential calculations does not exceed 10% in the whole temperature range considered.

4. Conclusions

Analytical expressions for the energy levels of an open-shell ^3P atom in an external electric field have been derived. The obtained expressions provide a simple and valuable means to calculate the long-range part of the interaction potential between an open-shell

^3P atom and a neutral molecule. They have been applied to the $O(^3P) + C_2H(^2\Sigma^+)$ reaction. The comparison between the rate coefficients obtained with the use of the analytical and *ab initio* potentials show very good agreement in the whole considered temperature range (5-500 K). This good agreement suggests that the theoretical predictions are of high accuracy.

Acknowledgments

This work was supported by NASA's Planetary Atmospheres Program through grant NNH09AK24I. Computational support was provided by the U. S. Department of Energy, Office of Basic Energy Sciences, Division of Chemical Sciences, Geosciences, and Biosciences under DOE Contract Number DE-AC02-06CH11357. We thank Larry Harding for valuable discussions.

References

Wakelam, V., Smith, I. W. M., Herbst, E., Troe, J., Geppert, W., Linnartz, H., Öberg, K., Roueff, E., Agúndez, M., Pernot, P., Cuppen, H. M., Loison, J. C., & Talbi, D. *Space Sci. Rev.* **2010**.

Miller, J. A. & Klippenstein, S. J. *J. Phys. Chem. A* **2006**, *110*, 10528.

Georgievskii, Y. & Klippenstein, S. J. *J. Chem. Phys.* **2005**, *122*, 194103.

Clary, D. C. *Ann. Rev. Phys. Chem.* **1990**, *41*, 61.

Troe, J. *Adv. Chem. Phys.* **1997**, *101*, 819.

Gentry, W. R. & Giese, C. F. *J. Chem. Phys.* **1977**, *67*, 2355.

Landau, L. D. & Lifshitz, E. M. *Quantum Mechanics;* Pergamon Press: New York, 1991.

T. H. Dunning, J. *J. Chem. Phys.* **1989**, *90*, 1007.

Celani, P. & Werner, H.-J. *J. Chem. Phys.* **2000**, *112*, 5546.

Molpro, version 2010.1, a package of *ab initio* programs. Werner, H.-J., Knowles, P. J., Manby, F. R., Schütz, M., Celani, P., Knizia, G., Korona, T., Lindh, R., Mitrushenkov, A., Rauhut, G., Adler, T. B., Amos, R. D., Bernhardsson, A., Berning, A., Cooper, D. L., Deegan, M. J. O., Dobbyn, A. J., Eckert, F., Goll, E., Hampel, C., Hesselmann, A., Hetzer, G., Hrenar, T., Jansen, G., Köppl, C., Liu, Y., Lloyd, A. W., Mata, R. A., May, A. J., McNicholas, S. J., Meyer, W., Mura, M. E., Nicklass, A., Palmieri, P., Pflüger, K., Pitzer, R., Reiher, M., Shiozaki, T., Stoll, H., Stone, A. J., Tarroni, R., Thorsteinsson, T., Wang, M., & Wolf, A. **2010**.

Hirschfelder, J. O., Curtiss, C. F., & Bird, R. B. *Molecular Theory of Gases and Liquids;* John Wiley & Sons: New York, 1967.

Georgievskii, Y. & Klippenstein, S. J. *J. Chem. Phys.* **2003**, *118*, 5442.

Discussion

Stephan Schlemmer: What is the influence of diabatic coupling on the results of calculation of $k(T)$, especially at low temperatures?

Stephen J. Klippenstein: The statistical treatment of the electronic states gives some indication of the maximum effect of the diabatic couplings. Sample calculations suggest that these effects are typically quite small: a few % or less.

E. F. van Dishoeck: Thanks for calculating all of these astrophysically interesting reactions. In interstellar space, the populations of the fine structure leves of, say, $O(^3P)$ or $C(^3P)$ are often subthermal. It would be very useful for astrochemists to "pull out" these state-specific rate coefficients explicitly from your calculation.

STEPHEN J. KLIPPENSTEIN: That can readily be done from our electronic state specific results and we will do so in the future.

IAN SMITH: In radical-radical reactions surfaces of several spin multiplicities correlate with the reactants. I would expect the surfaces of low multiplicity (where electrons have "paired up") to have no barrier. I am less sure about the surfaces of higher multiplicities. Have you checked to all that these surfaces have no barrier?

STEPHEN J. KLIPPENSTEIN: Yes, we have and this is one of the important steps in the rate constant calculation. Typically we found that for reactions with either $O(^3P)$ or $C(^3P)$ and a radical species in the doublet ($S = 1/2$) electronic state, the two doublet electronic surfaces and one quartet surface are barrierless while all other surfaces have significant barriers and, therefore, are not reactive.

The Molecular Universe
Proceedings IAU Symposium No. 280, 2011
José Cernicharo & Rafael Bachiller, eds.

doi:10.1017/S1743921311025130

Anions in Space and in the Laboratory

Veronica M. Bierbaum

Department of Chemistry and Biochemistry
Center for Astrophysics and Space Astronomy
University of Colorado
Boulder, Colorado 80309-0215
email: veronica.bierbaum@colorado.edu

Abstract. The astronomical detection of molecular anions has prompted our study of their chemical reactions with atomic species that are abundant in the interstellar medium. We have recently explored the chemistry of a variety of $C_xN_y{}^-$ anions with hydrogen atoms and determined their reaction rate constants and products using the flowing afterglow-selected ion flow tube technique. Computational studies allow characterization of the structures of reactants and products, as well as the energetics along the reaction pathway. For anions containing one or two nitrogen atoms, reactions with hydrogen atoms are facile, and proceed primarily by associative detachment. In contrast, anions containing three nitrogen atoms are unreactive with hydrogen atoms due to reaction barriers and unfavorable thermodynamics.

Keywords. Negative ions, gas phase reactions, associative detachment, reaction rate constants, flowing afterglow-selected ion flow tube

1. Introduction

Ions are found throughout nature in many environments. They are abundant in the earth's oceans and atmosphere, as well as in lightning and flames; ions are critical in the biochemical reactions of life, and they have been detected in many space environments. Positive ions have been known to exist in the interstellar medium (ISM) since the discovery of CH^+ in 1941 (Douglas & Herzberg 1941), but as recently as five years ago, scientists were debating whether negative ions could survive in these environments. Then in December 2006, McCarthy and co-workers reported the *Laboratory and Astronomical Identification of the Negative Molecular Ion C_6H^-* (McCarthy *et al.* 2006). A total of six anions have now been detected including three hydrogenated carbon chain anions, C_4H^-, C_6H^-, C_8H^-, and three cyano anions, CN^-, C_3N^-, C_5N^- (McCarthy *et al.* 2006; Brünken *et al.* 2007; Cernicharo *et al.* 2007, 2008; Remijan *et al.* 2007; Thaddeus *et al.* 2007; Agúndez *et al.* 2010). These negative ions represent a large fraction of recently detected chemical species, and certainly many more are awaiting discovery.

Our research program has focused on the gas phase chemistry of ions with neutrals that are abundant in the ISM (Snow & Bierbaum 2008), and the potential role of these reactions in the synthesis of interstellar molecules. This presentation will provide an overview of our studies of negative ion chemistry, and then describe our recent work on the nitrogen-containing carbanions, $C_xN_y{}^-$.

2. Experimental

We employ the versatile Flowing Afterglow-Selected Ion Flow Tube (FA-SIFT) Technique shown in Figure 1 (Van Doren *et al.* 1987). Ions are formed in a flowing afterglow source utilizing a variety of ionization methods, including dc discharge, electron impact,

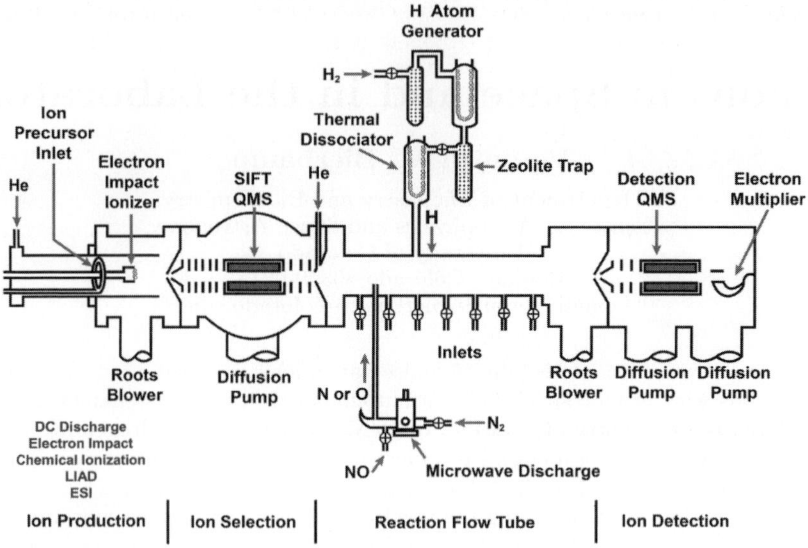

Figure 1. The Flowing Afterglow-Selected Ion Flow Tube (FA-SIFT).

and chemical ionization; these sources allow the formation and study of a wide array of ions. The desired reactant ion is mass-selected with a quadrupole mass filter, and injected through a venturi inlet into a stream of helium. The ions are thermalized by thousands of collisions with helium; stable molecular reactants are added through a manifold of inlets, and atomic reactants can also be introduced. Reaction occurs throughout the remainder of the flow tube; the reactant and product ions are monitored with a quadrupole mass filter, and kinetics are determined from the dependence of ion intensities on the flow rate of the neutral reactant. To generate H-atoms, molecular hydrogen is dissociated thermally; hydrogen is passed through liquid nitrogen traps and over a hot filament. The flow of H-atoms is proportional to the voltage applied to the filament. The method produces a few percent dissociation and the hydrogen atom density is measured with calibration reactions. Nitrogen atoms are generated by passing molecular nitrogen through a microwave discharge; oxygen atoms are formed by addition of NO, which reacts with nitrogen atoms. The concentrations of N-atoms and O-atoms are determined from the endpoint of the titration.

This method has several valuable features for studying gas phase ion chemistry; in particular, the ions are thermalized by many collisions with helium before reaction, so that the energy is well-defined; however, the energy can be varied by heating or cooling the flow tube, or applying a drift field; because distance and time are related in a flow tube, kinetic analysis, measurement of rate constants, is straightforward; there are high ion densities and good sensitivity – this has allowed coupling of the flowing afterglow with many other techniques over the years; perhaps the most valuable feature is the chemical versatility – a wide variety of ionic and neutral reactants can be formed and studied, including atomic reactants.

3. Results

Our studies of negative ions have included five categories: pure carbon chains (C_x^-), hydrogenated carbon chains (C_xH^-), organic anions ($H_wC_xN_yO_z^-$), hydride anion (H^-), and nitrogen-containing carbanions ($C_xN_y^-$).

Carbon chain anions (C_x^-). Pure carbon chain anions, with x = 1–11, can be formed in the ion production region using a cold cathode dc discharge; a graphite rod is raised to several kilovolts relative to the flow tube. Helium and argon are used as carrier gases, and the discharge is stabilized by water-cooling. An anion with a specific mass-to-charge ratio is selected with the SIFT quadrupole, and injected into the reaction flow tube for study. None of the carbon chain anions react with molecular hydrogen, but they react efficiently with hydrogen atoms by associative detachment. The larger carbanions also show simple addition to form the hydrogenated carbon chain anions. Typical rate constants are about 7×10^{-10} cm^3 s^{-1} (Barckholtz *et al.* 2001).

Reactions of the carbon chain anions with atomic oxygen occur rapidly by associative detachment and by the loss of a single carbon atom to form carbon monoxide and an anion with one fewer carbon atom. Reactions of carbon chain anions with nitrogen atoms are somewhat slower ($\sim 2 \times 10^{-10}$ cm^3 s^{-1}), but show a rich variety of pathways that involve formation of carbon-nitrogen bonds (Eichelberger *et al.* 2007; Yang *et al.* 2010a).

Hydrogenated carbon chain anions (C_xH^-). These anions do not react with molecular hydrogen but undergo rapid associative detachment reactions with hydrogen atoms (Barckholtz *et al.* 2001). With oxygen atoms, these anions show associative detachment, transfer of carbon to form carbon monoxide, as well as replacement of hydrogen by oxygen . The reactions with nitrogen atoms show pathways similar to those for the pure carbon chain anions, but the reactions are extremely slow; this lack of reactivity is attributed to problems of spin conservation for the reaction of the singlet anions with the quartet nitrogen atoms (Eichelberger *et al.* 2007).

Organic anions ($H_wC_xN_yO_z^-$). We have explored the reactivity of several organic anions formed by deprotonation of the neutral parents, including methyl cyanide, ethyl cyanide, iso-propyl cyanide, acetaldehyde, acetone, ethyl acetate, methanol, acetic acid and glycine (Snow *et al.* 2009; Yang *et al.* 2010b). Reactions of these anions with hydrogen atoms occur by associative detachment, and there is a direct correlation between reactivity and the reaction exothermicity. The low reactivity of the acetate and glycine anions reflects the high stability of these delocalized anions.

In all cases, the reactions of these organic anions with oxygen atoms have larger rate constants relative to the corresponding reactions with nitrogen atoms. Our computational studies indicate that spin-allowed reactions dominate the O-atom reactions, whereas slow spin-forbidden pathways occur for the N-atom reactions (Yang *et al.* 2010c).

Hydride anion (H^-). The reactions of hydride ion with a series of neutral reagents occur by a wide array of mechanisms (Martinez *et al.* 2010). Efficient proton abstraction processes are observed for hydride anion reacting with acetylene, water, methyl cyanide, methanol, acetone, and acetaldehyde. Facile associative detachment is observed for molecular oxygen, whereas extremely slow associative detachment is observed for carbon monoxide. Reaction of hydride with nitrous oxide forms the hydroxide ion, while reaction with carbon dioxide occurs by addition to produce formate anion. Reaction of hydride with methyl chloride occurs by nucleophilic substitution (S_N2), whereas reaction with tert-butyl chloride proceeds by elimination (E2); both of these processes generate chloride ion as the product. Finally, reactions of hydride ion with diethyl ether, benzene, and molecular deuterium were immeasurably slow.

Nitrogen-containing carbanions ($C_xN_y^-$). These species are generated in the ion production region with the cold cathode discharge source, using the graphite rod in the presence of molecular nitrogen (Yang *et al.* 2011).

A. Ions containing one nitrogen atom. We have mass-selected and studied anions containing a single nitrogen atom, and from one to six carbon atoms, as shown in Figure 2.

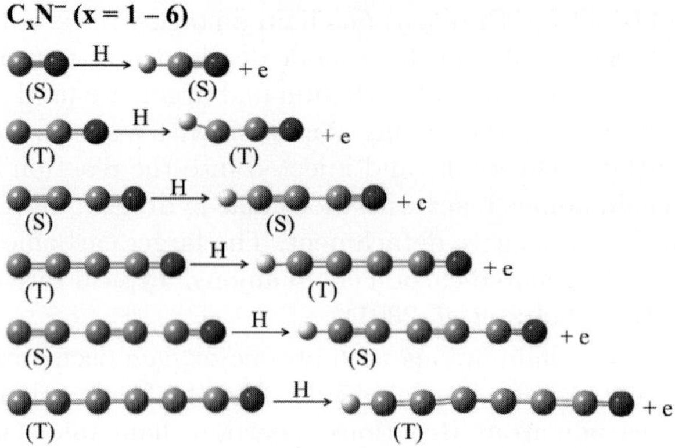

Figure 2. Associative Detachment Reactions of C_xN^-. The singlet ground state is designated by S, and the triplet ground state is designated by T.

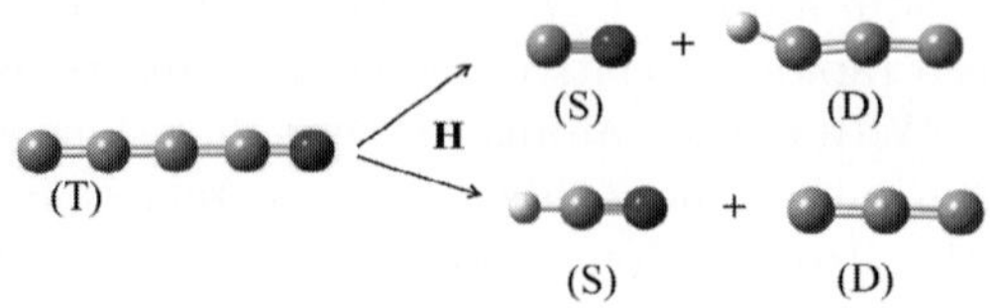

Figure 3. Reaction of C_4N^- with Hydrogen Atom to Form Ionic Products. S, D, and T designate singlet, doublet, and triplet, respectively.

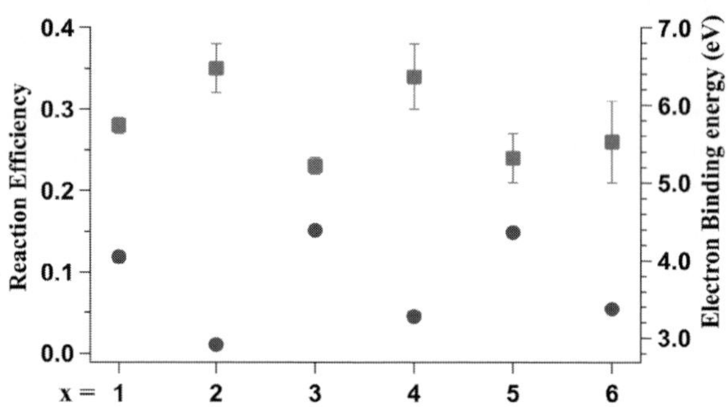

Figure 4. Reaction Efficiency (squares) for C_xN^- + H (x = 1–6) and the Electron Binding Energy of the Anions (circles).

The S and T designations indicate the singlet or triplet nature of the ground electronic state, as determined by our computations. The anions are unreactive with molecular hydrogen, but react readily with atomic hydrogen by associative detachment, to form for example, HCN and an electron. The larger anions react in a similar manner. Some of the anions, for example C_4N^-, also give ionic products; as shown in Figure 3, reaction with H-atom forms CN^- and HC_3, or products in which there is a proton transfer to form HCN and C_3^-. Although these reactions are relatively rapid, there is a slight alternation in reactivity. If reaction efficiency is plotted versus the number of carbon atoms (Figure 4), the ions with an even number of carbon atoms have slightly higher reactivity as indicated by the squares, but have lower electron binding energies as indicated by the circles.

B. Ions containing two nitrogen atoms. We have studied the reactions of the anions $C_xN_2^-$, where x = 1, 3, 4, and 5. These doublet negative ions react readily with H-atom by associative detachment; $C_4N_2^-$ and $C_5N_2^-$ also give ionic products.

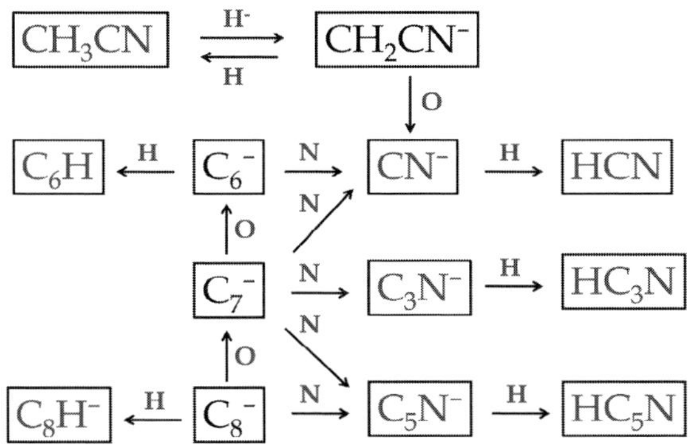

Figure 5. Partial Chemical Network for Interstellar Molecular Synthesis. This diagram illustrates some of the reactions studied in our laboratory.

C. Ions containing three nitrogen atoms. The $C_2N_3^-$ and $C_4N_3^-$ anions show no measurable reactivity with hydrogen atoms. Our computational studies indicate that there are high energy transition states along the reaction surface, and only endothermic channels exist. Thus, these ions will exhibit high stability in the ISM, even in regions of high H-atom density. The $C_4N_3^-$ anion has a symmetrical structure [$C(CN)_3^-$] and therefore no dipole moment. In contrast, the $C_2N_3^-$ anion is bent and polar; thus, this ion would be a good target for laboratory characterization of the rotational spectrum and for eventual astronomical detection.

4. Summary

Figure 5 shows nine molecules and anions that have been detected in the ISM, as well as four anions that have not been detected, but are likely present (CH_2CN^-, C_6^-, C_7^-, C_8^-). This scheme provides a glimpse of how ion chemistry can contribute to interstellar synthesis and interconnect the observed species. The arrows represent some of the chemical reactions with hydrogen, nitrogen, and oxygen atoms described in this presentation. Clearly this ion-neutral chemistry can provide efficient avenues for the generation and interconversion of these molecular ions and neutrals.

5. Future Directions

Many additional studies are desirable. In particular, we plan to quantify the product ratios for many of the reactions; these studies require that we account for mass discrimination, secondary reactions, and the loss of ion signal due to associative detachment. In addition, we plan to study the reactions of the nitrogen-containing carbanions with nitrogen and oxygen atoms.

Finally, we propose to extend our studies to larger ionic species, including polycyclic aromatic hydrocarbons and biomolecules. In preliminary studies, we have successfully generated deprotonated benzene ($C_6H_5^-$), naphthalene ($C_{10}H_7^-$), and anthracene ($C_{14}H_9^-$), and explored their ion chemistry. These singlet anions have moderate electron binding energies, and may be readily formed in molecular clouds; therefore, these anions should be included in the modeling of cloud chemistry. In contrast, the doublet molecular anions benzene ($C_6H_6^-$), naphthalene ($C_{10}H_8^-$), and anthracene ($C_{14}H_{10}^-$), formed by electron attachment to the parent molecules, are more weakly bound.

In order to extend these studies to larger compounds of lower volatitily, we are currently incorporating Laser Induced Acoustic Desorption (LIAD) and Electrospray Ionization (ESI) as new ion sources for our FA-SIFT. We hope that these modifications will enable the study of a broad range of chemical systems.

6. Acknowledgements

NASA and the National Science Foundation are gratefully acknowledged for their financial support of this research. I am grateful to my collaborators for their invaluable contributions to this work: Ted Snow, Zhibo Yang, Nicholas Demarais, Callie Cole, Nadine Wehres, Oscar Martinez, Nick Betts, Brian Eichelberger, and Cindy Barckholtz. I thank Zhibo Yang for help in preparing the figures, and Nicholas Demarais and Callie Cole for creating the Latex version of this manuscript.

References

Agúndez, M., Cernicharo, J., Guélin, M., Kahane, C., Roueff, E., Klos, J., Aoiz, F. J., Lique, F., Marcelino, N., Goicoechea, J. R., González García, M., Gottlieb, C. A., McCarthy, M. C., & Thaddeus, P. 2010, *A&A*, 517, L2

Barckholtz, C., Snow, T. P., & Bierbaum, V. M. 2001, *ApJ*, 547, L171

Brünken, S., Gupta, H., Gottlieb, C. A., McCarthy, M. C., & Thaddeus, P. 2007, *ApJ*, 664, L43

Cernicharo, J., Guélin, M., Agúndez, M., Kawaguchi, K., McCarthy, M., & Thaddeus, P. 2007, *A&A*, 467, L37

Cernicharo, J., Guélin, M., Agúndez, M., McCarthy, M. C., & Thaddeus, P. 2008, *ApJ*, 688, L83

Douglas, A. E. & Herzberg, G. 1941, *ApJ*, 94, 381

Eichelberger, B., Snow, T. P., Barckholtz, C., & Bierbaum, V. M. 2007, *ApJ*, 667, 1283

Martinez Jr., O., Yang, Z., Demarais, N. J., Snow, T. P., & Bierbaum, V. M. 2010, *ApJ*, 720, 173

McCarthy, M. C., Gottlieb, C. A., & Gupta, H., Thaddeus. P. 2006, *ApJ*, 652, L141

Remijan, A. J., Hollis, J. M., Lovas, F. J., Cordiner, M. A., Millar, T. J., Markwick-Kemper, A. J., & Jewell, P. R. 2007, *ApJ*, 664, L47

Snow T. P., Bierbaum, V. M. 2008, *Annu. Rev. Anal. Chem.*, 1, 229

Snow, T. P., Stepanovic, M., Betts, N. B., Eichelberger, B. R., Martinez Jr., O., & Bierbaum, V. M. 2009, *Astrobiol.*, 9, 1001

Thaddeus, P., Gottlieb, C. A., Gupta, H., Brünken, S., McCarthy, M. C., Agúndez, M., Guélin, M., & Cernicharo, J. 2007, *ApJ*, 677, 1132

Van Doren, J. M., Barlow, S. E., DePuy, C. H., & Bierbaum, V. M. 1987, *Int. J. Mass Spectrom. Ion Proc.*, 81, 85

Yang, Z., Snow, T. P., & Bierbaum, V. M. 2010a, *Phys. Chem. Chem. Phys.*, 12, 13091

Yang, Z., Eichelberger, B., Carpenter, M. Y., Martinez Jr., O., Snow, T. P., & Bierbaum, V. M. 2010b, *ApJ*, 723, 1325

Yang, Z., Eichelberger, B., Martinez Jr., O., Stepanovic, M., Snow, T. P., & Bierbaum, V. M. 2010c, *J. Am. Chem. Soc.*, 132, 5812

Yang, Z., Cole, C. A., Martinez Jr., O., Carpenter, M. Y., Snow, T. P., & Bierbaum, V. M., *ApJ*, in press 2011.

Discussion

HERBST: Can you study attachment reactions?

BIERBAUM: Yes, in principle, electron attachment can be studied in flow tube instruments by incorporating a Langmuir Probe. However, we have not pursued studies in this area.

SMITH: Can you confirm that you can observe and measure the reaction rates of C_3^-? I had the impression that it was missing from some of your slides.

BIERBAUM: Although C_3^- (m/z 36) is generated in the graphite discharge source, there are chloride ion impurities present at m/z 35 and 37. Therefore, it is difficult to mass select and inject a clean signal of C_3^- for studies of the chemistry.

SCHLEMMER: What are the routes to form C_x^- in space? (Reference to your chemical network diagram where such a route is missing.) Please comment.

BIERBAUM: The carbon chain anions are most likely formed in space by electron attachment to neutral carbon chains; radiative electron attachment to the larger carbon chains should be facile. The chemical network diagram (Figure 5) includes only reactions that we have studied, and represents only a sampling of our work.

HAMA: Could you please give us a comment on how you confirmed the structures of the reactant and products?

BIERBAUM: Many ionic structures are possible for a given m/z value. The reactant and product ions shown in the slides are the most stable isomers as determined by our computations. Structures could be probed by collision-induced dissociation experiments, by mobility studies, or possibly by spectroscopic experiments.

The Molecular Universe
Proceedings IAU Symposium No. 280, 2011
José Cernicharo & Rafael Bachiller, eds.

doi:10.1017/S1743921311025142

Solid State Pathways towards Molecular Complexity in Space

Harold Linnartz[1] Jean-Baptiste Bossa[1], Jordy Bouwman[1†], Herma M. Cuppen[1‡], Steven H. Cuylle[1], Ewine F. van Dishoeck[1,2], Edith C. Fayolle[1], Gleb Fedoseev[1], Guido W. Fuchs[1], Sergio Ioppolo[1], Karoliina Isokoski[1], Thanja Lamberts[1], Karin I. Öberg[1¶], Claire Romanzin[1‖], Emily Tenenbaum[1], Junfeng Zhen[1]

[1]Sackler Laboratory for Astrophysics, Leiden Observatory, University of Leiden, PO Box 9513, NL 2300 RA Leiden, the Netherlands
email: linnartz@strw.leidenuniv.nl

[2]Max-Planck-Institut für Extraterrestrische Physik, Giessenbachstrasse 1, D-85741 Garching, Germany

Abstract. It has been a long standing problem in astrochemistry to explain how molecules can form in a highly dilute environment such as the interstellar medium. In the last decennium more and more evidence has been found that the observed mix of small and complex, stable and highly transient species in space is the cumulative result of gas phase and solid state reactions as well as gas-grain interactions. Solid state reactions on icy dust grains are specifically found to play an important role in the formation of the more complex "organic" compounds. In order to investigate the underlying physical and chemical processes detailed laboratory based experiments are needed that simulate surface reactions triggered by processes as different as thermal heating, photon (UV) irradiation and particle (atom, cosmic ray, electron) bombardment of interstellar ice analogues. Here, some of the latest research performed in the Sackler Laboratory for Astrophysics in Leiden, the Netherlands is reviewed. The focus is on hydrogenation, *i.e.*, H-atom addition reactions and vacuum ultraviolet irradiation of interstellar ice analogues at astronomically relevant temperatures. It is shown that solid state processes are crucial in the chemical evolution of the interstellar medium, providing pathways towards molecular complexity in space.

Keywords. astrochemistry, infrared: Astrochemistry, ISM, ISM: atoms, ISM: molecules, molecular processes, ISM: abundances, infrared: ISM, methods: laboratory, techniques: spectroscopic

1. Introduction

At the low temperatures in space, micrometer sized dust grains act as small cryopumps onto which gas-phase molecules accrete. In dense cores, the approximate timescale at which gas-phase species deplete-out onto grains is about 10^5 years - shorter than the typical lifetime of a dense core - and hence, during the first stage of star formation virtually all species (except H_2) are expected to be frozen-out onto interstellar grains. The resulting ice coating provides a molecule reservoir, comprising tens of monolayers (ML)

† Present address: Lawrence Berkeley National Laboratory, 1 Cyclotron Road, MS 6R-2203 Berkeley, CA 94720-8226 USA

‡ Present address: Institute for Molecules and Materials (IMM), Radboud University Nijmegen, PO Box 9010, NL 6500 GL Nijmegen, The Netherlands

¶ Present address: Harvard-Smithsonian Center for Astrophysics, 60 Garden Street, MS 42, Cambridge, MA 02138 USA

‖ Present address: LPMAA, Université Pierre et Marie Curie, 4 place Jussieu, F 75252 Paris cedex, France of accreted molecules, and a heat sink capable of absorbing reaction excess energy, effectively catalyzing surface reactions. Basically, inter- and circumstellar surfaces offer an environment *where atoms and molecules accrete, meet and greet.* Grain surface chemistry is governed by the accretion rate of gas-phase species onto the grains, the surface migration rate, which sets the reaction network, and the desorption rate. *I.e.*, it depends on a number of physical parameters, such as ice temperature and ice morphology. External triggers are needed to overcome reaction barriers and to initiate chemical processes in the ice (*i.e.*, after accretion and before desorption), *e.g.*, by photon irradiation - particularly Ly-α and x-rays - thermal effects and particle bombardment (H, D, C, N, O, S atoms, free electrons as well as cosmic rays). In order to interpret the relevance of these processes in space, it is necessary to fully understand the physical and chemical processes that are at play and for this detailed laboratory experiments are a prerequisite. This is the topic of the present proceeding that is organized in the following way. It starts with the spectral features of interstellar ices and their laboratory analogues. An example is given for CO ice mixed with different constituents, and summarizing the findings from a very recent paper (Cuppen *et al.* 2011). It continues with surface processes induced by H-atom addition reactions and vacuum ultraviolet irradiation, and concludes with experiments studying molecular complexity in ice. The hydrogenation and UV irradiation part of this proceeding are based on highlights presented by Ioppolo *et al.* 2011c, and in a upcoming book chapter on solid state laboratory astrophysics.

2. Ice spectroscopy

Spectral features of ices in star-forming regions have been observed particularly in mid-infrared absorption studies of embedded objects or along the line of sight of a background star. Ground-based observations in the mid-IR spectral range are limited largely to the 3-5 micron range because of telluric absorptions, so airborne and space observations have been used to identify solid-state species at longer wavelengths. Observations with the *Infrared Space Observatory* (ISO) have improved our understanding of interstellar ices substantially. Because of its limited sensitivity, ISO observed mostly bright sources, such as high-mass young stellar objects (YSOs) and quiescent dense clouds toward luminous background stars (Gibb *et al.* 2000, 2004). More recently, the *Spitzer Space Telescope* characterized the molecular content of icy grain mantles in the 5 - 35 μm wavelength range towards more than 40 low-mass protostars within the c2d (cores to disks) program (*e.g.*, Boogert *et al.* 2008; Pontoppidan *et al.* 2008; Öberg *et al.* 2008; Bottinelli *et al.* 2010) and dozens more within other programs (Zasowski *et al.* 2009). The Spitzer data are complemented by ground-based 3-5 micron surveys of the same sources (e.g., Pontoppidan *et al.* 2003, van Broekhuizen *et al.* 2005). Following the outcome of dedicated laboratory studies (Hudgins *et al.* 1993), this has made possible the unambiguous identification of a series of solid state constituents, like H_2O, CO_2, CO, CH_4, CH_3OH, and NH_3, as well as their likely mixing ratios (see *e.g.*, Bouwman *et al.* 2009). As ice matrix interactions are different for different ice morphologies, peak positions, band widths and intensity ratios differ for these features and this allows quite accurate comparisons between astronomical and laboratory spectra (Tielens 1991). The latter are typically recorded in Fourier Transform InfraRed (FTIR) transmission experiments of ices deposited at different mixing ratios and studied in full dependence of the temperature. An interstellar ice feature inventory with abundances relative to water ice towards dark clouds, low- and high-mass young stellar objects is given by Öberg *et al.* (2011).

These are the basic ingredients taken for the laboratory recipes (described in the next sections) to form new molecules. Besides information on the actual chemical composition

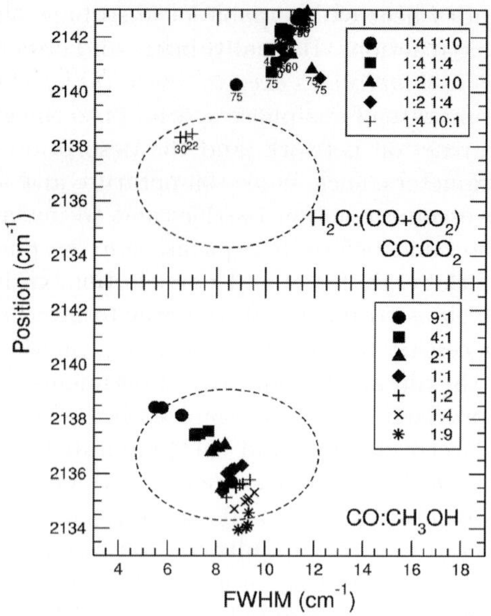

Figure 1. The FWHM and position of the CO absorption band in laboratory spectra of a range of different H_2O:CO:CO_2 mixtures (top panel) and CO:CH_3OH (bottom panel) ice mixtures at different temperatures. The circles indicate the observational data position within the panels. Figure from Cuppen *et al.* (2011).

of ices in space, accurate comparisons between astronomical and laboratory data also provides information on the chemical history of the ice. This is illustrated for a specific case in Section 2.1.

2.1. *CO ice mixed with CH_3OH ice: a proof for methanol ice residing in a water poor environment*

In observations towards icy sources, the CO stretching band comprises a prominent and broad component around 2136.5 cm^{-1}. This feature is generally attributed to solid CO mixed in a hydrogen-bonded environment like water, but CO:H_2O laboratory studies have not been able to fully reproduce this feature. Water-containing CO ice cannot reproduce the observed band position and bandwidth without simultaneously producing a shoulder at 2152 cm^{-1}. This band, believed to originate from the interaction of dangling-OH bonds with CO, is not observed in any of the astronomical spectra. It has been suggested that the 2152 cm^{-1} feature is suppressed in astronomical ices by blocking of the dangling OH-bonds by other species (Fraser *et al.* 2004). A recent spectroscopic study of H_2O:CO:CO_2 mixtures (Cuppen *et al.* 2011) - given the abundances listed by Öberg *et al.* (2011) a logical choice - shows that such an ice mixture does not offer an explanation for this discrepancy between astronomical and laboratory data. This is illustrated in the upper part of the Fig. 1. Even though the 2152 cm^{-1} band can be suppressed under low temperature conditions, it is not possible to reproduce peak position and band width of the 2136.5 cm^{-1} band at the same time. This is indicated by the circle that defines boundary conditions as derived from interpreting the astronomical data, and clearly the spectroscopic parameters for most of the studied mixing ratios are outside this circle. An ice mixture containing only CO and CH_3OH and no H_2O, however, does reproduce the spectra at low temperature, both in terms of peak position and band width of the CO component and the 2152 cm^{-1} band does not appear. This is illustrated for seven different mixing ratios in the lower panel of the figure. This provides additional proof that CO only freezes out after H_2O has formed on the grain, leading to at least two separate ice components; a H_2O rich and a CO rich one. In this scenario, the spectral characteristics of the CO central feature require a high fraction of CH_3OH to be present. This means that almost all methanol is present in the CO:CH_3OH component. This is in general agreement with astronomical observations of solid CH_3OH, also indicating that at least part of the methanol is in a water poor environment.

The observation that CH_3OH is predominantly mixed with CO in a H_2O-poor environment leads to the conclusion that methanol is mainly formed through hydrogenation of CO, as shown by Watanabe *et al.* (2004) and Fuchs *et al.* (2009) and not through UV or ion processing of a H_2O containing ice. Grain surface chemistry simulations of CO in a water-poor environment indeed show that methanol can be formed in this way and that CH_3OH is mainly present on the grain mantles in pure form or mixed with CO, and that it is not in a water-rich phase (Cuppen *et al.* 2009). Further details of this work are available from Cuppen *et al.* (2011). The main conclusion is that not all species are intimately mixed in the ice and that the composition of astrophysically relevant ice mixtures should therefore be treated with care. Within this context it is also relevant that ice segregation can take place. For separate studies see Öberg *et al.* (2009e) and Fayolle *et al.* (2011a).

3. Hydrogenation reactions in interstellar ice analogues

The first realistic gas-grain model was proposed by Tielens & Hagen (1982). Their astrochemical model includes a complex grain surface reaction network to explain molecule formation in quiescent dark clouds. According to this model in a first phase, H_2O ice can be produced by the sequential hydrogenation of O atoms landing on the grain. Atomic oxygen can also react with other O atoms to form O_2 and O_3. Reaction of O_3 with H reforms O_2 and OH. O_2 can be hydrogenated to form H_2O, while the OH radical can react with H or H_2 to form again H_2O, or it can form CO_2 and H_2 with the CO accreted on the surface. Since hydrogen is the most abundant component of the interstellar gas, this first phase of grain-surface chemistry results in the formation of a polar ice (water-rich) mantle onto the dust grains. During the second phase, when the density increases in the molecular cloud, CO freezes-out onto the grains (its accretion rate is higher than that of H) forming an apolar (water-poor) ice layer on top of the polar one. Under these conditions, the hydrogenation of CO ice leads to the formation of H_2CO and CH_3OH (Tielens & Hagen 1982; Charnley *et al.* 1997). CO_2 can be formed through the reaction CO + OH (Goumans *et al.* 2008). As suggested by Charnley *et al.* (2001), the ongoing reaction between CO or CO hydrogenation reaction products and elements heavier than H may lead to the formation of more complex molecules such as CH_3COH and C_2H_5OH (see reaction diagrams in Charnley & Rodgers 2005). These and other astrochemical reaction networks were based on chemical intuition and analogues from gas-phase routes. It took several decades before experimental techniques allowed laboratory astrochemists to put all these reactions to the test in the solid state.

Typical experiments are performed using Ultra High Vacuum (UHV) setups with pressures well below 10^{-10} mbar, such as our SURFace REaction Simulation DEvice (SURFRESIDE), which comprises a cryogenic unit where ices are grown for astronomically relevant temperatures and a H-atom beam line. A schematic view of the experimental apparatus is shown in Fig. 2.

Ices are monitored by means of Reflection-Absorption InfraRed Spectroscopy (RAIRS) using a Fourier Transform InfraRed (FTIR) spectrometer, which covers the range between 4000 and 700 cm^{-1} (2.5 - 14 μm). A spectral resolution between 1 and 4 cm^{-1} is used and several scans are co-added. Two different procedures are used; the ice is first deposited and then hydrogenated/deuterated (see *e.g.*, Fuchs *et al.* 2009; Ioppolo *et al.* 2011b, and

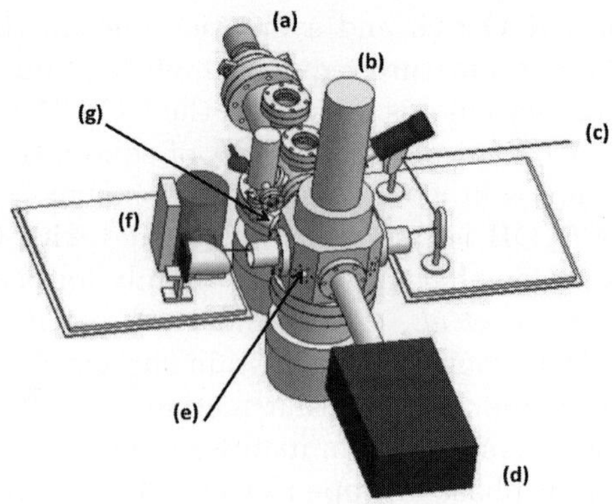

Figure 2. Schematic top-view of the solid-state experimental UHV setup (SURFRESIDE): [a] H-atom source; [b] cold finger; [c] IR beam; [d] mass spectrometer; [e] main chamber; [f] IR detector; [g] deposition line.

Romanzin *et al.* 2011) - in this case, RAIR difference spectra with respect to the initial deposited ice are acquired during H/D exposure - or molecules are co-deposited with H atoms and RAIR difference spectra are acquired with respect to the bare substrate during co-deposition (Cuppen *et al.* 2010). In all cases, newly formed solid species are monitored by using unique IR spectral signatures. Spectra are recorded at different stages during hydrogenation, providing time resolved information about the destruction (*i.e.*, use-up) of the precursor ice (the deposited ice layer) and the formation of new molecules that are identified through their spectral fingerprints. The intensity of a spectrum can be translated into a column density using a modified Lambert-Beer equation (Bennett *et al.* 2004). At the end of the H-atom addition a Temperature Programmed Desorption (TPD) experiment can be performed to constrain the spectroscopic results. Surface hydrogenation reactions of simple ices, like pure CO, O_2, O_3, and CO:O_2 mixtures are investigated for a full range of different laboratory conditions including H/D-atom fluxes, ice temperatures, ice thicknesses, ice structures, and mixture ratios. This makes it possible to unreveal the physics and chemistry of molecule formation and where applicable to examine in more detail the astronomical implications.

In the past the chemistry of inter- and circumstellar ice analogues has been studied using a top-down scenario: icy mixtures of astronomical constituents with more or less realistic mixing ratios were chemically triggered through UV/cosmic ray irradiation. The resulting residue was shown by several groups to contain complex organic compounds. The experimental results using this approach have been compared to interstellar ices generally in a more qualitative than quantitative way. More recently, a bottom-up approach has become experimentally possible, through the use of the aforementioned UHV setups in which individual reactions of simpler ices (*i.e.*, not the cumulative outcome of a chemical network in ice mixtures) can be studied *in situ* and in real time under fully controlled laboratory conditions. This approach makes it possible to derive fundamental and molecule specific parameters, like reaction rates and diffusion barriers, which can then be included in astrochemical models to simulate the ice evolution under much longer timescales (10^5 yr) than accessible in the laboratory (<1 day).

Here we follow a bottom-up approach and summarize a representative sample of relevant experiments mainly by our group (see *e.g.*, Fuchs *et al.* 2009; Ioppolo *et al.* 2008, 2010a, 2011a,b; Cuppen *et al.* 2010; Romanzin *et al.* 2011), but it should be explicitly stated that other groups - such as the groups of Watanabe, Lemaire and Dulie - are very active in this field as well. These experiments prove that species like H_2CO, CH_3OH

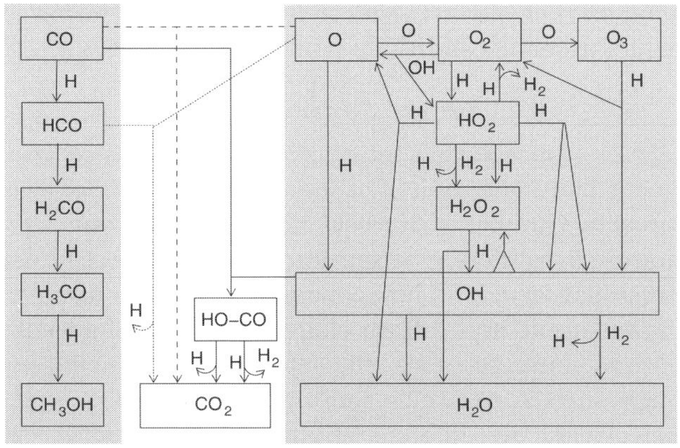

Figure 3. Schematic representation of the hydrogenation reactions in CO, $O/O_2/O_3$ and $CO{:}O_2$ mixed ice.

and H_2O can be formed at low temperatures by simple hydrogenation (*i.e.*, without the need for thermal, UV or cosmic ray processing) and provide the basic molecular data to simulate their formation on astronomical timescales (*e.g.*, Cuppen *et al.* 2009), even though the ice as a whole is not fully representative for a realistic astronomical ice.

Figure 3 shows the schematic representation of the reaction network that is discussed here. The formation of formaldehyde and methanol ice through the CO + H channel is shown on the left-side of the figure, while the water formation channels ($O/O_2/O_3$ + H) are plotted on the right-side. The possible CO_2 formation routes in a $CO{:}O_2$ mixed ice are shown in between the aforementioned channels.

3.1. *Surface formation of methanol*

The surface formation of formaldehyde (H_2CO) and methanol (CH_3OH) by hydrogenation of CO ice has been topic of several studies (*e.g.*, Hiraoka *et al.* 2002; Watanabe & Kouchi 2002; Fuchs *et al.* 2009). Methanol has been observed in interstellar ices and has been experimentally shown as a good starting point for the formation of more complex species (Öberg *et al.* 2009c). The hypothesis that interstellar methanol forms in the solid phase has been recently supported by experimental and modelling work, showing that the gas-phase route via ion neutral reactions is less efficient than previously assumed (Geppert *et al.* 2005; Garrod *et al.* 2006). However, the first studies on solid CO hydrogenation by two different groups yielded conflicting results: in one study (Hiraoka *et al.* 2002) only the formation of H_2CO was reported, whereas in the other study also CH_3OH was observed (Watanabe & Kouchi 2002). The prevailing discrepancy between these studies was experimentally shown by Fuchs *et al.* (2009) to be a consequence of different experimental conditions, most noticeable the adopted H-atom fluxes and they were also able to determine reaction rates from RAIR data for different ice temperatures and ice thicknesses, as well as H-atom fluxes.

The formation of H_2CO and CH_3OH in the ice was also confirmed mass spectrometrically. The experiments showed that the hydrogenation process is thickness independent for layers thicker than 4 monolayers, which represents the maximum penetration depth of the H atoms into the CO ice. For temperatures higher than 15 K, a clear drop in the production rate of methanol is observed. This is probably due to the reduced residence-time of H atoms in the ice at those temperatures, which decreases the probability that hydrogenation reactions occur. Since the rate of formation of molecules depends on a sequence of events: deposition of H atoms, diffusion and then reaction, reaction barriers can only be determined using a detailed model to fit the experimental data. These results can then be used in an astrochemical model to investigate the formation of formaldehyde and methanol under interstellar conditions (Cuppen *et al.* 2009).

3.2. *Surface formation of water*

Tielens & Hagen (1982) proposed that interstellar water forms on grain surfaces through three reaction channels: hydrogenation of atomic and molecular oxygen and ozone. Using a Monte Carlo approach, Cuppen & Herbst (2007) and Cazaux *et al.* (2010) showed that the contribution of the different formation channels strongly depends on the local environment in interstellar clouds. They concluded that the atomic oxygen channel is the main route in translucent and diffuse clouds, while the molecular oxygen channel, together with the ozone route, is more efficient in dense cold molecular clouds. Water ice formation through surface reactions following O + H, O_2 + H and O_3 + H has been studied extensively and by different groups (Miyauchi *et al.* 2008; Ioppolo *et al.* 2008; Matar *et al.* 2008; Oba *et al.* 2009; Mokrane *et al.* 2009; Ioppolo *et al.* 2010a; Cuppen *et al.* 2010). Dulieu *et al.* (2010) used TPD in the submonolayer regime as their main analysis technique; the hydrogenation of solid O_2 is the most extensively studied channel. Ioppolo *et al.* (2008, 2010a) investigated this reaction channel for a large range of astronomically relevant temperatures (12 - 28 K) (see Fig. 4). The main and surprising finding is that the initial formation rate of H_2O_2 and H_2O is much less temperature dependent than the analogous reactions for CO hydrogenation. Furthermore, O_2 hydrogenation results in a much larger yield than the few monolayers found for CO hydrogenation. This yield is strongly temperature dependent. Both effects are shown to be a direct consequence of the ability of H atoms to diffuse easily through an O_2 ice (Ioppolo *et al.* 2010a). The experiments show that the penetration depth of H atoms into the ice depends strongly on the ice composition. In Ioppolo *et al.* (2010a) O_2 hydrogenation is investigated extensively from a physical approach, *i.e.*, studying different ice thicknesses, ice temperatures, ice structures and H_2 concentrations in the atomic beam, whereas in a follow-up paper (Cuppen *et al.* 2010) the focus was on the reaction scheme with the intent to assess reaction routes and branching ratios. The latter work shows that the initially proposed reaction network by Tielens & Hagen (1982) of only three mainly isolated channels is too simple and that several of the channels are actually linked through additional reactions. The third water formation channel (the hydrogenation of solid O_3) was tested by Mokrane *et al.* (2009) and more recently by Romanzin *et al.* (2011). Since this channel is connected to the O_2 channel after the first reaction step, special care was taken to deposit a pure O_3 ice by keeping the substrate temperature between the O_2 and O_3 desorption temperature during deposition. If such a temperature is also kept during H-atom addition, the O_2 molecules formed upon O_3 hydrogenation will desorb from the surface of the ice. In this way the reaction of OH to form water via H or H_2 addition can be probed. The hydrogenation of O_3 is found to behave more similar to CO hydrogenation in the sense that only the top few monolayers of O_3 are hydrogenated. Moreover, the reaction OH + H_2 may be more efficient than the reaction OH + H: reaction OH + H_2 likely proceeds through tunneling, while reaction OH + H needs to dissipate 5.3 eV of excess energy with just one final product, which could be difficult. These experimental results complete the reaction scheme on water formation initially proposed. The general conclusion that the three channels ($O/O_2/O_3$ + H) are strongly linked, is of importance for astrochemical models focusing on water formation under interstellar conditions (Wakelam *et al.* 2010). The recent identification of H_2O_2 in the interstellar medium (Bergman *et al.* 2011) is important within this context.

3.3. *Surface formation of CO_2*

Ioppolo *et al.* (2011b) proved the surface formation of CO_2 at low temperatures through the reaction CO + OH and subsequent dissociation of the resulting HO-CO complex

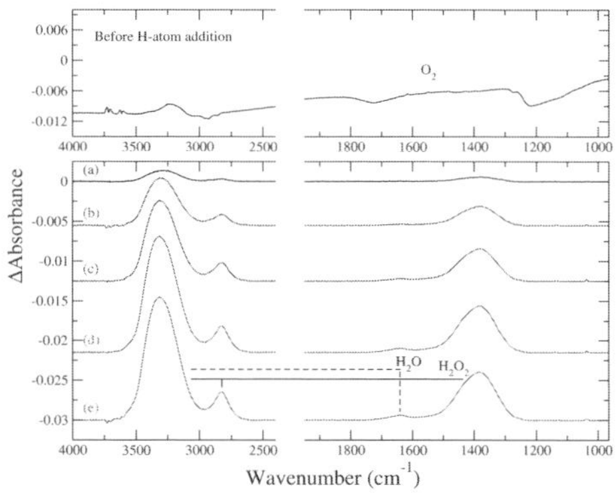

Figure 4. RAIR difference spectra of O_2 + H taken from (Ioppolo *et al.* 2010a). The top panel shows the O_2 ice spectrum at T = 15 K. The bottom panels shows the RAIR difference spectra with respect to the initial deposited ice after H-atom exposure at T = 25 K for increasing fluences (a-e). The formation of H_2O_2 and H_2O is clearly observed.

by hydrogenation of a CO:O_2 ice mixture. Such a binary ice is not fully representative for an interstellar ice, but the hydrogenation of O_2 ice produces OH radicals, which allows the investigation of the interstellar relevant CO + OH solid state reaction. As discussed above similar astrophysical ice studies have focused on the investigation of isolated surface reaction schemes, starting from the hydrogenation of pure ices, like solid CO and O_2. For such ices no CO_2 formation was observed upon H-atom exposure. The hydrogenation of binary ice mixtures allows to investigate the influence of the presence of other species in the ice on the pure ice reaction schemes. Mixtures of CO:O_2 are deposited on a substrate under UHV conditions at low temperatures (15 and 20 K) and are subsequently hydrogenated. The RAIRS shows that solid CO_2 is formed in all studied CO:O_2 mixtures. Within the experimental uncertainties no dependency on ice temperature or composition was observed. The competition between CO and O_2 hydrogenation in the mixed ice reveals that the penetration depth of H atoms into the ice depends strongly on the ice composition, and that the CO and O_2 channels influence each others final product yields. The formation rate for all the final products is found to be less sensitive on the mixture composition than the final yield. Therefore, the formation rates found for H_2CO, CH_3OH, H_2O_2 and H_2O in the isolated studies of the CO + H and O_2 + H channels are valid for use in astrochemical models. In addition, surface CO_2 formation without energetic input is found to be an important formation mechanism, which may explain the formation of CO_2 together with H_2O ice during the dense cold core phase prior to star formation. More details are available from Ioppolo *et al.* (2010b).

4. VUV irradiation ices: photo desorption and photo processing

The interstellar radiation field (ISRF) results in a background UV field of about 10^8 photons cm^{-2}s^{-1}. The UV field is however enhanced by orders of magnitude close to young stellar objects and even in environments protected from external UV (*e.g.*, in cloud cores and disk mid-planes) due to cosmic ray interactions with hydrogen. Photo-processing of ices may induce non-thermal desorption of surface molecules. It triggers the formation of complex species and thus offers a complimentary pathway towards molecular complexity in space. In fact, the photochemistry of icy grain mantles was suggested as a path towards chemical complexity more than three decades ago. Famous are the UV irradiation experiments on interstellar ice mixtures comprising of many different constituents and resulting in the formation of large organic compounds, among which are amino acids (Bernstein *et al.* 2002; Muñoz Caro *et al.* 2002; Ehrenfreund *et al.* 2001;

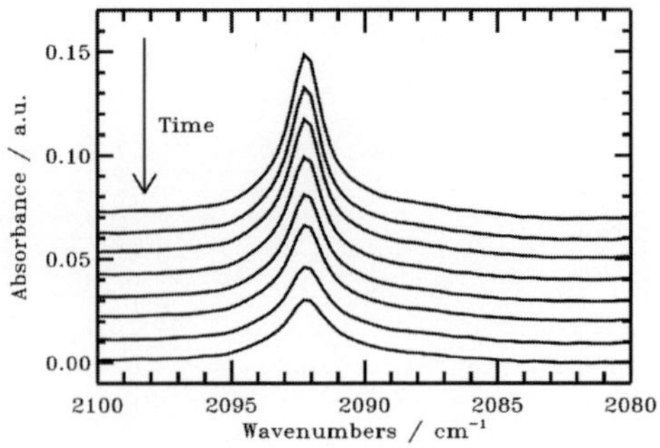

Figure 5. RAIR difference spectra of $C^{18}O$ $\nu = 1$-0 vibrational band at 2093 cm^{-1} acquired before irradiation of 8 ML of CO ice and subsequently after every hour of 8-hr irradiation. The drop in integrated absorbance of the CO ice band is linear with UV irradiation time. The photodesorption rate for $C^{16}O$ is the same within the experimental uncertainty. Figure from Öberg *et al.* (2007).

Charnley *et al.* 1992). The latter work has set the scene, but more controlled experiments are needed to study molecule formation *in situ* and in real time.

In the laboratory interstellar UV radiation is typically simulated using the output of broadband hydrogen-microwave discharge lamps. These peak around 121.6 nm (Ly-α) and cover 115 - 170 nm (7.5 - 10.5 eV) with a variable photon flux from roughly 1 to 10×10^{13} photons cm^{-2}s^{-1} at the ice surface. Since this flux is much higher than that in space, special care has to be taken to extrapolate laboratory results to processes that occur in the ISM. In Leiden the ices are prepared in another UHV setup CRYOgenic Photoproduct Analysis Device (CRYOPAD) (Fuchs *et al.* 2006) in an identical way as described in the previous paragraph for the H-atom addition reactions in SURFRESIDE. The ice is subsequently UV processed. Here we will discuss results using RAIRS to identify the formation and destruction of ice constituents and TPD as an additional tool to monitor desorption processes and ice evolution.

One of the major conclusions of a series of recent UV irradiation studies of CO, N_2, CO_2, H_2O and CH_3OH ice (Öberg *et al.* 2007, 2009a,b,c; Westley *et al.* 1995) is that photodesorption, *i.e.*, the process in which a photon induces a non-thermal evaporation of one or several molecules is more efficient than has been assumed in the past. This is an important result which may also explain why for example gaseous CO can still be found in regions in space where it should have been fully frozen onto dust grains (Bergin *et al.* 2001).

CO ice photodesorbs non-dissociatively; the process is illustrated in Fig. 5. Here the decreasing CO RAIRS signal is shown as a function of irradiation time. With the known UV flux at the surface, the loss of signal with time can be converted into a photodesorption yield, as the signal can be translated into a number density. It is possible to derive a photodesorption yield which is in the order of 10^{-3} desorbed molecules per incident UV photon for CO. Very recently it was shown by Fayolle *et al.* (2011b) that this process is wavelength dependent. Experiments using the DESIRS beam line at SOLEIL showed that the CO photodesorption rate follows the CO electronic absorption pattern. In contrast to CO, CO_2, H_2O and CH_3OH dissociate upon UV irradiation and the fragments can recombine or photodesorb. If the fragments are in the surface layers, they either desorb directly from the excess energy provided by the photon or react with other radicals, which may still result in indirect photodesorption due to the release of energy. If the dissociated fragments remain in the ice mantle, the radicals can either become frozen in the ice matrix, diffuse through the ice to react with other radicals or recombine reforming the parent molecule. To constrain the photodesorption of the ice and thus determine the loss of precursor molecules into the gas phase rather than into photoproducts in the

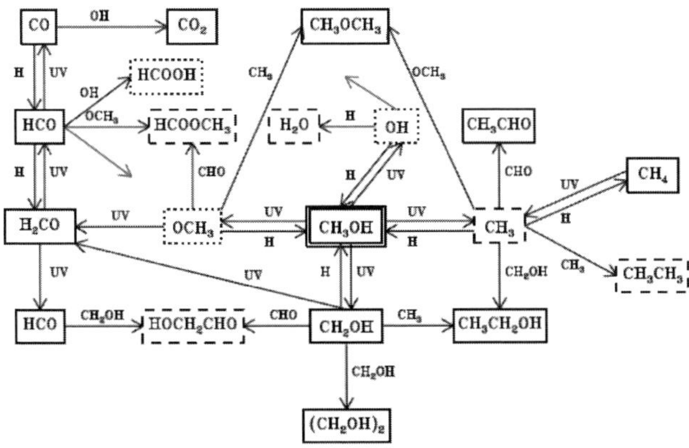

Figure 6. The proposed reaction scheme to form the observed products following UV-irradiation of pure CH_3OH ice (solid boxes), of CO or CH_4:CH_3OH ice mixtures (dashed boxes) and products whose production could only be constrained with upper limits (dotted boxes).

ice, a special method has to be used. Photodesorption from a multilayer ice is a zeroth order process with respect to photon fluence, since it only depends on the amount of molecules in the surface layers. The photodesorption yield will not change with fluence as long as the original ice is sufficiently thick. Ice photolysis, however, is a first order process, since it depends on the total amount of ice. Through simultaneous modeling of the ice loss with an exponential decay and a linear function, the two processes can be separated, resulting in an effective photodesorption yield (Öberg *et al.* 2009a). For CO_2 and H_2O again values of the order of 10^{-3} photon^{-1} have been found, for CH_3OH the photodesorption yield is about twice as large (Öberg *et al.* 2009c). The photodesorption of N_2 ice has a yield that is at least an order of magnitude lower; N_2 only co-desorbs with other molecules since it has no dipole-allowed transition within the energy range of the lamp. This observation is consistent with the recent wavelength dependent results found for CO (Fayolle *et al.* 2011b). It should be noted that not only the UV field, but also the ice thickness, ice temperature, and temperature dependent mean-free-path of the excited molecule play a role. Empirically derived photodesorption yields as function of these parameters are available from Wakelam *et al.* (2010).

In pure CH_3OH ices UV-induced photodissociation produces several different fragments - CH_3, CH_3O, CH_2OH, OH - which can react to form complex organic molecules such as ethanol from CH_3 and CH_2OH. This is illustrated in Fig. 6. Such reactions followed by ice desorption have been proposed to explain observations of gas-phase complex organic molecules in the warm inner regions of protostellar envelopes, so-called hot cores. In a series of dedicated RAIRS experiments, monitoring the UV photoproducts as a function of fluence, Öberg *et al.* 2009c produced all of the complex molecules commonly observed in hot cores (Bottinelli *et al.* 2007). Complementary TPD experiments have been used to further support identifications, since each product has a unique combination of mass signal and desorption temperature. The formation yields typically depend on ice temperature and ice composition as well as the UV flux and fluence. More details are available from Öberg *et al.* (2009d).

The role of the H_2O concentration on the photochemistry of more volatile species was studied as well (Öberg *et al.* 2010). It was found, among others, that increasing the water concentration also increases the destruction efficiency of the volatile mixture constituent with a substantially higher trapping efficiency of the formed radicals in the ice. The formation of more complex species in this way is interesting since solid methanol has been observed towards a range of young stellar objects, including solar-type protostars.

A UV driven methanol-ice chemistry may thus be a common pathway to chemical complexity in space. Complex molecule formation starting from methanol formation through CO hydrogenation and subsequent UV irradiation of methanol containing ices seems to provide a consistent reaction pathway for molecular complexity in space up to species with about 10 atoms. The presence of H_2O may be essential as a catalyst.

At this stage, it should be noted that H-atom bombardment and UV irradiation are not the only possibilities to trigger chemical reactions in an ice for astronomical conditions. Many efforts have been made, for example, to characterize the impact of ions on ice (Ioppolo *et al.* 2009; Palumbo 2006; Mennella *et al.* 2004; Baratta *et al.* 2002; Strazzulla *et al.* 2001). Thermal processing as well as the influence of electron and cosmic-ray impact on interstellar ice analogues have been studied (see *e.g.*, Bennett *et al.* 2010). In recent years much progress has also been made in simulating the dynamical processes that take place in the ice (see *e.g.*, Andersson & van Dishoeck 2008; Arasa *et al.* 2010). This provides additional and valuable information to interpret laboratory observations in terms of interacting surface species.

5. Complex molecules in ice

From a laboratory perspective it is clear that complex molecules can be formed under astronomical conditions. The physical and chemical parameters, derived from work as described in the previous paragraphs, implemented into astrochemical models, indicate that solid state processes are relevant for the formation of larger species in space. A remaining question is which level of complexity can be reached in the ice. An alternative approach to answer this question is to skip the formation process in a top-down approach and to implement large complex molecules, such as polycyclic aromatic hydrocarbons (PAHs) directly into the ice. PAHs are found to be common throughout the interstellar medium, as evidenced by their (mid-IR) emission features that originate from vibrational relaxation after UV excitation. Towards dense clouds, however, these mid-IR features are strongly quenched. In these regions, as discussed before, most volatile molecules are frozen out on grains, forming layers of ice, and less volatile molecules such as PAHs are expected to condense on interstellar grains as well.

Previous spectroscopic ice studies focused on infrared experiments and have resulted in a large amount of spectroscopic data, also providing dynamical information (see *e.g.*, Gudipati *et al.* 2003, 2006). In the specific case of PAHs, however, these have the disadvantage of spectral overlap; for different PAH members essentially the same vibrational modes are excited. This problem is partially overcome in the optical regime, since electronic energies will be different for different PAH species. Optical studies of photo-induced processes at play during the irradiation of PAH containing interstellar H_2O ice analogues are available only for a few species and it was shown that these PAHs are readily ionized and stabilized within the ice, suggesting that trapped ions may play important, but overlooked roles in cosmic ice processes. Beyond this, there is little information on UV induced photochemistry and photophysics of PAH containing H_2O-rich ices.

A new measuring principle allows the spectroscopic and photo-dynamic study of PAH containing ice analogues upon UV irradiation, *in situ* and in real time (Bouwman *et al.* 2009). A high vacuum chamber is used in which ice is grown as described before, as well as a microwave hydrogen discharge source generating UV light for the photoprocessing, and a special Xe-arc light source emitting broad-band light that is transmitted through the ice onto a fast and fully computer controlled monochromator equipped with a sensitive CCD camera. This configuration results in a spectral coverage of 270 to 1200 nm and a time resolution down to 5 ms per spectrum. The solid precursor PAH species is brought into the gas phase through resistive heating and mixed with H_2O vapor. A HeNe laser is used to determine the ice thickness of the mixture and in conjunction with optical spectroscopy on the PAH species this allows for an accurate concentration measurement. The innovative aspect of this setup is that it combines sensitive broadband spectroscopy with a resolution that is more than sufficient to monitor solid state features, *i.e.*, it is possible to cover large spectral ranges without loss of spectral details, and a fast time response which allows monitoring ice chemistry in real time.

So far the UV induced photochemistry of the PAHs anthracene ($C_{10}H_{12}$ or Ant), pyrene ($C_{16}H_{10}$ or Py), benzo[ghi]perylene ($C_{22}H_{12}$ or BghiP), and coronene ($C_{24}H_{12}$ or Cor) in a $\sim$ 1:10,000 diluted H_2O ice has been studied (Bouwman *et al.* 2009, 2010a,b, 2011a,b). The main conclusions of this work is that PAHs are likely to play a chemical role in the full solid state astrochemical solid state network.

6. Outlook

The richest inventory of complex organic molecules (COMs) has been found in a Galactic-center hot-core Sgr B2(N-LMH), including acetone, ethylene glycol, and glycolaldehyde. The temperature in these regions is >100 K and molecules are in the gas phase. One of the major questions today is whether these COMs are first generation evaporated species produced by pure grain surface chemistry or second generation molecules formed by high temperature gas phase chemistry (after ice desorption). While several models suggest that the latter does not adequately explain the observed abundances (Wakelam *et al.* 2006), little quantitative information is available on the actual physical and chemical processes. More than 30 years ago experiments were performed in which ices containing typical astrophysical precursors were photo-processed with hard UV radiation (Agarwal *et al.* 1985; Hagen *et al.* 1979). The resulting organic refractory synthesized under high-vacuum conditions was extracted at room temperature and analysed by gas chromatography showing a rich variety of highly complex molecules. Whether these compounds were formed during UV irradiation of the ice sample or during the relatively intrusive analysis procedure could not be discriminated. New laboratory techniques with an emphasis on more delicate and sensitive sampling methods are therefore needed to tackle the subtle and complex chemistry taking place in the interstellar ice analogues. Limitations in IR spectroscopy, evidently suitable for simpler ices, become apparent when studying more complex ice mixtures. Overlapping IR absorption features and low detection sensitivity due to decreased state densities make the analysis difficult (one reason for studying PAHs with optical spectroscopy as described in the previous paragraph). The TPD mass spectroscopic analysis discussed before, requires the ice sample to be thermally evaporated, which may affect the chemical composition of the ice, and indefinitely implies the destruction of the ice sample. The inevitable prerequisite for mass spectroscopic analysis is the evaporation of the ice sample. An alternative for thermal desorption used in TPD is soft laser ablation - a technique typically used for non-volatile samples. In laser ablation, a short ns-UV laser pulse is focused onto the sample surface, where the photon energy is transferred to the sample leading to local desorption. MATRI2CES (Mass Analytical Tool for Reactions in Interstellar ICES) is a newly constructed setup in Leiden in which this technique has been implemented for the study of interstellar ice analogues. MATRI2CES combines laser ablation with time-of-flight mass spectroscopy, and is comparable to a standard MALDI-TOF setup. The freedom from having to heat-up and evaporate the ice as a whole means that the reactions in the ice analogues can be monitored *in situ* and online under astronomical conditions. Moreover, pulsed repetitive operation enables time-resolved monitoring of the chemical composition as well as depth-profiling of the ice. The construction of MATRI2CES was recently completed.

To conclude: understanding the chemistry of complex organic molecules under interstellar conditions is particularly interesting due to their potential connection to pre-biotic chemistry. Is it possible to start with a realistic interstellar ice analogue and end up with an amino-acid? And if so, does this depend, for example, on temperature or the presence of specific constituents? How is this chemistry driven: atom bombardment, UV irradiation, a combination of both, or should other processes be considered? These are important questions that still have to be addressed, but in view of the progress obtained in the last years this is now work that can be dealt with in a systematic way in the nearby future.

Acknowledgement

The research topics highlighted in this proceeding have been obtained during several years of dedicated work by many (PhD) students, postdocs and visitors and whose names are listed in the list with references. We are very thankful for their many contributions and discussions that have improved our understanding of solid state astrochemical processes. Financial support by NOVA (the Dutch Research School for Astronomy), NWO (the Netherlands Research Council), FOM (the Foundation for Fundamental Research of Matter), and the European Community's Seventh Framework Programme (FP7/2007-2013) under grant agreement no 238258 (LASSIE) is gratefully acknowledged.

References

Agarwal, V. K., Schutte, W., Greenberg, J. M., Ferris, J. P., Briggs, R., Connor, S., van de Bult, C. P. E. M., & Baas, F. 1985, *Origins of Life and Evolution of Biosphere*, 16, 21

Andersson, S. & van Dishoeck, E. F. 2008, *A&A*, 491, 907

Arasa, C., Andersson, S., Cuppen, H. M., van Dishoeck, E. F., & Kroes, G. 2010, *JCP*, 132, 184510

Baratta, G. A., Leto, G., & Palumbo, M. E. 2002, *A&A*, 384, 343

Bennett, C. J., Jamieson, C. S., Mebel, A. M., & Kaiser, R. I. 2004, *PCCP*, 6, 735

Bennett, C. J., Jamieson, C. S., & Kaiser, R. I. 2010, *PCCP*, 12, 4032

Bergin, E. A., Ciardi, D. R., Lada, C. J., Alves, J., & Lada, E. A. 2001, *ApJ*, 557, 209

Bergman, P., Parise, B., Liseau, R., Larsson, B., Olofsson, H., Menten, K. M., & Güsten, R. 2011, *A&A* (Letters), 531,L8

Bernstein, M. P., Dworkin, J. P., Sandford, S. A., Cooper, G. W., & Allamandola, L. J. 2002, *Nature*, 416, 401

Boogert, A. C. A., Pontoppidan, K. M., & Knez, C. *et al.* 2008, *ApJ*, 678, 985

Bottinelli, S., Ceccarelli, C., Williams, J. P., & Lefloch, B. 2008, *A&A*, 463, 601

Bottinelli, S., Boogert, A. C. A., & Bouwman, J. *et al.* 2010, *ApJ*, 718, 1100

Bouwman, J., Paardekooper, D. M., Cuppen, H. M., Linnartz, H., & Allamandola, L. J. 2009, *ApJ*, 700, 56

Bouwman, J., Cuppen, H. M., Bakker, A., Allamandola, L. J., & Linnartz, H. 2010a, *A&A*, 511, 33

Bouwman, J. 2010b, *Spectroscopy and chemistry of interstellar ice analogues*, Ph.D. Thesis, Leiden Observatory, Leiden University, P.O. Box 9513, 2300 RA Leiden, The Netherlands

Bouwman, J., Mattioda, A. L., Linnartz, H., & Allamandola, L. J. 2011a, *A&A*, 525, 93

Bouwman, J., Cuppen, H. M., Steglich, M., Allamandola, L. J., & Linnartz, H. 2011b, *A&A*, 529, 46

van Broekhuizen, F. A., Pontoppidan, K. M., Fraser, H. J., & van Dishoeck, E. F. 2005, *A&A*, 441, 249

Cazaux, S., Cobut, V., Marseille, M., Spaans, M., & Caselli, P. 2010, *A&A*, 522, A74

Charnley, S. B., Tielens, A. G. G. M., & Millar, T. J. 1992, *ApJ* (Letters), 399, L71

Charnley, S. B., Tielens, A. G. G. M., & Rodgers, S. D. 1997, *ApJ* (Letters), 482, L203

Charnley, S. B., Rodgers, S. D., & Ehrenfreund, P. 2001, *A&A*, 378, 1024

Charnley, S. B. & Rodgers, S. D. 2005, in: D.C. Lis, G.A. Blake, E. Herbst (eds.), *Astrochemistry: Recent Successes and Current Challenges*, Proc. IAU Symposium No. 231 (Pacific Grove, California, USA), p. 237

Cuppen, H. M., van Dishoeck, E. F., Herbst, E., & Tielens, A. G. G. M. 2009, *A&A*, 508, 275

Cuppen, H. M., Ioppolo, S., Romanzin, C., & Linnartz, H. 2010, *PCCP*, 12, 12077

Cuppen, H. M., Pendeado, E. M., Isokoski, K. M., van der Marel, N., & Linnartz, H., 2011, *MNRAS*, in press

Cuppen, H. M. & Herbst, E. 2007, *ApJ*, 668, 294

Dulieu, F., Amiaud, L., Congiu, E., Fillion, J., Matar, E., Momeni, A., Pirronello, V., & Lemaire, J. L. 2010, *A&A*, 512, A30

Ehrenfreund, P., Bernstein, M. P., Dworkin, J. P., Sandford, S. A., & Allamandola, L. J. 2001, *ApJ* (Letters), 550, L95

Fayolle, E. C., Öberg, K. I., Cuppen, H. M., Visser, R., & Linnartz, H. 2011a, *A&A*, 529, A73

Fayolle, E. C., Bertin, M., Romanzin, C., Michaut, X., Öberg, K. I., Linnartz, H., & Fillion, J. H. 2011b, *ApJ* (Letters), in press

Fraser, H. J., Collings, M. P., Dever, J. W., & McCoustra, M. R. S., 2004, *MNRAS*, 353, 59

Fuchs, G. W., Acharyya, K., & Bisschop, S. E. *et al.* 2006, *FaDi*, 133, 331

Fuchs, G. W., Cuppen, H. M., Ioppolo, S., Bisschop, S. E., Andersson, S., van Dishoeck, E. F., & Linnartz, H. 2009, *A&A*, 505, 629

Garrod, R., Park, I. H., Caselli, P., & Herbst, E. 2006, *FaDi*, 133, 5

Geppert, W. D., Thomas, R. D., & Ehlerding, A., *et al.* 2005, *JPhCS*, 4, 26

Gibb, E. L., Whittet, D. C. B. & Schutte, W. A. *et al.* 2000, *ApJ*, 536, 347

Gibb, E. L., Whittet, D. C. B., Boogert, A. C. A., & Tielens, A. G. G. M. 2004, *ApJ* (Suppl.), 151, 35

Goumans, T. P. M., Uppal, M. A., & Brown, W. A. 2008, *MNRAS*, 384, 1158

Gudipati, M. S. & Allamandola, L. J. 2003, *ApJ* (Letters), 596, L195

Gudipati, M. S. & Allamandola, L. J. 2006, *ApJ*, 638, 286

Hagen, W., Allamandola, L. J., & Greenberg, J. M. 1979, *Ap&SS*, 65, 215

Hiraoka, K., Sato, T., Sato, S., Sogoshi, N., Yokoyama, T., Takashima, H., & Kitagawa, S. 2002, *ApJ*, 577, 265

Hudgins, D. M., Sandford, S. A., Allamandola, L. J., & Tielens, A. G. G. M. 1993, *ApJSS*, 86, 713

Ioppolo, S., Cuppen, H. M., Romanzin, C., van Dishoeck, E. F., & Linnartz, H. 2008, *ApJ*, 686, 1474

Ioppolo, S., Palumbo, M. E., Baratta, G. A., & Mennella, V. 2009, *A&A*, 493, 1017

Ioppolo, S., Cuppen, H. M., Romanzin, C., van Dishoeck, E. F., & Linnartz, H. 2010a, *PCCP*, 12, 12065

Ioppolo, S. 2011d, *Surface formation routes of interstellar molecules - A laboratory study*, Ph.D. Thesis, Leiden Observatory, Leiden University, P.O. Box 9513, 2300 RA Leiden, The Netherlands

Ioppolo, S., Cuppen, H. M., van Dishoeck, E. F., & Linnartz, H. 2011a, *MNRAS*, 410, 1089

Ioppolo, S., van Boheemen, Y., Cuppen, H. M., van Dishoeck, E. F., & Linnartz, H. 2011b, *MNRAS*, 413, 2281

Ioppolo, S., Cuppen, H. M., & Linnartz, H. 2011c, *Rend. Fis. Acc. Lincei*, DOI 10.1007/s12210-011-0135-3

Matar, E., Congiu, E., Dulieu, F., Momeni, A., & Lemaire, J. L. 2008, *A&A*, 492, 17

Mennella, V., Palumbo, M. E., & Baratta, G. A. 2004, *ApJ*, 615, 1073

Miyauchi, N., Hidaka, H., Chigai, T., Nagaoka, A., Watanabe, N., & Kouchi, A. 2008, *CPL*, 456, 27

Mokrane, H., Chaabouni, H., Accolla, M., Congiu, E., Dulieu, F., Chehrouri, M., & Lemaire, J. L. 2009, *ApJ* (Letters), 705, 195

Muñoz Caro, G. M., Meierhenrich, U. J., & Schutte, W. A. *et al.* 2002, *Nature*, 416, 403

Oba, Y., Miyauchi, N., Hidaka, H., Chigai, T., Watanabe, N., & Kouchi, A. 2009, *ApJ*, 701, 464

Öberg, K. I., Fuchs, G. W., Awad, Z., Fraser, H. J., Schlemmer, S., van Dishoeck, E. F., & Linnartz, H. 2007, *ApJ* (Letters), 662, L23

Öberg, K. I., Boogert, A. C. A., Pontoppidan, K. M., Blake, G. A., Evans, N. J., Lahuis, F., & van Dishoeck, E. F. 2008, *ApJ*, 678, 1032

Öberg, K. I., van Dishoeck, E. F., & Linnartz, H. 2009a, *A&A*, 496, 281

Öberg, K. I., Linnartz, H., Visser, R., & van Dishoeck, E. F. 2009b, *ApJ*, 693, 1209

Öberg, K. I., Garrod, R. T., van Dishoeck, E. F., & Linnartz, H. 2009c, *A&A*, 504, 891

Öberg, K. I. 2009d, *Complex processes in simple ices: laboratory and observational studies of gas-grain interactions during star formation*, Ph.D. Thesis, Leiden Observatory, Leiden University, P.O. Box 9513, 2300 RA Leiden, The Netherlands

Öberg, K. I., Fayolle, E. C., Cuppen, H. M., van Dishoeck, E. F., & Linnartz, H. 2009e, *A&A*, 505, 183

Öberg, K. I., van Dishoeck, E. F., Linnartz, H., & Andersson, S. 2010, *ApJ*, 718, 832

Öberg, K. I., Boogert, A. C. A., Pontoppidan, K. M., van den Broek, S., van Dishoeck, E. F., Bottinelli, S. Blake, G. A., & Evans, II, N. J. 2011, *preprint (astro-ph/1107.5825)*,

Palumbo, M. E. 2006, *A&A*, 453, 903

Pontoppidan, K. M., Fraser, H. J., Dartois, E., Thi, W.-F., van Dishoeck, E. F., Boogert, A. C. A., d'Hendecourt, L., Tielens, A. G. G. M., & Bisschop, S. E. 2003, *A&A*, 408, 981

Pontoppidan, K. M., Boogert, A. C. A., & Fraser, H. J. *et al.* 2008, *ApJ*, 678, 1005

Romanzin, C., Ioppolo, S., Cuppen, H. M., van Dishoeck, E. F., & Linnartz, H. 2011, *JCP*, 134, 084504

Strazzulla, G., Baratta, G. A., & Palumbo, M. E. 2001, *AcSpe*, 57, 825

Tielens, A. G. G. M. & Hagen, W. 1982, *A&A*, 114, 245

Tielens, A. G. G. M., Tokunaga, A. T., Geballe, T. R., & Baas, F. 1991, *ApJ*, 381, 181

Wakelam, V., Herbst, E., & Selsis, F. 2006, *A&A*, 451, 551

Wakelam, V., Smith, I. W. M., & Herbst, E. *et al.* 2010, *Space Sci. Revs*, 156, 13

Watanabe, N., Nagaoka, A., Shiraki, T., & Kouchi, A. 2004, *ApJ*, 616, 638

Watanabe, N. & Kouchi, A. 2002, *ApJ* (Letters), 571, L173

Westley, M. S., Baragiola, R. A., Johnson, R. E., & Baratta, G. A. 1995, *Nature*, 373, 405

Zasowski, G., Kemper, F., Watson, D. M., Furlan, E., Bohac, C. J., Hull, C., & Green, J. D. 2009, *ApJ*, 694, 459

Discussion

SMITH: I have two questions relating to the activation energies/barriers that you cited. 1) Are all degrees of freedom thermalized at the same temperature in these experiments? 2) Are the values of these barriers dependent on the nature of the ice and/or the way in which it is prepared?

LINNARTZ: 1) Yes. The best proof of this is that the H atoms being formed in a thermal cracking source, and subsequently thermalized to room temperature through collisions, instantaneously thermalize upon surface impact. Details of this process are available from Fuchs *et al.* (2009). 2) Yes. The values depend on ice morphology (e.g., amorphous vs. crystalline) but the values should not depend on the way the ice has been prepared. This is also confirmed by reproducible results for identical ice systems from different laboratories (i.e., different experimental setups).

OBA: Some reactions energetically unfavourable are included in the reaction network to the formation of H_2O, e.g., $HO_2 + H_2 \rightarrow H_2O_2$ (endothermic with huge barrier). What is the energy source to promote the reactions (in laboratory)?

LINNARTZ: A detailed and extensive answer to this question can be found in "Water formation by surface O_3 hydrogenation", C. Romanzin, S. Ioppolo, H. M. Cuppen, E. F. van Dishoeck, H. Linnartz, J. Chem. Phys. 134 (2011) 084504, in section IVc, where these reactions Eqs. 7 and 8) are discussed. See also Figure 5.

The Molecular Universe
Proceedings IAU Symposium No. 280, 2011
José Cernicharo & Rafael Bachiller, eds.

doi:10.1017/S1743921311025154

Water Ice Formation and the o/p Ratio

François Dulieu

LERMA, UMR 8112 du CNRS
Observatoire de Paris et Université de Cergy Pontoise
5, mail Gay Lussac, 95000 Cergy Pontoise, France
email: francois.dulieu@obspm.fr

Abstract. Water is observed in many astrophysical environments in both gas and solid phase. Water ice, for its specific properties, is probably the most important template that structures the gas-solid interaction. In cold environments, its synthesis is supposed to occur directly in the solid phase and then water acts as a catalytic matrix for subsequent synthesis of various molecules. When the medium begins to warm again, water sublimates and nourishes the gas phase, as occurs for example in comets or in star forming regions. Over the last four years, water formation on cold surfaces has been studied experimentally. Different precursors (O, O_2, O_3...) have been used to understand the complex mechanisms that take place. Although numerous questions remain unanswered, at present, it is clear that water is easily formed by different pathways, and that the ice formed has an amorphous structure. The recent observations of the ortho/para ratio of water with Herschel satellite have similarities with the previous o/p ratio observations of water in comets. Some experimental work have been recently reported in this domain, mostly rare gas matrix studies where nuclear spin conversion is measured even at 4.2 K. H_2 molecules adsorbed on amorphous solid water ice also exhibit a nuclear spin conversion in presence of a tiny fraction of O_2. Finally, I will discuss if microphysics properties of water desorption can explain the o/p ratio values observed.

Keywords. astrochemistry, molecular processes, methods: laboratory, ISM: molecules

1. Introduction

Hydrogen and oxygen are the two most abundant reactive atoms. H_2 is the most abundant molecule in the universe and is mainly present in the gas phase, even if the solid phase chemistry is thought to be crucial for its formation. Also, adsorbed H_2 may play an important role in the deuteration of species (Kristensen *et al.* (2011)).

Water is the most abundant molecule in the solid phase. It has been detected in gaseous or solid form in numerous astrophysical environments such as planets, comets, interstellar clouds and star forming regions (Ehrenfreund *et al.* (2003), Dartois (2005), Tielens (2005)). Amorphous water ice was directly detected in dark interstellar clouds through infra-red absorption bands Leger *et al.* (1979).

Water has the property to create H-bonds, and this has several kinds of consequences. The first consequence is that solid water ice has a high energy of sublimation (46.9 kJ per mole) (Speedy *et al.* (1996), Fraser *et al.* (2001)). This is why this molecule is often in its solid form, because the range of conditions where solid water ice can exist is wider than for the majority of other small molecules detected in space. The second main implication is that water shapes the morphology and geometry of the solid phase because it is the main compound of icy mantles and has rather strong interactions with the adsorbates. The morphology of the solid phase has important effect on gas-grain interactions, i.e, by changing the binding properties of adsorbed molecules (Fillion *et al.* (2009)), by changing the kinetics of reactions like in the case of H_2 formation (Hornekaer *et al.* (2003), Congiu *et al.* (2009)). Moreover, the gas-grain interaction is only the first step of molecular synthesis that occurs in solid phase. In addition, surface chemistry is of course influenced by the morphology of the ice, for example in the case of D diffusion in function of the thickness of the porous film (Matar *et al.* (2008)), or in the case of hydrogenation of CO (Hidaka *et al.* (2007)).

There is a wealth of models that predict that water should be synthesized on the surface of dust grains (e.g., Cuppen & Herbst (2007) and references therein). The first part of the present paper focuses mainly on experimental studies of the water formation itself on cold solid substrates . The neutral gas phase route is known have a low efficiency because the $OH + H \longrightarrow H_2O$ has an activation barrier. The main difference between the gas phase and the solid phase is that typical collision timescale in gas phase can be a day, whereas in solid phase, if the reactants are nearby, the frequency of attempts is typically of the order of the picosecond. Therefore, the main question in solid state is "Are the reactants nearby?" rather than "What is the height of the activation barrier of the reaction?". The diffusion properties, densities on the surface and the residence time compared to accretion time are probably of higher importance than the height of activation barriers. If the reaction can proceed in a time shorter than the timescale of the change in the environmental configuration of either reactant, the reaction occurs, no matter how short this time may be (picosecond, nanosecond, millisecond...).

The varieties of the surface conditions (surface densities of the reactants, temperature, fluxes, composition and morphology) studied in the recent experimental reports tend to make the comparison between different experimental results very hard and of difficult interpretation for the non-experimental astrochemists. The aim of the first part of this article is to show that some important and apparent trends can be drawn. One of them is that the water (and hydrogen peroxide) can be easily formed from any of the oxygen precursor deposited on a cold (10-20 K) substrate exposed to H atoms, and the that first step of hydrogenation is barrierless. The second important fact is that the water formed by surface reactions is amorphous and compact.

The second part of this article is devoted to the question of ortho-to-para ratio of molecules in connection with the solid state. Molecules having two protons that can be exchanged like in H_2, H_2O, H_2CO... have different spin isomers. The change of one isomer to the other requires a flip of nuclear spin. The nuclear spin conversion (NSC) may happen if interactions, such as for example strong magnetic gradients, are present. Nuclear spin conversions are also favored if the isomer states are energetically close. For a molecule in gas phase, it is believed that the nuclear spin conversion is not efficient because the interaction (collision) time between two molecules is too short. Of course, for the total population of the molecules, nuclear spin conversion can happen especially in the case where molecules are formed and destroyed. Accretion to solid phase and successive desorption to gas phase are processes that can drive, at least partly, the ortho-to-para ratio observed in different astrophysical environments. In the case of comets, where desorption plays an obvious role, the opr of water has been found to be between 1.8 and 3, with a maximum around 2.4, corresponding spin temperature of around 30 K (Bonev *et al.* (2007), Bockelée-Morvan *et al.* (2009)). This value does not correspond to the kinetic temperature, and it is also not the case for H_2, in shock regions (Neufeld *et al.* (2006)).

The Herschel satellite provides new input for interstellar medium regions Lis *et al.* (2010), Melnick *et al.* (2010)) and also for D_2O opr (Vastel *et al.* (2010)). The question of the origin of the observed diversity is nowadays examined as never before.

The aim of this paper is to present the recent progress made in the laboratory or theoretical field, and to try to see if the microphysics of the gas-grain interaction can have an effect on the opr ratio observed. I will first discuss the case of H_2 adsorbed on the surface of solid water ice, and show that a tiny fraction of paramagnetic molecules such as O_2 is able to produce NSC in a time measurable in the laboratory. I will present some results obtained in noble gas matrix that demonstrate the dependence of the NSC on the concentration of H_2O, suggesting that the NSC is accelerated with water-water interaction. I will recall the theoretical results obtained by Buntkowsky *et al.* (2008), that show that the interaction of 2 water molecules should produce NSC in the millisecond timescale. I will finally discuss the possibility that selective desorption of water in function of the spin isomer can produce a disequilibrium in the opr of the nascent molecules detected in the gas phase.

2. Experimental studies of water formation

During the last 5 years, water formation on cold surfaces has been studied experimentally. In 2006 our group presented the first successful attempt to synthesize water molecules on a cold substrate starting from oxygen and hydrogen. This experiment was carried out on a water substrate held at 10 K, mimicking thus the growth of water mantles (Dulieu *et al.* (2006)). However, the chosen system was by far the most complicated to study even if the qualitative results were clear, namely, production of water. (Dulieu *et al.* (2010)). The first complication was provided by the water template, which, though perfectly relevant to astrophysical purposes, increased greatly the difficulty of the analysis. The water substrate makes the analysis by infrared spectroscopy very hard, and the isotope exchange ($H_2O + D_2O \rightleftharpoons 2\ HDO$) occurring during the desorption of water makes the Temperature Programmed Desorption (TPD) analysis much more complex. The second intrinsic difficulty is that oxygen atoms were used as reactants together with D atoms. Under O and H irradiation of a cold surface, O atoms can as well react with themselves so the O+H, O_2+H and O_3+H pathways are simultaneously open. Therefore, the reactions network leading to the observed production of water is of maximal complexity. The first column of Table 1, presents 16 of the 18 reactions proposed as active in the formation of water in solid phase, starting from pure O or H species. Beside the complexity of this first attempt, the conclusions of this study are that oxygen atoms are included efficiently in the water formed, although it was not possible to determine unambiguously which chemical route led to water and its exact efficiency.

One claim of this work was that the O_2+H reaction was very efficient and it has been rapidly explicitly demonstrated by two independent group in Japan (Miyauchi *et al.* (2008)) and in the Netherlands (Ioppolo *et al.* (2008)). In contrast with the previous study, the most abundant compound formed was not water, but H_2O_2. In these two studies a thick O_2 layer was exposed to H (or D) atoms and the water formation was identified following the formation of hydrogen peroxide. Later, in other experiments where several H/O_2 ratios were investigated, water was found to become the major product as soon as the H/O_2 became very large (Oba *et al.* (2009)).

Ozone too was been used as a precursor at submonolayers coverages on a water substrate on which O_3, exposed to H, mostly leads to the formation of water (Mokrane *et al.* (2009)). In the case of a O_3 ice used as a substrate, water and also hydrogen peroxide are formed (Romanzin *et al.* (2011)). As a last part of a complex experimental puzzle, infrared spectroscopy allowed to isolate some reaction intermediates such as O_2H and OH, and to propose a complex chemical network (Ioppolo *et al.* (2010), Cuppen *et al.* (2010)).

One common result of all these studies is that water is always produced starting from O, O_2 or O_3 by addition of H atoms, regardless of the substrate. However, some obvious and important discrepancies appear such the H_2O_2 /H_2O ratio. These discrepancies are

Table 1. Proposed reactions for water formation on cold solid state from oxygen and hydrogen. Opinion of (○) H. Linnartz (Leiden Observatory), (∗) N. Watanabe (Okkaido University) and (†) F. Dulieu (Paris Observatory). ***: Certain, ** : Presumption, * : Opinion

Reaction	#	No Barrier	Barrier	No reaction
$O + O \longrightarrow O_2$	1	∗∗ ††		
$O_2 + O \longrightarrow O_3$	2	∗∗∗	†† ○○	
$O + H \longrightarrow OH$	3	∗∗ †† ○		
$O_2 + H \longrightarrow HO_2$	4	∗∗∗ †† ○○○		
$O_3 + H \longrightarrow O_2 + OH$	5	∗∗ ††† ○○○		
$HO_2 + H \longrightarrow H_2O_2$	6	∗∗∗ †○○○		
$HO_2 + H \longrightarrow OH + OH$	7	∗†○○○		
$HO_2 + H \longrightarrow H_2O + O$	8	†○○○		
$OH + H \longrightarrow H_2O$	9	∗∗†† ○○		
$OH + OH \longrightarrow H_2O_2$	12	∗∗∗† ○○○		
$OH + OH \longrightarrow H_2O + O$	13	∗∗∗†	†	
$H_2O_2 + H \longrightarrow H_2O + OH$	14		∗∗∗††○○○	
$O + H_2 \longrightarrow H_2O$	15			∗∗∗††
$O_2 + H_2 \longrightarrow H_2O + OH$	16			∗∗∗†††○○○
$OH + H_2 \longrightarrow H_2O + H$	17		∗∗○○	††
$O_3 + H_2 \longrightarrow H_2O + O_2$	18			††† ○○○

probably due to different experimental conditions, and most of all the accessibility to the reactants (surface density, mobility, diffusion...) which is of major importance as explained earlier.

In order to show that the experimental results present more similarities than differences, I asked the two leaders of the other group their opinion about the chemical reactions that occur on a cold surface at 10 K. Table 1 presents 16 reactions that can occur on a cold substrate in the water formation network. The reactions may have no noticeable barrier and appear to be very efficient. It is for example the case of O_3+H which occurs even at 50 K, a temperature at which the residence time of H on the substrate is extremely short. Reactions can have a barrier and therefore, some competing mechanism can prevent the reaction to occur or tend to minimize its role, or a certain product may take more time to develop with respect to others. No reaction, in Table 1, means that under these experimental conditions it is not possible to detect any effect such as the case of $O_2 + H_2$.

Table 1 shows that there is a large consensus on most of the reactions. For those reactions involving only oxygen (#1 and #2) there is still a question about the presence or not of an activation barrier in the formation of ozone by addition of O and O_2. On the contrary, it is clear that O, O_2 and O_3 undergo hydrogenation very efficiently.

The two reaction intermediates OH and O_2H promptly react with H, but also together, and basically will form H_2O or H_2O_2 at the end of the chain, but the branching ratios are still an open question.

H_2O_2 can react with H, but this reaction is slower than any oxygen-hydrogen reaction (typically one order of magnitude slower). This can be attributed to a barrier. However, as H_2O_2 is thought to react only with H, whenever it is accessible to H atoms, it will end up in water formation. This is mainly the reason why water appears always to be the major final compound in case of sub-monolayer reactions, that is when complete hydrogenation can be achieved.

Finally H_2, which is always present in our experimental conditions, as well as on icy dust grains in dark cloud, does not participate in the water formation except maybe for the $OH+H_2$ reaction. Yet, our group has never succeeded to demonstrate this possible reaction in a direct way. It is, however, one of the key reaction that should regulate the deuterium enrichment of water (Tielens *et al.* (1983), Kristensen *et al.* (2011)).

Another important aspect of the experimental study of water formation is that the morphology of the ice produced has been determined to be compact (Oba *et al.* (2009), Accolla thesis (2010)), and even if porosity can arise from the adsorption of gaseous water on cold dust grains, the molecular hydrogen formation is able to reduce dramatically the porosity of the ice. This is likely to be due to the fraction of the formation energy released which acts as a local heating of the ice network, and thus reduces the porous nature of water ice (Accolla *et al.* (2011)).

The experimental study of water formation has been extensively studied over the last 5 years, and has demonstrated that the water formation on dust grains is plausible. Thanks to the variety of precursors and of experimental conditions, the chemical networks involved have been determined even if some open questions still remain.

There are also some challenging matters like the determination of the fraction of compounds that could desorb directly during the formation phase, due to the large energy released. The study of the first layer of water on silicates (see poster by Vidali *et al.*) or carbonaceous material has to be investigated and probably the fraction of ejected molecules upon formation is even more important in such a context.

3. Nuclear spin conversion on cold surfaces

Because their protons have two interchangeable identical positions, molecules like H_2, H_2O or H_2CO have spin isomers. The spin isomers are sometimes called by extension ortho and para molecules, according to their nuclear spin states. The total wave function of such molecules shall be antisymmetric, so for a given nuclear spin parity exists a ro-vibrational wave function of opposite parity. For the case of H_2, in the v =0 state, the odd J rotational numbers correspond to ortho population (I=1, triplet) and the even J states correspond to the para population (I=0, singlet).

If such a molecule is perfectly isolated and we consider that the nuclear spin has no interaction with the rest of the molecule, the nuclear spin cannot flip and therefore the conversion between the two spin isomers is not possible. This is why one can consider this spin isomers as two independent population of molecules.

One of the new capabilities opened by the Herschel satellite is to be able to measure, in some environments, the ortho-to-para (opr) ratio of water (e.g. Lis *et al.* (2010)). In different astrophysical context, especially in comets, it is already well established that the ratio between the two spin isomer populations can define a spin temperature that does not correspond to the kinetic temperature (Croviser (1998)). At high temperature ($T>300$K) the opr is equal to the statistical weight of the nuclear spin degeneracy $\simeq$3 (3/4 for ortho molecules and 1/4 for para molecules).

Let us consider the case of H_2. Figure 1 shows the distribution of the population considering two extreme cases of no nuclear spin conversion, hence two separate populations (right panel) and a fast spin conversion which allows the two spin isomers to be considered as a unique population (central panel). The two cases are presented for two different temperatures. At 30 K (and below) the molecules are in the lowest possible rotational states. If conversion is possible, all the molecules are in the J=0 state. In the case of two

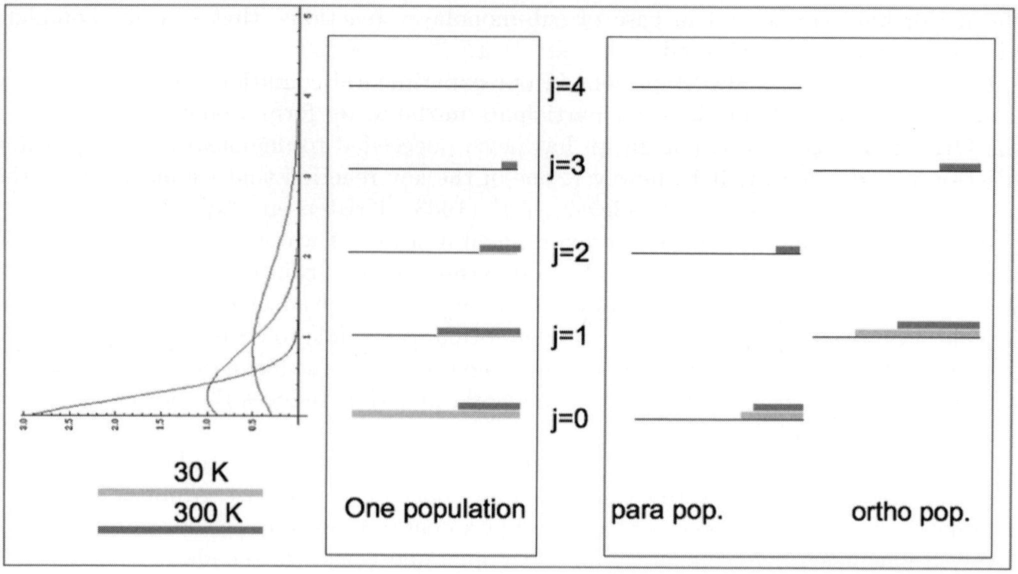

Figure 1. Relative population of H_2 molecules for a gas at 30K (blue bars), and 300 K (orange bars) of the v=0 state in function of the J rotational states. The left panel represents the normalized continuous Boltzmann distribution (at 30K, 100 K and 300K) as a guide for the eye. The central panel represents the case of a unique spin isomer population (obtained if NSC is very efficient), and the right panel corresponds to two different isomer populations populated with a opr =3.

separate populations, 1/4 of the molecules is in the J=0 state, and 3/4 is in the $J = 1$ state which is the lowest state of para molecules. At 300 K, the molecules are spread over the 4 first levels. Under either assumption on nuclear spin conversion (NSC), opr is equal to 3. In the case of a unique population, we can notice that at very low temperature opr=0, and when the temperature increases the opr grows to reach the asymptotic value of 3 at T>250K. If the opr is comprised in the [0,3] range, it corresponds to a spin temperature. In numerous astrophysical observations, this spin temperature does not correspond to the kinetic temperature.

On water ice surfaces, H_2 desorbs below 30 K (Amiaud *et al.* (2006) Fillion *et al.* (2009)). Let's then consider the simple experiment consisting in adsorbing H_2 at 10 K and heating the water ice sample until H_2 desorbs (the so called Temperature Programmed Desorption). If nuclear spin conversion happens very rapidly at 10 K, all the molecules should desorb in the $J = 0$ rotational state. If nuclear spin conversion does not occur, therefore 3/4 of the molecules should desorb in the $J = 1$ state and 1/4 in the $J = 0$. Using a tunable laser, it is possible to detect selectively a quantum state (Selected Quantum State TPD). In the case of H_2 (or D_2) it is clear that the second scenario, i.e. no NSC, corresponds to the experiments. We have measured that molecules at 300K (having an opr=3), deposited at 10 K, desorb below 30 K and have a population of both $J = 0$ and $J = 1$ states.

We have also noticed that the dynamic of the desorption is slightly different because of an energy difference in the adsorption energy of ortho and para states (Amiaud *et al.* (2008)). For D_2 this difference has been estimated to be 1.4 ± 0.3 meV. We must note that the $J = 1$ molecules are more bound to the surface than the $J = 0$ ones, which is a little counterintuitive, because one could think that rotation may favor the desorption process. This underlines the fact that the D_2 molecules can rotate almost freely on the water ice substrate. The case of H_2 adsorbed on a water surface is very similar to a gas phase problem, in the sense that the ro-vibrational wave function is clearly defined and therefore the spin isomers keep fully their meaning. The great difference with the gas phase is that the time of interaction with the local surroundings can be very long.

We have, therefore, performed another set of experiments, basically the same as above, but where we varied the time (from one minute to one hour) between the adsorption and the desorption phase for H_2 and D_2 molecules to see if some conversion happens during the wait time. We were able to detect a NSC in the case where we added O_2 molecules on the substrate (Chehrouri *et al.* (2011)). We observed a typical NSC time of one hour in the case where 3/10000 adsorption sites where occupied by O_2 molecules and an occupation of >1/6 of D_2 molecules. The NSC efficiency increased with the O_2 coverage and with the D_2 coverage. The conclusion of this study is that even trace quantities of O_2 (or other paramagnetic molecules) present on the surface is able to induce NSC at the astrophysical time scale.

Infra red spectroscopy allows to detect ortho and para states of water imbedded in a rare gas matrix. Even at 4.2 K, NSC occurs (Pardanaud *et al.* (2008)), but in this case with a typical timescale of one day. Particularly, the NSC increases with the concentration of water molecules demonstrating that even at long range, water-water interaction is sufficient to induce the conversion. Once again, the case of rare gas matrix is not exactly the case of water in solid phase, because the water molecule in the rare gas cage can rotate with small constraints. The nature of the rare gas case affects the NSC timescale (Pardanaud thesis (2007)). The ortho and para definition of molecules in these systems is therefore close to the one in gas phase.

But in solid phase, which means in the case of water in solid ice, water molecules form H bonds, and the rotation is widely impeded. Buntkowsky *et al.* (2008) have studied the case of two water molecules from a theoretical point of view. They have shown that the frustration of the rotation reduces considerably the difference in energy of the rotational levels. This implies that NSC in water ice should happen in the millisecond range.

Considering that the NSC on solid water ice is accelerated by the presence of paramagnetic species, or simply by the frustration of rotation, it is reasonable to think that for one molecule the NSC is faster that the desorption process in an astrophysical context. Figure 2 represents a schematic view of water desorption taking into account the ortho and para water molecules.

We have proposed a simple model of water desorption that includes the possibility of spin conversion in solid phase, and takes into account a differential desorption of the ortho and para molecules and no efficient spin conversion in the gas phase Dulieu *et al.* in prep. We assume a rapid (compared to desorption) NSC in the solid phase, hence one population can be transferred to another, if one of the two populations evaporates more efficiently. If we consider that the rotational energy difference in the gas phase induces differences in the desorption energy, therefore one can see that the para molecules (which include the 000 state) have on average a lower desorption energy and as a consequence a higher desorption flux. We have estimated that in the 100-120 K range, the opr ~ 2.4 just after desorption. Para molecules have a higher desorption probability, and that their reduction in number (due to the desorption) is compensated by rapid NSC in solid phase. This is why the selective desorption favors the para population and reduce apparently the spin temperature. The conclusion is that the desorption process could be one of the reason why the observed spin temperature does not correspond to the kinetic temperature. New experiments are required to validate this proposition.

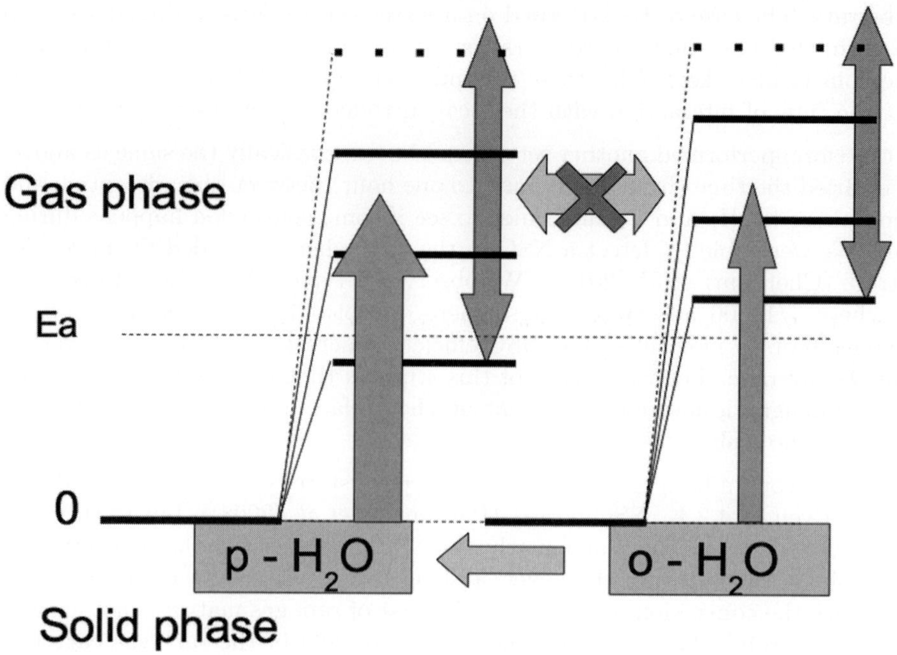

Figure 2. Schematic representation of the model of selective desorption. Large gray arrow: Nuclear Spin conversion; Green arrows: desorption; Salmon arrow: thermal equilibrium induced by gas phase collisions.

4. Conclusion

Numerous experimental studies during the last years have demonstrated that the water formation on cold solid surface is a efficient mechanism, mainly because no activation barrier has been measured for the first step of hydrogenation of any pure oxygen molecules (O, O_2, O_3). The reaction network is, however, complex but a rather good accord exists in the community. Nevertheless, some very interesting questions should be addressed in the next years. Among these questions, the fraction of water molecules that can directly go into the gas phase upon their formation will be probably one of the most stimulating.

Also the ortho and para topic in solid state has made some interesting progress, especially for the H_2/H_2O system and for rare gas matrix experiments. There is for the moment no experimental evidence of how the spin isomers evolve when the solid-gas phase transition occurs. However, we have proposed a simple explanation why spin temperature and kinetic temperature should not be the same after a simple thermal desorption.

Acknowledgement

I would like to thanks Dr Michaut and Pr Fillion (Université Pierre et Marie Curie), Pr Linnartz (Leiden Observatory) and Pr Watanabe (Okkaido University) for our exchanges before the conference, and Dr Congiu for his help after the conference.

References

Mario Accolla PHD thesis, 2010.

Accolla, M., Congiu, E., Dulieu, F., Manicò, G., Chaabouni, H., Matar, E., Mokrane, H., Lemaire, J. L., & Pirronello, V., 2011, *Phys. Chem. Chem. Phys.*, 13 (17), 8037 - 8045

Amiaud, L., Fillion, J. H., Baouche, S., Dulieu, F., Momeni, A., & Lemaire, J. L. 2006, *J. Chem. Phys.*, 124, 094702

Amiaud, L., Momeni, A., Dulieu, F., Fillion, J. H., Matar, E., & Lemaire, J. L., 2008, *Phys. Rev. Lett.*, 100, 056101.

Bockelée-Morvan D. Woodward, C. E., Kelley, M. S., & Wooden D. H., 2009, *ApJ* 696, 2737

Bonev B. P., Mumma, M. J., Villanueva, G. L., Disanti M. A. *et al.*, 2007, *ApJ*, 661, L97

Buntkowsky G., Limbach H.-H., Walaszek B., Adamczyk A., Xu Y., Breitzke H., Schweitzer A., Gutmann T., Warchtler M., Frydel J., Emmler Th., Amadeu N., Tietze D., Chaudret B., 2008, *Z. Phys. Chem.* 222 10491063

Cherhouri M., Fillion J-H, Chaabouni, H. and Mokrane *et al.Phys. Chem. Chem. Phys.*, 13, 2172

Congiu E., Matar E., Kristensen L. E., Dulieu F. and Lemaire J. L., 2009, *Mon. Not. R. Astron. Soc.* 397, L96

Cuppen, H. M. & Herbst, E. 2007, *ApJ*, 668, 294

Cuppen, H. M., Ioppolo, S., Romanzin, C., van Dishoeck, E. F., & Linnartz, H, 2010, *Phys. Chem. Chem. Phys.*, 12, 12077

Croviser, J., 1998, *Faraday discuss.* 109, 437.

Dartois, E. 2005, *Space Science Reviews*, 119, 293

Workshop: Molecular databases for Herschel, ALMA and SOFIA from 6-8 Dec 2006, Lorentz center, Leiden, The Netherlands.

Dulieu, F., Amiaud, L., Fillion, J.-H.., Matar, E., Momeni, A., Pirronello, V., & Lemaire, J. L. 2007, in Molecules in Space and Laboratory, Ed. J.L. Lemaire, F. Combes., p. 79

Dulieu, F., Amiaud, L., Congiu, E., Fillion, J.-H., Matar, E., Momeni, A., Pirronello, V., & Lemaire, J. L. 2010, *A&A*, 512, A30

Dulieu F., Congiu E., Fillion J.H., Michaut X. in preparation.

Ehrenfreund, P., Fraser, H. J., Blum, J., Cartwright, J. H. E., Garcìa-Ruiz, J. M., Hadamcik, E., Levasseur-Regourd, A. C., Price, S., Prodi, F., & Sarkissian, A. 2003, *Planetary and Space Science*, 51, 473

Fraser, H. J., Collings, M. P., MacCoustra M. R. S., & Williams, D. A., 2001, *Mon. Not. R. Astron. Soc.* 327, 11651172

Fillion, J.-H., Amiaud, L., Congiu, E., *et al.* 2009, *Phys. Chem. Chem. Phys.*, 11, 4396

Hidaka, H., Kouchi, A., & Watanabe, N., 2007, *J. Chem. Phys.*, 126, 204707

Hornekaer, L., Baurichter A ., Petrunin, V. V., Field. D. & Luntz, A., C., 2003 *Science*, 302, 1943

Hiraoka, K., Miyagoshi, T., Takayama, T., Yamamoto, K., & Kihara, Y. 1998, *ApJ*, 498, 710

Ioppolo, S., Cuppen, H. M., Romanzin, C., van Dishoeck, E. F., & Linnartz, H. 2008, *ApJ* 686 1474

Ioppolo, S., Cuppen, H. M., Romanzin, C., van Dishoeck, E. F., & Linnartz, H. 2010, *Phys. Chem. Chem. Phys.*, 12, 12065

Kristensen, L. E., Amiaud, L., Fillion, J.-H., Dulieu, F., & Lemaire, J.-L., 2011, *A&A* , 527, A44

Oba Y., Miyauchi N., Hidaka H., Chigai T., Watanabe N. and Kouchi A., Astrophys. J., 2009, 701, 464.

Lis D. C., Phillips, T. G., Goldsmith, F., & Neufeld, D. A., *et al.* 2010, *A&A*, 521, L26,

Leger, A., Klein, J., de Cheveigne, S., Guinet, C., Defourneau, D., & Belin, M. 1979, *A&A* 79, 256

Matar, E., Congiu, E., Dulieu, F., Momeni, A., & Lemaire, J. L. , 2008, *A&A*, 492, L17-L20

Mokrane H., Chaabouni H., Accolla M., Congiu E., Dulieu F., Chehrouri M., Lemaire J. L., 2009, *ApJ* ,705, L195

Melnick, G. J., Tolls, V., Neufeld D. A., Bergin E. A., *et al.*, 2010, *A&A*, 521, L27

Miyauchi, N., Hidaka, H., Chigai, T., Nagaoka, A., Watanabe, N., & Kouchi, A. 2008, *Chemical Physics Letters*, 456, 27

Neufeld, D. A., Melnick G. J., Sonnentrucker, P., Bergin, E. A. *et al.*, 2006, *ApJ*, 649, 816.

Cedric Pardanaud PHD thesis, 2007, http://tel.archives-ouvertes.fr/tel-00259961/fr/

Pardanaud, C., Vasserot, A.-M., Michaut, X., and Abouaf-Marguin, L., (2008), *Journal of Molecular Structure* 873, 181

Romanzin, Ioppolo, S., C., Cuppen, H. M., van Dishoeck, E. F., & Linnartz, H., 2011, *J. Chem. Phys*, 134, 084504

Speedy, R. J., Debenedetti, P. G., Scott Smith R., Huang, C., & Kay, B. D., 1996, *J. Chem. Phys.* 105, 240.

Tielens, A. G. G. M., Hagen, W., & Greenberg, J. M. 1983, *J. Phys. Chem.*, 87, 4220

Tielens, A. G. G. M. 2005, The Physics and Chemistry of the Interstellar Medium (Cambridge University Press)

van Dishoeck, E. F. 2004, *Annu. Rev. Astron. Astrophys.*, 42, 119

Vastel, C., Ceccarelli, C., Caux, E., Coutens, A., *et al.*, 2010, *A&A*, 521, L31

Discussion

L. HORNEKAER: Chairwoman

ERIC HERBST: Why does H_2 o/p conversion require O_2 whereas H_2O does not?

Changing the concentration of O_2 on the substrate is a way to control experimentally the o/p conversion time of H_2 and makes it measurable in an easier way under our experimental conditions. The O_2 present on the substrate probably also affects the o/p conversion of H_2O, but it has not been yet measured in ice. In absence of paramagnetic species, the difference between H_2 and H_2O adsorbed on the substrate is mostly due to the fact that H_2 can rotate almost freely at the surface whereas water form H bonds and therefore has very important constraints. In one case the rotational energy is almost unchanged, and in the other case, the rotational energy levels are almost degenerated, and this strongly favors the NSC.

NAOKI WATANABE: Why does the tiny addition of O_2 accelerate conversion of H_2? For the ratio H_2/O_2 10^{-5}, it is quite rare to be close to O_2 on the surface.

O_2 offers a magnetic gradient that provokes NSC. The conversion is complete in the case of D_2 which means that these molecules are mobile at 10 K and are able to adsorb (at least temporarily) close to one O_2 molecule. If molecules were not mobile only very little conversion would occur.

NAOKI WATANABE: Have you measured the surface-temperature dependence of o/p conversion of H_2?

No we haven't, but it is an excellent idea, not necessary easy to realize. If we increase the temperature, the adsorbed molecules can desorb during the wait time, so we cannot do the same experiments at a temperature higher than 12-14 K probably. By reducing the temperature, we shall at least partly reduce the mobility of D_2, except when it is already dominated by tunnel diffusion. We would maybe measure an increase in the NSC time at very low temperature.

EDWIN BERGIN: The o/p ratio of NH_3, CH_4 and H_2O seen in comets are different but reflect a common spin temperature. If thermal evaporation determines the o/p ratio, how could this be replicated ?

Though we did not calculated specifically the case of NH_3, I expect that it gives about the same kind of results than for water. The case of CH_4 is different because the evaporation temperature range is rather different, there is less rotational constraints, so I don't know what could happen.

DAVID NEUFELD: If I understood you correctly, the opr on the grain surface is always 3, regardless of the formation temperature, but the para-water vaporizes slightly more rapidly, leading to an opr of 2.4. But what opr is expected when water ice are released by sputtering or photodesorption?

You understood perfectly! We did not considered sputtering or photodesorption processes, so I'd rather not answer. I can just say that if the sputtering eject randomly a unique molecule, the opr should be close to 3, but if a small cluster of water ice is sputtered and thereafter it sublimates, then I expect more the value of 2.4, but this is just my opinion.

E. F. van Dishoek: Just a follow-up on the question of David Neufeld. Photodesorption can change the opr from that in the ice. There are two basic mechanisms for H_2O photodesorption (cf Andersson & van Dishoek 2008, Arasa *et al.* 2010, 2011):

i) H_2O dissociates in H + OH and then recombines and escapes

ii) the energetic H resulting from photodissociation of H_2O kicks out a neighboring water molecule.

In process i) the opr will change, in process ii) not. The relative importance of these processes varies with ice monolayer, but are both significant.

The Molecular Universe
Proceedings IAU Symposium No. 280, 2011
José Cernicharo & Rafael Bachiller, eds.

doi:10.1017/S1743921311025166

Recent Results of Solid-State Spectroscopy

Cornelia Jäger[2], Thomas Posch[3], Harald Mutschke[1], Simon Zeidler[1], Akemi Tamanai[4] and Bernard L. de Vries[5]

[1]Astrophysikalisches Institut, Friedrich-Schiller-Universität Jena, Schillergässchen 2-3, D-07745 Jena, Germany

[2]Laboratory Astrophysics and Cluster Physics Group, Max Planck Institute for Astronomy and Institute of Solid State Physics, Friedrich-Schiller-Universität Jena, Max-Wien-Platz 1, D-07745 Jena, Germany, Email: cornelia.jaeger@uni-jena.de

[3]Institut für Astronomie, Türkenschanzstraße 17, A-1180 Vienna, Austria

[4]Kirchhoff-Institut für Physik, Universität Heidelberg, Im Neuenheimer Feld 227, D-69120 Heidelberg, Germany

[5]Instituut voor Sterrenkunde, Katholieke Universiteit Leuven, Celestijnenlaan 200D, 3001 Leuven, Belgium

Abstract. Solid state spectroscopy continues to be an important source of information on the mineralogical composition and physical properties of dust grains both in space and on planetary surfaces. With only a few exceptions, artificially produced or natural terrestrial analog materials, rather than 'real' cosmic dust grains, are the subject of solid state astrophysics. The Jena laboratory has provided a large number of data sets characterizing the UV, optical and infrared properties of such cosmic dust analogs. The present paper highlights recent developments and results achieved in this context, focussing on 'non-standard conditions' such as very low temperatures, very high temperatures and very long wavelengths.

Keywords. methods: laboratory; ISM: dust, extinction; stars: AGB and post-AGB; circumstellar matter

1. Introduction

Solids are strangers in space. By far the largest part of the baryonic matter in the universe is in the gas or plasma state. Less than 1% of cosmic matter consists of solids, such as ices, dust grains or terrestrial planets. Yet, many of the molecules known from astrochemical investigations would not have formed without the catalytic role that the surfaces of solids play in various environments (e.g. in circumstellar shells, in the interstellar medium, protoplanetary disks, etc.).

Proper understanding of solids in space requires studies on analog materials in terrestrial laboratories. Here we shall focus on the *infrared spectra* of cosmic dust analogs (CDAs), which are a small, but crucial subset of CDA properties. An obvious goal of infrared spectroscopy of CDAs is to create databases of dielectric functions enabling us to identify cosmic dust species from their observed astronomical spectra. One such database is the Jena Database of Optical Constants of Solids:
`http://www.astro.uni-jena.de/Laboratory/Database/databases.html`.
For most of the solids discussed here, data files of their optical constants can be retrieved from this website. It should be noted that while a precise analytical characterization has been performed for all materials that are present in the Jena Database, the respective detailed results of physico-chemical analyses are not contained there, but only in the pertinent scientific articles.

Our paper is structured as follows. Section 2 is on mid-infrared spectroscopy of solids, while section 3 is on far-infrared spectroscopy. In both sections, we focus on recent measurements, a substantial part of which were performed at either cryogenic temperatures (especially in the case of the FIR-measurements) or at high temperatures, namely in order to explore conditions under which cosmic dust doubtlessly exists. At some points, a comparison to astronomical infrared spectra is made, as a part of the above mentioned, still ongoing effort to identify the main components of circumstellar, interstellar and in protoplanetary dust.

2. Mid-Infrared Spectroscopy

2.1. *General considerations*

As for the identification of cosmic dust species on the basis of their vibrational bands, the mid-infrared region (5-50 μm) turned out to be the most fruitful wavelength range.

For some species of solids – e.g. crystalline silicates and highly ionic crystalline oxides in which intense dipole oscillations can be excited – MIR powder spectra are strongly influenced by the shapes, sizes, and possible agglomeration of the respective particles (e.g. Salisbury *et al.* 1992). Crystalline grains which are *not* small compared to the examined wavelengths, as well as powders containing grains strongly deviating from spherical symmetry, show markedly different spectra (band positions, band profiles) than very small spherical grains. For amorphous solids, the influence of the particle shapes on the infrared (powder) spectra is less pronounced.

There are several ways to overcome or 'control' the grain-shape-dependence of infrared spectra. One method is to determine optical constants, n and k – e.g. from reflectance measurements of polished surfaces of bulk material – and to use these quantities for the calculation of small particle spectra for arbitrary grain shapes (e.g. Bohren & Huffman 1983). Compared to mineral powder spectra – for which the actual grain shapes are often unknown or difficult to disentangle –, calculated small particle spectra are better suited for a comparison with the spectra of cosmic dust, since the grain shape can then be varied in the calculations in a controlled way. Another method to get grip on grain-shape-dependence is aerosol spectroscopy; exemplary results are presented in the following subsection.

2.2. *Effects of grain shapes on band profiles examined by aerosol spectroscopy*

The Jena laboratory astrophysics group has adopted the aerosol technique and performed IR spectroscopic measurements in a matrix-free environment (Tamanai *et al.* 2006, 2009a, 2009b). With this method, it is possible to directly study the influence of particle morphologies on IR spectra. Any spectral influence of a matrix material used in powder spectroscopy (e.g. potassium bromide, KBr) is ruled out in aerosol spectroscopy, simply because there is no matrix substance in this case (the particles are free floating).

Images of two forsterite (Mg_2SiO_4) samples with different shapes obtained by a scanning electron microscope (SEM) are shown in Fig. 1. The image of 'Forsterite a' (left) is dominated by roundish grains. On the other hand, 'Forsterite b' particles (right) are irregularly shaped and show sharp edges. X-ray diffraction has been used to examine the crystal structure of both samples, which turned out to be identical. As shown in Fig. 2, the band profiles measured for forsterite particles with ellipsoidal (roundish) shapes clearly differ from those with shapes including sharp edges. The peaks of the resonances undergo a blueshift of up to 0.22 μm for the rounded grains (Tamanai *et al.* 2006). Although it is difficult to demonstrate that the peak shift is purely caused by the grain shape, a blueshift of the rounded grains is expected, since the resonances of a sphere

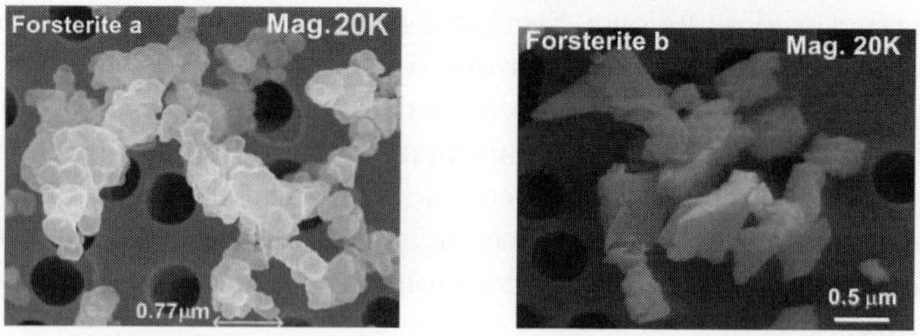

(a) (b)

Figure 1: SEM images of forsterite particles. (a) 'Forsterite a', with nearly spheroidal and ellipsoidal (rounded) grain shapes. (b) Irregular shapes with sharp edged particles dominating in 'Forsterite b'. Note: the black dots are the holes of the polyester membrane filter which have a diameter of $0.5\,\mu$m.

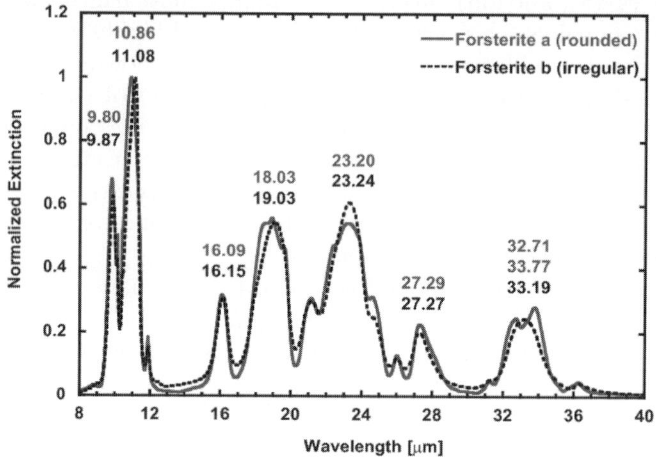

Figure 2: Normalized extinction vs. wavelength for two different shapes of forsterite particles. The grey solid line refers to the 'Forsterite a' (rounded), while the black dotted line refers to the 'Forsterite b' (irregular shapes).

arise at shorter wavelengths, which can be predicted by means of theoretical simulations (Bohren & Huffman 1983). It can be predicted that resonances of particles far from a spherical shape, such as needle-like ones, are located predominantly at the transverse optical (TO) frequency, which is located at longer wavelengths than the resonance of spherical grains (see also Koike *et al.* 2010). Another interesting difference between the IR band profiles of rounded and of irregular shaped particles is that rounded grains may produce double peaked (rectangular) profiles (see e.g. the peaks at 19, 23, $33\,\mu$m). This tendency has been observed in oxide particle investigations as well (see more details in Tamanai *et al.* 2009a, b).

The particle size and agglomeration state may partially affect the spectral features as well. Thus it is essential to do theoretical modeling in order to distinguish grain shape effects from grain size and agglomeration effects.

2.3. *High temperature measurements of spinel and corundum and comparison with the observed 'astronomical' $13\,\mu m$ band*

A second 'hot topic' in solid-state astrophysics is to fully cover the range of temperatures that dust grains reach in space. In the case of circumstellar dust, this may imply the necessity to heat the solid samples up to 1000 K.

Two different cases of high temperature measurements have to be distinguished:

- *Annealing*: In this case a sample undergoes some persistent structural transition upon heating and keeping it hot for a certain time. The effect of the respective structural change on the dielectric properties, crystal structure etc. can be measured after cooling the sample to room temperature.
- *'In situ'* high-T-measurements: These have to be applied if reversible changes (of infrared properties etc.) are expected to occur for a cosmic dust analog, such that for instance its IR reflectance continuously changes upon heating. IR spectra consequently have to be taken while the sample is still hot, which may require sophisticated experimental setups.

While annealing experiments have been carried out for cosmic dust analogs in a number of laboratories, in situ high-T-spectra are so far available for a few substances only.

We performed in situ high-T measurements of the IR properties of corundum (α-Al_2O_3) and spinel ($MgAl_2O_4$), since these dust species are indeed expected to condense at high temperatures (above 1000 K) in stellar envelopes (see Gail 2010) and protoplanetary disks.

Both mineral species have been discussed as potential carriers of the so-called 13 μm band – a noteworthy infrared feature detected in oxygen-rich AGB stars with low mass loss rates (Posch *et al.* 1999, Fabian *et al.* 2001, Sloan *et al.* 2003). Our measurements set new constraints on the carrier substance of this band and on its possible range of temperatures, as will be discussed below.

For our laboratory measurements, a high-temperature high-pressure (HTHP) cell was used. This water-cooled cell allowed us to carry out the temperature-dependent measurements up to 1073 K. The HTHP cell (type Specac P/N 5850) was integrated into a Bruker 113v Fourier-Transform-Infrared spectrometer. We measured reflection spectra of crystallographically oriented minerals heated to 573, 773, and 973 K, relative to the reflectance of a gold mirror (kept at room temperature). From the reflectance spectra, the complex refractive indices of the minerals have been calculated using a Lorentz oscillator model. In the case of spinel, we used a Lorentz model with five oscillators, while for corundum, 11 and 9 oscillators for the crystallographic orientations E||c and E⊥c were needed, respectively. With increasing temperature, the oscillator

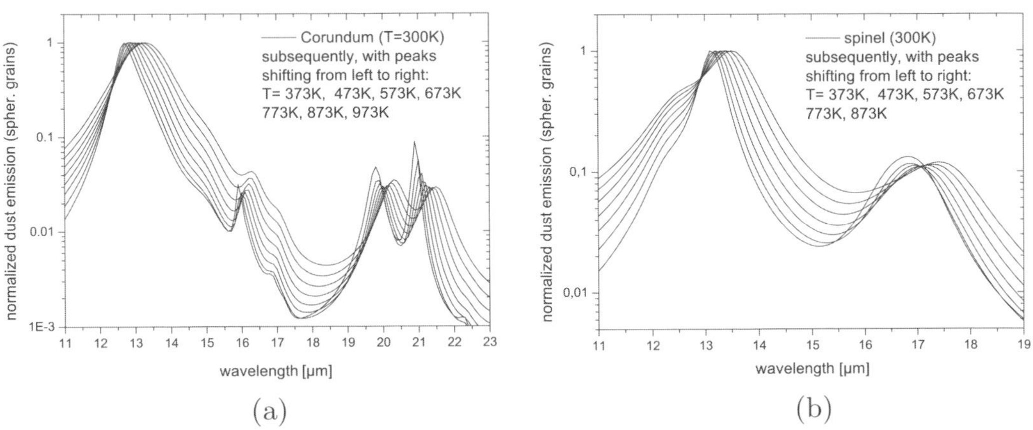

Figure 3: Normalized emission spectra of small spherical corundum (a) and spinel grains (b) for different temperatures ranging from 300 to 973 K. Orientational averaging has been performed as required by the anisotropy of α-Al_2O_3. Note the shift of the peaks to longer wavelengths and the increasing bandwidth as the temperature increases.

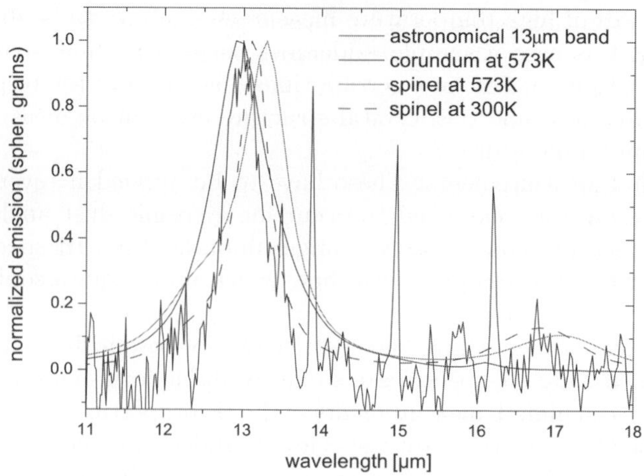

Figure 4: Comparison between the average profile of the astronomical 13 μm emission band – as derived by Fabian *et al.* (2001) for a sample of oxygen-rich AGB stars – with the normalized emission spectra of corundum and spinel grains at temperatures of 300 K and 573 K.

frequencies generally shift to longer wavelengths and the damping parameters increase strongly, while the oscillator strength changes only slightly. This behavior is expected for resonances in an anharmonic vibrational potential (e.g. Henning & Mutschke 1997).

In order to derive the complex refractive indices (n+i k) – also for intermediate temperatures – we fitted the temperature dependence of the oscillator parameters by second order polynomials. In consequence, we obtained analytical expressions for the temperature-dependent optical constants of the measured mineral for each crystallographic axis.

From the optical constants, small-particle spectra $Q(\lambda)$ were calculated for the simplest case of spherical grains in the Rayleigh size limit (particle size $a \ll \lambda$). The emitted flux at a given temperature T finally results from multiplying $Q(\lambda)$ with the corresponding T-dependent blackbody function $B_\nu(T)$. The respective emission spectra for temperatures ranging from 300 K to 973 K are shown in Fig. 3.

In Fig. 4, we compare the small-spherical-grain emission spectra of spinel and corundum to the average profile of the above mentioned 13 μm band as derived by Fabian *et al.* (2001).

The comparison shows that
i) both spinel and corundum grains – if mainly radiating at a 'suitable' narrow range of temperatures – can account for the average profile of the 13 μm band: spinel at about 300 K and corundum at about 573 K;
ii) at high grain temperatures, the resulting integrated emission both of spinel and of corundum grains becomes too broad as to fit the observed astronomical band (see esp. the spinel spectrum for 573 K; cf. Fig. 3a for corundum at T > 600 K).

As for corundum, it should be noted that only our high temperature measurements enable a successful fitting to the 13 μm band. However, the problem remains that corundum has additional weaker bands around 16 μm and 20–21 μm which have no clear counterparts in the observed astronomical spectra.

2.4. *Spectral properties of carbonaceous material*

We now turn to the spectral properties of carbonaceous materials, another major component of cosmic dust, and start this subsection with a brief introduction into the secrets of carbonaceous structures prepared by gas-phase condensation and with a short overview on possible formation processes of carbonaceous solids in astrophysical environments. The structure of particulate carbon materials can be characterized on different length scales describing the short-, medium-, and long-range order of the material. The long-range order describes the size and shape of the primary particles and their agglomeration state. The medium-range order characterizes the arrangement, shape and size of the structural subunits and can be derived by analytical characterization such as high-resolution transmission electron microscopy. The hybridization state of carbon, the nature of the bonds between carbon atoms as well as the incorporation of hydrogen or heteroatoms in the carbon grains can be specified by the short-range order of the material (Schnaiter *et al.* 1998, Jäger *et al.* 1999). The short- and medium-range order of carbon grains can be well described by the terms 'amorphous carbon' (AC) or 'hydrogenated amorphous carbon' (HAC). HACs are frequently used to describe the carbonaceous dust component in astrophysical studies (Jones *et al.* 1990, Duley 1994, Pendleton & Allamandola 2002, Dartois 2011). In AC and HAC, the carbonaceous structure can be characterized by aromatic islands of different sizes, which are linked by aliphatic structures including sp, sp^2, and sp^3 carbon atoms. In the astrophysical literature, the term HAC is preferentially employed for structures containing aromatic units of 2-8 rings (Jones *et al.* 1990).

In recent laboratory studies on the gas-phase condensation of carbonaceous material, a low- and a high-temperature formation process has been found (Jäger *et al.* 2009). At condensation temperatures higher than 3500 K, fullerene fragments or complete fullerenes build up the nucleating particles. Fullerene-like carbon grains and fullerene compounds are formed. Low-temperature condensation ($T \leqslant 1700$ K) favors a nucleation and growth process with PAHs as precursors and particle-forming elements resulting in carbonaceous grains revealing well developed planar or slightly bent graphene layers in their interior and a mixture of PAHs as side products.

The MIR spectra of gas-phase condensed carbonaceous solids that may consist either of a mixture of grains and soluble components (PAHs or fullerenes) or of carbonaceous grains exclusively, without a soluble component, show aromatic IR bands (AIBs) as well as aliphatic IR band, partly superimposed on broad plateau features around 8 and 12 μm (see Fig. 5). Aliphatic $-CH_x$ absorptions can be observed at 3.4, 6.8, and 7.25 μm. In fullerene-like carbonaceous grains, saturated aliphatic $-CH_x$ groups are mainly responsible for the links between the fullerene fragments (Jäger *et al.* 2008). In this material, signatures related to the presence of $-C{\equiv}C-$ triple bonds can be observed in the in-situ recorded IR spectra at 3.0 and 4.7 μm, which points to the presence of polyynes. For the formation of fullerene-like soot grains, polyynes are discussed as intermediates. They finally form up the fullerene snatches that are incorporated into the grains. Additional bands due to –C–C– stretching, –C–H, and –C–C– twisting, wagging, and rocking bands can be expected in the range between 7.4 and 10 μm. In the 11-15 μm range, out-of-plane –C–H bending bands can be observed caused by –C–H groups at the periphery of the fullerene snatches.

The more aromatic carbonaceous material produced by low-temperature condensation of hydrocarbons in laser pyrolysis (Jäger *et al.* 2009) show a series of aromatic bands including 3.3, 6.3, 8.6, 11.3, 12.3, and 13.3 μm as well as aliphatic IR bands at 3.4, 6.8, and 7.25 μm (–C–H stretching and deformation bands). IR features observed beyond 6.7 μm (6.8, 7.25, 7.93, 8.47, 9.28, 9.72, 10.5 μm) mainly arise from –C–C– stretching, C-H deformation, and other bands due to –C–H twisting, wagging, rocking, and –C–C– deformation bands. In this spectral range, a firm assignment of the IR features to a specific functional group is rather difficult even for carbonaceous grains without adsorbed

PAHs, which may contain a lot of different functional groups with single and double bonds integrated in varying chemical environments leading to a distribution of vibrational bands in this spectral range.

Bands of more diagnostic relevance can be found at wavelengths larger than 11 μm, which can be assigned to aromatic =C–H out-of-plane bending vibrations (see Fig. 5). The 11.3 μm feature can be assigned to a single =C–H band, whereas the band at around 12.3 μm is caused by 2–3 and the feature at 13.27 μm by 3–4 adjacent H atoms in the periphery of PAH molecules. The 13.56 and 14.3 μm bands can be attributed to around 4 adjacent H atoms bound to the aromatic ring. The observed IR features in carbonaceous condensates are typical for both non-substituted PAHs or PAHs containing methyl or methylene groups, as well as for carbonaceous grains containing aromatic and aliphatic structures.

Sometimes, also oxygen-containing functional groups such as –C=O (carbonyl group) are incorporated into the carbonaceous structures which can be identified at wavelengths between 5.8–6.0 μm. In addition, higher fullerenes can show vibrational modes in this range (Mordkovich 2000).

Carbonaceous condensates consisting of grain/PAH blends show spectral characteristics that are comparable to the observed IR spectra of post AGB stars and protoplanetary nebulae (Kwok *et al.* 2001, Hony *et al.* 2003, Hrivnak *et al.* 2007). A comparison shown in Fig. 5 demonstrates that nearly all of the IR bands observed in the spectrum of the post-AGB-star HD 56126 are also present in the spectrum of the laboratory low-temperature condensate. Differences in band ratios, apparent for the =C-H out-of-plane bands between 11 and 14 μm, are due to a higher abundance of small PAHs, or larger PAHs containing outer rings with up to 3 to 4 H atoms compared to larger, compactly condensed species which are thought to be very stable under astrophysical conditions. Carbonaceous particles built up by large PAHs in the growth process and purified from the adsorbed, smaller PAHs by extraction show a better coincidence with the observed bands (see gray solid curve in Fig. 5). However, in order to obtain a dominance of the 11.3 μm out-of-plane band due to solo =C–H frequently observed in PPNs and PNs, very large ($N_C \geqslant 100$) and rather elongated PAH molecules are required (Hony *et al.* 2001). The comparison reveals that the low-temperature condensate is a promising dust analog for carbonaceous materials produced in carbon-rich AGB stars.

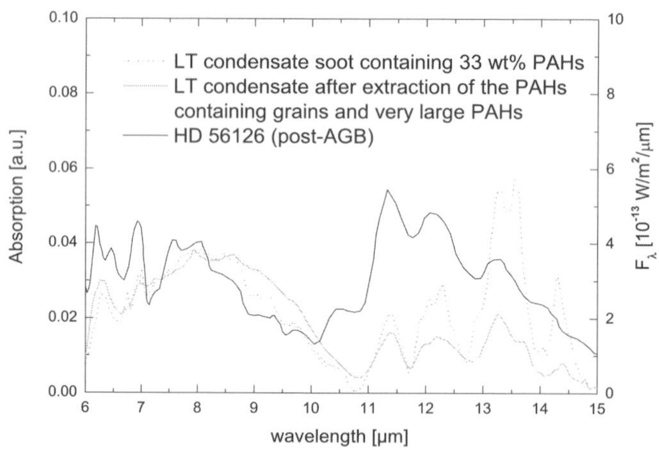

Figure 5: Comparison between the IR spectral properties of a carbonaceous condensate that consists of a mixture of solid carbon particles and PAHs and the observed spectrum of a post AGB star adapted from Hony *et al.* (2003).

3. Far-Infrared Spectroscopy

3.1. *Selected FIR bands of solids: carbonates, phyllosilicates and the 69 micron forsterite band*

A number of crystalline dielectric solids show phonon bands in the far-infrared range (λ>50 μm), where the relatively cold dust located at larger distances from stars (e.g. in planetary nebulae or protoplanetary systems) is emitting thermal radiation. If observed, these bands would be of high diagnostic value to assess the mineralogy of this dust and to draw conclusions about its history. The profiles of these bands have been demonstrated to be quite sensitive to temperature, i.e. the bands get sharper and shift to shorter wavelengths at low temperatures. Consequently, it is necessary to perform laboratory spectroscopy at these low temperatures for providing comparison spectra, or for the determination of low-temperature optical constants as input for calculations, respectively (see Sect. 2.2 for the analogous case of high temperatures). Such low-temperature spectroscopic data have been produced by laboratory work in the past years. E.g., temperature-dependent optical constants have been measured for forsterite (Suto *et al.* 2006) and for carbonates (Posch *et al.* 2007). These optical constants are very valuable because of their flexibility and independence from the measurement conditions (see Sect. 2.1). Other studies have determined the changes of band positions and band widths with temperature in absorption measurements based on the pellet technique. E.g., the long-wavelength bands of several crystalline silicates have already been measured at low temperatures about ten years ago (Bowey & Adamson 2002, Chihara *et al.* 2002).

A more complete investigation of the low-temperature absorption spectra of olivine minerals with different iron contents was presented by Koike *et al.* (2006). The quantification of the nonlinear peak shift of the 69 μm band and of the band broadening with temperature allows to use these bands as thermometers under certain conditions such as at known iron content of the silicate. As an example of how temperature dependent optical constants can be used, Fig. 6 shows the 69 μm band in the Herschel/PACS spectrum of the post-AGB star HD161796 (obtained by the MESS consortium, Groenewegen *et al.* 2011; see also de Vries *et al.* 2011). A one-temperature fit of the 69 μm band indicates that the forsterite grains in this system have a temperature of the order of 124 K. For our derivation of the temperature, we used the laboratory measurements of forsterite's optical constants at different temperatures by Suto *et al.* (2006).

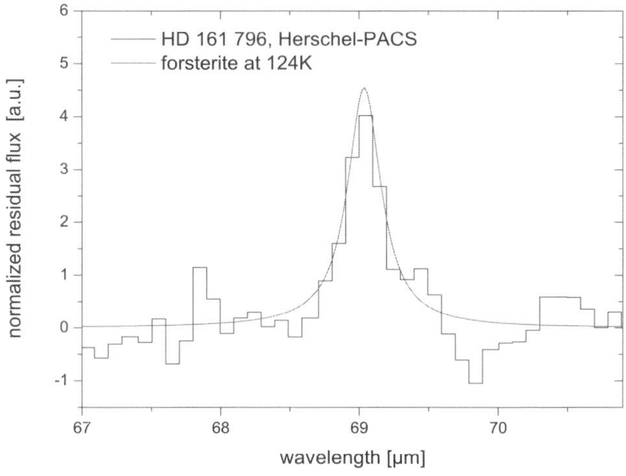

Figure 6: The 69 μm band in the (binned) Herschel-PACS spectrum of the post-AGB star HD161796. The 'smooth' line shows the best fit, obtained with forsterite at T=124 K.

Apart from olivines, several other crystalline silicates show clear FIR bands in the long-wavelength range as well. In Fig. 7, bands of diopside ($CaMgSiO_3$) and of the hydrous silicates serpentine ($(Mg,Fe)_3Si_2O_5(OH)_4$) and talc ($Mg_3[Si_4O_{10}(OH)_2]$) are shown for example. Diopside has a very strong 66 μm band (note the five-times reduction of the peak intensity in our plot). It is an abundant constituent of early condensates in the solar nebula (see, e.g. Posch *et al.* 2007a), but the band has not yet been identified in the spectra of protoplanetary disks. By contrast, it may in fact contribute to a band seen at this wavelength in planetary nebulae (Kemper *et al.* 2002, Chihara *et al.* 2007).

Hydrous silicates are weathering products of silicate minerals including olivines and pyroxenes and they are abundant in some classes of primitive meteorites. Hydrous silicates were also discussed to be a carrier of a broad 100 μm emission band detected in Herbig Ae/Be stars (Malfait *et al.* 1999). Far-infrared absorption spectra of hydrous silicates have been measured by different laboratory astrophysics groups (Koike *et al.* 1982, Koike and Shibai 1990, Hofmeister and Bowey 2006, Mutschke *et al.* 2008). The spectra show very interesting features such as the longest-wavelength phonon bands known to our knowledge (e.g. one at about 270 μm for minerals of the chlorite group). However, far-infrared emission features could not be successfully assigned to bands of hydrous silicates so far (see Fig. 8).

Finally, broad far-infrared bands observed in the emission spectrum of the planetary nebula NGC 6302 have been proposed to be caused by carbonate dust (Kemper *et al.* 2002). Carbonates have extremely strong far-infrared bands (note the ten-times reduction of the peak intensity in Fig. 7), so that already small amounts of these minerals, which may be formed by non-equilibrium reactions of silicates with a H_2O-CO_2-rich gas (Toppani *et al.* 2005), should be detectable. Low-temperature measurements of the optical constants of carbonates by Posch *et al.* (2007) largely confirmed this possible assignment for NGC 6302, but did not give support to an assignment to a 100 μm emission feature observed in protostars (Chiavassa *et al.* 2005).

3.2. *Dust continuum opacities in the submillimeter wavelength range*

3.2.1. *Wavelength and temperature dependence*

Apart from the far-infrared phonon bands, which may provide information about the composition and temperature of cold, mostly crystalline dust, the underlying continuum

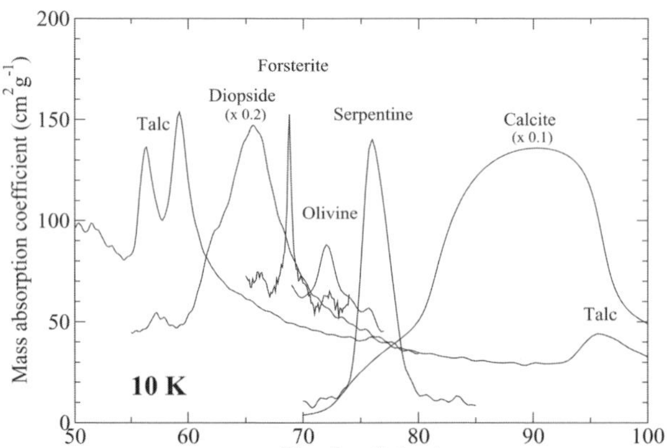

Figure 7: Far-infrared bands of a variety of minerals measured at a temperature of 10 K. The calcite spectrum has been calculated from the optical constants by using the CDE model, whereas all other spectra have been measured on submicron-sized particles embedded in polyethylene pellets.

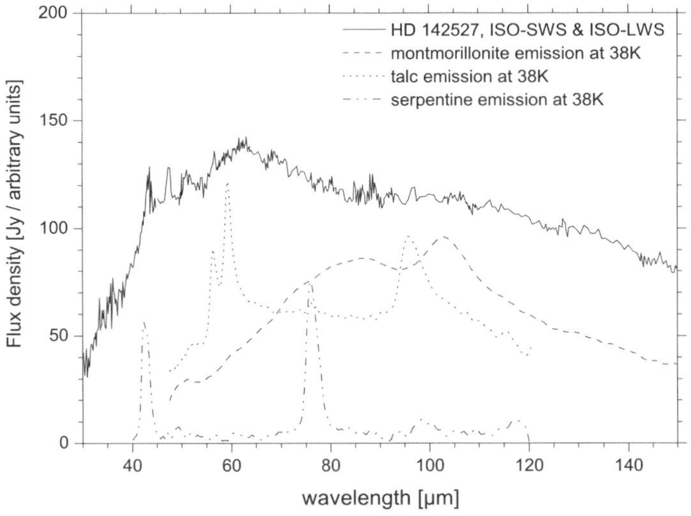

Figure 8: Comparison of simulated emission of cold hydrous silicate dust with the thermal emission of the Herbig AeBe object HD 142527 (after Mutschke *et al.* 2008).

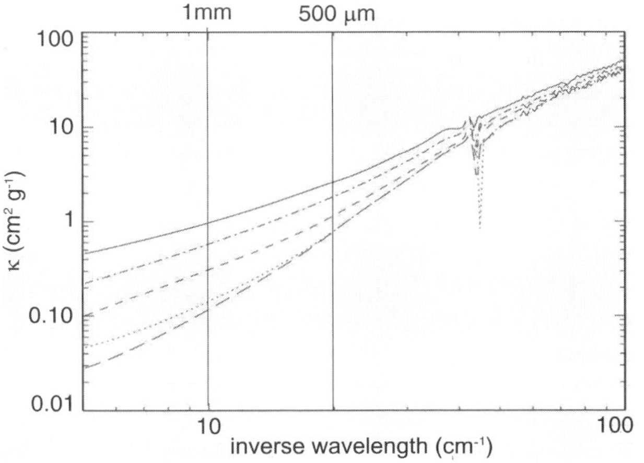

Figure 9: Mass absorption coefficient for MgSiO3, synthetized by sol-gel reaction at 300 K (solid line), 200 K (dot-dashed line), 100 K (short dashed line), 30 K (dotted line), and 10 K (long-dashed line) (Boudet *et al.* 2005).

dust opacity is of great importance. For example, one of the big uncertainties in determining the amount of cold dust in interstellar space arises indeed from the uncertainty in the emissivity of the dust particles at these wavelengths. Furthermore, models of the spatial dust distribution of circumstellar dust, e.g. in planetary systems (debris disks) depend on the wavelength and temperature dependence of the dust opacity.

In contrast to the far-infrared bands, the continuum opacity of dust in most astrophysical environments is determined by that of amorphous solids, especially amorphous silicates, because their emissivity is generally higher at far-infrared wavelengths. The reason for that is the possibility of additional low-energetic transitions due to structural disorder.

Table 1: Continuum opacities of dust species at low temperatures (4-24 K). The values are given as mass absorption coefficients at 100 μm (read from figures – approximate) and 1 mm wavelength, plus the power-law slope value β at 1mm wavelength (for forsterite, we refer to the slope at 40-80 μm). Room-temperature values are given in parentheses.

Dust species	κ (100 μm) [$cm^2 g^{-1}$]	κ (1 mm) [$cm^2 g^{-1}$]	β	Reference
Am. Fe-silicate ("FAYA")	95(110)	0.86(5.0)	2.04(1.35)	Mennella *et al.* (1998)
Am. Mg-Silicate ($MgSiO_3$ glass)	45(50)	0.22(0.75)	2.14(1.58)	Boudet *et al.* (2005)
Am. Mg-Silicate ($MgSiO_3$ sol-gel)	40(50)	0.12(0.98)	2.74(1.44)	Boudet *et al.* (2005)
Am. Mg-Silicate ($MgSiO_3$ sol-gel)		0.28	$\sim$ 2.0	Agladze *et al.* (1996)
"Astronomical silicate" (spheres)	29	0.30	1.6	Li & Draine (2001)
Forsterite	2.1(20)		5.0(2.6)	Chihara *et al.* (2001)
Forsterite	80(120)	0.16(0.43)	2.32(2.04)	Mennella *et al.* (1998)
Forsterite (unheated)		0.22		Agladze *et al.* (1996)
Am. Carbon ("BE")	300(500)	21(72)	1.15(0.76)	Mennella *et al.* (1998)

Among them, the disorder-induced phonon transitions give a temperature-independent contribution which has been shown by Schlömann (1964) to lead to a λ^{-2} wavelength dependence at shorter and a λ^{-4} dependence at longer wavelengths (see also Meny *et al.* 2007). A strongly temperature-dependent behavior is, however, introduced by additional low-energetic tunneling transitions of structural groups in the amorphous network.

First studies of the temperature-dependent absorption coefficient of amorphous silicates in the sub-millimeter and millimeter continuum have been performed by Agladze *et al.* (1996), Henning & Mutschke (1997) and Mennella *et al.* (1998). Recently, these studies have been extended to a larger set of amorphous silicate materials measured in a wide wavelengths and temperature range (Boudet *et al.* 2005, Coupeaud *et al.* 2010; see also their poster at this conference). In these studies, the power-law spectral index β of absorption was found to increase with decreasing temperature, especially at wavelengths larger than about 300 μm. For amorphous Mg silicates and SiO_2, this anti-correlation of T and β is demonstrated in Fig. 9 (values taken from Boudet *et al.* 2005). The spectral index varies between 1 and 3 for temperatures between 300 and 10 K, which confirms the expectations from theory.

In Tab. 1 we summarize opacity values and power-law indices for a number of amorphous and crystalline Mg- and Fe-silicates, taken from four of the mentioned publications (apologies for being not complete). Additionally, we compare with values measured for an amorphous carbon by Mennella *et al.* (1998) and with the numbers calculated for spherical grains composed of the model material "astronomical silicate (astrosil)" (Li & Draine 2001).

At a wavelength of 1 mm, the low-temperature opacity values of amorphous magnesium silicates seem to concentrateconverge to values of 0.1–0.3 $cm^2 g^{-1}$, which is close to the "astrosil" value. However, we have to mention that Agladze *et al.* (1996) have reported higher values up to 3.7 $cm^2 g^{-1}$ for some of their sol-gel produced silicate samples as well. The reason for that is not clear, but may have to do with a high porosity of these samples. Fe-silicates seem to have somewhat higher opacities, and a somewhat stronger temperature dependence, but this is only based on one measurement so far (Mennella *et al.* 1998). For mixed Mg/Fe silicates, investigations are still lacking. The power-law exponent around 1 mm wavelength has typically values between 2 and 3 for amorphous silicates. The "astrosil" value has been reduced in order to better match galactic dust emission data. Such values seem to correspond to certain data in the Agladze *et al.* measurements, but just these have unusually high opacity values at the same time.

For amorphous carbon, a similar temperature dependence of the opacity as for amorphous silicates has been reported by Mennella *et al.* (1998), although at a strongly enhanced opacity level (see Tab. 1). The excitation mechanisms behind this could actually be totally different because contributions from free charge carriers might be dominating (see following subsection).

For crystalline silicates, the measured values do not seem to be in perfect agreement so far. For forsterite, Mennella *et al.* (1998) and Agladze *et al.* (1996) found low-temperature opacity and β values, which were similar to those of the amorphous silicates, although for the same composition κ values lower by a factor of four were reported (Agladze *et al.* 1996) and the temperature dependence was considerably weaker. Chihara *et al.* (2001), however, measured values at 100 μm wavelengths, which were only one order of magnitude higher, and therefore in disagreement with a $\beta \sim 2$ behavior. This discrepancy certainly deserves further investigation.

3.2.2. *Grain shape dependence and carbonaceous materials*

For carbonaceous materials, the FIR-behavior is not only influenced by the internal structure, but also strongly by the morphology of the carbon grains. It is well known that the clustering of carbon grains decreases the spectral index β very strongly (Stognienko *et al.* 1995, Michel *et al.* 1996). The influence of the internal structure on the far-infrared behavior of differently structured carbon materials was already studied by Koike *et al.* (1995), Mennella *et al.* (1998), and Papoular *et al.* (1996). These measurements were performed in transmission using polyethylene pellets. However, one has to keep in mind that this method cannot disentangle the structural and morphological effects on the FIR absorption.

For the interpretation of observational FIR data provided by ground-based or satellite-borne observations such as with the Herschel satellite, the opacity of small grains in vacuum is required, which can be calculated from the optical constants n and k for spherical shapes by the Mie theory. In addition, shape effects can roughly be simulated by a continuous distribution of ellipsoids with the same probability of all shape parameters (CDE) in the Rayleigh limit (Bohren & Huffman 1983). In Fig. 10, the calculated

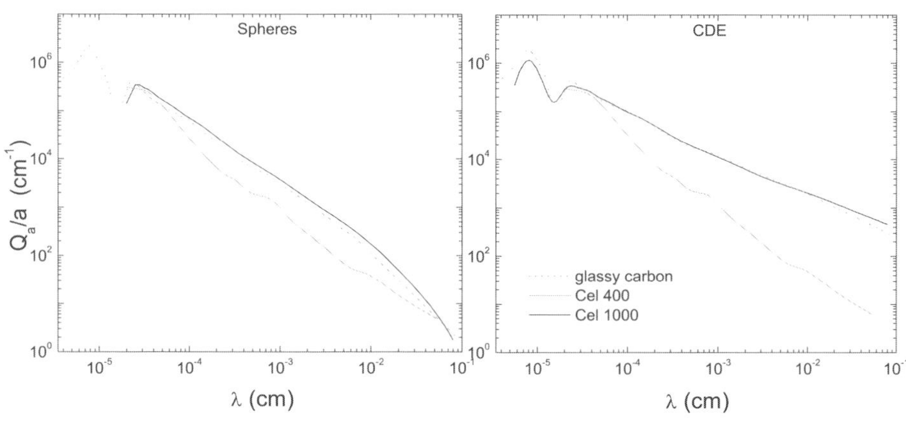

Figure 10: Absorption efficiency divided by particle radius calculated for spherical particles and ellipsoidal grains in vacuum from the optical data of pyrolized cellulose materials.

normalized absorption efficiencies of two carbon materials ('cel 400' and 'cel 1000') are compared with those of glassy carbon. The two materials represent a sp^3 hybridization dominated non-conducting material ('cel 400') and an aromatic carbonaceous material with a high number of free charge carriers ('cel 1000'; see Jäger *et al.* 1998).

Whereas in case of the aliphatic-dominated carbon material 'cel 400' the spectral index β in the FIR region ($\lambda \geqslant 100$ μm or 10^{-2} cm) is about 1.25 irrespective of the particle model, for the aromatic sample carbonized at 1000 °C it is 2.28 for spherical particles, but only 0.71 for agglomerated (CDE) particles. The latter value is very close to the one measured by Mennella *et al.* (1998) for amorphous carbon particles embedded in a polyethylene pellet (see Table 1), although the absolute absorption coefficient is in between those calculated for spherical and CDE particles. This may indicate that in the pellet the carbon particles are strongly aggregated. We conclude that for cosmic particles formed of conducting materials such as highly graphitized amorphous carbon, the far-infrared emissivity can be very high, but this will strongly depend on the grain morphology.

References

Agladze, N. I., Sievers, A. J., Jones, S. A., Burlitch, J. M., & Beckwith, S. V. W., 1996, *ApJ*, 462, 1026

Begemann, B., Dorschner, J., Henning, Th., *et al.* 1997, *ApJ*, 476, 199

Bohren, C. F. & Huffman, D. R. 1983, Absorption and Scattering of Light by Small Particles (New York: John Wiley)

Boudet, N., Mutschke, H., Nayral, C., Jäger, C., Bernard, J.-P., Henning, Th., & Meny, C. 2005, *ApJ*, 633, 272

Bowey, J. E. & Adamson, A. J., 2002, *MNRAS*, 334, 94

Chiavassa, A., Ceccarelli, C., Tielens, A. G. G. M., Caux, E., & Maret, S., 2005, *Astron. Astrophys.*, 432, 547

Chihara, H., Koike, C., Tsuchiyama, A., Tachibana, S., & Sakamoto, D., 2002, *Astron. Astrophys.*, 391, 267

Chihara, H., Koike, C., & Tsuchiyama, A., 2007, *Astron. Astrophys.*, 464, 229

Coupeaud, A., Demyk, K., Meny, C., & Nayral, C., 2010, in: SF2A-2010: Proceedings of the Annual meeting of the French Society of Astronomy and Astrophysics, Eds. S. Boissier, M. Heydari-Malayeri, R. Samadi and D. Valls-Gabaud, p. 205

Dartois, E., 2011, EAS Publications Series 46, 381

De Vries, B. L., Klotz, D., Lombaert, R., *et al.*, 2011, in: F. Kerschbaum, Th. Lebzelter, B. Wing (eds.), Why Galaxies Care About AGB Stars II, *ASP Conf. Ser.*, **vol. 445**, in press

Duley, W., 1994, *ApJ*, 430, L133

Fabian, D., Jäger, C., Henning, Th., Dorschner, J., & Mutschke, H. 2000, *Astron. Astrophys.*, 364, 282

Fabian, D., Posch, Th., Mutschke, H., Kerschbaum, F., & Dorschner, J., 2001, *Astron. Astrophys.*, 373, 1125

Gail, H.-P., 2010, Formation and Evolution of Minerals in Accretion Disks and Stellar Outflows, in: Th. Henning (ed.), Astromineralogy, 2nd ed., Lecture Notes in Physics (Berlin and Heidelberg: Springer Verlag)

Groenewegen, M. A. T., Waelkens, C., Barlow, M. J., Kerschbaum, F., Garcia-Lario P., *et al.*, 2011, *Astron. Astrophys.*, 526, A162

Henning, Th. & Mutschke, H. 1997, *Astron. Astrophys.*, 327, 743

Hofmeister, A. M. & Bowey, J. E., 2006, *MNRAS*, 367, 577

Hony, S., Van Kerckhoven, C., Peeters, E., Tielens, A. G. G. M., Hudgins, D. M., & Allamandola, L. J., 2001, *Astron. Astrophys.*, 370, 1030

Hony, S., Tielens, A. G. G. M., Waters, L. B. F. M., & de Koter, A. 2003, *Astron. Astrophys.*, 402, 211

Hrivnak, B. J., Geballe, T. R., & Kwok, S. 2007, *ApJ*, 662, 1059

Hoppe, P., 2009, *ASP Conf. Ser.*, 414, 149

Jäger, C., Mutschke, H., & Henning, Th., 1998, *Astron. Astrophys.*, 332, 291

Jäger, C., Henning, Th., Spillecke, N., & Schlögl, R., 1999, *J. Non-Cryst. Solids* 258, 161

Jäger, C., Mutschke, H., Henning, Th., & Huisken, F., 2008, *ApJ* 689, 249

Jäger, C., Huisken, F., Mutschke, H., Jansa, I. L., & Henning, Th., 2009, *ApJ* 696, 706

Jones, A. P., Duley, W. W., & Williams, D. A., 1990, *Q. Jl. R. Astr. Soc.* 31, 567

Kemper, F., Jäger, C., Waters, L. B. F. M., Henning, Th., Molster, F. J., Barlow, M. J., Lim, T., & de Koter, A., 2002, *Nature*, 415, 295

Koike, C., Hasegawa, H., & Hattori, T., 1982, *Astrophys. Space Sci.*, 88, 89

Koike, C. & Shibai, H., 1990, *MNRAS*, 246, 332

Koike, C., Kimura S., Kaito C., 1995, *ApJ* 446, 902

Koike, C., Mutschke, H., Suto, H., Naoi, T., Chihara, H., Henning, Th., Jäger, C., Tsuchiyama, A., Dorschner, J., & Okuda, H., 2006, *Astron. Astrophys.*, 449, 583

Koike, C., Imai, Y., Chihara, H., Suto, H., Murata, K., Tsuchiyama, A., Tachibana, S., & Ohara, S., 2010, *ApJ*, 709, 983

Kwok, S., Volk, K., & Bernath, P. 2001, *ApJ*, 55, L87

Li, A. & Draine, B. T., 2001, *ApJ*, 554, 778

Malfait, K., Waelkens, C., Bouwman, J., de Koter, A., & Waters, L. B. F. M., 1999, *Astron. Astrophys.*, 345, 181

Mennella, V., Brucato, J., Colangeli, L., Palumbo, P., Rotundi, A., & Bussoletti, E., 1998, *ApJ*, 496, 1058

Meny, C., Gromov, V., Boudet, N., Bernard, J.-P., Paradis, D., & Nayral, C., 2007, *Astron. Astrophys.*, 468, 171

Michel, B., Henning, Th., Stognienko, R., & Rouleau, F., 1996, *ApJ* 468, 834

Mordkovich, V. Z., 2000, *Chem. Mater.* 12, 2813

Mutschke, H., Zeidler, S., Posch, Th., Kerschbaum, F., Baier, A., & Henning, Th., 2008, *Astron. Astrophys.*, 492, 117

Papoular, R., Conard, J., Guillois, O., Nenner, I., Reynaud, C., Rouzaud, J.-N., 1996 *Astron. Astrophys.* 315, 222

Pendleton, Y. J. & Allamandola, L. J., 2002, *Astrophys. J. Suppl. Ser.* 138, 75

Posch, Th., Kerschbaum, F., & Mutschke, H., *et al.*, 1999, *Astron. Astrophys.*, 352, 609

Posch, Th., Mutschke, H., Trieloff, M., & Henning, Th., 2007, *ApJ* 656, 615

Posch, Th., Baier, A., Mutschke, H., & Henning, Th., 2007, *ApJ*, 668, 993

Salisbury, J. W., *et al.*, 1992, Infrared (2.1–25μm) Spectra of Minerals (Baltimore and London: The John Hopkins University Press), p. xv

Schlömann, E., 1964, *Phys. Rev.*, 135, A413

Schnaiter, M., Mutschke, H., Dorschner, J., Henning, Th., & Salama, F., 1998, *Astrophys. J.* 498, 486

Sloan, G. C., Kraemer, C., Goebel, J., & Price, S. 2003, *ApJ*, 594, 483

Stognienko, R., Henning, Th., & Ossenkopf, V., 1995, *Astron. Astrophys.* 296, 797

Suto, H., Sogawa, H., Tachibana, S., Koike, C., Karoji, H., Tsuchiyama, A., Chihara, H., Mizutani, K., Akedo, J., Ogiso, K., Fukui, T., & Ohara, S., 2006, *MNRAS*, 370, 1599

Tamanai, A., Mutschke, H., Blum, J., & Meeus, G., 2006, *ApJ*, 648, 147

Tamanai, A., Mutschke, H., & Blum, J., 2009a, *Astronomical Society of the Pacific Conference Series*, 414, 438

Tamanai, A., Mutschke, H., Blum, J., Posch, Th., Koike, C., & Ferguson, J. W., 2009b, *Astron. Astrophys.*, 501, 251

Toppani, A., *et al.*, 2005, *Nature*, 437, 1121

Discussion

GOUMANS: You mentioned that if you have the optical constants, one can get the spectrum of grains of any shape and size. What about very small grains, or clusters, that do not behave like bulk material any more? Is spectroscopy on those desirable and possible?

POSCH: Studies of clusters, including infrared spectroscopy, can certainly help to clarify the nature of some unidentified broad infrared bands that were hitherto assigned to mineral species. However, IR spectroscopy of clusters requires very sophisticated techniques and will therefore take even more time than solid state spectroscopy. Additionally, the spectra critically depend on the cluster size, so experiments must be carried out for a wide range of possible cluster sizes.

The Molecular Universe
Proceedings IAU Symposium No. 280, 2011
José Cernicharo & Rafael Bachiller, eds.

doi:10.1017/S1743921311025178

How Can We Use Complete Experimental Catalogs in the Complex Spectra Limit?

Frank C. De Lucia[1], Sarah M. Fortman[1], Ivan R. Medvedev[2], and Christopher F. Neese[1]

[1]Department of Physics, Ohio State University
[2]Department of Physics, Wright State University

Abstract. There is a broad consensus that many, if not most, of the unidentified spectral lines in astrophysical spectra are due to transitions in excited vibrational states of a relatively small number of molecules, the astrophysical weeds. For these unidentified lines, it is somewhat less well understood that the spectroscopic effort required to characterize them in the traditional quantum mechanical catalog approach is substantially larger because of significant perturbations. We have previously discussed a new experimental approach that addresses this challenge. This approach is based on the analysis of many complete, intensity-calibrated spectra taken over a range of temperatures. However, the spectroscopic completeness of this approach results in a much larger database. These data can be transfer to the astrophysical community in a variety of ways, but because an order of magnitude larger number of lines is included, consideration must be given to implementation strategies.

1. Introduction

We have previously proposed and demonstrated a method to record complete, intensity calibrated spectra as a function of temperature and to use these results to provide complete astrophysical catalogs or complete point-by-point simulations as a function of temperature (Medvedev & De Lucia (2007), Fortman *et al.* (2010a, 2010b, 2010c, 2011). Table 1 shows an example of a small portion of a resultant catalog and Figure 1 shows an example of a small region in the point-by-point simulation.

Table 1. A small portion of a catalog derived from an experimental catalog.

Frequency (MHz)	$S_{ij}\mu^2$ (D^2)	Energy cm^{-1}	Figures of Merit
227820.916	197	1056	W, G, T
227821.777	600	1196	W
227826.268	335	1106	W
227827.710	315	1093	W, T
227834.675	630	1257	W, G, T
227838.149	9	395	W, G, T
227850.297	703	1244	W
227857.065	429	1146	W
227858.221	310	637	
227866.026	532	825	
227870.968	304	654	
227872.144	415	1145	W
227892.603	485	1229	W, G, T
227897.612	635	157	
227906.700	694	150	

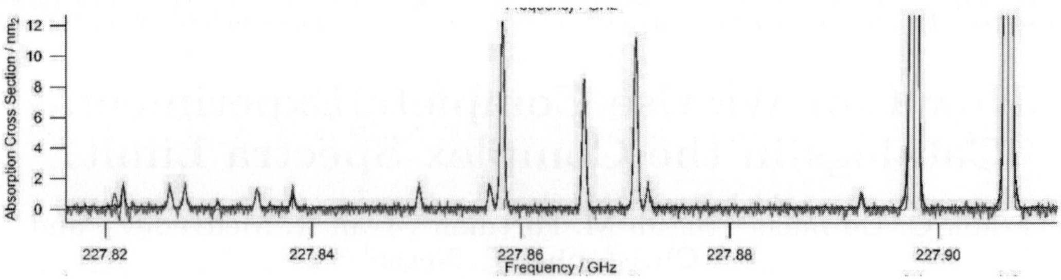

Figure 1. A point-by-point simulation for a small spectral region. The green is a single experimental trace at 300 K. The black is a simulation of this spectrum based on an analysis of ~400 different spectral scans. The red sticks are lines included in the QM catalog.

In Figure 1 the black line, which represents the spectrum calculated from a fit to all ~400 temperature scans, is of much higher sensitivity than the single green experimental scan because of statistical redundancy and contains many lines which are unobservable on the scale of the figure. The red sticks represent the predictions of the quantum mechanical (QM) model. The figure illustrates that many relatively weak lines (e. g. the line near 227.84 GHz) are in the catalogs, whereas much stronger lines (e. g. the line near 227.87 GHz) are not. Additionally, this spectrum shows that the weak line predicted near 227.86 GHz is overlapped by a much stronger line not included in the QM catalog.

One of the challenges for the experimental catalog approach is blended lines because their shapes vary, often significantly, with temperature. To meet this challenge we developed and tested what we refer to as a point-by-point analysis that uses a method similar to (but augmented by knowledge of the Gaussian profile associated with laboratory Doppler limited spectra) our line-by-line catalog approach, but which treats each (~10^6) frequency point independently. Figure 2 illustrates the application of this approach to blends.

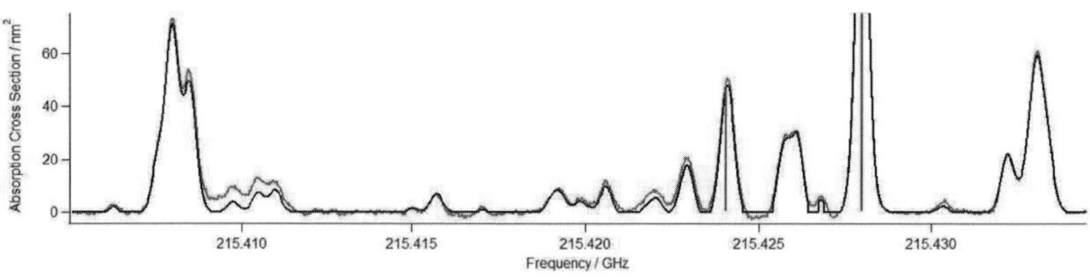

Figure 2. A blow up of a small portion of blended spectra. The black is a simulation of this spectrum based on an analysis of ~400 different spectral scans. The red sticks are lines included in the QM catalog.

In this figure the green trace is a single experimental scan near 300 K, and the black trace is a simulation from the point-by-point analysis. Inspection of the feature near 215.410 GHz shows that the simulation faithfully reproduces the complex lineshape. Additionally, because of the statistical redundancy of the fit to ~400 spectral scans, the noise and baseline issues are both significantly reduced.

In this paper we will consider the use of these complete experimental catalogs in an astrophysical context. Figure 3 shows on a very compressed scale both the astrophysical data and five laboratory spectra.

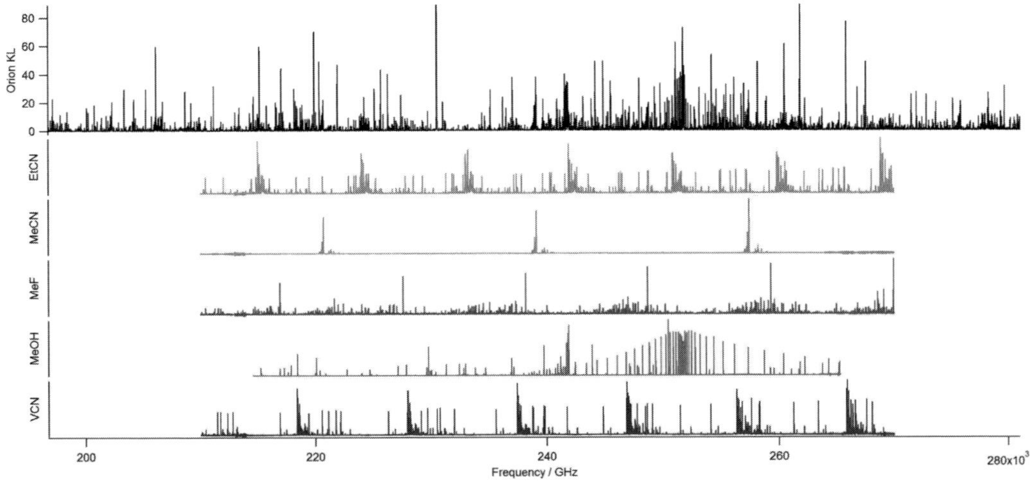

Figure 3. Astrophysical data (upper panel) and intensity calibrated laboratory data (lower five panels, top to bottom: ethyl cyanide, methyl cyanide, methyl formate, methanol, and vinyl cyanide) on highly compressed scales. Astrophysical data from IRAM, courtesy of Jose Cernicharo and Belen Tercero.

2. Quantum mechanical catalogs and catalogs/spectra derived from experimental spectra

Quantum mechanical catalogs and catalogs/spectra derived from experimental data provide complimentary information. Table 2 summarizes this.

Table 2. Comparison of QM catalogs and intensity calibrated experimental catalogs.

	Quantum Mechanical Catalogs	**Experimental Spectra**
Line Frequencies	Model redundancy Model extrapolation error	Directly measured
Lower State energy	Very accurate from model	Accurate enough to give ∼1% intensity error
Completness	Only for analayzed vibrational states and for rotational states within cutoffs	Currently down to ∼1/1000 of strongest line for species
Quantum numbers	Known	Unknown

2.1. *Line frequencies*

In the QM catalogs the frequencies are calculated via a Hamiltonian model from a large number of assigned lines, whereas in the experimental catalogs each measured frequency stands on its own. The QM frequencies have the advantage that the statistical redundancies of the fit can provide increased accuracy. A comparison for vinyl cyanide in the 210-270 GHz band is shown in Fig. 4, which compares the experimental and QM catalog frequencies. This figure shows that for lines for which there are not QM model errors, the agreement is on the order of 5 kHz. This is better than we expected for the experimental frequencies and results from the fact that the varying temperature randomizes baseline effects that can be traced to standing waves. These standing wave effects can be one of the limitations on frequency measurement accuracy. The larger systematic differences are due to model errors in the QM analyses.

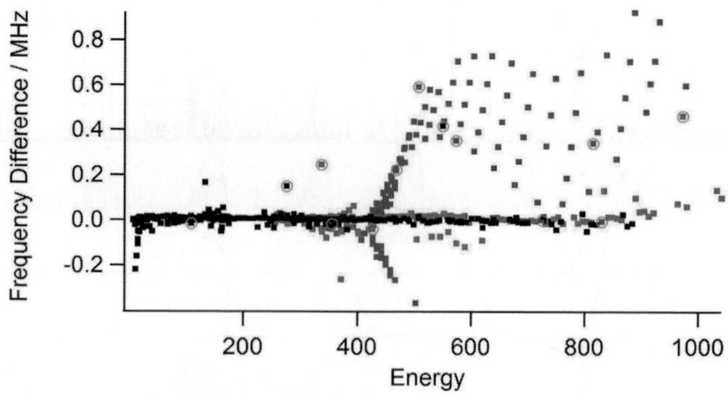

Figure 4. A comparison between QM catalog frequencies and experimentally measured frequencies for vinyl cyanide. The black points are ground vibrational state lines, the red points are first excited state lines, and the blue points are lines in the second excited vibrational state. The systematic differences are presumably due to model errors caused by perturbations in the QM analysis.

The shifts in higher frequency bands are considerably larger and it is not possible to make a figure similar to that of Fig. 4 because it can be hard to match experimental lines with QM model predictions. Figure 5 shows an example of shifts in a small frequency region. In Fig. 5 the calibrated intensities greatly aid the spectral assignments shown.

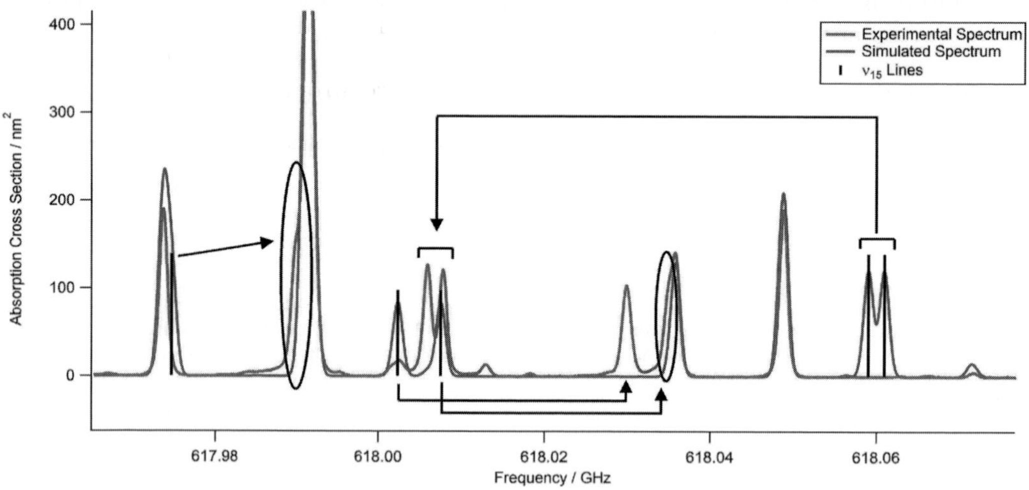

Figure 5. Frequency shifts in the calculated spectrum of vinyl cyanide.

2.2. *Lower state energies/calculated intensities*

QM analyses provide extremely accurate lower state energies, very small fractions of a wave number, whereas the uncertainties associated with the lower state energies of the experimental fits are much greater, $\sim$10 cm^{-1} for unblended lines. For blended lines, the calculated lower state energies have no meaning. In fact, the calculated lower state energies for blended lines are best considered as fitting parameters for calculation of the spectral line intensities as a function of temperature. The accuracy of these calculated intensities is typically 1%, with greater fractional uncertainties for weaker lines.

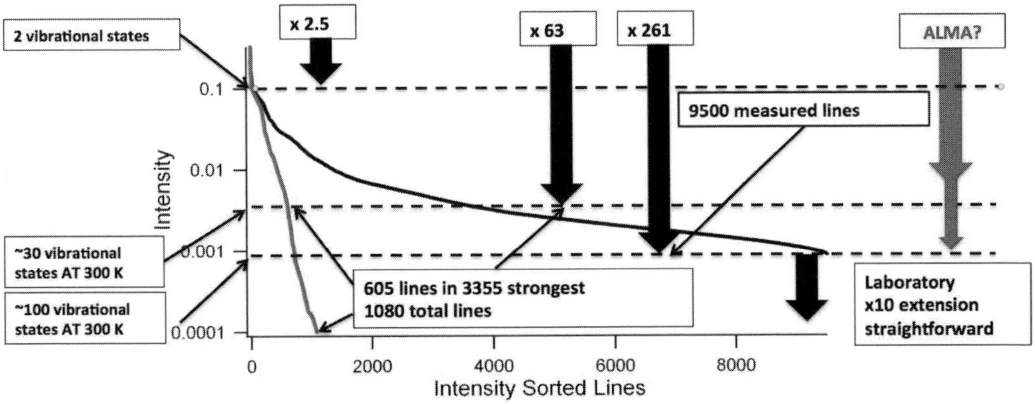

Figure 6. Completeness of the ethyl cyanide QM catalog. The red points (which are so numerous as to appear as lines) are the QM catalog intensities, sorted by strength and the black points are the experimental intensities sorted by strength. These diverge at the strength of the first experimental line that is not included in the catalog.

2.3. *Completeness*

The most important feature of the experimental method is that it is complete. We have discussed this issue in quantitative detail (Fortman *et al.* 2010c). Figure 6 shows this for ethyl cyanide and Figure 7 shows an overview for eight astrophysical weeds.

QM catalogs are complete in the sense that for every vibrational state that has been analyzed as many rotational transitions as desired can be calculated, but with varying accuracy according to model extrapolation and the significance of perturbations. The experimental catalogs only include those lines that are strong enough to observe and are thus intensity incomplete. However, this is not a practical limitation because the experimental simulations include lines that are weaker by $\sim 10^3$–10^4 of the strongest lines in their spectrum. In contrast, the QM catalogs are incomplete because they only include lines for which there has been an analysis. It is this incompleteness that is believed to be the source of many, if not most, of the unidentified astrophysical lines.

2.4. *Quantum numbers*

The experimental approach provides a means for the calculation of spectra as a function of temperature. While its completeness, intensity calibration, and known lower state energies can greatly aid a QM assignment (Fortman *et al.* 2010d), it does not provide assigned quantum numbers. There are many astrophysical circumstances for which the knowledge of the quantum numbers would be advantageous. These include vibrational temperatures that are not in equilibrium with rotational temperatures and state populations that cannot be described by a temperature.

3. Astrophysical Impact

In this section we will use a comparison between our experimental spectra and astrophysical spectra to show their impact as well as challenges in making use of them. We are grateful to Jose Cernicharo and Belen Tercero for making prepublication IRAM data for Orion KL available to us for these preliminary comparisons.

Because of the large amount of laboratory experimental data involved, it was our goal to do these comparisons and analyses in as automated a fashion as possible. This led us to fitting procedures similar to those that we have developed for laboratory analytical

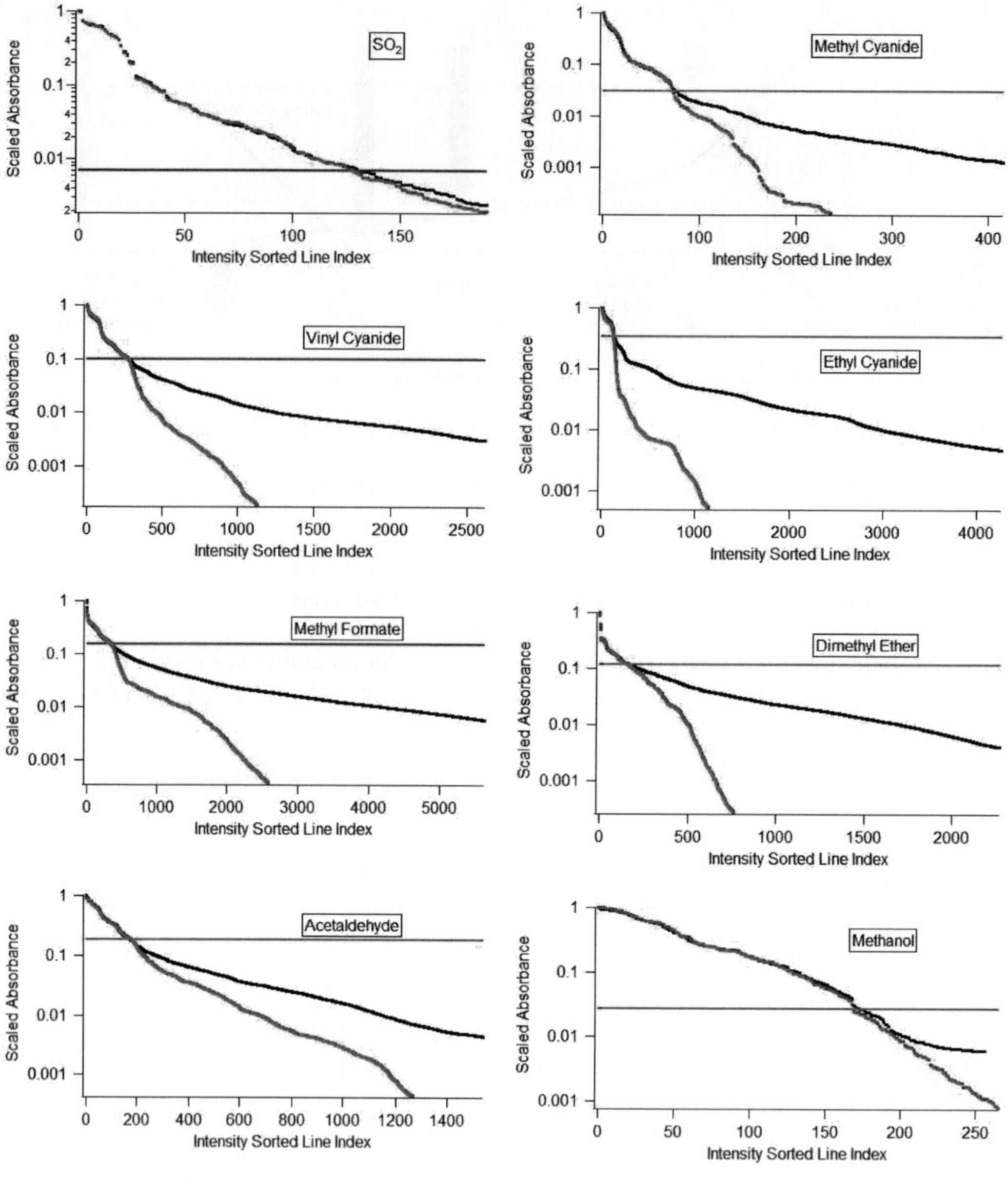

Figure 7. Overview of the completeness for eight astrophysical weeds. The red points (which are so numerous as to appear as lines) are the QM catalog intensities, sorted by strength and the black points are the experimental intensities sorted by strength. These diverge at the strength of the first experimental line that is not included in the catalog. The expected location of this divergence is determined by the Boltzmann factor of the lowest excited vibrational state not included in the analysis. In each figure this is indicated by the horizontal red line.

chemistry, but with the recognition that the interstellar medium is a much more complex environment.

To begin, we needed to convolve our laboratory data with appropriate astrophysical lineshapes, one for each species. Indeed, a fundamental question is whether or not this is possible. To do this, we chose a number of assigned lines for each of the species and

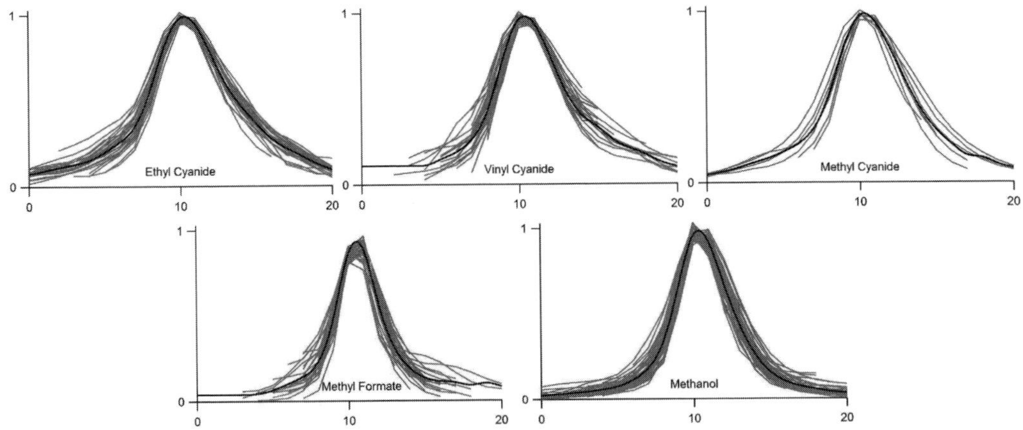

Figure 8. Lineshapes for five of the observed species in the astrophysical data. The red lines are from fits to relatively isolated astrophysical lines. The black lines are the averages of the astrophysical lines that are used for convolution with the astrophysical data.

overlaid the lineshapes derived from the astrophysical data as shown in Fig. 8. This figure shows that to within the bin resolution of the astrophysical data and some overlap on the wings of the spectra, that it was possible to obtain a single lineshape for each species that was representative of the astrophysical data. These lineshapes are considerably broader than those observed by interferometers, and it is inferred that beam averaging plays a significant role.

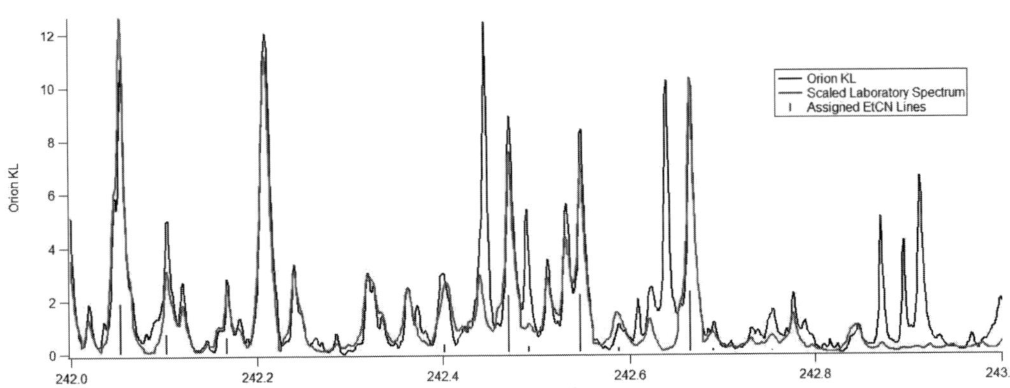

Figure 9. Fit of an experimental simulation of 190 K ethyl cyanide to the Orion KL spectrum.

Above in Figs. 6 and 7, we showed catalog completeness in statistical terms for eight astrophysical weeds. Figure 9 shows this in spectral detail. This figure shows the results of a fit of an experimental simulation of a 190 K spectrum of ethyl cyanide to the astrophysical data, with the abundance as the only free variable. This fit was performed iteratively with any residual that resulted from the astrophysical data being stronger removed from the fit under the assumption that the extra intensity is from another species. There are many similar examples in the astrophysical data.

These results are very encouraging. In this spectral region there are 10 features that can be ascribed to QM cataloged line, but more than 15 additional lines are predicted

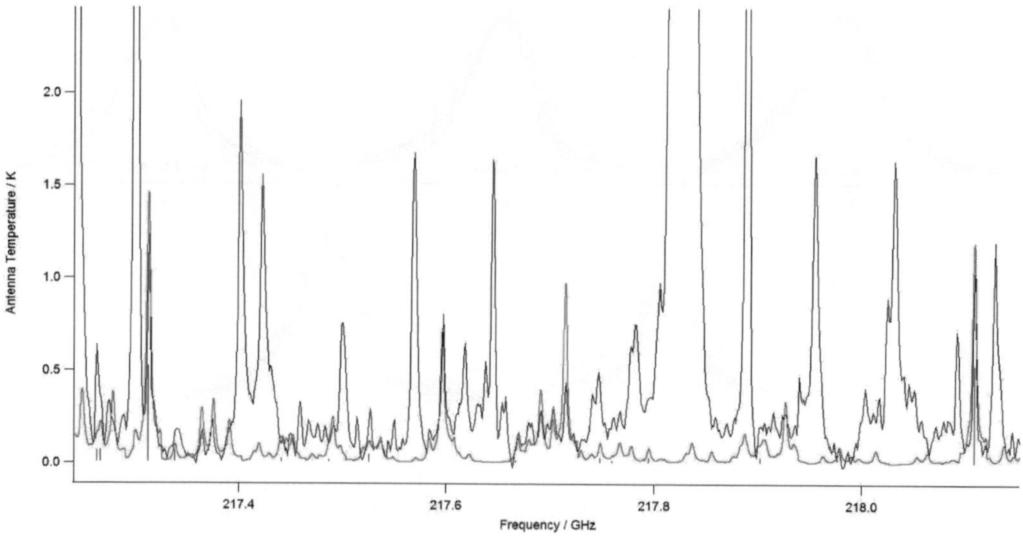

Figure 10. A comparison of the astrophysical spectrum of Orion KL (black trace) and the 190 K spectrum of methyl formate (green trace). The predictions of the QM catalog are shown as red sticks.

by the experimental spectrum. For a number of these additional lines, the experimental spectrum provides additional detail for the structure of the blends that make up these features.

Figure 10 shows a similar comparison for methyl formate in a relatively crowed region of the astrophysical spectrum. Again, there are many additional methyl formate lines that can be assigned from the methyl formate spectrum, many due to lines that are considerably stronger than other lines that are include in the QM catalog. However, in this preliminary comparison the intensity matches in some cases are not as good as those shown in Fig. 9.

4. Observations and conclusions

From these comparisons we make several preliminary observations:

(*a*) The experimental simulation accounts for many features that are not in the QM catalog, including intensities and details of blends. Thus, these astrophysical features are not available for assignment to "new" species.

(*b*) The experimental simulation gets the intensity right in regions where the QM calculation is a clear underestimate-presumably due to a missing line in the catalog. See the example near 242.4 GHz in Fig. 9.

(*c*) The experimental simulation does not predict any lines that are not in the astrophysical spectrum.

(*d*) These one temperature fits vary in the accuracy with which they reproduce intensities. It will be interesting to see if multiple temperature fits or interferometer data from arrays such as ALMA that reduce beam averaging improve the intensity fit.

(*e*) The experimental fits are based on terrestrial isotopic abundances. Either experimental fits on samples with different abundances or the use of ground state QM analyses will be needed to accurately account for astronomical isotopic abundances.

References

Fortman, S. M., Medvedev, I. R., Neese, C. F., & De Lucia, F. C., 2010a, *ApJ*, 714, 476
Fortman, S. M., Medvedev, I. R., Neese, C. F., & De Lucia, F. C., 2010b, *ApJ*, 725, 1682
Fortman, S. M., Medvedev, I. R., Neese, C. F., & De Lucia, F. C., 2010c, *ApJL*, 725, L11
Fortman, S. M., Medvedev, I. R., Neese, C. F., & De Lucia, F. C., 2010d, *Chem. Phys. Lett.*, 493, 212
Fortman, S. M., Medvedev, I. R., Neese, C. F., & De Lucia, F. C., 2011, *ApJL*, in press
Medvedev, I. R. & De Lucia, F. C., 2007, *ApJ*, 656, 621

Discussion

R. Mauersberger: In star-forming regions, visible mm lines can contribute 30-50% to the total flux (line + continuum). Is there any guess of the contribution of blended weed lines to the "continuum"?

F. De Lucia: We do not have an estimate at this time, but since the experimental spectra are complete down to a relatively low level, we should be able to make such an estimate. It would be interesting to compare this estimate with an estimate that one would get from a calculated vibrational partition function.

K. Menten: 1) What do you mean when you sy you toolo 400 spectra? 2) How many species did you do? did you do methanol?

F. De Lucia: 1) We scan 50 to 100 GHz in about 30 seconds (120 scans/hr). Each scan is a complete spectrum. While we scan for about 4 hours, we ramp the temperature by about 200 K. Thus, the temperature varies by about 0.5 K over each scan. 2) In the 210-270 GHz region we have done 8 molecules, including methanol. In the 575-650 GHz region, we have done 2. We expect do the remaining 6 in the next few months.

The Molecular Universe
Proceedings IAU Symposium No. 280, 2011
José Cernicharo & Rafael Bachiller, eds.

doi:10.1017/S174392131102518X

Analysis Tools for Spectral Surveys

Peter Schilke[1], Rainer Rolffs[1,2] and Claudia Comito[2]

[1]I. Physikalisches Institut der Universität zu Köln,
Zülpicher Str. 77, 50937 Köln, Germany
email: schilke@ph1.uni-koeln.de

[2]MPIfR, Auf dem Hügel 69, 53121 Bonn, Germany
email: rrolffs,ccomito@mpifr.de

Abstract. Spectral surveys in the past were a hobby of a few, usually restricted to strong, line-rich and close-by sources which were considered templates for source classes, e.g. Orion KL for hot cores, IRC+10216 for AGB stars, and CRL618 for protoplanetary nebulae. Not any more, since with the large bandwidths and high sensitivities of modern instruments, notably ALMA, all but a few sources will show many lines from many molecules at every observations. So (involuntary) line surveys will be the norm rather than the exception. A common strategy is to ignore all lines but the few one is interested in. Since all data will be available through the archive, this does not mean that the data are lost, since eventually the information will be extracted. Another strategy is to take the bull by the horns, and try to analyze all or at least a large portion of the spectrum. This includes the steps of line identification, source modeling and linking to physical and chemical models. With the data volumes at hand doing it the traditional, pedestrian, way is somewhere between impractical and impossible, semi-automatic methods need to be employed.

Keywords. Astrochemisty, line:identification, radiative transfer, methods:data analysis, stars: formation, ISM: molecules

1. Introduction

Spectral line surveys used to be a very specialized field, targeting very few template sources and trying to get a complete inventory of the molecular content (Schilke *et al.* 2001, Comito *et al.* 2005 and references therein). There are a variety of reasons why this was the case. There were technical difficulties, receiver sensitivities were relatively low, receiver and backend bandwidths were narrow, which meant frequent retunings to scan a given frequency range. Particularly in the submillimeter wavelength range observations are difficult from most sites that were available a few years ago, which meant that for line surveys of even a strong source like Orion some nights of observing time were needed, which, given the weather uncertainties, easily translated in a few weeks at a telescope. But then, the data were there, and needed to be processed.

Sideband deconvolution often had to be performed, and relative calibration or pointing offsets needed to be dealt with, but the end product was, after months of work, a full band spectrum of a single source. This then needed to be analyzed, which meant first line identification, often done in a pedestrian fashion from line lists provided by catalogs like JPL (Pickett *et al.* 1998) or CDMS (Müller *et al.* 2001, 2005). In line rich sources one often has line blending, and the number of unidentified features was large, due to incompleteness of the catalogs. After identification, rotation temperatures and column densities needed to be extracted, often by means of rotation diagrams, which often are plagued by unknown line opacties, and give completely erroneous and misleading results if high opacties are not taken into account. Many analyses ran out of steam after that, and no more sophisticated data modeling was done, which effectively means that a lot of information contained in the data was not extracted. The legacy value of line surveys was perceived to be little more than road maps, so that people observing other sources could determine what all the little bumps in their spectrum were. Very useful, but not very exciting.

With the advent of powerful, sensitive and broadband instruments, such as APEX, SMA, HIFI and, soon, ALMA, or the upgrade of existing instruments such as the IRAM 30m. Mopra or ATCA, spectra containing a multitude of lines or, in the case of interferometers, data cubes containing a multitude of line maps, will become the rule rather than the exception for many Galactic (High-Mass and Low-Mass Star forming regions, Protoplanetary disks, AGB stars and Supergiants, Planetary and Protoplanetary nebula) and nearby Extragalactic sources (Galactic Nuclei, Starburst Regions. Martin, this volume). Also, high sensitive extremely wide-band line surveys such as the ones obtained in the HIFI Key Programs HEXOS (Bergin *et al.* 2010) and CHESS (Ceccarelli *et al.* 2010) are available now. Even though the original proposers and owners of the data in many cases may not be interested in all their data, concentrating on one or a few lines perhaps, most (unfortunately not all) data will end up in publicly available archives after a proprietary period, and can then be analyzed by other groups (e.g. Shi *et al.* 2010). But not only the quantity, also the quality of the data has improved, high sensitivity combined with high spectral resolution allows to investigate line shapes in detail, and combined with high spatial resolution can give a detailed picture of the source structure, both physical and chemical, if analyzed properly.

So, while data taking certainly has taken a major leap forward, we have to look further downstream in the data analysis, which always was a bottleneck. If this bottleneck will not been widened, and widened significantly, a growing fraction of the acquired data stream will go directly to the archive, with only a tiny rivulet branching off to publications. In this article, we will have a critical look at the tools available now, and make suggestions on the further course of developments.

2. Line Identification

The foundation of line identification are molecular line catalogs (JPL†, CDMS‡, Toyama¶, for an enhanced compilation see Splatalogue‖). As will become clear, all this is done by computer analysis, so in a certain sense molecular line data which are not in one of the available databases do not exist. They can be used for dedicated searches and analyses of a handful of lines, but for the analysis of hundreds or thousands of lines they have to be in digital form. There are a number of application programs that make use of these data, and allow direct access to them in programs that can be used to display and manipulate the astronomical data themselves. These are, to the best of our knowledge, XCLASS†† **??**(Schilke *et al.* 2001, Comito *et al.* 2005)]schilke2001,comito2005, CASSIS‡‡(Caux *et al.*, this volume) and WEEDS (Maret *et al.* 2011). They differ in details, but are similar in many respects. Because we know it best, we in the following describe the method of XCLASS only, and refer the readers to the publications and websites of CASSIS and WEEDS for informations on these programs.

† http://spec.jpl.nasa.gov
‡ http://www.astro.uni-koeln.de/cdms
¶ http://www.sci.u-toyama.ac.jp/phys/4ken/atlas
‖ http://www.splatalogue.net
†† https://www.astro.uni-koeln.de/projects/schilke/XCLASS
‡‡ http://cassis.cesr.fr

The line density in the catalogs is high, if all entries are considered, so the naive way of line identification, listing all lines within a certain range in the catalog, is impractical, as typically 20-30 lines are listed, for ranges of 10 MHz. While for reasonably strong lines, it often is possible to make an educated guess, for weaker lines this is not necessarily the case. For very weak lines this procedure does not work with any amount of accuracy even for very experienced specialists. It also depends on an educated guess, which may be easy for people who have spend years of their life analyzing surveys, but hard for graduate students just starting with this topic. So, clearly, a different line identification method is needed.

We have found that the most efficient method is to simulate molecular spectra for line identification purposes. In XCLASS, this is done using the LTE assumption. One can assume several components, whose intensities are just added (i.e. they are assumed to be non-interacting), and each component is calculated to be the emission of a clump of a certain size (to take beam coupling into account), assuming a certain excitation temperature, column density, line width and line offset, and a Gaussian line shape. All molecules have to be modeled simultaneously to take account of blending, even if one is interested in one species only. It has been found (Crockett *et al.*, this volume) that this method manages to model typical hot core spectra with a surprising amount of accuracy, although clearly unrealistic assumptions are made – constant temperature and LTE for each component, i.e. no source structure and deviations from LTE are taken into account. The assumption of LTE is, for many species existing only in the very densest part of the hot core, valid, but fails completely for species that extend to low density envelopes, for example for abundant linear rotors. The source structure to some degree is approximated by multiple components.

This seems to be a lot of effort for mere line identification. It is useful though for identification of all but the most abundant species, as it takes opacity effects and line blending into account. To identify new species with weak lines, this procedure is essential, since in this case one not only has to demonstrate that all predicted lines of the species in the observed range are compatible with observations, i.e. either detected at the predicted strengths or blended with other features, but also has to demonstrate a good overall knowledge of the spectral features. A claimed detection of a new species is more convincing if the amount of unidentified features is 10% rather than 90%. This method has been convincingly demonstrated by Belloche *et al.* (2008, 2009), and is also being used in the identification of the most line-rich surveys coming out of HIFI, the Orion-KL, SgrB2(N) and (M) and NGC6334 spectra.

This method, however, it is not very well suited to describe the chemical structure of a source. One often finds that, if the parameters have been optimized to give the best fit, the components of different molecules have different temperatures and also different line widths, which does not give a coherent picture of the sources. In reality, is due to different distribution of molecules, e.g. a lower excitation temperature will suggest that a molecule has its peak abundance in the outer regions of centrally heated sources, while higher excitation temperatures point to a more central distribution. There is also the complete failure to describe more complex line shapes of lines with very high opacities. This is a common feature of all programs mentioned above. While CASSIS can, for certain molecules, do LVG predictions, it still does not take any source structure into account. To compare with chemical and physical models, one has to take source structure into account in a more explicit fashion, which is the topic of the next section.

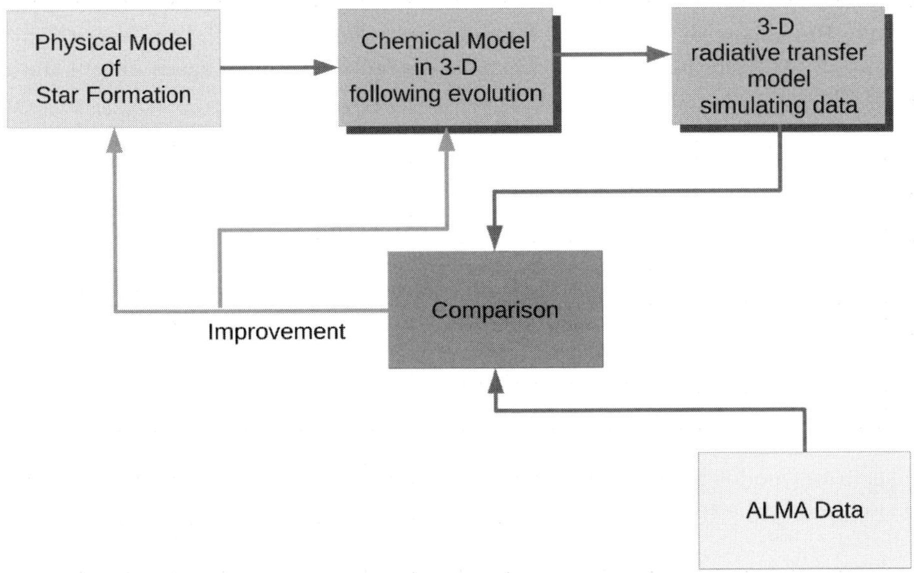

Figure 1. Scheme for modeling ALMA data. To minimize the number of free parameters, it is best to start from physical and chemical models, which then may have to be modified based on the comparison. The comparison itself needs to be done automatically, in a fashion that gives not only the best fit solution, but also information about the quality of the fit, the existence or not of secondary minima, and confidence intervals of parameters.

3. Source Modeling

3.1. *Spherical Modeling*

Systematic radiative transfer modeling of molecular cores in spherical symmetry has been pioneered by the Leiden group (see e.g. van Dishoeck & van der Tak 2000, van der Tak *et al.* 2000), and its basic approach remains valid even today. To minimize the (potentially) large number of free parameters for the source structure, it is parametrized, guided by physical models, e.g. with a power law in density, then first the dust data (maps, fluxes, SEDs) are fitted, preferably using a self-consistent way of calculating the temperatures, given the heating sources (DUSTY (Nenkova *et al.* 2000) has been frequently used for this purpose). Then the structure is frozen, and molecular line fluxes are calculated, using assumptions on the molecular abundances, either ad hoc, or guided by chemical models. Usually, fairly simple abundance laws have been used, either constant abundance, or abundance jumps at some critical temperature, e.g. for ice evaporation. The RATRAN program (Hogerheijde & van der Tak 2000) has been used extensively for this purpose (see also van der Tak, this volume), and can be used also to model line shapes (Chavarría *et al.* 2010, Rolffs *et al.* 2010, 2011), which then also constrain the velocity field. While spherical models are used for sake of simplicity, and because often no constraints on the source structure exists – many sources, particularly hot cores, are fairly small and therefore unresolved in single-dish observations. It becomes in many cases very obvious that deviations from spherical symmetry are present, and cannot be modeled by spherical models. As an example, Rolffs *et al.* (2011) finds cases of centrally heated sources where HCN lines, in spite of very high opacities, are not self-absorbed, which would be inevitable for spherically symmetric models. In this case, stating that the source is non-spherical is where one has to stop, since without additional constraints from high-resolution observations the set of models that would reproduce such a behavior is very large. In cases with known non-spherical symmetry, such as disks, e.g. (Panić *et al.* 2009), two-dimensional models have been successfully used, but again only if the models could be constrained by high-resolution observations.

Deviations from spherical symmetry are noticeable mostly when very high opacities are involved. For optically thin lines, the intensity is an integral over the line-of-sight through the gas, and the exact distribution of abundances and temperatures along the line of sight does not matter. For very optically thick lines, one always looks at a surface, a photosphere. Where this photosphere is, along the line of sight, depends on the temperature, density and abundance distribution, and one observes the excitation temperature at this photosphere. Since the opacity varies across the line, different velocity bins have different opacities, and therefore trace different depths into the cloud, and it is here that deviations from spherical structures are most noticeable. This comes both as a curse and a blessing, since while one has to go through a lot more effort, and construct a much more elaborate model to reproduce the observations, one in the end also has a much more accurate model of the source structure, and it allows access to the third spatial dimension, along the line of sight, which normally is otherwise inaccessible.

3.2. *3-d Modeling*

Hence, spherical modeling is to be abandoned, and 3-d models have to be employed. This comes, as already mentioned, at the price of an even higher complexity, and an even larger number of free parameters. We still feel that the time is ripe for it, since ALMA will give us data of very high quality. That includes high spectral and spatial resolution, high sensitivity and high image fidelity, and this will enable us to constrain many of the free parameters. Another factor is that ALMA will open the submillimeter window for interferometers much more than the SMA has done, which will make multi-line studies feasible and, ALMA TAC willing, the rule rather than the exception. The power of this kind of study to overcome the limitations of having only a 2-d projection of a 3-d structure has been demonstrated by Rolffs *et al.* (2011).

Still, many free parameters remain, and one has to base the models on a set of preconceived source structure, based on physical models of star formation, which form the building blocks of a star-forming region. For example, the source is assumed to be composed to a collection of spherical clumps rather than a single one, each of them having a power law or some other structure that can be parametrized or, instead of clumps, disks. To this one can add outflows, filaments, and an embedding medium. Heating sources are added, the temperature structure is calculated, and the continuum fluxes and shapes are modeled. This then has to be compared with observations, and iterated until a model or a set of models that fit the observations is found. Then, the structure is frozen, and molecules are added, their abundances either based on simple considerations (jump models) or chemical models calculated with the real source structure. This kind of model then has a morphology fixed by observations (positions of clumps, opening angle of flows), and a finite set of free parameters, which have to be optimized (Fig. 1).

In the past, this kind of optimization often has been done manually (Rolffs *et al.* 2011) or, for simple cases, using χ^2 methods and parameter grids (van der Tak *et al.* 2000). This is a problem for multi-parameter models, because it can often be too computationally expensive to run grids spanning the whole multi-dimensional parameter range, and fitting by-eye can never be sure to find the global minimum or indeed, a minimum at all. Ideally, one would like to know the shape of the solution space – are there multiple

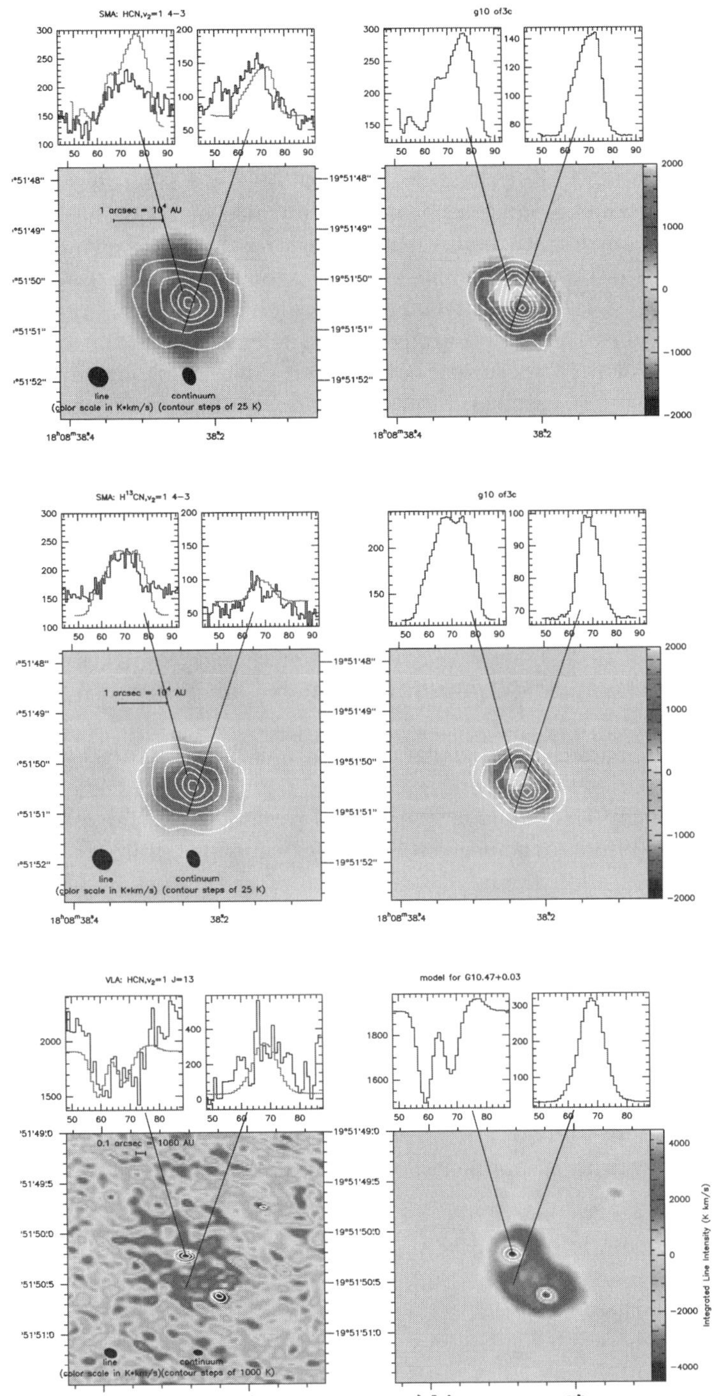

Figure 2. `radmc-3d` model (right) of observed lines (left) of vibrationally excited HCN(4–3) (top), $H^{13}CN$(4–3) (center), and the $J = 13$ ℓ-type line of HCN in G10.47.

minima, other local minima – and the size of confidence intervals for the free parameters. Often, parameters are degenerate, creating shallow minima diagonal to the parameter axes. The model as such may be inappropriate, so even the best fit parameters may not give a good fit – e.g. if one tries to fit an externally heated source with an internally heated source model. Hence, the use of automatic fitting routines (e.g. MAGIX†, Bernst *et al.* 2010) which can provide all this information is advocated.

We have noted the need for 3-d models several times, but have not mentioned any that can be used. The oldest 3-d model we are aware of is the Accelerated Lambda Iteration (ALI) code `MOLLIE`, written by Eric Keto (Keto 1990, Keto *et al.* 2004, Keto & Caselli 2010). Another model based on Monte Carlo methods has been described by Juvela (1998). We concentrate here on two more recent codes, which are available on request to interested parties, and come with an extensive documentation: `radmc-3d`‡ by Kees Dullemond, and LIME¶ by Brinch & Hogerheijde (2010).

`radmc-3d` by origin is a continuum radiative transfer method, so it can self-consistently calculate the temperature structure of cores, given the density distribution and distribution of heating sources. It uses the Monte Carlo method, and an adaptive grid. Its line radiative transfer possesses the ability to do multi-line LTE and LVG approximations, a full radiative transfer package is planned for the near future. Since for many species collision rates are not available, nor seem to be necessary (see discussion of validity of LTE above, and van der Tak, this volume), the then only possible LTE method is a valid choice for modeling many molecules.

LIME does not calculate the temperature structure, so this is an input parameter at present. It also uses a Monte Carlo method, but unstructured Delaunay latices instead of an adaptive grid. It does exact radiative transfer though, so it is at present the model of choice for cases where collision rates are available, and where LTE or even LVG are supposed to fail.

An example of an `radmc-3d` model of SMA observations can be seen in Fig. 2. The model employed here has two heating sources (seen as UCHII regions in the lower figure, at 7 mm). The density distribution consists of a distribution of clumps with random velocities. Both the clumps themselves and the clump distribution follow a Plummer-profile, the temperature is calculated self-consistently. An outflow which points toward the observer has been added. The clumps themselves are spatially unresolved, but manifest themselves in the line shapes of the rotational HCN lines in the v_2 vibrational state which are, in spite of very high optical depths, not self-absorbed. This so far is based on the analysis of just the vibrationally excited lines, because we used only LTE, which is not appropriate for the ground state lines, but it gives a glimpse of the potential that lies in this kind of modeling.

4. Outlook

Our goal is it to cast the flow of Fig. 1 into a pipeline which runs without much human intervention, apart from of course defining the input structure and the free parameters, and the judging of the output. The building blocks of this pipeline exist, but they have to be connected together. We feel the need for a semi-automatic pipeline because ALMA will produce hundreds of data cubes in a short time, and in each source one will have to fit many species. This will allow to put star formation research on a much more reliable statistical basis. We will be able to extract the fragmentation structure, the velocity field – how does the global infall of a cluster break up into the individual sources? – density, temperature and chemical structure for hundreds of sources spanning a large range

† http://www.astro.uni-koeln.de/projects/schilke/MAGIX
‡ http://www.ita.uni-heidelberg.de/ dullemond/software/radmc-3d/
¶ http://www.strw.leidenuniv.nl/ brinch/website/limecode.html

in masses, environmental conditions and age, and isolate commonalities and individual features.

4.1. *Caveats*

We have advocated the use of prefabricated building blocks, physical entities of known, or at least parameterizable structure, and the same is true for velocity fields. For the observer and for modeling purposes, such a method is easy to apply. However, modern star formation models are mostly not analytical. They use MHD calculations (e.g. Krumholz *et al.* 2009, Wang *et al.* 2010, Peters *et al.* 2010), and are thus not directly comparable with observations of a specific region. One has to find a metric to do this comparison, and one way is to use the building blocks for observations, and analyze the models, folded with chemical models and processed through the same radiative transfer models, in the same way. Depending on how good the fits are, one might want to enlarge the set of primitive building blocks. An alternative would be to derive some metric that is independent on building blocks, such as Probability Distribution Functions (PDFs) of column densities or velocity centroids, or spatial correlation functions. Growing experience with modeling ALMA data will converge on the best comparison methods.

References

Belloche, A., Menten, K. M., Comito, C., Müller, H. S. P., Schilke, P., Ott, J., Thorwirth, S., & Hieret, C. 2008, *A&A*, 482, 179

Belloche, A., Garrod, R. T., Müller, H. S. P., Menten, K. M., Comito, C., & Schilke, P. 2009, *A&A*, 499, 215

Bergin, E. A., *et al.* 2010, *A&A*, 521, L20

Bernst, I., Schilke, P., Möller, T., Panoglou, D., Ossenkopf, V., Röllig, M., Stutzki, J., & Muders, D. 2010, Astronomical Society of the Pacific Conference Series, 442, 505

Brinch, C. & Hogerheijde, M. R. 2010, *A&A*, 523, A25

Ceccarelli, C., *et al.* 2010, *A&A*, 521, L22

Chavarría, L., *et al.* 2010, *A&A*, 521, L37

Comito, C., Schilke, P., Phillips, T. G., Lis, D. C., Motte, F., & Mehringer, D. 2005, *ApJS*, 156, 127

Hogerheijde, M. R. & van der Tak, F. F. S. 2000, *A&A*, 362, 697

Juvela, M. 1998, *A&A*, 329, 659

Keto, E. R. 1990, *ApJ*, 355, 190

Keto, E. & Caselli, P. 2010, *MNRAS*, 402, 1625

Keto, E., Rybicki, G. B., Bergin, E. A., & Plume, R. 2004, *ApJ*, 613, 355

Krumholz, M. R., Klein, R. I., McKee, C. F., Offner, S. S. R., & Cunningham, A. J. 2009, *Science*, 323, 754

Maret, S., Hily-Blant, P., Pety, J., Bardeau, S., & Reynier, E. 2011, *A&A*, 526, A47

Müller, H. S. P., Schlöder, F., Stutzki, J. & Winnewisser, G. 2005 *J. Mol. Struct.* 742, 215

Müller, H. S. P., Thorwirth, S., Roth, D. A., & Winnewisser, G. 2001, *A&A*, 370, L49

Nenkova, M., Ivezić, Ž., & Elitzur, M. 2000, *Thermal Emission Spectroscopy and Analysis of Dust, Disks, and Regoliths*, 196, 77

Panić, O., Hogerheijde, M. R., Wilner, D., & Qi, C. 2009, *A&A*, 501, 269

Peters, T., Klessen, R. S., Mac Low, M.-M., & Banerjee, R. 2010, *ApJ*, 725, 134

Pickett, H. M., Poynter, R. L., Cohen, E. A., Delitsky, M. L., Pearson, J. C., & Müller, H. S. P. 1998 *J. Quant. Spectrosc. & Rad. Transfer 60, 883*

Rolffs, R., *et al.* 2010, *A&A*, 521, L46

Rolffs, R., Schilke, P., Wyrowski, F., Menten, K. M., Güsten, R., & Bisschop, S. E. 2011, *A&A*, 527, A68

Schilke, P., Benford, D. J., Hunter, T. R., Lis, D. C., & Phillips, T. G. 2001, *ApJS*, 132, 281

Shi, H., Zhao, J.-H., & Han, J. L. 2010, *ApJ*, 710, 843

van der Tak, F. F. S., van Dishoeck, E. F., Evans, N. J., II, & Blake, G. A. 2000, *ApJ*, 537, 283

van Dishoeck, E. F. & van der Tak, F. F. S. 2000, *From Molecular Clouds to Planetary Systems*, Proc. IAU Symposium No. 197, p. 97

Wang, P., Li, Z.-Y., Abel, T., & Nakamura, F. 2010, *ApJ*, 709, 27

Discussion

GUÈLIN: I fully understand that we need to turn to automatic reduction routines in view of the large number lines present in the spectra. Yet I am worried about the residual of automatic line fitting and removal. Line shapes vary with level energy and with optical depth an there will remain ghosts that may be assigned by mistake to astronomical valuable information.

SCHILKE: There obviously is some danger associated to a black box approach and results should not be taken blindly, but with careful visual control by humans. Some safeguards can also be built into the fitting programs so that e.g. the significance of results obtained in the vicinity of strong lines can be weighted down, so that some residual outflow wings will not be interpreted as new spectral features.

The Molecular Universe
Proceedings IAU Symposium No. 280, 2011
José Cernicharo & Rafael Bachiller, eds.

doi:10.1017/S1743921311025191

Radiative Transfer and Molecular Data for Astrochemistry

Floris van der Tak

SRON Netherlands Institute for Space Research, Landleven 12, 9747 AD Groningen, The Netherlands
email: vdtak@sron.nl

Abstract. The estimation of molecular abundances in interstellar clouds from spectroscopic observations requires radiative transfer calculations, which depend on basic molecular input data. This paper reviews recent developments in the fields of molecular data and radiative transfer. The first part is an overview of radiative transfer techniques, along with a "road map" showing which technique should be used in which situation. The second part is a review of measurements and calculations of molecular spectroscopic and collisional data, with a summary of recent collisional calculations and suggested modeling strategies if collision data are unavailable. The paper concludes with an overview of future developments and needs in the areas of radiative transfer and molecular data.

Keywords. radiative transfer, atomic data, molecular data

1. Introduction

At the heart of observational astrochemistry is the estimation of molecular abundances from telescope data. Extracting the information out of the data requires radiative transfer calculations which need molecular input data, and these two topics are discussed in this paper. In many cases, measuring the chemical composition of interstellar gas is coupled to the estimation of its physical conditions, in particular the kinetic temperature and the volume density. In addition, molecular abundances usually need interpretation with a chemical model, to estimate parameters such as the age of the cloud or its ionization rate by cosmic rays. These two important related topics are not covered here.

Before starting the review, it is useful to consider the limitations of radiative transfer calculations. Some limitations are of observational nature, such as resolution mismatch between observations at different wavelengths and the presence of unresolved substructure in the data. In addition, the lines in a spectrum may not all sample the same gas, for instance when some lines appear in emission and others in absorption. Furthermore, the observations may only cover part of the source, and only a few molecular energy levels. Finally the radiative transfer calculations may be limited by the availability of molecular data: either spectroscopic information or collisional excitation probabilities. When interpreting observations, it is important to keep these limitations in mind and to choose the model such to minimize their impact.

The field of molecular data and radiative transfer is large, and due to space limitations, this paper cannot treat every aspect in depth. This review focuses on the analysis of spectral lines; see Pascucci *et al.* (2004) for a discussion of continuum radiative transfer and Shirley *et al.* (2011) for a recent update about dust opacities. In the area of line radiative transfer, see Peraiah (2001) for an introduction and Pavlyuchenkov *et al.* (2007) for recent developments. For detailed accounts on laboratory spectroscopy, see Pickett *et al.* (1998) and Müller *et al.* (2005) and references therein. The book by Flower (2003) reviews the field of interstellar molecular collisions. Finally, the estimation of physical parameters of interstellar clouds is reviewed by Evans (1999). Note that throughout this review, the term 'molecules' includes mono-atomic and ionic species.

2. Analyzing molecular line spectra

The goal of radiative transfer calculations is to estimate physical parameters from astronomical observations. In the case of astrochemistry, the prime application of radiative transfer is the estimation of molecular abundances from line spectra at infrared and (sub)millimeter wavelengths. This review focuses on the analysis of radio and far-infrared data, but most of the mentioned techniques are also applicable into the mid-infrared. The near-infrared, visible and ultraviolet spectra of circumstellar envelopes and disks are also of interest in astrochemistry and make additional demands on molecular data and radiative transfer; however, these issues are beyond the scope of this paper.

There is no single best method for radiative transfer calculations; rather, the amount and nature of available data and other constraints determine the appropriate level of analysis. For example, the use of non-LTE models is often limited by the availability of molecular collision data (§ 3.2). Also, applying a two-dimensional model with 1000 grid points is clearly overkill if only a handful of spectral lines have been observed in a single direction. On the other hand, the increasing data volumes from modern telescope facilities demand ever faster and more sophisticated models for the data analysis, and developing such tools is an active area of work. This section reviews radiative transfer models at three different levels, ranging from fairly basic LTE models to fairly complex non-local models in 2D or even 3D.

2.1. *Single excitation temperature ('LTE') models*

The simplest way to estimate a molecular column density from an observed line intensity is to avoid radiative transfer altogether, and the first step in doing so is the estimation of the molecular column density in the initial state of the observed line (upper state for emission lines, lower state for absorption lines). The conversion from line intensity to optical depth is straightforward for absorption lines, but involves assuming a beam filling factor if the lines appear in emission. Observations at higher angular resolution help to constrain this factor if they are available; otherwise a point source and a very extended source provide useful limiting cases.

The second step is to account for the other, unobserved states of the molecule, and the common approach is to assume that the energy levels are populated according to the Boltzmann distribution. The characteristic 'excitation temperature' ($T_{\rm ex}$) is a priori unknown, but is often taken to equal $T_{\rm kin}$, which is the case of local thermodynamic equilibrium (LTE) which applies at high densities when collisions determine the excitation. At low densities, radiative decay competes with collisions, so that generally $T_{\rm ex} < T_{\rm kin}$, but even in this case, this method is often (misleadingly) called the LTE method.

If several lines from a range of energy levels of the same molecule have been observed, their relative intensities can be used to constrain $T_{\rm ex}$. The most direct way to do so is to fit a straight line to the line intensities as a function of upper level energy. When rotational lines are probed, the popular name 'rotation diagram' is correct, but in general, the alternative name 'Boltzmann plot' is more appropriate. For low line optical depths, this method yields reliable estimates of $T_{\rm ex}$ and molecular column density, both for submillimeter emission lines (e.g., Kristensen *et al.* 2010) and for visual/infrared absorption lines (e.g., Kaźmierczak *et al.* 2010). However, at high column densities, the derived $T_{\rm ex}$ is inaccurate because of optical depth effects, and correction factors must be applied

(Goldsmith & Langer 1999). The advantage of this improved method (sometimes called the 'population diagram') is that it also provides an estimate of the size of the emitting region. An example of combining the two methods is the analysis of CH_3OH line images of the Cep A region, where classic rotation diagrams work well for the diffuse extended gas, while optical depth correction are needed close to the central protostar (Torstensson *et al.* 2011).

Many programs to perform 'LTE' calculations exist, but one popular program is CASSIS which is publicly available†. Besides multi-line models for single species, this program is capable of synthesizing wide-band spectra based on models for many molecules, each with their own set of excitation conditions.

2.2. *Non-LTE models*

The LTE method is fast and applies to any molecule with a known spectrum, but its assumption of a Boltzmann distribution for the level populations often does not hold, and measuring $T_{\rm ex}$ is often difficult. More sophisticated non-LTE methods retain the assumption of a local excitation, but solve explicitly for the balance between excitation and de-excitation of molecular energy levels. Both collisional and radiative processes contribute to this balance, and these 'non-LTE' models require molecular collision data as input, in addition to spectroscopic data. This requirement limits the use of non-LTE models as collision data do not exist for all astrophysically relevant species. However, the advantage of non-LTE models is that not only the column density but also the kinetic temperature and the volume density of the gas can be constrained from multi-line observations.

There are two main ways in which non-LTE models treat the effect of nonzero line optical depth. One way is the Sobolev approximation, where the cloud is assumed to have a significant velocity gradient, of the order of $1\,{\rm km\,s^{-1}\,pc^{-1}}$ or more. The resulting Doppler shift of the line profile decouples the excitation at one end of the cloud from radiation from the other, and has the effect of localizing the radiative transfer problem. Such models match observed spectra, but the reliability of the derived parameters is unclear since most clouds do not have systematic velocity gradients (e.g., Ossenkopf 1997). A somewhat better treatment of optical depth effects is the 'escape probability' formalism, in which the chance for photons to leave the cloud is based on the current local estimate for the optical depth and an assumed geometry. Both treatments require iteration because the molecular excitation and the radiation intensity are mutually dependent, but on modern computers, convergence is fast.

Many programs to perform non-LTE calculations exist, but one popular program is RADEX (Van der Tak *et al.* 2007) which is publicly available‡. This program exists as an on-line calculator for quick estimates of line intensities from assumed column densities or vice versa, as a module inside the CASSIS package, and as a stand-alone program for extensive calculations, for instance over a grid of kinetic temperatures and volume densities. An example of this latter use is the work by Ginsburg *et al.* (2011) who constrain the densities of molecular gas close to ultracompact H II regions with non-LTE models of H_2CO line ratios.

Multi-zone non-LTE models divide a cloud into grid cells and calculate the escape probabilities for photons from one cell into each of the others. These programs are much faster than true non-local radiative transfer programs but are often limited to plane-parallel geometry (Elitzur & Asensio Ramos 2006) or particular molecules (Poelman &

† `http://cassis.cesr.fr/`
‡ `http://www.sron.rug.nl/~vdtak/radex/index.shtml`

Spaans 2005). Versions of the multi-zone escape probability method in spherical geometry exist as well (Maloney 1987; see Van Dishoeck *et al.* 1991), and the results appear to be in good agreement with those of non-local methods (Yun *et al.* 2009).

2.3. *Non-local models*

The most sophisticated radiative transfer programs use non-local methods: they drop the assumption of a local excitation and solve for the molecular excitation as a function of position in the cloud. To estimate the local radiation field at all line frequencies due to radiation from everywhere else in the cloud, two main methods exist. Accelerated Lamda Iteration (ALI) program discretize the cloud onto a grid of points and split off the local contribution from the full radiation field at each point. Monte Carlo (MC) methods divide the cloud into grid cells and send photon packages in random directions from each cell. The performance and convergence of both types of programs has been tested by Van Zadelhoff *et al.* (2002) for HCO^+ and by Van der Tak *et al.* (2005) for H_2O. The high accuracy of these models comes at a price: for complicated model geometries, large gradients in excitation conditions, and molecules with many energy levels, the calculations can take several hours.

Several programs to perform non-local radiative transfer calculations exist, but one popular program is RATRAN (Hogerheijde & van der Tak 2000). The 1D version of this Monte Carlo program is publicly available† and the 2D version is available upon request. Recently, the program LIME has appeared (Brinch & Hogerheijde 2010), which is capable of doing 3D non-local calculations. This program is not publicly available, but the authors provide a copy upon request‡.

Programs such as RATRAN and LIME are useful to constrain the physical (temperature, density and velocity) structure of astrophysical sources, and also to estimate molecular abundances as a function of position in a cloud. An example of the former is the work by Panić *et al.* (2009), who model interferometric images of IM Lup in the CO J=2–1 line to constrain the 2D density structure of its protoplanetary disk. An example of the latter is the work by Chavarría *et al.* (2010), who model multi-line Herschel-HIFI observations of H_2O toward the high-mass protostar W3 IRS5 to constrain the distribution of H_2O within the protostellar envelope.

3. Molecular input data

The interpretation of interstellar line spectra with radiative transfer calculations generally requires two kinds of molecular input data: spectroscopic data (energy levels, statistical weights, line frequencies, and Einstein A coefficients) and collision data (for all but the simplest LTE models). This section describes how such data are obtained, what the results are, and which limitations exist.

3.1. *Spectroscopic data*

Characterizing the spectrum of a molecule begins with measuring its line frequencies in the laboratory. Many species are available commercially; many others can be produced in the reaction between two or more species. The molecules are let into a gas cell, which is typically a few meters long and made of glass. The gas cell is pumped to high vacuum, a radiation source is put at one end and a spectrometer at the other.

The bandwidth over which the line frequencies are recorded depends on the measurement technique, and is limited by atmospheric transmission. Ideally, spectra are measured

† `http://www.sron.rug.nl/~vdtak/ratran/`
‡ `http://www.strw.leidenuniv.nl/~brinch/website/limecode.html`

in the microwave, (sub)millimeter, and near- and mid-infrared ranges. The resolution is usually limited by Doppler broadening of the lines, which amounts to $R = \Delta\nu/\nu_0 \sim 10^{-6}$ or 0.1 MHz at 100 GHz for air at 300 K. Special techniques to achieve sub-Doppler resolution exist but are not always available.

The second step is to apply a Hamiltonian model and to assign quantum states to the measured lines. Even for linear molecules, this step is not trivial, since absorption spectra at room temperature usually contain several vibrational states, and sometimes multiple isotopologues may contribute. For large molecules with many vibrational modes, the number of parameters in the Hamiltonian becomes large; see Xu *et al.* (2004) for an example with internal rotation with ≈50 parameters.

The standard program to perform Hamiltonian model fits is `spfit` by Herb Pickett; for certain specialized situations, other programs exist. Besides line frequencies and assignments, such programs need the molecular dipole moment, which is often calculated ab initio, or sometimes measured directly (e.g., Botschwina & Oswald 2008). The output from these programs is a list of the complete molecular line spectrum, indicating where this prediction deviates from the measurements. These deviations may then be used to change some assignments in order to optimize the overall quality of the fit.

The results of the spectroscopic work are collected in databases: JPL† (Pickett *et al.* 1998) and CDMS‡ (Müller *et al.* 2005) are the leading efforts at submillimeter wavelengths and HITRAN¶ (Rothman *et al.* 2009) at infrared wavelengths. As of Summer 2011, the submillimeter databases contain well over 500 entries, including many species, isotopologues, and vibrational states. The HITRAN database covers fewer species than JPL and CDMS, but many more vibrational states, as this database is aimed at atmospheric research.

Although the completeness and accuracy of the current spectroscopic work is sufficient for many astrophysical applications, areas exist where astrochemical work is limited by spectroscopy. One example are the many unidentified features in spectral line surveys, especially at frequencies >1 THz (e.g., Bergin *et al.* 2010). In this regime, line frequencies cannot be measured directly in the lab, so that their identification relies on Hamiltonian predictions which are sometimes quite uncertain.

For large molecules with significant internal structure, a Hamiltonian fit to the laboratory spectra and assigning each of the many lines from the various bending/torsional/vibrational states is impractical. Instead, Fortman *et al.* (2010) have measured the spectra of several such molecules at various temperatures so that features in spectral line surveys can be assigned to particular species, if not to specific transitions. Such efforts are essential for the reliable identification of new molecular species in crowded astronomical spectra (e.g., Belloche *et al.* 2008).

In some cases, astronomical observations are useful to characterize molecular spectra. The intrinsic line width of cold, quiescent interstellar cloud cores is smaller than achievable in the laboratory, so that higher spectral resolution can be obtained. In addition, certain species are difficult to produce on Earth or very short-lived under terrestrial conditions. Examples include the spectra of N_2H^+ and N_2D^+ (Pagani *et al.* 2009a), DCO^+ (Caselli & Dore 2005) and DNC (Van der Tak *et al.* 2009).

3.2. *Collision data*

Besides spectroscopic data, modeling interstellar line spectra requires knowledge of the probabilities for (de-)excitation of the molecule upon collision with other molecules. In the

† `http://spec.jpl.nasa.gov/ftp/pub/catalog/catform.html`
‡ `http://www.astro.uni-koeln.de/cdms/`
¶ `http://www.cfa.harvard.edu/hitran/`

case of interstellar clouds, the most common collision partner is H_2, but collisions with H, He, H^+ and electrons are also sometimes relevant. Collisional excitation probabilities are usually cast in the form of rate coefficients (with dimension $cm^3\,s^{-1}$) which are collisional cross-sections (cm^2) integrated over velocity ($cm\,s^{-1}$). These collision data are the results of quantum-mechanical calculations, with the occasional experiment for verification. Only downward collisions need to be calculated: the upward rates follow from detailed balance.

The calculation of molecular collision data is a two-step process: first the potential energy surface of the collisional system is calculated, and second the dynamics of the interaction are evaluated, taking the Maxwell average of the velocity distribution. These calculations are time-consuming, which is why early work (before ≈2000) concentrated on closed-shell molecules, used He as collision partner, and adopted simplified methods for the calculation of the potential energy surface and the dynamics. Thanks to recent advances in computer power and generous funding from the European Union, modern calculations usually consider H_2 as collision partner, use accurate potential energy surfaces and close-coupling level dynamics, while treatment of open-shell molecules such as CN is now also possible.

Experimental verification of collisional calculations is sometimes possible, albeit for a limited number of states. Besides pressure boadening experiments (see § 3.4) which measure the quality of the potential energy surface as a whole, detailed state-to-state experiments are valuable tests of quantum-mechanical computations. One recent example is the study of Asvany *et al.* (2009) on the H_2D^+ – H_2 system, where both inelastic and reactive collisions play a role.

Due to the computational limitations mentioned above, the field of molecular collisions is not nearby as advanced as that of spectroscopy. As of Summer 2011, the two leading databases, LAMDA† (Schöier *et al.* 2005) and BASECOL‡ (Dubernet *et al.* 2006b) contain collision data for 32 molecules. This number is only a fraction of the ≈160 molecules species known to exist¶ in interstellar and circumstellar media (H_2O_2 being the latest addition; Bergman *et al.* 2011). Since the detection rate of interstellar molecules is similar to the calculation rate of collision data (a few molecules per year), this gap is unlikely to close – in fact it may even widen with time. Moreover, coverage of isotopologues and vibrational states is usually limited to the most common species in the ground state.

3.3. *Recent collision calculations*

Since the previous review on this topic (Van der Tak & Hogerheijde 2007), many new calculations of molecular collision cross sections have appeared in the literature. Below is a list of calculations from the past 3 years, which may not be complete.

The calculations for H_2D^+ and D_2H^+ by Hugo *et al.* (2009) are a major step forward, even if they are based on a statistical approach rather than quantum scattering calculations. These ions are crucial for the study of cold and dense pre-stellar cloud cores (Caselli *et al.* 2008; Parise *et al.* 2011). Some of the state-to-state rates have been verified by experiment, which is important for the study of the ortho-para ratio of H_2, for example (Pagani *et al.* 2009b).

New collision rates for H_2CO with H_2 have been published by Troscompt *et al.* (2009), which are valuable because H_2CO line ratios are often used to measure the kinetic temperature (and sometimes the volume density) of interstellar clouds (e.g., Mühle *et al.* 2007). The new rates are however incomplete as only the ortho species is covered; hopefully the rates for para-H_2CO will follow soon.

† `http://www.strw.leidenuniv.nl/~moldata/`
‡ `http://basecol.obspm.fr/`
¶ `http://www.astro.uni-koeln.de/cdms/molecules`

The calculation of collision rates of HCN and HNC with He (Sarrasin *et al.* 2010; Dumouchel *et al.* 2010) are useful because the HCN/HNC ratio is often used as a chemical diagnostic of temperature (Loenen *et al.* 2008; Roberts *et al.* 2011). Previously, HCN rates were used for HNC, which turns out not to be accurate at all. Recent observations of the $H^{13}CN/HN^{13}C$ ratio in a sample of dense cores are in good agreement with the newly calculated collision rates (Padovani *et al.* 2011b).

The collision rates for O_2 with He by Lique (2010) have appeared just in time for their use to interpret the definitive detection of interstellar O_2 with Herschel (Goldsmith & the HOP team 2011). This result is an important confirmation of the earlier tentative detection with Odin (Larsson *et al.* 2007). Although the excitation of O_2 is usually not far from LTE, the new rates are useful to quantify this effect.

New results have also appeared for the CN radical (Lique & Kłos 2011), which is often used as a probe of dissociating ultraviolet radiation (e.g., Pérez-Beaupuits *et al.* 2009), especially together with HCN. The new collision rates are especially useful because they include the hyperfine structure of CN, which allows quantitative estimates of the optical depths of its emission lines.

Models for high-J lines of CO have become much more accurate with the collision calculations by Yang *et al.* (2010). These data are important for example to model Herschel observations of the CO emission from young stellar objects which heat their surroundings to several 100 K by UV irradiation and shocks (e.g., Van Kempen *et al.* 2010).

One key interstellar organic molecule is CH_3OH, and its lines are a diagnostic of both the kinetic temperature and the volume density of interstellar clouds (Leurini *et al.* 2007). The new collision calculations by Rabli & Flower (2010) bring models of CH_3OH excitation to higher accuracy. The coverage of the first torsionally excited state also enables to constrain the intensity of the far-infrared radiation field within astrophysical sources. Torsionally excited emission has already been observed in some objects (e.g., Torstensson *et al.* 2011) and will become routine with ALMA.

The excitation of interstellar H_2O is a long-standing problem in astrophysics. While the main isotope is treated in the following section, we note here the recent calculations for HDO by Wiesenfeld *et al.* (2011b) and for D_2O by Scribano *et al.* (2010). The HDO spectrum is of particular interest as a probe of the depletion of molecules onto grain surfaces (e.g., Parise *et al.* 2005), whereas D_2O is only detectable under extreme conditions (Butner *et al.* 2007; Vastel *et al.* 2010).

Other noteworthy collision calculations from recent years are those for NH–He (Toboła *et al.* 2011), CH^+–He (Turpin *et al.* 2010), SO_2–H_2 (Cernicharo *et al.* 2011), and NHD_2–H_2 (Wiesenfeld *et al.* 2011a).

In regions with strong radiation fields such as PDRs, collisions with electrons may contribute to the excitation of molecules. The calculation of collisional excitation probabilities with electrons is a specialized field with its own approximations and pitfalls, as reviewed by Faure *et al.* (2009). Examples of recent calculations are H_3^+–e^- (Kokoouline *et al.* 2010), CS–e^- (Varambhia *et al.* 2010), H_2O–e^- (Zhang *et al.* 2009), SiO–e^- (Varambhia *et al.* 2009) and the calculations by Faure *et al.* (2007b) for the excitation of HCN, HNC, DCN and DNC in collision with electrons.

3.4. *H_2O as an example*

The excitation of interstellar H_2O is a long-standing problem in astrophysics, which remains to be solved today. At this moment, several calculations exist for its collision rates, which apply to different situations. At the low temperatures ($\lesssim 20$ K) as occur for instance in dense pre-stellar cores and the outer zones of protoplanetary disks, the detailed quantum-mechanical calculations by Dubernet *et al.* (2006a) are the best rates to use. For higher temperatures (up to $\sim$300 K), such as in modeling protostellar envelopes and photon-dominated regions, the quasi-classical rates by Faure *et al.* (2007a) are to be preferred, which at low temperatures join smoothly with the full quantum results. Finally at high temperatures ($\gtrsim$ 300 K) where vibrational excitation becomes important, the rates by Faure & Josselin (2008) should be used.

The existence of one set of collision rates for H_2O that applies to all temperatures is obviously desirable, and work is ongoing towards this goal (Dubernet *et al.* 2009; Daniel *et al.* 2010). Currently the quantum calculations for H_2O at high temperature have only been published for collisions of ortho-H_2O with para-H_2 and para-H_2O with ortho-H_2, so that 50% of the rates is unavailable. Publication of the remaining rates would be very useful for the modeling of H_2O observations, in particular with Herschel (e.g., Van Dishoeck *et al.* 2011). Use of the older H_2O-He rates by Green *et al.* (1993) is not recommended, as these numbers lie factors 3–4 below the corresponding rates with H_2, even after scaling for the different masses.

The H_2O molecule is one of the rare cases where collision calculations have been verified in a laboratory experiment. While direct verification of state-to-state collision rates is limited to a few special circumstances (see § 3.2), pressure broadening experiments offer a way to measure the total inelastic collision rates to and from a given molecular energy level. Thus, measuring the derivative of the line width as a function of gas pressure is a way to test the accuracy of the potential energy surface that is also used to calculate collision rates.

Dick *et al.* (2009, 2010) have measured the pressure broadening of the ground-state lines of ortho- and para-H_2O in collisions with H_2 gas as a function of temperature. They find that at temperatures above $\approx$80 K, the slope of the pressure broadening curve as a function of temperature agrees well with theoretical predictions using the 5D potential energy surface by Valiron *et al.* (2008), and that the vertical level of the curve indicates an ortho-para ratio of $\approx$3 for the H_2 gas. However, at temperatures below $\approx$80 K, the experiments indicate a sharp drop of the pressure broadening coefficient, which is in apparent contradiction with the theoretical predictions.

As Wiesenfeld & Faure (2010) point out, the disagreement between theory and experiment may only be apparent, because the interpretation of the experimental data depends critically on the approximation that the molecules do not rotate during their collision. However, new experiments by Drouin *et al.* (2011) show that this so-called 'impact approximation' holds even in the case of collisions with H_2 at low temperatures, and that instead, conversion of ortho- into para-H_2 in the gas cell has caused the apparent decrease in the measured pressure broadening coefficients at $T < 80$ K.

State to state experiments provide an even stronger test of a collisional potential energy surface. In the case of the H_2O–H_2 system, Yang *et al.* (2011) have achieved a confirmation of the quantum mechanical calculations using a crossed molecular beam experiment where the angular velocity distributions of fully specified rotationally excited final states are obtained using a technique called velocity map imaging. The experimental data agree with the theoretical predictions in impressive detail.

3.5. *What if collision data do not exist?*

Astronomers are often in the situation that collision data are unavailable for the molecular line that they have observed, so that they need workarounds. One option is to assume LTE and accept the limited accuracy of the model prediction, but often it is possible to use existing collision data with minor modifications.

For isotopologues, adopting the collision data for the main species is a common strategy. This method works well for species like $C^{34}S$, which have a similar structure as the main species, so that the potential energy surface of the collision is the same. The mass difference with the main species $C^{32}S$ is also small, so that the collision dynamics are similar. On the other hand, neglecting isotope effects is not an option for species like HDO, which lacks the ortho/para moieties of its main species H_2O, and where the mass effect is also larger.

Collision data for vibrationally excited states exist for very few molecules only. The collisional excitation of vibrational modes is therefore often neglected, which is a good approximation if the radiative decay from these levels is fast, as is often the case. For low-lying vibrational states with small Einstein A coefficients, LTE models are the only option if collision data do not exist.

For certain molecules, it is possible to borrow the collision data from other species with a similar structure. Such 'substituted' rates are generally good enough for non-LTE estimates of molecular column densities, but not good enough to constrain the kinetic temperature or the volume density of the parent gas. Cases where substitution works well include H_2S and H_2CS, where H_2O and H_2CO data are used respectively, after scaling the collision rates for the difference in mass (see Schöier *et al.* 2005 for details). Similarly, it should be possible to use scaled HF rates (Reese *et al.* 2005) for HCl. The same approach is sometimes applied to cases like HOC^+, where HCO^+ rates are used, but the result is less accurate because the dipole moments of the two species differ significantly.

Frequently, collision data only exist for low-excitation lines of a molecule, for a limited range of temperatures. To model observations of high-excitation lines of this molecule, or to model conditions outside the range for which collision data exist, extrapolation of the existing data is the most common strategy. Schöier *et al.* (2005) make recommendations on how to perform such extrapolations for small molecules, and Faure *et al.* (2008) do so for larger molecules.

Finally, cases exist where only collision data with He as partner are available. The common approach is to scale these rates to the different mass of the H_2 molecule, which only takes care of the dynamics part of the interaction. Dedicated studies show that the potential energy surfaces of X-He and X-H_2 sometimes differ substantially, especially if collisions with ortho-H_2 with its non-zero quadrupole moment are included (e.g., Cernicharo *et al.* 2011). Collisional excitation rates with H_2 appear to be larger by factors of 3–10 than expected by scaling He rates, as indicated by several calcuations and also some experiments (Mengel & De Lucia 2000).

4. Future developments

In the coming years, the field of radiative transfer and molecular data will undergo rapid changes to keep up with the increasing data rates and higher precision of new generations of telescopes and instruments. Comparison of observations and models by hand will become impossible, and automated routines will appear which will search large pieces of parameter space for the best match to a set of observations. While such tools exist already at the individual level, the development of a parameter search engine for general use is actually a major undertaking. One such attempt is the CATS project (see the contribution by Schilke to these proceedings).

A second, more ambitious project is ARTIST (Padovani *et al.* 2011a), which besides a parameter search engine aims to develop a coupled radiative transfer program for dust and gas, where the thermal balance calculation is integrated with the radiative transfer. Like CATS, a link to solving chemical networks is foreseen, which is important for highly reactive species such as OH^+ and H_2O^+. Third, ARTIST aims to develop continuum radiative transfer tools with polarization capability, which is of interest for ALMA observations which will be sensitive to polarization.

Another future development is the merging and streamlining of the various databases. A few years ago, the Splatalogue† has appeared as a useful front-end to the JPL and CDMS spectroscopic databases, allowing for easy searching and switching between various units. One useful addition would be the option to create a state list and to calculate partition functions at any temperature. At some point, a comprehensive database of spectroscopic and collision information may appear as the VAMDC (Dubernet *et al.* 2010). While such meta-databases certainly make the access to molecular data easier for non-expert users, close ties between database developers and data sources are essential to ensure that the information in the database is up-to-date, and to ensure that the spectroscopists, physicists, and chemists who produce the data get proper credit for their important work.

Last but not least, the future will see new calculations of molecular collision rates. Commonly observed molecules for which no collision data currently exist include CH and C_2H; for the latter, calculations are ongoing in Meudon and Le Havre. The collisional excitation of the hyperfine transitions of NH and ND is also being studied in Le Havre, while the Grenoble group is working on the H_2O isotopologs. Other species for which collision calculations would be useful are OH^+ and H_2O^+, especially to model the observation of their emission lines in some extragalactic systems (Van der Werf *et al.* 2010). For 'large' molecules (with more than $\approx$4) atoms, the quantum-mechanical calculation of collision potentials will likely remain impractical for the foreseeable future, and developing recipes or approximations for their collisional excitation probabilities would be very useful for the interpretation of spectral line surveys with Herschel and ground-based telescopes. See Faure *et al.* (2011) for the example of $HCOOCH_3$. Calculations of the excitation of atomic fine structure lines in collisions with H_2 would be helpful to understand the physical structure of photon-dominated regions and 'warm dark gas' (Langer *et al.* 2010): for O, C, N^+ and C^+, only fairly old ($\approx$1990) data exist, while for S, no calculations exist at all. Finally, inclusion of electronic and vibrational energy levels and radiative transitions into molecular databases will be crucial to quantify the radiative excitation rotational lines through infrared, optical and ultraviolet pumping.

Acknowledgement

This paper is dedicated to the memory of Fredrik Larsen Schöier, the initiator of the LAMDA database, who passed away on 14 January 2011 at age 41. The author thanks K.-S. Wang, C. Brinch and Z. Nagy for input, V. Ossenkopf, P. Goldsmith, F. de Lucia, S. Schlemmer, J. Horáček and E. van Dishoeck for useful discussions, and F. Lique, H. Müller, L. Wiesenfeld, A. Faure and J. Black for comments on the manuscript.

References

Asvany, O., Hugo, E., Wahed, S., & Schlemmer, S. 2009, Journal of Physics Conference Series, 192, 012010
Belloche, A., Menten, K. M., Comito, C., *et al.* 2008, A&A, 482, 179
Bergin, E. A., Phillips, T. G., Comito, C., *et al.* 2010, A&A, 521, L20
Bergman, P., Parise, B., Liseau, R., *et al.* 2011, A&A, 531, L8
Botschwina, P. & Oswald, R. 2008, International Journal of Mass Spectrometry, 277, 180
Brinch, C. & Hogerheijde, M. R. 2010, A&A, 523, A25
Butner, H. M., Charnley, S. B., Ceccarelli, C., *et al.* 2007, ApJ, 659, L137

† `http://www.splatalogue.net/`

Caselli, P. & Dore, L. 2005, A&A, 433, 1145
Caselli, P., Vastel, C., Ceccarelli, C., *et al.* 2008, A&A, 492, 703
Cernicharo, J., Spielfiedel, A., Balança, C., *et al.* 2011, A&A, 531, A103
Chavarría, L., Herpin, F., Jacq, T., *et al.* 2010, A&A, 521, L37
Daniel, F., Dubernet, M.-L., Pacaud, F., & Grosjean, A. 2010, A&A, 517, A13
Dick, M. J., Drouin, B. J., & Pearson, J. C. 2009, JQSRT, 110, 619
Dick, M. J., Drouin, B. J., & Pearson, J. C. 2010, Phys. Rev. A, 81, 022706
Drouin, B. J., Pearson, J. C., Wiesenfeld, L., & Faure, A. 2011, in 66th International Symposium On Molecular Spectroscopy
Dubernet, M. L., Boudon, V., Culhane, J. L., *et al.* 2010, JQSRT, 111, 2151
Dubernet, M.-L., Daniel, F., Grosjean, A., *et al.* 2006a, A&A, 460, 323
Dubernet, M.-L., Daniel, F., Grosjean, A., & Lin, C. Y. 2009, A&A, 497, 911
Dubernet, M.-L., Grosjean, A., Flower, D., *et al.* 2006b, Journal of Plasma Research SERIES, Volume 7, p. 356–357, 7, 356
Dumouchel, F., Faure, A., & Lique, F. 2010, MNRAS, 406, 2488
Elitzur, M. & Asensio Ramos, A. 2006, MNRAS, 365, 779
Evans, II, N. J. 1999, ARA&A, 37, 311
Faure, A., Crimier, N., Ceccarelli, C., *et al.* 2007a, A&A, 472, 1029
Faure, A. & Josselin, E. 2008, A&A, 492, 257
Faure, A., Josselin, E., Wiesenfeld, L., & Ceccarelli, C. 2008, in IAU Symposium, Vol. 251, IAU Symposium, ed. S. Kwok & S. Sandford, 137–138
Faure, A., Szalewicz, K., & Wiesenfeld, L. 2011, J. Chem. Phys., 135, 024301
Faure, A., Tennyson, J., Kokoouline, V., & Greene, C. H. 2009, Journal of Physics Conference Series, 192, 012016
Faure, A., Varambhia, H. N., Stoecklin, T., & Tennyson, J. 2007b, MNRAS, 382, 840
Flower, D. 2003, Molecular Collisions in the Interstellar Medium (Cambridge University Press)
Fortman, S. M., Medvedev, I. R., Neese, C. F., & De Lucia, F. C. 2010, ApJ, 725, L11
Ginsburg, A., Darling, J., Battersby, C., Zeiger, B., & Bally, J. 2011, ApJ, in press
Goldsmith, P. F. & Langer, W. D. 1999, ApJ, 517, 209
Goldsmith, P. F. & the HOP team. 2011, ApJ, in press
Green, S., Maluendes, S., & McLean, A. D. 1993, ApJS, 85, 181
Hogerheijde, M. R. & van der Tak, F. F. S. 2000, A&A, 362, 697
Hugo, E., Asvany, O., & Schlemmer, S. 2009, J. Chem. Phys., 130, 164302
Kaźmierczak, M., Schmidt, M. R., Bondar, A., & Krełowski, J. 2010, MNRAS, 402, 2548
Kokoouline, V., Faure, A., Tennyson, J., & Greene, C. H. 2010, MNRAS, 405, 1195
Kristensen, L. E., van Dishoeck, E. F., van Kempen, T. A., *et al.* 2010, A&A, 516, A57
Langer, W. D., Velusamy, T., Pineda, J. L., *et al.* 2010, A&A, 521, L17
Larsson, B., Liseau, R., Pagani, L., *et al.* 2007, A&A, 466, 999
Leurini, S., Schilke, P., Wyrowski, F., & Menten, K. M. 2007, A&A, 466, 215
Lique, F. 2010, J. Chem. Phys., 132, 044311
Lique, F. & Kłos, J. 2011, MNRAS, 413, L20
Loenen, A. F., Spaans, M., Baan, W. A., & Meijerink, R. 2008, A&A, 488, L5
Maloney, P. R. 1987, PhD thesis, Arizona Univ., Tucson.
Mengel, M. & De Lucia, F. C. 2000, ApJ, 543, 271
Mühle, S., Seaquist, E. R., & Henkel, C. 2007, ApJ, 671, 1579
Müller, H. S. P., Schlöder, F., Stutzki, J., & Winnewisser, G. 2005, Journal of Molecular Structure, 742, 215
Ossenkopf, V. 1997, New Astr., 2, 365
Padovani, M., Jørgensen, J. K., Bertoldi, F., *et al.* 2011a, in IAU Symposium, Vol. 270, IAU Symposium, ed. J. Alves, B. G. Elmegreen, J. M. Girart, & V. Trimble, 451–454
Padovani, M., Walmsley, M., Tafalla, M., & Hily-Blant, P. 2011b, in IAU Symposium, Vol. 280
Pagani, L., Daniel, F., & Dubernet, M.-L. 2009a, A&A, 494, 719
Pagani, L., Vastel, C., Hugo, E., *et al.* 2009b, A&A, 494, 623
Panić, O., Hogerheijde, M. R., Wilner, D., & Qi, C. 2009, A&A, 501, 269

Parise, B., Belloche, A., Du, F., Güsten, R., & Menten, K. M. 2011, A&A, 526, A31
Parise, B., Caux, E., Castets, A., *et al.* 2005, A&A, 431, 547
Pascucci, I., Wolf, S., Steinacker, J., *et al.* 2004, A&A, 417, 793
Pavlyuchenkov, Y., Semenov, D., Henning, T., *et al.* 2007, ApJ, 669, 1262
Peraiah, A. 2001, An Introduction to Radiative Transfer (Cambridge University Press)
Pérez-Beaupuits, J. P., Spaans, M., van der Tak, F. F. S., *et al.* 2009, A&A, 503, 459
Pickett, H. M., Poynter, R. L., Cohen, E. A., *et al.* 1998, JQSRT, 60, 883
Poelman, D. R. & Spaans, M. 2005, A&A, 440, 559
Rabli, D. & Flower, D. R. 2010, MNRAS, 406, 95
Reese, C., Stoecklin, T., Voronin, A., & Rayez, J. C. 2005, A&A, 430, 1139
Roberts, H., van der Tak, F. F. S., Fuller, G. A., Plume, R., & Bayet, E. 2011, A&A, 525, A107
Rothman, L. S., Gordon, I. E., Barbe, A., *et al.* 2009, JQSRT, 110, 533
Sarrasin, E., Abdallah, D. B., Wernli, M., *et al.* 2010, MNRAS, 404, 518
Schöier, F. L., van der Tak, F. F. S., van Dishoeck, E. F., & Black, J. H. 2005, A&A, 432, 369
Scribano, Y., Faure, A., & Wiesenfeld, L. 2010, J. Chem. Phys., 133, 231105
Shirley, Y. L., Huard, T. L., Pontoppidan, K. M., *et al.* 2011, ApJ, 728, 143
Toboła, R., Dumouchel, F., Kłos, J., & Lique, F. 2011, J. Chem. Phys., 134, 024305
Torstensson, K. J. E., van der Tak, F. F. S., van Langevelde, H. J., Kristensen, L. E., & Vlemmings, W. H. T. 2011, A&A, 529, A32
Troscompt, N., Faure, A., Wiesenfeld, L., Ceccarelli, C., & Valiron, P. 2009, A&A, 493, 687
Turpin, F., Stoecklin, T., & Voronin, A. 2010, A&A, 511, A28
Valiron, P., Wernli, M., Faure, A., *et al.* 2008, J. Chem. Phys., 129, 134306
Van der Tak, F. & Hogerheijde, M. 2007, ArXiv Astrophysics e-prints, astro-ph
Van der Tak, F., Neufeld, D., Yates, J., *et al.* 2005, in ESA Special Publication, Vol. 577, ESA Special Publication, ed. A. Wilson, 431–432
Van der Tak, F. F. S., Black, J. H., Schöier, F. L., Jansen, D. J., & van Dishoeck, E. F. 2007, A&A, 468, 627
Van der Tak, F. F. S., Müller, H. S. P., Harding, M. E., & Gauss, J. 2009, A&A, 507, 347
Van der Werf, P. P., Isaak, K. G., Meijerink, R., *et al.* 2010, A&A, 518, L42
Van Dishoeck, E. F., Black, J. H., Phillips, T. G., & Gredel, R. 1991, ApJ, 366, 141
Van Dishoeck, E. F., Kristensen, L. E., Benz, A. O., *et al.* 2011, PASP, 123, 138
Van Kempen, T. A., Kristensen, L. E., Herczeg, G. J., *et al.* 2010, A&A, 518, L121
Van Zadelhoff, G.-J., Dullemond, C. P., van der Tak, F. F. S., *et al.* 2002, A&A, 395, 373
Varambhia, H. N., Faure, A., Graupner, K., Field, T. A., & Tennyson, J. 2010, MNRAS, 403, 1409
Varambhia, H. N., Gupta, M., Faure, A., Baluja, K. L., & Tennyson, J. 2009, J. Phys. B, 42, 095204
Vastel, C., Ceccarelli, C., Caux, E., *et al.* 2010, A&A, 521, L31+
Wiesenfeld, L. & Faure, A. 2010, Phys. Rev. A, 82, 040702
Wiesenfeld, L., Scifoni, E., Faure, A., & Roueff, E. 2011a, MNRAS, 413, 509
Wiesenfeld, L., Scribano, Y., & Faure, A. 2011b, Physical Chemistry Chemical Physics (Incorporating Faraday Transactions), 13, 8230
Xu, L.-H., Müller, H. S. P., van der Tak, F. F. S., & Thorwirth, S. 2004, Journal of Molecular Spectroscopy, 228, 220
Yang, B., Stancil, P. C., Balakrishnan, N., & Forrey, R. C. 2010, ApJ, 718, 1062
Yang, C.-H., Sarma, G., Parker, D. H., Ter Meulen, J. J., & Wiesenfeld, L. 2011, J. Chem. Phys., 134, 204308
Yun, Y. J., Park, Y.-S., & Lee, S. H. 2009, A&A, 507, 1785
Zhang, R., Faure, A., & Tennyson, J. 2009, Phys. Scr., 80, 015301

Discussion

HORÁČEK: My group has measured the rovibrational and electronic spectra of HF and HCl, which may be useful in include in the astrophysical data bases.

VAN DER TAK: Yes, those data would be very useful.

The Molecular Universe
Proceedings IAU Symposium No. 280, 2011
José Cernicharo & Rafael Bachiller, eds.

doi:10.1017/S1743921311025208

Conference Summary

John H. Black

Department of Earth & Space Sciences, Chalmers University of Technology,
Onsala Space Observatory, 439 92 Onsala, Sweden
email: John.Black@chalmers.se

Abstract. By any measure, IAU Symposium 280 has been an outstanding success: more than 400 participants represented at least 30 countries with 74 presentations and more than 300 posters. Beyond these numbers, it is evident that the cross-disciplinary field of astrochemistry is flourishing with excellent prospects for growth in the future. We have enjoyed the excitement of new, unexpected results from the Herschel Space Observatory and eagerly await new opportunities and facilities that will arise in the coming months and years.

Keywords. astrochemistry, molecular processes, molecular data, stars: formation, planets: formation, ISM: molecules, circumstellar matter

---

The task of summarizing a very full week of almost 400 presentations – oral and posters – is, to quote Ian Smith, "...challenging, by which I mean impossible". In particular, I cannot give due credit to all authors and their work. Instead, I will try to identify a selection of important themes in astrochemistry with emphasis on areas that have progressed rapidly since the last symposium and areas that show special promise of development during the next few years.

Astrochemistry is concerned with all aspects of cosmic evolution. At the present time much effort and many resources are devoted to understanding how stars and planets form. This chemical and physical evolution takes place within the context of a universe that has been expanding for some 13.7 gigayears starting with an initial baryonic composition of hydrogen and helium nuclei. Our molecular universe is interesting because so much of it is out of equilibrium. This is precisely why astrochemistry is such a rich and essential research area: astronomical phenomena can be fully understood only with detailed knowledge of myriad atomic and molecular processes.

Let me offer a few general observations about what we have seen during this week.

• The large number of young scientists here suggests that astrochemistry is healthy. The new facilities and large projects related to astrochemistry offer hope that the new generation of researchers can remain employed and stimulated.

• Very interesting research is being carried out effectively through large, international collaborations.

• Databases and data archives are expanding. Free and effective access to data has long been one of the moral glories of astronomy. It is gratifying to see the growth of databases for molecular spectra and processes, as well as public-domain computer codes for modeling. However, databases make it easy for users to overlook the original sources of the data. We can continue feeding our inexhaustible appetite for molecular data only if we support the efforts of colleagues who measure, compute, and publish the data by giving proper credit to original sources and by stressing the importance of laboratory astrophysics to funding agencies. Fortunately, many of the things that are of interest in astrochemistry lie close enough to the state of the art in molecular physics and physical chemistry that there are strong and basic justifications for the work.

• Observations are rapidly outrunning theory and models, especially in three-dimensional detail at high resolution.

The thrill of discovery has been in the air around Toledo this week, especially with the flood of new results from Herschel Space Observatory during the last year, as the spectrum at submillimeter and far-infrared wavelengths has been opened to extremely sensitive observation. Commissioning data from the Atacama Large Millimeter/submillimeter Array (ALMA) have just been released. The looming deadline for ALMA early-science proposals will surely make it difficult for authors and editors to meet the publication deadline of this book! What is new?

• Fullerenes C_{60} and C_{70} have been unambiguously identified in space.

• New and important interstellar molecules have been reported, in some cases with surprisingly high abundances: OH^+, H_2O^+, SH^+, H_2Cl^+.

• The large bandwidths of heterodyne receivers are revolutionizing spectroscopy in radio and submm-wave astronomy. This will increase the demands on molecular data and on our ability to analyze spectra.

• Molecular anions have been added to the inventory of interstellar and circumstellar molecules since the last IAU symposium on astrochemistry. This has stimulated new laboratory experiments and efforts to include more anions in chemical models.

Themes

Panchromatic astronomy. Although the submillimeter and infrared parts of the electromagnetic spectrum have predominated here, other regions are important. X-ray absorption spectroscopy makes it possible to measure total abundances of elements in both gaseous and solid forms. In the ultraviolet, there is a new capability with the sensitive Cosmic Origins Spectrograph on Hubble Space Telescope. This has already revealed interesting CO fluorescence in disks of young stellar objects. Those who construct models of such disks should keep in mind that some T Tau stars are directly observed in the ultraviolet; therefore, it is possible to use empirical UV spectra as input to photochemistry, rather than to depend on inappropriate model atmospheres of more evolved stars. In the future, we will certainly see more interplay between molecular astrophysics and high-energy astrophysics as molecular tracers of cosmic-ray and X-ray ionization (e.g. H_3^+, OH^+, H_2O^+) lead us to a better understanding of buried sources of energetic particles and photons. We will also see more cases where electronic and vibrational transitions in molecules affect their rotational excitation.

Models. The Herschel/HEXOS project has provided a dramatic illustration of the problems involved in doing a spectroscopic analysis of some 10^5 lines in the submm-wave spectrum of Orion. Models of photon-dominated regions (PDR) are now able to include hydrodynamics, ice processing, grain growth, stochastic surface chemistry, and even fractal structures. Model-makers of exoplanet atmospheres should be reminded to benchmark their models against a one Jupiter-mass planet at 5 astronomical units from a G2 V star – remember that the UV/EUV spectrum of the sun is well observed and its intensity varies.

"Water, water everywhere, nor any drop to drink". The plain of La Mancha can be rather arid, but apparently nothing like low-mass young stellar objects, whose cores are quite devoid of H_2O. The question – why is there so little water? – will probably still be with us for a few years to come. Likewise the origin of the water on Earth is still an unsolved problem. The Cassini space mission has revealed remarkable transport of water between Saturn's atmosphere and its system of satellites and rings. The ortho and para

nuclear-spin symmetry species of the water molecule seem to reflect a characteristically low spin temperature in comets. More data on water and its singly deuterated isotopologue HDO in protoplanetary disks suggest long-term mixing of warm and cold regions. Water exists in both gaseous and solid form in space and its phase changes are an important part of the chemical evolution of planet-forming disks.

New insight from our own solar system. Fascinating new results have come both from traditional meteoritics and from space probes that capture solid particles in interplanetary space and return them to terrestrial laboratories. For example, the Japanese HAYABUSA mission that returned samples from an asteroid after a seven-year journey is a thrilling story of space travel, aside from the chemical insights that will soon emerge from detailed analysis of the samples. The Stardust spacecraft that sampled Comet 81P/Wild 2 has revealed heterogeneous distributions of both minerals and organic materials that are mostly protosolar rather than pre-solar. There is evidence of extensive mixing between the inner and outer parts of the protosolar nebula, without significant processing within the comet.

Exoplanets and their atmospheres. At the time of the symposium more than 500 planets had been discovered around other stars, with a rapidly growing list of new identifications and candidates from the CoRoT and Kepler space missions, which search for photometric evidence of transiting planets. Spectroscopic signatures of atmospheres have been reported for a few exoplanets. The structure and atmospheric chemistry of a gas-giant planet located very close (less than 0.1 astronomical unit) to its parent star pose interesting problems for theory and models. Although such hydrogenic atmospheres may seem similar to photon-dominated regions of interstellar clouds at first glance, they are likely to have dynamical outer boundaries and to suffer large effects of transport and mixing that static PDR lack. The search for habitable, Earth-like planets has intensified, and along with it the need for a better understanding of remotely detectable signatures of life.

Evolved stars. Both carbon-rich and oxygen rich circumstellar envelopes show extremely rich spectra at submm and infrared wavelengths. Red giant stars in the asymptotic giant branch (AGB) phase of evolution tend to show the largest rates of mass loss in winds, which produce extended molecular envelopes. Some of them process gas into very large molecules and solid dust particles. Aromatic infrared bands are not seen in AGB stars, but do appear in a later evolutionary stage of young planetary nebulae after the envelope has detached from the central core of the star and stands exposed to intense ultraviolet radiation of the hot core. Whether the appearance of aromatic bands in the proto-planetary nebulae is mainly a chemical effect or a result of UV excitation is not certain yet. Spatially resolved circumstellar envelopes show stratified molecular distributions. The extreme supergiant star VY CMa is not even spherically symmetric. A recent interferometric spectrum survey (270-355 GHz) has been carried out in close cooperation with laboratory spectroscopy, both to identify possible new molecules and to determine abundances. This is one of the few cases where H_2O is found to be the second most abundant molecular after H_2, rather than CO. Herschel observations of CW Leo (IRC $+10^o216$) reveal that H_2O molecules arise deep inside despite the predominance of carbon in its outer envelope. Arcs of dust are detected out to 300 arcseconds around CW Leo and thus represent a record of mass-losing events over thousands of years. The most extreme forms of mass loss occur in supernova explosions. Molecules like CO and SiO have been detected in the spectra of supernovae. Chemistry and dust processing evidently continue in the expanding remnants of these explosions, although it is still difficult to distinguish molecules and dust particles formed in supernovae remnants from those that have been swept up from the surrounding interstellar medium.

PDR, XDR, and TDR. Photon-dominated regions (PDR) of molecular clouds attract attention because they tend to be the brightest thermal sources in galaxies at submm and far-infrared wavelengths. An important variation on this theme is the X-ray-dominated region (XDR). Renewed interest in both PDR and XDR has come from Herschel observations of star-forming galaxies and gas surrounding active galactic nuclei: although it is difficult to distinguish between PDR and XDR at mm wavelengths, models show that the spectroscopic diagnostics become more distinct in molecular transitions in the submm and far-infrared parts of the spectrum. Turbulence is an important property of the interstellar medium on all scales from the dimensions of giant molecular clouds down to the microscopic scale (dissipation length). In recent years the statistical properties of turbulence in the diffuse interstellar medium have become better characterized by observation. In parallel, magnetohydrodynamical simulations gas clouds and theoretical models of small vortices have helped to elucidate the many physical and chemical implications of interstellar turbulence. Turbulent dissipation regions (TDR) associated with small-scale structures of high vorticity appear able to explain the observed products of high-temperature chemistry (e.g. interstellar CH^+ and SH^+) as well as excess emission in pure rotational lines of H_2.

Chemistry of the early universe and the first stars. We can be impressed that cosmology is now claimed to be a precise science, with well constrained values for the most important cosmological parameters. However, we should be alarmed to be informed that the rates of fundamental chemical reactions, like three-body association of hydrogen

$$H + H + H \rightarrow H_2 + H,$$

are sufficiently uncertain to limit theoretical simulations of accretion and fragmentation in primordial protostars. At the same time, it is gratifying that cosmologists take such a keen interest in the details of atomic and molecular processes that affect the evolution of structure in the Universe. The chemistry of star-forming clouds at early epochs may differ significantly from that at the present time, with hydrides like OH predominating over heavier molecules like CO.

Molecules in galaxies at low and high redshift. Even though molecular clouds remain spatially unresolved in all but nearby galaxies, it is now becoming possible to do more quantitative chemistry. The broad bandwidths now available for mm/submm-wave spectroscopy of emission lines in galaxies make it possible to estimate both molecular abundances and average physical conditions in line-forming regions. A recent, unbiased absorption line survey at 7 mm wavelength of the interstellar gas in a galaxy at redshift $z = 0.89$ identified 34 different molecules plus isotopic variants. Although the number of suitable mm-wave background sources is very limited, such measurements are an important complement to high-resolution optical absorption spectroscopy of distant quasi-stellar objects, which can be used to measure redshifted ultraviolet lines of H, H_2, HD, and CO at high redshift. We can expect to see more and more sensitive cosmological tests of rolling fundamental constants and of the z-dependence of the cosmic background radiation temperature. Out of more than 35000 catalogued quasars, there are now approximately 30 with detected CO emission lines.

Laboratory astrochemistry. After twenty-five years, the PAH hypothesis (polycyclic aromatic hydrocarbons as the carriers of strong, broad emission bands in the mid-infrared) is alive and well. This hypothesis has stimulated a wide range of experimental work ranging from spectroscopy of gas-phase PAH molecules to investigation of their chemical interactions and behavior in both anionic and cationic charge states. Many groups are struggling valiantly to meet the increasing demands for spectroscopic data of molecules, for reaction rates at low temperatures, and for inelastic collision cross sections and rates.

There is a parallel development of databases, which are gradually being incorporated into the international virtual observatory and into the pipelines for proposals and data at real observatories. Among the many interesting posters were several quantum chemical studies relating to reaction dynamics and experimental studies of neutral atom-molecule reactions at temperatures as low as 10 K. Ion traps are being used in various ways to investigate processes of astrophysical interest. It is especially gratifying to see results of projects that combine such disparate techniques as molecular dynamics computations, phase-space theory, and astrochemical modeling in order to investigate problems like the survival of PAH clusters in PDR.

The promise and peril of ALMA. The Atacama Large Millimeter/submillimeter Array (ALMA) will go into partial operation this year. ALMA will surely stimulate even more activity in astrochemistry. At the same time, ALMA may overwhelm our capacity to interpret observations. Owing to the large bandwidths of the receivers, every spectroscopic observation will represent a line survey, with a terrific demand on analysis tools to identify, measure, and model line intensities. The high angular resolution (of the order of arcsec or better) over a relatively small instantaneous field of view (of the order of arcmin) will make it difficult to see the details in relation to their larger-scale environments. Fortunately, this symposium has demonstrated that all aspects of astrochemistry are thriving and maturing, so I am hopeful that we will cope with the next wave of discovery. In particular, informal and formal networks of collaboration growing and becoming more effective, I think.

Thanks to the Local Organizing Committee and all the sponsors for a wonderful meeting in a lovely environment. Thanks to all the participants for a splendid symposium!

The Molecular Universe
Proceedings IAU Symposium No. 280, 2011
José Cernicharo & Rafael Bachiller, eds.

doi:10.1017/S174392131102521X

Appendix: Additional Contributions

The number of contributions to the Symposium was so high that only the review and invited talks have found place, in the form of articles, in this volume. This Appendix lists all these additional contributions (oral and posters) which are not present as articles. The abstracts of all contributions were published in a booklet produced by the Local Organizing Committee and are available at the SAO/NASA Astrophysics Data System (ADS). More information on these contributions (PowerPoint presentations and/or articles) have been made public in the Internet web site of the conference *(http://cab.inta-csic.es/molecular_universe/)*.

Oral Presentations

Exoplanetary atmosphere observations
G. Tinetti

Exoplanetary atmosphere models
E. Hebrard

Galactic Center clouds and complex molecules
J. Martín-Pintado, M.A. Torres-Requena

The D/H ratio of water ice in star formation
Jeong-Eun Lee, Edwin Bergin

Herschel introduction
G. Pilbratt

Molecules in protostellar shocks: the CHESS view on L1157-B1
Bertrand Lefloch, M. Benedettini, S. Cabrit, E. Caux, C. Ceccarelli, J. Cernicharo, C. Codella, T. Giannini, B. Nisini, B. Parise, M. Salez, M. Vasta, S. Viti, and the CHESS team

Herschel observations of EXtra-Ordinary Sources (HEXOS): Analysis of the HIFI 1.2 THz Wide Spectral Survey toward Orion KL
Nathan R. Crockett, Edwin A. Bergin, Tom Bell, Geoffrey Blake, Jose Cernicharo, Martin Emprechtinger, Harshal Gupta, Dariusz Lis, Steven Lord, John Pearson, Rene Plume, Peter Schilke, Floris van der Tak, Shiya Wang, Shanshan Yu

WISHes coming true: low-mass protostars as chemical fountains
Lars E. Kristensen, Ewine van Dishoeck, Umut Yıldız, Ruud Visser, Greg Herczeg, Jes Jørgensen, Tim van Kempen, Michiel Hogerheijde, and the WISH team

The chemistry of water in the ultracompact HII region (UC HII) Mon R2

Asunción Fuente, Paolo Pilleri, José Cernicharo, Olivier Berné, José Ricardo Rizzo, Manuel González-García, Javier R. Goicoechea, Volker Ossenkopf, Maryvonne Gerin, and the WADI KP team

The atmospheres of Titan and Saturn in the infrared from Cassini: the interplay between observation and laboratory studies

Donald E. Jennings, Conor A. Nixon, F. Michael Flasar, Virgil G. Kunde, Athena Coustenis

Herschel observations of comet 103P/Hartley 2: D/H in a Jupiter family comet

D. C. Lis, P. Hartogh, D. Bockelée-Morvan, M. de Val-Borro, N. Biver, M. Küppers, M. Emprechtinger, E.A. Bergin, J. Crovisier, M. Rengel, R. Moreno, G.A. Blake, S. Szutowicz

Detecting the cold water reservoir in a protoplanetary disk

Michiel Hogerheijde, Edwin Bergin, Christian Brinch, Lauren Cleeves, Jeffrey Fogel, Geoffrey Blake, José Cernicharo, Carsten Dominik, Darek Lis, Gary Melnick, David Neufeld, Olja Panić, John Pearson, Lars Kristensen, Umut Yıldız, Ewine van Dishoeck

Time-dependent anion chemistry in the CSE IRC+10216

Michel Guelin, Marce Agundez, Jose Cernicharo, Carl Gottlieb, Mike McCarthy, Patrick Thaddeus

Formation and destruction processes of carbonaceous interstellar dust

Cesar S. Contreras, Claire L. Ricketts, Farid Salama

New chemical models for new era observations: a multiphase Monte Carlo model of gas-grain chemistry

Anton I. Vasyunin, Eric Herbst

Herschel oxygen project observations of molecular oxygen in Orion

Paul F. Goldsmith, René Liseau, Tom Bell, Darek Lis, Jo-Hsin Chen, Ron Snell, Di Li, Michael Kaufman, Edwin Bergin, Gary Melnick, and the HOP Team

Chemical enrichment of the interstellar medium through the mass loss of evolved stars

Leen Decin, on behalf of the HIFISTARS and MESS GTKP consortia

Modelling the gas and the dust of protoplanetary disks in the Herschel-GASPS sample

Wing-Fai Thi, and the GASPS Team

Where is the molecular gas in low metallicity dwarf galaxies?

Suzanne C. Madden, Diane Cormier, Vianney Lebouteiller, SPIRE SAG 2 consortium, and PACS GT Consortium

An interferometric 270–355 GHz spectral line survey of the red supergiant VY CMa

Karl M. Menten, Ken H. Young, Nimesh A. Patel, Carl A. Gottlieb, Patrick Thaddeus, Michael C. McCarthy, Mark A. Gurwell, Arnaud Belloche, Tomasz Kaminski, Lies Verheyen, Leen Decin, Sandra Bünken, Holger S. P. Müller

DISCS: A Disk Imaging Survey of Chemistry with the SMA

Edwin Bergin, Karin Öberg, Chunhua Qi, David J. Wilner, Jeffery K. J. Fogel, Sean M. Andrews, Lauren Cleeves, Catherine Espaillat, Ilaria Pasucci

The far-ultraviolet molecular spectrum of protoplanetary disks: new views from Hubble

Kevin France

Competing mechanisms in the formation of molecular hydrogen on silicates in conditions pertinent to the ISM

J.L. Lemaire, G. Vidali, S. Baouche, M. Chehrouri,

Molecular clouds at the reionization epoch

Amiel Sternberg, Alex Dalgarno, Eric Herbst, Yezhe Pei

The chemistry of exoplanet atmospheres

Catherine Walsh, T. J. Millar

Nobeyama 45 m telescope legacy project: Line survey

Shuro Takano, Yuri Aikawa, Vivien Chen, Naomi Hirano, Masaaki Hiramatsu, Tomoya Hirota, Kazuhisa Kamegai, Kaori Kobayashi, Kotaro Kohno, Yi-Jehng Kuan, Sheng-Yuan Liu, Taku Nakajima, Hideko Nomura, Nagayoshi Ohashi, Masatoshi Ohishi, Hiroyuki Ozeki, Nami Sakai, Takeshi Sakai, Shoichi Shiba, Yu-Nung Su, Mika Sugimura, Shigehisa Takakuwa, Tomofumi Umemoto, Kuo-Song Wang, Masako Yamada, Takahiro Yamaguchi, Satoshi Yamamoto, Qi-Zhou Zhang

Overview of data bases

T. J. Millar

Laboratory investigations of the formation of superhydrogenated PAHs

John Thrower, Emil Friis, Bjarke Jørgensen, Louis Nilsson, Saoud Baouche, Richard Balog, Andrew Cassidy, Liv Hornekær

H_2CCC - A diffuse interstellar band carrier

Ranjini Raghunandan, John P. Maier, Gordon A. H. Walker, David A. Bohlender, Fabio J. Mazzotti, Jan Fulara, Iryna Garkusha, Adam Nagy

Spectroscopy of PAHs with carbon side chains

Gaël Rouillé, Mathias Steglich, Yvain Carpentier, Friedrich Huisken, Thomas Henning

Exploring the Central Molecular Zone of the Galaxy with H_3^+ and CO

T. R. Geballe, T. Oka, M. Goto, N. Indriolo, B. J. McCall

Herschel/PACS detection of far-IR OH emission towards the Orion Bar PDR

J. R. Goicoechea, C. Joblin, A. Contursi, O. Berné, J. Cernicharo, M. Gerin, J. Le Bourlot, E.A. Bergin, and the HEXOS-GT-KP team

The ubiquitous nature of HF

T. G. Phillips, R. Monje, P. Sonnentruker

Probing the diffuse interstellar medium with hydroxyl cations and water cations

David Neufeld, David Hollenbach, Michael Kaufman, Mark Wolfire, Javier Goicoechea, Maryvonne Gerin, and the PRISMAS team

Carbon chemistry in transitional clouds from the GOT C+ Survey of CII 158 micron emission in the Galactic Plane

William D. Langer, Thangasamy Velusamy, Jorge Pineda, Karen Willacy, Paul F. Goldsmith

Herschel observation of C_3 in star forming regions

B. Mookerjea, T. Giesen, J. Stutzki, J. Cernicharo, J. Goicoechea, J. Black

Posters: Session 1

I.01. AKARI observations of ice absorption bands towards edge-on YSOs

Yuri Aikawa, Dai Kamuro, Itsuki Sakon, Yoichi Itoh, Jennifer A. Noble, Klaus M. Pontoppidan, Helen J. Fraser, Hiroshi Terada, Motohide Tamura, Ryo Kandori, Akiko Kawamura, Munetaka Ueno

I.02. Thermal structure of a protoplanetary disk around HD163296: A study of vertical temperature distribution by CO emission lines

Eiji Akiyama, Munekate Momose, Hiroyuki Hayashi, Yoshimi Kitamura

I.03. Deuterium chemistry in protoplanetary disks

Tobias Albertsson, Dmitry Semenov, Thomas Henning

I.04. Complex molecules in protostellar outflows

H. Arce, J. Santiago-Garcia, D. Mardones, J. K. Jorgensen, G. Garay, M. Tafalla, R. Bachiller

I.05. X-rays in protoplanetary disks: their impact on the thermal and chemical structure, a grid of models.

Giambattista Aresu, Inga Kamp, Rowin Meijerink, Peter Woitke, Wing Fai Thi, Marco Spaans

I.06. New probes of the chemistry in the inner regions of planet-forming disks

Jeanette Bast, Avi Mandell, Ewine van Dishoeck, Daniel Harsono, Simon Bruderer

I.07. Tracing Xray and FUV radiation in the embedded phase of star formation

Arnold O. Benz, Simon Bruderer, Susanne F. Wampfler, Carolin Dedes, Ewine F. van Dishoeck

I.08. Herschel/HIFI-HRS observation of CH absorption in IRAS16293-2422

Sandrine Bottinelli, Emmanuel Caux, Valentine Wakelam, Cecilia Ceccarelli, Claudine Kahane, and the CHESS team

I.09. Principles and promise of Fabry-Perot resonators at THz frequencies

Rogier Braakman, Geoffrey A. Blake,

I.10. A VLT-CRIRES 4.7 micron survey of CO emission from young protoplanetary disks

Joanna Brown, Klaus Pontoppidan, Ewine van Dishoeck, Gregory Herczeg

I.11. High-J CO emission in young stellar objects: Disks and Outflow walls! FUV or Shocks?

Simon Bruderer, Ewine F. van Dishoeck, S.D. Doty

I.12. New constraints on the origin of protostellar jets: The SiO abundance in HH212

Sylvie Cabrit, Claudio Codella, Frederic Gueth, Vianney Taquet

I.13. Star forming regions towards stellar wind bubbles: the ring nebulae RCW 52 and RCW 78

Cristina Cappa, Gisela A. Romero, Mónica Rubio, M.C. Martín

I.14. Negative ions induced processes in proto-planetary atmospheres and ISM

F. Carelli, F. Sebastianelli, F. A. Gianturco

I.15. Construction of a high-resolution terahertz cavity ringdown spectrometer

P. Brandon Carroll, Brett A. McGuire, Susanna L. Widicus Weaver

I.16. The role of OH in the chemical evolution of protoplanetary disks

G. Chaparro, I. Kamp

I.17. Water in massive star-forming regions with Herschel Space Observatory

Luis Chavarria, Fabrice Herpin, Sylvain Bontemps, Thierry Jacq, Alain Baudry, Jonathan Braine, Floris van de Tak, Friedrich Wyrowski, Ewine van Dishoeck

I.18. Chemistry in Very Low Luminosity Objects (VeLLOs)

Jo-Hsin Chen, Neal Evans, Jeong-Eun Lee, Michael M. Dunham, Hyo Jeong Kim

I.19. Chemical signposts in transition disks

Ilse Cleeves, Edwin Bergin, Jeffrey Fogel

I.20. The SEVO experiment onboard NASA's O/OREOS Small Satellite

Amanda Cook, Andrew Mattioda, Nathan Bramall, Kathryn Bryson, Julie Chittenden, Pascale Ehrenfreund, Giovanni Minelli, Richard Quinn, Antonio Ricco

I.21. Organic chemistry of Cha-MMS1 and IRAS 15194-5115 probed by microwave spectroscopy

Martin Cordiner, Steven Charnley

I.22. Molecular anions in protostars, prestellar cores and dark clouds

Martin Cordiner, Steven Charnley, Jane Buckle, Catherine Wash, Tom Millar

I.23. Study of deuterated water in the low-mass protostar IRAS16293-2422

Audrey Coutens, Charlotte Vastel, Emmanuel Caux, Cecilia Ceccarelli, and the CHESS team

I.24. HNC in the C-rich envelope of the AGB star IRC+10216

Fabien Daniel, Marcelino Agundez, Elvire de Beck, Jose Cernicharo, Leen Decin, Robin Lombaert

I.25. Detection of iron containing crystalline olivines in the debris disk of β Pictoris

B.L. de Vries, B. Acke, B. Vandenbussche, L. B. F. M. Waters, J. A. D. L. Blommaert, C. Waelkens

I.26. The photodissociation of formaldehyde in comets

Paul D. Feldman

I.27. The relation between deuteration and evolution in massive star formation

Francesco Fontani, Aina Palau, Paola Caselli, Alvaro Sanchez-Monge, Jonathan Tan, Michael J. Butler, Izaskun Jimenez-Serra

I.28. An improved equation of state for protoplanetary simulations

Marina Galvagni, Tristen Hayfield, Lucio Mayer, Prasenjit Saha

I.29. On the formation of fullerenes in H-rich circumstellar environments

Anibal Garcia-Hernandez

I.30. Spitzer spectral line mapping of protostellar outflows: H_2 emission in L1448, BHR71, and NGC2071

Teresa Giannini, Brunella Nisini, David Neufeld, Yuan Yuan, Simone Antoniucci, Antoine Gusdorf

I.31. New observations of molecular absorptions toward GV Tau N

Erika Gibb, Matthew Troutman

I.32. SAFARI: A Far Infrared Imaging FTS-Spectrometer for SPICA

J.R. Goicoechea, P.R. Roelfsema, W. Jellema, B.M. Swinyard

I.33. Molecular emission in the IC443 supernova remnant

Antoine Gusdorf, Rolf Güsten, Yuan Yuan, David Neufeld, and the WADI team

I.34. PAH Anions in circumstellar and interstellar environments

Mark Hammonds, Alessandra Candian, Peter Sarre

I.35. Modelling molecular emission from young embedded disks

Daniel Harsono, Ruud Visser, Ewine van Dishoeck, Lars Kristensen, Simon Bruderer, Christian Brinch, Michiel Hogerheijde

I.36. Warm water in Herschel/PACS observations of NGC 1333 IRAS 4B: the outflow, not the disk!

Gregory J. Herczeg, Agata Karska, Lars E. Kristensen, Ewine F. van Dishoeck, Ruud Visser, Jes Jorgensen, Simon Bruderer, Umut Yildiz, and the WISH Herschel Key Program Team

I.37. H_2 energetics and gravitational fragmentation in the overlap region in the Antennae. "Early stages of star formation"

Cinthya Herrera, Francois Boulanger, Nicole Nesvadba

I.38. Chemical evolution of atmospheres of H_2O dominant comet nuclei

Subhon Ibadov

I.39. Modelling the chemistry of a gravitationally unstable protoplanetary disc

John Ilee, Aaron Boley, Paola Caselli, Richard Durisen, Tom Hartquist, Jonathan Rawlings

I.40. Chemical segregation in hot cores: SMA imaging of the AFGL2591 star forming region

Izaskun Jimenez-Serra, Qizhou Zhang, Jesus Martin-Pintado, Serena Viti, Willem-Jan de Wit

I.41. The structure and appearance of irradiated protoplanetary disks: the role of chemistry

Inga Kamp, Giambattista Aresu, German Chaparro, Peter Woitke, Wing-Fai Thi

I.42. Herschel/PACS observations of Class 0/I low-mass young stellar objects.

Agata Karska, Gregory Herczeg, Ewine van Dishoeck, Lars E. Kristensen, and WISH team

I.43. The search for evidence of episodic mass accretion: CO_2 features in low-luminosity, embedded protostars

Hyo Jeong Kim, Neal J. Evans II, Michael M. Dunham, Jeong-Eun Lee

I.44. Epic changes in the IRS 46 mid-infrared spectrum; an inner disk chemistry study

Fred Lahuis, Inga Kamp, Wing Fai Thi, Ewine van Dishoeck, Peter Woitke

I.45. The molecular universe as seen by JWST-MIRI

Fred Lahuis, Ewine van Dishoeck, Gillian Wright, George Rieke, and the MIRI Team

I.46. Class I methanol maser observations at 44 GHz in the direction of some SNRs and SFRs.

G. M. Larionov, I. D. Litovchenko, I. E. Val'tts, A. V. Alakoz

I.47. OH 1720-MHz observations toward northern class I methanol masers with 70-m Ukrainian Telescope.

G. M. Larionov, I. D. Litovchenko, O. S. Bayandina, I. E. Val'tts, A. V. Alakoz, D. V. Mukha, A. S. Nabatov, A. A. Konovalenko, V. V. Zakharenko, E. V. Alekseev, V. S. Nikolaenko, V. F. Kulishenko, S. A. Odincov

I.48. Non-LTE infrared emission from protoplanetary disk surfaces

Alexandra Lockwood, Geoffrey Blake

I.49. PACS molecular spectroscopy of OH/IR stars

R. Lombaert, B. L. de Vries, L. Decin, J. A. D. L. Blommaert, P. Royer, E. De Beck, A. de Koter, L. B. F. M. Waters

I.50. Modeling photodissociation of CO isotopologues in a turbulent solar nebula

James Lyons

I.51. Emission spectra of Geminid Fireballs from 3200 Phaeton: Preferential depleation of volatile phases

Jose Maria Madiedo, Josep Maria Trigo-Rodriguez

I.52. Water profiles of intermediate mass YSOs from HIFI

Carolyn MCoey, Samuel Tisi, Doug Johnstone, Michel Fich, Tim A. van Kempen, Asuncion Fuente, Paola Caselli, Jose Cernicharo, Lars E. Kristensen, Ewine F. van Dishoeck

I.53. Molecules in protoplanetary HAEBE discs as seen with Herschel.

Gwendolyn Meeus

I.54. The Herschel Orion Protostar Survey: Probing protostellar evolution with molecules, ice and dust

Tom Megeath, Manoj Puravankara, Dan Watson, Charles Poteet, Will Fischer, Babar Ali, Roland Vavrek, and the HOPS team

I.55. Centimeter, millimeter, and submillimeter observations of Comet 103P/Hartley 2

Stefanie N. Milam, Steven B. Charnley, Yo-Ling Chuang, Yi-Jehng Kuan, Iain M. Coulson, Anthony J. Remijan

I.56. Multiwavelength observations of volatiles in comets

Stefanie N. Milam, Steven B. Charnley, Yi-Jehng Kuan, Yo-Ling Chuang, Michael A. DiSanti, Boncho P. Bonev, Anthony J. Remijan

I.57. Can PAH clusters survive in PDRs?

Julien Montillaud, Christine Joblin, Dominique Toublanc, Aude Simon, Pascal Parneix

I.58. Sulphur-bearing molecules in the ultracompact HII region G10.6-0.4

Bhaswati Mookerjea, Massimo De Luca, Tom Bell, Audrey Coutens, Maryvonne Gerin, Harshal Gupta, Raquel Monje, Charlotte Vastel

I.59. Chemical evolution of protoplanetary disks: The effects of viscous accretion, turbulent mixing, and disk winds

Hideko Nomura, Dominikus Heinzeller, Catherine Walsh, Tom Millar

I.60. Observations of near-infrared line ratios of molecular hydrogen emission to diagnose dust evolution in protoplanetary disks

Hideko Nomura, Yuhei Takagi, Yoichi Itoh, Koji Sugitani, Makoto Watanabe, Jeff Bary, David Weintraub, Yuri Aikawa, Masahiro Tsujimoto, Yoshitsugu Nakagawa, Tom Millar

I.61. Carbon fractionation in photon-dominated regions

Volker Ossenkopf, Markus Röllig, Asunción Fuente, Robert Simon, Carsten Kramer, Edwin Bergin, the whole WADI key project team, the whole HEXOS key project team

I.62. Molecular complexity in the Cep E protostellar outflow

Susana Pacheco Vazquez, Bertrand Lefloch, Cecilia Ceccarelli

I.63. Disks, outflows, and hot cores in the mm range at subarcsecond angular resolution

Aina Palau, Asunción Fuente, Jérémie Boissier, Vincent Piétu, Álvaro Sánchez-Monge, Luis Zapata, Francesco Fontani, Josep Miquel Girart, Roberto Neri, Gemma Busquet, Qizhou Zhang, Paul T. P. Ho, Robert Estalella, Marc Audard, Tomas Alonso-Albi

I.64. Chemical evolution in protoplanetary disks around T Tauri stars

Andrey Paska, Andrew Markwick

I.65. Origin of water around deeply embedded low-mass protostars

Magnus V. Persson, Jes K. Jørgensen, Ewine F. van Dishoeck

I.66. Production of small hydrocarbons in photo-dissociation regions

Paolo Pilleri, Christine Joblin, Maryvonne Gerin, Jerome Pety, Julien Montillaud, Francois Boulanger, Asuncion Fuente

I.67. Herschel/PACS observations of young sources in Taurus: the far-infrared counterpart of optical jets

Linda Podio, & the GASPS team

I.68. Carbon chemistry in oxygen rich planetary nebulae

Lizette Ramirez

I.69. Stellar winds in hot water

Anita M S Richards, Khudhair Assaf, Moshe Elitzur, Sandra Etoka, Malcolm Gray, Graham Harper, Liz Humphreys, Jeremy Lim, Din van Trung, Jeremy Yates

I.70. Thermal emission of CO and isotopomers from the envelopes of water fountain stars

J. Ricardo Rizzo, J. F. Gómez, M. Osorio, O. Suárez, L. F. Miranda

I.71. The search for the magnetic precursor of C-type shocks in young molecular outflows

Julia Roberts, Izaskun Jiménez-Serra, Jesus Martín-Pintado

I.72. Radio spectroscopy of circumstellar molecular masers

Georgij M. Rudnitskij, Pierre Colom, Evguenij E. Lekht, Mikhail I. Pashchenko, Vladimir A. Samodurov, Il'nur A. Surbaev, Alexander M. Tolmachev

I.73. Deuterium fractionation in massive clumps in early evolutionary stages of high-mass star formation

Takeshi Sakai, Nami Sakai, Kenji Furuya, Yuri Aikawa, Tomoya Hirota, Satoshi Yamamoto

I.74. Linking high-J CO emission from low- to high-mass protostars with Herschel–HIFI

Irene San José-García, Lars E. Kristensen, Umut A. Yıldız, Ewine F. van Dishoeck, and the WISH team

I.75. Molecular complexity in O-rich circumstellar envelopes around evolved stars: IK TAU and OH 231.8+4.2

C. Sánchez Contreras, L. Velilla Prieto, J. Cernicharo, J. Alcolea, J.R. Pardo, M. Agúndez, V. Bujarrabal, F. Herpin, K.M. Menten, F. Wyroswky

I.76. Hot molecular cores in young stellar objects of different luminosities

Álvaro Sánchez-Monge, Aina Palau, Robert Estalella, Stan Kurtz

I.77. Observing water in low-mass proto-stellar outflows: the case of L1448

Gina Santangelo, Brunella Nisini, Simone Antoniucci, Teresa Giannini, Milena Benedettini, Claudio Codella, René Liseau, Andrea Lorenzani, Mario Tafalla, Magda Vasta, Ewine van Dishoeck, Lars Kristensen

I.78. Characterizing CO fourth positive emission in proto-planetary disks

Eric Schindhelm, Kevin France, Eric Burgh, Alex Brown, Greg Herczeg

I.79. Cool molecular environment of stellar mergers V838 Mon and V4332 Sgr. Observations of the circumstellar ammonia 1_0–0_0 lines in carbon-rich AGB stars by the Herschel/HIFI

Mirosław R. Schmidt, Romuald Tylenda, Tomasz Kamiński, M.R. Schmidt, D. A. Neufeld, R. Szczerba, J. He, N. Siódmiak, and the HIFISTARS consortium

I.80. Effect of metallicity on the chemical properties of ices around embedded young stellar objects

Takashi Shimonishi, Takashi Onaka, Daisuke Kato, Yoshifusa Ita, Itsuki Sakon, Akiko Kawamura, Hidehiro Kaneda

I.81. Spectroscopic analysis of cool galactic R CrB Star candidates

Olesja Smirnova, Laimons Začs, Yakiv V. Pavlenko, Bogdan Kaminsky

I.82. Observational signatures of ^{12}CO–^{13}CO partitioning in ice and gas towards local young stellar objects and molecular clouds

Rachel L. Smith, Klaus M. Pontoppidan, Edward D. Young, Mark R. Morris

I.83. Cosmic origins spectrograph observations of dense translucent clouds

Theodore P. Snow, E. B. Burgh, J. P. Destree

I.84. Periodic outbursts in the 6.7GHz methanol maser line towards G22.356+0.066

M. Szymczak, P. Wolak, A. Bartkiewicz

I.85. A molecular survey of outflow gas: velocity-dependent shock chemistry and the peculiar composition of the extremely high-velocity (EHV) gas

Mario Tafalla, Joaquín Santiago-García, Alvaro Hacar, Rafael Bachiller

I.86. Molecular chemistry as a probe of pre-stellar core dynamics

Konstantinos Tassis, Karen Willacy, Harold Yorke, Neal Turner

I.87. Herschel/HIFI view on massive evolved stars: the HIFISTARS sample of supergiant and yellow hypergiant envelopes

David Teyssier, Anthony Marston, Javier Alcolea, Valentin Bujarrabal, and the HIFISTARS consortium

I.88. Modelling CH^+ in the protoplanetary disk HD100546

Wing-Fai Thi, F. Ménard, G. Meeus, C. Martin-Zaïdi, P. Woitke, E. Tatulli, M. Benisty, I. Kamp, I. Pascucci, C. Pinte, C. A. Grady, S. Brittain, G.J. White, C. D. Howard, G. Sandell, C. Eiroa

I.89. Modelling the gas and dust of protoplanetary disks in the Herschel-GASPS sample

Wing-Fai Thi, and the GASPS Team

I.90. Low-temperature kinetic studies of OH radical reactions relevant to planetary atmospheres

Thomas M. Townsend, Maria Antiñolo, Bernabe Ballesteros, Elena Jimenez, Andre Canosa

I.91. The pathways of C: from AGB stars, to the Interstellar Medium, and finally into the protoplanetary disk

Josep M. Trigo-Rodriguez, D. Anibal Garcia-Hernandez

I.92. The extent of aqueous alteration in C-class asteroids, and the survival of presolar isotopic signatures in chondrites

Josep M. Trigo-Rodriguez

I.93. Searching for a spectro-astrometric signal from molecules in circumstellar disks

Matthew Troutman, Erika Gibb

I.94. Complex organic molecules toward low-mass protostars and outflows

Nienke van der Marel, Karin I. Öberg, Lars Kristensen, Ewine F. van Dishoeck

I.95. Herschel spectroscopy of planetary nebulae

P.A.M. van Hoof, M.J. Barlow, K.M. Exter, R. Wesson, G.C. Van de Steene, M.A.T. Groenewegen

I.96. L1157-B1: water as a unique tracer of low velocity shocks

Serena Viti, I. Jimenez-Serra, J. Yates, C. Codella, M. Vasta, P. Caselli, B. Lefloch, C. Ceccarelli

I.97. Herschel HIFI observation of HCl emission in the low mass protostar IRAS16293-2422

Valentine Wakelam, Emmanuel Caux, Sandrine Bottinelli, Cecilia Ceccarelli, Claudine Kahane, and the CHESS team

I.98. An expanding disk around the young massive star AFGL 2591

Kuo-Song Wang, Floris van der Tak, Michiel Hogerheijde

I.99. Spectral line survey toward RCrA IRS7B in the 345 GHz window with ASTE

Yoshimasa Watanabe, Nami Sakai, Johan Lindberg, Jes Jørgensen, Suzanne Bisschop, Satoshi Yamamoto

I.100. The $^{12}C/^{13}C$ ratio as a chemistry indicator

Eva Wirström, Wolf Geppert, Carina Persson, Steven Charnley

I.101. Sub-mm spectroscopy and infrared interferometry of molecular layers of evolved stars

Markus Wittkowski, E. M. L. Humphreys, M. D. Gray, D. A. Boboltz, C. de Breuck

I.102. High-J CO survey of low-mass protostars observed with Herschel-HIFI

Umut A. Yıldız, Lars E. Kristensen, Ewine F. van Dishoeck, Jes K. Jørgensen, Ruud Visser, Irene San José-García, and the WISH Team

I.103. Molecular chemistry in protostellar disk winds and observational predictions for Herschel and ALMA

W. Yvart, S. Cabrit, G. Pineau des Forêts, P.J.V. Garcia, J. Ferreira, F. Casse

I.104. Molecules in the atmosphere and circumstellar shell of proto-planetary nebula IRAS22272+5435

Laimons Zacs, Aija Laure, Julius Sperauskas

Posters: Session 2

II.01. The effects of grain size and grain growth on the chemical evolution of cold dense clouds

Kinsuk Acharyya, Eric Herbst

II.02. Revisited nitrogen isotopic ratio in molecular clouds

Gilles Adande, Lucy Ziurys

II.03. Astronomical identification of CN^-, the smallest observed molecular anion

Marcelino Agúndez, José Cernicharo, Michel Guélin, Claudine Kahane, E. Roueff, Jacek Kłos, Francisco Javier Aoiz, François Lique, Nuria Marcelino, Javier R. Goicoechea, Manuel González García, Carl A. Gottlieb, Mike C. McCarthy, Patrick Thaddeus

II.04. Photoionization models of the H_2 emission of the Narrow Line Region of AGNs

Isabel Aleman, Ruth Gruenwald

II.05. SO and SO_2 line observations and chemical evolution of Orion KL

Gisela B. Esplugues, José Cernicharo, Serena Viti, Belén Tercero, Nuria Marcelino, Javier R. Goicioechea

II.06. Probing the physical conditions in Orion KL with methyl cyanide

T. A. Bell, J. Cernicharo, and the HEXOS team

II.07. Exciting Results from the HEXOS Herschel GT Key Program

Edwin A. Bergin, and the HEXOS team

II.08. Herschel insights into the evolution of very small dust particles and the heating of interstellar gas

O. Berné, C. Joblin, P. Pilleri, J. Goicoechea, F. Salgado, Y. Okada, M. Gerin, V. Ossenkopf, E. Bergin

II.09. The molecular gas rich galaxy NGC 3934: a multiwavelenght view

Daniela Bettoni, Giuseppe Galletta, Roberto Rampazzo, Antonietta Marino, Paola Mazzei, Lucio Buson

II.10. Nitrogen isotopic fractionation in cold quiescent clouds: first results for L1544

Luca Bizzocchi, Paola Caselli, Luca Dore

II.11. DR21(OH) - a high-mass star cluster in formation observed with Herschel-HIFI, and the IRAM PdBI

S. Bontemps, T. Csengeri, F. Herpin, N. Schneider, F. Motte, L. Chavarria, A. Baudry

II.12. On the relative abundance of LiH and LiH+ molecules in the early universe: new results from quantum reactions

Stefano Bovino, Mario Tacconi, Franco Gianturco

II.13. Dense gas and star formation along the major axis of M33 (HERM33ES)

Christof Buchbender, Carsten Kramer, Erik Rosolowsky

II.14. On the behavior of the NH_3/N_2H+ ratio in high-mass star forming regions

Gemma Busquet, Robert Estalella, Aina Palau, Qizhou Zhang, Paul T.P. Ho, Serena Viti

II.15. Analysis of velocity components along the line of side towards SgrB2(M)

Denis Büchel, Peter Schilke, Claudia Comito, Sheng Li Qin, Edwin A Bergin, Darek C Lis

II.16. Molecular gas and star formation in Sh2-196 and Sh2-206

Cristina Cappa, M.C. Martín, S. Cichowolski, Javier Vasquez, Ricardo Zinn

II.17. Water observations with Herschel/HIFI toward AFGL 2591

Yunhee Choi, Floris van der Tak, Ewine van Dishoeck

II.18. The water content of Sgr B2(M)

Claudia Comito, Peter Schilke, Darek Lis, Edwin Bergin, *et al.*

II.19. Radiative cooling functions for primordial molecules

Carla Maria Coppola, Lorenzo Lodi, Jonathan Tennyson

II.20. Vibrational level population of H_2 and H_2^+ in the early Universe

Carla Maria Coppola, Savino Longo, Mario Capitelli, Francesco Palla, Daniele Galli

II.21. Diffuse interstellar bands in the Andromeda and Triangulum galaxies

Martin Cordiner, Nick Cox, Keith Smith, Christopher Evans, Carrie Trundle, Karl Gordon, Peter Sarre

II.22. Gas properties in dwarf galaxies from Herschel FIR spectroscopic observations

Diane Cormier, Suzanne Madden, Vianney Lebouteiller, Frederic Galliano, Sacha Hony

II.23. A mm spectral survey of the Orion Bar Photo Dissociation Region

J. R. Goicoechea, S. Cuadrado, A. Fuente, G. B. Esplugues, J. Cernicharo, S. García-Burillo, A. Usero, N. Marcelino,

II.24. The molecular UCHII region W31C seen by Herschel-HIFI

Massimo De Luca, Tom Bell, Audrey Coutens, Maryvonne Gerin, Harshal Gupta, Raquel Monje, Bhaswati Mookerjea, Charlotte Vastel

II.25. Studying cooling mechanisms in the massive star forming region IRAS 12326-6245

Carolin Dedes, Fabrice Herpin, Luis Chavarria, Susanne Wampfler, Friedrich Wyrowski, Floris van der Tak, Arnold Benz, Simon Bruderer, Edward Polehampton, Martin Melchior

II.26. Large area mappings of formaldehyde at 6-cm toward Giant Molecular Clouds

Jarken Esimbek, Jianjun Zhou, Chuanpeng Zhang

II.27. Gas and dust in the Galactic Centre - ISO LWS, submillimetre line and continuum observations.

Mireya Etxaluze, Glenn J. White, Howard A. Smith, Eduardo Gonzalez-Alfonso, Antony A. Stark, Gordon J. Stacey, Sarah J. Leeks, Ian Gatley, Jacqueline Fisher, Douglas Pierce-Price, John S. Richer, Tim W. Grundy, Ed T. Polehampton

II.28. $HCOOCH_3$ as a probe of temperature and structure of Orion-KL

Cecile Favre, Didier Despois, Nathalie Brouillet, Alain Baudry, Francoise Combes, Michel Guelin, Al Wootten, Georges Wlodarczak

II.29. NGC4945 and NGC 253 line survey

Paola Fiadino, Julia Roberts

II.30. Water in the Milky Way: Absorption lines along the PRISMAS lines of sight

Nicolas Flagey, Paul Goldsmith, David Neufeld, Darek Lis, and the PRISMAS team

II.31. Chemical differentiation in magnetized young starless cores in the Pipe Nebula

Pau Frau, Josep M. Girart, Maria T. Beltrán, Óscar Morata

II.32. Carbon Isotope and Isotopomer Fractionation in Dense Molecular Cloud Cores

Kenji Furuya, Yuri Aikawa, Nami Sakai, Satoshi Yamamoto

II.33. Molecular evolution of a first core in 3 dimensional hydrodynamic Calculations

Kenji Furuya, Yuri Aikawa, Tomoaki Matsumoto, Kengo Tomida, Kazuya Saigo, Kohji Tomisaka, Franck Hersant, Valentine Wakelam

II.34. Dense molecular gas tracers and star formation rate in galaxies near and far

Yu Gao

II.35. Large-scale molecular shocks in galaxies: the imprint of mechanical dominated regions (MDR)

Santiago Garcia-Burillo, Antonio Usero, Asuncion Fuente, Javier Gracia-Carpio, Andrew J Baker, Jesus Martin-Pintado

II.36. Formation of carbonaceous molecules in the ISM

Thomas F. Giesen, Volker Lutter, Imke Gottbehüt, Jürgen Krieg, Christian Endres, Sven Thorwirth, Stephan Schlemmer

II.37. Tunnelling in space

T.P.M. Goumans, S. Andersson, J. Kaestner

II.38. The Horsehead Nebula: a template for extragalactic high density tracers studies ?

Pierre Gratier, Jérôme Pety, Maryvonne Gerin, Julien Montillaud, Viviana Guzmán, Javier R. Goicoechea,

II.39. Detection of OH^+ and H_2O^+ towards Orion KL

Harshal Gupta, Paul Rimmer, John C. Pearson, Eric Herbst, Shanshan Yu, Edwin A. Bergin, and the HEXOS team

II.40. H_2CO in the Horsehead nebula

Viviana Guzmán, Jerôme Pety, Javier R. Goicoechea, Maryvonne Gerin, Evelyne Roueff

II.41. Physical conditions and chemical evolution of the gas towards the Orion Bar photodissociation region using Herschel/PACS and SPIRE spectro-imaging observations

Emilie Habart, Jeronimo Bernard-Salas, Emmanuel Dartois, Heddy Arab, Alain Abergel, and the SPIRE consortium

II.42. Model of molecular abundance in an AGN disk of NGC 1068

Nanase Harada, Todd Thompson, Eric Herbst

II.43. Physical/chemical structure and kinematics of an extremely filamentary Infrared Dark Cloud

Jonathan D. Henshaw, Paola Caselli, Izaskun Jiménez-Serra, Francesco Fontani, Jonathan Tan

II.44. Oxygen depletion in dense molecular clouds: a clue to a low O_2 abundance?

Ugo Hincelin, Valentine Wakelam, Franck Hersant, Stéphane Guilloteau, Jean-Christophe Loison, Pascal Honvault, Jurgen Troe

II.45. PAHs in the LMC and comparison to other galaxies

Hony and SAGE-Spec team.

II.46. Spitzer spectroscopy of young stellar object candidates in the central molecular zone - Determing an average star formation rate for the center of our Galaxy

Katharina Immer, Frederic Schuller, Karl M. Menten

II.47. Spectral survey of the star-forming region W51 e1/e2 in the 3-mm wavelength range

Sergei Kalenskii, Lars E. B. Johansson

II.48. The effects of subsurface chemistry in the grain mantles on the deuterium chemistry in the molecular clouds

Juris Kalvans, Ivar Shmeld

II.49. The Production of O_2 in molecular shocks: HOP observations of Orion H_2Peak 1

Michael Kaufman, Paul Goldsmith, and the HOP team

II.50. Barely detectable molecules in translucent interstellar clouds

Jacek Krełowski, Maja Kaźmierczak

II.51. Investigation of dust properties within the Ophiuchus region with multi-band observations

Dae-Hee Lee, Woong-Seob Jeong, Jeonghyun Pyo, Kwang-Il Seon

II.52. Turbulent formation and excitation of molecules in diffuse media.

Pierre Lesaffre, Benjamin Godard, Guillaume Pineau des Forêts, François Boulanger, Édith Falgarone, Maryvonne Gerin, Pierre Guillard

II.53. The distribution of warm gas in the G327.3-0.6 star forming region

Silvia Leurini, Friedrich Wyrowski, Floris van der Tak, Fabrice Herpin, and the WISH team

II.54. The impact of strong radiation fields on protostars in the R CrA region

Johan E. Lindberg, Jes K. Jørgensen, and the DIGIT team

II.55. A Molecular line survey of IRDC star forming clumps with the Nobeyama 45M telescope

Sheng-Yuan Liu, Yu-Nung Su, and the NRO 45 m survey team

II.56. Star formation and black hole accretion in nearby (U)LIRGs.

Edo Loenen, Paul van der Werf, Rowin Meijerink, and the HerCULES consortium

II.57. o-H_2, p-H_2 and HD vibrational kinetics in the Early Universe

S. Longo, R. D'Introno, A. Panarese, C. M. Coppola, J. Tennyson

II.58. SiO outflows in high-mass star forming regions: a potential chemical clock

Ana López-Sepulcre, Malcolm Walmsley, Riccardo Cesaroni, Claudio Codella, Frédéric Schuller, Leonardo Bronfman, Sean J. Carey, Karl Menten, Sergio Molinari, Alberto Noriega-Crespo

II.59. Investigating the molecular content of lensed submillimeter galaxies

Roxana Lupu, and the Z-Spec team

II.60. Large scale CO emission in Orion A: Star formation feedback on the molecular gas

Nuria Marcelino, Olivier Berné, José Cernicharo

II.61. Warm deuteration of hydrogen cyanide in Orion

Nuria Marcelino, Belén Tercero, José Cernicharo, Evelyne Roueff, Aina Palau, Javier R. Goicoechea, Edwin Bergin, and the HEXOS team

II.62. Hot Core modelling with stochastic grain surface modelling

Daniel McElroy, Tom Millar

II.63. Observations of nitrogen isotope fractionation in prestellar cores

Stefanie N. Milam

II.64. Can we trust CO emission as a probe of the densities and temperatures of molecular clouds?

Faviola Molina, Simon Glover, Christoph Federrath

II.65. Herschel/HIFI observations of hydrogen fluoride along a strip towards Sagittarius B2 (M)

Raquel R. Monje, Darek Lis, Paul F. Goldsmith, Thomas G. Phillipas, David Neufeld

II.66. Dense molecular gas in starburst galaxies: Warmer than expected?

Stefanie Mühle, Christian Henkel, Tahlia de Maio, Ernie R. Seaquist

II.67. The quest for complex molecules in space. Searches for cyanides related to *n*-propyl cyanide in Sgr B2(N)

Holger S. P. Müller, Arnaud Belloche, Karl M. Menten, Audrey Coutens, Adam Walters, Jens-Uwe Grabow, Stephan Schlemmer

II.68. Orion S: a test for Oxygen chemistry

Zsófia Nagy, Floris van der Tak, Gary Melnick, Paul Goldsmith, Rene Plume, Edwin Bergin, Volker Tolls, and the HEXOS and HOP teams

II.69. Line survey project of external galaxies with NRO 45-m Telescope

Tac Nakajima, Shuro Takano, Kotaro Kohno, Hirofumi Inoue, and the NRO 45-m survey team

II.70. Methyl formate in star forming regions

A. Occhiogrosso, S. Viti, P. Modica, M.E. Palumbo

II.71. Nitrogen-bearing molecules in prestellar cores

Marco Padovani, Malcolm Walmsley, Mario Tafalla, Pierre Hily-Blant

II.72. ortho-H_2 and the age of dark clouds and prestellar cores

Laurent Pagani, Pierre Lesaffre, Pascal Honvault, Alexandre Faure, Mohammad Jorfi, Tomas Gonzales-Lezana

II.73. A Comprehensive survey of hydrogen chloride in the galaxy

Ruisheng Peng, Hiroshige Yoshida, Richard A. Chamberlin, Thomas G. Phillips, Dariusz C. Lis, Maryvonne Gerin

II.74. Acetone in Orion: high-resolution images of a special oxygen-bearing molecule

Tzu-Cheng Peng, Nathalie Brouillet, Cécile Favre, Dider Despois, Alain Baudry, Anthony Remijan, Tom Wilson, Alwyn Wootten

II.75. Structure and dynamics of an AGN torus: 3D hydrodynamical simulations for ALMA

J.P. Perez-Beaupuits, K. Wada, M. Spaans, R. Guesten

II.76. Nitrogen hydrides in interstellar gas towards G10.6-0.4 (W31C) and W49N

C. M. Persson, M. De Luca, B. Mookerjea, M. Gerin, J. H. Black, T. A. Bell, B. Godard, J. Goicoechea, G. Hassel, E. Herbst, P. Hily-Blant, K. M. Menten, H. S. P. Müller, A. O. H Olofsson, J. C. Pearson, S. Yu, and the PRISMAS team

II.77. The Role of environment in star formation: Young clusters forming in isolation

Dawn Peterson, Rob Gutermuth, Tyler Bourke

II.78. CO (1-0) mapping of local interacting Luminous Infrared Galaxies

Andreea Petric, Joe Mazzarella, Lee Armus, Jason Surace, Jin Koda, Aaron Evans, Dave Frayer

II.79. Physical condition of molecular gas at the centre of the active galaxy NGC 1097

Nuria Piñol Ferrer, Kambiz Fathi, Andreas Lundgren, Glenn van de Ven

II.80. Probing ISM dust through X-RAY spectroscopy

Ciro Pinto, Jelle S. Kaastra, Elisa Costantini, F. Verbunt

II.81. A Direct determination of the C18O column density in Orion KL

Rene Plume, E. A. Bergin, T. G. Phillips, D. C. Lis, and the HEXOS team

II.82. A High-resolution study of the near-infrared diffuse interstellar bands

Mark Rawlings, Andy Adamson, Ben McCall, Tom Kerr

II.83. The APEX spectral line surveys, more and more excited

Miguel Angel Requena Torres, Rolf Güsten, Jesus Martín-Pintado, Sergio Martín, Rebeca Aladro, Axel Weiss, Bernd Klein, Stephan Heyminck

II.84. Three possible explanations for transitional molecules in Orion KL

Paul Rimmer, Eric Herbst

II.85. The overall systematic trends in the kinematics of massive star forming regions. Observations of HC_3N^* in hot cores

Víctor M. Rivilla, Jesús Martín-Pintado, Izaskun Jiménez-Serra, Pablo de Vicente

II.86. A search for PAHs in the ISM: High-resolution UV observations confronted with laboratory spectra

Gaël Rouillé, Roland Gredel, Yvain Carpentier, Mathias Steglich, Friedrich Huisken, Thomas Henning

II.87. Herschel-SPIRE spectroscopy of nearby Seyfert galaxies

Nicola Sacchi, and the SPIRE Sag2 consortium

II.88. Polycyclic aromatic hydrocarbons and the diffuse interstellar bands. A survey

Farid Salama, Gazinur Galazutdinov, Jacek Krelowski, Ludovic Biennier, Yuri Beletsky, In-Ok Song

II.89. Polycyclic aromatic hydrocarbon and emission line ratios in Active Galactic Nuclei and Starburst Galaxies

Dinalva A. Sales, Miriani Pastoriza, Rogério Riffel

II.90. Global collapse of the DR21 filament

N. Schneider, T. Csengeri, S. Bontemps, F. Motte, R. Simon, P. Hennebelle, C. Federrath, R. Klessen

II.91. Chemical and dynamical evolution of infrared dark clouds, massive protostars and proto star clusters

Jonathan Tan, Audra Hernandez, Michael Butler, Bo Ma, Yichen Zhang, Peter Barnes, Stefan O'Dougherty, Paola Caselli, Francesco Fontani, Izaskun Jimenez-Serra

II.92. A new model for the formation of grain mantles in Prestellar Cores

Vianney Taquet, Cecilia Ceccarelli, Claudine Kahane

II.93. A line confusion limited millimeter survey of Orion KL

B. Tercero, J. Cernicharo, J. R. Pardo, J. R. Goicoechea

II.94. Herschel observations of extra-ordinary sources: Sulfur carbon chains and silicon bearing species in Orion KL

B. Tercero, J. Cernicharo, N. Marcelino, and the HEXOS team

II.95. A multispecies survey of the active galaxy NGC1068

Antonio Usero, Santiago García-Burillo, Asunción Fuente, Susanne Aalto, Roberto Neri, Melanie Krips

II.96. Water emission towards the chemical rich outflow L1157: the WISH spectral line survey.

Magda Vasta, Claudio Codella, Andrea Lorenzani, Gina Santangelo, Brunella Nisini, Teresa Giannini, Mario Tafalla, Rene Liseau, Lars Kristensen, Edwine van Dischoeck

II.97. Chemical composition of infrared dark clouds: observations vs modeling

Tatiana Vasyunina, Eric Herbst, Hendrik Linz, Thomas Henning, Igor Zinchenko, Henrik Beuther, Maxim Voronkov

II.98. C+/CO transitions in the diffuse ISM: Transitional Cloud Sample from the GOT C+ Survey of [CII] in the inner Galaxy at l = -30 to 30

T. Velusamy, J. L. Pineda, W. D. Langer, K. Willacy, P. F. Goldsmith

II.99. Prestellar cores with ALMA

Malcolm Walmsley

II.100. First hyperfine structure resolved OH FIR spectrum of a star-forming region

S. F. Wampfler, S. Bruderer, L. E. Kristensen, E. A. Bergin, A. O. Benz, E. F. van Dishoeck, G. J. Herczeg, F. F. S. van der Tak, J. R. Goicoechea, S. D. Doty, F. Herpin

II.101. Chemical and thermal structure of infrared dark cloud cores

Dmitri Wiebe, Yaroslav Pavlyuchenkov

II.102. The first Herschel FTS look of the M83 FIR spectrum

Ronin Wu, Suzanne Madden, Pasquale Panuzzo, and the Herschel SAG2 Very Nearby Galaxy team

II.103. APEX telescope observations of hydrides

Friedrich Wyrowski, Karl Menten, Rolf Güsten, Arnaud Belloche, Thomas Klein, Christian Leinz

II.104. CO(3-2),CO(2-1) and CO(1-0) observations toward the Galactic bubbles N54, N73, N74, N82 and N130

Jian-jun Zhou, Jarken Esimbek, Wei-guang Ji

II.105. Structure and chemistry of the high mass star forming region S255 on small scales

Igor Zinchenko, Stan Kurtz, Sheng-Yuan Liu, Devendra Ojha, Yu-Nung Su

II.106. A search for HeH+ and CH in a high redshift QSO

Igor Zinchenko, Victor Dubrovich, Christian Henkel

Posters: Session 3

III.01. A triplet ground state for cationic polyaromatic hydrocarbons after hydrogen loss

Hector Alvaro Galue, Jos Oomens

III.02. Photolysis and Radiolysis of Formic Acid: Relevances to Astrophysical Ices.

Diana Andrade, Alicja Domaracka, Enio Frota da Silveira, Herman Rothard, Philippe Boduc

III.03. Molecular dynamics simulations of CO_2 formation in interstellar ices

Carina Arasa, Stefan Andersson, Ewine F. van Dishoeck, Geert-Jan Kroes

III.04. Molecular dynamics simulations of D_2O ice photodesorption

Carina Arasa, Stefan Andersson, Herma Cuppen, Ewine F. van Dishoeck, Geert-Jan Kroes

III.05. Interstellar chemistry of atomic nitrogen: low temperature kinetics of the N + OH and N + NO reactions.

Astrid Bergeat, Julien Daranlot, Kevin M. Hickson

III.06. Differential adsorption of complex organic molecules isomers at interstellar ices surfaces

Mathieu Bertin, Marie Lattelais, Hakima Mokrane, Françoise Pauzat, Jerôme Pilmé, Claire Minot, Yves Ellinger, Claire Romanzin, Xavier Michaut, Pascal Jeseck, Jean-Hugues Fillion, Henda Chaabouni, Emanuele Congiu, François Dulieu, Saoud Baouche, Jean-Louis Lemaire

III.07. Gas phase chemical kinetics at high temperature of carbonaceous molecules: application to circumstellar envelopes

Ludovic Biennier, Aline Gardez, Ghassen Saidani, Robert Georges, Bertrand Rowe, K.P.J. Reddy

III.08. A new geometry-free astrochemistry code for treating HII/PDR complexes

Thomas G. Bisbas, Tom A. Bell, Serena Viti, Jeremy Yates, Mike Barlow, Barbara Ercolano

III.09. Studies of sulphur containing model interstellar ices

Wendy A. Brown, Daren J. Burke, John L. Edridge

III.10. Analyzing astronomical observations with the NASA Ames PAH database.

Jan Cami, Christiaan Boersma, Els Peeters, Charles Bauschlicher Jr., Douglas Hudgins, Louis Allamandola

III.11. Emission model of AIBs: the case of the 11.2 λm band

Alessandra Candian, Peter Sarre

III.12. Observing PAH hydrogenation with scanning tunneling microscopy

Andrew M. Cassidy, Louis Nilsson, Richard Balog, John Thrower, Bjarke Jørgensen, Liv Hornekær

III.13. Grain surface polarization: ferroelectric mantles?

Andrew M. Cassidy, Richard Balog, David Field, Nykola Jones, Oksana Plekan

III.14. CASSIS, a software package to analyse high spectral resolution observations

Emmanuel Caux, Sandrine Bottinelli, Charlotte Vastel, Jean-Michel Glorian

III.15. A unified Macro Monte Carlo Simulation of gas phase chemistry and micro Monte Carlo simulation of grain surface chemistry

Qiang Chang, Eric Herbst

III.16. Stochastic models of molecule formation on dust

Steven Charnley, Eva Wirstroem

III.17. Ice, dust and extinction in dense clouds

Jean Chiar, Adwin Boogert, Caltech, Claudia Knez, Xander Tielens

III.18. Low temperature FIR and submm opacity of interstellar silicate dust analogs: experimental variation of the emissivity spectral index with the wavelength and temperature

Anne Coupeaud, Karine Demyk, Claude Mény, Céline Nayral

III.19. Nuclear spin dependence of hydrogenic plasmas in the laboratory and the diffuse interstellar medium

Kyle N. Crabtree, Nick Indriolo, Holger Kreckel, Carrie A. Kauffman, Brian A. Tom, Eftalda Becka, Brett A. McGuire, Benjamin J. McCall,

III.20. Photodesorption and product formation in UV-irradiated ices under ultra-high-vacuum conditions.

Gustavo A. Cruz Díaz, Antonio Jiménez Escobar, Guillermo M. Muñoz Caro

III.21. Spectroscopic networks

Attila G. Császár, Tibor Furtenbacher

III.22. Simulations of grain surface chemistry under interstellar conditions

Herma Cuppen

III.23. UV induced chemistry of polycyclic aromatic hydrocarbons in NH_3-containing interstellar ice analogues

Cuylle, S.H., Tenenbaum, E.D., Bouwman, J., Linnartz, H.

III.24. Tentative detection of HCl^+ in diffuse clouds

Massimo De Luca, Harshal Gupta, David Neufeld, Maryvonne Gerin, David Teyssier, John Pearson, Darek Lis, and the PRISMAS team

III.25. Electronic absorption spectra of diacetylene cation and comparison with diffuse interstellar bands

Rainer Dietsche, Satrajit Chakrabarty, Corey A. Rice, Fabio J. Mazzotti, John P. Maier

III.26. Formation of complex organics by gas phase and intracluster ion-molecule reactions involving acetylene and hydrogen cyanide

Samy El-Shall, Ahmed Hamed, Abdel-Rahman Soliman, Paul O. Momoh

III.27. VAMDC: A new home for molecular data for astronomy

Christian Endres, Thomas Giesen, Holger Müller, Peter Schilke, Jürgen Stutzki, Stephan Schlemmer

III.28. Electronic absorption spectra of small carbon chain oxides in 6 K neon matrices.

Sonia M. Erattupuzha Joseph, Jan Fulara, Adam Nagy, Iryna Garkusha, John P. Maier

III.29. Photodesorption of interstellar ice analogues: a wavelength-dependent study

Edith Fayolle, Mathieu Bertin, Claire Romanzin, Xavier Michaut, Karin Öberg, Harold Linnartz, Jean-Hugues Fillion

III.30. The chemistry of high mass star forming regions with "Chemical Differentiation": Orion KL, W75N, & W3

Douglas N. Friedel, Susanna Widicus Weaver

III.31. Experimental studies of the formation of molecular hydrogen through superhydrogenation of PAHs

Emil E. Friis, John D. Thrower, Bjarke Jørgensen, Saoud Baouche, Alan Luntz, Liv Hornekær

III.32. Infrared spectroscopy of D_2O in amorphous and crystalline water environments

Óscar Gálvez, Belén Maté, Rafael Escribano, Víctor J. Herrero

III.33. Electronic transitions of protonated PAHs in 6 K neon matrices

Iryna Garkusha, Jan Fulara, Adam Nagy, University of Basel, Sonia M. Erattupuzha Joseph, John P. Maier

III.34. Modeling the formation of interstellar CO_2, CO and water ice

Robin Garrod, Tyler Pauly

III.35. Modeling complex molecular chemistry in astrophysically-relcvant laboratory ices

Robin Garrod, Karin Oberg

III.36. Sulfur and carbon chemistries driven by turbulent dissipation in the diffuse ISM

Benjamin Godard, Edith Falgarone, Guillaume Pineau des Forêts, Maryvonne Gerin, Massimo De Luca

III.37. The influence of cosmic rays on the 3.4 microns interstellar absorption band

Marie Godard, Géraldine Féraud, Marin Chabot, Thomas Pino, Rosario Brunetto, Yvain Carpentier, Jean Duprat, Cécile Engrand, Philippe Bréchignac, Louis d'Hendecourt, Emmanuel Dartois

III.38. Visible spectra of titanium dioxide

Varun Gupta, Ramya Nagarajan, John P. Maier, Xiujuan Zhuang, Anh Le, Timothy C. Steimle

III.39. Cavity ring down spectroscopy of molecular ions in the 3 micron region

Joseph Guss, Harald Verbraak, Harold Linnartz

III.40. Diffusion mechanism of hydrogen and deuterium atoms and the spin temperature of molecules on the surface of water ice at 8–15 K

Tetsuya Hama, Naoki Watanabe, Yuki Kimura, Akira Kouchi, Takeshi Chigai, Valerio Pirronello

III.41. Ammonium and formate ions in interstellar ice analogues. Ice morphology and the elusive 6.85 μm band

Víctor. J. Herrero, Belén Maté, Óscar Gálvez, Delia Fernández-Torre, Miguel A. Moreno, Rafael Escribano

III.42. FTIR measurements of ammonia formation by the successive H-atom addition to N-atom in an N_2 matrix at 10K

Hiroshi Hidaka, Motohiro Watanabe, Akira Kouchi, Naoki Watanabe

III.43. Formation routes of interstellar molecules through surface reactions at low temperature

Sergio Ioppolo, Gleb Fedoseev, Thanja Lamberts, Herma M. Cuppen, Claire Romanzin, Ewine F. van Dishoeck, Harold Linnartz

III.44. MATRIICES - Mass Analytical Tool for Reactions in Interstellar ICES

Karoliina Isokoski, Jean-Baptiste Bossa, Harold Linnartz

III.45. Photoprocessing of ice containing H_2S in circumstellar icy grain mantles

Antonio Jiménez-Escobar, Guillermo M. Muñoz Caro

III.46. Solid-state spectroscopy: silicates & carbonaceous grains
Cornelia Jäger, Harald Mutschke, Thomas Henning, Friedrich Huisken

III.47. Associative detachment reaction H- + H: Ion trap study at 10-100 K
Pavol Jusko, Štěpán Roučka, Illia Zymak, Dmytro Mulin, Sergii Opanasiuk, Radek Plašil, Dieter Gerlich, Juraj Glosík

III.48. UV irradiation of hydrogenated amorphous carbon (HAC) as a carrier candidate of the interstellar UV bump at 4.6 μm^{-1}
K. Gadallah, H. Mutschke, C. Jäger

III.49. Pyrolysis of simple chiral aromatic alcohols. Survivability and preservation of chirality on minerals of astrophysical interest: a case study
Yeghis Keheyan

III.50. Protonation of molecular oxygen probed in a cold 22-pole ion trap
Lars Kluge, I. Physikalisches Institut, Universität zu Köln, Sabrina Gärtner, I. Physikalisches Institut, Universität zu Köln, Sandra Brünken, I. Physikalisches Institut, Universität zu Köln, Oskar Asvany, I. Physikalisches Institut, Universität zu Köln, Stephan Schlemmer, I. Physikalisches Institut, Universität zu Köln

III.51. Laboratory rotational spectrum of singly ^{13}C-substituted dimethyl ether up to 1.5 THz and interstellar detection of $^{13}CH_3O^{12}CH_3$ - a fruitful interplay between laboratory work and inter
Monika Koerber, Suzanne Bisschop, Christian Endres, Frank Lewen, Stephan Schlemmer,

III.52. Laboratory and observational studies of methyl ethyl ketone
Jay A. Kroll, Steven Shipman, Susanna L. Widicus Weaver

III.53. The distributions of complex organic molecules in the Orion KL molecular core
Yi-Jehng Kuan, Yu-Sen Hsu, Steven B. Charnley, Kuo-Song Wang

III.54. Methanol photodissociation and its effects on complex chemistry in the ISM
Jacob Laas, Susanna Widicus Weaver

III.55. The revised version of class I methanol maser catalog
G.M. Larionov, I.D. Litovchenko, I.E. Val'tts, A.V. Alakoz,

III.56. Atom - H_2 reactions at very low temperatures and astrophysical implications
Sébastien D. Le Picard, Meryem Tizniti, Coralie Berteloite, André Canosa, Ian R. Sims

III.57. Competing mechanisms in molecular hydrogen formation on silicates in conditions relevant to the interstellar medium (up to 70K)

Jean Louis Lemaire, Gianfranco Vidali, Saoud Baouche, Mourad Chehrouri, Henda Chaabouni, Hakima Mokrane

III.58. Applying the Meudon PDR code on dense structures from MHD simulations of the ISM

Francois Levrier, Franck Le Petit, Patrick Hennebelle, Pierre Lesaffre, Maryvonne Gerin, Edith Falgarone

III.59. Effects of reagent rotation and vibration on H + OH (v, j)-¿ O + H_2

X. Li, Leiden Observatory, E. F. van Dishoeck, Leiden Observatory, M. C. van Hemert, Leiden Institute of Chemistry, C. Arasa

III.60. State-specific rate coefficients for dissociation of $H_2(v, j)$ + $H_2(v', j')$

Margot Mandy

III.61. An infrared study of glycine in astrophysical ices.

Belén Maté, Yamilet Rodríguez-Lazcano, Óscar Gálvez, Isabel Tanarro, Rafael Escribano

III.62. Residual gas analysis of samples formed from the UV irradiation of astrophysical ice analogs

Christopher K. Materese, Michel Nuevo, Scott A. Sandford

III.63. Progress towards the terahertz rotational spectrum of H5+ and its isotopologues

Brett A. McGuire, Yimin Wang, Joel Bowman, Susanna L. Widicus Weaver

III.64. Molecular hydrogen formation on aromatic carbon structures

Vito Mennella, Liv Hornekaer, John Thrower, Mario Accolla

III.65. A solid state physics approach to the interaction between organic molecules and interstellar dust grains: (C_{60}) on SiC

Pablo Merino, Jose Angel Martin-Gago, Jose Cernicharo

III.66. PAH bombardment by energetic particles: models and astrophysical implications

Elisabetta Micelotta, Anthony Jones, Alexander Tielens

III.67. Effects of H_2 coating of grains on depletion

Oscar Morata, Tatsuhiko Hasegawa

III.68. Contrasting [CII], methylidynium ion CH+, and methanol physical and chemical tracers around Orion KL

Pat Morris, John Pearson, David Neufeld, Harschal Gupta, and the HEXOS team

III.69. QUB low energy ion-ices irradiation experiment

Andra Muntean, Tom Field, Adam. Hunniford, Bob McCullough, Jorge Konanoff, Tom Millar

III.70. Electronic absorption spectra of mass-selected hydrocarbon cations in solid neon: $C_nH_4^+$ (n=5–8,10,12)

Adam Nagy, Jan Fulara, Iryna Garkusha, John P. Maier

III.71. Electron irradiation of interstellar ice analogues

B. G. Nair, N. J. Mason

III.72. Formation of nucleobases and other prebiotic species from the UV irradiation of pyrimidine in astrophysical ices

Michel Nuevo, Scott A. Sandford, Stefanie N. Milam, Christopher K. Materese, Jamie E. Elsila, Jason P. Dworkin

III.73. Sequential hydrogenation of molecular oxygen on cold surfaces: reaction kinetics and structure of formed H_2O ice.

Yasuhiro Oba, Naoya Miyauchi, Hiroshi Hidaka, Takeshi Chigai, Naoki Watanabe, Akira Kouchi

III.74. Neutral radical-molecule reactions CO + OH on cold interstellar ices

Yasuhiro Oba, Naoki Watanabe, Akira Kouchi, Tetsuya Hama, Valerio Pirronello

III.75. Formation of CO_2 and OCS after cosmic ion irradiation of icy grain mantles

M. Elisabetta Palumbo, Giuseppe Baratta, Daniele Fulvio, Mario Garozzo, Sergio Ioppolo, Zuzana Kanuchova, Giuseppe Leto, Ivana Sangiorgio, Giovanni Strazzulla

III.76. About the role of ice in the reconstruction of ice

F. Pauzat, P. Redondo, A. Markovits, Y. Ellinger

III.77. A submillimeter study of CH in Orion KL

John C Pearson, Harshal Gupta, Patrick Morris, Maryvonne Gerin, Edwin Bergin

III.78. A hybrid method of moment equations and rate equations to modeling gas-grain chemistry

Yezhe Pei, Eric Herbst

III.79. An approach to global rovibrational analysis based on anharmonic ladder operators: Application to hydrogen selenide and hydrogen sulfide

F. Pérez Bernal, M. Carvajal, O. Álvarez-Bajo

III.80. Secondary fragmentation routes of glycine in ice under irradiation

A. Pernet, J. Pilmé, Y. Ellinger,

III.81. Near-infrared to visible spectral lines of cold H_3^+

Annemieke Petrignani, Max Berg, Dennis Bing, Florian Grussie, Andreas Wolf

III.82. Possible evidence for crystallization of astrophysical ice analogs by heavy and energetic cosmic rays

Sergio Pilling, Eduardo Seperuelo Duarte, Enio F. da Silveira, Hermann Rothard, Alicja Domaracka, Philippe Boduch

III.83. Formation of unsaturated hydrocarbons by cosmic ray analogs in interstellar ices

Sergio Pilling, Diana P. P. Andrade, Enio F. da Silveira, Hermann Rothard, Alicja Domaracka, Philippe Boduch

III.84. Chiral molecules in the ISM: the best candidates

J. Pilmé, G. Marloie, M. Lattelais, F. Pauzat, Y. Ellinger

III.85. An FTIR study on the catalytic effect of water molecules on the reaction of CO successive hydrogenation at 3 and 10K

Claire Pirim, Lahouari Krim

III.86. Surface science studies of ethene containing model interstellar ices

Fabrizio Puletti, Mark Whelan, Wendy A. Brown

III.87. Reaching the line confusion limit: new spectral analysis software and its application to a molecular line survey of Orion-KL

Mary L. Radhuber, Matthew C. Sumner, Jay A. Kroll, Jonas Zmuidzinas, Georey A. Blake, Susanna L. Widicus Weaver

III.88. Chemical and physical conditions at the disk-halo interaction places in the Galactic center region

Denise Riquelme, M. Aranzazu Amo-Baladron, Jesus Martin-Pintado, Rainer Mauersberger, Sergio Martin, Leonardo Bronfman

III.89. H_2 formation pumping signatures: a modellist's view

Evelyne Roueff, Jacques Le Bourlot, Franck Le Petit, Simon Casassus

III.90. Ab initio calculations for the spectral analysis of dimethyl ether (CH_3OCH_3) and their isotopologues

M.L. Senent, M. Villa, R. Domínguez-Gómez, O. Álvarez-Bajo, M. Carvajal

III.91. H atom diffusion on amorphous ice at low temperature

Bethmini Senevirathne, Gunnar Nyman, S. Andersson

III.92. A Spitzer survey of an isolated globule: DC314.8-5.1

S.S. Shenoy, D.C.B. Whittet, Y.J. Pendleton, C. Boersma, L.J. Allamandola, D. Horne, P.A. Mayeur

III.93. Deuterium fractionation of massive star forming clumps in infrared dark clouds

Yu-Nung Su, Sheng-Yuan Liu, Vivien H.-R. Chen, Qizhou Zhang

III.94. Dissociative electron attachment to hydrocarbons. A laboratory study

E. Szymanska, N. J. Mason

III.95. A theoretical study of the electronic dissociative recombination of an ion of astrophysical interest : N_2H+

Dahbia Talbi, David Kashinski, Peet Hickman

III.96. SUCCESS: A SUbmm Catalogue of Circumstellar Envelope of StarS with Herschel/HIFI

David Teyssier, Javier Alcolea, Valentin Bujarrabal, Arantxa Castro-Carrizo, Jose Cernicharo, Pedro Garcia-Lario, Anthony Marston, Hans Olofsson, Christophe Risacher, Fredrik Schoeier, Eva Verdugo

III.97. Kinetic studies on low-temperature solid-state reactions in interstellar ices

Patrice Theule, Fabrice Duvernay, Florent Mispelaer, Vassilissa Vinogradoff, Jean-Baptiste Bossa, Fabien Borget, Gregoire Danger, Thierry Chiavassa

III.98. Infrared spectroscopy of carbon- and carbon-silicon clusters

Sven Thorwirth, Jürgen Krieg, Volker Lutter, Imke Gottbehüt, Stephan Schlemmer, Thomas F. Giesen

III.99. Laboratory studies of astronomically relevant molecules harboring second-row elements

Sven Thorwirth, Valerio Lattanzi, Leonie A. Mück, Dewayne T. Halfen, Sandra Brünken, Carl A. Gottlieb, Patrick Thaddeus, Lucy M. Ziurys, Jürgen Gauss, Michael C. McCarthy

III.100. Isotope effect on charge transfer in collisions of H with He^+ and He^{2+}

Nathalie Vaeck, Jérôme Loreau, Sergey Ryabchenko

III.101. Recent development of the photoionization / PDR code Cloudy

P.A.M. van Hoof, R.L. Porter, G.J. Ferland

III.102. Formation of water on dust grains

Gianfranco Vidali, Dapeng Jing, Jiao He, Paul Frank, John Brucato, Antonio De Sio, Lorenzo Tozzetti

III.103. Chemical evolution of the interstellar ice analogs: HMT formation mechanism

Vassilissa Vinogradoff, Fabrice Duvernay, Grégoire Danger, Patrice Theulé, Fabien Borget, T. Chiavassa

III.104. Photodissociation and radiative stabilization of HeH^+ in its a and b $^3\Sigma^+$ states

Stéphane Vranckx, Jérôme Loreau, Michèle Desouter-Lecomte, Nathalie Vaeck

III.105. The KInetic Database for Astrochemistry (KIDA)

V. Wakelam and the KIDA team

III.106. New molecules in hydrogen precipitates

Mark Walker, Leaf Lin, Andrew Gilbert, Peter Gill

III.107. Deuterated glycoaldehyde: laboratory measurements, analysis and proposed astrophysical research

Aurélia Bouchez, Adam Walters, Laurent Margulés, Roman Motiyenko, Jean-Claude Guillemin, Sandrine Bottinelli, Cecilia Ceccarelli, Claudine Kahane

III.108. Laboratory measurements and analysis for two molecules of astrophysical interest: ^{13}C-ethanol and n-butyl cyanide

Adam Walters, Matthias Ordu, Aurélia Bouchez, Holger Müller, Marc Nuñez, Frank Lewen, Sandrine Bottinelli, Stephan Schlemmer

III.109. Isolation of OH radicals in water ice and study of their reactivity

Emilie-Laure Zins, Prasad Joshi, Lahouari Krim,

III.110. T BASECOL and VAMDC: A new way to access and manipulate molecular collisional data for interstellar applications

F. Daniel, M. L. Dubernet, M. Doronin, L. Nenadovic, J. Bureau

III.111. The chemistry of AGB stars at low metallicity

Paul Woods, Ciska Kemper, Martin Cordiner

III.112. Amino Acids and Sugars in the Gas Phase: Microwave Data for Astrochemistry

Santiago Mata, Carlos Cabezas, Marcelino Varela, Isabel Peña, Cristobal Pérez, Susana Blanco, María E. Sanz, Juan C. López, José L. Alonso

III.113. Commissioning and Science Verification of ALMA

R. Mauersberger, E. Villard, A. B. Peck, S. Asayama, D. Barkats, P. Calisse,S. Corder, J. Cortes, P. Cortes, I. de Gregorio, W. R. F. Dent, E. Fomalont, D. Fulla, D. Garcia-Appadoo, D. Gunawan, A. Hales, R.E. Hills, T. Kamazaki, L. Knee, R. Kneissl, S. Komugi, S. Leon, R. Lucas, A. Lundgren, G. Marconi, S. Matsushita, P. Planesas, M. Rawlings, A. Richards, T. Sawada, R. Simon, K. Tachihara, T. van Kempen, T. Wiklind

III.114. The ORGANIC Experiment on the ISS EXPOSE-R

K. Bryson, Z. Peeters, F. Salama, B. Foing, P. Ehrenfreund, E. Jessberger, A. Bischoff, M. Breitfellner, W. Schmidt

III.115. Experimental results obtained with the Interstellar Astrochemistry Chamber (ISAC) for the simulation of ice processes

G. M. Muñoz Caro, A. Jimenez Escobar, G.A. Cruz Diaz, A. Ciaravella, C. Cecchi-Pestellini

III.116. Composition of the grain mantle: A Monte Carlo Study

A. Das, S. K. Chakrabarti

III.117. Chemical Evolution around star forming region: A time dependent study

L. Majumdar, A. Das, S.K. Chakrabarti, S. Chakrabarti

Author Index

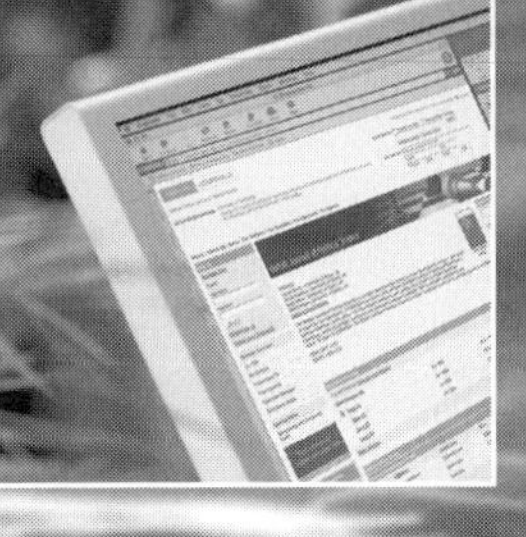